WITHDRAWN

ANNUAL REVIEW OF NUCLEAR AND PARTICLE SCIENCE

EDITORIAL COMMITTEE (1983)

GORDON A. BAYM
JAMES D. BJORKEN
GERALD T. GARVEY
HARRY E. GOVE
ERNEST M. HENLEY
J. D. JACKSON
LEE G. PONDROM
ROY F. SCHWITTERS

Responsible for the organization of Volume 33
(Editorial Committee, 1981)

CHARLES BALTAY
GORDON A. BAYM
HARRY E. GOVE
ERNEST M. HENLEY
J. D. JACKSON
ROY F. SCHWITTERS
G. M. TEMMER
JAMES D. BJORKEN

Production Editor   MARGOT PLATT
Indexing Coordinator   MARY GLASS

ANNUAL REVIEW OF NUCLEAR AND PARTICLE SCIENCE

Volume 33, 1983

J. D. JACKSON, *Editor*
University of California, Berkeley

HARRY E. GOVE, *Associate Editor*
University of Rochester

ROY F. SCHWITTERS, *Associate Editor*
Harvard University

ANNUAL REVIEWS INC.   4139 EL CAMINO WAY   PALO ALTO, CALIFORNIA 94306 USA

ANNUAL REVIEWS INC.
Palo Alto, California, USA

COPYRIGHT © 1983 BY ANNUAL REVIEWS INC., PALO ALTO, CALIFORNIA, USA. ALL RIGHTS RESERVED. The appearance of the code at the bottom of the first page of an article in this serial indicates the copyright owner's consent that copies of the article may be made from personal or internal use, or for the personal or internal use of specific clients. This consent is given on the condition, however, that the copier pay the stated per-copy fee of $2.00 per article through the Copyright Clearance Center, Inc. (21 Congress Street, Salem, MA 01970) for copying beyond that permitted by Sections 107 or 108 of the US Copyright Law. The per-copy fee of $2.00 per article also applies to the copying, under the stated conditions, of articles published in any *Annual Review* series before January 1, 1978. Individual readers, and nonprofit libraries acting for them, are permitted to make a single copy of an article without charge for use in research or teaching. This consent does not extend to other kinds of copying, such as copying for general distribution, for advertising or promotional purposes, for creating new collective works, or for resale. For such uses, written permission is required. Write to Permissions Dept., Annual Reviews Inc., 4139 El Camino Way, Palo Alto, CA 94306 USA.

International Standard Serial Number: 0163-8998
International Standard Book Number: 0-8243-1533-2
Library of Congress Catalog Card Number: 53-995

Annual Review and publication titles are registered trademarks of Annual Reviews Inc.

Annual Reviews Inc. and the Editors of its publications assume no responsibility for the statements expressed by the contributors to this *Review*.

PRINTED AND BOUND IN THE UNITED STATES OF AMERICA

Annual Review of Nuclear and Particle Science
Volume 33, 1983

CONTENTS

UPSILON RESONANCES, *Paolo Franzini and Juliet Lee-Franzini*	1
GAUGE THEORIES AND THEIR UNIFICATION, *P. Ramond*	31
PROGRESS AND PROBLEMS IN PERFORMANCE OF e^+e^- STORAGE RINGS, *R. D. Kohaupt and G.-A. Voss*	67
NUCLEAR MATTER THEORY: A Status Report, *A. D. Jackson*	105
PHYSICS WITH THE CRYSTAL BALL DETECTOR, *Elliott D. Bloom and Charles W. Peck*	143
SUM RULE APPROACH TO HEAVY QUARK SPECTROSCOPY, *M. A. Shifman*	199
BAG MODELS OF HADRONS, *Carleton E. DeTar and John F. Donoghue*	235
FUSION REACTIONS BETWEEN HEAVY NUCLEI, *J. R. Birkelund and J. R. Huizenga*	265
ELEMENTAL AND ISOTOPIC COMPOSITION OF THE GALACTIC COSMIC RAYS, *J. A. Simpson*	323
MUON SCATTERING, *J. Drees and H. E. Montgomery*	383
CHANNELING RADIATION, *J. U. Andersen, E. Bonderup, and R. H. Pantell*	453
COSMIC-RAY RECORD IN SOLAR SYSTEM MATTER, *R. C. Reedy, J. R. Arnold, and D. Lal*	505
MEASUREMENT OF CHARMED PARTICLE LIFETIMES, *Ronald A. Sidwell, Neville W. Reay, and Noel R. Stanton*	539
INELASTIC ELECTRON SCATTERING FROM NUCLEI, *J. Heisenberg and H. P. Blok*	569
INTERNAL SPIN STRUCTURE OF THE NUCLEON, *Vernon W. Hughes and Julius Kuti*	611
GRAND UNIFIED THEORIES AND THE ORIGIN OF THE BARYON ASYMMETRY, *Edward W. Kolb and Michael S. Turner*	645
INDEXES	
Cumulative Indexes of Contributing Authors, Volumes 23–33	697
Cumulative Index of Chapter Titles, Volumes 23–33	699

ANNUAL REVIEWS INC. is a nonprofit scientific publisher established to promote the advancement of the sciences. Beginning in 1932 with the *Annual Review of Biochemistry*, the Company has pursued as its principal function the publication of high quality, reasonably priced *Annual Review* volumes. The volumes are organized by Editors and Editorial Committees who invite qualified authors to contribute critical articles reviewing significant developments within each major discipline. The Editor-in-Chief invites those interested in serving as future Editorial Committee members to communicate directly with him. Annual Reviews Inc. is administered by a Board of Directors, whose members serve without compensation.

1983 Board of Directors, Annual Reviews Inc.

Dr. J. Murray Luck, Founder and Director Emeritus of Annual Reviews Inc.
  Professor Emeritus of Chemistry, Stanford University
Dr. Joshua Lederberg, President of Annual Reviews Inc.
  President, The Rockefeller University
Dr. James E. Howell, Vice President of Annual Reviews Inc.
  Professor of Economics, Stanford University
Dr. William O. Baker, *Retired Chairman of the Board, Bell Laboratories*
Dr. Winslow R. Briggs, *Director, Carnegie Institution of Washington, Stanford*
Dr. Sidney D. Drell, *Deputy Director, Stanford Linear Accelerator Center*
Dr. Eugene Garfield, *President, Institute for Scientific Information*
Dr. Conyers Herring, *Professor of Applied Physics, Stanford University*
Mr. William Kaufmann, *President, William Kaufmann, Inc.*
Dr. D. E. Koshland, Jr., *Professor of Biochemistry, University of California, Berkeley*
Dr. Gardner Lindzey, *Director, Center for Advanced Study in the Behavioral Sciences, Stanford*
Dr. William D. McElroy, *Professor of Biology, University of California, San Diego*
Dr. William F. Miller, *President, SRI International*
Dr. Esmond E. Snell, *Professor of Microbiology and Chemistry, University of Texas, Austin*
Dr. Harriet A. Zuckerman, *Professor of Sociology, Columbia University*

Management of Annual Reviews Inc.

John S. McNeil, Publisher and Secretary-Treasurer
Dr. Alister Brass, Editor-in-Chief
Mickey G. Hamilton, Promotion Manager
Donald S. Svedeman, Business Manager
Richard L. Burke, Production Manager

ANNUAL REVIEWS OF
Anthropology
Astronomy and Astrophysics
Biochemistry
Biophysics and Bioengineering
Earth and Planetary Sciences
Ecology and Systematics
Energy
Entomology
Fluid Mechanics
Genetics
Immunology
Materials Science
Medicine
Microbiology
Neuroscience
Nuclear and Particle Science
Nutrition
Pharmacology and Toxicology
Physical Chemistry
Physiology
Phytopathology
Plant Physiology
Psychology
Public Health
Sociology

SPECIAL PUBLICATIONS

Annual Reviews Reprints:
  Cell Membranes, 1975–1977
  Cell Membranes, 1978–1980
  Immunology 1977–1979
Excitement and Fascination
  of Science, Vols. 1 and 2
History of Entomology
Intelligence and Affectivity,
  by Jean Piaget
Telescopes for the 1980s

A detachable order form/envelope is bound into the back of this volume.

SOME RELATED ARTICLES IN OTHER *ANNUAL REVIEWS*

From the *Annual Review of Astronomy and Astrophysics*, Volume 21 (1983):

Scientists I Have Known and Some Astronomical Problems I Have Met, Bengt Strömgren

Galactic Gamma-Ray Sources, G. F. Bignami and W. Hermsen

From the *Annual Review of Earth and Planetary Sciences*, Volume 11 (1983):

Radioactive Nuclear Waste Stabilization: Aspects of Solid-State Molecular Engineering and Applied Geochemistry, Stephen E. Haggerty

Recent Developments in the Dynamo Theory of Planetary Magnetism, F. H. Busse

From the *Annual Review of Energy*, Volume 8 (1983):

Recent Changes in U.S. Energy Consumption: What Happened and Why, Eric Hirst, Robert Marlay, David Greene, and Richard Barnes

Prospects for the U.S. Nuclear Reactor Industry, John F. Ahearne

The Development of Breeder Reactors in the United States, Herbert Kouts

Energy and Security in the 1980s, E. William Colglazier, Jr., and David A. Deese

From the *Annual Review of Fluid Mechanics*, Volume 15 (1983):

Contributions of Ernst Mach to Fluid Mechanics, H. Reichenbach

From the *Annual Review of Materials Science*, Volume 13 (1983):

Solitons in Solids, Tarik Ö. Ogurtani

From the *Annual Review of Physical Chemistry*, Volume 34 (1983):

Laser Microchemistry and Its Application to Electron-Device Fabrication, R. M. Osgood, Jr.

UPSILON RESONANCES

Paolo Franzini

Department of Physics, Columbia University, New York, New York 10027

Juliet Lee-Franzini

Department of Physics, SUNY at Stony Brook, Stony Brook, New York 11794

CONTENTS

1. INTRODUCTION 2
   1.1 *Summary and Plan of the Paper* 2
   1.2 *Discovery of the Υ's and χ_b's* 3
   1.3 *Production of the Upsilons in e^+e^- Annihilations* 3
   1.4 *Transitions and Decays of the Bound Upsilons* 3
   1.5 *Decay of the Υ'''* 4
   1.6 *Outlook for the Study of Fine and Hyperfine Structure* 4
2. DISCOVERY OF THE UPSILON STATES 4
   2.1 *Discovery of the Υ's* 4
   2.2 *Discovery of the χ_b's* 6
3. PRODUCTION OF THE UPSILONS IN e^+e^- ANNIHILATIONS 7
   3.1 $\sigma_{\mu\mu}, \sigma_{had}, \Gamma_{ee}, B_{\mu\mu}, \Gamma_{tot}$ 7
   3.2 *Measurements of R, Γ_{ee}, and Resonance Masses* 9
   3.3 *Limits on the Production of Other Resonances* 10
   3.4 *Comparison of Measurements and Theoretical Models* 12
4. TRANSITIONS AND DECAYS OF BOUND UPSILONS 15
   4.1 *Decay of the Ground-State Υ* 15
   4.2 *Decays of the Υ' and the Υ''* 16
   4.3 *Partial Rates for Υ's Decays* 21
   4.4 *Comparison with Theory* 21
   4.5 *Search for Rare Decays of the Υ* 24
5. DECAY OF THE Υ''' 24
   5.1 *b-Flavor Threshold and $B^* \to B + \gamma$* 25
6. OUTLOOK FOR THE STUDY OF SPIN-ORBIT AND SPIN-SPIN INTERACTIONS 25
   6.1 *Fine Structure of the Triplet P-Wave States* 25
   6.2 *Search for Singlet $b\bar{b}$ States* 25
7. CONCLUSIONS 27

1. INTRODUCTION

The latest addition to the apparently endless list of strongly interacting particles is a family of vector mesons called the upsilons, Υ. They are the heaviest hadrons known to date, with masses around 10 GeV, and are especially rich in confrontation opportunities between experiments and the continuously developing theories of elementary particles. Some of the salient consequences of the existence of the Υ's are listed in the following. (a) Prior to their discovery, all known hadrons were understood as bound or quasibound states of fundamental, point-like constituents, the quarks, which came in four "flavors": u, d, s, c. The discovery of the Υ's required the existence of a fifth quark, the b-flavor quark, b. (b) Upsilons are bound states of a $b\bar{b}$ pair whose relative motion is sufficiently slow that relativistic effects are quite negligible, in contrast to light hadrons and even the heavier psions, J/ψ. Schroedinger's equations can be used to accurately describe the level structure and thus we can learn about the interquark forces responsible for the binding of these states, which range in size from 0.2 to 1 fermi. (c) The bound upsilons (Υ, Υ', Υ'') decay primarily through the annihilation of the $b\bar{b}$ pair into three "gluons," the quanta of the gauge field coupled to the "color charge" carried by all the quarks. Gluons turn ultimately into hadrons; thus properties of the invisible gluons and of the gluon-quark coupling can be gleaned through the study of the Υ's final states and their decay rates. (d) The fourth upsilon is quasibound, i.e. it decays with much higher rate into a pair of b-flavor mesons, the B mesons, each containing a b quark and a light, d or u, quark. B mesons provide means to study the weak interaction of the b quark. (e) Finally the massiveness of the b quark implies a nonnegligible coupling to the elusive Higgs scalars, making the Υ's, at present, the best hunting grounds for light Higgs bosons.

This article reviews the experimental work on upsilon spectroscopy, i.e. the study of the level structure, the transitions between states, and the decays of these states. We present only the minimal amounts of theoretical background and results that are necessary to understand the way measurements are performed and the discussion of the results. The reader is referred to the article on the "Sum Rule Approach to Heavy Quark Spectroscopy," by M. A. Shifman in this volume for a detailed theoretical exposition and references to the extensive theoretical reviews.

1.1 *Summary and Plan of the Paper*

The following subsections contain a brief summary of the contents of each section. We first, however, explicitly list some topics of relevance to the subject of this article that will not be reviewed. (a) Some B-meson properties are mentioned, but not their decays. (b) The fragmentation mechanisms of gluons and quarks, mentioned in the discussion of hadronic decays, are not explained or justified. In particular we do not present data on charged and neutral multiplicities, and production of strange particles and baryons. (c) While a great deal of interest and speculation exists concerning the mechanism of hadroproduction of flavored and unflavored heavy particles, this subject is not discussed since, except for the original experiment that discovered the upsilons and similar ones thereafter, there is almost no experimental information on the subject.

1.2 Discovery of the Υ's and χ_b's

In 1977, at Fermilab the first three Υ's were discovered, unresolved, as an enhancement in the mass spectrum of muon pairs, produced in collisions of 400-GeV protons on beryllium (1, 2). The first two upsilons (Υ and Υ') were observed fully resolved as resonances in e^+e^- annihilations into hadrons, in 1978 at the DORIS e^+e^- storage ring at the DESY laboratory in Hamburg (3–5), and the third one (Υ'') in 1979 at CESR (Cornell Electron Storage Ring) (6, 7). The fourth Υ (Υ'''), which decays into muon pairs with a branching ratio of the order of 10^{-5}, was discovered in 1980 at CESR as a resonance much broader than the three previous ones (8, 9). The χ_b's (scalar, axial vector and tensor $b\bar{b}$ states) are not produced directly in e^+e^- collisions but are reached via photon emission from the Υ' and Υ''. They were first observed at CESR in 1982–1983 (10–12).

1.3 Production of the Upsilons in e^+e^- Annihilations

The masses, leptonic widths, and other properties of the upsilons have been obtained mostly from the study of e^+e^- annihilations into hadrons. The first three Υ's are observed as narrow enhancements in the hadronic cross section with widths determined by the machine energy spread. In the interpretation of the Υ's as $b\bar{b}$ bound states, the mass differences correspond to the excitation energies and the leptonic widths are related to the wave function at the origin. The comparison between measurements and calculations in "potential models," leads to the identification of the four upsilons as the first four triplet S-wave states. The charge of the b quark is determined in this way to be $\frac{1}{3}e$. The Υ''' is a quasibound state lying above the free b-flavor threshold. At energies above the mass of the Υ''', the ratio of the hadronic cross section to the muon pair cross section is observed to increase by $\frac{1}{3}$, which confirms that the b-quark charge is $\frac{1}{3}e$ (13).

1.4 Transitions and Decays of the Bound Upsilons

Experimental observation of a three-jet structure in the final state of the lightest Υ is direct evidence that the decay occurs via annihilations of the $b\bar{b}$

pair into three gluons (10, 14–16); this confirms the existence and nonobservability of color and gives tangible proof for the original explanation of the narrowness of the widths of both psions and upsilons. The first two excited ϒ's can decay to the ground state, emitting two pions. The partial rates for these transitions compared to those for the ψ' give one of the most convincing proofs of the vector nature of the gluon (17–19). The χ_b's, which are identified as triplet P-wave $b\bar{b}$ states, are reached via electric dipole transition from the excited ϒ's. The measured center of gravity of these states gives additional proof of the color independence of the interquark forces (20). Searches for the Weinberg-Wilczek axion in ϒ's decays gives strong evidence against the existence of such particles (21).

1.5 Decay of the ϒ'''

The fourth upsilon has a total width of ~20 MeV, whereas the first three have total widths in the range of tens of kilovolts. This suggests that ϒ''' can decay into B mesons, a fact confirmed by the observation of a strong signal corresponding to the weak decay of a particle of ~5-GeV mass (22, 23), the B meson. Excited B mesons, B*'s, are not produced in ϒ''' decays as proved by a search for photons from B* → B+γ. The B-meson mass is therefore bounded as $5.263 < M_B < 5.278$ GeV (24).

1.6 Outlook for the Study of Fine and Hyperfine Structure

The potential models that so successfuly describe the level structure of the $b\bar{b}$ bound states cannot unambiguously predict the fine structure of the ^3P states, nor the hyperfine splitting between $^3S_1$ (ϒ's) and $^1S_0$ (η_b's) states. Experimental information is necessary to establish the form of the effective spin-spin and spin-orbit interactions responsible for these effects. The next generation of experiments both at CESR and at DORIS will concentrate in this area, and will require very fine resolution and very high machine intensities.

2. DISCOVERY OF THE UPSILON STATES

2.1 Discovery of the ϒ's

The upsilons were discovered in 1977 by Lederman's group (Columbia–FNAL–Stony Brook collaboration), in the continuing study of muon-pair production in hadron collisions, at Fermilab, with proton energies up to 400 GeV, thus extending the reachable muon-pair mass to above 15 GeV (1, 2). A plot of the muon-pair cross section versus dimuon mass obtained by this group is shown in Figure 1 (25). A prominent peak, ~500 MeV wide (of the order of the experimental mass resolution), is strikingly visible over a sharply falling continuum. The continuum-subtracted signal, shown in

Figure 2, is in fact consistent with the production of two and possibly three states spread over a mass range of about 1 GeV, centered around 10 GeV (25). While the three states were not resolved in the Fermilab experiment, the observed mass spectrum was consistent with the natural width of the states being much smaller than their separation. Their interpretation as bound states of a new flavored heavy quark, of mass around 5 GeV, was almost immediate, after the J/ψ experience of only a couple of years earlier (25a).

The Fermilab discovery prompted the DESY Laboratory to push the maximum energy of the DORIS e^+e^- collider up to 5 GeV per beam, which allowed the observation of the first two upsilons fully resolved. The machine energy spread of DORIS at these energies is around 20 MeV, and the experiments at DORIS obtained the first accurate values for the Υ and Υ' masses as 9.46 and 10.01 GeV, respectively (3–5). The Υ'' was not resolved until the autumn of 1979 (6, 7) when CESR began operation. Shortly thereafter a fourth state, the Υ''', was discovered (8, 9). The Υ'' and Υ''' masses are 10.35 and 10.58 GeV, respectively. This last state is the first Υ with a natural width of tens of MeV, indicative of the fact that it is just above the open flavor threshold. The Υ''' will probably remain for some time the heaviest state in the Υ family to be observed. Figure 3 shows the four resonances as they are seen at CESR.

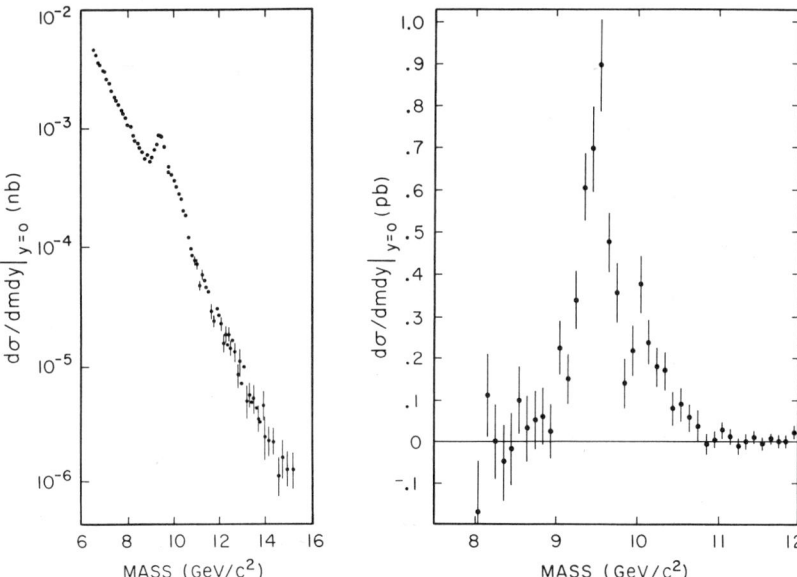

Figure 1 Dimuon mass spectrum observed at Fermilab (25).

Figure 2 Dimuon mass spectrum after background subtraction (25).

Searches at CESR for additional resonances were extended up to ~11.6 GeV, with negative results (13). However, a step in R ($\equiv \sigma_{\text{had}}/\sigma_{\mu\mu}$) was observed around a total energy $W = 10.58$ GeV, of magnitude approximately one third, which confirmed the crossing of the threshold for the production of a pair of new quarks of charge $\pm\frac{1}{3}e$ (13). The value of R around the Υ''' is shown in an expanded scale in Figure 4.

The four Υ's, copiously produced in e^+e^- hadronic annihilations, are vector mesons with quantum numbers identical to those of the photon. With the aid of potential models these states are identified as S-wave triplet states, with the $\Upsilon(9.46)$ being the ground state and the other three Υ's being the $n = 2, 3, 4$ radially excited states. In addition to the good agreement between measured and calculated excitation energies of the four Υ's, further evidence that they are bound states of the same b quark comes from the observation of hadronic transitions between the various levels. The transition between Υ' and Υ was observed at DORIS and CESR in 1980 (17–19). Transitions from Υ'' to Υ were observed at CESR in 1981 and to Υ' in 1982 (26, 27).

2.2 Discovery of the χ_b's

Bound $b\bar{b}$ states other than $^3S_1$ cannot be directly produced in e^+e^- annihilation either because they have wrong J^{PC} values or because the wave function vanishes at the origin. Both hadronic and electromagnetic

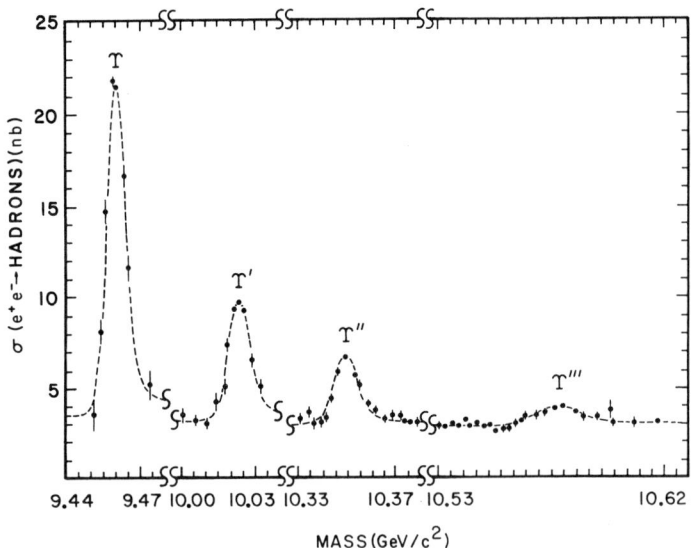

Figure 3 Cross section for e^+e^- annihilations into hadrons at CESR (CUSB data).

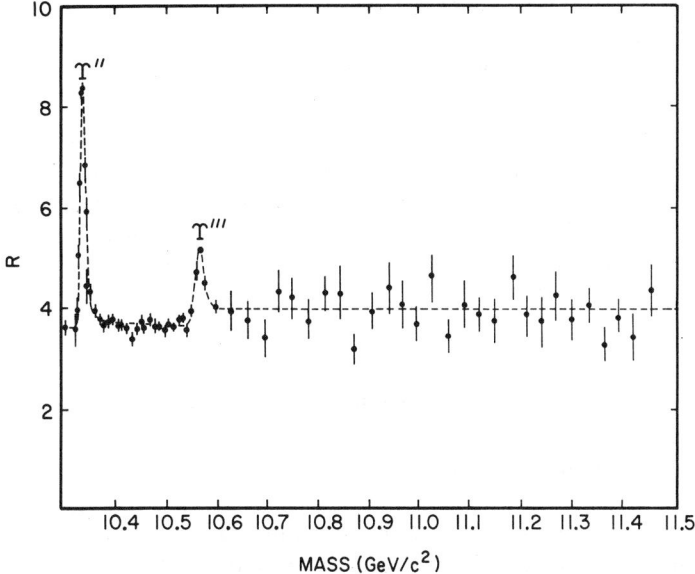

Figure 4 R versus center-of-mass energy around the b-flavor threshold (CUSB data).

transitions can, however, feed-down from the first three ϒ's to other states. In particular the ^3P_J states (χ_b's) can be reached via electric dipole (E1) transitions from the ^3S states. Indirect evidence for the decay of ^3S to ^3P states was obtained in 1981 by the CUSB collaboration at CESR (10) (see Section 4.2.2). In 1982 CUSB observed the photons from the transition ϒ" → $\chi'_b + \gamma$ in the inclusive ϒ" decay photon spectrum as well as the photons from decays of the χ'_b to ϒ' and ϒ in exclusive channels. This established for the first time the existence of P-wave bound $b\bar{b}$ states (11, 12).

3. PRODUCTION OF THE UPSILONS IN e^+e^- ANNIHILATIONS

3.1 $\sigma_{\mu\mu}$, σ_{had}, Γ_{ee}, $B_{\mu\mu}$, Γ_{tot}

Positron-electron collisions result in, to lowest order in QED, four processes: (*a*) elastic or Bhabha scattering; (*b*) annihilation into two photons; (*c*) lepton-pair production; and (*d*) hadron production, via production of a quark-antiquark pair. Bhabha scattering is used to measure the "luminosity" L of e^+e^- colliders, defined by $N = TL\sigma$, where T is the time of the measurement, σ the cross section, and N the number of reactions. The two-photon annihilation cross section provides an independent measurement of the luminosity. The cross section for muon-pair production well above the threshold is given by

$$\frac{d\sigma}{d\Omega} = \frac{\alpha^2}{4W^2}(1+\cos^2\theta),\qquad 1.$$

where α is the fine structure constant and θ is the center-of-mass angle between electron and negative muons. The same result applies to the production of any pair of point-like, spin-$\frac{1}{2}$ fermions of charge e. It is therefore assumed to apply for quarks, taking properly into account charge and color. Integrating over the polar angle, we obtain

$$\sigma(e^+e^- \to \mu^+\mu^-) = \frac{4\pi\alpha^2}{3W^2} = \frac{86.8}{W^2}\,\text{nb}$$

(with W in GeV), and for N quark flavors of charge q_i with three colors

$$\sigma(e^+e^- \to \text{hadrons}) = 3(86.8/W^2)\sum_i^N q_i^2\,\text{nb}$$

assuming, according to the quark-parton model, that quarks evolve into hadrons with unit probability. Note that both these cross sections scale as $1/W^2$ as expected for point-like particles. It is convenient to define the dimensionless ratio $R = \sigma_{\text{had}}/\sigma_{\mu\mu}$, which is expected to have no energy dependence except if a new threshold is reached. Below the b-flavor threshold we know of the u, d, s, and c quarks of charges $|q_i| = \frac{2}{3}, \frac{1}{3}, \frac{1}{3}$, and $\frac{2}{3}$ respectively. R is thus predicted to be $\frac{10}{3} = 3.33$. The experimental value is around 3.6. Note that the inclusion of the color factor of 3 is quite necessary for good agreement. The remaining difference is almost entirely accounted for by higher order QCD corrections to the quark-pair cross section (28).

If the total energy W is very close to the mass M of a $J^{PC} = 1^{--}$ bound $q\bar{q}$ state, such as a vector meson V, then the total cross section acquires additional resonant contributions given (29) by

$$\sigma_i = \frac{3\pi}{M^2}\frac{\Gamma_{ee}\Gamma_i}{(W-M)^2+\Gamma_{\text{tot}}^2},\qquad 2.$$

where σ_i is the cross section to channel i, Γ_{ee} and Γ_i are respectively the partial widths for V decay to e^+e^- and to channel i, and Γ_{tot} is the total decay width of the meson V. Integrating this additional contribution and summing over all channels, one obtains a relation between Γ_{ee} and the total resonant cross section:

$$\int \sigma_{\text{tot,res}}\,dW = \frac{6\pi^2}{M^2}\Gamma_{ee}.\qquad 3.$$

We wish, however, to express the leptonic width in terms of the resonant hadronic cross section only. Defining $B_{\mu\mu} = \Gamma_{\mu\mu}/\Gamma_{\text{tot}}$ and assuming lepton universality, i.e. $B_{ee} = B_{\mu\mu} = B_{\tau\tau}$, we can write

$$\Gamma_{tot} = \Gamma_{had} + 3B_{\mu\mu}\Gamma_{tot} = \frac{\Gamma_{had}}{(1-3B_{\mu\mu})}$$

and thereby obtain

$$\Gamma_{ee} = \frac{M^2}{6\pi^2} \frac{\int \sigma_{had,res}\, dW}{1-3B_{\mu\mu}}. \qquad 4.$$

From measurements of $\sigma_{had}(W)$ and $B_{\mu\mu}$ we can therefore obtain Γ_{ee} and $\Gamma_{tot} = \Gamma_{ee}/B_{\mu\mu}$.

3.2 Measurements of R, Γ_{ee}, and Resonance Masses

Two types of detectors are used to study the properties of the upsilons: (a) magnetic spectrometers such as PLUTO (3), DASP-II (4), and ARGUS at DORIS and CLEO at CESR (30); and (b) calorimetric spectrometers such as LENA (15) and the Crystal Ball (see the article by E. D. Bloom and C. W. Peck in this volume) at DORIS and CUSB at CESR (30). Magnetic detectors have tracking in magnetic field, which provides momentum analysis. Calorimetric detectors emphasize precise measurements of electromagnetic energy (e's or γ's) by the use of segmented arrays of active converters such as NaI or lead glass. Tracking is provided with various degrees of accuracy before the electromagnetic calorimeter. These two classes of detectors are complementary in their main aims but can both measure many of the quantities of interest.

An extensive mapping of R versus energy has been obtained in the ϒ region. Figure 4 shows the measurements of R obtained with the CUSB detector. The reported values of R are summarized in Table 1. We recall

Table 1 Measurements of R in the ϒ region

Energy (MeV)	R^a	Experiment	
9.4	3.67±0.23±0.29	PLUTO	(31)
9.5	3.73±0.16±0.28	DASP-II	(32)
7.4–9.4	3.37±0.06±0.28	LENA	(33)
10.4–10.5	3.63±0.06±0.37	CUSB	(13, 34)
10.4–10.5	3.77±0.06±0.26	CLEO	(35)
ΔR across flavor threshold			
	0.36±0.09±0.03	CUSB	(13, 34)
	0.34±0.09±0.11	CLEO	(35)

[a] The first error is statistical and the second is an estimate of the systematic uncertainty.

that above the b-flavor threshold R is expected to be higher by $3(\frac{1}{3})^2 = 0.33$ for $q_b = \frac{1}{3}e$, in good agreement with observation.

At e^+e^- colliders the masses of the vector mesons are determined by the total energy W of the colliding beams at which the resonances are observed. The energy of the stored beams is usually estimated from measurements of the magnets providing the guide field. In this way DORIS obtained for the Υ a mass of 9460 MeV with an estimated uncertainty of 10 MeV (3–5). At CESR a value of 9433 MeV was obtained, with an uncertainty of about 30 MeV (6, 7). Recently at VEPP-4 the $g-2$ depolarizing resonances have been observed at energies close to the Υ mass; this improves better than tenfold the accuracy with which the Υ mass is determined (36). Their result, $M(\Upsilon) = 9459.7 \pm 0.6$ MeV, in close agreement with the original DORIS value, is therefore used to rescale all the CESR mass measurements and is the only entry for the upsilon mass in the table.

The determination of $B_{\mu\mu}$ is a much harder task, requiring precise knowledge of σ_{had} at the Υ peak and good measurements of the μ-pair yield on and off peak. The first good signal for $\Upsilon \to \mu^+\mu^-$ was obtained by LENA in 1980. CLEO has measured $B_{\mu\mu}$ directly and by comparing the two-pion decay of the Υ' in inclusive and exclusive channels. There is a strong tendency for $B_{\mu\mu}$ to decrease with time. Only these last three values are reported here.

In Table 2 we give the mass, leptonic width, and $B_{\mu\mu}$ for the four resonances. The latest determinations of these parameters are given first, followed by the average values. Systematic uncertainties are not given. They typically are 2 to 5 MeV for masses and $\sim 10\%$ for Γ_{ee} and $B_{\mu\mu}$. Since potential models compute excitation energies and give more accurate predictions for ratios of leptonic widths, these are also given. The fourth upsilon has, at CESR, a width that is clearly wider than the machine energy spread but its shape is not known because of the lack of a detailed scan of the Υ''' peak. This fact is reflected in the difference between the values reported by CLEO and CUSB (34, 35) for mass, Γ_{ee} and Γ_{tot}.

3.3 Limits on the Production of Other Resonances

The existence of other narrow resonances, besides the three upsilons, in $\sigma(ee \to hadrons)$ has been suggested in various contexts (see Section 3.4.4). LENA at DORIS has established that in the energy intervals 7.4–7.5 GeV and 8.6–9.4 GeV there are no resonances with leptonic widths larger than 233 eV (41). CUSB (13, 34) and CLEO (35) at CESR have established the following limits:

$10.27 < W < 10.34$ GeV, $\Gamma_{ee} < 20$ eV (CUSB) and $\Gamma_{ee} < 20$ eV (CLEO)
$10.36 < W < 10.55$ GeV, $\Gamma_{ee} < 20$ eV (CUSB) and $\Gamma_{ee} < 40$ eV (CLEO).

Table 2 Measured parameters of the ϒ resonances

Parameter	ϒ	ϒ'	ϒ''	ϒ'''	Experiment	Reference
M (MeV)	9459.7±0.6				VEPP-4	(36)
		10016.8±1.5			DASP-II	(37)
		10013.6±1.2			LENA	(38)
		10021.2±0.7	10352.1±0.7	10575.0±1.1	CLEO	(35)
		10023.4±0.7	10350.0±0.7	10578.0±3.0	CUSB	(34)
Γ_{ee} (keV)	1.35±0.11	0.61±0.11			DASP-II	(37)
	1.13±0.09	0.53±0.07			LENA	(38)
	1.14±0.05	0.50±0.03	0.35±0.03	0.21±0.05	CUSB[a]	(34)
	1.30±0.05	0.52±0.03	0.42±0.04	0.32±0.03	CLEO	(35)
$B_{\mu\mu}$ (%)	3.8±1.5				LENA	(39)
	3.5±0.8				CLEO	(18)
	2.7±0.3	1.9±1.3	3.3±1.3		CLEO	(40)
Averages						
M (MeV)	9459.7±0.6	10020.5±0.7	10350.0±0.7	10576.0±3.0		
$M - M_\Upsilon$ (MeV)		560.8±0.4	890.3±0.4	1116.5±3.0		
Γ_{ee} (keV)	1.22±0.03	0.52±0.02	0.38±0.02	0.29±0.03		
$\Gamma_{ee}/\Gamma_{ee}(\Upsilon)$		0.42±0.02	0.31±0.02	0.24±0.03		
$B_{\mu\mu}$ (%)	2.83±0.28	1.9±1.3	3.3±1.3	$\sim 10^{-5}$		

[a] From a reanalysis of the data in Reference 34 (P. M. Tuts, paper in preparation).

Higher excitations of the ^{3}S b$\bar{\text{b}}$ states are also expected, with the 5^3S and 6^3S states a few hundred MeV above the Υ''' and with natural widths most likely in the 100-MeV range. Searches at CESR did not have enough sensitivity for the detection of such states. A scan in the energy region 10.6 < W < 11.6 GeV excludes the existence of resonances with natural widths of 20 MeV and Γ_{ee} > 0.3 keV (CLEO) and 40 MeV with Γ_{ee} > 0.2 keV (CUSB).

3.4 Comparison of Measurements and Theoretical Models

3.4.1 POTENTIAL MODELS In the potential model description of heavy q$\bar{\text{q}}$ bound states, the Schroedinger equation is solved using an effective central potential $V(r)$. The energy eigenvalues E_n give the excitation energies and, if the wave function can be calculated reliably at the origin, one can obtain the leptonic width Γ_{ee} from the Van Royen & Weisskopf (42) formula:

$$\Gamma_{ee} = 16\pi\alpha^2 q_q^2 \frac{|\Psi(0)|^2}{M_n^2}, \qquad 5.$$

where α is the fine structure constant, q_q is the quark charge, and M_n is the mass. It is more often the case that Equation 5 is solved the other way around to obtain a constraint on the wave function. Equation 5 is valid to zeroth order in quantum chromodynamics, QCD, and in addition one should include relativistic corrections because even for the lowest Υ state $\langle v^2/c^2 \rangle \cong 0.08$. It is reasonable to assume that most of these corrections cancel whenever ratios of widths are considered.

In the last few years, over a dozen potentials have been proposed to simultaneously describe c$\bar{\text{c}}$ and b$\bar{\text{b}}$ system, with considerable success, thus satisfying and proving a fundamental requirement of QCD, namely that the interquark forces be flavor independent. These potentials are virtually indistinguishable at distances of 0.2 to 1 fermi, the region probed by the ψ and the Υ families. For a comparison with data we have chosen four models, each representative of a particular approach in the choice of the functional dependence of V on r and the determination of its parameters. Eichten et al (43) used a linear combination of a confining potential and a Coulomb term, $V(r) = -K/r + r/a^2$, where a and K are determined by the spacing of the first two levels and the quark masses are free parameters. Krasemann & Ono (44) incorporated "asymptotic freedom" in the Bhanot & Rudaz (45) version of the above potential. Defining $\alpha_s = g^2/4\pi$, where g is the strength of the fundamental QCD quark-gluon vertex, asymptotic freedom refers to the fact that α_s vanishes logarithmically for $q^2 \to \infty$. Büchmuller, Grundberg & Tye (46) include higher order QCD corrections in a modified version of the Richardson potential (47). Martin (48) assumes a simple power law potential of the form $V(r) = A + Br^\alpha$. Data from Υ's, ψ's, and ϕ's are used in a global fit to determine the parameters A, B, and α as well as the quark masses m_b, m_c, and m_s. Similar results were obtained by Quigg & Rosner (49), who used $V(r) = \lambda r^v$ as well as the inverse scattering method (50).

3.4.2 QCD SUM RULES This approach to the problem of bound states of heavy quarks does not require a potential but derives the "current" quark mass and the net effect of long-range gluon field fluctuation, as measured by the vacuum expectation value of the gluon field squared $\langle G^2 \rangle$ from experiments. Using these two parameters the masses of other ground states and the leptonic width of the 1^3S state can be calculated (51).

3.4.3 COMPARISON OF CALCULATIONS AND MEASUREMENTS Table 3 lists the calculated quantities for the five examples chosen, and the measured values, for the first four ^3S states, labelled in the table as nS.

The agreement between data and calculations is excellent for the first two spacings, which confirms the identification of the Υ' and Υ'' as the 2^3S_1 and 3^3S_1 $b\bar{b}$ states. It is much harder to obtain accurate values for the excitation energy of states above threshold; therefore even an approximate agreement between experiment and calculations for the Υ''' confirms the identification of this last state as the 4^3S_1 $b\bar{b}$ radial excitation. Quigg & Rosner (49) have predicted on general grounds, in the context of potential models, that there should be just three bound $b\bar{b}$ states below the flavor threshold.

3.4.4 OTHER NARROW STATES The ^3D states are predicted by potential models to lie some 150 MeV below the corresponding ^3S states. While ^3D states have $J^{PC} = 1^{--}$ as the photon, they cannot be directly produced in e^+e^- annihilations since their wave function vanishes at the origin. They could, however, be observed if there is significant mixing with a nearby S state. In the $b\bar{b}$ case the second D state could be produced via mixing with the Υ'''. This mixing is, however, expected to be very small (52).

In theories of quark confinement, excitations of the color-flux tube are expected in addition to the potential model levels. The lowest of these modes, referred to as vibrational states, with $J^{PC} = 1^{--}$ has been predicted to exist with excitation energies between 910 and 1000 MeV and a leptonic width between 70 and 270 eV (53). The results of searches at CESR (given in Section 3.3) appear to exclude the existence of these states for the predicted width values, as well as any unexpected production of D states.

3.4.5 THE CHARGE OF THE b QUARK Combining results from potential models with Equation 5, one can in principle obtain the charge of the bound quarks. Quigg & Rosner (49) have given lower bounds for the leptonic widths of the Υ's in terms of those of the ψ's and the quark charges. Comparison of these bounds with the leptonic width of the Υ' establishes

Table 3 Comparison of prediction and measurements for the Υ's

Quantity	Eichten et al (43)	Krasemann & Ono (44)	Büchmuller, Grunberg & Tye (46)	Martin (48)	Voloshin (51)	Experiment
$M(2S) - M(1S)$ (MeV)	898	862	890	565	—	561
$M(3S) - M(1S)$ (MeV)	1170	1108	1180	900	—	890
$M(4S) - M(1S)$ (MeV)			1142	1142	—	1116
$\Gamma_{ee}(1S)$ (keV)		1.05	1.07		1.15 ± 0.20	1.22 ± 0.03
$\Gamma_{ee}(2S)/\Gamma_{ee}(1S)$	0.39	0.43	0.44	0.41	—	0.42 ± 0.02
$\Gamma_{ee}(3S)/\Gamma_{ee}(1S)$	0.27	0.31	0.32	0.35	—	0.31 ± 0.02
$\Gamma_{ee}(4S)/\Gamma_{ee}(1S)$	0.22	0.25	0.26	0.27	—	0.24 ± 0.03

that the b quark has charge $|q_b| < \frac{2}{3}e$. (From the leptonic width of the Υ, one obtains a less stringent bound.) Thus the charge of the b quark is $\frac{1}{3}e$, in agreement with the observed change in R.

4. TRANSITIONS AND DECAYS OF BOUND UPSILONS

4.1 *Decay of the Ground-State Υ*

One of the outstanding properties of the first three Υ's is their width of tens of keV, while their mass is in the 10-GeV range. This fact was first explained in the framework of QCD by Appelquist & Politzer (54) for the J/ψ meson. The decay of a heavy vector meson into hadrons is supposed to proceed mostly via annihilation of the heavy quark antiquark pair into three gluons, in analogy to the decay of triplet positronium into three photons, followed by evolution of the gluons into hadrons with unit probability. Properly accounting for the fact that the three gluons must be in a color singlet state, one obtains (54)

$$\Gamma_{\text{ggg}} = \frac{160}{81} \alpha_s^3 (\pi^2 - 9) \frac{|\Psi(0)|^2}{M^2}, \qquad 6.$$

which, using current values of α_s and the value of $|\Psi(0)|^2/M^2$ derived from Γ_{ee}, gives $\Gamma_{\text{ggg}} \approx 30$ keV in agreement with observation (see Section 4.4).

A direct proof that vector mesons decay into three gluons has been obtained for the first time by observing the three-jet structure in Υ decay. The three-jet structure has been observed at DORIS and CESR. While many global parameters may be employed to characterize the structure of a many-body final state, the one used in various forms by all groups and most amenable to calculations in QCD is the quantity called thrust, defined as

$$T = \text{Max} \left(\sum_i \frac{|\mathbf{p}_i \cdot \mathbf{n}|}{|\mathbf{p}_i|} \right),$$

where $\mathbf{p}_i$ are the particle momenta (energy clusters in a calorimeter) and the unit vector $\mathbf{n}$ is rotated until a maximum is found. For a collinear two-jet final state, $\mathbf{n}$ is the jet axis; for a three-jet event, $\mathbf{n}$ is close to the axis of the most energetic jet. The thrust distribution dN/dT can be calculated for two-quark and three-gluon (55) final states.

The thrust axis angular distribution, for three gluons from vector mesons also depends on the gluon spin. The angular distribution for the polar angle of the unit vector $\mathbf{n}$ is given by $1 + \rho \cos^2 \theta$, where $\rho = 1.0$ for two-quark jets (hadronic events from continuum e^+e^- annihilations), $\rho = 0.39$ for Υ decays into three spin-1 gluons, and $\rho = -1.0$ for decays into three spin-0 gluons (56). The thrust distributions obtained by PLUTO (14), LENA (15), CLEO (16), and CUSB (10) change significantly between continuum ($q\bar{q}$) and Υ (mostly three gluons), and the data agree well with predictions for three-gluon decays. The parameter ρ has been measured at DORIS (41) and by CLEO (16), who reports the most accurate value of 0.32 ± 0.11. Averaging with the other results gives $\rho = 0.32 \pm 0.09$, in good agreement with the value for decay to three spin-1 gluons.

4.2 Decays of the Υ' and the Υ''

Additional decay channels are available to the first two excited Υ's, namely transitions to lower-lying $b\bar{b}$ states. Because of the smallness of the annihilation width, processes with very low Q-value and/or due to electromagnetic interactions can compete favorably with annihilation into three gluons. This is well known for the ψ', whose total width is about three times larger than that of the J/ψ. Half of this width is due to the transition $\psi' \to \psi \pi \pi$ (57) and about 26% to electric dipole (E1) transitions to P-wave $c\bar{c}$ states (58). In the Υ family, these transitions are suppressed with respect to the charm case. For E1 transitions, one expects a suppression of a factor 4 because of the b-quark charge and of a factor ~ 3 from the smaller size of the $b\bar{b}$ system in the approximation that the E1 matrix element $|\langle i|r|f \rangle|^2$ scales like $\langle r^2 \rangle$. A similar argument had been advanced by Gottfried (59) for the pion transitions between upsilons, which can be understood in a multipole expansion of the color-electric field. In particular, for spin-1 gluons the lowest contribution to the emission of two pions is a double "E1" transition, thus being suppressed for the upsilon case by a factor of $\langle r^2 \rangle^2$ or ~ 10.

4.2.1 TWO-PION TRANSITIONS BETWEEN Υ'S   In the upsilon family there are three two-pion transitions between ^3S states: (a) $\Upsilon' \to \Upsilon \pi \pi$, (b) $\Upsilon'' \to \Upsilon \pi \pi$, and (c) $\Upsilon'' \to \Upsilon' \pi \pi$. Reaction (a) is the easiest to observe and was detected in 1980 both at DORIS and at CESR. In calorimetric detectors these reactions can only be observed in the decay cascade chain $\Upsilon^n \to \Upsilon^{n-1} \pi \pi$, $\Upsilon^{n-1} \to e^+e^-$ or $\mu^+\mu^-$. For the case of reaction (a), the LENA group (19) found 5 $\pi^+\pi^-e^+e^-$ and 2 $\pi^+\pi^-\mu^+\mu^-$ events; the CUSB group (18) found 23 $\pi^+\pi^-e^+e^-$ events in the decay of about 10,000 Υ', and CLEO (18) found 17 $\pi^+\pi^-e^+e^-$ events. These three results combined give $B_{\pi^+\pi^-\ell\bar{\ell}} = 0.0066 \pm 0.0010$. CLEO (18) also detects the two-pion decay of the Υ' to Υ by computing the mass recoiling against any pair of opposite-sign pions. They observe a clear peak at the Υ mass (Figure 5) above a large combinatorial background and obtain $B(\Upsilon' \to \Upsilon \pi^+ \pi^-) = 0.191 \pm 0.031$. Combining all results and assuming that $\Gamma(\Upsilon' \to \Upsilon \pi^+ \pi^-) = 2\Gamma(\Upsilon' \to \Upsilon \pi^0 \pi^0)$, one finally obtains $B_{\Upsilon \pi \pi}(\Upsilon') = 0.29 \pm 0.04$.

The two-pion decays of the Υ'' have much smaller widths; the reaction (b) is suppressed because $\Delta n = 2$, whereas the decay to the Υ' is suppressed by phase space because the Q value for the transition is only 47 MeV. Results for the decay to the ground state have been obtained by both CLEO (26) and CUSB (27). A few decays to the Υ' have been observed by CUSB. These results were obtained from the analysis of a sample of about 50,000 Υ''. CLEO (26) also observes the decay to the ground state in the recoiling mass spectrum. In addition, CUSB (27) observed emission of two π^0's by detecting four photons and the rates were consistent with being one half of those for emission of $\pi^+\pi^-$ as expected from isospin invariance. The results reported by the two groups are $B(\Upsilon'' \to \pi^+\pi^-\Upsilon) = 0.049 \pm 0.009$ (CLEO), $B(\Upsilon'' \to \pi^+\pi^-\Upsilon) = 0.039 \pm 0.012$ (CUSB), and $B(\Upsilon'' \to \pi^+\pi^-\Upsilon') = 0.031$

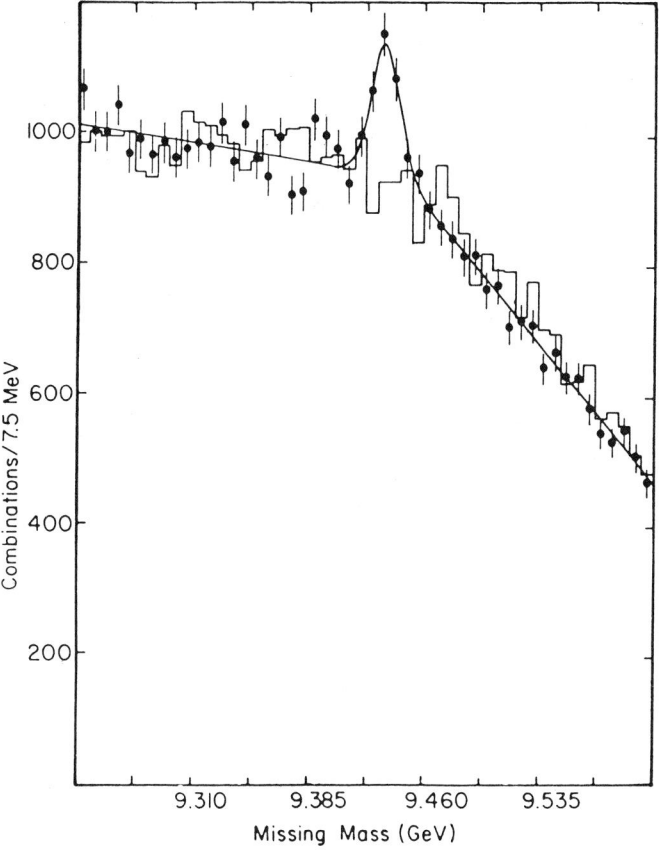

Figure 5 Spectrum of the mass recoiling against any $\pi^+\pi^-$ pair in Υ' decays (*data points*) and against same-sign π pairs (*solid line histogram*) (18).

±0.020 (CUSB). Again, combining all the results and using isospin invariance, we obtain $B_{\Upsilon\pi\pi}(\Upsilon'') = 0.069 \pm 0.011$ and $B_{\Upsilon'\pi\pi}(\Upsilon'') = 0.046 \pm 0.030$. CLEO has also searched in the recoiling mass spectrum for evidence of two-pion transitions to singlet P-wave $b\bar{b}$ states. They do not observe any signal and establish an upper limit of 3% for this decay (60).

4.2.2 PHOTON TRANSITIONS TO P-WAVE $b\bar{b}$ STATES The study of these transitions in the $b\bar{b}$ system is still in its infancy, partly owing to the low production rates of upsilon resonances thus far. While two million ψ' have been studied in one single experiment at SPEAR (58), the largest equivalent upsilon sample, up to 1982, consisted of about 37,000 Υ''.

Until 1981 the prospects for the search for a signal from E1 transitions were rather dim. Predictions for the branching ratios were in the 5 to 20% range, while the total accumulated samples of Υ' and Υ'' were only 10,000 and 7,000 respectively. A study of the Υ' and Υ'' showed an excess of two-jet event topologies in their decays (10). An explanation of this fact is that a fraction of the Υ's decay via E1 transitions to ^3P_J. These states have $J^{PC} = 2^{++}, 1^{++}$, and 0^{++} and therefore are expected to decay mostly by annihilation of the $b\bar{b}$ pair into two gluons, which leads to a two-jet final state. The experimental observation leads to an estimate of the branching ratio for E1 transition of the order of 10% for the Υ' and 30% for the Υ''. That this branching ratio should be larger for Υ'' is reasonable because most other decay channels are suppressed while the E1 rate is enhanced by the larger value of $\langle r^2 \rangle$ of the Υ''.

Encouraged by these expectations, CESR ran in 1981–1982 with increased luminosity for three months at the Υ'' peak. Some 65,000 hadronic events were observed by CUSB for a total time-integrated luminosity of 14 pb^{-1}, of which about 37,000 were Υ'' decays and the remainder were continuum events. The inclusive photon spectrum from these events shows a clear enhancement centered around 100 MeV (11). This is shown in Figure 6a and, after background subtraction, the net excess signal is shown in Figure 6b. This signal is attributed to the decay $\Upsilon'' \to 2^3P_J + \gamma$, where 2 stands for the second P-wave state. The ^3P states are expected to show fine structure, with splittings predicted to be in the 15–40-MeV range. The fine structure is not resolved in the spectrum of Figure 6b; however, the observed enhancement is significantly broader than the resolution for photons around 100 MeV, which in CUSB is ~ 17 MeV full width at half maximum (FWHM). The observed spectrum is therefore fitted to three lines with shapes given by the resolution function and with free position and intensity. This fit is shown in Figure 6b.

An alternative way to search for E1 transition to P-wave states is to observe the decay chains:

1. $\Upsilon'' \to 2^3P + \gamma_1 \to \Upsilon' + \gamma_2 + \gamma_1 \to \gamma_1 + \gamma_2 + \mu^+\mu^-$ or e^+e^-
2. $\Upsilon'' \to 2^3P + \gamma_1 \to \Upsilon + \gamma_2 + \gamma_1 \to \gamma_1 + \gamma_2 + \mu^+\mu^-$ or e^+e^-
3. $\Upsilon'' \to 1^3P + \gamma_1 \to \Upsilon + \gamma_2 + \gamma_1 \to \gamma_1 + \gamma_2 + \mu^+\mu^-$ or e^+e^-.

The number of events that one can find in this way is very small since it is proportional to the product of three small branching ratios and the fractional solid angle enters at least cubed in the acceptance. However, the background is almost nonexistent and the events can be fitted to satisfy kinematical constraint; this results in better accuracy for the photon energies.

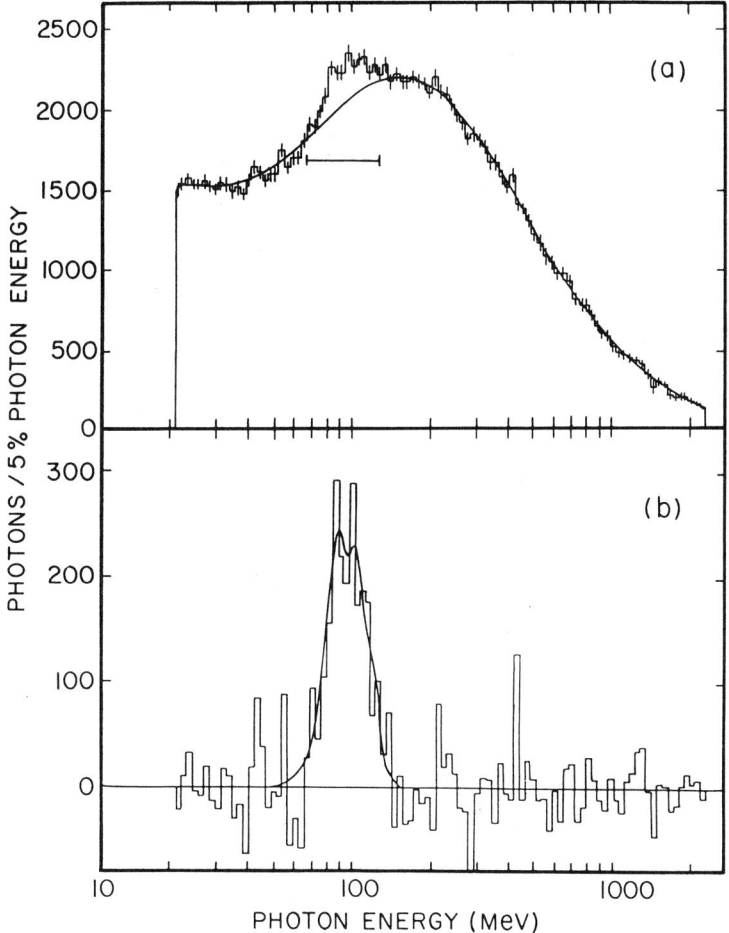

Figure 6 (a) the observed inclusive photon spectrum from Υ'' decays, the solid line is a fit to the background; (b) the background-subtracted photon signal in Υ'' decays.

Fourteen examples of Reaction 1 and 15 examples of Reaction 2 were observed (12) during the same run described above, in the CUSB detector, with an estimated background of 1.7 and 1.3 events, respectively. One candidate is observed for Reaction 3. The lower energy photons for both Reactions 1 and 2 cluster around 100 MeV, which thus establishes that the 2^3P level lies about 100 MeV below the Υ'', and that the signal observed in the inclusive photon spectrum from Υ'' decays is due to the direct transition to the 2^3P state. The energies of the photons, as extracted from fitting both the inclusive and exclusive spectra, are 84.4 ± 2.0, 99.5 ± 3.2, 117.2 ± 5.0 MeV, and 84.0 ± 3.0, 99.0 ± 2.0, 119.0 ± 5.0 MeV, respectively. The relative intensity of each of the three lines in the inclusive spectrum is consistent with being proportional to $k^3(2j+1)$, as expected for E1 transitions, where k is the photon energy. The overall picture presented by these results is quite a convincing proof that the first excited $b\bar{b}$ P-wave states have indeed been observed.

It is usual to characterize triplet P-wave states by their center of gravity ("cog") and their fine structure. Using the fitted photon energies above, one obtains $\cos(2^3P) \simeq M(\Upsilon'') - 93$ MeV $= 10256$ MeV including a correction of ~ 0.5 MeV for recoil. The fine structure splittings are 15 and 18 MeV, with errors of the order of 3 to 7 MeV accounting for systematic uncertainties in addition to the statistical errors. The ratio of these splittings,

$$r = \frac{M(J=2) - M(J=1)}{M(J=1) - M(J=0)},$$

is often used to compare calculations and experiments. The data give $r(2P) = 0.83 \pm 0.3$. Finally, from the observed enhancement in the inclusive photon spectrum, the CUSB group obtains a branching ratio for $3^3S \to 2^3P$ E1 transitions of 0.34 ± 0.03, with an estimated additional systematic uncertainty of 0.03. From the observed numbers of two-photon events, CUSB also obtains the following sums of products of branching ratios for decays of the Υ'' to the Υ' and Υ, via the 2^3P_J states (12):

$$\sum_i B(\Upsilon'' \to \gamma\chi'_{bi})B(\chi'_{bi} \to \gamma\Upsilon') = 0.059 \pm 0.021,$$

$$\sum_i B(\Upsilon'' \to \gamma\chi'_{bi})B(\chi'_{bi} \to \gamma\Upsilon) = 0.036 \pm 0.012;$$

for transitions via the 1^3P_J to the Υ an upper limit is given (12):

$$\sum_i B(\Upsilon'' \to \gamma\chi_{bi})B(\chi_{bi} \to \gamma\Upsilon) < 0.03 \ (90\% \ \text{C.L.}).$$

Recently, transitions from the Υ' to the first P-wave $b\bar{b}$ level have also been observed by CUSB at CESR, both in the inclusive photon spectrum and in exclusive channels. The cog of the 1^3P_J has a mass of ~ 9901 MeV, the fine structure splittings are 20 and 21 MeV corresponding to a value of $r(1P) \approx 0.9$, and the branching ratio for the decay $\Upsilon' \to 1^3P_J + \gamma$ is $\sim 0.15 \pm 0.05$ (61).

4.3 Partial Rates for Υ's Decays

As described in Section 3.1, a knowledge of $B_{\mu\mu}$ is necessary to obtain Γ_{ee}. The value of $B_{\mu\mu}$ is known at present with reasonable accuracy only for the Υ. For the other two narrow Υ's it is possible, however, to obtain Γ_{tot} indirectly, and therefore $B_{\mu\mu}$, by using the assumption that $\Gamma_{annihilation}/\Gamma_{ee}$ is the same for all three states, together with measurements of the branching ratios for decays without annihilation of the $b\bar{b}$ pair. In addition to the annihilation of the $b\bar{b}$ pair into three gluons, one must also include annihilation into a virtual photon, leading to lepton pairs and quark pairs. For three lepton flavors and by definition of R, the partial width for decays via a virtual photon is given by $(3+R)\Gamma_{ee}$. We can therefore write $\Gamma_{tot} = \Gamma_{ggg} + (3+R)\Gamma_{ee} + \Gamma_{other}$, where Γ_{other} is the decay width for all channels not requiring annihilation (Γ_{ggg} as defined here includes a $\sim 4\%$ (62) contribution from $\Upsilon \to \gamma gg$). We thus obtain

$$\Gamma_{tot}^n = \frac{\Gamma_{ee}^n}{B_{\mu\mu}(\Upsilon)(1 - B_{other}^n)},$$

where n refers to the nth Υ, and B_{other} is the branching ratio for all decays without annihilation. Using the values given in Sections 4.2.1 and 4.2.2, we derive $B_{other}(\Upsilon') = 0.49 \pm 0.06$ and $B_{other}(\Upsilon'') = 0.46 \pm 0.04$. [Searches for other decay modes of the Υ' and Υ'' have given limits of $< 3\%$ (41, 60) for their branching ratios.] The results in Table 4 are derived using the outlined procedure. The directly measured quantities are given first, followed by the derived ones.

4.4 Comparison with Theory

4.4.1 HADRONIC TRANSITIONS The two-pion transitions between the Υ's not only confirm the assumptions that these states are composed of the same new quark but also offer insight into the mechanism for such decays, especially in comparison with the ψ' case. The partial width for the decay $\psi' \to \psi\pi\pi$ is ~ 107 keV (57), while in the Υ' case the width for the $\pi\pi$ transition is only ~ 10 keV, although the Q value for the two decays is very similar. That this should be the case had already been surmised from the fact that production of the Υ' is quite obvious in proton collisions on nuclei, while ψ' production is hardly observed in similar experiments; this implies that the

Table 4 Measured and derived rates and branching ratios for the bound ϒ's

Parameter	ϒ	ϒ'	ϒ''
Inputs			
Γ_{ee} (keV)	1.22 ± 0.03	0.52 ± 0.02	0.38 ± 0.02
$B_{\mu\mu}$ (%)	2.83 ± 0.28		
$B_{\pi\pi}$		0.29 ± 0.04	0.115 ± 0.033
B_{E1}		0.15 ± 0.05	0.34 ± 0.03
Derived			
Γ_{tot} (keV)	43.1 ± 4.3	32.8 ± 4.6	24.6 ± 3.4
Γ_{ggg} (keV)	35.0 ± 4.3	14.9 ± 1.8	10.9 ± 1.3
$\Gamma_{\pi\pi}$ (keV)		9.5 ± 1.4	2.8 ± 0.9
Γ_{E1} (keV)		4.9 ± 1.8	8.4 ± 1.4
$B_{\mu\mu}$ (%)		1.58 ± 0.28	1.54 ± 0.22

$\pi\pi$ feed-down mechanism should be suppressed for the ϒ's (59). Following an original suggestion of Gottfried (59), Yan (63), by a multipole expansion in the "color electric" field, has computed that the rate for $\Upsilon' \to \Upsilon\pi\pi$ should be $\sim \frac{1}{16}$ the rate for $\psi' \to \psi\pi\pi$, in good agreement with observation. For spinless gluons the $\pi\pi$ decay width should be proportional to the color charge and therefore identical for ϒ' and ψ'. The strong suppression of the $\pi\pi$ transition from ϒ' to ϒ is a particularly beautiful proof that the gluons carry spin 1. Kuang & Yan (64) have also estimated the widths for the $\pi\pi$ transitions of the ϒ'', which, within the present limited statistical accuracy, are consistent with observation. Finally, predictions have been made for the $\pi\pi$ mass spectrum for these transitions. Both CLEO (26) and CUSB (27) results for the ϒ' are in agreement with the simplest expected shape described by Brown & Cahn (65) and by Yan (63). In the decay $\Upsilon'' \to \Upsilon\pi\pi$ the enhancement of high mass $\pi\pi$ pairs observed for the ϒ' does not appear to be present, the spectrum being essentially consistent with phase space.

4.4.2 P-WAVE STATES AND PHOTON TRANSITIONS The discovery of the P-wave b$\bar{\text{b}}$ states brings a wealth of new information to be compared with predictions of potential models as well as with results of QCD sum rules and bag model calculations. We can compare three quantities: the center of gravity (cog) of the level masses, the E1 transition rates and the fine structure splitting. The cog's are perhaps easier to predict, because of our good knowledge of the binding potential. The transition rates and splittings are sensitive to relativistic effects [they have not yet been correctly predicted for the ψ' case (58)] and there is no satisfactory theory of the effective spin dependence of the forces in quarkonium. While all these problems are expected to be less severe for the b$\bar{\text{b}}$ case, the predictions for the ratio r for the χ'_b range from 0.48 to 1 and the splittings themselves range from 14 to 24 MeV for the two upper levels and from 17 to 40 MeV for the two lower ones. The spread of the predictions is even larger for the χ_b case. Since r is poorly determined at present, we compare only measurements and calculations for masses and transition rates. Similarly, the products of branching ratios given in Section 4.2.2 are consistent with expectations (12, 64) although the experimental accuracy is very limited. Results of many authors are shown in Table 5 for the ^{3}P level masses and transition rates. Since the rate for E1 transitions is proportional to k^3, where k is the photon energy, we have rescaled all E1 widths using the experimental value $k_{\text{cog}}(1P) = 119$ MeV and $k_{\text{cog}}(2P) = 93$ MeV.

The agreement between measurements is in fact impressive for potential models, poor for the bag model calculations (70), and extremely poor for the QCD sum rules predictions of the lowest ^{3}P state mass (71, 72).

4.4.3 ^{3}P STATES' CENTER OF GRAVITY AND FLAVOR INDEPENDENCE Quigg & Rosner (49) have pointed out that ratios of level spacings can be directly related to the power v of a potential of the form λr^v. Thus from the ratio $[M(3S)-M(2S)]/[M(2S)-M(1S)]$ they derive $v(\psi) = 0.2$ and $v(\Upsilon) \approx 0.33$. Using the 2S, 1P, and 1S levels gives $v(\psi) \approx 0.15$. From the new information available about the $b\bar{b}$ P-wave states, one obtains $v(\Upsilon) \approx -0.22$ using the 2S, 1P, and 1S levels (61), and $v(\Upsilon) \approx -0.15$ using the 3S, 2P, and 2S levels (20). The fact that all these values are close to zero confirms the flavor independence of the interquark forces and suggests that the potential becomes more Coulomb-like at small distances.

Table 5 Comparison of predictions and measurements for ^{3}P levels

Author	1^3P cog (MeV)	2^3P cog (MeV)	$\Gamma(2^3S \to 1^3P)^a$ (keV)	$\Gamma(3^3S \to 2^3P)^a$ (keV)
Büchmuller, Grunberg & Tye (46)	9890	10250	4.2	6.1
Büchmuller (66)			4.1	6.8
Eichten et al (43, 64)	9924	10271	4.4	6.1
Martin (48)	9861	10242		
Quigg & Rosner (50)	9888	10244	4.3	7.2
Krasemann (44, 67)	9936	10271	3.1	4.8
Gupta et al (68)	9898	10256		
McClary & Byers (69)	9923	10267	4.4	7.6
Baacke et al (70)	9971	10312	5.2	5.2
Voloshin et al (71)	9835 ± 30			
Bertlmann (72)	9803 ± 10			
Experiment	9901 ± 5	10256 ± 5	4.9 ± 1.8	8.4 ± 1.4

[a] Adjusted using experimental cog's (see text).

4.5 Search for Rare Decays of the Υ

Current models of the electroweak interactions require the existence of at least one pseudoscalar Higgs particle (73). The Higgs particle couples to quarks proportionally to the square of the quark mass, which results (74) in the prediction $BR(\Upsilon \to \gamma + H) = 2.6 \times 10^{-4}(1 - M_H^2/M_\Upsilon^2)$. Models with two Higgs fields modify this result by a factor $(v_1/v_2)^2$ for charge-$\frac{1}{3}$ quarks (Υ) while the corresponding ratio for the ψ is multiplied by $(v_2/v_1)^2$, where v_1 and v_2 are the vacuum expectation values of the two Higgs fields (75).

The second model applies to the Weinberg (76) and Wilczek (77) axion, a, which is a light ($M < 1$ MeV), long-lived, and semiweakly interacting pseudoscalar particle, distinguished by its "invisibility." Since v_1/v_2 is not known, one cannot predict the branching ratios for either Υ or $\psi \to \gamma + a$. However, the product of the two branching ratios is predicted to be 1.6×10^{-8}. The CUSB group found no example of this decay in an equivalent sample of $\sim 60{,}000$ Υ decays, giving an upper limit of 1.2×10^{-4} for the branching ratio. Combining this result with the limit of Edwards et al (78) for $\psi \to \gamma + a$ of 1.4×10^{-5}, CUSB quote a limit of 0.6×10^{-9} for the product of the two branching ratios at 90% C.L., an unambiguous result arguing against the existence of the axion (21).

In the case of heavier axions the CUSB group has searched for the reaction $\Upsilon \to H + \gamma$, where the Higgs particle decays into hadrons. While these searches are not at present sensitive to the expected rate for the single Higgs model, they can exclude a large region in the $v_1/v_2, M_H$ plane, the upper boundary of which is defined by values of $(v_1/v_2, M_H)$ such as: (2,4), (3,6), (4,7), and (6,8), with M_H in GeV (J. Lee-Franzini, personal communication).

5. DECAY OF THE Υ'''

If the mass of the ^3S $b\bar{b}$ state is larger than twice the mass of the lightest bound state of a b quark with a light antiquark such as $\bar{u}$ or $\bar{d}$ (B mesons), the Υ''' can decay into $B\bar{B}$ pairs with a rate more typical of conventional strong interactions, possibly limited by the available phase space. That the Υ''' has a natural width of the order of 20 MeV [CLEO gives $\Gamma(\Upsilon''') = 32 \pm 7$ MeV (35) and CUSB reports 20 ± 3 MeV (34); see Section 3.2 for comments] suggests that the free b-flavor threshold has in fact been reached. Support for this assumption was obtained at CESR, from the observation of a sharp increase at the peak of the Υ''' in the yield of leptons of energies up to ~ 2.5 GeV as expected from the weak decays of a B meson of mass ~ 5 GeV (22, 23).

5.1 b-Flavor Threshold and $B^* \to B + \gamma$

Of general and practical interest is the knowledge of the B meson mass. An obvious bound is $M(\Upsilon'')/2 < M(B) < M(\Upsilon''')/2$ since the first state is below threshold and the Υ''' is above. However, if the mass of the B is close to the lower bound, the Υ''' would copiously decay into B*'s (24). From scaling arguments it follows that the B* is about 50 MeV heavier than the B meson, in which case its dominant decay is $B^* \to B + \gamma$ (43). Decays of the Υ''' into B* should be detectable by the presence of monochromatic photons around 50 MeV, in the decays of the Υ'''.

A search for these photons has been carried out by the CUSB group at CESR with negative results (24). Their negative result can be expressed as an upper limit for the branching ratio for $\Upsilon''' \to BB^*$, which they give as $BR < 0.09$, at 90% C.L., for a B*-B mass difference between 45 and 65 MeV. From this limit the same authors (24) establish bounds on the mass of the B meson: $5263 < M_B < 5278$ MeV.

6. OUTLOOK FOR THE STUDY OF SPIN-ORBIT AND SPIN-SPIN INTERACTIONS

6.1 Fine Structure of the Triplet P-Wave States

The fine structure of the 2^3P_J states is not at present resolved because their splittings of ~ 15 MeV are very close to the CUSB resolution at ~ 100 MeV. CUSB is in the process of adding an array of bismuth germanate crystals that will improve resolution by a factor of two. The Crystal Ball detector, at DORIS since mid 1982, has approximately 1.5 times better resolution than CUSB. CESR and DORIS have plans for increasing their luminosities, which at present are 550 and 250 nb^{-1} per day, respectively. Good measurement of the fine structure splittings and angular correlations will be performed within the coming year and will allow us to choose among the various ansatz proposed to describe the effective spin-orbit interaction in the $b\bar{b}$ system.

6.2 Search for Singlet $b\bar{b}$ States

The search for the singlet states, ^1P and ^1S or η_b, is a much harder proposition. We recall that the lowest ^1P state has not been found in charmonium (79) and that the search for the η_c's had a long and complicated history. The positions of these states are, however, crucial to the understanding of the spin-spin interaction, which is expected to be a short-range effect arising from hard gluon exchange. The singlet-triplet hyperfine splittings for P-wave states are expected to be extremely small

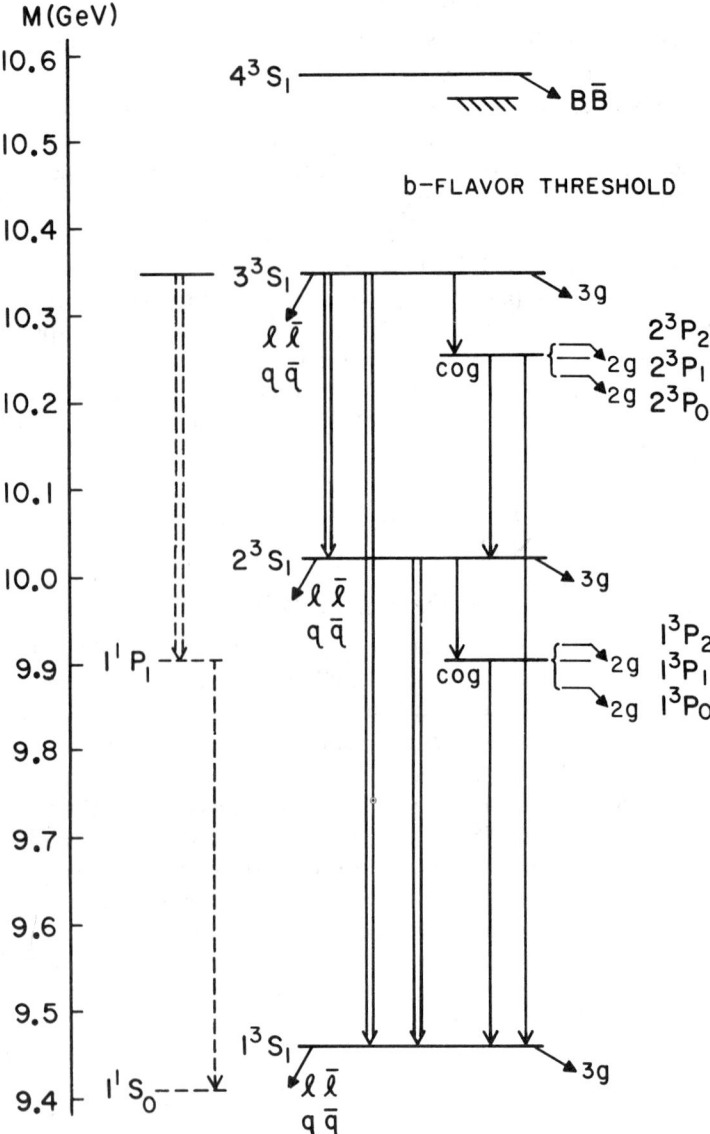

Figure 7 A level diagram of the $b\bar{b}$ bound system; double lines indicate $\pi\pi$ transitions and single lines indicate γ transitions. Solid lines indicate observed levels and transitions (see text).

since P waves vanish at the origin. The hyperfine splittings for S-wave states are also predicted to be small, resulting in branching ratios for the M1 transition $\Upsilon \to \eta_b + \gamma$ of $\sim 5 \times 10^{-4}$ (43, 67, 80). However, in the Υ system there appears to be an alternate path to reach both the ^{1}P and the η_b that is experimentally promising, as shown on the left in Figure 7. This is the decay chain $\Upsilon'' \to \pi\pi 1^1\text{P}$, $1^1\text{P} \to \eta_b + \gamma$. Kuang & Yan (64) estimate the first process to have a branching ratio of $\sim 2\%$, and potential models predict for the second transition, an E1 transition with a photon energy of ~ 470 MeV, branching fractions of 0.35 to 0.5. A modus operandi would therefore be to locate the 1^1P state via the two-pion transition, a task mostly for magnetic spectrometers, and to search for a monochromatic photon from the $1^1\text{P} \to \gamma + \eta_b$, a task for electromagnetic calorimeters. Very good energy resolution and high statistics are required because of the many nearby lines due to Υ's and χ_b's E1 decays. It is estimated that samples of 10^6 Υ'' and Υ' each would yield a definite signal for the η_b in the upgraded CUSB detector.

7 CONCLUSIONS

The present experimental status of upsilon spectroscopy is best summarized by the level diagram of Figure 7, where the observed levels and transitions are illustrated (*solid lines*). We have also reviewed the phenomenal success with which potential models are able to describe most features of the spectrum. We have indicated directions for future forays lying outside the central potential phenomenology. The study of the Υ's will continue to provide us with valuable information about properties of quarks, gluons, and perhaps Higgs particles.

Literature Cited

1. Herb, S. W., et al. 1977. *Phys. Rev. Lett.* 39:252
2. Innes, W. R., et al. 1977. *Phys. Rev. Lett.* 39:1240
3. Berger, Ch., et al. 1978. *Phys. Lett.* 76B:243
4. Dardeen, C. W., et al. 1978. *Phys. Lett.* 76B:246
5. Bienlein, J. K., et al. 1978. *Phys. Lett.* 78B:360
6. Andrews, D., et al. 1980. *Phys. Rev. Lett.* 44:1108
7. Böhringer, T., et al. 1980. *Phys. Rev. Lett.* 44:1111
8. Finocchiaro, G., et al. 1980. *Phys. Rev. Lett.* 45:222
9. Andrews, D., et al. 1980. *Phys. Rev. Lett.* 45:219
10. Peterson, D., et al. 1982. *Phys. Lett.* 114B:277
11. Han, K., et al. 1982. *Phys. Rev. Lett.* 49:1612
12. Eigen, G., et al. 1982. *Phys. Rev. Lett.* 49:1616
13. Rice, E., et al. 1982. *Phys. Rev. Lett.* 48:906
14. Berger, Ch., et al. 1979. *Phys. Lett.* 82B:449; also 1981. *Z. Phys.* C8:101
15. Niczyporuk, B., et al. 1981. *Z. Phys.* C9:1
16. Cabenda, R. C. 1982. PhD thesis, Cornell University (unpublished)
17. Mageras, G., et al. 1981. *Phys. Rev. Lett.* 46:1115
18. Mueller, J., et al. 1981. *Phys. Rev. Lett.* 46:1181
19. Niczyporuk, B., et al. 1981. *Phys. Lett.* 100B:95
20. Lee-Franzini, J. 1983. In *Proc. Summer Institute on Particle Physics*, ed. A.

Mosher, Stanford Univ. Press, Calif. In press
21. Sivertz, M., et al. 1982. *Phys. Rev.* D26:717
22. Bebeck, C., et al. 1981. *Phys. Rev. Lett.* 46:84; also Chadwick, K., et al. 1981. *Phys. Rev. Lett.* 46:88
23. Spencer, L. J., et al. 1981. *Phys. Rev. Lett.* 47:771
24. Schamberger, R. D., et al. 1982. *Phys. Rev.* D26:720
25. Lederman, L. M. 1979. In *Proc. 19th Int. Conf. on High Energy Physics, Tokyo, Japan*, ed. S. Homma, M. Kawaguchi, H. Miyazawa, p. 706. Tokyo: Phys. Soc. Jpn.
25a. Chinowsky, W. 1977. *Ann. Rev. Nucl. Sci.* 27:393
26. Green, J., et al. 1982. *Phys. Rev. Lett.* 49:617
27. Mageras, G., et al. 1982. *Phys. Lett.* 118B:453
28. Barnett, R., Dine, M., McLerran, L. 1980. *Phys. Rev.* D22:594
29. Blatt, J. M., Weisskopf, V. F. 1952. *Theoretical Nuclear Physics*, p. 423. New York: Wiley
30. Franzini, P., Lee-Franzini, J. 1982. *Phys. Rep.* 81:239–291; also Andrews, D., et al. 1982. *Cornell Rep. CLNS 82/538* (unpublished)
31. Berger, Ch., et al. 1979. *Phys. Lett.* 81B:410
32. Weseler, S. 1981. Diplomarbeit, University of Heidelberg (unpublished, in German).
33. Niczyporuk, S., et al. 1983. *Z. Phys.* C15: In press
34. Rice, E. 1983. PhD thesis, Columbia University (unpublished)
35. Plunket, R. 1983. PhD thesis, Cornell University (unpublished)
36. Artamonov, A. S., et al. 1982. *J. Phys.* 43:C3–789
37. Albrecht, H., et al. 1980. *Phys. Lett.* 93B:500
38. Niczyporuk, B., et al. 1981. *Phys. Lett.* 99B:169
39. Niczyporuk, B., et al. 1981. *Phys. Rev. Lett.* 46:92
40. Andrews, D., et al. 1983. Submitted to *Phys. Rev. Lett.*
41. Bienlein, J. K. 1981. In *Proc. 1981 Int. Symp. on Lepton and Photon Interactions at High Energies*, ed. W. Pfeil, p. 190. Universität Bonn, West Germany
42. Van Royen, R., Weisskopf, V. F. 1967. *Nuovo Cimento* 50A:617
43. Eichten, E., et al. 1980. *Phys. Rev.* D21:203
44. Krasemann, K. H., Ono, S. 1979. *Nucl. Phys.* B154:283
45. Bhanot, G., Rudaz, S. 1978. *Phys. Lett.* 78B:119
46. Büchmuller, W., Grunberg, G., Tye, S.-H. H. 1980. *Phys. Rev. Lett.* 45:103, 587 (E); also Büchmuller, W., Tye, S.-H. H. 1981. *Phys. Rev.* D24:132
47. Richardson, J. L. 1979. *Phys. Lett.* 82B:272
48. Martin, A. 1980. *Phys. Lett.* 93B:338; also 1981. *Phys. Lett.* 100B:511
49. Quigg, C., Rosner, J. L. 1979. *Phys. Rep.* 56:167
50. Quigg, C., Rosner, J. L. 1981. *Phys. Rev.* D23:2625; also Quigg, C., Thacker, H. B., Rosner, J. L. 1980. *Phys. Rev.* D21:234
51. Voloshin, M., Zakharov, V. I. 1980. *Phys. Rev. Lett.* 45:688
52. Eichten, E. 1980. *Phys. Rev.* D22:1819
53. Büchmuller, W., Tye, S.-H. H. 1980. *Phys. Rev. Lett.* 44:850
54. Appelquist, T., Politzer, H. D. 1975. *Phys. Rev. Lett.* 34:43
55. Koller, K., Walsh, T. F. 1978. *Nucl. Phys.* B140:449
56. Koller, K., Krasemann, K. H. 1979. *Phys. Lett.* 88B:119
57. Particle Data Group. 1982. *Phys. Lett.* 111B
58. Gaiser, J. E. 1982. In *Proc. Moriond Workshop on New Flavor*, ed. J. Tran Thanh Van, L. Montanet, p. 11. Gif-sur-Ivette: Edition Frontiere
59. Gottfried, K. 1977. In *Proc. Int. Symp. on Lepton and Photon Interactions at High Energies, Hamburg, Germany*, ed. F. Gutbrod, p. 667. Hamburg: DESY
60. Gilchriese, M. G. D. 1982. In *Proc. 2nd Int. Conf. on Physics in Collision: High Energy ee/ep/pp Interactions, Stockholm*. New York: Plenum
61. Lee-Franzini, J. 1983. In *Proc. 7th Int. Conf. on Experimental Meson Spectroscopy*, ed. S. U. Chung, S. J. Lindembaum. To be published
62. Chanowitz, M. 1975. *Phys. Rev.* D12:918
63. Yan, T.-M. 1980. *Phys. Rev.* D22:1652
64. Kuang, Y.-P., Yan, T.-M. 1981. *Phys. Rev.* D24:2874
65. Brown, L. S., Cahn, R. N. 1975. *Phys. Rev. Lett.* 35:1
66. Büchmuller, W. 1982. See Ref. 58, p. 91
67. Krasemann, K. H. 1981. *CERN Rep. TH.3036*
68. Gupta, S. N., Redford, S. F., Repko, W. W. 1982. *Phys. Rev.* D26:3305
69. McClary, R., Byers, N. 1982. *UCLA Rep. UCLA/82/TEP/12*; also McClary, R. 1982. PhD thesis. Univ. Calif., Los Angeles (unpublished)
70. Baacke, J., Igarashi, Y., Kasperidus, G. 1981. *Dortmund Rep. DO-TH81/10* (unpublished)

71. Voloshin, M., et al. 1980. *ITEP Rep. ITEP-21*; also Voloshin, M., Zakha, V. 1980. *DESY Rep. F15-80/03* (unpublished)
72. Bertlmann, R. A. 1981. *CERN Rep. TH-3192* (unpublished)
73. Ellis, J., Gaillard, M. K., Girardi, G., Sorba, P. 1982. *Ann. Rev. Nucl. Part. Sci.* 32:443
74. Wilczek, F. 1977. *Phys. Rev. Lett.* 39:1304
75. Peccei, R. D., Quinn, H. R. 1977. *Phys. Rev. Lett.* 39:1440
76. Weinberg, S. 1978. *Phys. Rev. Lett.* 40:223
77. Wilczek, F. 1978. *Phys. Rev. Lett.* 40:220
78. Edwards, C., et al. 1982. *Phys. Rev. Lett.* 48:903
79. Porter, F. C., et al. 1982. See Ref. 58, p. 27
80. Barik, N., Jena, S. N. 1982. *Phys. Rev.* D26:618

GAUGE THEORIES AND THEIR UNIFICATION

P. Ramond[1]

Physics Department, University of Florida, Gainesville, Florida 32611

CONTENTS

1. INTRODUCTION AND SUMMARY 31
2. THE STANDARD MODEL 41
   2.1 Quantum Electrodynamics: QED 41
   2.2 The Strong Interactions: QCD 42
   2.3 The Electroweak Theory 47
   2.4 Physics at 10^{-16} cm 52
3. GRAND UNIFICATION 54
   3.1 The SU_5 Model 54
   3.2 GUT Generalizations 61
4. CONCLUSIONS 64

1. INTRODUCTION AND SUMMARY

The subject matter of this review is a physics equivalent of the famous epic novel *Roots*. Particle physicists think (not for the first time) that they are on the verge of understanding the common ancestry of all interactions identified to date: gravitational, electromagnetic, weak, and strong interactions. Just as in the macroscopic world where different lifeforms were found to share the same underlying molecular structure, one finds that the basic interactions resemble one another more and more as the distance over which they are probed diminishes. In the following we describe how this metamorphosis comes about.

At first sight, of the four basic interactions, electromagnetism and gravity bear the greatest resemblance because they both are long-range forces, and thus were the first to be identified. However, the similarity ends there: their

---

[1] Work supported in part by the US Department of Energy under contract no. DE-AS05-81-ER40008.

interaction strengths are vastly different; one is purely attractive, the other both attractive and repulsive. Yet, immediately after the formulation of general relativity, there were many attempts at unifying Einstein's and Maxwell's theories; they all relied on underlying classical (non-quantum-mechanical) geometrical principles. These theories, to which the names of Weyl, Kaluza-Klein, Einstein, et al, are associated, did not bear fruit (that is, no useful predictions), but there is no stigma attached to this early failure since we still lack a working theory that unites gravity to the other interactions. We only have guesses (for example, supergravity), but they are not the main subject of this review.

Rather, this review concerns itself with the unification of the electromagnetic, weak, and strong interactions. At first sight, this would seem to be a ludicrous proposition for these interactions manifest themselves so very differently in the laboratory. Indeed, their unification is made possible through the magic of quantum field theory, to be contrasted with the aforementioned attempts, which were only classical in nature. However, as experimental knowledge of the weak interactions increased, one similarity with Maxwell's theory emerged: both had repulsive and attractive channels or, in the language of quantum field theory, both behaved as if they were caused by the exchange of a spin-1 (vector) particle. In the case of Maxwell's theory (quantum electrodynamics, QED), this spin-1 particle is the electrically neutral photon and the long-range of the force requires it to be massless. For weak interactions the equivalent of this particle, if it existed, would have to be massive, so as to explain the short range of the weak force, and charged, to explain the fact that weak interactions transfer electric charges from one particle to another, as in neutron beta decay. In addition, weak interactions violate parity while QED does not, which means that this hypothetical mediator of weak interactions would have to couple to matter differently.

The reader might be puzzled with our focus on such a slender similarity in the midst of all these differences but it is a very fundamental one: the vector nature and the long range of electromagnetism can be deduced from an invariance principle called gauge invariance. From this principle one can (almost) derive Maxwell's equations. This is very satisfying because it makes Maxwell's theory rather unique. Of course, Einstein's general relativity provides the seminal example of a theory derived on the basis of a (different) grand invariance principle. As it turns out, gauge invariance can be generalized to more complicated theories called Yang-Mills (1) theories in which the vector nature of the interaction is linked to the underlying gauge principle. Thus it became natural to some daring theorists (2) to hypothesize that weak interactions were a more complicated version of QED, thus predicting the existence of a massive charged vector particle called the $W^\pm$. (As this review was being completed, it was announced that the $W^\pm$ has been "seen" in the $p\bar{p}$ collider at CERN) (2a, 2b).

The next step was to see if one could not smooth out the remaining differences between QED and weak interactions: while the photon is massless, electrically neutral, and couples to charged matter diagonally (i.e. a particle does not change into another particle by radiating a photon) in a left-right symmetric way, the hypothesized $W^\pm$ is massive, electrically charged, and couples to matter nondiagonally (i.e. a particle changes into another particle by radiating a $W^\pm$) in a parity-violating way.

These differences appear in two forms, one involving the coupling of the vector (gauge) particles and the other in the mass of the $W^\pm$. Neglecting the mass of the $W^\pm$ for a moment, it is possible to incorporate the $W^\pm$ and γ (photon) couplings (without introducing any new matter particles) within the framework of a Yang-Mills theory. The price for this formal unification is the prediction of a new type of parity-violating weak interaction, mediated by a massive neutral vector (gauge) particle, called the Z^0. The manifestation of this new interaction has now been found in the laboratory—the so-called neutral current interaction—although the Z^0, just like the $W^\pm$, still awaits direct observation. The fact that the $W^\pm$ and the Z^0 have to be massive looked at first to be an essential complication because a nonzero mass for the gauge particle breaks the gauge invariance from which its existence is deduced! However, it was found soon afterward that there was a (sneaky) way to introduce a mass for gauge particles and yet keep the gauge invariance in the fundamental equations that describe the theory. While a mass certainly breaks gauge invariance, it does not say where it is to be broken, for there are two ways to break a symmetry: explicit breaking of the symmetry in the equations of motion, or breaking of the symmetry in the choice of ground state for the system, while keeping the equations of motion invariant; this latter way is called spontaneous breaking. The Heisenberg ferromagnet, described by a rotationally invariant Hamiltonian, provides an example of spontaneous symmetry breaking if its ground-state wave function is taken to be asymmetrical. In a similar way, a Yang-Mills theory can be constructed to display spontaneous breaking of its symmetry, at the expense of introducing spinless (scalar) particles whose configuration of minimum energy breaks the symmetry. In this fashion the equations still retain their gauge invariance although it is broken spontaneously. This mechanism goes under the generic name of the Higgs mechanism (3), and the requisite spinless particles are called Higgs particles.

The incorporation of the Higgs mechanism into the Yang-Mills theory that unifies the weak and electromagnetic forces produced a very compact description of these forces (4). It yields the electroweak theory based on the gauge symmetry $SU_2 \times U_1$ where the SU_2 is the weak isospin group and

U_1 is the hypercharge. In this language Maxwell's theory is described by a Yang-Mills theory based on the gauge symmetry U_1, to be identified with the electric charge. Here U_1, refers to a phase symmetry and we say it is identified with electric charge when the value of that phase for a given field is equal to the charge of its corresponding particle. The electroweak theory has two coupling constants, one associated with the U_1 symmetry and the other with the SU_2. In order to reproduce the beta-decay rate, one has to arrange that the $SU_2 \times U_1$ symmetry be spontaneously broken to the electromagnetic U_1 at a distance of 10^{-16} cm. This means that at distances larger than 10^{-16} cm, the residual symmetry is that of Maxwell's equations, U_1, while at smaller distances, the underlying symmetry is $SU_2 \times U_1$. The two independent coupling constants of the theory can be identified in terms of the electromagnetic coupling constant α, and a parameter called the Weinberg angle, which measures the amount of parity violation in the neutral current interactions. Thus the electroweak theory does not yet provide a complete unification of weak and electromagnetic interaction.

So far in our description we have only used quantum field theory in a superficial way, to indicate the equivalence between fields and particles. QED provided the first great triumph of quantum field theory, the discipline that unites quantum mechanics and classical field theory. This union did not come without its inherent divergences but they were overcome, at least in QED, by the renormalization process invented by Feynman, Schwinger, and Tomonaga. Gauge invariance was an essential ingredient to the success of the renormalization procedure, as was the smallness of the electromagnetic coupling "constant" α, which allowed for a perturbative evaluation of QED. As is well known, the spectacular agreement between theory and experiment is a great testimony to the soundness of QED. There were, however, some theoretical misgivings concerning QED. For one, the electromagnetic coupling strength was found to be *increasing* at shorter distances and was in fact calculated to leave the perturbative regime at an absurdly short distance (eventually blowing up at the mythical Landau singularity). In QED, thus, the effective coupling strength decreased as the distance increased but so slowly that this effect hardly drew any attention. This scale-dependent coupling constant phenomenon is due to pair creation and will turn out to be of fundamental importance in the unification with strong interactions. In passing, note that the decrease of the effective coupling with distance makes the notion of electric charge meaningful at large distances.

The electroweak theory had to wait to be deemed renormalizable in order to serve as *the* candidate theory. It was shown by G. 't Hooft (5) that Yang-Mills theories with unbroken gauge symmetries and those with spontaneously broken symmetries were indeed renormalizable—in particular, the electroweak theory of Glashow, Weinberg, Salam, and Ward (2,

4). Henceforth in the mind of most theorists there was no more doubt that this theory provided the correct way to unify weak and electromagnetic theory. It is only now that accelerators capable of verifying its many predictions are coming on-line. Only the new p-p̄ collider at CERN is, so far, capable of identifying through their distinctive decays both charged and neutral vector bosons of the weak interactions ($W^\pm$, Z^0). In fact, the $W^\pm \to e\nu$ mode has just been observed (2a). It would be a tremendous surprise if the expectation of the electroweak theory were not experimentally verified. Complete verification of the theory will demand experimental discovery of the vector gauge bosons $W^\pm$ and Z^0, at least one neutral Higgs particle, and another charge-2/3 quark, called the top quark. The author hopes that by the time this review is published both the Z^0 and the $W^\pm$ will have manifested themselves in the laboratory. Should experiment fail to verify the electroweak theory, it would put in question in a fundamental way our understanding of the fabric of the universe, especially our gradual realization that gauge theories are the mainstay of the physical world at low energies. This belief is heightened by the most amazing development of all which provides the missing link in our unified view of the interactions: the strong interactions themselves are a gauge theory in disguise!

This is even more surprising than the previous identification between weak and electromagnetic interactions. At least they both involved vector forces, but, as everyone knows, strong interactions are essentially attractive, corresponding to the exchange of a spinless particle, the pion (with still a more complicated structure at shorter distances described by vector particles ρ, ω,...exchange). Furthermore the strength of the strong interaction is very large. This bears no resemblance to what is naively expected from gauge theories. Indeed, in order to understand how one can make such an apparently crazy identification, we have to appeal to several pieces of the strong interaction puzzle: (a) the quantum numbers of the strongly interacting particles, which led to the notion of quarks; (b) the experimental probing of nucleons at short distances, which led to the concept of partons; and (c) the marvelous property of Yang-Mills theories called asymptotic freedom, which enabled theorists to understand how a nucleon can be made up of quasi free quarks/partons that can never leave the nucleon on their own. It is precisely the force that holds the quarks together that has the characteristics of a gauge interaction.

The precise nature of this gauge theory was revealed indirectly by the hadron spectrum. The varieties of strongly interacting particles (hadrons) were first classified by Gell-Mann & Ne'eman (7) under the aegis of SU_3, which generalized Heisenberg's SU_2 isospin to include strangeness. However, it was noticed soon afterwards by Gell-Mann (8) and Zweig (9) that the description could be further simplified if one assumed that the quantum numbers of the hadrons were carried by smaller units, called quarks by Gell-Mann. These arrange themselves in such a way that baryons are made up of three quarks, and scalar or vector mesons are made up of a quark and antiquark with their spin antialigned or aligned, respectively. According to this description there would be three quarks flavors—the u (up), d (down), s (strange) quarks with charges 2/3, −1/3, −1/3—all subject to weak interactions. Furthermore if the quark existed as a particle it would have to have spin 1/2, thus presumably obeying Fermi statistics. Next, by combining the quark spin with the quark flavors, one arrived at a unified description of the observed vector mesons and pseudoscalar mesons.

The same success could be duplicated in the baryon sector by combining three 1/2 spins to form a spin 3/2, only if one assumed that the quarks violated the Pauli exclusion principle even though they appeared to have spin 1/2 (10)! For instance, the state $u_\uparrow u_\uparrow u_\uparrow$ with the spins of all three up quarks aligned, corresponding to a spin-3/2 Δ^{++}, is known to exist; yet if quarks obey the Pauli exclusion principle, it should be forbidden. One way out of this dilemma was to assume that there were three different up quarks (11) and that the Δ^{++} combination involved each new species only once, with $\Delta^{++} \sim u_\uparrow^R u_\uparrow^W u_\uparrow^B$; here the indices R, W, B (red, white, blue) label the quark colors. Then the Pauli exclusion principle is finessed as it does not operate between different species. We are thus led to a tripling of the numbers of quarks: each quark flavor (up, down, strange, ...) occurs in three colors, call them R, W, B. A baryon state would then appear to be a combination of three quarks, antisymmetrized over their color indices in accord with the Pauli principle. The meson states would be combinations of quarks and antiquarks summed over their colors. Thus as far as quantum numbers are concerned, hadrons behave as if they were made up of several flavors of quarks, each appearing in three colors.

A bit later, measurements of electron-nucleon inelastic scattering were performed. (For a review, see 12.) These experiments provide a way to probe by means of (virtual) photons the nucleons at distances that are small with respect to nucleon sizes. The measurements showed a behavior akin to that found much earlier by Rutherford for atoms: protons, on close observation, do not behave like jelly, but rather act as if they were composed of point-like particles. Feynman called these point-like scattering centers partons (13). It became clear that these partons were the hypothesized quarks, which now acquired a reality of their own. Nucleons *do* behave, upon close examination, as if they were made up of quarks. How then are quarks held together in the hadron? Nambu (14) hypothesized early on that the quark-force was in fact a vector force of the Yang-Mills type, and since there were three colors the corresponding gauge symmetry was SU_3, not to be confused with the earlier flavor SU_3 of the eightfold way. This Yang-Mills theory with color SU_3 gauge invariance, according to which there would be eight vector particles called gluons mediating the interquark interaction, was called by Gell-Mann quantum chromodynamics (QCD). This hypothesis on the nature of the quark force became compelling when it was realized that Yang-Mills theories exhibited "asymptotic freedom."

We have alluded to the fact that the coupling constant appearing in gauge theories is distance dependent. In QED, for instance, the effect of this phenomenon is to make the repulsion between two electrons increase faster than r^{-2} as they are brought closer together; alternatively their repulsion becomes ever so weaker as they are separated. Now, for Yang-Mills theories, the situation is exactly the opposite: as quarks are brought together, the strength of the force between them decreases, and as they are separated the attractive force between them increases. That is presumably the reason quarks do not leave the proton, and also the reason why quarks appear as point particles to short-distance probes: quarks appear to be free as the energy of the probe increases: asymptotic freedom (15). However, quarks in certain combination have no net strong charge on which this relentless force can act over large distances. These "colorless" combinations are the only ones that can leave nuclear matter and thus be observed in the laboratory—these are the meson "$q\bar{q}$" and the nucleons "qqq" and their antiparticles. In addition QCD demands that quarkless states made up of two or three gluons, called glueballs, appear in the laboratory, but estimates of their masses and widths are for the moment theoretically imprecise (for a review, see 16). If the color symmetry is not broken, free quarks are not expected to appear in the laboratory.

At present, experimental work continues on the verification of the QCD hypothesis. In high-energy machines, quarks and gluons flaring out of nuclear matter manifest themselves in the form of jets, and it is hoped that a close analysis of gluon jets will allow for a determination of the gluon spin. Unfortunately, it is very difficult to have a direct verification of QCD with present energy machines, but all circumstantial evidence is in the favor of QCD (see, for example, 17). Also, the study of the properties of heavy quark bound states such as charmonium or bottomonium, further favors the QCD picture (for a review, see 18). Three quark colors are necessary to achieve agreement with experiment in the e^+e^- total cross section (19) and in the decay rate of π^0 into two photons. QCD even provides a reason for the heaviness of the flavor singlet η' meson (20).

The link between QCD and the hadronic world has not yet been resolved at the dynamical level, because theorists have been unable to compute reliably with theories that involve strong coupling. Approximations of QCD on the lattice yield very promising results, reproducing the systematics of the hadron spectrum, but they are a long way from reproducing strong interaction parameters such as the proton's mass or magnetic moment with the desired accuracy (21). QCD presumably involves only one dimensional parameter (called Λ_{QCD}) in terms of which all hadronic parameters should eventually be expressed.

To be fair, it should be said that QCD does not come without its problems; the theory inherently violates CP invariance, albeit with unknown strength! This effect produces a nonvanishing electric dipole moment for the neutron (measured so far to be absent within the experimentally accessible distances). However, faith in QCD has motivated theorists to seek ingenious solutions to explain either the smallness or the absence of QCD-induced CP violation. The introduction of a quark-chiral global symmetry (22) can obviate the problem, at the expense of introducing a hypothetical neutral particle called the axion (23).

We have now arrived at the threshold of the grand unified description of the interactions.

At 10^{-16} cm, the world now looks much simpler. All interactions, except gravity, appear to be of the Yang-Mills type. Matter is made up of spin-1/2 leptons such as the electron, the muon, the tau, and their neutrinos, which interact only via the electroweak gauge theory with coupling constants of the order of $\alpha_I \sim 1/30$ and $\alpha_Y \sim 1/95$, their ratio being fixed by the Weinberg angle. There are also quarks [u, d, s, c (charm), b (bottom), t (top)?,...], which interact not only electroweakly, but also through the color force generated by the gluons. The quark-gluon coupling constant is much larger than the others but still much less than one and decreasing fast as distances shorten. In this context, then, the unification of quarks and leptons into a single Yang-Mills theory (24) seems to be a very natural next step: Grand Unification! The remarkable thing is that there are *two* apparently independent indications that point towards grand unification:

1. The quantum numbers (i.e. electroweak and color charges) of the quarks and leptons fit so snugly inside such a larger gauge structure that, by fixing the neutrino to be electrically neutral, the relative charges of the quarks and leptons are reproduced (25). This constitutes the first understanding of the equality of the proton and positron charges. In this classification the emergence of (at least) three known "families" of elementary fermions occurs: the first family, the lightest one, includes the electron, its left-handed neutrino, and the u and d quarks; the second family consists of the muon, its left-handed neutrino, and the c and s quarks; and the third family, the heaviest, is made up of the τ-lepton, its left-handed neutrino, and the t and b quarks. The third family is not yet complete for the t quark is presumably too massive to have been detected in the laboratory.

2. As we have just described, there are three gauge coupling constants, corresponding to color, weak isospin, and hypercharge. The first two are decreasing with decreasing distance, the third is increasing. Their variation with distance is determined within the framework of these three gauge

theories by the renormalization group, and it happens (26) that they all meet at the same point: all three coupling constants merge into one at a distance of $\sim 10^{-28}$ cm!

Conversely, by demanding unification, one can say that the ratio of the two electroweak coupling constants is fixed, thus yielding the hitherto undetermined Weinberg angle: it turns out that the most appealing grand unified theory (GUT) based on the gauge group SU_5 (25) yields a value for the Weinberg angle that is compatible with experiment, a great triumph of the grand unification idea. In this picture, there is only one theory at 10^{-28} cm with one coupling constant $\alpha_5 \sim 1/40$, and all spin-1/2 particles part of the same theoretical structure. In particular it means that a quark can turn into lepton by radiating a new type of vector particle called a leptoquark, and in the simplest GUT the same leptoquark gauge particles can turn a quark into an antiquark!

These new gauge particles are part of the unification package, and, because they violate baryon number, they cause the proton to decay (24, 25); it is a good thing for us that these new forces are most efficient only at 10^{-28} cm. This great disparity of scales enables the theory to predict proton decay lifetimes just a shade longer than the experimental bound in the minimal SU_5 model. However, the proton lifetime can easily be lengthened by adding new spin-1/2 and spin-0 particles with masses corresponding to distances between 10^{-16} and 10^{-23} cm, the very region where simplest GUTs predict a "desert" of interactions and particles. In the desert their effect is to slow down the descent of the QCD coupling and postpone the meeting of the couplings to a scale less than 10^{-28} cm. (As of this writing, a large fraction of the experimental community has gone underground to look for proton decay.) In fact, by pushing a little bit beyond the most elementary models (see 27 and 28 for models such as SO_{10} and E_6, respectively), even lepton number is violated, and neutrinos are understood to be very light because the unification scale is so much smaller than the weak scale (29). Thus grand unification implies that the strong, weak, and electromagnetic interactions merge into one weak force at $\sim 10^{-28}$ cm, and approach the altar of quantum gravity ($\sim 10^{-33}$ cm) as a single gauge theory.

The prototype theory is SU_5, although there are many variations on the market. In this picture, there are no new gauge interactions to be met between 10^{-16} cm and 10^{-28} cm, resulting in a so-called desert: after the $W^\pm$ and Z^0 vector bosons are discovered, the next vector (gauge) bosons to be discovered will weigh at least 10^{15} GeV! Thus at distances shorter than 10^{-28} cm the gauge symmetry is, say, SU_5. (Actually our claim to understanding applies only to distances larger than the Planck length, $\sqrt{G\hbar/c^3} \sim 10^{-33}$ cm, which parameterizes the beginning of our ignorance or the end of our knowledge.) Between 10^{-28} and 10^{-16} cm, the symmetry is broken down to SU_3 for color, SU_2 for weak isospin, and U_1 for hypercharge. At distances greater than 10^{-16} cm it is broken down further to the color SU_3 and U_1 for electromagnetism.

Before discussing the special new problems created by this description, let us note in passing two unexpected dividends the theory gives to our picture of the early universe. Under the hypothesis of the Big Bang, our journey to smaller distances is equivalent to retracing the steps of our evolution. Thus GUT provides cosmology with baryon-number-violating processes in the early universe that contribute to an understanding of the predominance of matter over antimatter (30), and also a first-order phase transition that can be used to alleviate long-standing problems of standard cosmology (31). On the other hand, magnetic monopoles (32) can be produced in the early universe by the GUT transition (33), but their predicted abundance today is linked with the horizon problem of standard cosmology (see the article by R. Kolb and E. Turner in this volume).

Ambitious as GUT models may be they do not provide answers to the observed multiplicity of fermions [Rabi's famous "Who ordered the muon?" is now replaced by "Who ordered the muon family?"], nor to the values of their masses. In fact, GUT models are singularly unpredictive as far as fermion masses and mixing angles such as the Cabibbo angle are concerned, despite some success for the heavier fermions. This can all be traced to the fact that fermion masses are closely tied to the symmetry-breaking mechanisms, and therefore to the Higgs particles. There is no compelling guiding principle that restricts the type and interactions of these spinless particles, and no hints from experiments since none have been observed to date. Yet GUT relies on them to provide the scales at which the gauge symmetries are broken, as well as the pattern of fermion masses and mixing angles. In fact the requisite ratio (10^{-12}) between electroweak and GUT breaking is very hard to understand in the absence of symmetries capable of taming the unruly behavior of the Higgs; this leads to a GUT-generated difficulty called the gauge hierarchy problem (34). Recently, the introduction of a new type of symmetry, which necessitates scalar particles, called supersymmetry (for a review, see 35) has had some successes in that direction, but not enough to give GUT predictive power in determining fermion masses and multiplicities.

Hence the theory still appears to be incomplete, which is not surprising since gravity is yet to be included. Recall that GUT unification occurs just a few orders of magnitude below the Planck length. So it is likely, after all, that particle physics will be successfully understood only after inclusion of gravity, thus going full circle back to the early geometrical unification attempts. This time around, however, the ultimate unification will no doubt involve supergravity (36), the generalization of gravity that includes supersymmetry. At the time of this review, the union has not yet been consummated, owing to diverging interests of one of the parties involved. Still some feel we are at the threshold of understanding the very origins of our universe, and that all the remaining problems will be solved by the union of gravity and the GUT interaction under the watchful eye of quantum mechanics. It is more likely, however, that many surprises still await us since none of the supergravity models, pretty as they may be, can be made to mesh even with the GUT description of the nongravitational interactions. In particular, the supergravity models do not share the chiral structure of the "low" energy interactions.

2. THE STANDARD MODEL

In this section we review in a more technical way the three nongravitational interactions and develop the tools necessary in their description and unification. We arrive at a consistent description based on a Yang-Mills theory with $SU_3^c \times SU_2^{I_w} \times U_1^Y$ gauge symmetry (c = color, I_w = weak isospin, Y = hypercharge), all with only 19 undetermined parameters! We start with a review of the mother theory: QED.

2.1 Quantum Electrodynamics: QED

In the context of quantum field theory (QFT), the electromagnetic force is due to the exchange of a massless particle, the photon. The charge of a particle or antiparticle is just the strength of its coupling to the photon and, for elementary particles, it is sufficiently small (in units of $\sqrt{\hbar c}$) to allow for a perturbative treatment of the photon interaction. In view of the remarkable success of this theory, let us review its structure and use its familiar context to introduce some novel notions. The classical Lagrangian describing QED is given by

$$\mathscr{L}_{QED} = -\frac{1}{4}F_{\alpha\beta}F_{\alpha\beta} + \psi_L^\dagger \sigma_\alpha(\partial_\alpha + ieA_\alpha)\psi_L + \psi_R^\dagger \bar\sigma_\alpha(\partial_\alpha + ieA_\alpha)\psi_R$$
$$+ m(\psi_L^\dagger \psi_R + \psi_R^\dagger \psi_L),$$

where $F_{\alpha\beta} = \partial_\alpha A_\beta - \partial_\beta A_\alpha$ is the Maxwell field strength given in terms of the photon field A_α. Its coupling to a matter field is written in two-component (Weyl) notation; ψ_L and ψ_R are two-component fields representing a prototype charged particle, and σ_α and $\bar\sigma_\alpha$ are 2 × 2 matrices. In the limit of zero mass they would represent a spin-1/2 particle with left- or right-helicity. The advantage of this notation is to clarify an important point: the photon couples to ψ_L and ψ_R in the same way, thus leaving parity intact. Photon interactions preserve the chiral structure while the mass term breaks it. The point of this complicated description is to emphasize that a gauge field can couple independently to ψ_L and ψ_R. It is only the left-right symmetry of the electromagnetic interactions that dictates the same photon coupling to ψ_L and ψ_R. The QED Lagrangian has two input parameters, the dimensionless charge e and the mass m of the field ψ.

This Lagrangian has another remarkable property, it is invariant under the following gauge transformation:

$$\psi_{L(R)}(x) \to e^{i\Lambda(x)} \psi_{L(R)}(x), \qquad A_\alpha(x) \to A_\alpha(x) - \frac{1}{e}\partial_\alpha \Lambda(x),$$

where $\Lambda(x)$ is an arbitrary function of the space-time point x. It is easily shown that this symmetry requirement on the ψ field alone necessitates the introduction of a photon field. Note that any mass term for the A_α field, of the form $M^2 A_\alpha A_\alpha$, would break this symmetry. In this case the gauge invariance appears as a U_1 phase transformation on the ψ field.

The corresponding QFT is beset with inherent divergences due to both short- and long-distance effects. The divergences in long-distance effects (infrared) are due to our difficulty in distinguishing between a particle state with or without very low-energy photons; the resulting divergences can be absorbed by a careful analysis of the measuring process. On the other hand the short-distance (ultraviolet) divergences can be absorbed in the input parameters and fields of the theory. The gauge invariance of QED plays a crucial role in this renormalization process. As a result the parameters of the theory, in this case the mass and charge of a particle, have to be specified at some time or distance scale for they are no longer absolute parameters, but vary with the scale at which they are measured. In QED, the electromagnetic coupling is measured by the Thomson cross section, which is (for us) essentially a static measurement, and it is found to be numerically small at that scale. It turns out that the electromagnetic coupling "constant" increases slowly with inverse distance and will leave the perturbative region only at a scale much shorter than the Planck length! This very slow growth of the photon's coupling strength is not significant in measurements made on scales larger than or equal to atomic distances. Physically, the effect can be viewed as a consequence of pair creation: an energetic photon will excite many particle-antiparticle pairs out of the vacuum. This will result in a clumping of the opposite-charge partner around the test charge, leading to an increase in net charge. The more energetic the photon, the more the net charge increases. To conclude, the essential features of QED are (a) U_1 gauge invariance under a local phase transformation, which results in a neutral massless photon, (b) left-right symmetric coupling of the photon to matter, and (c) renormalizability and a very slowly varying coupling strength, increasing with energy. We will use some of these characteristics to describe QCD and the electroweak theory.

2.2 The Strong Interactions: QCD

At first glance, strong interactions bear no resemblance to QED. They appear to be due to the exchange of a multitude of particles, dominated by pseudoscalar pion exchange, with large coupling constant. This bars the way to a perturbative evaluation as in QED. However, in this case appearances proved to be misleading, for now it is thought that the strong interactions we have just referred to are due to a gauge theory much like QED that becomes recognizable as such only to probes with wavelengths shorter than 1 fermi. This surprising development was spearheaded by two independent approaches: the classification of the plethora of hadron states, which led to the notion of quarks as constituents of hadrons, and the probing of nucleons at short distances by electromagnetic currents, which showed that the nucleon's electric charge is concentrated in point-like partons. The latter gave reality to the quarks as the constituents of hadrons. The force holding the quarks together inside a nucleon is now believed to be due to a gauge theory, a more complicated version of QED, called QCD (quantum chromodynamics) (for a review, see 37). This view is strengthened by the fact that QCD exhibits "asymptotic freedom," which explains why quarks appear as free particles to short-distance probes and yet strongly bound to large-distance ($\gtrsim 1$ fermi) probes. This view is totally vindicated by the existence and properties of long-lived massive quark-antiquark bound states, such as $c\bar{c}$, $b\bar{b}$.

By the middle of the 1960s, hadron spectroscopy could be very neatly summarized in the following way. Baryons appear as if made up of three spin-1/2, fractionally charged constituents called quarks, while integer-spin hadrons look like quark-antiquark pairs. It was necessary to introduce three flavors of quarks, up (u), down (d), and strange (s), to account for the known particles at that time. It was soon realized that three flavors of quarks were not enough; for instance, one needed three up quarks with their spins aligned to reproduce the spin-3/2 Δ^{++} resonance, but this could not be done (in an S state) if the quarks obeyed Fermi statistics. It was suggested that the quarks making up the Δ^{++} were all different, although they have the same flavor. That is, each quark flavor appears in three different colors. In this way the Δ^{++} still contains three quarks all with the same flavor, but each with a different color. Color, then, is a quark quantum number and it serves to differentiate quarks from the colorless leptons.

This view of the constituents of hadrons was hypothesized from a consideration of quantum numbers, but it was soon to be corroborated by experiment, such as deep inelastic scattering where an energetic electron scatters off nuclear matter. According to QED this process is understood in terms of a virtual photon probing the nucleon. With sufficient energy available, experimenters (6) were able to achieve resolution less than a nuclear size, and they found their cross section exhibited "scaling" (38), which resulted in many large-angle scatterings. This was recognized by Feynman as indicative of quasi-free point-like scattering centers inside the nucleus, which he called partons. These must be the quarks whose existence was deduced earlier on totally different grounds. One also notices that all three quark colors are needed to reproduce the observed cross section for $e^+e^- \to$ hadrons.

Now that hadrons appear to be quark composites, there remained to identify the force strong enough to hold quarks together and yet weak enough to allow quarks to be quasi-free scattering centers. Just as the electric charge is the charge for the electromagnetic force, Nambu theorized that the three quark colors were the charges for this interquark force, based on a generalization of gauge theories to include different charges, due to Yang and Mills. In order to proceed, we have to understand its salient features.

In Yang-Mills theories (1), (a) the U_1 (phase) gauge invariance (as in QED) is generalized to any Lie group G; (b) the spin-1/2 or spin-0 fields of the theory transform as representations of G, and thus their number is limited by the representations of G; (c) there are as many spin-1 gauge bosons as there are parameters in the group G; (d) the gauge invariance produces nonlinear interactions among the gauge fields and requires them to be massless; and (e) the gauge fields couple to spin-1/2 and spin-0 fields by means of currents, just as in QED. Specifically let us illustrate the construction for spin 1/2. Let $\psi_L(x)$ [or $\psi_R(x)$] be any d_r-component spinor field transforming as a d_r-dimensional representation, r, of G:

$$\psi_L(x) \to U(x)\psi_L(x), \quad U(x) = \exp[i\omega^A(x)T_r^A].$$

Here T_r^A are the $d_r \times d_r$ matrices that represent the group G in the representation r; they satisfy the Lie algebra

$$[T_r^A, T_r^B] = if^{ABC}T_r^C,$$

where f^{ABC} are the structure constants of G, and the indices A, B, C run over the d parameters of G. The Yang-Mills Lagrangian is given by

$$-\frac{1}{4}F_{\alpha\beta}^B F_{\alpha\beta}^B + \psi_L^\dagger \sigma_\alpha [\partial_\alpha + igT_r^B A_\alpha^B(x)]\psi_L,$$

where we have introduced the d gauge fields $A_\alpha^B(x)$; g is the dimensionless gauge coupling constant, and $F_{\alpha\beta}^B$ are the Yang-Mills field strengths

$$F_{\alpha\beta}^B = \partial_\alpha A_\beta^B - \partial_\beta A_\alpha^B - gf^{BCD}A_\alpha^C A_\beta^D.$$

This Lagrangian is invariant under the combined gauge transformations

$$A_\alpha \to -iU(x)\partial_\alpha U^\dagger(x) + U(x)A_\alpha(x)U^\dagger(x), \quad \psi_L \to U(x)\psi_L$$

where we have set $A_\alpha \equiv A_\alpha^B T_r^B$. A similar construction for spinless fields is easily done by generalizing all derivatives to the covariant derivative $(\partial_\alpha + igA_\alpha)$.

If the three quark colors are to serve as the Yang-Mills charges, one is led to take each quark flavor to transform as a triplet of a new SU_3 that causes transitions between quarks of the same flavor but different colors. An SU_3 Yang-Mills theory dictates the introduction of eight new spin-1 gauge particles, called gluons. Furthermore, since strong interactions do not violate parity, it seems reasonable to demand that the gluons couple to left-handed and right-handed quarks in the same way. One arrives in this way at quantum chromodynamics (QCD).

We give as an illustration the coupling of the gluon to one flavor of quark (in this case the u quark)

$$u_L^\dagger \sigma_\alpha \left(\partial_\alpha + ig_3 \frac{\lambda}{2} \cdot \mathbf{A}_\alpha \right) u_L + u_R^\dagger \bar\sigma_\alpha \left(\partial_\alpha + ig_3 \frac{\lambda}{2} \cdot \mathbf{A}_\alpha \right) u_R,$$

where g_3 is the gluon coupling constant, and we have suppressed all quark color indices. The theory includes eight gluons transforming as 8^c of color while quarks (antiquarks) transform as 3^c ($\bar{3}^c$). The gluons have no charge, just color. The gluons couple only to states with nonzero color. If gluons are indeed responsible for a strong static interquark force, only color singlet states would be allowed to appear as free states in the laboratory. Group theory indicates that color singlets appear in the fully antisymmetrized product of three triplets (antitriplets), i.e. qqq, in the product of a triplet and as antitriplet, i.e. q$\bar{\text{q}}$, and also in the product of two and three octets. The first two combinations correspond to half-integer-spin baryons and integer-spin mesons, while the last one yields neutral quarkless states made of gluons alone, called glueballs. So far, there is no evidence for glueball states, and there is no precise theoretical determination of their masses and widths, although they are expected to be heavier than the lighter quark composites.

The property that makes QCD overwhelmingly compelling as the candidate theory of strong interaction is asymptotic freedom. 't Hooft proved that Yang-Mills theories, like QED, are renormalizable (up to possible chiral anomalies). In QED we have described how the coupling constant varies with distance due to pair creation. In QCD one expects a similar phenomenon to take place for the color charge, but with the important difference that the gauge particles themselves have color. Thus a very short wavelength probe of a color charge will create gluon pairs in addition to quark-antiquark pairs, which, as in QED, tend to increase the net charge. It turns out that the cloud of gluon pairs thus created more than offsets the effects of the quark pairs, resulting (as long as there are not too many flavors of quarks) in a net decrease of the coupling strength as the wavelength of the probe is shortened! Thus the interquark force becomes weaker at shorter distances, and stronger at larger distances. This is exactly the desired property. From the theoretical point of view it means that at short distances QCD can be treated perturbatively while at larger distances

($\gtrsim 1$ fermi) the quark force becomes so strong it cannot be treated perturbatively. One understands the typical strong interaction scale of 1 fermi to be the scale at which the QCD coupling strength leaves the perturbative regime.

At present there are no dynamically precise ways of obtaining standard low-energy hadron physics from QCD because we cannot yet tame field theory in the nonperturbative regime. However, approximation techniques such as the lattice approximation have proven themselves remarkably successful in reproducing the systematics of the hadron spectrum. Eventually one should be able to express all hadronic quantities, such as the proton mass and its magnetic moment, in terms of the one fundamental scale appearing in QCD. In the short-distance regime, QCD calculations can already be reliably performed since the coupling there is perturbative, but direct comparison with experiment is complicated by two factors. First, the quark coupling becomes weak logarithmically slowly so that present machine energies do not provide probes of sufficiently short wavelength. Second, measurements always involve physical particles, which are quark bound states, making it difficult to separate the perturbative from the nonperturbative part of QCD. Nonetheless, QCD predicts definite corrections to the free quark (scaling) picture, and all evidence points to QCD, albeit in a circumstantial way. In high-energy experiments, quarks and gluons flare out of nuclear matter, creating jets of particles as they fragment into hadrons.

The existence of new heavy quark flavors such as the charm (c) quark and the bottom (b) quark has further corroborated the existence of weak coupling in the quark system. These new flavors appear as heavy $c\bar{c}$ and $b\bar{b}$ bound states in e^+e^- annihilation experiments, with lifetimes very long compared with typical hadronic lifetimes. This is due to the weakness of the gluon coupling at scales defined at the masses of these bound states. Again, their phenomenology is totally consistent with the quark QCD picture even though a precise QCD derivation of the mechanism that produces these bound states is still lacking.

To conclude this quick description of QCD, one must mention that the theory predicts CP violation in the strong interactions of an undetermined strength, due to nonperturbative instanton effects. This would presumably lead, for instance, to a nonzero electric dipole moment for the neutron, which is now limited by experiment to be very small. It is not clear at present how to resolve this problem, but two attitudes seem prevalent: ignore the problem, leaving it to a grander theory to explain the smallness of the CP violation, or introduce a global quark-chiral symmetry, which, when broken spontaneously by quark masses and explicitly by instanton effects, will lead to a pseudo Nambu-Goldstone boson called the axion. Such a particle has been searched for in accelerators but not found. If it is very light astrophysical considerations are required to put limits on its properties. On the other hand, the same QCD instanton effects provide a reason for the large η' mass. To complete the description of the low-energy world, we need to discuss the weak interactions (6).

2.3 The Electroweak Theory

A long time ago, by analogy with QED, Fermi almost derived the form of the beta-decay Hamiltonian. Twenty-five years later his picture was slightly altered to include parity violation by saying that since the weakly interacting neutrino (antineutrino) is purely left- (right-) handed, it can be represented solely by one two-component field v_L. (Henceforth we use the particle symbol to represent its corresponding field.) The charged current weak interactions can be summarized by the effective Hamiltonian

$$H_w = \frac{4G_F}{\sqrt{2}} (J_\alpha^\ell + J_\alpha^q)(J_\alpha^\ell + J_\alpha^q)^\dagger + \text{h.c.},$$

where G_F is the dimensionful Fermi constant. The leptonic vector current J_α^ℓ (including the τ lepton and its neutrino) is given by

$$J_\alpha^\ell = e_L^\dagger \sigma_\alpha v_{eL} + \mu_L^\dagger \sigma_\alpha v_{\mu L} + \tau_L^\dagger \sigma_\alpha v_{\tau L} + \cdots.$$

The hadron current J_α^q is given in terms of quarks fields by

$$J_\alpha^q = (s_L^\dagger \sin \theta_c + d_L^\dagger \cos \theta_c)\sigma_\alpha u_L + (s_L^\dagger \cos \theta_c - d_L^\dagger \sin \theta_c)\sigma_\alpha c_L + \cdots,$$

where θ_c is the Cabibbo angle. We have included for illustration the charm c quark, which, through the GIM mechanism (39), explains the suppression of flavor-changing processes. We have omitted the t and b quark whose presence provides a likely mechanism for the experimentally observed CP violation of the weak interaction (40). Also not shown are the additional quark mixing angles and color indices. This current-current interaction could be due to exchange of a charged spin-1 boson $W_\alpha^\pm$, which causes transitions between v_{eL} and e_L, $v_{\mu L}$ and μ_L, d_L and u_L, s_L and c_L, etc.

The next step consists of incorporating the charged Ws into a Yang-Mills theory. Because the $W^\pm$ causes transitions between two left-handed fields at a time, one is led to consider an SU_2 Yang-Mills theory with the two fields transforming as the components of a doublet. This SU_2 is called the weak isospin. Hence left-handed matter is arranged in weak doublets

$$\begin{pmatrix} v_e \\ e \end{pmatrix}_L, \begin{pmatrix} v_\mu \\ \mu \end{pmatrix}_L, \begin{pmatrix} v_\tau \\ \tau \end{pmatrix}_L; \quad \begin{pmatrix} u \\ d' \end{pmatrix}_L, \begin{pmatrix} c \\ s' \end{pmatrix}_L, \begin{pmatrix} t \\ b' \end{pmatrix}_L.$$

The prime on the quark fields means that the mixing angles have been incorporated. Because the right-handed partners of the quarks and the charged leptons suffer no charge-changing weak interactions, they are assigned to weak singlets. Since SU_2 is a three-parameter group, the Yang-Mills construction introduces three gauge bosons—a charged one and its antiparticle, and a neutral one. The charged boson can be identified with the $W_\alpha^\pm$ for it is designed to couple just to the desired current $J_\alpha^q + J_\alpha^\ell$. Note that in this theory universality of the gauge coupling between leptons and (mixed) quarks is automatic. The third gauge boson is neutral and its current, corresponding to the third component of weak isospin, is just

$$J_\alpha^3 = \tfrac{1}{2}(v_{eL}^\dagger \sigma_\alpha v_{eL} - e_L^\dagger \sigma_\alpha e_L + u_L^\dagger \sigma_\alpha u_L - d_L^\dagger \sigma_\alpha d_L) + \cdots,$$

Exchange of the third gauge boson corresponds to a new neutral interaction that violates parity and involves the neutrinos. Note that there are no flavor-changing terms like $d_L^\dagger \sigma_\alpha s_L$ in this current, thanks to the GIM mechanism. It is useful to compare it to the electromagnetic current

$$J_\alpha^\gamma = -(e_L^\dagger \sigma_\alpha e_L + e_R^\dagger \bar\sigma_\alpha e_R) + \frac{2}{3}(u_L^\dagger \sigma_\alpha u_L + u_R^\dagger \bar\sigma_\alpha u_R)$$
$$-\frac{1}{3}(d_L^\dagger \sigma_\alpha d_L + d_R^\dagger \bar\sigma_\alpha d_R) + \cdots,$$

which clearly breaks weak isospin since members of the same doublet have different electric charges. On the other hand, the difference between these two neutral currents, the hypercharge current

$$J_\alpha^Y \equiv J_\alpha^\gamma - J_\alpha^3$$

has the simpler form

$$J_\alpha^Y = -\frac{1}{2}(v_{eL}^\dagger \sigma_\alpha v_{eL} + e_L^\dagger \sigma_\alpha e_L) - e_R^\dagger \sigma_\alpha e_R + \frac{1}{6}(u_L^\dagger \sigma_\alpha u_L + d_L^\dagger \sigma_\alpha d_L)$$
$$+ \frac{2}{3} u_R^\dagger \sigma_\alpha u_R - \frac{1}{3} d_R^\dagger \sigma_\alpha d_R + \cdots.$$

The hypercharge current has the property that each member of the weak doublet has the same hypercharge value, $-1/2$ for lepton doublets, $1/6$ for quark doublets. It is thus independent of the SU_2 and can be gauged separately. The gauge structure of the electroweak theory is then taken to be $SU_2^{Iw} \times U_1^Y$. The interaction part of the Lagrangian is

$$\mathscr{L}_{int} = g_1 B_\alpha J_\alpha^Y + g_2 W_\alpha^3 J_\alpha^3 + \frac{g_2}{\sqrt{2}}[W_\alpha^+ (J_\alpha^\ell + J_\alpha^h) + W_\alpha^- (J_\alpha^\ell + J_\alpha^q)^\dagger],$$

where $B_\alpha(x)$ is the hypercharge gauge boson with coupling g_1, and the three $W_\alpha(x)$ are the weak isospin gauge particles, with gauge coupling g_2: the theory contains one charged boson W^+ and its antiparticle, W^-, and two neutral gauge particles, W^3 and B. At this stage they are all massless. The photon field A_α is a combination of these two neutral gauge fields, mixed by the Weinberg angle θ_w,

$$A_\alpha = \cos\theta_w B_\alpha + \sin\theta_w W_\alpha^3,$$

with another neutral gauge boson field

$$Z_\alpha = -\sin\theta_w B_\alpha + \cos\theta_w W_\alpha^3$$

as the orthogonal linear combination. Now we can re-express $\mathscr{L}_{\text{int}}$ in terms of the A_α and Z_α fields. Since the neutrino is electrically neutral, we must identify

$$\tan\theta_w = g_1 g_2^{-1}.$$

We find for the neutral-current part of $\mathscr{L}_{\text{int}}$, after algebraic rearrangement,

$$\mathscr{L}_{\text{int}}^{\text{NC}} = g_2 \sin\theta_w A_\alpha J_\alpha^\gamma + \frac{g_1}{\sin\theta_w} Z_\alpha (J_\alpha^3 - \sin^2\theta_w J_\alpha^\gamma).$$

The charged-current part is unaltered. From the above we identify the electric coupling constant to be

$$e = g_2 \sin\theta_w = g_1 g_2 (g_1^2 + g_2^2)^{-1/2}.$$

We thus reproduce the familiar electromagnetic interaction at the cost of introducing a new neutral current interaction mediated by Z_α.

In order to make contact with nature, we have to give a mass to both the charged $W^\pm$ and the neutral Z gauge bosons, while the photon, of course, must remain massless. These masses, M_W and M_Z break the $SU_2 \times U_1$ electroweak gauge symmetry down to the U_1 of electromagnetism.

The required electroweak breaking can be obtained most elegantly in the following way (4): introduce a complex two-component scalar field $\Phi(x)$ that transforms as a weak doublet and has hypercharge value of $+1/2$, chosen so that one of its components is electrically neutral. The electroweak Lagrangian density is augmented by

$$\mathscr{L}_s = \left|\left(\partial_\alpha + ig_2\frac{\tau}{2}\cdot \mathbf{W}_\alpha + \frac{i}{2}g_1 B_\alpha\right)\Phi\right|^2 - V(\Phi),$$

where $V(\Phi)$ is the scalar field potential, built to be invariant under $SU_2 \times U_1$. Its most general form, compatible with renormalizability, is chosen to be

$$V(\Phi) = \lambda(\Phi^*\Phi - v^2)^2,$$

where λ is a positive dimensionless parameter, and v is a constant with the dimension of mass. It is arranged so that this potential assumes its minimum value for a nonzero value of Φ, $|\Phi|^2 = v^2$. In this manner, the symmetry carried by Φ is spontaneously broken by its configuration of minimum energy, not in the equations of motion. The extremum $|\Phi| = v$ defines the physical vacuum. It is convenient to expand the scalar field about its vacuum value as

$$\Phi(x) = \exp\left[\frac{i}{2v}\tau \cdot \omega(x)\right][\rho(x) + v]\begin{pmatrix} 0 \\ 1 \end{pmatrix}.$$

In this form, the four parameters of the field are separated into the three Nambu-Goldstone bosons, $\omega(x)$, and a neutral Higgs field $\rho(x)$. As per the Higgs mechanism, the three $\omega(x)$ are eaten by the three gauge fields and become their longitudinal degrees of freedom. The neutral ρ field remains behind as a real particle, predicted by the electroweak theory.

The scalar Lagrangian at minimum yields

$$\mathcal{L}_s = \frac{v^2}{4}g_2^2|W_\alpha^1 + iW_\alpha^2|^2 + \frac{v^2}{4}(g_2 W_\alpha^3 - g_1 B_\alpha)^2 + \cdots,$$

which gives the $W^\pm$ and the Z^0 their requisite masses

$$M_W^2 = \frac{1}{2}v^2 g_2^2; \quad M_Z^2 = M_W^2 \cos^{-2}\theta_w.$$

The Z^0 is more massive than the $W^\pm$ and the vacuum value of Φ, v, is fully determined in terms of G_F. The relation between the $W^\pm$ and Z^0 masses fixes the strength of the neutral and charged currents to be the same; it is a characteristic of the way the symmetry has been broken.

On the other hand, the mass of the neutral Higgs field, given by

$$M_H^2 = 8\lambda v^2,$$

is undetermined. In addition the scalar field has Yukawa couplings to the fermions. For the leptons, they are of the form

$$\left[m_e\begin{pmatrix} v_e \\ e \end{pmatrix}_L^\dagger e_R + m_\mu \begin{pmatrix} v_\mu \\ \mu \end{pmatrix}_L^\dagger \mu_R + m_\tau \begin{pmatrix} v_\tau \\ \tau \end{pmatrix}_L^\dagger \tau_R\right]\frac{\Phi}{v},$$

such that, at minimum, the charged lepton fields become massive. The Higgs interactions therefore break universality, with strength proportional to the mass of the fermion they couple to. Note that the electroweak Lagrangian does not allow the neutrinos to have (or even acquire radiatively) Majorana mass terms of the form $v_L^T \sigma_2 v_L$. Their absence is a sign that the electroweak theory conserves lepton number.

The quark masses and mixing angles are generated in the same way through couplings of the form

$$a \begin{pmatrix} u \\ d' \end{pmatrix}_L^\dagger d_R \Phi + b \begin{pmatrix} u \\ d' \end{pmatrix}_L^\dagger u_R \tilde\Phi + \cdots$$

appearing in all possible combinations that preserve $SU_2 \times U_1$ [$\tilde\Phi \equiv \tau_2 \Phi^*$]. There are in this sector ten Yukawa coupling constants, corresponding to the masses of the six quark flavors and the four mixing angles including CP violation. Again, the Higgs field coupling is proportional to the mass of the quark it couples to. The Higgs sector of the electroweak theory thus contains at least 14 unknown parameters: the Higgs mass, the fermion masses, and mixing angles. In spite of its apparent complexity, this particular breaking mechanism is very beautiful: it breaks the gauge symmetry while keeping renormalizability, it automatically leaves U_1^γ invariant, and it provides masses to the fermions. On the negative side it sheds no light on the values of the fermion masses and mixing angles. At this level, they appear as free parameters in the theory.

Experimentally, the electroweak model is well on its way to being verified. First of all, the old (V-A) effective Hamiltonian of the charge-changing interaction is reproduced with

$$G_F = \frac{g_2^2}{4\sqrt{2}m_w^2} = \frac{1}{2\sqrt{2}v^2},$$

so that $v \approx 188$ GeV. Now for the new predictions: the neutral current part of H_w is given by

$$H_w^n = \frac{1}{2} \frac{g_1^2}{\sin^2\theta_w m_Z^2} (J_\alpha^3 - \sin^2\theta_w J_\alpha^\gamma)^2,$$

and its strength is the same as that of the charged interactions, with this particular $\Delta I_w = 1/2$ Higgs breaking. It is responsible for totally new interactions such as elastic processes $v_\mu e \to v_\mu e$, $\bar v_\mu e \to \bar v_\mu e$, $v_\mu p \to v_\mu p$, $\bar v_\mu p \to \bar v_\mu p$, inelastic processes $v_\mu N \to v_\mu N \pi^0$, inclusive processes $v_\mu N \to v_\mu X$, and others. All have been measured to occur at the predicted electroweak rate. In inclusive hadron experiments one finds (6) a consistent value for the Weinberg angle

$$\sin^2\theta_w = 0.216 \pm 0.010 \quad \text{(world average)},$$

as well as the relative strength between charged and neutral currents processes

$$\frac{m_W^2}{m_Z^2 \cos^2\theta_w} \equiv \rho = 0.999 \pm 0.025 \quad \text{(world average)}.$$

Polarization experiments (41) have directly verified the presence of parity violation in the electron neutral current. The electroweak theory predicts at least three new particles, the $W^\pm$, the Z^0, and the spinless Higgs ρ, all with definite decay mechanisms (41a). The $W^\pm$ and Z^0 masses can be determined in terms of the Weinberg angle, the electromagnetic coupling constant, and the Fermi constant. Since the most precise determinations of $\sin^2\theta_w$ come from hadronic experiments, corrections due to quark dynamics (QCD) must be carefully included, leading to the values (6, 42)

$$M_W = 83^{+3.0}_{-2.8} \text{ GeV} \qquad M_Z = 93.8^{+2.5}_{-2.4} \text{ GeV},$$

with the major uncertainties coming from our imprecise knowledge of the value of the QCD coupling constant. It is within the capability of the CERN $p\bar{p}$ collider to detect the decay of the W^- into $\bar{v}_e e$ as well as the Z^0 decay into lepton pairs. The Z^0 does not couple as strongly as the $W^\pm$ and one expects ten $W^- \to e\bar{v}_e$ events for each $Z^0 \to \ell^+\ell^-$ event, although the Z^0, with its decay into two charged particles, is more detectable than the $W^\pm$, which loses half of its energy into v_e. It is rumored that the CERN collider experiment has indeed yielded $W^\mp$ decays at the requisite rate. Another expected effect of the Z^0 is to make itself felt in corrections to electromagnetic processes due to photon-Z^0 interference.

The third predicted particle is the neutral Higgs. Unfortunately its coupling to a fermion of mass m is depressed over the gauge coupling by m/M_W. Hence the best hope for observing it lies in the heavy particle sector. Up to now, experimental searches have set a limit of $M_H \gtrsim 6$ GeV. When the top quark is found, it is expected that the toponium $\bar{t}t$ bound state will decay into the Higgs by emitting a photon whose spectrum will signal the Higgs particle (43).

2.4 Physics at 10^{-16} cm

We can now summarize the resulting picture as follows: at the scale of 10^{-16} cm, the three interactions are of the Yang-Mills type, with $SU_2^{lw} \times U_1^Y \times SU_3^C$ gauge symmetry. The strong interaction has become QCD, with a large, but perturbative coupling constant α_3 estimated to be between 0.10 and 0.15, and decreasing relatively fast with inverse length l^{-1}, according to the renormalization group equation

$$\alpha_3^{-1}(l^2) = \alpha_3^{-1}(l_0^2) + \frac{1}{6\pi}(33-2n_f)\ln(l_0 l^{-1}),$$

where n_f is the number of quark flavors (at present $n_f = 5$, without the t quark) and l_0 is an arbitrary length scale. The electroweak theory has two independent coupling constants α_1 and α_2, with their ratio given by the Weinberg angle. The hypercharge coupling constant, α_1, is increasing with inverse length l^{-1} according to

$$\alpha_1^{-1}(l^2) = \alpha_1^{-1}(l_0^2) - \frac{n_f}{3\pi}\ln(l_0 l^{-1}),$$

and the weak isospin coupling constant α_2 is decreasing with inverse length l^{-1} according to

$$\alpha_2^{-1}(l^2) = \alpha_2^{-1}(l_0^2) + \frac{1}{3\pi}(11 - n_f)\ln(l_0 l^{-1}).$$

At this stage, several remarks are in order. First, the knowledge of the electromagnetic constant at small energy ($\alpha_{em} = 1/137$) together with the knowledge of the Weinberg angle, enables one to determine the value of α_2 and α_1 at 10^{-16} cm. One finds them both to be very small, and, more significantly, one finds α_2 to be larger than α_1, which allows them to meet at a much shorter scale, where they are both small (perturbative). Second, the QCD coupling constant α_3 is much larger at 10^{-16} cm than is α_2, as determined from experiment, and it is decreasing much faster with energy than α_2, so that it will also meet α_2 in the perturbative region. Thus both α_3 and α_1 will cross α_2 at a yet smaller scale. Do α_1, α_2, and α_3 all meet at the same point?

Furthermore, the Yang-Mills charges according to $SU_2^{Iw} \times U_1^Y \times SU_3^C$ can be used to classify all the spin-1/2 quarks and leptons. One finds the lowest-lying particles (within each charged sector) to have the quantum numbers

$$(2, 1^C)_{-1/2} + (1, 1^C)_1 + (1, \bar{3}^C)_{1/3} + (1, \bar{3}^C)_{-2/3} + (2, 3^C)_{1/6},$$

$$\begin{pmatrix} \nu_{eL} \\ e_L \end{pmatrix} \quad \bar{e}_R \quad \bar{d}_R \quad \bar{u}_R \quad \begin{pmatrix} u_L \\ d_L \end{pmatrix}$$

where the first entry in the parentheses denotes the weak isospin representation, the second entry the color representation, and the subscripted number the hypercharge value. All fields are left-handed so that right-handed fields appear conjugated. We have neglected the quark mixing angles. This quantum number pattern is repeated with particles of slightly higher mass, forming the second family

$$\begin{pmatrix} \nu_{\mu L} \\ \mu_L \end{pmatrix} \quad \bar{\mu}_R \quad \bar{s}_R \quad \bar{c}_R \quad \begin{pmatrix} c_L \\ s_L \end{pmatrix},$$

and the yet heavier third family with the tentative assignment

$$\begin{pmatrix} \nu_{\tau L} \\ \tau_L \end{pmatrix} \quad \bar{\tau}_R \quad \bar{b}_R \quad \bar{t}_R \quad \begin{pmatrix} t_L \\ b_L \end{pmatrix}.$$

Of these, the charge-2/3 t quark still awaits discovery. We include it here because observed pattern of b-containing matter seems to indicate the same family structure. On the theoretical side, one knows that some new particle must exist, because the τ, ν_τ, and b by themselves have a chiral anomaly that has to be cancelled to avoid nonrenormalizability. The neutrino partner of the τ^- lepton has, of course, not been detected directly but kinematics indicate that when the left-handed τ^- decays, it emits a light particle of mass $\lesssim 200$ MeV.

The apparent triplication of this $SU_3^C \times SU_2^{l_W} \times U_1^Y$ quantum number pattern stands as one of the great mysteries confronting theorists. As discussed below, grand unification offers no explanation for it. Having described how matter and its interactions appear at 10^{-16} cm, we are now ready to explore the possibility that at yet shorter distances these three gauge interactions unify into one grand unified theory (GUT).

3. GRAND UNIFICATION

Once the strong interactions were recognized to be a gauge theory in disguise, theorists could not resist the urge to unify the three "low-energy" gauge theories into one (for reviews and extensive references, see 44). The earliest attempt of Pati & Salam (24) unified quarks and leptons by generalizing SU_3^C to SU_4 where the leptons provide the fourth color. Leptons escape confinement when the SU_4 symmetry is broken to SU_3^C while its coupling constant is still very weak. However, the most economical way to mesh the $SU_3 \times SU_2 \times U_1$ symmetry into one is provided by the SU_5 model of Georgi & Glashow (25), which we now proceed to describe.

3.1 The SU_5 Model

The gauge theory that unifies the three low-energy gauge theories must have a gauge symmetry at least as large as $SU_3 \times SU_2 \times U_1$ and must reproduce the observed quantum number pattern for the light spin-1/2 particles. Of the many symmetry groups in which $SU_3 \times SU_2 \times U_1$ can fit, the most economical group with the complex representations necessary to describe the left-handed spinors is SU_5 (25, 45). Recall that a Yang-Mills theory with a gauge symmetry such as SU_5 is characterized by one coupling constant and by spinor fields that transform as SU_5 representations. The representations of SU_5 are all obtained by taking products of its fundamental complex five-dimensional representation **5** and its conjugate **5̄**. For our purposes it suffices to list a few such representations, for example, the real adjoint representation **24** obtained in $5 \times \bar{5} = 1 + 24$, and the complex decuplet **10** obtained in the antisymmetric product of two quintets $(5 \times 5)_A = 10$. Thus SU_5 is a 24-parameter group and we expect 24 gauge bosons among which eight must be the QCD vector gluons, three the weak isospin bosons, and one the hypercharge boson; the remaining twelve new bosons predicted by the theory will cause new interactions, such as proton decay.

The low-mass left-handed spinors fit beautifully into two representations of SU_5:

$$\bar{5} = (2, 1^c) + (1, \bar{3}^c); \quad 10 = (1, 1^c) + (2, 3^c) + (1, \bar{3}^c),$$

$$\begin{pmatrix} \nu_{eL} \\ e_L \end{pmatrix} \quad \bar{d}_R \qquad \bar{e}_R \quad \begin{pmatrix} u_L \\ d_L \end{pmatrix} \quad \bar{u}_R$$

obtained by decomposing the $\bar{5}$ and 10 representations of SU_5 into its $SU_3 \times SU_2 \times U_1$ subgroups. This assignment automatically reproduces the charges of quarks and leptons, especially the factor of 1/3 relating quark and lepton charges; as such it provides an understanding for the equality between proton and positron charges. This is because the electric charge is an SU_5 quantum number so that its sum over a representation vanishes. For instance, in the 10, we note that, while the charge of u_L cancels that of $\bar{u}_R$ (because electromagnetism conserves parity), the charge of $\bar{e}_R$ must be opposite to that of the three colors of d_L. In this way the factor of 3 is related to the number of colors and to the parity-conserving nature of electromagnetism.

In the SU_5 theory the other families are obtained by triplicating this $\bar{5} + 10$ pattern: this simple GUT provides no clue to the triplication of fermion families except that it does tend to arrange the particles according to their masses. Hence, ambitious as they may seem, GUTs still appear to be incomplete theories.

The remarkable success of SU_5 in fitting the quantum numbers of known particles into its representations is matched by an analysis of the way in which the three low-energy gauge coupling constants α_1, α_2, and α_3 mesh into one (26). Recall that when last seen these coupling constants were heading for one another, with α_1 increasing, α_2 larger than α_1 and decreasing, and α_3 larger than α_2 and decreasing faster than α_2 with energy. It is astounding that, given the experimental value of the Weinberg angle, the approximately known value of α_3, and the electromagnetic coupling at, say, 10^{-16} cm, the renormalization group equations lead these three coupling constants to meet all at the same point!

Alternatively we can turn the argument around and say that the SU_5 theory predicts a value for the Weinberg angle that is in agreement with experiment, given the experimental value of the QCD and QED couplings at 10^{-16} cm. The reasoning goes as follows. According to SU_5 all three couplings α_1, α_2, and α_3 must meet at one point. Thus boundary values for

α_1 and α_3 fix that point, which determines α_2 and therefore the ratio $\alpha_2\alpha_1^{-1}$ at lesser energy. The output is the Weinberg angle and the scale at which the couplings meet. One finds that scale to be approximately $V \simeq 10^{15}$ GeV. The occurrence of such an enormous scale in a theory designed to classify phenomena appearing in the laboratory is perhaps the most profound aspect of grand unification. In a single stroke it predicts no new fundamental gauge interactions between 10^{-16} and 10^{-28} cm. At this very short scale it predicts new interactions that could have been directly visible only in the very early universe, and it brings a unified world of elementary particles at the threshold of gravity whose characteristic scale is given by the Planck length $l_p \approx 10^{-33}$ cm. On the theoretical side it creates the brand new problem of understanding how to handle a theory with widely different scales: the scale at which the SU_5 symmetry is broken, $\sim 10^{15}$ GeV, and the scale at which the electroweak theory is broken, $\sim 10^2$ GeV. This enormous disparity of scales cannot be easily understood nor maintained in perturbation theory (34), a difficulty known as the gauge hierarchy problem.

The SU_5 model predicts new gauge bosons that can cause new effective interactions among the known low-mass particles in much the same way the electroweak $W^{\pm}$ generate the effective V-A theory of beta decay. These new gauge bosons have the quantum numbers of the adjoint representation of SU_5, which decomposes as

$$\mathbf{24} = (\mathbf{1}, \mathbf{8}^c)_0 + (\mathbf{3}, \mathbf{1}^c)_0 + (\mathbf{1}, \mathbf{1}^c)_0 + (\mathbf{2}, \mathbf{3}^c)_{-5/6} + (\mathbf{2}, \bar{\mathbf{3}}^c)_{5/6}$$

under $SU_2 \times SU_3 \times U_1$. The new spin-1 particles of the SU_5 theory are called the X and Y bosons, and their antiparticles $\bar{X}$, $\bar{Y}$. Both are color triplets with respective electric charge of $-4/3$ and $-1/3$ and form a weak isodoublet. According to the Yang-Mills construction, one can read off the currents they couple to:

$$J_\alpha^X = e_L^\dagger \sigma_\alpha \tilde{d}_L + \tilde{u}_L^\dagger \sigma_\alpha u_L + \cdots$$

$$J_\alpha^Y = v_{eL}^\dagger \sigma_\alpha \tilde{d}_L + \tilde{u}_L^\dagger \sigma_\alpha d'_L + \cdots,$$

where the tilde indicates $\tilde{\psi}_L \equiv \sigma_2 \psi_R^*$ for any field.

The X and Y bosons are seen to mediate both lepton-quark and quark-antiquark transitions. For this reason they are commonly called leptoquarks. Their effect among low-mass particles is to generate an effective Hamiltonian of the form

$$\frac{g_5^2}{8m_X^2} J_\alpha^{X\dagger} J_\alpha^X + \frac{g_5^2}{8m_Y^2} J_\alpha^{Y\dagger} J_\alpha^Y,$$

where g_5 is the SU_5 coupling constant and m_X, m_Y are the masses of the leptoquarks. However, since the SU_5 symmetry is broken at very large energy, their masses are necessarily very large

$$m_{X,Y} \approx g_5 V,$$

so that the strength of the four-fermi interaction induced by X or Y exchange has an effective coupling G_{GG} of the order of 10^{-30} $(\text{GeV})^{-2}$ [to be compared with $G_F \approx 10^{-5}$ $(\text{GeV})^{-2}$ for beta decay]. This is just as well since X and Y exchange causes proton decay by changing two quarks into an antiquark-antilepton pair! The prototype decay modes are

$$p \to e^+ \pi^0 \text{ (X exchange)} \qquad p \to \bar{\nu}_e \pi^+ \text{ (Y exchange)}.$$

In the minimal SU_5, the predicted lifetime for the proton (46),

$$\tau_p \simeq 10^{30 \pm 1} \text{ years} \qquad \text{(minimal } SU_5\text{)},$$

turns out to be slightly larger than the experimental bound, 3×10^{29} years. Note that we have quoted an error in the exponent (just as in astrophysics); it is due to experimental uncertainties in determining α_3 and to theoretical uncertainties concerning the wavefunctions of quarks inside the proton. We should note that this particular value for the proton lifetime is a prediction of the minimal SU_5 model. As this review was being completed, a more stringent bound, 6.5×10^{31} years, on the decay mode $p \to e^+ \pi^0$ was announced by the IMB collaboration (46a); this would seem to put the minimal SU(5) model in conflict with this bound. It can be increased by adding new particles to the theory. Also, in more complicated models, other interactions such as Higgs exchange can become the dominant mechanism for proton decay. If the proton decays through this latter type of mechanism, it will display lack of universality, modulo the quark mixing angles. The closeness of the theoretical prediction to the existing experimental bound has motivated many experimental groups to go underground (to cut down on background events) and monitor large chunks of matter. So far these searches have not yet established the existence of proton decay.

Any description of the SU_5 theory would be incomplete if it did not include a discussion of the Higgs particles, which are responsible for breaking the SU_5 symmetry (45). The symmetry must be broken in two stages: stage I when SU_5 is broken down to $SU_3^C \times SU_2^{I_w} \times U_1^Y$ at about 10^{-28} cm; stage II when $SU_3^C \times SU_2^{I_w} \times U_1^Y$ is broken down to $SU_3^C \times U_1^\gamma$ at about 10^{-16} cm. We have already seen how to achieve the second breaking (stage II) within the electroweak theory by including Higgs that transform as weak doublets. In SU_5, the stage I symmetry breaking is most economically achieved by introducing a spinless field transforming as the **24** of SU_5. Its coupling to the SU_5 gauge bosons is dictated by the Yang-Mills construction and it can have in addition self-interactions made up of

SU_5-invariant terms at most quartic to avoid nonrenormalizability. Such a potential can depend on at least three parameters, a mass M and two dimensionless coupling constants. It is convenient to represent the twenty-four Higgs fields H^A in traceless matrix form $H \equiv H^A T_r^A$ where T_r^A are properly normalized SU_5 matrices in its r representation. The potential is taken to be

$$V_{24}(H) = -\frac{1}{2}M^2 \operatorname{Tr}(H^2) + \frac{1}{4}a[\operatorname{Tr}(H^2)]^2 + \frac{1}{2}b \operatorname{Tr}(H^4).$$

A simple calculation shows that V_{24} has an absolute minimum corresponding to the required stage I breakdown as long as the coupling constants satisfy the inequality $b > 0$, $a > -7b/15$. Then at the minimum of V_{24}, the matrix H can be written as

$$H = \frac{h}{\sqrt{60}}\begin{pmatrix} 2 & & & & \\ & 2 & & & \\ & & 2 & & \\ & & & -3 & \\ & & & & -3 \end{pmatrix}; \quad h = M\left(\frac{30}{15a+7b}\right)^{1/2},$$

which clearly achieves the desired stage I breaking. This results in equal masses for the X and Y bosons

$$m_X^2 = m_Y^2 = \frac{5}{12}g_5^2 h^2.$$

Of the 24 spinless particles, twelve are used to become the longitudinal degrees of freedom of the X and Y bosons. The remaining twelve with quantum numbers of the $SU_3 \times SU_2 \times U_1$ gauge bosons, i.e. $(\mathbf{1},\mathbf{8}^c)+(\mathbf{3},\mathbf{1}^c)+(\mathbf{1},\mathbf{1}^c)$, all acquire very large masses. Since the H fields cannot couple to fermions in an SU_5-invariant way, their effect on the low-energy theory is negligible.

The electroweak (stage II) breaking is achieved by means of spinless fields Φ transforming as $\mathbf{5}$ or $\bar{\mathbf{5}}$ under SU_5, so that, in addition to the usual Weinberg-Salam doublets, SU_5 invariance introduces color triplet, iso-singlet Higgs fields. The most general renormalizable potential for this field alone depends on two parameters, a mass and a coupling constant:

$$V_5(\Phi) = -\frac{1}{2}m^2\Phi^\dagger\Phi + \frac{\lambda}{2}(\Phi^\dagger\Phi)^2.$$

However, this is not the whole story, for renormalizability allows terms in the potential that contain both Φ and H. Actually such terms are necessary in order to assure that the stage II breaking does break SU_2^{lw} and not, for instance, SU_3^C. These cross-terms are of the form

$$V_{5-24} = c\Phi^\dagger \Phi \, \text{Tr}\, H^2 + d\Phi^\dagger H^2 \Phi.$$

The d coupling constant is chosen negative so as to minimize the value of the potential when Φ breaks SU_2. The full potential depends on two masses and five coupling constants. Its parameters must be carefully chosen so as to reproduce the required breaking scales. The minimization of the full potential $V_{24} + V_5 + V_{5-24}$ yields the required pattern with the vacuum value of H slightly altered by the electroweak breaking to include an isovector-breaking component, with strength of the order of v^2/V, i.e. much weaker than electroweak breaking. One can easily arrange the ratio v/V to be of the required order of 10^{-12} if the input parameters are carefully adjusted. At present there is no understanding of the systematics behind such an adjustment. The problem is made worse by radiative corrections, which at each order of perturbation theory detune the delicate cancellations that must take place to reproduce the 10^{-12} gauge hierarchy. One is free, of course, to give up control of the gauge hierarchy and leave it to a future theory to explain why the renormalized parameters obey such precise relations. Otherwise one can appeal to a symmetry that allows control of the radiative corrections, and thus alleviates the technical problem but still leaves unanswered the reasons behind the hierarchy. Supersymmetry is such a symmetry; it relieves the technical pains but still does not explain the hierarchy.

Just as in the electroweak theory, the Higgs that break the electroweak symmetry have the right quantum numbers to couple to the fermions. The quantum numbers of the fermion masses are bilinear in the fermion fields, thus belonging to the product of fermion representations

$$\bar{5} \times 10 = 5 + 45; \qquad 10 \times 10 = \bar{5} + \overline{45} + 50.$$

The minimal SU_5 theory introduces only Higgs particles transforming as the **5** of SU_5, which, as we have seen, serve to break the electroweak theory. With this minimal Higgs structure, the masses of the charge-(-1) leptons are related to those of the charge-$(-1/3)$ quarks, because they both lie in the $\bar{5} \times 10$ combination. The mass of the charge-$2/3$ quark of each family lies in the 10×10 combination. At a scale where the theory is exact ($\sim 10^{-28}$ cm), having only a Higgs **5** (and $\bar{5}$) implies that the mass of the charge-(-1) lepton is exactly equal to the mass of the charge-$(-1/3)$ quark. However, as SU_5 is broken, this mass ratio gets affected by radiative correction since the quark masses get corrected by the electroweak force and by the much stronger color force while the lepton mass gets its corrections only from electroweak forces. The net result is that at laboratory energies, the ratio gets enhanced roughly by a factor of 3. This relation turns out to be very good for the third family, yielding (45)

$$\left.\frac{m_b}{m_\tau}\right|_{q^2=4\,\text{GeV}^2} \simeq 3,$$

and disastrous for the two lighter families, where it gives

$$\frac{m_d}{m_s} \simeq \frac{m_e}{m_\mu},$$

off by a factor of 10 from experiment. We have combined two results from the two lighter families into one that involves only quark mass ratios, which are much better known from current algebra than the quark masses themselves. Thus the minimal Higgs assignment is desirable only for the heaviest family, and certainly unacceptable for the lightest family. This means that the coupling of scalar to fermion of the lighter families has to be altered by adding to the theory new Higgs fields transforming according to the **45** or **50** of SU_5. This provides another indication of the incompleteness of the theory.

In SU_5, the fermions of each family arrange themselves in the reducible representation $\bar{5}+10$. This creates a global phase symmetry resulting after electroweak breaking in conservation of baryon number minus lepton number, $B-L$. This conservation has two results: the proton decay modes will of course respect this symmetry and the neutrinos remain massless in the theory. This is because a left-handed neutrino can have a Majorana mass that has lepton number 2 and transforms as an isotriplet. While the isotriplet can be generated in higher orders of perturbation theory, $B-L$ conservation prevents the Majorana mass from ever appearing in the theory. In the simplest generalizations of SU_5 that introduce new spin-1/2 particles, $B-L$ is no longer conserved and the neutrinos are expected to get tiny masses since they are no longer protected by $B-L$.

At this point, a new feature of the SU_5 GUT transition should be mentioned. When SU_5 breaks into $SU_3^C \times SU_2^{I_w} \times U_1^Y$, the equations of motion of the theory have magnetic monopole solutions. These arise within the framework of Yang-Mills theories whenever a simple group is broken into a subgroup containing a phase U_1 symmetry. As pointed out long ago by Dirac, magnetic monopoles imply charge quantization, which is exactly what we found earlier by direct quantum number matching. The existence of such solutions implies that magnetic monopoles were produced when the SU_5 symmetry was broken; these monopoles are very heavy and their abundance is severely limited by cosmological considerations. Particle theory offers no suggestions as to their abundance although it is reasonable to assume that they were created at the rate of one per causally related region of space-time because monopoles exist only within regions where the orientation of the Higgs fields can be correlated. Their abundance is therefore linked to the number of horizons that have evolved into our visible universe. We should also note that the GUT symmetry breaking is thought to cause a first-order phase transition in the very early universe, an idea that has revolutionized the standard cosmological models. Lastly, the existence of baryon-number-violating processes in the very early universe has provided a conceptional framework for the understanding of the predominance of matter over anitmatter in our present universe.

It is perhaps appropriate at this point to take stock of the salient features of grand unified theories. Given that the three nongravitational interactions have revealed themselves to be gauge theories, the impulse to unify them into one is irresistible. As a result of such a unification, the description of the quantum numbers of the known fermions is greatly simplified, reproducing the quark-to-lepton electric charge ratio. However, fermions are still not completely unified into one entity: three families of fermions emerge, one almost massless and the other two with increasing masses. In the absence of new particles the three low-energy gauge coupling constants mesh naturally into one at the very small scale of at least $\sim 10^{-28}$ cm, roughly reproducing the experimental value of the Weinberg angle. The appearance of two vastly different scales in one theory, namely 10^{-28} cm and 10^{-16} cm, poses entirely new problems. One is purely technical and addresses itself to the mechanics of maintaining such a small ratio of scales in the face of detuning radiative corrections. It can be resolved by appealing to supersymmetry. The second problem is to understand how such a small ratio comes about. Within the framework of perturbative field theory, the scales and the ways in which gauge symmetries are broken, the fermion masses and mixing angles, are all consequences of the Higgs sector of the theory. Thus it can be said that while GUT provides unifying principles for the vector gauge couplings, all its shortcomings can be blamed on the spinless sector of the theory. No elementary spinless particles have been experimentally observed to date, much less their interactions, and therein lie all the uncertainties associated with GUTs. We should also mention that attempts at understanding symmetry breakings as a result of fermion condensates [technicolor (47)] have so far proved wanting and in fact have served to underline the efficiency of the Higgs particles. Thus while a good part of natural phenomena can be explained by gauge spin-1 interactions, we are at present dependent on spinless Higgs particles to explain the rest. The lack of an organizing principle for the Higgs sector of gauge theories is the Achilles' heel of GUTs.

3.2 GUT Generalizations

There are, of course, many ways to unify the three low-energy gauge interactions—we have presented only the simplest one. However, all require more elementary fermions than the minimal SU_5 does and none succeeds in explaining the fermion multiplicities, masses, and mixing angles in a convincing way. Still, these models give different predictions for the proton lifetime, predictions that tend to be longer than 10^{31} years, and feature neutrino oscillations. Thus they are interesting in their own right.

Perhaps the nicest extension of the minimal model is the SO_{10} model (27). Since SU_5 is a subgroup of SO_{10}, it suffices to discuss the SO_{10} model in SU_5 language. Here the world is described by an SO_{10}-invariant Yang-Mills theory with three fermion families, each transforming as a complex 16-dimensional spinor representation. Under SU_5 it decomposes as

$$\mathbf{16 = \bar{5} + 10 + 1},$$

so that each SU_5 family is unified at the cost of introducing an extra neutral spinor particle, which has no electroweak interactions. As a result, each family contains two neutral particles, v_L and N_L, where v_L has weak isospin 1/2 and N_L is isoscalar. Then it is easy to see that any mechanism for giving a mass to the charge-2/3 quark gives the same Dirac mass to the neutrino, treating $\tilde{N}_L$ as the right-handed partner of v_L. However, disastrous consequences are avoided when, according to the Gell-Mann-Ramond-Slansky mechanism (29), it is realized that N_L itself can have a Majorana mass of the same magnitude as the inverse scale at which SO_{10} is broken. Then the neutral mass matrix looks like

$$\begin{pmatrix} 0 & m \\ m & M \end{pmatrix},$$

where m is of order of electroweak breaking and M is of order of SO_{10} breaking. As a result, the usual left-handed neutrino develops only a tiny mass of $\sim m(m/M)$, depressed from the usual quark masses by the ratio of the electroweak to the SO_{10}-breaking scale. It also suggests that the lightest neutrinos belong to the lightest family and that the best hope for detection of the concomitant neutrino oscillations lies in the $v_\tau - v_\mu$ sector.

Generalizations of the minimal SU_5 model fall in at least two categories: those that increase the size of the gauge symmetry and those that impose supersymmetry on any of the existing gauge models. As we have already mentioned, one does not seem to gain very much by extending the gauge symmetry. There exists no model where the fermion mass patterns and multiplicities can be understood as part of one symmetry: the problem of family triplication still remains although there have been attempts to link the number of families to a family group (48). In some models, fermions and spinless particles are predicted to exist between the electroweak and GUT scale, with the result that the unification point is pushed to yet shorter distances. In those models, the proton lifetime is lengthened over the minimal SU_5 value. However, there is no compelling reason for any of these models, and since they offer no new testable predictions at low energy, we simply refer the reader to the literature (e.g. 49).

There exists a generalization of GUT models (e.g. 50) that may be sufficiently outrageous to be true: SuperGUT.

Supersymmetry, which was first conceived within the framework of string models (51), is a generalization of the space-time symmetry of special relativity (52), and is therefore different from the (internal) gauge symmetries we have considered so far. Its characteristics are to relate particles of different spins, and in fact exact supersymmetry demands an equal number of bosons and fermions with the same mass. As this is clearly not the case in nature, supersymmetry can only occur in broken form. On the positive side, it turns out that theories with supersymmetry display a remarkable resilience to radiative corrections; so much so in fact that some supersymmetric toy theories in four dimensions have no ultraviolet divergences!

Within the context of GUT, supersymmetry can solve the technical aspect of the gauge hierarchy problem. In a supersymmetric theory, it suffices to fix parameters at the Lagrangian level without worrying about the radiative corrections, now domesticated by supersymmetry. The reasons behind the delicate relations among the input parameters are still as mysterious as ever. The price for supersymmetry is the doubling of the number of elementary particles: for every known particle one has to add a supersymmetric partner, i.e. for each spin-1 add a spin-1/2, for each spin-1/2 (0), add a new spin-0 (1/2).

Although it is difficult to present in a model-independent way the experimental consequences of supersymmetric GUTs, the salient features are as follows. (a) The supersymmetric partners of the known particles cannot be too heavy if supersymmetry is to help with the gauge hierarchy problem. Hence it is likely that their effect will be observed first indirectly, as small deviations from the standard electroweak theory, and then directly by future accelerators probing the TeV region. (b) Proton decay takes place the fastest through mechanisms involving spinless particles. Since spinless particles seem to couple strongest to the heaviest particles, any mechanism induced by their exchange would tend to favor proton decay into the heaviest flavor allowed by energy conservation: strangeness. However, in supersymmetric theories the decay rates are not well known at all, and theorists are clearly at the mercy of experimentalists. (c) The X,Y-boson-induced proton decays are relatively suppressed because the gauge couplings meet at smaller distances than in usual GUT theories. (d) There is also the exciting possibility of having nonzero neutrino masses even at the minimal SuperGUT level, and of relating certain proton decay modes to neutrino oscillations.

It must be said, however, that the inclusion of supersymmetry into grand unification has not (so far) produced any tangible results except in assuaging theorists' concerns about the gauge hierarchy problem. Still, many theoretical factors make supersymmetry appealing to the theorist, one being that it allows for a consistent effective theory below the Planck mass.

4. CONCLUSIONS

We have tried to present the reader with an overview of the subject of grand unification. While there are convincing arguments for the three low-energy interactions to be gauge theories, it is by no means necessary for them to merge into one. Yet, given what we know, it is a most compelling scenario. The fact that it can be done with the known set of particles is an indication that it is not so far-fetched an idea. Of course, experiment will tell. Grand unification means that global symmetries such as baryon number and lepton number cannot be exact and one should expect, at some level, proton decay and neutrino oscillations. In the theoretical development of the idea, a fundamental role for Higgs particles has emerged. While no fundamental spinless particles seem to exist in the low-energy world, they are a theoretical necessity for describing gauge theories with broken symmetries. However, just as for spinors, we do not know the reason for their existence. In the absence of any obvious organizing principle in the Higgs sector, it is tempting to hypothesize that the Higgs particles are not fundamental. For instance, the peculiar fact that the unbroken symmetries are all vector-like (QCD and QED) and the broken symmetries are chiral, would seem to indicate a closer relationship between spin-1/2 and spin-0 particles. At present, however, one can only postulate the Higgs structure needed to reproduce the low-energy world. Perhaps another type of symmetry, such as (extended) local supersymmetry, will some day explain why the Higgs sector appears to be so unprincipled. This may be related to the fact that the theories we have described do not include gravity, and in this sense, are to be regarded as incomplete theories. Particle physicists have been led to consider matter and its interactions at very short scales (10^{-28}–10^{-33} cm) as a result of the unification of the three known nongravitational interactions. Let us hope that experimentalists will enlighten their search by detecting proton decay and other unforeseen phenomena!

ACKNOWLEDGMENT

I would like to thank Prof. P. Sikivie for his reading of the manuscript and critical comments.

Literature Cited

1. Yang, C. N., Mills, R. L. 1954. *Phys. Rev.* 96:191
2. Glashow, S. L. 1961. *Nucl. Phys.* 22:579; Salam, A., Ward, J. C. 1964. *Phys. Lett.* 13:168
2a. Arnison, G., et al. 1983. *Phys. Lett.* 122B:103–16
2b. Banner, M., et al. 1983. *Phys. Lett.* 122B:476–85
3. Anderson, P. W. 1963. *Phys. Rev.* 130:439; Higgs, P. W. 1964. *Phys. Lett.* 12:132; Englert, F., Brout, R. 1964. *Phys. Rev. Lett.* 13:321; Guralnik, G. S., Hagen, C. R., Kibble, T. W. B. 1964. *Phys. Rev. Lett.* 13:585
4. Weinberg, S. 1967. *Phys. Rev. Lett.* 19:1264; Salam, A. 1968. In *Elementary Particle Theory*, ed. N. Svartholm, p. 367. Stockholm: Almquist & Wiksells
5. 't Hooft, G. 1971. *Nucl. Phys.* B35:167; *Phys. Lett.* 37B:195
6. Bilenky, S. M., Hosiak, J. 1982. *Phys. Rep.* 90(2):73
7. Gell-Mann, M., Ne'eman, Y. 1964. *The Eightfold Way*. New York: Benjamin
8. Gell-Mann, M. 1964. *Phys. Lett.* 8:214
9. Zweig, G. 1964. CERN Rep. No. TH401. Unpublished
10. Gürsey, F., Radicati, L. 1964. *Phys. Rev. Lett.* 13:173; Sakita, B. 1964. *Phys. Rev.* 136B:1765
11. Greenberg, O. W. 1964. *Phys. Rev. Lett.* 13:598
12. Kirkby, J. 1979. *Proc. Int. Symp. on Lepton and Photon Interactions at High Energies*, ed. T. Kirk, H. Abarbanel. Batavia, Ill: Ferm: Natl. Accel. Lab.
13. Feynman, R. P. 1969. *Phys. Rev. Lett.* 23:1415
14. Nambu, Y. 1966. In *Preludes in Theoretical Physics*, ed. A. de Shalit. Amsterdam: North-Holland; Fritzsch, H., Gell-Mann, M. 1972. *Proc. XVI Int. Conf. on High Energy Physics*, ed. J. D. Jackson, A Roberts, vol. 2, p. 35. Batavia, Ill: Fermi Natl. Accel. Lab.
15. Politzer, H. D. 1973. *Phys. Rev. Lett.* 30:1346; Gross, D., Wilczek, F. 1973. *Phys. Rev. Lett.* 30:1343
16. Meshkov, S. 1982. *Sixth Workshop on Current Problems in High Energy Particle Theory, Firenze, Italy, June 1982*. Baltimore: Johns Hopkins Univ.; also UCI Rep. 82-35. Univ. Calif., Irvine
17. Buras, A. J. 1981. *Lepton-Photon Conf., Bonn, Germany*. Univ. Bonn Phys. Inst.; also *Fermilab-Conf-81/69*. Batavia, Ill: Fermi Natl. Accel. Lab.
18. Quigg, C., Rosner, J. L. 1979. *Phys. Rep.* 56:167
19. Richter, B. 1977. *Rev. Mod. Phys.* 49:251
20. 't Hooft, G. 1976. *Phys. Rev. Lett.* 37:8
21. Hasenfratz, P. 1982. See Ref. 16
22. Peccei, R., Quinn, H. 1977. *Phys. Rev. Lett.* 38:1440; *Phys. Rev.* D16:1791
23. Weinberg, S. 1978. *Phys. Rev. Lett.* 40:223; Wilczek, F. 1978. *Phys. Rev. Lett.* 40:279
24. Pati, J., Salam, A. 1973. *Phys. Rev.* D8:1240
25. Georgi, H., Glashow, S. L. 1974. *Phys. Rev. Lett.* 32:438
26. Georgi, H., Quinn, H., Weinberg, S. 1974. *Phys. Rev. Lett.* 33:451
27. Fritzsch, H., Minkowski, P. 1975. *Ann. Phys.* 93:193; Georgi, H. 1975. In *Particles and Fields, 1974 (APS/DPF Williamsburg)*, ed. C. E. Carlson. New York: Am. Inst. Phys.
28. Gürsey, F., Ramond, P., Sikievie, P. 1975. *Phys. Lett.* B60:177
29. Gell-Mann, M., Ramond, P., Slansky, R. 1979. In *Supergravity*, ed. P. Van Nieuwenhuizen, D. Z. Freedman. New York: North-Holland; and Ramond, P. 1979. Sanibel Talk, February. CALT 68-709, unpublished; Yanagida, T. 1979. *Proc. Workshop on Unified Theory and the Baryon Number of the Universe* KEK
30. Yoshimura, M. 1978. *Phys. Rev. Lett.* 41:28; Sakharov, A. D. 1967. *JETP Lett.* 5:24
31. Guth, A. 1981. *Phys. Rev.* D23:347
32. 't Hooft, G. 1974. *Nucl. Phys.* B79:276; Polyakov, A. M. 1976. *Pis'ma Zh. Eksp. Teor. Fiz.* 20:430
33. Preskill, J. P. 1979. *Phys. Rev. Lett.* 43:1365
34. Gildener, E. 1976. *Phys. Rev.* D14:1667
35. Fayet, P., Ferrara, S. 1977. *Phys. Rep.* 32C:249
36. Ferrara, S., Friedman, D. Z., Van Nieuwenhuizen, P. 1976. *Phys. Rev.* D13:3214; Deser, S., Zumino, B. 1976. *Phys. Lett.* 62B:335
37. Marciano, W., Pagels, H. 1978. *Phys. Rep.* 36C:137
38. Bjorken, J. D. 1966. *Phys. Rev.* 148:1467
39. Glashow, S. L., Iliopoulos, J., Maiani, L. 1970. *Phys. Rev.* D2:1285
40. Kobayashi, M., Maskawa, T. 1973. *Prog. Theor. Phys.* 49:652
41. Prescott, C. Y. et al. 1978. *Phys. Lett.* 77B:347
41a. Ellis, J., Gaillard, M. K., Girardi, G., Sorba, P. 1982. *Ann. Rev. Nucl. Part. Sci.* 32:443–97
42. Marciano, W. J., Sirlin, A. 1981. *Phys. Rev. Lett.* 46:163
43. Wilczek, F. 1977. *Phys. Rev. Lett.* 39:1304
44. Langacker, P. 1981. *Phys. Rep.* 72C:185;

and Ross, G. G. 1981. *Rep. Prog. Phys.* 44:655
45. Buras, A. J., Ellis, J., Gaillard, M. K., Nanopoulos, D. V. 1978. *Nucl. Phys.* B135:66
46. Goldman, T. J., Ross, D. A. 1979. *Phys. Lett.* 84B:208; Marciano, W. J. 1979. *Phys. Rev.* D20:274
46a. Bionta, R. M., et al. 1983. *Phys. Rev. Lett.* 51:27
47. Susskind, L. 1979. *Phys. Rev.* D20:2619; Weinberg, S. 1976. *Phys. Rev.* D13:974
48. Wilczek, F., Zee, A. 1979. *Phys. Rev. Lett.* 43:1571; Ramond, P. 1979. Sanibel Talk, CALT 68-709, unpublished.
49. Zee, A. 1982. *Unity of Forces in the Universe.* Singapore: World Sci.
50. Dine, M. 1982. See Ref. 16
51. Ramond, P. 1971. *Phys. Rev.* D3:2413
52. Gol'fand, Yu. A., Likhtman, E. P. 1971. *JETP Lett.* 13:323; Wess, J., Zumino, B. 1974. *Nucl. Phys.* B70:39

PROGRESS AND PROBLEMS IN PERFORMANCE OF e+e− STORAGE RINGS

R. D. Kohaupt and G.-A. Voss

Deutsches Elektronen-Synchrotron DESY, Notkestrasse 85, D-2000 Hamburg 52, West Germany

CONTENTS

1. INTRODUCTION 68
    1.1 Energy 68
    1.2 Luminosity 69
    1.3 Energy Spread 71
    1.4 Background 72
    1.5 Polarization 72
2. ELECTRON AND POSITRON ACCUMULATION 73
    2.1 Bunch Coalescing Through Vernier Phase Space Compression 73
    2.2 Positron Accumulator Rings 74
3. SINGLE-BEAM LIMITATIONS 76
    3.1 Satellite Resonances 76
    3.2 Higher-Order Mode Losses 77
    3.3 Coherent Single-Beam Instabilities 78
4. SPACE CHARGE LIMITATIONS 83
    4.1 General Description 83
    4.2 Observations 84
    4.3 Theory and Computer Tracking Results 85
    4.4 Ways to Improve the Space Charge Limited Luminosity 86
    4.5 Variation of the Number of Bunches in Each Beam 87
    4.6 Adjustment of the Horizontal Emittance 87
    4.7 Low-beta Interaction Regions 89
    4.8 Chromaticity Problems 90
    4.9 Mini-beta Interaction Regions 91
    4.10 Micro-beta Interaction Regions 93
    4.11 Space Charge Compensation 93
5. POLARIZATION 94
    5.1 History and Observations 94
    5.2 Polarization and Depolarization 95
    5.3 Spin Rotators 98
6. CONCLUSION AND OUTLOOK 98

1. INTRODUCTION

Electron-positron storage rings have been in operation for high-energy physics since 1967. But even before 1967 much pioneering work had been done on the electron-positron ring ADA at Frascati (1), VEPP I at Novosibirsk (2), and the two intersecting Princeton-Stanford electron rings (3). It was the 380-MeV ring VEPP II at Novosibirsk where the first e^+e^- annihilation events were observed in 1966 (4); soon after that events were also observed at the ACO storage ring at Orsay (5).

Since then electron-positron storage rings have played an ever increasing role in elementary particle physics. The simplicity of the initial state of an electron-positron annihilation event, the fact that (given sufficient energy) one can produce any particle together with its antiparticle as long as they have electromagnetic coupling, has made electron-positron storage rings one of the most powerful tools in modern high-energy physics. This is particularly true with storage rings getting into the energy range where weak neutral currents play a significant role.

Figure 1 lists the center-of-mass energies actually obtained for high-energy physics experimentation in all e^+e^- projects built so far. Table 1 gives the operating parameters of those storage rings presently doing e^+e^- physics.

For elementary particle physics certain aspects of electron-positron storage rings are of great importance; these are discussed in the next sections.

1.1 *Energy*

Fifteen years ago it was assumed that the largest energy of practical importance for e^+e^- storage rings would be about 2×3 GeV. This was because form factors for production of strongly interacting particles were throught to decrease with a high power of the center-of-mass energy E_{cms}, and cross sections for QED processes decreasing with E_{cms}^{-2} would also become impractically small at $E_{cms} > 6$ GeV. In particular, the first point has been shown to be wrong. Production of strongly interacting particles proceeds mainly via production of quarks with different flavors. With the opening of new channels at higher energies, cross sections decrease more slowly than the QED cross sections. Progress in storage ring technology resulted in ever increasing event rates (luminosity) and now makes physics with colliding electron-positron beams with center-of-mass energies up to 200 GeV not only feasible but also very desirable.

The physics of electron storage rings is dominated by energy losses of the circulating beams due to synchrotron radiation. These losses make the installation of large and expensive radiofrequency accelerating systems

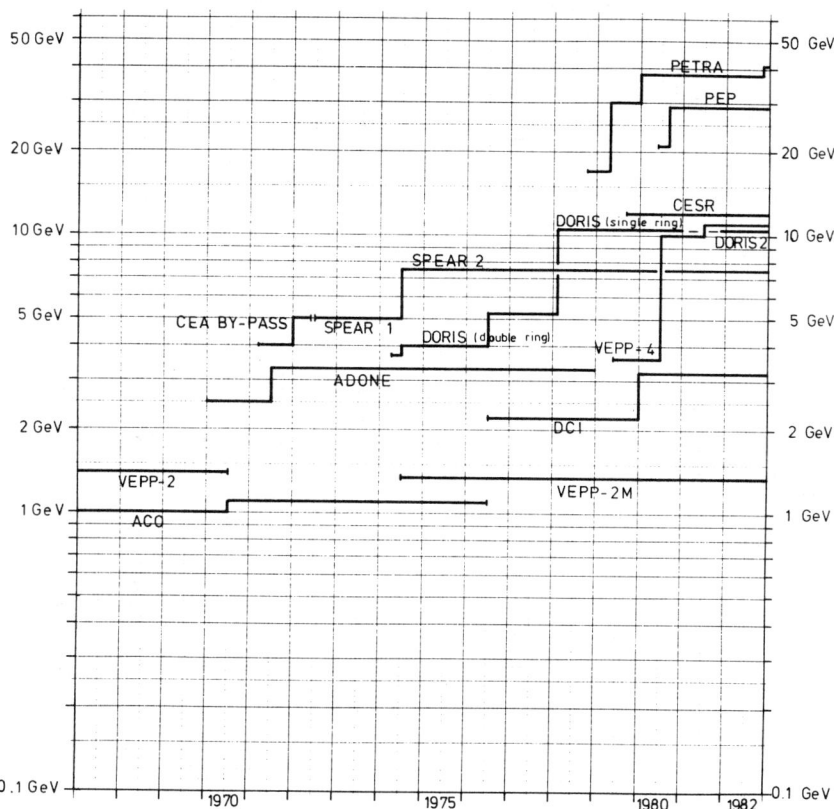

Figure 1 Maximum available center-of-mass energies for e^+e^- colliding-beam physics.

necessary. Cost optimization (6) makes the size and cost of electron and positron storage rings grow quadratically with particle energy. This law sets an upper limit for the energy of electron-positron storage rings at center-of-mass energies around 300 GeV, at which point linear colliders may become more economical (7). Superconducting radiofrequency systems, under development at various laboratories, may push the economical energy limit for electron storage rings somewhat higher; but they do not change the quadratic cost-vs-energy law.

1.2 Luminosity

The luminosity L is defined as the event rate n for a particular reaction divided by the cross section σ for that reaction, $L = n/\sigma$ cm^{-2} s^{-1}. (The units cm^{-2} s^{-1} are assumed for luminosity throughout this article.) Peak luminosities in early storage rings were in the 10^{27} to 10^{28} range. Since then

Table 1 Colliding-beam parameters of e^+e^- storage rings presently in operation (December 1982)

	Circumference (m)	Maximum useful collision energy achieved for high-energy physics (GeV)	Maximum luminosity achieved ($cm^{-2} s^{-1}$)	Maximum luminosity at the energy (GeV)	Currents for maximum luminosity (mA)
VEPP-2M (Novosibirsk)	17.9	2×0.67	5×10^{30}	2×0.6	2×20
DCI (Orsay)	94.6	2×1.56	1.4×10^{30}	2×1.55	2×125
SPEAR 2 (Stanford)	234	2×3.7	1.6×10^{31}	2×3.7	2×20
DORIS 2 (Hamburg)	288	2×5.1	1.0×10^{31}	2×4.7	2×30
VEPP-4 (Novosibirsk)	366	2×5.5	2×10^{30}	2×4.7	6×6
CESR (Cornell)	768	2×6.0	1.6×10^{31}	2×5.5	2×17
PEP (Stanford)	2200	2×14.5	1.1×10^{31}	2×14.5	6×6
PETRA (Hamburg)	2304	2×20.03	1.7×10^{31}	2×17	4×6

several machines have reached values in the 10^{31} range. How this was achieved is described in the following sections. At the highest energies of a storage ring, luminosities are usually limited by the available rf power, which limits the possible synchrotron radiation power loss by the beams and thereby the beam currents. At low energies, luminosities are limited by the *beam-beam interaction* forces (Section 4). At intermediate energies, luminosities may be limited by the amount of current that can be accumulated and brought to collision. Single-current limitations may be due to *beam instabilities* or *intensity-dependent resonances* (Section 3).

Currents stored in storage rings decay because of the interaction with the residual gas. At average pressures of 10^{-7} torr the typical lifetime due to bremsstrahlung losses is 30 minutes. Ultra high vacua are difficult to obtain in electron-positron storage rings because of gas desorption due to synchrotron radiation (8). Progress in vacuum technologies has resulted in an average pressure smaller than 10^{-8} torr with corresponding lifetimes of several hours for most storage rings now. Since the luminosity is proportional to the product of both currents, it will decrease faster than the currents. For colliding-beam experiments the important number is the *time-averaged luminosity* as given by peak luminosity, luminosity decay time, and the frequency and time needed for refilling the ring with electrons and positrons (Section 2). Average luminosity may range from only 20 up to 90% of peak luminosities, depending on positron production and accumulation rates, on whether injection can take place at the operating energy or if acceleration of newly injected currents to the operating energy is necessary, and on the luminosity decay time.

1.3 Energy Spread

Counterrotating currents in electron storage rings are contained in short single bunches. The normal current distribution is Gaussian in all three dimensions and given by the quantum nature of the synchrotron radiation emission. The stochastic excitation of particle oscillations by synchrotron radiation increases with the fourth power of particle energy, while an integral damping effect increases with the third power. Both together are responsible for the Gaussian shape of the bunches, which in all dimensions increase linearly with energy (height, width, and relative energy spread). For different machines, but with the same energy, the energy spread is inversely proportional to the square root of the bending magnet radius (9). The relative energy spread at maximum energy for each listed storage ring is of the order of 10^{-3} to 10^{-4}. Its narrowness is ideally suited for analyzing narrow particle resonances. But at low energy this Gaussian energy distribution can be disturbed by the turbulent *bunch lengthening*, an effect that increases the energy spread in low-energy storage rings by factors up to

4 or 5 (Section 3) and decreases particle production rates in very narrow resonances by the same factor.

1.4 Background

Interaction regions in electron-positron storage rings have an extremely small volume because of the short bunch length and the very small transverse beam dimensions. The number of reactions from e^+e^- collisions originating from this small volume exceeds the number of reactions with the residual gas and cosmic ray events by a very large factor. In general, discrimination of one type of these events against the others does not present any difficulty. A greater problem is the general radiation background in the detectors. This background originates from particles lost near the detector region due to bremsstrahlung or large-angle scattering. The problem is aggravated by the experimenter's desire for small-diameter beam pipes at the interaction point in order to improve the resolution of vertex detectors. Particle tracking programs are now used to understand better the source of the high-energy background and to design countermeasures (10). Another serious source of background can be the high-energy tail of the synchrotron radiation spectrum. Making the bending magnets nearest the interaction point very weak, the SPEAR group decreased this kind of background significantly (11), a technique now used in all electron-positron storage rings. Monte Carlo programs are used to calculate synchrotron radiation background and to design suitable absorbers (12, 13).

1.5 Polarization

Electron and positron beams can be transversely polarized. This is because, for the two possible spin states in the magnetic guide field, the synchrotron radiation processes associated with a spin flip have different probabilities. Hence, a polarization of the initially unpolarized beams may gradually build up. The mechanism was predicted by Sokolov & Ternov (14) and has since been observed in several storage rings. Apart from the physics aspects of such a polarization and the additional handle it gives on interpreting colliding-beam events, there are also important machine physics aspects. Measurement of depolarizing resonance frequencies makes it possible to determine the energies of stored beams with unparalleled accuracy ($\Delta E/E < 10^{-6}$). Rotation of the transverse spin direction into the direction of motion at the collision points will be of great importance in very high-energy machines in order to investigate weak interaction effects. Unfortunately, polarization is very sensitive to the smallest machine imperfections. Polarized beams are more difficult to obtain in the very high-energy machines for which polarization aspects become most interesting. In Section 5 we summarize the present state of the art.

In the following an attempt is made to describe some of the progress made in the performance of e^+e^- storage rings and some of the more fundamental limits as seen today. While the energy that can be reached with a new storage ring project depends mostly on the available budget, luminosities are limited for a number of reasons. Some of the problems can be defined in an analytical way, but much of our present "understanding" comes from extended simulation and beam tracking programs made possible by modern computer technology. At the end of this article a glossary defines some of the symbols and special terms (marked with an asterisk) used by us. For the reader who wants to learn more about the physics of electron-positron storage rings we recommend the excellent introduction written by M. Sands (9).

2. ELECTRON AND POSITRON ACCUMULATION

In electron-positron storage rings, particles must be accumulated in a few small bunches. Radiation damping* permits particles to be injected off-axis. The injection repetition frequency is typically between 4 and 100 Hz, corresponding to damping times of the order of 100 to 4 ms. The effect of radiation damping together with the stochastic excitation of betatron* and phase oscillations* through synchrotron radiation eliminates all injection peculiarities and concentrates all particles eventually in the same small phase volume. Electron-positron storage rings are quite different in this respect from proton storage rings in which radiation excitation and damping effects at energies presently achieved have been quite negligible so far owing to the large proton mass. Injection problems arise in conjunction with positron production and accumulation.

Since positron injection rates can be very small, they usually determine the time it takes to refill the storage ring and thereby the average luminosity. Positrons are produced in a double conversion process by bombarding tungsten targets with high-energy electrons. Depending on the energy of the electrons and the details of the converter systems, positron yields may vary between 0.1 and 15%. There are two limitations to the positron accumulation rate: The short electron bursts required to produce the short positron bunches are limited in their peak current, and the repetition frequency of the injector has to match the injection rate into the storage ring as given by radiation damping times. Two new systems have been developed during the last few years to overcome these limitations.

2.1 *Bunch Coalescing Through Vernier Phase Space Compression*

In the CESR project (15) a linear accelerator (linac) produces a train of 60 positron bunches per injection pulse. Since in the electron part of the linac the peak current is limited, one now gets 60 times as many positrons as in a single bunch. These positrons are postaccelerated in a second linac to an energy of 200 MeV and in a synchrotron to 8 GeV. At that energy the storage ring has sufficiently strong radiation damping to permit injection and accumulation at a rate of 60 Hz in the storage ring. After the desired number of positrons have been accumulated, the 60 bunches are put on top of each other by a novel vernier phase space compression scheme: The storage ring CESR is arranged concentric to the synchrotron and has a 2.4% larger circumference. By ejecting one bunch at a time and injecting it back into the synchrotron and allowing it to catch up for a few turns with the beam in the storage ring, one can, after transfer back into the storage ring, put each of the original 60 positron bunches into the same bunch location. The practical overall gain factor for this coalescing scheme compared to a direct injection of single bunches into the storage ring is about 15. With this the positron accumulation time is 15 minutes; otherwise it would have been 3.5 hours.

2.2 *Positron Accumulator Rings*

Injection energy at the PETRA storage ring (16) is limited to 7 GeV because of the available injector synchrotron. At that energy, radiation damping in PETRA limits injection rates to 8 Hz. To overcome this handicap and at the same time the problem of peak current limitation in the positron-producing electron linac, an accumulating ring PIA (positron intensity accumulator) (Figure 2) was built (17). This small storage ring with a circumference of 28.8 m employs very strong magnets with a bending radius of 1 m, such that radiation damping at the operating energy of 450 MeV permits injection and accumulation at 50 Hz. The linac injects a bunch train of positrons with a length of 30 m. Radiation damping in the accumulator ring collects these positrons in one 1-m long bunch (the accelerating rf system works on the circumferential frequency of the particles). After 10 linac pulses have been accumulated, this bunch is further compressed by a higher harmonic rf acceleration system. It is then transferred to the injector synchrotron where it is accelerated as a single bunch to the injection energy for PETRA. Such positron accumulator rings make the positron accumulation rate in e^+e^- storage rings independent of the radiation damping times at injection; they become particularly useful for very large rings where the injection energy usually is much lower than the maximum energy. The PIA ring produces eight positron bunches per second, each containing 2×10^9 positrons. With this, injection times for PETRA are reduced to the order of one minute.

e⁺e⁻ STORAGE RINGS   75

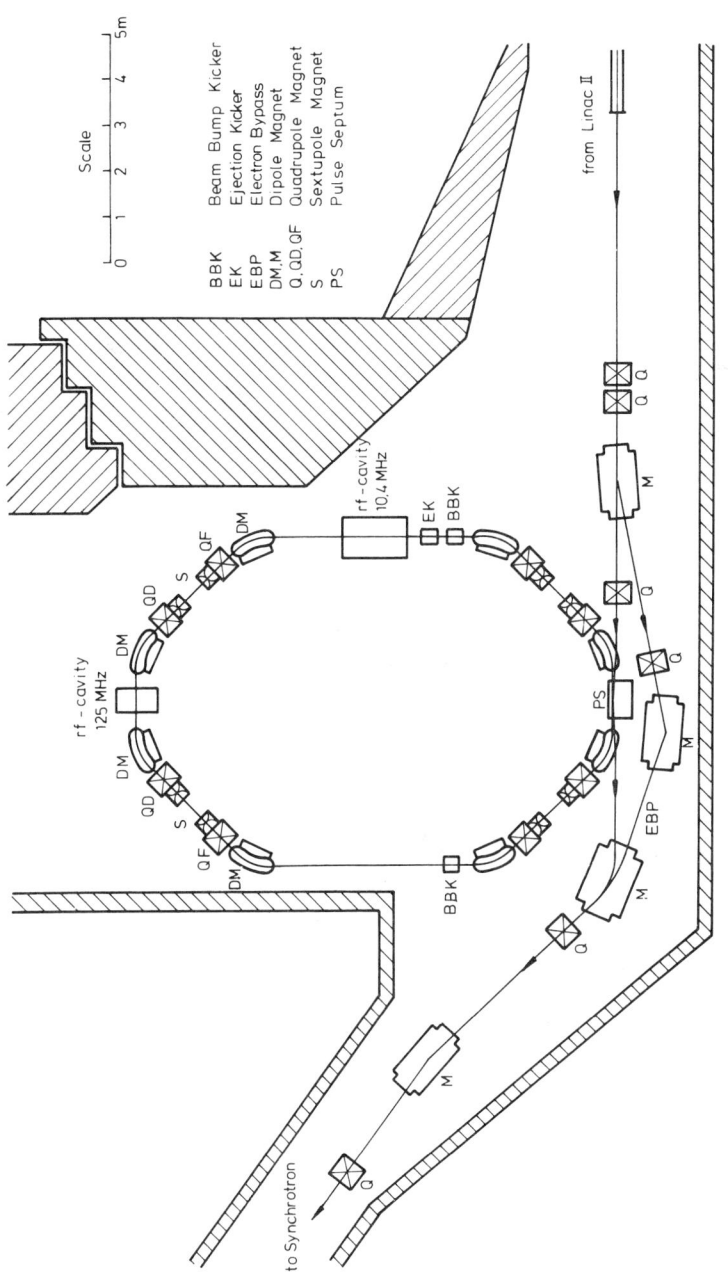

Figure 2 Positron intensity accumulator (PIA).

3. SINGLE-BEAM LIMITATIONS

One of the most severe problems of particle accumulation in e^+e^- storage rings is the limitation of single-beam currents due to intensity-dependent phenomena. Since the circulating particles are focussed in the longitudinal and transverse directions, the beams form an oscillatory system in all three space directions. At high intensities of the stored beam, the circulating particle bunches generate strong electromagnetic fields that can excite electromagnetic resonances in the elements of the vacuum chamber (bellows, kicker, tanks, etc) and particularly in the accelerating rf resonators. The electromagnetic forces induced by these interactions can produce various instabilities of the oscillating particles, so that the intensity of the stored beam will be limited. The interaction of the stored beam with the "environment" is most dangerous for very high-energy machines in which the injection energy is relatively small. These machines need a large number of accelerating resonators for operation at high energies. But these resonators already interact with the beam at low energies where instabilities easily arise because of the lower stiffness of the beam. Much work has gone into the investigation of these current-dependent phenomena.

3.1 *Satellite Resonances*

Among the fundamental perturbations of the stability of particle motion are the current-dependent resonances.

3.1.1 EXCITATION OF SATELLITES BY SPURIOUS DISPERSION IN THE CAVITIES When a particle bunch passes through an rf resonator the individual particle experiences an energy change depending on its longitudinal position within the bunch. In the presence of dispersion* in the resonator, the energy change leads to a displacement of the equilibrium closed orbit*. For particles off this new orbit, this leads to an excitation of betatron oscillations*. Since the energy change depends on the longitudinal position within the bunch and this position changes according to synchrotron oscillations*, the excitation of betatron oscillations is modulated with the synchrotron frequency. As a consequence new "stop bands"* appear as side bands (satellites) of the integer stop bands. With increasing intensities in the beam, the bunches also excite highly nonlinear longitudinal fields in the resonators, such that the longitudinal motion of the individual particles becomes highly nonlinear. This generates higher harmonics of the longitudinal oscillation and produces higher-order satellite stop bands that are current dependent and spaced around the integer stop bands by multiples of the synchrotron frequency. The width of the stop bands increases with current and may considerably reduce the range of possible machine tunes (18–21).

3.1.2 EXCITATION OF SATELLITES BY TRANSVERSE FIELDS   If a particle bunch passes through an rf resonator off-axis, it excites transversely acting fields. These fields contain high-frequency components such that the field strength varies within the bunch. With this the transverse forces on the individual particles depend on the longitudinal position, which changes with time. As in the case of spurious dispersion (i.e. unavoidable small values of dispersion due to imperfections of the storage ring), the modulated transverse forces produce current-dependent satellite stop bands around the integer stop bands. Both types of satellite-producing mechanisms have been observed, and although satellite resonances have been investigated with great effort (22–24), they still severely limit single-bunch currents. These satellite effects can be reduced by the following procedures, which lead to a better performance of the storage rings:

1. precise control of transverse and longitudinal oscillation frequencies,
2. minimization of horizontal and vertical dispersion in the rf resonators, and
3. precise control of closed orbit position.

3.1.3 EXCITATION OF SATELLITE RESONANCES AT NONVANISHING CHROMATICITY   If the chromaticity* is not completely compensated, the betatron frequency depends on the particle's energy deviation from the ideal synchronous particle. But this deviation changes for each particle with the synchrotron motion. The result is a frequency modulation of the betatron oscillation leading to satellite side bands to resonances. The higher-order satellites become current dependent if the synchrotron motion becomes nonlinear due to the excitation of nonlinear longitudinal electric fields (25, 26).

Satellite resonances belong to the class of incoherent phenomena: Although the higher-order satellites are induced by collective action of all particles in the bunch, the particles become unstable individually because of resonance conditions. Another important example of this class of incoherent current-dependent effects is connected with the higher-order mode losses.

3.2 *Higher-Order Mode Losses*

When a high-intensity particle bunch passes through an rf resonator (or through a cavity-like object in the ring), it excites an infinite number of resonator modes that differ in frequency and field configuration. A fraction of the energy gained by passing through the accelerating cavity is immediately lost through excitation of higher cavity modes by the bunch (27–36). This energy loss increases with the bunch current and is equivalent to a reduction of the accelerating voltage, such that individual particles may leave the stable phase focussing regime. This in turn may lead to a reduction of beam lifetime. At high storage ring energies, which normally are limited by the available rf power, higher-order mode losses cause an intensity limitation. The simplest way to avoid this limitation is to provide sufficient rf power. However, recent progress in the application of computational methods using modern computer technology allows the design of rf resonators for optimum conditions. In existing machines (e.g. PEP, PETRA) great effort was made to keep the vacuum chambers as smooth as possible in order to reduce the higher-order mode losses.

3.3 *Coherent Single-Beam Instabilities*

While satellite resonances and the reduction of lifetime due to higher-order mode excitation are examples of incoherent phenomena, the coherent instabilities represent another class of severe current limitations. Coherent instabilities are accompanied by an unstable coherent motion of particle groups in the beam.

3.3.1 COUPLED-BUNCH INSTABILITIES In the simplest case of coherent instabilities, the bunches of the beam oscillate as a whole. For a given number of bunches B there are B eigenmodes possible differing in the phase between adjacent bunches. An unstable growth of the oscillation amplitude can occur if the oscillating bunches excite parasitic resonator modes with a decay time longer than the time between bunch passages. Because of this long decay time all bunches are coupled and a phase shift can arise between the oscillating bunches and the forces acting back on the beam. This phase shift then can lead to a coupled-bunch instability. Since the bunches are coupled, the currents of all individual bunches contribute. Thus the total current is limited by this kind of instability, which can occur in the longitudinal or transverse direction depending on the field configuration of the excited resonator modes (37–39). One of the basic problems in the case of coupled-bunch instabilities is the identification of the excited resonator modes. Once the offending modes have been identified, "absorbing systems" can be built to match the frequency and the field configuration of the mode. These absorbing systems reduce the "coupling impedance" so that the instability be cured. In the multibunch machine DORIS I, the rf spectrum of the unstably oscillating beam was observed and analyzed. This led to a mode identification. Accordingly constructed and installed absorbers improved considerably the DORIS performance (40). For machines with a small number of bunches where the time-of-flight interval between the bunch passage is sufficiently long (0.5 μs), a "feed-back" system is the most efficient cure for coupled-bunch instabilities. Transverse and longitudinal feed-back systems have been constructed for several machines

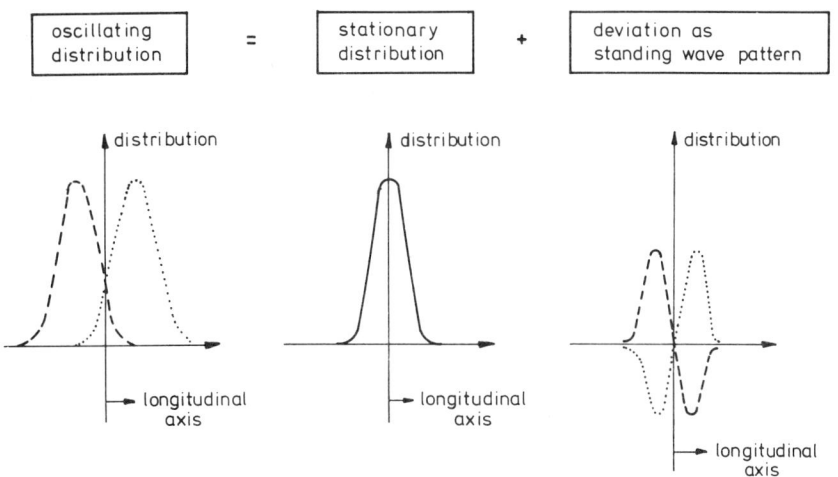

Figure 3 Classification of longitudinal modes.

and are successfully operated to improve the storage ring performance (41–44).

The principal mechanism of a feed-back system is simple: An electrical signal generated from the oscillating bunches will be detected by a pick-up station that distinguishes between individual bunches. This signal passes an amplifier that provides sufficient gain and controls a deflecting device acting on the individual bunches. With an appropriate choice of the phase advance between pick-up and feed-back station, the bunch system can be strongly damped.

3.3.2 SINGLE-BUNCH INSTABILITIES Besides satellite resonances, coherent single-bunch instabilities are the source of the severest limitations of storage ring performance. This kind of instability is characterized by the fact that the bunches of the beam become unstable individually. A single bunch is described by a three-dimensional particle distribution, and in such a distribution an infinite number of internal longitudinal and transverse oscillation modes can exist. Those modes leading to a center-of-mass motion of the bunch form only one special configuration.

For the longitudinal motion, the time-dependent particle distribution is quantitatively described by its deviation from the stationary distribution. This deviation is analyzed in terms of standing wave patterns for the classification of modes. Figure 3 shows the case where the bunch oscillates as a whole.

The transverse configurations so far observed in e^+e^- storage rings are characterized by the fact that particles at a fixed longitudinal position within the bunch oscillate with the same phase. Between particles at different longitudinal positions, however, any phase shift may exist. Accordingly, the transverse modes are classified by analyzing the transverse deformation of the bunch along the longitudinal axis in terms of standing wave patterns (Figure 4).

In principle any longitudinal or transverse mode can become unstable. However, modes with a high mode number can be driven only by fields varying strongly within the bunch. These fields contain high-frequency components that normally are less strongly excited. The excitation of electromagnetic fields by the beam is described in terms of two fundamental quantities, the longitudinal and transverse impedance function*. If these functions are known for all objects in the ring, the strength of single-bunch instabilities can be calculated in the framework of a modern formalism developed by F. Sacherer (45–48). The theoretical determination of the impedance functions is one of the basic problems in this field. Recently it has become possible to find numerical solutions for the impedance functions using new computational methods and large computers (49).

3.3.3 BUNCH LENGTHENING From a general point of view the concept of single modes is only a mathematical tool. For weak interactions (low influences), the single modes have a physical meaning because they are eigenstates of the dynamic system. If the intensity increases, the isolation of single modes becomes physically meaningless. The coupling of longitudinal modes at high intensities leads to a severe single-bunch instability known as "turbulent bunch lengthening" (50–58).

Because of the unstable motion of particles within the bunch, the energy spread of the bunch increases and therefore the bunch length also increases. As a consequence of an increased energy spread, the lifetime can be considerably reduced. Bunch lengthening with increasing single-bunch current was observed in nearly all existing e^+e^- storage rings (59, 60). The

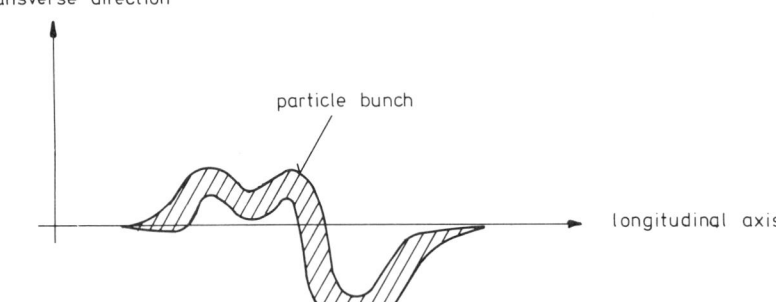

Figure 4 Transverse deformation of particle bunch due to internal transverse oscillation.

investigations of this effect were performed with great effort because of its importance for the large storage rings under construction. Attempts were made to derive "scaling laws" that would allow one to scale bunch lengthening from one machine to another without the explicit knowledge of the longitudinal impedance function. The progress in the numerical computation of resonator fields as part of longitudinal tracking programs led to a better understanding of the observed effects (61).

3.3.4 THE HEAD-TAIL EFFECT The head-tail effect is the classical transverse single-bunch instability that has been observed in nearly all storage rings and was first analyzed for the ADONE ring (62, 63). The mechanism of this effect can be easily demonstrated if the particle distribution of a bunch is approximated by two rigid particle "collectives." If one of the collectives of the circulating bunch is leading and the other is trailing, the leading collective having a small transverse displacement leaves a wake field behind that acts transversely on the trailing collective. Since particles (collectives) differing from the equilibrium longitudinal position perform synchrotron oscillations, the local constellation of two collectives changes with time, and after half a synchrotron period the two collectives have exchanged their longitudinal positions. The two collectives drive each other via this exchange.

The synchrotron oscillation is accompanied by a change of the energy, so that the two collectives have different energies during the exchange of their positions. If the chromaticity of the machine is not compensated, different energies effect different betatron frequencies. Because of this a phase shift between the collectives driving each other causes the head-tail instability. If the chromaticity is compensated, the instability can be avoided. Therefore the head-tail effect does not cause a severe current limitation in e^+e^- storage rings, provided the chromaticity can be compensated. However, the investigation of this effect is of great importance. Since transverse instabilities are governed by the transverse impedance function, the observation of the head-tail instability as a function of various machine parameters (current, chromaticity, bunch length, etc) gives detailed experimental information about this fundamental quantity. Besides an instability the head-tail mechanism causes an observable current-dependent shift of the betatron frequency. The experimental and theoretical investigations of both the head-tail instability and the coherent frequency shift have contributed to a better understanding of transverse effects.

3.3.5 TRANSVERSE "TURBULENCE" After PETRA began to operate the first high-current experiments showed that the single-bunch currents were limited by a vertical instability occurring even at compensated chromaticity. This single-bunch instability seemed to be in disagreement with all mechanisms of known transverse instabilities. The effect turned out to be a severe limitation of the PETRA performance when the number of accelerating resonators was increased. Theoretical and experimental studies showed that this instability could be explained—in analogy to the longitudinal case—by transverse mode coupling (transverse "turbulence"). The theory predicted an increase of the maximum single-bunch current with increasing bunch length, with increasing synchrotron frequency, and with a reduced amplitude function in the rf sections (64, 65). These predictions were confirmed experimentally (66). For the luminosity performance of PETRA the instability can be avoided by an artificial increase of the bunch length during injection and by using a special injection optics with reduced amplitude functions in the rf sections. In the meantime the new instability has also been observed at PEP (67, 68). Since this instability so severely limits single-bunch intensities in large storage rings, further effort is needed to find new concepts for controlling it in future large e^+e^- storage rings such as LEP (69).

A most effective system that can cure various types of instabilities was proposed a long time ago (70): the higher harmonic rf system. The idea of applying such an additional rf system is based on the fact that the gradient of the accelerating voltage can be compensated at the bunch center without affecting those parts of the rf voltage that determine the beam lifetime. This compensation effects a considerable reduction of the small amplitude synchrotron frequency without loss of beam lifetime (71, 72). As a consequence of the reduced synchrotron frequency, the bunch length increases without a change in energy spread.

Since the restoring force of the synchrotron motion becomes strongly nonlinear, the higher harmonic system contributes sufficient Landau damping* (73) to cure longitudinal coupled-bunch instabilities. The reduction of the small amplitude synchrotron frequency shifts the current-limiting satellites out of the range of normal machine tunes.

A higher harmonic rf system was installed in SPEAR (74), and the satellite resonances were completely suppressed. In PETRA the influence of a higher harmonic rf system operating on the second harmonic was studied in detail at the injection energy. In agreement with the observations at SPEAR, the satellite resonances disappeared. In addition the turbulent transverse instability was suppressed up to single-bunch currents far above the values needed for good luminosity performance.

As a result of these intensive investigations of various current-dependent phenomena, the current limitations in all existing e^+e^- storage rings are understood at least qualitatively. However, we are left with the problem of quantitative predictions. These predictions are of great importance for the proper design of the new machines. The problems may be solved by an

4. SPACE CHARGE LIMITATIONS

4.1 *General Description*

The most important luminosity limitation in storage rings is given by incoherent beam-beam interaction at the collision points. Often this effect is also referred to as space charge limitation. Space charge forces are proportional to $1/\gamma^2$ (γ = energy divided by rest energy). In a single beam, they are totally negligible in high-energy electron-positron storage rings. Space charge forces between the beam and its image currents in the resistive wall of the vacuum chamber can also be neglected, but the electric and magnetic fields of an opposing beam add up in their effects. If electrons and positrons travel on the same orbit (as is almost always the case in a single electron-positron storage ring and in the absence of transverse electric fields), the effect of these space charge forces on particles close to the axis is that of a focussing lens. This focussing lens increases the horizontal and vertical betatron wave numbers* Q_x and Q_z (betatron oscillations per turn) by ΔQ_x and ΔQ_z (75), (ΔQ is the wave number change per interaction region):

$$\Delta Q_x = \frac{r_e N_B \beta_{x\mathrm{ip}}}{2\pi\gamma\sigma_x(\sigma_x+\sigma_z)} \qquad 1.$$

$$\Delta Q_z = \frac{r_e N_B \beta_{z\mathrm{ip}}}{2\pi\gamma\sigma_z(\sigma_x+\sigma_z)} \qquad 2.$$

where r_e is the classical electron radius; N_B is the number of particles in the opposing bunch; $\beta_{x,z\mathrm{ip}}$ are the beta functions* at the interaction point; and $\sigma_{x,z\mathrm{ip}}$ are the beam dimensions at the interaction point (standard deviation).

Such a change of the number of betatron oscillations per turn can move the operating point in the Q_x-Q_z diagram onto a resonance* and cause beam loss. More important though is the fact that this focussing lens is linear, i.e. independent of the transverse position, only in the area where the current density of the opposing beam is constant. But since particle densities have Gaussian distributions, the space charge lens as given by the opposing beam is highly nonlinear, creating on its own a large number of new nonlinear resonances. The strength of these resonances increases with the opposing current. Many of these resonances are still too weak or may be avoided by the right choice of Q values at small currents. As the currents increase, one reaches a point where stable storage is no longer possible. This point is characterized by a maximum linear tune shift ΔQ for particles close to the axis. But it must be stressed that this linear tune shift is not the cause of the problem. Its effect could be compensated by a readjustment of storage ring parameters. The ΔQ serves only as a convenient way to describe the strength of the nonlinear effects, which are proportional to the linear effects and are the real cause of current limitations.

4.2 Observations

The most recent summary of observations on incoherent beam-beam interaction in electron-positron storage rings and its effect on luminosity was given in a special session on beam-beam effects at the International Accelerator Conference 1980 in Geneva (76). The observations at different storage rings are more or less in agreement and can be summarized in the following way:

1. As the currents in the opposing beams are increased, the vertical beam size, which normally is small as compared to the horizontal beam size, increases. This reduces the specific luminosity L_{sp} ($L_{sp} = L/i^+i^-$).
2. As the currents are further increased, the luminosity itself may reach a maximum and drop again. Eventually the beam lifetime will become very short. In some cases the point of short beam lifetime is reached before the luminosity has gone through a maximum.
3. The point of highest luminosity is characterized by vertical ΔQ values of usually between 0.025 and 0.06 (e.g. 77). This space charge limit ΔQ increases somewhat with the radiation damping of betatron oscillations, i.e. in a given storage ring with the beam energy. Proton-antiproton storage rings with no radiation damping have reached ΔQ values of 0.0035 (78).
4. If there is more than one bunch of each kind in the ring, the space charge limit per interaction region is usually lower, but there are also cases where the limit, with up to three bunches in the beam, was not smaller than that with only one bunch of each kind. With the larger bunch number, a luminosity three times as large could be obtained (79).
5. The space charge limits are higher if the vertical betatron frequency is slightly above an integral number (77). But only a few storage rings can be operated in this region.
6. If the opposing bunches have different currents, the stronger one is much less affected by beam-beam forces. Unequal beam currents result in unequal vertical blow-ups and in smaller specific luminosities. It is therefore important to make both currents as equal as possible. Sometimes it is difficult to prevent an unequal increase of vertical beam size even with identical currents [flip-flop effect (80)].

4.3 *Theory and Computer Tracking Results*

The theoretical understanding of this space charge effect has proven to be exceedingly difficult and has resulted only in qualitatively correct predictions. According to Chirikov (81) the width of nonlinear resonances created by the space charge forces is proportional to ΔQ. If there were only one degree of freedom for particle motion, stability would exist as long as the widths of the resonances were smaller than their spacing (Chirikov criterion). If resonances overlap, stochastic or chaotic motion will result with subsequent beam blow-up. But even in the case of narrow resonances there are stochastic regions that in the case of multidimensional motion (for instance, two degrees of betatron motion and one degree of synchrotron motion) may be connected such that a gradual blow-up may result (Arnold diffusion).

Quantitative agreement between experiment and the Chirikov theory is not very good. One problem may be that small imperfections in the storage ring optics (slightly different beam size at the different interaction points, small differences in the betatron phase advance* between interaction points, differences in dispersion or small spurious vertical dispersion at the interaction points, and many others) may have a profound effect on the strength of certain nonlinear resonances but are very difficult to measure.

During the last few years it has become possible to simulate beam-beam interactions with large computers (82, 83). Such simulations take all conceivable effects into account. The kick a particle receives through the electromagnetic fields of the opposing bunch will lead to a slightly different betatron motion in each turn. Simulations assume the proper charge distribution at the interaction points (already affected by beam-beam interactions) and include also phase motion of the particles. The stochastic excitation of particle oscillations due to synchrotron emission as well as the radiation damping are included. Assumptions about machine errors of the kind mentioned above are found to be very important. Simulations allow the separation of the different effects to such an extent that the effectiveness of each machine error in lowering the space charge limit can be judged. A comparison of computer predictions with observations indicates that, with reasonable assumptions of machine errors, good agreement can be reached. Computer simulations seem to describe all the observed features of incoherent beam-beam interactions correctly: The time dependence of the blow-up when both beams are brought to collision, the difference between strong-weak beam and strong-strong beam effects and the energy dependence of the maximum ΔQ shifts (84).

4.4 Ways to Improve the Space Charge Limited Luminosity

The incoherent space charge interaction limits the currents that can be brought to collision and thereby the luminosity. For Gaussian current distributions one can derive the equation for the luminosity to be

$$L = \frac{N_1 N_2 f_c}{4\pi \sigma_x \sigma_z B}.\qquad\qquad 3.$$

Here $N_{1,2}$ are the numbers of electrons (positrons) in each beam, B is the number of bunches in each beam, f_c is the circumferential frequency, and $\sigma_{x,z}$ are the standard deviations of the beam dimensions at the interaction points. If one eliminates from Equation 3 the quantities N_1 and N_2 by using Equations 1 and 2, one gets an expression for the space charge limited luminosity:

$$L = \frac{\pi B f_c \gamma^2}{r_e^2} \frac{(1+K)^2 \varepsilon_x \Delta Q^2}{\beta_{zip}}.\qquad\qquad 4.$$

In this expression ε_x is the horizontal emittance* of the beam and K is the "coupling factor" describing the vertical emittance ($\varepsilon_z = K\varepsilon_x$). The beam dimensions* have been expressed by $\sigma = \sqrt{\varepsilon \beta}$. It is assumed that the limiting $\Delta Q_z = \Delta Q_x$ and that both ΔQ have been made the same by the right choice of β_{xip} at the interaction points ($\beta_{xip} = \beta_{zip}/K$).

Equation 4 indicates all possibilities for improving the space charge limited luminosity in electron-positron storage rings. For given ΔQ and ε_x the luminosity increases quadratically with γ, i.e. with energy. But since ε_x itself increases quadratically with energy (for a given storage ring optics), the luminosity normally would increase with the fourth power of energy. Since ΔQ also may have a small energy dependence (85), the overall energy dependence of the space charge limited luminosity is usually stronger than the fourth power. The luminosity is proportional to the number of bunches B in each beam, provided a larger B does not decrease the space charge limit ΔQ (see above). The coupling factor K is not a free parameter; K normally is determined by betatron coupling in imperfectly aligned quadrupoles and by vertical betatron excitation from synchrotron radiation in conjunction with spurious vertical dispersion (also due to machine imperfections). In a well-aligned machine with small currents, K may be as small as a few percent. The first indication of beam-beam interaction is an increase of K, which subsequently allows for larger currents and thereby larger luminosities. If the vertical beam size in both beams would increase by the same amount, the luminosity would increase until the vertical beam dimensions reach a limiting aperture. At that moment the lifetime would be drastically reduced. The only truly free parameters that can be adjusted for high luminosity are the horizontal emittance ε_x, the vertical beta function β_{zip}, and the bunch number B.

4.5 Variation of the Number of Bunches in Each Beam

In a single ring the minimum number of bunches B in each beam is determined by the desired number of beam-beam collision points, which is $2B$. An increase in B will only lead to higher luminosities if the space charge limit per interaction point Q is not drastically reduced with the now larger number of interaction points. To overcome this problem one could build two separate rings for the electron and positron beams such that they intersect each other only at a small number of interaction points. In such a setup a very large number of bunches in each beam should be permissible, since each bunch would only meet an opposing bunch a few times per turn. This was the rationale for the original design of the DORIS I storage rings: In two independent rings up to 480 bunches could be accumulated. The two rings intersected each other at two points. One would assume that the space charge limited luminosity is now 480 times as large as that of a single ring with only one bunch per beam. At low energies DORIS indeed had larger luminosities than comparable single rings. But the factor 480 was never realized. It turned out that the space charge limits for crossing-beam geometries are an order of magnitude lower than those for head-on collisions (86). Further, at higher energies the total current was limited by multibunch instabilities. A luminosity advantage of such a two-ring arrangement as compared to a single ring existed only for beam energies smaller than 2 GeV. In single rings a way of increasing the number of bunches B in each beam is by avoiding unwanted beam-beam collisions with electrostatic separator fields. This technique is planned for LEP (87), where in the so-called Phase I four electron bunches are to collide with four positron bunches at four interaction points. The other four potential beam-beam collisions are avoided by vertical beam separation. Similar techniques are being prepared for the storage ring CESR.

4.6 Adjustment of the Horizontal Emittance

According to Equation 4, the space charge limited luminosity is proportional to the horizontal emittance. For a given storage ring optics, i.e. for a given lattice configuration and for given focus strengths of the quadrupole magnets in the ring, the emittance is proportional to γ^2. At a certain maximum energy the circulating beam will use up the acceptance of the machine as determined by vacuum chamber cross sections and orbit distortions or by the effect of nonlinear resonances. At energies below that limiting energy, the luminosity could be increased by an artificial increase of beam emittance and a simultaneous increase of beam current such as to stay at the space charge limit. Such emittance control can be achieved in three different ways.

4.6.1 EMITTANCE CONTROL THROUGH WIGGLER MAGNETS Wiggler magnets (88) consist of a series of dipole magnets with alternating polarity such that the overall deflection of the beam is zero. They are arranged in straight sections of the storage ring. If the magnetic field strength in the wigglers dominates that in the rest of the machine, stochastic excitation of betatron and phase oscillations together with an increased integral damping effect due to the larger average synchrotron radiation energy losses will result in larger beam emittances. The result is a set of beam parameters similar to those of the machine at higher energy. The disadvantage of emittance control through wigglers is the unavoidable increase in energy spread and the increased overall synchrotron radiation losses.

4.6.2 EMITTANCE CONTROL THROUGH VARIABLE OPTICS The beam emittance depends on the storage ring optics. Lattices with weaker focussing result in larger beam emittances. By choosing the optics that will maximize beam emittance at a given energy, the space charge limited luminosity can be maximized. The energy spread is not changed with this change in optics because it depends only on the beam energy and the radius of curvature in the magnetic guide field. A drawback of this method is that it does not permit a continuous emittance variation. By varying the focussing strength, one also changes the frequency of the betatron oscillations* and would soon lose the beams on a resonance.

4.6.3 EMITTANCE CONTROL THROUGH VARIATION OF DAMPING PARAMETERS The emittance is determined by the equilibrium between radiation damping and stochastic excitation of betatron and phase oscillations through the emission of synchrotron radiation. According to a theorem formulated by Robinson (89), the sum of the three damping rates for horizontal and vertical betatron oscillations and phase oscillations is a constant, dependent only on machine radius and beam energy. But the partition of total damping rate onto the three modes of oscillation depends on the lattice and can be changed. Horizontal betatron oscillations can be less strongly damped by transferring some of that damping to phase oscillations; this will increase the horizontal emittance. In large storage rings such a change of damping rates is extremely simple. The partition of damping rates depends critically on the average beam position in the quadrupole elements. A small change in orbit radius as it can be easily facilitated by a small change of frequency of the accelerating cavities can change the damping rate for horizontal betatron oscillations to the point where they are antidamped (Figure 5). The corresponding change in orbit radius is, for example, only 2.9 mm for the PETRA storage ring.

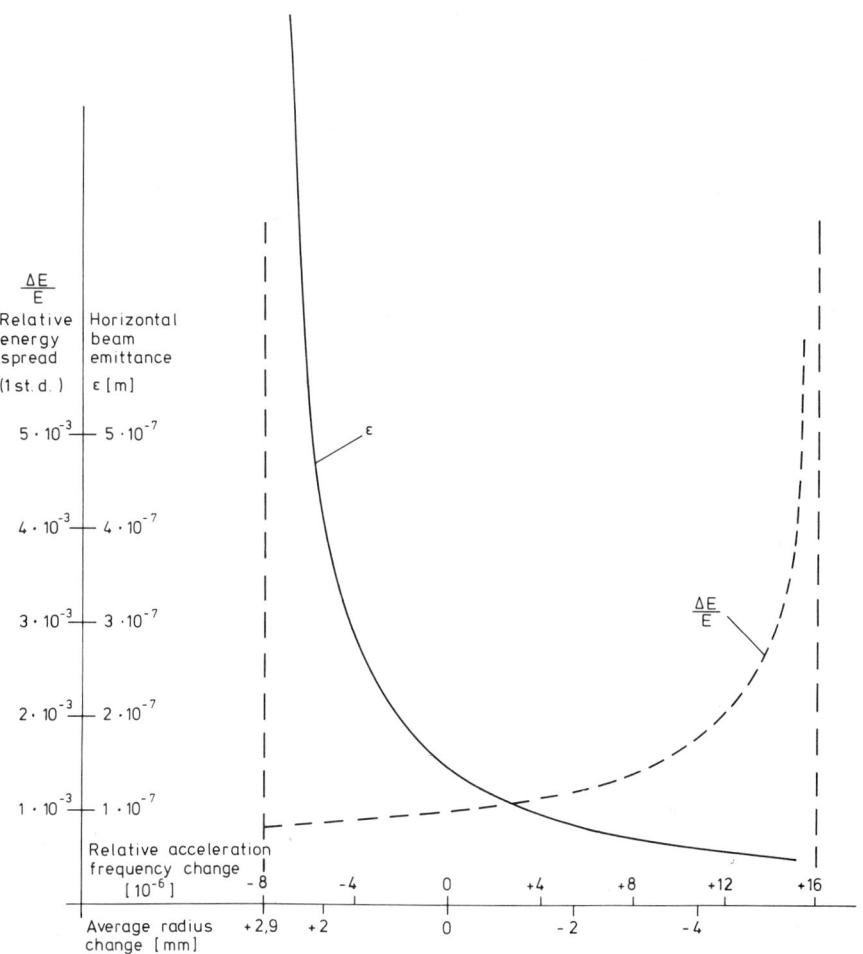

Figure 5 Variation of damping parameters by variation of orbit radius in PETRA.

Wiggler magnets are presently in use at the PEP facility (79). Variable optics and the damping rate adjustment technique are being used in the PETRA storage ring (16), and all three techniques are planned for LEP (90).

4.7 Low-beta Interaction Regions

The most obvious and promising way of increasing the space charge limited luminosity is by decreasing the value of the vertical beta function at the interaction points (see Equation 4). This technique was first proposed for the CEA-bypass project (91) and also tried out on that machine (92). By arranging quadrupole magnets near the interaction point one could focus the beam to an extremely small cross section (0.5 mm wide, 0.07 mm high). The corresponding beta functions were $\beta_x = 7$ cm, $\beta_z = 22$ cm. This "low-beta" interaction region increased the luminosity by about two orders of magnitude as compared to an interaction point in the normal storage ring lattice. Low-beta interaction regions have since been incorporated in most storage rings. The limit to how small beta can be made is given by chromatic effects and the problems of how to correct them (next section).

4.8 Chromaticity Problems

For a given beam emittance* the beam divergence at the interaction point is inversely proportional to the beam size. Very small beam cross sections at the interaction points imply large angular divergences and consequently large beam cross sections at the adjacent quadrupoles. The beta function at the first lens is given by

$$\beta(s) = \beta_{ip} + \frac{s^2}{\beta_{ip}}, \qquad 5.$$

with s being the distance to the interaction point. Large beta functions at these lenses cause large chromatic errors: Particles with energies deviating from the central energy will be focussed differently. This in turn will lead to a different betatron wave number. The energy dependence of the betatron wave number is called chromaticity:

$$\xi = \gamma \frac{dQ}{d\gamma}. \qquad 6.$$

The chromaticity contribution of a single quadrupole magnet to the total chromaticity is

$$\Delta \xi = -\frac{\beta}{4\pi f}, \qquad 7.$$

where f is the focal length of the quadrupole magnet. Large beta functions at the quadrupole cause large chromaticity contributions. The overall natural chromaticity of a storage ring is negative. Low-beta interaction regions contribute a large portion to the overall chromaticity. This chromaticity must be corrected to values close to zero for several reasons:

1. The strongest coherent single-beam instability is the so-called head-tail instability (see Section 3). For the most important mode of this instability, the strength is directly proportional to the negative chromaticity. By correcting ξ to zero this instability is suppressed.
2. The dependence of the betatron wave number on energy will cause satellite resonances (see Section 3). These resonances are side bands to

the main linear and nonlinear resonances and are caused by the frequency modulation of betatron oscillations by phase oscillations. The strength of these resonances is directly proportional to the chromaticity.
3. The sensitivity of the storage ring to orbit errors is much larger for a chromatically uncorrected machine (93).
4. The acceptance for off-momentum particles is greatly reduced if the chromaticity does not equal zero.

For these reasons it is desirable, even necessary, to correct the machine chromaticity. Such a correction can only be done with sextupole magnets at locations of beam dispersion, i.e. where particles of different momentum travel on different orbits. For horizontal beam displacements sextupoles act like quadrupoles with a strength depending on orbit position. To particles travelling on different orbits due to different energy deviation, they can give the necessary extra focussing (or defocussing) strength to make the wave number independent of energy. These new elements produce, on the other hand, very bad side effects: For particles with vertical displacement they act like quadrupoles rotated by 45°, giving rise to a coupling of horizontal to vertical betatron oscillations. They also give rise to nonlinear betatron resonances for particles with large betatron amplitudes. The result is a limitation of the possible beam emittance that can be stably stored. Large chromaticities need strong sextupole corrections and consequently lead to small machine acceptances. Many investigations have been made to find the best chromatic corrections which preserves the largest machine acceptance (see, for example, 94–96). It is obvious that the desire for a very small beta function at the interaction points has to be balanced against the difficulty of correcting the chromaticity. In order to reach the largest space charge limited luminosity, one must make a compromise between these conflicting demands. This then determines the smallest value of β in a low-beta interaction region. (Other limitations may be given by the bunch length, see Section 4.10.)

4.9 Mini-beta Interaction Regions

The focal length of the quadrupoles near the interaction region is proportional to the distance between quadrupole and interaction point. The chromaticity contribution (Equations 5 and 7) of these quadrupoles can therefore be rewritten as

$$\Delta \xi \approx \frac{s^2/\beta_{ip} + \beta_{ip}}{4\pi s} \approx \frac{s}{\beta_{ip}} \qquad \text{for } s \gg \beta_{ip}. \qquad 8.$$

If one wants smaller beta values without increasing the chromaticity, s should be made as small as possible. An optical arrangement in the storage

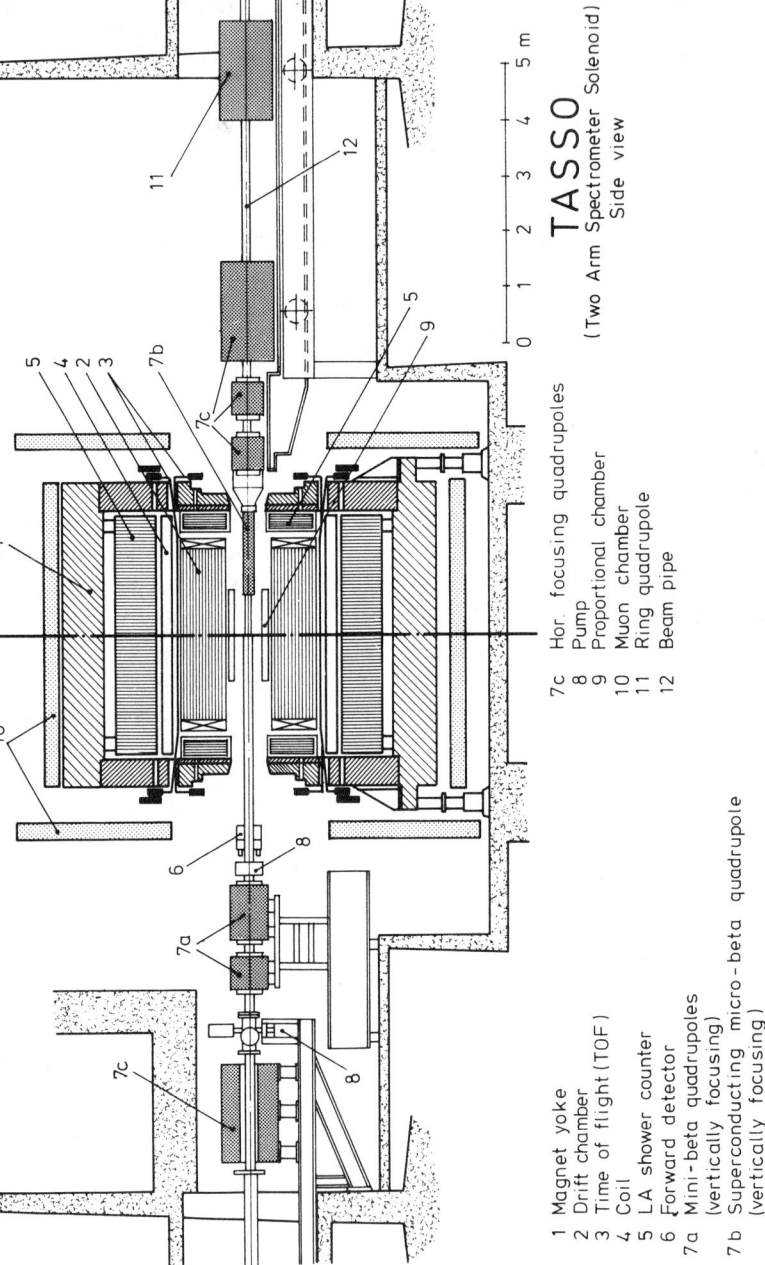

Figure 6 (*Left*) Mini-beta arrangement with TASSO detector (in operation). (*Right*) Micro-beta arrangement with TASSO detector (as planned).

1 Magnet yoke
2 Drift chamber
3 Time of flight (TOF)
4 Coil
5 LA shower counter
6 Forward detector
7a Mini-beta quadrupoles (vertically focusing)
7b Superconducting micro-beta quadrupole (vertically focusing)
7c Hor. focusing quadrupoles
8 Pump
9 Proportional chamber
10 Muon chamber
11 Ring quadrupole
12 Beam pipe

ring PETRA, in which the quadrupoles were moved as close to the detector as possible, increased luminosity by a factor of 3 (97, 98). This arrangement was dubbed "mini-beta interaction region." Similar reductions of the quadrupole distance from the interaction point were subsequently introduced in the PEP and CESR storage rings. Figure 6 (*left*) shows a PETRA "mini-beta interaction region."

4.10 *Micro-beta Interaction Regions*

The limit of beta in a "mini-beta" region is reached when the quadrupoles forming the final focus are just outside the detector. If one wants to further decrease β_{ip}, quadrupoles would have to move into the detector. Most of the detectors use longitudinal magnetic fields that are incompatible with a steel quadrupole magnet. This, and the fact that when the focussing quadrupole moves ever closer to the interaction point the focal strength must get larger and larger, led to the proposal of air core superconducting quadrupoles inside the magnetic detector very close to the interaction point (99). An arrangement as shown in Figure 6 (*right*) is proposed for the PETRA storage ring. It is expected that beta functions as small as $\beta_{xip} = 70$ cm and $\beta_{zip} = 4$ cm can be reached, which would further increase luminosity by a factor of 2 as compared to the "mini-beta" solution. A similar arrangement is planned for the LEP storage ring at CERN.

What now constitutes the limit of beta reduction in a "micro-beta" region? In storage rings with very high energy it is the maximum gradient in superconducting quadrupoles. For LEP a minimum distance of 3.5 m between interaction point and front face of the first quadrupole seems to be a practical limit if one wants to stay within tolerable levels of chromaticity. This then results in a β_{zip} of 10 cm. For storage rings with smaller energies it was shown by computer tracking (83) that the space charge limit ΔQ is independent of beta only if beta is large compared with the bunch length. The maximum luminosity is reached when $\beta_{zip} = 2\sigma_L$ ($\sigma_L =$ one standard deviation of the bunch length).

4.11 *Space Charge Compensation*

An interesting attempt to overcome the space charge limitation is being made with the DCI storage ring at Orsay (100). Space charge forces are made to vanish with a second bunch of the same shape and intensity but with opposite charge travelling in the same direction. Beam collisions now take place between four beams, an electron and a positron beam coming from both sides (Figure 7). So far space charge compensation with four beams has not resulted in larger luminosities. An explanation may be that small differences in the cross sections of the four beams create nonlinear fields strong enough to cancel the benefits of any partial compensation.

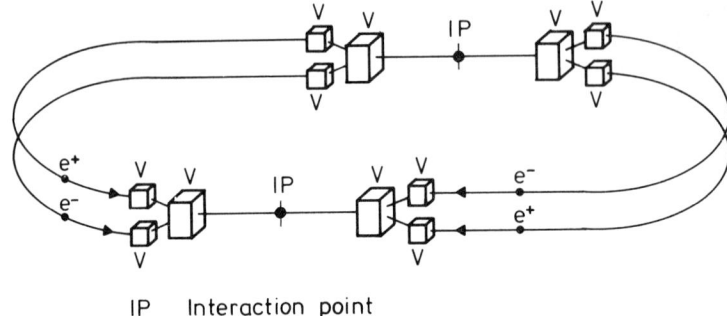

IP Interaction point
V   Vertical deflecting magnet (10°)

Figure 7 The four-beam space charge compensation scheme (schematic). Each of the two rings contains one electron and one positron bunch.

There also seem to be coherent beam-beam instabilities between the four beams.

5. POLARIZATION

5.1 *History and Observations*

Observations of spin polarization in electron storage rings were first made in 1971 independently by the Orsay (101) and Novosibirsk (102) groups. Polarization was measured by the spin-dependent intrabeam elastic scattering (Touschek scattering). The observed build-up times and the observed degree of polarization agreed well with the Sokolov-Ternov predictions, and evidence for various types of depolarizing resonances was found. The main resonances occur when the spin precession f_{sp} is an exact multiple n of the circumferential frequency f_c:

$$f_{sp} = a\gamma f_c = nf_c$$

with

$$a = \frac{g-2}{2} = 1.16 \times 10^{-3},$$

where g is the gyromagnetic moment of the electron.

This means that the main imperfection resonances occur at energy intervals of 440 MeV. By applying longitudinal or transverse magnetic fields with a frequency matched to f_{sp}, the beams can be depolarized at a given energy γ. This made it possible for the Novosibirsk group to determine the exact masses of the ϕ, $K^\pm$, K^0, ψ, and ψ' mesons and recently the Y resonance with an accuracy of 3×10^{-5} (103). At PETRA an accurate measurement of the frequency that depolarizes the beam allowed an absolute energy measurement at an energy of 16 GeV with an accuracy of 3 × 10^{-6} (104). In 1975 the SPEAR group made use of the polarization: By measuring the distribution of the azimuthal angle for the jet production in e^+e^- collision experiments it was possible to determine the spin of the initial state quarks to be $-1/2$ (120). Also, a new method for measuring the polarization was developed at SPEAR: Circularly polarized laser photons were Compton-backscattered from the stored beam. Since the vertical distribution of the high-energy backscattered photons depends on the spin direction of the electrons, this is an effective way for measuring the degree of polarization (105). This method is now generally used and is also applicable to new large storage rings such as HERA or LEP (106, 108). Shátunov (107) applied longitudinal magnetic fields on the spin precession resonance in such a way as to produce a complete spin flip of the electrons without appreciable depolarization. This method may have great significance for very large storage rings in which individual bunches could now have different spin states. This together with a spin rotation into the longitudinal direction at the interaction points would be an excellent means of investigating weak interaction effects (108).

5.2 *Polarization and Depolarization*

The time constant τ for the synchrotron radiation induced build-up of transverse polarization is, according to Sokolov & Ternov (14),

$$\tau = 98.7 \frac{\rho^3}{E^5} \frac{R}{\rho},$$

where ρ is the radius of curvature in the guide field measured in meters, E is the particle energy in GeV, R is the average radius of the machine, and τ is measured in seconds.

The theoretical maximum polarization that can be reached in the absence of depolarizing effects is 92%. The principal mechanism of depolarization is that of spin diffusion: In an ideal plane machine the equilibrium spin will always be perpendicular to the plane of motion. Particles not 100% polarized will, classically speaking, precess about this equilibrium spin direction with the frequency $f_{sp} = a\gamma f_c$. If the machine is disturbed by imperfections or by special inserts, a new equilibrium spin direction can be defined: a spin motion that closes upon itself after one turn. This alone would reduce polarization because as soon as the equilibrium spin no longer points in the direction of the magnetic guide field the polarizing mechanism is lessened. In such a disturbed machine, the spins of particles with different momenta do not necessarily have the same direction at each point along the circumference. The change of this vector with momentum is called spin dispersion in close analogy to orbital motion. But particles with large betatron oscillations will also have a different spin motion. At the instant when a particle emits a photon through synchrotron radiation, the off-momentum closed orbit and betatron amplitude change abruptly. The spin has not changed its direction (except for the rare case of a spin flip), but with respect to the new closed orbit and betatron oscillation amplitude the spin direction no longer corresponds to that of a fully polarized particle. After the excited betatron and phase oscillations have damped down (which happens in a short time as compared to the polarization time) the spin will have changed as compared to the time just before the photon emission. This change is a vector, which may be designated **d** and which should be called a "spin orbit coupling vector." Its magnitude determines the rate of spin diffusion that sets in because of the uncorrelated photon emissions. If **d** is known everywhere in the machine, the degree of polarization may be calculated using an equation derived by Derbenev & Kondratenko (109):

$$P = 0.92 \frac{\langle \rho^{-3} \mathbf{B}(\mathbf{n} - \mathbf{d}) \rangle}{\langle \rho^{-3} [1 - \frac{2}{9}(\mathbf{n}\boldsymbol{\beta})^2 + \frac{11}{18}(\mathbf{d})^2] \rangle},$$

where ρ is the radius of curvature, **B** is the magnetic field direction, **n** is the equilibrium spin direction, and $\boldsymbol{\beta}$ is the particle direction.

If the machine distortions are known, **d** and thereby the polarization can be numerically calculated, e.g. with the SLIM program developed by A. W. Chao (110). SLIM accurately predicts the polarization, as long as one is not right on a resonance, where linear approximations are not valid. The main imperfection resonances ($f_{sp} = k_0 f_c$) and the betatron side-band resonances [$f_{sp} = (k_0 \pm Q)f_c$] together with synchrotron side-band resonances [$f_{sp} = (k_0 \pm Q_s)f_c$] are accurately described, but not the effect of nonlinear resonances. The only way in which nonlinear resonances excited by beam-beam interaction or by imperfect magnetic fields may be evaluated in their effect on polarization is through tracking programs (111). Figure 8 shows a typical prediction for the polarization in the HERA storage ring. Because machine imperfections and beam-beam effects turn the spin by an amount proportional to $a\gamma$, their effectiveness grows linearly with energy. Special efforts are necessary to maintain polarization in large machines.

One way of eliminating the effect of resonances is to make the machine "spin transparent." Originally this method was invented by Chao & Yokoya (112) to cancel the depolarizing effects of spin rotators. For such an insert, ten integral conditions can be defined to assure that the spin orbit coupling **d** remains zero. Steffen (113) extended these spin transparency conditions to improve the performance of imperfect machines. Rossmanith & Schmidt (114) showed through SLIM calcu-

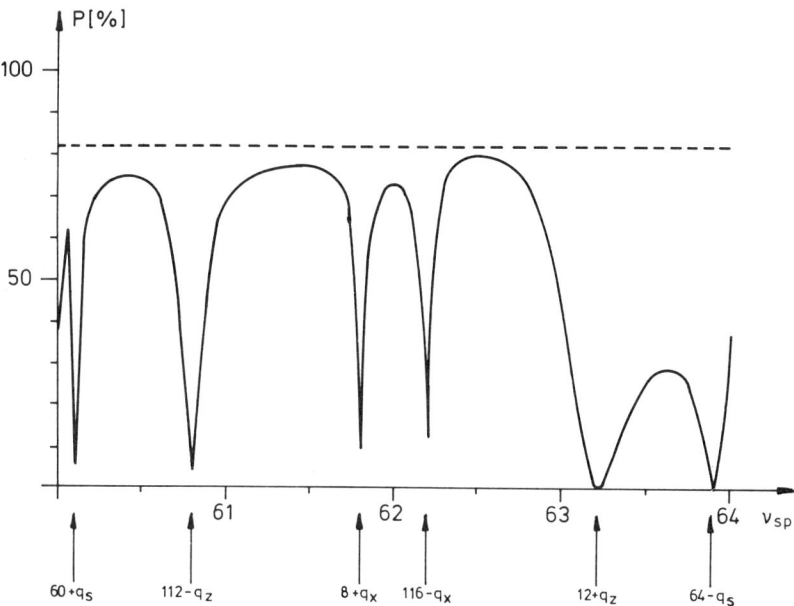

Figure 8 Polarization in a storage ring with an antisymmetric spin rotator (predictions by the SLIM program).

lations that, by cancelling the driving forces of those resonances closest to the working point, depolarization should be greatly reduced. Such cancellation is relatively easily done with the help of a few vertical steering coils. The experiment performed in PETRA (115) showed indeed a dramatic improvement of polarization, from 20 to 80%, when such empirical corrections were applied.

The increasing destructiveness of machine errors with increasing energy is not the only reason why high-energy electron storage rings have difficulties with polarization: The absolute spread in energy gets quite large in these rings and this means that the spread in spin precession frequencies grows to the point where it can no longer fit between the main imperfection resonances, which are spaced at intervals of 440 MeV. The solution to this problem may be the so-called Siberian Snakes (109): spin rotators that turn the spin such that in one half of the ring it is opposite to that in the other half. In this way the spin precession frequency is made to be 1/2, independent of particle energy. The spread in spin precession frequency is reduced to zero, and staying away from harmful resonances is no longer a problem. But at the same time the Sokolov-Ternov mechanism of polarization is reduced to zero. Additional wigglers with unequal field strength in the alternating poles may restore some of the polarization.

Even if polarization can be maintained for single beams under such adverse conditions it will be of no use if beam-beam effects depolarize the beams. Polarization has to be maintained under operating conditions with luminosity. The analytical treatment of this problem is extremely difficult. Estimates (116) make it likely that depolarization through beam-beam interaction may become a strong effect when the space charge forces begin to blow up the beams, i.e. when the specific luminosity begins to drop. In this connection, measurements on PETRA (117) that seem to confirm these estimates are of great interest. With colliding beams characterized by a space charge interaction of $\Delta Q_z = 0.023$, a polarization of 80% was measured; with more intense beams, polarization dropped to 20%. The best values were 80% polarization at a luminosity of 4×10^{30} cm^{-2} s^{-1}.

5.3 Spin Rotators

Most of the interest in polarization with very high-energy storage rings is connected with weak interaction physics. For this, longitudinal polarization at the interaction points with both helicities is essential. That way it may be possible, for example, to discover effects of right-handed neutral currents, should they exist. For this, different kinds of spin rotators have been designed (e.g. 118), but are not built yet. Figure 9 shows a spin rotator for the proposed e-p storage ring HERA (119). Most rotators have the drawback that they only work for a particular energy unless they permit some flexibility in their beam geometry, which is quite limited even in the best cases. Designing a rotator that maintains a high degree of polarization is a formidable task. Optics requirements and spin transparency form a set of twelve or more matching conditions that need to be simultaneously fulfilled.

Polarization represents a new dimension of storage ring technology, as yet not fully exploited. There are no reasons known why polarized electron-positron beam collisions should not be possible even at the highest storage ring energies, provided the necessary precautions are taken. And such beam-beam collisions should even be possible with longitudinally polarized beams. But all this requires a high degree of sophistication and complete mastering of the art of accelerator technology. At this moment one is only beginning to get into this field.

6. CONCLUSION AND OUTLOOK

During the last 15 years electron-positron storage rings have evolved into one of the most important tools in experimental high-energy physics. Their energy has increased from 2×500 MeV in the ACO storage ring to 2×59 GeV for the LEP (Phase I) storage ring presently under construction at

e^+e^- STORAGE RINGS 99

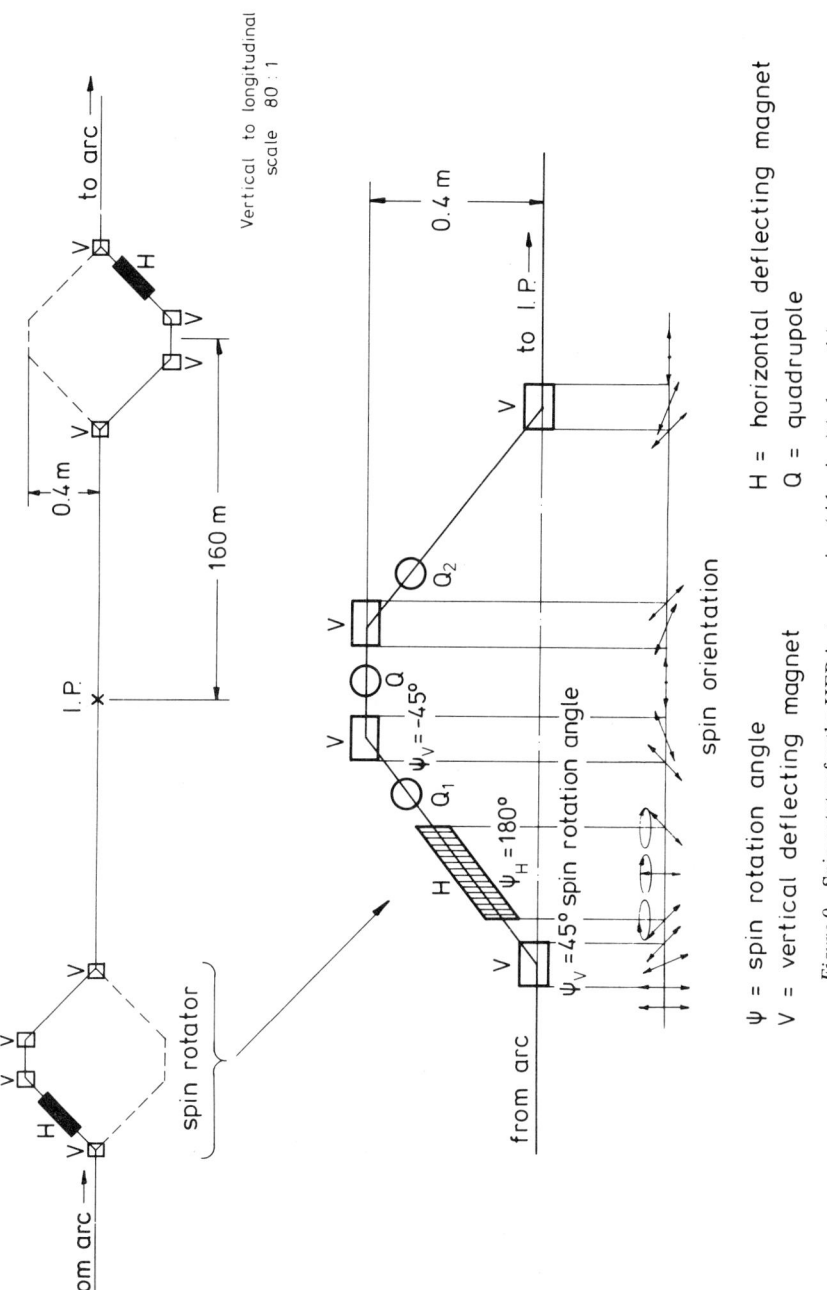

Figure 9 Spin rotator for the HERA storage ring (side view) (schematic).

CERN. At the same time the luminosity of these devices has increased by 3 to 4 orders of magnitude to values in the 10^{31} cm^{-2} s^{-1} range.

This remarkable development was made possible through a major machine physics effort reflected in a very large number of publications in this new field. The main directions of research were in the areas of single-beam stability and beam-beam interaction. Understanding of the effects of the "environment" (vacuum chambers, radiofrequency cavities, etc) on the stability of high-density current bunches has matured to the point where predictions for machines not yet built can be made with confidence. Although the analytical treatment of these problems is very difficult, quantitative descriptions of the relevant effects have become possible through numerical methods for calculating the electromagnetic fields excited by the circulating beams together with tracking programs to determine the interplay between fields and charge distribution in the beams. Similar methods have been very successful in describing the effects of beam-beam interaction. From all this the importance of certain machine features has become evident: storage rings must be built with smooth vacuum chambers to avoid the excitation of fields that can act back on the beams, and with strong focussing at the places of radiofrequency cavities to minimize the effect of beam induced fields on the beam. For high luminosity an extremely strong focussing at the interaction points is essential.

Storage ring technology has improved to the point where major breakthroughs can perhaps no longer be expected. There are only two areas in which we can look forward to major developments. (a) Superconducting rf systems may make storage rings cheaper (by perhaps as much as 50%?) and improve their performance somewhat. Because of the higher gradients in superconducting cavities, their required number will be smaller and this will result in a smaller impedance as seen by the beams. (b) Polarization of the beams, and particularly longitudinal polarization at the interaction points, is a distinct possibility, although much work is still necessary to make this a practical tool.

ACKNOWLEDGMENTS

We would like to thank our colleagues from DESY and other laboratories for helping us with the material for this article. We also wish to thank Drs. D. Degèle, K. Steffen, and Professor L. Hand for helpful comments regarding the manuscript.

GLOSSARY OF SOME OF THE SPECIAL TECHNICAL TERMS

Betatron oscillations
Beta function or amplitude function

Betatron wave number

Betatron phase advance: Owing to the transverse focussing, particles deviating from the closed orbit perform transverse oscillations about the closed orbit. If s is the coordinate along the closed orbit and y the deviation from the closed orbit, the betatron oscillation can be described as an amplitude- and phase-modulated harmonic oscillation:

$$y = A\sqrt{\beta(s)} \cos[\phi(s) + \phi_0].$$

The term $\beta(s)$ is called the *amplitude function* or *beta function*, $\phi(s)$ is the *phase advance*, and ϕ_0 and A are constants, where A is the invariant amplitude of the *betatron oscillations*. If C is the circumference of the closed orbit one has

$$\phi(s+C) = \phi(s) + 2\pi Q.$$

Here Q is the *betatron wave number* (number of betatron oscillations per revolution). Typical values for large e^+e^- storage rings are $15 < Q < 70$.

Chromaticity: Change of betatron wave number with energy deviation

$$\xi = \frac{\Delta Q}{\Delta E/E}.$$

Closed Orbit: That particular particle trajectory which, at equilibrium energy, closes upon itself after one revolution.

Dispersion: Deviation of a particle with relative energy deviation $\Delta E/E_0 = 1$ from the closed orbit.

Emittance

Beam dimensions: Emittance is the area in phase space filled with particles. It is customary to define

ε_x as the area in the horizontal $(x, dx/ds)$ plane
ε_z as the area in the vertical $(z, dz/ds)$ plane.

The emittance ε is independent of s, and the *transverse beam dimensions* along the closed orbit are described by $\sqrt{\varepsilon \beta(s)}$.

Impedance function: The action of the e.m. forces on the particles is measured in terms of electric potentials. These potentials are proportional to the beam current. The factor of proportionality has the dimension of an impedance and depends on the frequency spectrum of the current. It is called the impedance function.

Landau damping: If the single particles of a particle distribution have different (longitudinal or transverse) frequencies, the individual particles excited by a common perturbation run out of phase after a time that can be characterized by the Landau damping time. This causes a damping of collective motion.

Radiation damping: Synchrotron radiation in electron-positron storage rings damps the betatron and synchrotron oscillations of individual particles with damping rates that are of the order of the energy loss per turn divided by particle energy. The radiation damping rates increase with the third power of the beam energy. Typical damping times τ in large e^+e^- storage rings are $1 < \tau < 200$ ms.

Resonances

Stop bands: Since any imperfection in the magnetic guiding fields of the ring causes a periodic perturbation for the circulating particles, certain values of the wave number Q lead to resonant conditions so that the oscillations of the particles become unstable. Such resonant conditions can occur for the horizontal (Q_x) and vertical (Q_z) direction; they effect in the Q_x, Q_z plane forbidden resonance lines (*stop bands*), on which the particle motion becomes unstable.

Synchrotron oscillations or

Phase oscillations: Particles deviating from the equilibrium energy E_0 by an amount ΔE perform *synchrotron* or *phase oscillations*. Because of these oscillations the energy changes periodically within the interval $\pm \Delta E$. The energy oscillations are accompanied by a periodic change of the time-of-arrival (phase) at the accelerating rf resonators. The time-of-arrival oscillations and the energy oscillations are out of phase by 90°. The number of synchrotron oscillations Q_s along one revolution is the *synchrotron wave number* Q_s (typical values for large e^+e^- machines are $0.03 < Q_s < 0.1$).

Literature Cited[1]

1. Bernardini, C., et al. 1962. *Nuovo Cimento* 23:202
2. Bayer, V. N. 1963. In *Proc. Int. Conf. High-Energy Accel.*, ed. A. A. Kolomenski, p. 274. Moscow: Dubna
3. Barber, W. C., et al. 1966. In *Proc. Int. Symp. on Electron Positron Storage Rings*, ed. E. Cremien-Alcan. Paris: Saclay
4. Budger, G. I. 1966. See Ref. 3
5. Marin, P. 1967. In *Proc. Symp. on Electron Photon Interactions at High Energies*, p. 376. Stanford: AEC (Atomic Energy Commission)
6. Richter, B. 1976. *Nucl. Instrum. Methods* 136:473
7. Voss, G.-A. 1982. In *Proc. of ECFA-RAL Conf.* Geneva: CERN/ECFA
8. Kouptsidis, J. S. 1977. In *Proc. 7th Int. Vacuum Congress and 3rd Int. Conf. on Solid. Surfaces*, p. 93, ed. R. Dobrozemsky. Vienna
9. Sands, M. 1971. In *Proc. Int. Sch. Phys. "Enrico Fermi", Course XLVI 1969*, ed. B. Touschek. New York, London: Academic
10. Bartel, W., Wrulich, A. 1981. *DESY F11-81/03*. Hamburg: DESY
11. *SLAC Annual Report*. 1975. Stanford: SLAC
12. Scholz, K.-U. 1978. *DESY PET-78/03*. Hamburg: DESY

[1] Much of the material has been published in laboratory reports. The best way of obtaining those reports may be by requesting them from the corresponding laboratory libraries.

13. Potter, K. 1981. Gen. Meet. on LEP, *ECFA 81/54*. Geneva: ECFA
14. Sokolov, A. A., Ternov, I. M. 1964. *Sov. Phys. Dokl.* 8 1203:91
15. Tigner, M. 1977. *IEEE Trans. Nucl. Sci.* NS24:1849
16. PETRA Storage Ring Group. 1980. In *Proc. 11th Int. Conf. on High Energy Accel.*, p. 16. Basel: Birkhäuser
17. Febel, A., Hemmie, G. 1980. *DESY M-80/17*. Hamburg: DESY
18. Crowley-Milling, M. C., Rabinowitz, I. I. 1971. *IEEE Trans. Nucl. Sci.* NS18:1052
19. Donald, M. H. R. 1973. *RHEL/M/Nim 18*. Chilton: Rutherford High Energy Lab.
20. Piwinski, A., Wrulich, A. 1976. *DESY 76/07*. Hamburg: DESY
21. Chao, A. W., Piwinski, A. 1977. *DESY PET-77/09*. Hamburg: DESY
22. Vinokurov, N. A., et al. 1977. In '*Proc. 10th Int. Conf. on High-Energy Accel.*, p. 254, ed. Yu. M. Ado. Protvino: IFWE
23. Sundelin, R. M. 1979. *IEEE Trans. Nucl. Sci.* NS26:3604
24. Piwinski, A. 1980. See Ref. 16, p. 638
25. Orlov, Yu. F. 1957. *Sov. Phys. JETP* 5:45
26. Robinson, K. W. 1958. *CEA 54*. Cambridge: CEA
27. Liboff, R. L. 1970. *J. Math. Phys.* 11:1295
28. Keil, E. 1972. *Nucl. Instrum. Methods* 100:419
29. Keil, E., et al. 1975. *Nucl. Instrum. Methods* 127:475
30. Chao, A. W. 1975. *PEP 105/SPEAR 198*. Stanford: SLAC
31. Chao, A. W. 1975. *PEP 119*. Stanford: SLAC
32. Zotter, B. 1979. *PEP-310*. Stanford: SLAC
33. Bassetti, M., et al. 1979. *DESY 79/07*. Hamburg: DESY
34. Halbach, K. 1976. *Part. Accel.* 7:213
35. Wilson, P. B. 1977. *SLAC PUB 1908/PEP 240*. Stanford: SLAC
36. Weiland, T. 1980. *CERN ISR-TH/80-07*. Geneva: CERN
37. Pellegrini, C. 1979. *Nuovo Cimento* 64A:477
38. Robinson, K. W. 1969. *CEATM-183*. Cambridge: CEA
39. Karliner, M. M. 1970. *IYa F Report 45-70*. Novosibirsk: Institut Yadernoi fisiki SOAN SSSR
40. Kohaupt, R. D. 1975. *IEEE Trans. Nucl. Sci.* NS22:1456
41. Pruett, C. H., Otte, R. A., Mills, F. E. 1965. In *Proc. 5th Int. Conf. on High-Energy Accel.*, ed. M. Guilli, p. 343. Rome: Comitato Nazionale Energia Nucleare
42. Pellegrin, J. L. 1975. *IEEE Trans. Nucl. Sci.* NS22:1500
43. Heins, D., et al. 1979. *DESY PET-79/06*. Hamburg: DESY
44. Wille, K. 1979. *DESY PET-79/01*. Hamburg: DESY
45. Sacherer, F. J. 1972. *CERN/SI,BR/72.5*. Geneva: CERN
46. Sacherer, F. J. 1973. *IEEE Trans. Nucl. Sci.* NS20:825
47. Sacherer, F. J. 1974. In *Proc. 9th Int. Conf. on High Energy Accel.*, p. 346. Stanford: SLAC
48. Sacherer, F. J. 1977. *IEEE Trans. Nucl. Sci.* NS24:1393
49. Weiland, T. 1982. *DESY 82-015*. Hamburg: DESY
50. Keil, E., Schnell, W. 1969. *CERN-TH-RF/69-48*. Geneva: CERN
51. Sessler, A. M. 1973. *PEP Note-28*. Stanford: SLAC
52. Channell, P. J., Sessler, A. M., 1976. *Nucl. Instrum. Methods* 136:473
53. Messerschmid, E., Month, M. 1976. *BNL-22411*. Brookhaven: BNL
54. Renieri, A. 1976. *LNF-76/11(R)*. Rome: Comitato Nazionale Energia Nucleare
55. Chao, A. W., Gareyte, J. 1976. *SPEAR-197/PEP-224*. Stanford: SLAC
56. Sacherer, F. J. 1977. *CERN/PS/BR/77-5*. Geneva: CERN
57. Ruggiero, A. 1977. *IEEE Trans. Nucl. Sci.* NS24:1205
58. Hardt, W., Kohaupt, R. D. 1977. *DESY 77/20*. Hamburg: DESY
59. Wilson, P. B., et al. 1977. *IEEE Trans. Nucl. Sci.* NS24:1211
60. Kohaupt, R. D. 1977. *DESY 77/66* Hamburg: DESY
61. Weiland, T. 1981. *DESY 81-088*. Hamburg: DESY
62. Pellegrini, C. 1969. *Nuovo Cimento* 64A:447
63. Sands, M. 1969. *SLAC-TN-69-8*. Stanford: SLAC
64. Kohaupt, R. D. 1980. *DESY 80/22*. Hamburg: DESY
65. Kohaupt, R. D. 1980. *DESY M-80/19*. Hamburg: DESY
66. Kohaupt, R. D. 1980. See Ref. 16, p. 582
67. Satoh, K., Chin, Y. 1982. *KEK 82-2*. Tsukuba: KEK
68. Chao, A. W., et al. 1982. To be published. Stanford: SLAC
69. Zotter, B. 1982. *LEP-Note 402*. Geneva: CERN
70. Averill, R., et al. 1973. *IEEE Trans. Nucl. Sci.* NS20:3
71. Gram, R., Morton, P. 1967. *SLAC TN-67-30*. Stanford: SLAC
72. Bramham, P., et al. 1977. *IEEE Trans. Nucl. Sci.* NS24:1490

73. Voss, G.-A. 1976. In *Proc. Int. Sch. Part. Accel. Erice*. Geneva: CERN
74. Wiedemann, H. 1977. See Ref. 22, p. 430.
75. Amman, F., Ritson, D. 1961. In *Proc. Int. Conf. on High-Energy Accel.* Brookhaven: BNL
76. Panel Discussion on Beam Beam Effects. 1980. See Ref. 16, p. 741
77. Amman, F. 1971. In *Proc. 8th Int. Conf. on High-Energy Accel.*, p. 63. Geneva: CERN
78. Hofmann, A., Vos, L., Zotter, B. 1980. See Ref. 16, p. 713
79. Paterson, J. M. 1980. See Ref. 16
80. Donald, M. H. R., Paterson, J. M. 1979. *IEEE Trans. Nucl. Sci.* NS26: 3580
81. Chirikov, B. V. 1979. *Phys. Rep.* 52: 263
82. Piwinski, A. 1980. *DESY 80/131*. Hamburg: DESY
83. Meyers, S. 1981. *CERN-ISR-RF/81-08*. Geneva: CERN
84. Piwinski, A. 1981. *IEEE Trans. Nucl. Sci.* NS28: 2440
85. Piwinski, A. 1981. *DESY 81-066*. Hamburg: DESY
86. DORIS Storage Ring Group. 1979. *IEEE Trans. Nucl. Sci.* NS26: 3135
87. Meyers, S. 1982. *LEP-Note*. Geneva: CERN
88. Paterson, J. M., et al. 1975. *PEP-125*. Stanford: SLAC
89. Robinson, K. W. 1948. *Phys. Rev.* 75: 1912
90. LEP Study Group. 1979. *CERN/ISR-LEP/79-33*. Geneva: CERN
91. Robinson, K. W., Voss, G.-A. 1966. See Ref. 3
92. Averill, R., et al. 1971. See Ref. 77, p. 163
93. Piwinski, A. 1976. *PETRA Note NR.79*. Hamburg: DESY
94. Autin, B. 1976. *CERN/ISR-LTD/76-37*. Geneva: CERN
95. Wiedemann, H. 1976. *PEP-220*. Stanford: SLAC
96. Zyngier, H. 1977. *LAL 77/35*. Paris: LAL
97. Steffen, K. G. 1979. *DESY PET-79/07*. Hamburg: DESY
98. PETRA Storage Ring Group. 1981. *IEEE Trans. Nucl. Sci.* NS28: 2025
99. Steffen, K. G., Voss, G.-A., Wolf, G. 1981. *DESY M-81/20*. Hamburg: DESY
100. Marin, P. 1974. See Ref. 47
101. Serednyakov, S. I., et al. 1976. *Sov. Phys. JETP* 44: 1063
102. LeDuff, J., et al. 1973. *Orsay 4-73*. Orsay: LAL
103. Artanonov, A. S., et al. 1982. *Novosibirsk Preprint 82-94*. Novosibirsk: Institut Yadernoi fisiki SOAN SSSR
104. Neumann, R., Rossmanith. R. 1982. *Nucl. Instrum. Methods* 204: 29
105. Gustavson, D. B., et al. 1979. *Nucl. Instrum. Methods* 165: 177
106. Montague, B. W. 1982. *DESY-82/09*. Hamburg: DESY
107. Shátunov, Ju. 1982. *DESY M-82/09*. Hamburg: DESY
108. Schwitters, R. 1982. *DESY M-82/09*. Hamburg: DESY
109. Derbenev, Ya. S., Kondratenko, A. M. 1973. *Sov. Phys. JETP* 37: 968
110. Chao, A. W. 1981. *Nucl. Instrum. Methods* 180: 29
111. Kewisch, J. 1982. *DESY M-82/09*. Hamburg: DESY
112. Chao, A. W., Yokoya, K. 1981. *KEK 81-7*. Tsukuba: KEK
113. Steffen, K. G. 1982. *DESY HERA 82-02*. Hamburg: DESY
114. Rossmanith, R., Schmidt, R. 1982. *DESY M-82/09*. Hamburg: DESY
115. Rossmanith, R., Schmidt, R. 1982. *DESY 82-026*. Hamburg: DESY
116. Courant, E. 1982. *DESY M-82/09*. Hamburg: DESY
117. Bremer, H. D., et al. 1982. In *Proc. of High-Energy Spin Phys.* Brookhaven: BNL
118. Buon, J. 1982. *SLAC Rep. 250*. Stanford: SLAC
119. HERA Proposal. 1981. *DESY HERA 81/10*. Hamburg: DESY
120. Schwitters, R. F., et al. 1975. *Phys. Rev. Lett.* 35: 1609

NUCLEAR MATTER THEORY: A Status Report

A. D. Jackson

Department of Physics, State University of New York, Stony Brook, New York 11794

CONTENTS

1. INTRODUCTION 105
2. BETHE-BRUECKNER-GOLDSTONE THEORY 112
3. VARIATIONAL CALCULATIONS 120
4. TOWARD A SUITABLE NUCLEAR HAMILTONIAN 128
5. CONCLUSIONS 138

1. INTRODUCTION

Initially, study of the infinite system of homogeneous nuclear matter was intended as a rapid and simple check of the firmly held conviction that the properties of finite nuclei could be determined microscopically starting from a two-body potential describing the elastic scattering of free nucleons. That such calculations are apt to be simple is seen immediately from Hartree-Fock calculations. In nuclear matter the homogeneity of the system indicates that the single-particle wave functions resulting from the solutions of the Hartree-Fock equations must be plane waves so that the calculation of the energy of the system is reduced to the evaluation of a few integrals. This is not the case in finite nuclei, where the solution of the Hartree-Fock equations requires a certain application. In the absence of direct information on this infinite system, the binding energy was assumed to be 16 ± 1 MeV per particle on the basis of the volume term in the usual semiempirical mass formula. The equilibrium density of nuclear matter is less accessible since it must be determined from extrapolations of the central densities of finite nuclei. Aside from the usual concern that nuclei are mostly surface, such determinations are clouded by the fact that central densities are sensitive to assumptions about high momentum transfer results for electron scattering, which are not always available, and to certain underlying assumptions regarding the nature of the electromagnetic interactions between electrons and nucleons in the presence of the strong interaction between nucleons. Specifically, the relationship of mesonic exchange current corrections to electromagnetic operators is not yet understood well enough to represent an unambiguous ingredient in the analysis of electron scattering data. The best estimate for the density of nuclear matter is 0.156 ± 0.015 F^{-3}, which is roughly consistent with estimates based on the Coulomb term in the semiempirical mass formula.

The information contained in nucleon-nucleon (NN) elastic phase shifts and deuteron properties is not sufficient to determine the underlying potential uniquely even if one makes the restrictive assumption that the potential is local. There are abundant theoretical grounds for believing that the NN interaction is both nonlocal and explicitly energy dependent. Composite (for example, quark) structure in the nucleon leads to nonlocal interactions. The elimination of inelastic channels (for example, two nucleons plus pions) leads to energy-dependent interactions. Thus, a secondary aim of nuclear matter calculations was to help select the correct NN interaction. Such hope was short-lived since none of the existing interactions was able to reproduce the observed binding energy and equilibrium density of nuclear matter. This is the "nuclear matter problem." The situation is illustrated in Figure 1, which shows the empirical values of

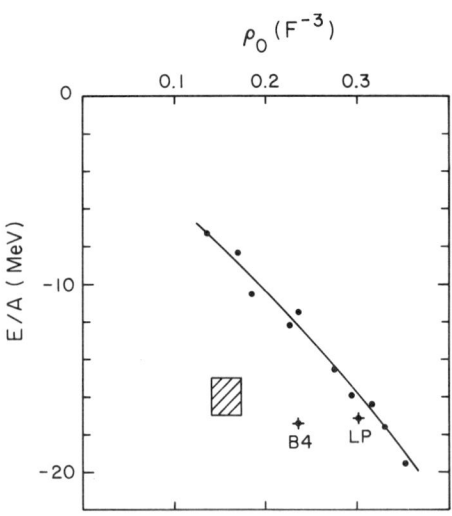

Figure 1 The Coester line showing equilibrium densities and energies for a variety of NN potentials.

binding energy and equilibrium density along with the *Coester line* (1), the line that summarizes the results of calculations in lowest-order Bethe-Brueckner-Goldstone (BBG) theory. If a given potential provides the correct equilibrium density, it underestimates the binding energy. If it provides the correct binding energy, it overestimates the equilibrium density. A qualitatively reliable measure of where a given potential will lie on this line is the strength of the tensor force in two-nucleon states with isospin zero, which, in turn, is measured by the D-state admixture in the deuteron wave function. Potentials with a stronger tensor force tend to saturate at lower densities and, hence, with smaller binding energy.

Obviously, there has been considerable interest in the origin of this discrepancy. One possibility is that none of the NN interactions studied to date is close enough to the true interaction. This explanation seems unlikely since some of the potentials considered, such as the Paris potential (2), represent the best theoretical understanding of the NN interaction currently available. The Paris potential explicitly includes the exchange of known mesons with acceptable coupling constants and describes the important two-pion-exchange pieces of the interaction with dispersion relation techniques firmly tied to πN scattering data and $\pi\pi$ phase shifts.

Two reservations can be expressed regarding the best mesonic theoretic potentials. The first is that they are phenomenological at short distances (for example, nucleon separations less than 0.8 F in the case of the Paris potential). It is generally acknowledged that meson theoretic arguments are not capable of providing more than qualitative guidance in this region, and one exploits this freedom to adjust the short-range interaction in such a manner as to provide a quantitative fit to NN scattering data. At some later stage in our understanding of the NN interaction, it seems clear that this phenomenological freedom will be replaced by models that account for the three-quark nature of the nucleon. In this regard a model such as the little bag model (3), which has explicit coupling of the nucleon bag to external pions, seems most appealing. Because of this pion coupling, such models will be relatively easy to graft on to mesonic theoretic models, which have been remarkably successful in describing peripheral NN phase shifts and are likely to provide a reliable description of the NN interaction at intermediate and large distances.

In the little bag model, the pion field provides an external pressure that tends to reduce the bag radius to a manageable 0.5 F. Preliminary studies of the NN interaction in the little bag model—which suggest that the increase in the pion field surrounding a single nucleon in the presence of a second nucleon further reduces the bag radius—provide additional support for the conventional picture that meson exchange dominates the NN interaction at distances greater than 0.8 F (4). Of course, if the pion presence outside nucleon bags is substantially less than assumed in the little bag model, as suggested by the MIT version of the bag model (5), one can anticipate overlapping nucleon bags at internucleon distances of as much as 2 F; this would require a major rethinking of the NN interaction. Bag models of the NN interaction will be confronted with a difficult task when their development allows a quantitative attack on the NN interaction. Specification of one- and two-pion-exchange pieces of the interaction places very strong constraints on the form of the interaction at short distances, where bag model modifications are to be expected, if a quantitative description of elastic NN scattering data is to be maintained. If bag model modifications can quantitatively describe NN scattering data, it is unlikely that they will move nuclear matter significantly away from the Coester line.

A second minor reservation regarding meson theoretic potentials is that, in the interests of technical simplicity, they tend to suppress the energy dependence, which is a natural ingredient in, for instance, isobar contributions to the two-pion-exchange interaction. Since such realistic energy dependence has not been given a thorough test in nuclear matter, there remains the possibility that its correct inclusion might help solve the problem.

A more fruitful line of attack during the past decade has been to argue that the origin of the discrepancy lies in the treatment of the many-body problem. Specifically, BBG theory, which in lowest order leads to the Coester line, is a hole-line expansion. In lowest order, one considers only the repeated scattering of each pair of particles in the nuclear medium. The effects of the medium are contained in the Pauli principle and dispersion effects. The Pauli principle prohibits scattering into occupied states with single-particle momenta less than the Fermi momentum. Dispersion effects account for the fact that each nucleon is bound in the medium and, thus, individual nucleon energies are shifted from their free space values. Except for these effects, lowest-order BBG results are close in spirit and execution to the summation of two-particle "ladder" diagrams describing the free scattering of two nucleons. In such a scheme, deficiencies in fitting NN phase shifts have immediate consequences in nuclear matter. This is the origin of one of the cardinal rules of the nuclear matter game: Interactions considered must describe NN phase shifts.

Over the years, a number of criticisms of the BBG theory have been advanced. The strongest attack began about a decade ago when "variational" techniques were applied to the nuclear matter problem. In developments parallel to those in the BBG approach, Jastrow (6) proposed a simple trial wave function aimed at realizing two-body correlations in nuclear matter. He suggested a function that was a product of two-body correlation functions and the usual Slater determinant of noninteracting plane waves. The energy of nuclear matter could be calculated using this trial wave function, and the energy minimized by varying the form of the two-body correlation function. The first step in this procedure is not trivial and requires a substantial numerical investment to proceed from the correlation function to the two-body distribution function, $g(\mathbf{r}_{12})$, which represents the probability that there is a nucleon at $\mathbf{r}_2$ given the fact that there is a nucleon at $\mathbf{r}_1$. This task was greatly simplified by the use of the hypernetted chain (HNC) approximation, which provides an approximate but straightforward relation between these quantities through an integral equation (7). With this technical improvement, the Jastrow/HNC scheme offered a powerful alternative to BBG calculations. Preliminary results using this method indicated large differences from BBG results. Indeed, the BBG approach is not well suited to systems of high density and, for this reason, has not been applied with conspicuous success to the systems of liquid ^{3}He and its simpler boson counterpart, liquid ^{4}He. The variational approach, which effectively incorporates the long-range correlations of phonon exchange, does not suffer from this drawback, and a variety of interesting and useful calculations for these atomic liquids have been performed in the guise of homework problems aimed at refining many-body techniques for use in the nuclear matter problem.

A number of concerns can also be expressed regarding the variational approach. First, in spite of the name and the variational framework, the hypernetted chain approximation does not respect the variational principle and the resulting energies do not provide an upper bound on the ground-state energy. The diagrams forming the basis for the lowest-order BBG calculations are included only approximately in the Jastrow/HNC approach. In particular, the Jastrow ansatz for the wave function explicitly indicates that the correlation between each pair of particles is the same regardless of their single-particle quantum numbers. This approximation is directly responsible for the inability of such simple trial functions to describe the effective mass in liquid ^{3}He. This is not the case in the BBG approach. Variational calculations also present certain linguistic barriers. While it is clear that they have a diagrammatic content richer than BBG theory, it has not always been easy to see precisely what this content is. Finally, the variational approach is at its best when dealing with the correlations induced by purely central forces. The existence of a strong tensor force is one of the least ambiguous statements meson theory can make regarding the NN interaction since it is associated with one-pion exchange. A reliable description of the resulting tensor correlations between nucleons is a necessary ingredient in nuclear matter calculations not easily obtained in the variational approach.

The last decade has seen substantial technical improvements in both types of calculations and considerable improvement in the agreement between results for the same interactions in the two approaches. The common difference between recent results and the results of lowest-order BBG theory is now larger than their internal disagreements. For example, lowest-order BBG results for the Reid potential (8) yield a binding energy of -10.5 MeV per particle at an equilibrium density of 0.183 F^{-3}. More elaborate BBG results (9) yield $E/A = -17.3$ MeV and $\rho_0 = 0.237$ F^{-3}, which can be compared with the values $E/A = -17.1$ MeV and $\rho_0 = 0.303$ F^{-3} obtained from variational calculations (10). Results of such elaborate calculations are available for a limited selection of potentials. Although they indicate a considerable shift along the Coester line, there is much less tendency to move off the Coester line. The discrepancy between these calculations and empirical nuclear matter parameters remains. We address some of the specifics of these calculations in the following sections.

If the solution to the nuclear matter problem lies neither in the two-body interaction nor in the methods used to solve the many-body problem, the only possibility remaining is genuine many-body forces. Such forces would have no effect on the free two-body system and could free nuclear matter from the constraints of the Coester line. The question arising is which many-body forces to consider first and, more generally, how to order many-body forces.

One obvious ordering principle is to consider first those three-body forces of longest range. Two examples of long-range, three-body forces are shown in Figures 2a and 2b. This is precisely the ordering principle that has been applied with success in the two-body problem. It is less obvious that it is valid in the present case. Integrating over one of the three nucleon lines, these three-body forces assume the form of density-dependent modifications of the two-body interaction, as indicated in Figures 2c and 2d. These diagrams, when supplemented by the remaining members of the obvious geometrical series, represent changes in the pion mass and propagator, respectively. The pion, owing to its pseudoscalar nature, leads to components in the NN interaction having only the spin/isospin structure $(\sigma_1 \cdot \sigma_2)(\tau_1 \cdot \tau_2)$ and $S_{12}(\tau_1 \cdot \tau_2)$. To lowest-order in the potential, neither term contributes to the energy of either nuclear matter or neutron matter. Thus, the longest-range three-body force can be regarded as a correction to a two-body force term that is itself zero.

Significant contributions from these three-body forces come from the iteration of this pion-range three-body force with the ordinary one-pion-exchange force as shown in Figure 2e. In this case we are dealing with the operator

$$(\tau_1 \cdot \tau_2)^2 (S_{12})^2 = [3 - 2(\tau_1 \cdot \tau_2)][6 + 2(\sigma_1 \cdot \sigma_2) - 2S_{12}]. \qquad 1.$$

Thus, this interaction can give rise to a central force that can be made to account for the discrepancy in nuclear matter.

Another approach to the problem of many-nucleon forces breaks some of the cardinal rules and has led a somewhat independent existence. A number of authors (11–13) have considered more manifestly field theoretic models of nuclear matter based on a chiral symmetric Lagrangian dominated by scalar mesons and nucleons. Calculations usually begin with a mean field approximation. Even when solved exactly, such model Lagrangians are not capable of providing the kind of fit to NN scattering data that is a precondition for playing the nuclear matter game. Further, the mean field approach represents an approximation not easily introduced in the scattering problem. In short, this approach distances itself from the intimate connection between NN scattering and nuclear matter in which lowest-order BBG theory would have us believe. As soon as one looks at the numbers, it is easy to appreciate this point of view.

In Figure 4a we show the leading three-body force that arises in such theories. It is the obvious analogue of Figures 2a and 2c. This diagram

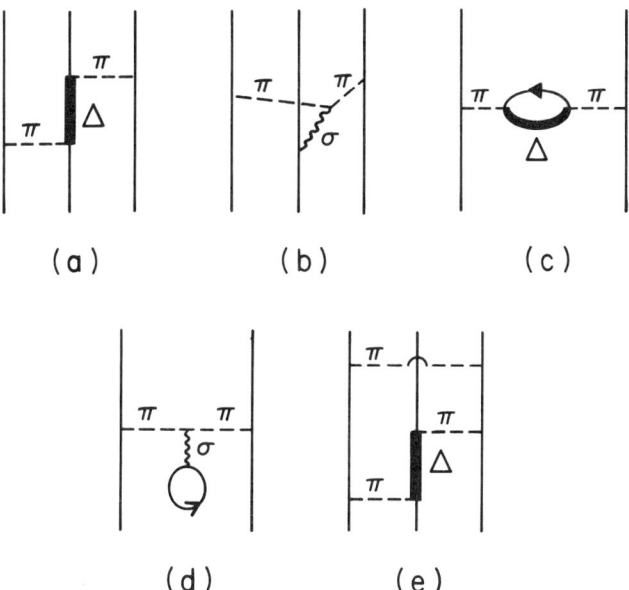

Figure 2 Conventional three-nucleon forces.

increases the binding energy of nuclear matter by something on the order of 14 MeV. In fact, such mean field theories can lead to a tightly bound, abnormal phase of nuclear matter in which the nucleons have zero mass unless prevented by vacuum fluctuation corrections. Individual vacuum fluctuation corrections can be extremely large and the notions of chiral symmetry can provide only limited guidance in how these terms should be grouped. Nonetheless, the physics upon which such chiral models are based is the solid physics of πN scattering and the two-pion-exchange force between nucleons. The many-body forces that arise in such models are the natural extension of this physics into the many-nucleon system. Such models and the nonperturbative way in which they are treated in mean field theories offer a systematic approach to the introduction of many-body forces not readily obtained in more conventional approaches. In Section 4 we consider both conventional and chiral descriptions of many-body forces and we offer some simple ways to embed the major effects of these forces in a more careful treatment of two-body forces.

Nuclear matter has been much reviewed; yet another review raises the prospect of filling a much-needed gap in the literature. For the pre-1972 status of nuclear matter theory, one cannot do better than consult Bethe's review in this series (14). In Section 2 we consider the BBG approach to the nuclear matter problem. Day's review (15) offers many details ignored here, while the review by Zabolitsky et al (16) provides a thorough discussion of the "exponential S" method. In Section 3 we discuss variational calculations. Feenberg's book (17) provides an excellent introduction to the subject. The use of correlation operators is described in the review by Pandharipande & Wiringa (18), while the spirit of the correlated basis function techniques is nicely, if pessimistically, stated by Ripka (19). The diagrammatic interpretation of variational calculations has been discussed at some length by Ripka (20) and by the present author (21). Section 4 is devoted to a discussion of many-nucleon forces both from a conventional point of view and using chiral Lagrangians. The latter point of view is discussed in greater detail in Walecka's review (11) and in References (12) and (13). The conclusions, relegated to Section 5, will have to stand without external support.

2. BETHE-BRUECKNER-GOLDSTONE THEORY

The Bethe-Brueckner-Goldstone (BBG) approach to the ground-state energy of nuclear matter is based on the Goldstone linked-cluster expansion (22). One starts by artificially breaking the full Hamiltonian,

$$H = \sum_i -\frac{\hbar^2}{2m}\nabla_i^2 + \sum_{i<j} V_{ij}, \qquad 2.$$

into two pieces:

$$H_0 = \sum_i \left[-\frac{\hbar^2}{2m} \nabla_i^2 + U(k_i) \right] \qquad 3.$$

and

$$H_1 = \sum_{i<j} V_{ij} - \sum_i U(k_i). \qquad 4.$$

Ignoring H_1, we observe that the solution to the Schrödinger equation for H_0,

$$H_0 \Psi = E\Psi, \qquad 5.$$

is trivial for any choice of $U(k_i)$. It leads to a ground-state wave function that is a Slater determinant of single-particle plane wave states, $\phi_\mathbf{k}$, where $|\mathbf{k}|$ is less than the Fermi momentum, k_F. For spin- and isospin-symmetric matter, the Fermi momentum and density are related by

$$\rho = \frac{2}{3\pi^2} k_F^3. \qquad 6.$$

The energy of each single-particle state is simply

$$E(k_i) = \frac{\hbar^2 k_i^2}{2m} + U(k_i), \qquad 7.$$

and the energy of the ground state involves summing this single-particle energy over the occupied states within the Fermi sphere. The single-particle potential, $U(k)$, is in principle arbitrary and is to be chosen to make a suitable perturbation theory in H_1 converge as rapidly as possible. The role played by $U(k)$ in enhancing convergence is not trivial, and the optimum prescription for its determination remains a matter of discussion.

Because of the strong short-range repulsion that characterizes realistic NN interactions, there is no reason to expect that a simple perturbation theory in H_1 will converge. It is generally acknowledged that the first sequence of terms to be included involving H_1 is that summed by the Bethe-Goldstone equation

$$G = V - V(Q/e)G, \qquad 8.$$

which sums the particle-particle ladder diagrams familiar from free particle scattering. This reaction matrix differs from the free scattering sums because a Pauli operator, Q, is present and prohibits scattering into occupied states with momentum less than k_F, and because a modified energy denominator,

$$e = E(p_1) + E(p_2) - E(h_1) - E(h_2), \qquad 9.$$

is also present. When the NN interaction contains strong short-range repulsion, this equation cannot be truncated. The presence of Q and the modification of e are not numerically innocent. They prevent the reduction of the Bethe-Goldstone equation to an integral equation in one scalar variable (i.e. the magnitude of the relative momentum) following a transformation to relative coordinates and a partial wave projection. In the interest of simplicity it has been conventional to adopt certain angle-averaged approximations based, again, on the assumed dominance of high-energy intermediate states as a consequence of short-range repulsion.

At this point it is easy to introduce the conventional choice of $U(k)$:

$$U(k) = \sum_{p<k_F} \langle \mathbf{kp}|G|\mathbf{kp}\rangle \qquad k < k_F$$
$$= 0 \qquad k > k_F. \qquad 10.$$

This choice of $U(k)$ is manifestly discontinuous at the Fermi surface. There are, however, strong reasons to believe that $U(k)$ should be continuous at the Fermi surface (23).

Characteristically, lowest-order BBG calculations yield a kinetic energy of $+25$ MeV per particle and a potential energy of roughly -35 MeV per particle for realistic NN interactions. The plot of energy versus density for the Reid potential is given in Figure 3 as the curve B2. It is at this level of sophistication that one obtains the Coester line of Figure 1, which indicates a global failure to reproduce the binding energy and equilibrium density of nuclear matter simultaneously.

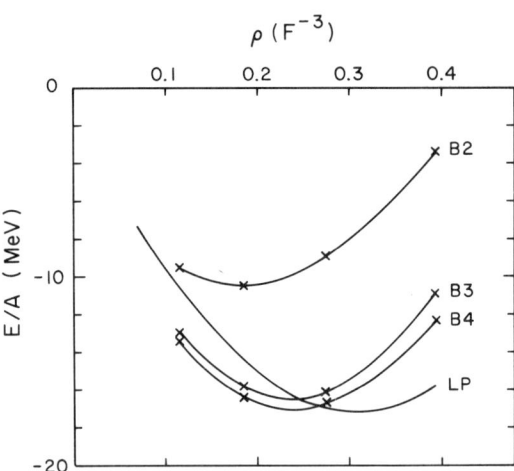

Figure 3 Saturation curves for BBG (B2–B4) and variational (LP) calculations using the Reid potential.

A minimum rearrangement of the terms in the Goldstone expansion might be aimed at providing a perturbation theory in powers of G. Such an expansion is still not convergent, and one must turn instead to an expansion in the number of independent hole lines for which the dominance of the effects of short-range repulsion again provides support. The number of independent hole lines in any given energy diagram is, as the name implies, the number of lines with momentum less than k_F whose momenta can be specified independently after the constraints of momentum conservation have been imposed. All diagrams contributing to the reaction matrix of Equation 8 have precisely two independent hole lines. Brandow (24) offered an argument to suggest that the contribution to the energy of a diagram with n independent hole lines should be of order $\kappa^{n-1}\rho\langle G_n\rangle$, where $\langle G_n\rangle$ is some average reaction matrix whose energy denominator reflects the excitation of all n independent hole lines. The quantity κ is essentially the probability that there is an unoccupied state within the Fermi sea. It is the "small" dimensionless parameter that provides a measure of the convergence rate of the independent hole-line expansion. In lowest order, κ may be estimated from the two-body correlations as

$$\kappa = \rho\left\langle \int [\phi_{mn}-\psi_{mn}]^2 \, d^3r \right\rangle, \qquad 11.$$

where $\langle \ \rangle$ represents an average over the Fermi sea of the hole-line labels m and n. Here, ϕ_{mn} is the uncorrelated two-body wave function (a simple product of plane waves) while ψ_{mn} is the correlated wave function coming from the solution of the Bethe-Goldstone equation:

$$\psi_{mn} = \phi_{mn} - (Q/e)G\phi_{mn}. \qquad 12.$$

The quantity κ is on the order of 1/6 at the empirical equilibrium density, which provides encouragement for the utility of a hole-line expansion (9). The magnitude of κ also reminds us that the equilibrium density of nuclear matter is not determined primarily by the range of the short-range repulsion, as it is in liquid ^{3}He where κ is of order one and hole-line expansions are of limited value. A more important mechanism is the reduction of the substantial attraction coming from the iterated one-pion-exchange tensor force due to the imposition of the Pauli principle in the intermediate state.

Brandow's simple result depends on the dominance of short-range correlations that render $(\psi_{mn}-\phi_{mn})$ rather insensitive to the occupied state labels m and n. As we shall see, this same notion of the approximate state independence of two-body correlations (for the same reasons) plays an important role in variational calculations based on Jastrow trial wave functions. Brandow was careful to point out that his analysis did not accurately reflect the convergence rate of long-range correlations (such as are embodied in the ring diagrams describing phonon exchange) for which there are even grounds to suspect an asymptotic divergence. [See Nyman et al (25) in this context.] These cautionary words have frequently been ignored by devotees of the BBG approach.

The leading diagrams of third order in G or U involving three independent hole lines are shown in Figure 4. Using the self-consistent prescription for $U(k)$ of Equation 10, we find diagrams a and b cancel precisely with analogous cancellations in higher orders. Diagram c is attractive and of order -0.5 MeV (9). The standard choice of $U(k)$ ensures that diagram d vanishes. This choice could be modified to yield cancellation between diagrams d and e. It is not clear that this would be useful. Diagrams e and f represent the leading terms in an infinite sequence of three-hole-line diagrams. Considerable cancellations are expected within this class, and the redefinition of $U(k)$ to allow cancellation of diagrams d and e is of unproven value.

The evaluation of the remaining three-hole-line diagrams has commanded considerable attention in recent years. The strong analogy between the Bethe-Goldstone equation and the Lippmann-Schwinger equation suggests that these diagrams can be summed through a version of the Faddeev equations modified to include the Pauli operator and suitable energy denominators. This point was made by Bethe (26). To perform this summation, one first picks three hole lines and solves the (off-shell) Bethe-

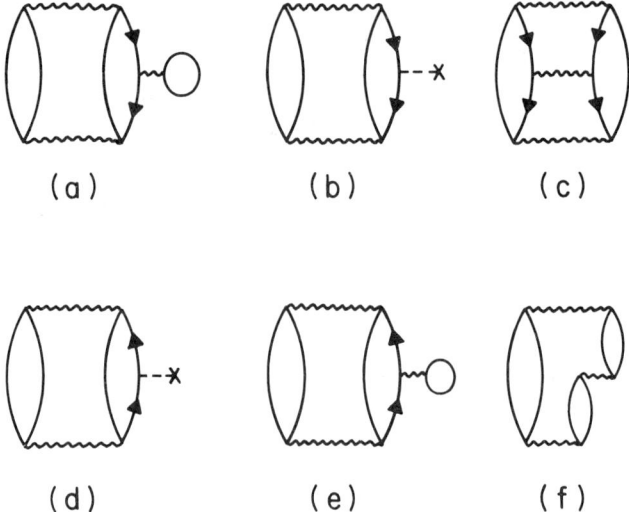

Figure 4 Leading diagrams with three independent hole lines.

Goldstone equation to describe the repeated scattering of a pair of particles in the presence of an excited third particle. The remaining diagrams are obtained by summing all sequences of pairwise interactions in G between all pairs of particles in any order. Of course, the consecutive interactions of a single pair are already included in G and must be excluded from this sum. This sum of diagrams is given by the Bethe-Faddeev equation:

$$\mathcal{G}_i = G_i - \sum_{j \neq i} G_i(Q/e)\mathcal{G}_j, \qquad 13.$$

where the indices run over particles 1 to 3 and G_i denotes a reaction matrix in which particle i is passive. The diagrams included in fourth order in G are shown in Figure 5. The sum $\sum_i \mathcal{G}_i$ represents the totality of three independent hole-line diagrams.

The reliable solution of the Bethe-Faddeev equations is an extremely difficult numerical task. It is substantially more difficult to solve than the three-particle equations in free space because of the presence of the Pauli operator and modified energy denominators, which again preclude elimination of the centre-of-mass coordinate. Day (9) has provided such careful solutions for the Reid potential. The contribution to the binding energy for $k_F = 1.4 \text{ F}^{-1}$ is found to be -5.4 MeV per particle with an uncertainty of 1 MeV. Adding these three-hole-line contributions to the two hole-line energies obtained from the reaction matrix directly yields the saturation curve B3 in Figure 3. The effect of three-hole-line terms is to increase both the binding energy and the equilibrium density for the Reid potential. Because of their extreme difficulty, such calculations have not been performed for the wide variety of potentials considered at the two hole line level. The calculations that have been performed with this degree of sophistication [including the Paris (2) and Bonn (27) potentials in addition to the Reid potential] suggest that the Coester line may be shifted somewhat but do not suggest that the nuclear matter problem will be solved by such improved calculations. These three potentials are among the most realistic NN interactions currently available. A successful calculation with

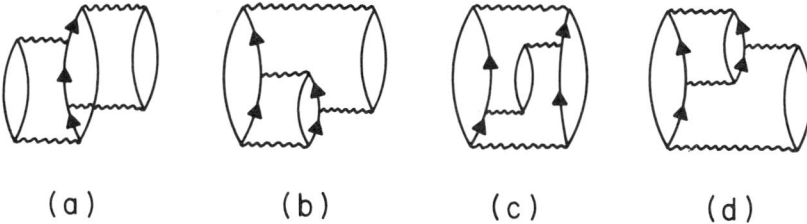

Figure 5 Fourth-order terms in the Bethe-Faddeev sum.

a substantially less realistic potential would in no sense provide a satisfactory resolution to the problem.

Day (28) has also extended his calculations to include an approximate evaluation of four-hole-line diagrams. At the four hole line level, the number of diagrams is bewilderingly large. Day has divided them into no less than 16 distinct classes and has offered simple scaling approximations to estimate their values. We do not even attempt the enumeration of these diagrams let alone their estimation. The reader is invited to consult Reference (28) for details. A full evaluation of these diagrams would require computer resources and human patience beyond that currently available.

The result of these approximate calculations suggests substantial cancellation and a net contribution to the binding energy from four hole line processes that is small and attractive. At $k_F = 1.4$ F^{-1} the Reid potential yields an additional binding energy of 0.5 MeV per particle leading to a total energy of -16.4 MeV per particle with an uncertainty of 1.1 MeV (9). These results are shown in Figure 3 as the curve B4. The relatively small difference between curves B3 and B4 can be viewed as evidence for the validity of the hole line expansion and as an *a posteriori* justification for the approximate evaluation of four-hole-line contributions. The B4 results suggest an equilibrium binding energy of -17.3 MeV per particle with an uncertainty of 1.5 MeV at an equilibrium density of 0.237 F^{-3}. Day assigns the rather conservative uncertainty of ± 0.047 F^{-3} to the density by assuming that there are no correlations between errors at the four densities actually considered. An optimistic estimate assuming that the errors at different densities are totally correlated leads to the smaller uncertainty of 0.009 F^{-3}. Even with his more conservative estimates, Day is led to conclude that "it is unlikely that any two-body potential fitted to scattering data and deuteron properties can account quantitatively for nuclear saturation. An additional effect must be found that gives more binding below the empirical saturation density and less binding at higher densities" (29). We return to this second observation in Section 4.

The BBG approach is an approximation scheme aimed at solving the coupled cluster equations that represent one form of the (exact) many-body Schrödinger equation. The Bochum group has advocated another truncation scheme aimed at the same problem. [The description of their "exponential S" method merits more discussion than space permits. The reader should consult Reference (16).] Their scheme does not rely on a rigorous hole line expansion but rather advocates the inclusion of other diagrams, such as the generalized rings, some of which are shown in Figure 6. Day has included the generalized ring diagrams in his calculations with specific concern for convergence rates. The ring diagram of Figure 6*a* is included in the Bethe-Faddeev sum while the four-hole-line diagram (Figure

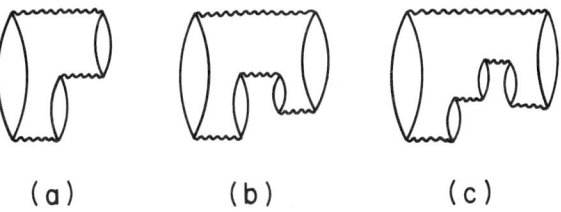

Figure 6 Some ring diagrams.

6b) is included in Day's B4 results. At $k_F = 1.4$ F^{-1} Day finds that ring diagrams of fifth order and higher make an attractive contribution to the energy of -0.32 MeV per particle. This relatively small contribution is taken as evidence for rapid convergence of the ring diagrams and, hence, of the basic validity of the BBG approach as well as an indication of the absence of the long-range correlations such phonon processes would represent.

Although appealing, this conclusion seems suspect. In considering the stability of any system it is useful to study the response of the system to infinitesimal perturbations. In a homogeneous system, such as nuclear matter, these should damp out exponentially (in r) with a scale determined by the compressibility if the system is stable. The compressibility is defined as

$$C = 9 \frac{d}{d\rho} \left(\rho^2 \frac{dE}{d\rho} \right) \qquad 14.$$

at all densities.[1] As C tends toward zero, small disturbances propagate further in the medium. When C passes through zero, as it does at densities sufficiently below the equilibrium density, small disturbances propagate with increasing amplitude and the system is unstable against collapse into a nonhomogeneous phase where only a fraction of the volume is filled with the fluid at equilibrium density. If the energy is precisely quadratic near equilibrium density, this instability will occur at 2/3 the equilibrium density. (Fitting Day's B4 results with a cubic polynomial suggests $C = 0$ at a density of 0.152 F^{-3}, which is 0.644 of his calculated equilibrium density.)

Thus, given Day's results, it should not be possible to find a locally stable homogeneous solution to the nuclear matter problem at or below the empirical equilibrium density for nuclear matter. In practice, this collapse should be generated by the ring diagrams and indicates that the energy of a phonon with wave number zero is negative. The "puddling" of the liquid

---

[1] The factor of 9 stems from the unfortunate convention adopted in nuclear matter that the compressibility at equilibrium represents the curvature with respect to k_F.

represents a condensate of phonons of infinite wave length. In variational calculations in which the effects of phonons are treated in a consistent fashion, it is not possible to find homogeneous solutions for precisely this reason (30). Since Day finds such solutions, it seems that the full effects of ring diagrams (which do not admit a perturbative treatment in this region) have not been incorporated. This raises the question of whether related estimates of convergence rates are reliable at the equilibrium density, which is only 50% higher than the density for which $C = 0$. It is precisely for such reasons that the two-pronged attack on the nuclear matter problem through both BBG and variational techniques is so useful.

It is worth mentioning that phonons can also play a role at higher densities where crystallization (and the attendant long-range order) can also be regarded as a phonon condensate. In this case the energy is lowered by the creation of phonons with wave numbers corresponding to the reciprocal lattice vectors in the crystal. This is a problem of importance in liquid helium where the density for solidification is only 50% more than the equilibrium density of the liquid. The effect is surely less important in nuclear matter where the repulsive core is not strong enough to support a solid. Nonetheless, one can wonder whether the "intermediate-range" order implied by slow convergence of the ring diagrams is adequately represented in BBG calculations.

These cautionary and personal concerns in no way detract from the value of Day's calculations. Rather, they represent the kind of questions that can be illuminated by comparing the BBG results with the variational calculations, to which we now direct our attention.

3. VARIATIONAL CALCULATIONS

Another approach to the treatment of short-range (two-body) correlations was offered by Jastrow; he proposed the use of a trial wave function of the form

$$\Psi = \prod_{i<j} f(r_{ij})\Phi, \qquad \qquad 15.$$

where Φ is the usual plane wave Slater determinant describing the noninteracting ground state (6). Here, short-range correlations correspond to the vanishing of $f(r)$ for small r. This wave function was intended for use in a variational calculation of the energy of nuclear matter. Although Jastrow suggested that $f(r)$ might be a correlation operator, most calculations have been performed with $f(r)$ as a simple function. Thus, the correlation between all pairs of particles is independent of their single-particle states. This corresponds to neglecting the state dependence in the energy denominators of Equation 8 and is expected to be reliable if short-range repulsion (and high-energy intermediate states) are dominant. Even with such simplification, it is not simple to determine the expectation value of the energy. This is most easily seen in Bose systems where the Slater determinant, Φ, is replaced by 1. In high-density systems like liquid helium, the difference between the Fermi system of ^{3}He and the Bose system of ^{4}He is not so great as to render this exercise useless.

For Bose systems it is easiest to express the energy in terms of the two-body distribution function, $g(r)$, defined as

$$g(r_{12}) = \frac{A(A-1)}{\rho^2} \frac{\int d^3r_3 \ldots d^3r_A \Psi^*\Psi}{\int d^3r_1 \ldots d^3r_A \Psi^*\Psi}.$$ 16.

With the distribution function, the energy becomes

$$\frac{E}{A} = \frac{1}{2}\rho \int d^3r g(r) \left[V - \frac{1}{4}\nabla^2 \ln f(r)^2 \right].$$ 17.

It is obvious that it is not simple to perform the integrals in Equation 16 to establish the connection between $g(r)$ and $f(r)$. It is conventional to rely on the hypernetted chain (HNC) approximation to obtain a tractable alternative to Equation 16 (7). This involves replacing f^2 by the short-range operator $(1+F)$ and performing a formal cluster expansion.

Two classes of terms are retained to all orders in this expansion. First, terms in which coordinates r_1 and r_2 are joined by a chain of Fs such as $F(r_{13})F(r_{34})F(r_{45})F(r_{52})$. Second, one retains terms in which $\mathbf{r}_1$ and $\mathbf{r}_2$ are connected by any number of parallel and independent chains. Performing such summations self-consistently leads to an integral equation relating f^2 and g:

$$\ln[g(r)/f^2(r)] = \rho \int d^3\mathbf{r}' [g(|\mathbf{r}-\mathbf{r}'|) - 1] \{ g(r')$$
$$- 1 - \ln[g(r')/f^2(r')] \}.$$ 18.

Equation 18 may be used in Equation 17 for an approximate determination of the energy. The HNC approximation (Equation 18) was obtained without regard for the variational principle, and the resulting estimate of the energy is no longer guaranteed to provide an upper bound on the ground-state energy.[2] Using Equations 17 and 18, we can minimize the

---

[2] Indeed, for sufficiently pathological choices of f, the kinetic energy can be made arbitrarily negative (30), which suggests that the variational principle is not at its best. The severity of this problem is affected by the form in which the kinetic energy operator is cast. Integration of Equation 17 by parts yields a variety of forms of the energy equivalent before the HNC approximation but different after.

energy by varying f or, more simply, g. This variation is intrinsically dangerous and should only be performed with the understanding that it must be supplemented by a second-order variational calculation to verify that any extremum found is an energy minimum. This has been done in practice (30) and it has been found that, in all cases where solutions to the resulting Euler-Lagrange equation have been found, they do represent energy minima.

With due regard for the fragility of the results, one can minimize the energy to find (31)

$$-\nabla^2 \sqrt{g} + V\sqrt{g} + W\sqrt{g} = 0 \qquad 19.$$

with

$$\tilde{W}(k) = -\frac{k^2(2S+1)(S-2)^2}{4\rho S^2}. \qquad 20.$$

Equation 19 has the form of a zero-energy Schrödinger equation for $\sqrt{g}$ that, from Equation 16, is wave-function-like. The potential in this equation is the sum of the two-body potential, V, and an additional potential, W, which is due to the polarization of the medium. In Equation 20, $\tilde{W}$ is the Fourier transform of W and S is the liquid structure function defined in Equation 21 below.

Some authors prefer to pick a functional form for f and vary built-in parameters. Near equilibrium density there is little difference in the resulting energies, although such calculations do not generally display collapse when the compressibility is negative. While such variational calculations have been applied with conspicuous success to liquid helium and nuclear matter, their language complicates interpretation in terms of the more familiar Goldstone diagrams of the previous section. Such interpretation is made easier when $f(r)$ is determined by the variational problem and can be expressed purely in terms of the two-body interaction. To see this we consider Equations 19 and 20 for Bose systems in two limiting cases.

First, for each order of the potential, we retain only the lowest power of the density. This corresponds to dropping the term W in Equation 19 and leads precisely to the sum of ladder diagrams constituting the two-hole-line terms in the BBG approach. In the Fermi version of the Euler-Lagrange equation, one obtains the ladder diagrams with the approximation of a state-independent propagator. This is, of course, a reflection of the fact that Equation 15 employs a state-independent correlation function.

The other interesting limit is the *uniform limit* in which both g and $g^{1/2}$ are approximately equal to one. This limit is realized by retaining only the highest power of the density for each order of the interaction in Equation 19. In this limit the Euler-Lagrange equation is algebraically soluble for the liquid structure function defined as

$$S(k) = 1 + \rho \int d^3r \, e^{i\mathbf{k}\cdot\mathbf{r}} [g(r) - 1]. \qquad 21.$$

One obtains

$$S(k) = k[k^2 + 4\rho \tilde{V}(k)]^{-1/2}. \qquad 22.$$

The energy contribution obtained from inserting Equation 22 into Equation 17 is the exact sum of ring diagrams describing the exchange of phonons between the particles. The analogous limit in Fermi systems reveals the sum of ring diagrams again with the use of state-independent average propagators. The remaining terms in Equation 19 have a form that invites diagrammatic interpretation. Specifically, Equations 19 and 20 approximate the sum of *parquet diagrams*, which represent the self-consistent sum of particle-particle ladder diagrams and ring diagrams, which in this context are most conveniently thought of as particle-hole ladder diagrams. Thus, the spirit of these variational calculations is to invoke rather strenuous approximations in the evaluation of specific diagrams in the interests of summing a much broader class of diagrams.

Let us now consider the parquet diagrams for bosons. [This topic is discussed more completely in Reference (21).] One can formally decompose the four-point function, Γ, into the bare potential, ladder diagrams (L) that are particle-particle reducible, and chain diagrams (C) that are particle-hole reducible. The ladder diagrams can be obtained from a linear integral equation, analogous to the Bethe-Goldstone equation, in which the driving term includes all particle-particle irreducible pieces (i.e. V and C). Similarly, the chain diagrams can be obtained from a linear integral equation in which the driving terms are particle-hole irreducible (i.e. V and L). These equations have the following simple form:

$$L = (V+C)G_{pp}(V+L+C) \qquad 23.$$

$$C = (V+L)G_{ph}(V+L+C) \qquad 24.$$

$$\Gamma = V+L+C. \qquad 25.$$

Here, G_{pp} and G_{ph} represent particle-particle and particle-hole propagators, respectively, and are suitable products of two single-particle propagators. These equations are shown diagrammatically in Figure 7. When supplemented by suitable counting rules these contributions to the vertex can be converted into contributions to the energy of the system by closing the external lines. The parquet contributions through fifth order in V are shown

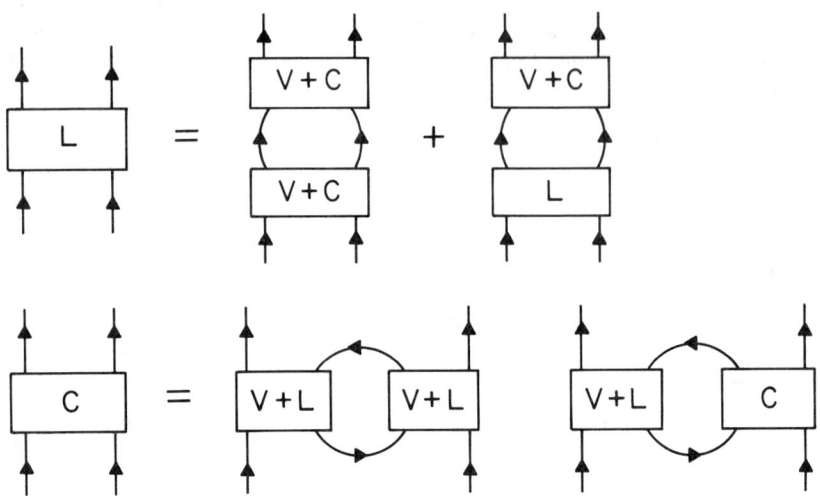

Figure 7 A schematic representation of the parquet equations.

in Figure 8. The class of diagrams thus summed includes a subset of the planar diagrams in which all diagrams contributing to Γ are *completely* two-particle reducible. (By this we mean that each diagram can be split into two disjoint pieces by cutting either two particle lines or one particle line and one hole line. Each of the resulting pieces is similarly reducible et cetera.) The self-consistent evaluation of rings and ladders is illustrated, for example, by Figure 8k, a ladder diagram in which one rung is a chain, and by Figure 8r, a chain diagram in which one link is a ladder.

The parquet equations can be cast in the following instructive form

$$\Gamma = V + \sum_i \Gamma(1 + G_i\Gamma)^{-1} G_i \Gamma, \qquad 26.$$

where the sum i is extended over both particle-particle and particle-hole channels. Equation 26 emphasizes that the parquet diagrams possess a particle-particle/particle-hole "crossing symmetry." This is no accident since the parquet equations were originally intended to provide a crossing symmetric alternative to the Bethe-Salpeter equation in the more familiar s-, t-, u-channel crossing symmetry of particle physics (32). Crossing symmetry in the many-body sense seems a useful way to retain both the effects of two-body correlations at short distances and the phonon effects at small momenta. It is also interesting to note that the arguments leading from Equations 23–25 to Equation 26 are topologically equivalent to the arguments leading to the HNC approximation summarized in Equation 18 (21).

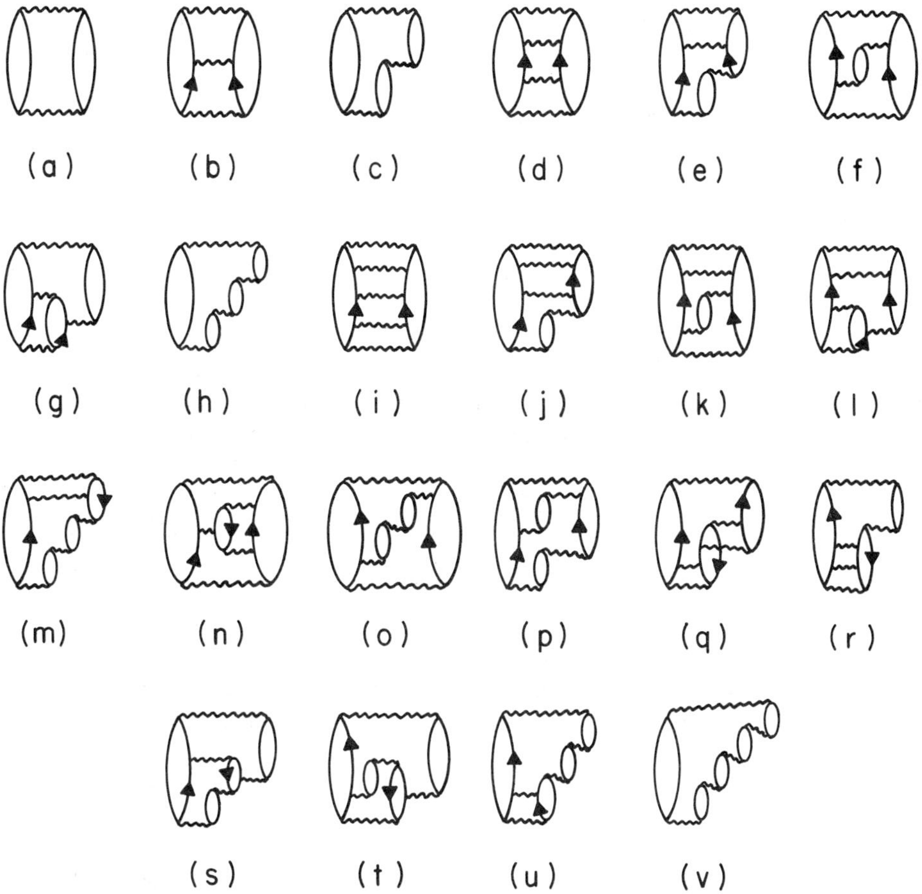

Figure 8 Parquet contributions to the energy through fifth order.

In general, Γ is a function of six scalar invariants, and the nonlinear parquet equations are not easily solved numerically. With guidance from the Euler-Lagrange equation (Equation 19), it is tempting to simplify these equations by making local approximations to the propagators G_{pp} and G_{ph}. Such approximations, along with the locality of V, will guarantee that the resulting equations for Γ involve only the magnitude of the momentum transfer. There is a precise correspondence between the terms arising in this approximate parquet sum and the terms appearing in an expansion in the Euler-Lagrange equation. All functional forms are equivalent with differences confined to overall numerical factors in diagrams that are neither pure rings nor pure ladders. Numerical calculations (21) suggest that these remaining differences are not significant. This is probably as close as one can come to a diagrammatic interpretation of the HNC variational calculations for bosons. Similar results are expected for fermions. The extension of the parquet equations to the Fermi systems is most easily made in Equation 26 by the simple extensions of the sum to include the remaining channel (i.e. the second particle-hole channel). We note that the parquet techniques can be extended by including in V all diagrams that are two-particle irreducible. The first such diagrams is the four-hole-line diagram of Figure 5a. It is difficult, however, to see how significantly broader classes of diagrams can be summed without resorting to some kind of perturbation theory.

The wave function (Equation 15) is efficient for the description of the purely central correlations induced by σ- and ω-meson exchanges. It is not able to follow the substantial state dependence that characterizes realistic NN interactions. Such state dependence falls loosely into three categories. First, the discrete state dependence, which indicates different correlations for pairs of particles whose spins/isospins are parallel or antiparallel. Discrete state dependence is also required to exploit two-body tensor correlations induced primarily by the strong tensor component in the one-pion-exchange interaction. As noted above, the iterated one-pion-exchange interaction gives rise to a substantial attraction in nuclear matter and the operation of the Pauli principle in intermediate states provides an important mechanism leading to saturation. Such effects can be incorporated by a suitable generalization of the f appearing in Equation 15 to include correlation operators such as $\sigma_1 \cdot \sigma_2$, $\tau_1 \cdot \tau_2$, and the usual tensor operator S_{12}. Owing to the commutation properties of these operators, such an extension strains the Jastrow/HNC approach and requires both additional approximations of a nonvariational nature and considerable numerical ingenuity. Finally, there is the genuine state dependence, which reflects the fact that, given the relative spin and isospin orientation of a pair of particles, their correlation also depends on the single-particle momenta of the states involved. It has been known for many years that such genuine state dependence is crucial in describing the effective mass of a ^{3}He impurity in liquid ^{4}He (33). In the absence of the "backflow" that such correlations describe, calculated effective masses in liquid ^{3}He or of ^{3}He impurities in liquid ^{4}He are found equal to the bare masses, which are more than a factor of two less than the empirical effective masses.

This problem is largely removed by the use of the following form in the generalized correlation operator (18)

$$f(\mathbf{r}_{ij}) + \eta(r_{ij})\mathbf{r}_{ij} \cdot \mathbf{\nabla}_{ij}. \qquad 27.$$

Calculations embodying these discrete and continuum operator generalizations of the Jastrow correlation function, as well as the effects of explicit three-body correlations beyond those implicit in the Jastrow trial wave function, have been performed by a number of authors (10, 34). For a more detailed review of such calculations, see Reference (18). In Figure 3 we show results obtained by Lagaris & Pandharipande (35) with their V12 variant of the Reid potential.[3] They obtain an energy of -17.11 MeV per particle at an equilibrium density of 0.303 F^{-3}. The crucial point here is that the results of Lagaris & Pandharipande are in remarkably good agreement with the best (B4) results of the BBG calculations of Day.

Clark and co-workers (36, 37) have adopted a slightly different approach to the improvement of the simple Jastrow trial function. Starting from that choice of $f(r)$ that minimizes the Jastrow/HNC energy, they define a complete but nonorthogonal set of states

$$\prod_{i<j} f(r_{ij}) | [m] \rangle, \qquad 28.$$

where $|[m]\rangle$ denotes a general Slater determinant of plane waves with the quantum numbers $[m]$ indicating changes in occupation from the ground state. These correlated basis functions (CBF) can be used to construct the perturbative improvements of the original Jastrow wave function. It is important to recognize that this CBF perturbation theory is not a slowly (or non-) convergent series in powers of the bare interaction but rather a rapidly convergent expansion in the difference between the "true" effective interaction and its hole line average. After careful formal arguments what emerges is a perturbation theory of familiar form whose convergence rate is governed by a factor formally analogous to the BBG κ of Equation 11. This parameter measures the difference between the state-dependent, correlated wave function and the hole line averaged, state-independent, correlated wave function. It is substantially smaller than the BBG κ. In neutron matter, where tensor forces are of little importance, this convergence parameter is on the order of 0.01 at $k_F = 1.5$ F^{-1} (38). In nuclear matter, where roughly half of the BBG κ is due to tensor correlations that must now be included perturbatively, this factor is rather larger and on the order of 0.10 near the empirical equilibrium density. Nonetheless, even the lowest-order CBF corrections to the binding energy of nuclear matter yield results in substantial agreement with those of other variational improvements of

---

[3] Using the six spin operators 1, $(\boldsymbol{\sigma}_1 \cdot \boldsymbol{\sigma}_2)$, S_{12}, $\mathbf{L} \cdot \mathbf{S}$, L^2, and $L^2(\boldsymbol{\sigma}_1 \cdot \boldsymbol{\sigma}_2)$ and the two isospin operators 1 and $(\boldsymbol{\tau}_1 \cdot \boldsymbol{\tau}_2)$, Lagaris & Pandharipande (35) could reproduce the Reid potential in the twelve NN channels with $J \leq 2$ in which it was defined and could continue the potential to higher channels as required by variational calculations. The V6 and V8 variants of the Reid potential (employing only the first three and first four spin operators, respectively) have also been used in nuclear matter calculations.

the Jastrow wave function. The virtue of the CBF scheme is that it is completely systematic and does not require the careful introduction of operator generalizations of $f(r)$ that are both physically suitable and numerically tractable. It is also possible to sum certain classes of diagrams within the CBF scheme (39). (An example is the RPA diagrams leading to an improved description of the ring diagrams already present in Jastrow/HNC calculations in a cruder form.)

We have seen that Jastrow/HNC calculations may be thought of as a sum of parquet diagrams with the use of hole line averaged, two-particle propagators. It is straightforward to construct a perturbative scheme to improve the evaluation of parquet diagrams where the perturbation is the difference between the true and approximate propagators. This is essentially the content of the CBF scheme, which is driven by the difference between exact two-body correlations and their hole line average. Both schemes allow for the systematic inclusion of terms not parquet in nature. The analogy between these approaches is strong and should prove to be useful.

Ambitious though it is, the BBG approach aims to evaluate a relatively small number of Goldstone diagrams with considerable precision and relies fairly heavily on the rapid convergence of the hole line expansion for its legitimacy. With the insight provided by the parquet diagrams, one can think of the Jastrow/HNC variational calculations as largely abandoning hole line ordering of diagrams in favor of the far less precise summation of a far broader class of diagrams. Given the relatively large differences in points of view, the agreement between variational and BBG calculations is far more striking than their remaining differences. (Perhaps the only substantial difference is financial. The computer time required for BBG and calculations is measured in hours. For variational calculations it is measured in seconds.) While the remaining differences merit continued attention, Pandharipande & Wiringa (18) observe that "we must conclude that a suitable nuclear Hamiltonian has not yet been found." This observation is strikingly similar to that which closed Section 2.

4. TOWARD A SUITABLE NUCLEAR HAMILTONIAN

In broad terms all calculations of the binding energy and equilibrium density of nuclear matter agree in their inability to reproduce empirical results. It seems reasonable to assume that the problem lies not in the solution of the many-body problem but in the choice of the nuclear Hamiltonian. It can be argued that, even though the range of NN interactions considered is broad, it has not been broad enough and fundamentally new classes of two-body interactions should be considered. Another possibility is that genuine many-body forces play an important role in determining the properties of nuclear matter. The construction of such forces requires field theoretic guidance supplemented by simple tests in $A = 3$ and $A = 4$ nuclei (since the two-nucleon system evidently has no insight to offer). Both possibilties merit consideration.

We first consider one prototypical modification of the two-body interaction. The intermediate-range central attraction in the NN interaction can be loosely associated with the exchange of a scalar (σ) meson of a mass on the order of $5\mu_\pi$. There is, in fact, no such meson. This attraction is to be understood as the exchange of two pions in a relative S wave. The dominant processes summarized by such an effective σ meson are shown in Figure 9. The pion rescattering term (Figure 9a) is most faithfully approximated as a scalar meson (with a distributed mass). The isobar contributions of Figures 9a and 9b require more attention. Simple perturbation theory arguments suggest that these processes are attractive below the production threshold for isobars and are explicitly energy dependent. In nuclear matter, both isobars and nucleons are subject to dispersion corrections, which tend to reduce the magnitude of this attraction, and to exclusion principle effects on the intermediate nucleon, which have a similar effect. Such effects are simply not included in the kinds of phenomenological NN interactions usually considered. They will result in additional repulsion in nuclear matter calculations provided that the fit to scattering data is preserved. By and large this conclusion, that more microscopic consideration of the NN interaction leads to additional repulsion in nuclear matter, is fairly general.

The three-body forces most frequently investigated are intimately related to the modifications of the NN interaction associated with Figure 9. These pion range forces are shown in Figures 2a and 2b. In the process of calculating the binding energy of nuclear matter it is necessary to close the hole lines in this diagram. Closing one hole lines leads to the processes of Figures 2c and 2d, which emphasize that these processes behave like self-

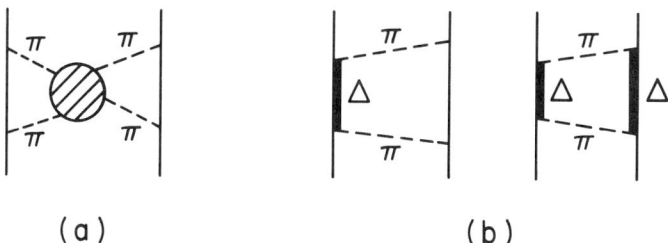

Figure 9 Contributions to the two-pion-exchange interaction.

energy and propagator modifications of the pion. Since the one-pion-exchange interaction in the absence of such corrections makes no contribution to the binding energy of nuclear matter due to spin and isospin symmetry, one expects little from these terms. Indeed, substantial contributions are only obtained when such corrections are introduced into the iterated one-pion-exchange term, which, as noted above, has a substantial central term. Thus, the isobar contribution of Figure 2e provides additional attraction in nuclear matter. The importance of high-energy intermediate states in this iteration renders the results less sensitive to this mass change than might be expected. The results of such calculations depend fairly sensitively on the way in which the three-body force is introduced. In general, however, inclusion of process 2e without simultaneous inclusion of dispersion and Pauli corrections to the related terms in the two-body interaction makes it difficult to obtain saturation of the binding energy at any density.

Thus, Carlson et al (40) have constructed a model of the three-body force based on isobar processes supplemented by a phenomenological term intended to simulate the dispersion corrections to the two-body process of Figure 9b. Parameters in these forces can be adjusted to provide a reasonable simultaneous description of the binding energy of $A = 3$ and $A = 4$ nuclei and equilibrium energy and density of nuclear matter without doing great violence to analytic continuations of the πN scattering amplitude. While such calculations indicate a possible resolution to the nuclear matter problem, more careful calculations will require a greater consistency between the description of two- and three-nucleon forces. For example, the two-pion-exchange force can be expressed as the t-channel iteration of the $N\bar{N} \to \pi\pi$ pseudophysical amplitude. As shown in Figure 10, the three-body force has a similar description. This connection, exploited in Reference (41), should be maintained along with the requirement that the pseudophysical amplitude chosen should be consistent with

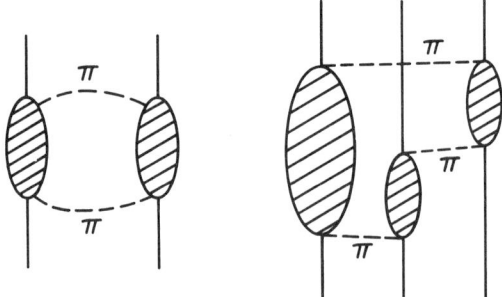

Figure 10 Relation between two- and three-nucleon forces.

πN data. In particular, some models of the πN amplitude (42) would suggest that ππ rescattering is relatively more important than the isobar processes, which are assumed in Reference (40) to be dominant. These remain questions for future study.

Returning to Figure 2, we may reasonably ask whether we should introduce similar self-energy and propagator modifications to those elements of the NN interaction that *do* account for substantial contributions to the energy of nuclear matter even in lowest order (i.e. σ and ω exchange). These corrections for the σ meson are shown in Figures 11a and 11c. The first of these created some excitement in nuclear matter circles and was identified as a process providing sufficient attraction to cause a phase transition in nuclear matter (43). At normal nuclear densities, this process was estimated as contributing a massive -15 MeV per particle to the energy of nuclear matter (44). Such contributions were discarded not because they were too small but but because they were too large. The clear inference from Reference (44) is that there must be other mechanisms tending to cancel this large effect. Given the magnitude of this effect, it was clear that such cancellations would be delicate and that essentially any binding energy could be obtained unless a clear and convincing ordering scheme were established.

The existence of such large effects along with equally dramatic cancellations is a familiar phenomenon in πN scattering. The πNN coupling constant is large ($g^2/4\pi = 14$). That the usual one-pion-exchange interaction between nucleons is relatively small is a consequence of the pseudoscalar nature of the pion. Each πNN vertex involves a factor of $g\bar{\psi}(p')\gamma_5\tau\psi(p)$ where ψ is a nucleon spinor. For the usual NN interaction, both $\psi(p')$ and $\psi(p)$ describe positive-energy spinors. In the nonrelativistic limit, γ_5 coupling connects large and small components of these spinors leading to a factor of

$$u^+(s')\frac{\sigma\cdot(\mathbf{p}-\mathbf{p}')}{2M}u(s) \qquad\qquad 29.$$

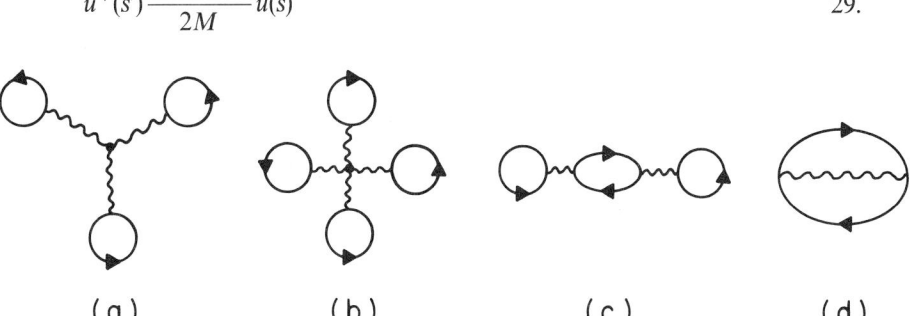

Figure 11 Many-body forces in the sigma model (*a–c*) and the missing Fock term (*d*).

at each vertex, where u is now a nonrelativistic, two-component spinor. In coordinate space these factors of $\boldsymbol{\sigma}\cdot(\mathbf{p}-\mathbf{p}')/M$ lead to the usual spin-spin and tensor forces characterizing one-pion exchange and to a factor of (μ_π^2/M^2), which is roughly 1/50. Such a small factor would not occur, for example, in the potential resulting from the exchange of a scalar meson. Neither does it arise in the calculation of πN scattering or in the construction of the two-pion-exchange potential.

Specifically, the process shown in Figure 12a can proceed through a negative-energy intermediate state so that the γ_5 coupling connects the large components of positive-energy spinors to large components of negative-energy spinors. This leads to a value of the isospin-averaged πN scattering length two orders of magnitude larger than the empirical value. This problem was overcome by incanting "pair suppression" and throwing the offending terms away. A more satisfying device was to introduce an intermediate σ meson, as shown in Figure 12b with coupling constants chosen to cancel Figure 12a. This was one of the reasons for the development of chiral models in which the π and σ fields form the components of a four-vector. (We do not consider here other reasons such as the Goldberger-Trieman relation and the Adler-Weisberger relation.) Although introduced in a relatively arbitrary fashion, this σ meson can be related to the more familiar scalar meson that provides the intermediate-range attraction in the NN interaction. An elementary discussion of these points can be found in Reference (45).

In addition to providing a simple mechanism for intermediate-range attraction through a σNN coupling, such a model provides for $\pi\sigma\sigma$, $\sigma\sigma\sigma$, and $\sigma\sigma\sigma\sigma$ vertices in a relatively parameter-free fashion. As seen from Figures 11a and 11b, these vertices lead to many-body forces. The sigma model thus offers the possibility of introducing many-body forces (not just three-body forces) between nucleons in a consistent manner. [In this regard see References (11), (12), and (13).] Since it has been constructed to enforce

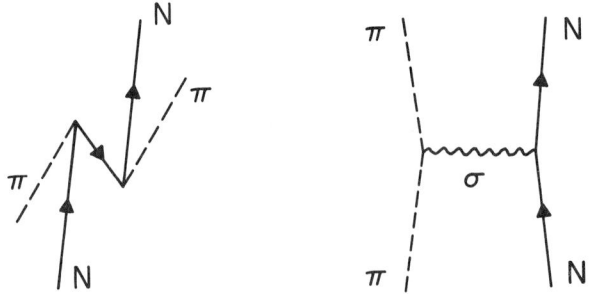

Figure 12 Processes whose cancellation leads to "pair suppression."

pair suppression in πN scattering, one can hope that the sigma model will also lead to the anticipated cancellations between various many-body forces. The standard line of attack is to treat the π and σ fields as classical fields constant in space. With such an approximation, the field theoretic Lagrangian describing the sigma model reduces to an essentially classical potential-energy surface whose minima yield the expectation values of the π and σ fields. When this is done the (initially massless) nucleon acquires a mass, and the σ field obtains a nonzero expectation value while the π field remains at expectation value zero. The parameters in this model can be adjusted to reproduce the empirical nucleon mass and the empirical πNN coupling constant. The only remaining parameter is the mass of the σ meson which is conventionally (and quite arbitrarily) set at or near the nucleon mass, M.

In a medium of finite nucleon density, the mean σ field is shifted from its vacuum expectation value by an amount ϕ and the nucleon acquires an effective mass

$$\bar{M} = M + g\phi. \qquad 30.$$

These shifts are responsible for the binding of nuclear matter in the mean field approximation. Specifically, one finds

$$E = \frac{1}{4}\lambda^2(4\sigma_0^2\phi^2 + 4\sigma_0\phi^3 + \phi^4) + \frac{2}{\pi^2}\int_0^{k_F} d^3\mathbf{k}\sqrt{\bar{M}^2 + k^2}, \qquad 31.$$

where the parameters λ and σ_0 may be related to the (known) nucleon mass and (unknown) σ-meson mass through

$$M = g\sigma_0 \quad \text{and} \quad m_\sigma = \sqrt{2\lambda}\sigma_0. \qquad 32.$$

The coupling constant, g, is approximately 13. The actual energy of nuclear matter is obtained by minimizing this expression with respect to ϕ. The lowest-order terms in ϕ contain the usual Hartree term describing the exchange of a single σ meson between nucleons. The term of order ϕ^3 describes the three-nucleon force of Figure 11a while the ϕ^4 term describes Figure 11b. The propagator correction of Figure 11c is implicit in the last term in Equation 31, as we demonstrate below. Before considering the results of such a mean field calculation, note that essentially all of the complicated manipulations with two-body potentials performed in preceding sections are reduced to the trivial Hartree term (and the Fock term not yet included). The remaining terms in Equation 31 describe the variety of genuine many-body forces contained in the sigma model.

We have previously seen how variational calculations sacrifice precision in the evaluation of individual diagrams in the interest of extending the class of Goldstone diagrams that can be summed. Mean field theories carry this point to the extreme of rendering the two-body interaction trivial in order to concentrate on many-body forces. Such approximations are not without justification at high densities, but at or near normal nuclear matter density they are surely not quantitatively reliable. Mean field theories are of considerable value in providing semiquantitative estimates of the role of many-body forces in the nuclear matter problem.

In the limit of small density we can replace the relativistic energy appearing in Equation 31 by $\bar{M}$ plus the nonrelativistic kinetic energy. Minimization with respect to ϕ then yields

$$\phi \approx -\frac{\rho g}{m_\sigma^2} \qquad 33.$$

and, retaining terms through $O(\phi^2)$,

$$\frac{E}{\rho} = M + \frac{3}{10}\frac{k_F^2}{M} - \frac{\rho g^2}{2[m_\sigma^2 + \frac{3}{5}\rho k_F^2 g^2/M^3]}\left(1 - \frac{3}{10}\frac{k_F^2}{M^2}\right)^2. \qquad 34.$$

The terms in Equation 34 all have an obvious physical interpretation. The first term is the bare nucleon mass. The second term is the usual kinetic energy for a nonrelativistic Fermi gas. The leading contribution from the third term, $-\rho g^2/2m_\sigma^2$, is the Hartree term describing σ-meson exchange. The propagator correction (Figure 11c) is contained implicitly in m_σ. In a nuclear medium this term must be modified to account for the effects of the Pauli principle on the intermediate nucleon. This is precisely the mass modification appearing in the denominator of Equation 34. For $g = 13$ and $\lambda^2 = 125$ we find an increase in m_σ of 0.5% at the empirical equilibrium density from this effect. The inclusion of the explicit ϕ^3 terms in Equation 31 leads to an additional contribution to the energy in Equation 34 of

$$-\frac{\rho^2 g^4}{2m_\sigma^4 M}, \qquad 35.$$

which represents an increased binding energy of roughly 12 MeV per particle at normal equilibrium density. This is precisely the three-body force of Figure 11a. When supplemented by other tree diagrams, this term may also be considered as a correction to m_σ^2. Including both mass corrections, we find that

$$\frac{E}{\rho} \approx M + \frac{3}{10}\frac{k_F^2}{M} - \frac{\rho g^2}{2m_\sigma^2(\rho)}, \qquad 36.$$

where

$$m_\sigma^2(\rho) = m_\sigma^2 + \frac{3}{5}\rho\frac{k_F^2}{M^3}g^2 - \frac{\rho g^2}{M}. \qquad 37.$$

The last term in Equation 37 represents a 7% decrease in m_σ at ρ_0. The virtue of writing the energy in the form of Equation 36 lies in the fact that it looks like the results from the usual two-body Hartree term with a density-dependent m_σ. To the extent that this replacement is reliable, this suggests that many-body forces can be incorporated in a more careful treatment of the two-body interaction with the conceptually and technically simple introduction of density-dependence in the region of the intermediate-range attraction in the NN interaction.

The results of mean field theories as they are commonly applied seem appealing at small densities but must be viewed with caution since they lead to phase transitions at physically unrealistic low densities. Above some critical density, on the order of $2\rho_0$, Equation 31 predicts a phase transition to a tightly bound abnormal phase of nuclear matter in which the effective mass of the nucleon is zero. The introduction of vacuum fluctuation corrections to the mean field results eliminates this phase transition (or at least displaces the critical density to inaccessibly large values) and, incidentally, to provide a mechanism for compensating the substantial attraction arising from the three-body force of Figure 11a (13). As we discuss below, individual vacuum fluctuation terms are extremely large, and it is essential to have a reliable ordering procedure for their evaluation.

The requirement that chiral symmetry be preserved rigorously in each stage of approximate calculations provides a partial answer. Specifically, chiral symmetry requires that the "one-loop" terms in Equation 31 be supplemented by the "one-loop" vacuum fluctuations whose leading terms are shown in Figures 13a and 13b. (Higher-order terms involve more σ lines attached to the central nucleon and sigma loops, respectively. Terms with only two σ lines are implicit in the $\bar{M}$ appearing in Equation 31 following renormalization, as indicated for example by Figure 11c.) These leading

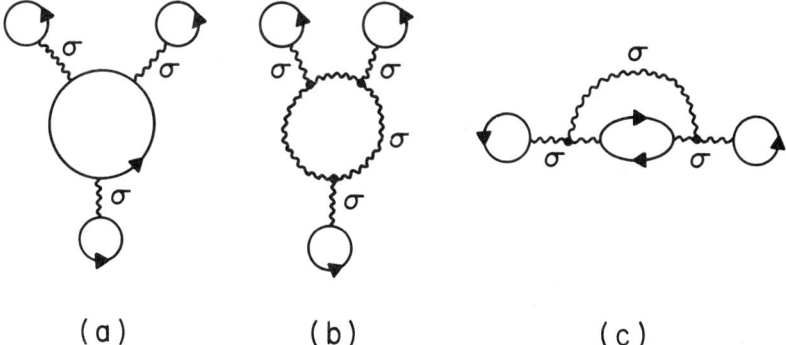

Figure 13 One-loop vacuum fluctuations for (a) nucleon and (b) sigma loops and (c) the two-loop propagator correction.

terms make contributions of

$$\frac{M\rho^2 g^6}{3\pi^2 m_\sigma^6} \quad \text{and} \quad -\frac{9\rho^2 g^6}{64\pi^2 M^3 m_\sigma^2} \qquad 38.$$

to E/ρ. The nucleon loop term is extremely repulsive ($+110$ MeV per particle at ρ_0) but the sigma loop term is almost equally attractive (-102 MeV per particle at ρ_0). This cancellation is clearly a delicate function of m_σ but, for the parameters adopted, persists when all one-loop terms are included.

In the calculations of Matsui & Serot (46) only the nucleon loop term is retained. The resulting repulsion renders it unnecessary for them to supplement the sigma model with vector-meson exchange in order to reproduce the binding energy and equilibrium density of nuclear matter. Lee & Margulies (12) and Nyman & Rho (13) rely more heavily on chiral invariance and advocate the retention of both classes of terms. Chiral invariance, which is realized for any value of m_σ, offers no guidance as to whether it is better to assume that these two vacuum fluctuation terms cancel exactly or that the sum of these vacuum fluctuations plus the three-body force of Figure 11a is zero. This important question requires further investigation, but we shall adopt the former prescription and neglect these vacuum fluctuations.

The sigma loop term is not mathematically well behaved as the density is increased. This is due to the neglect of sigma propagator corrections of the kind shown in Figure 11c in the evaluation of the sigma loop diagrams. Including such propagator corrections approximately along lines suggested by Lee & Margulies (12) leads to a modification of the sigma loop term. The leading term in this modification is shown in Figure 13c. This term has precisely the density dependence of Figure 11c and may, again, be thought of as providing a change in the mass of the σ meson:

$$\Delta m_\sigma^2 = \frac{27\rho k_F^2 g^4 m_\sigma^2}{160\pi^2 M^5}. \qquad 39.$$

This term is considerably larger than the propagator correction of Equation 37 owing to the size of g. It leads to a 3% increase in m_σ near ρ_0. To preserve chiral invariance, this two-loop diagram should be combined with all other two-loop diagrams including the Fock term of Figure 13d, which has not been included in our mean field calculations but which is, of course, included in any more detailed treatment of the two-body interaction.

We now have a simple picture for the density dependence of m_σ. At low densities the process of Figure 11a is dominant. It reduces m_σ and increases the binding energy. At higher densities process Figure 13c is dominant. It increases m_σ and decreases the binding energy. The perturbative argument leading to Equation 39 is not quantitatively reliable. A more careful calculation suggests that these effects cancel at a density of approximately 0.25 F^{-3}.

We could continue in this vein and add a simple ω-meson Hartree term to the sigma model in hopes of providing a more quantitative description of the properties of nuclear matter. We do not do this, but it is worth mentioning that the self-energy and propagator corrections that we considered for the σ meson need not be included for ω mesons. There is no evidence indicating the necessity for an $\omega\sigma\sigma$ coupling, and ω^3 coupling should vanish because of G-parity conservation. Thus, self-energy corrections to the ω meson analogous to those of Figure 11a should not be important. Propagator corrections analogous to Figure 11c should vanish in the relevant limit of zero ω-meson momentum for the same reason that such corrections do not modify the mass of the photon in electrodynamics.

Rather, we suggest that the salient features of many-body forces can be incorporated in any two-body calculation through the simple expedient of introducing a density-dependent, intermediate-range attraction of the form

$$m_\sigma(\rho) = m_\sigma(1 - \alpha\rho + \beta\rho^{5/3}). \qquad 40.$$

The present mean field results would suggest $\alpha \approx 0.5\ F^3$ and $\beta \approx 1.2\ F^5$. A cursory inspection of either the LP or B4 results of Figure 3 suggests that such a modification of the intermediate-range attraction in the Reid potential could restore agreement with the empirical binding energy and density. In this case we would expect the ratio α/β to remain essentially unaltered but the individual values to be substantially reduced because of the longer range of the central attraction in the Reid potential and the effects of short-range correlations. This is indeed the case. Calculations of the change in energy due to this modification of the intermediate-range attraction in the v8 variant of the Reid potential have been performed by Jackson et al (47) and the parameters α and β chosen to enforce agreement between the BBG results B4 and data. The parameters $\alpha = 0.05\ F^3$ and $\beta = 0.178\ F^5$ do admirably, as can be seen in Figure 14.

The small modifications in the NN interaction indicated by Equation 40 and the large energy changes resulting from them, shown in Figure 14, emphasize that the elimination of the remaining discrepancy between theoretical and empirical values of nuclear matter observables is a matter of extreme delicacy. Many-body forces can provide the answer, but more work remains to be done to refine both the conventional picture of many-body forces discussed earlier in this section and chiral models. In the interim, the nuclear matter problem can be solved by *fiat* in either picture. Since it seems unrealistic to deny the existence of many-body forces of the required magnitude, and implausible to assume that they can be calculated

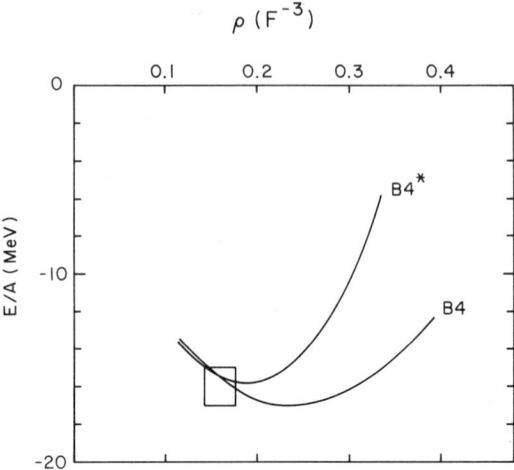

Figure 14 B4 results for the Reid potential modified to include a density-dependent sigma mass (B4*).

reliably, this approach has its merits. Its adoption will require the consideration of other nuclear matter "observables" for confirmation of theoretical models of nuclear matter.

5. CONCLUSIONS

Recent years have brought considerable improvement in the quality of both diagrammatic and variational many-body techniques. There appears to be general agreement that realistic NN potentials give a reasonable binding energy but an equilibrium density that is some 75% larger than the empirical value. Both conventional and chiral models of genuine many-body forces offer mechanisms for removing this discrepancy with each providing a simple picture for the required additional attraction for $\rho < \rho_0$. Neither approach is presently able to make *a priori* estimates of the magnitude of this effect with the delicacy required by the nuclear matter problem. Nor is it clear how to merge these pictures. We should not anticipate quick answers to these questions. One suitable interim strategy is to assume that many-body forces of the form suggested by either conventional or chiral pictures are responsible for the remaining discrepancy and to adjust parameters in such model many-body forces to restore agreement between theory and experiment. This might seem to be a summary dismissal of the standard nuclear matter problem. It should rather be regarded as revealing the next and richer layer of the nuclear matter problem.

This allows attention to be focussed on the various curvatures of the energy surface at ρ_0. If this extremum is a physically interesting energy minimum, small distortions of the Fermi surface will give rise to an increase in the energy of the system, which is necessarily second order in the distortion. A spherical distortion of the Fermi surface (a simple change in the density) will provide a measure of the ordinary compressibility of the system. A similar change in k_F in which protons and neutrons are precisely out of phase will correspond to the symmetry energy. In general, nonspherical distortions can also be considered. The spin-isospin symmetric $l = 1$ distortion provides a measure of the effective mass of a nucleon at the Fermi surface.

All generalized curvatures can be decomposed into two components. The first is a kinetic energy contribution that is necessarily positive. The second comes from the interaction between a particle and hole at the Fermi surface and is simply related to the usual Landau parameter. This term can have either sign but must not be so negative as to destroy the minimum nature of the energy extremum. This is why present results, such as the B4 and LP results of Figure 3, must be supplemented by many-body forces. As noted in Section 2, there are elementary grounds to expect a physically nonsensical negative compressibility for densities sufficiently far below the equilibrium density in any liquid. Both B4 and LP calculations yield negative compressibilities at the empirical equilibrium density of nuclear matter, and potentially useful comparisons with experiment are precluded from the start. This is not the case with the results B4* shown in Figure 14, which included a parameterized many-body force and do reproduce the empirical equilibrium density.

Fortunately, these energy curvatures and, hence, the Landau parameters themselves are moderately accessible experimentally. The symmetry energy is known from the semiempirical mass formula. The effective mass is related to the density of single-particle states in the vicinity of the ground state and is thus accessible, provided one is prepared to perform suitable A-averages to eliminate shell structure effects. Other Fermi liquid parameters can be extracted from the energies of giant resonances. In a classical liquid at finite temperature the compressibility determines the velocity of ordinary sound; this provides a linear relation between the wave number of an excitation and its energy. In a finite system boundary conditions quantize the allowed wave numbers so that the sound velocity (and compressibility) determines the energy of resonances. In more quantum mechanical language, these low-lying excitations of the bulk medium are coherent superpositions of particle-hole excitations or phonons. In the zero temperature limit, appropriate for nuclear matter, there are no collision terms and one must consider the propagation of zero rather than ordinary (i.e. first) sound.

Again, the Landau parameters determine the velocity. The corresponding excitations in finite nuclei are the well-known giant monopole (and giant quadrupole) resonances (48). Since these giant resonances are superpositions of large numbers of elementary particle-hole excitations, there are grounds to hope that the details of finite nuclear size (which determine the energy of any given particle-hole excitation) will be washed away in this coherent excited state. This expectation is supported by calculations of giant resonance energies with a variety of schematic nuclear forces (49). In such calculations the energy of the giant monopole resonance provides a clean measure of the compressibility for a given NN interaction.

One can imagine that the next few years will see considerable theoretical activity aimed at the reliable calculation of Landau parameters for nuclear matter and their application to the determination of giant resonance energies and widths in finite nuclei. This raises the hope that the next round of nuclear matter calculations will have an intimate connection with experiment, a connection that has not been a notable feature of such calculations in the past.

ACKNOWLEDGMENTS

This work was supported in part by the United States Department of Energy under contract number DE-AC02-76ER13001. The author would like to thank E. Krotscheck for providing the numerical results of Section 4, M. Rho for tuition in chiral symmetry, G. E. Brown for a number of lively discussions, and A. Lande for a critical reading of the manuscript.

Literature Cited

1. Coester, F., Cohen, S., Day, B., Vincent, C. M. 1970. *Phys. Rev.* C1:769
2. Lacombe, M., Loiseau, B., Richard, J. M., Vinh Mau, R., Cote, J., Pires, P., de Tourreil, R. 1980. *Phys. Rev.* C21:861
3. Brown, G. E., Rho, M. 1979. *Phys. Lett.* 82B:177; Brown, G. E., Rho, M., Vento, V. 1979. *Phys. Lett.* 84B:383
4. Brown, G. E. 1983. *Aust. J. Phys.* In press
5. Chodos, A., Jaffe, R. L., Johnson, K., Thorn, C. B. 1974. *Phys. Rev.* D10:2599; De Grand, T., Jaffe, R. L., Johnson, K., Kiskis, J. 1975. *Phys. Rev.* D12:2060
6. Jastrow, R. 1955. *Phys. Rev.* 98:1479
7. van Leeuwen, J. M. J., Groeneveld, J., de Boer, J. 1959. *Physica* 25:792; Fantoni, S., Rosati, S. 1974. *Nuovo Cimento* A20:179
8. Reid, R. V. 1968. *Ann. Phys.* 50:411
9. Day, B. D. 1981. *Phys. Rev.* 24C:1203
10. Lagaris, I. E., Pandharipande, V. R. 1981. *Nucl. Phys.* A359:349
11. Walecka, J. D. 1974. *Ann. Phys.* 83:491; Chin, S. A. 1977. *Ann. Phys.* 108:301
12. Lee, T. D., Margulies, M. 1975. *Phys. Rev.* D11:1591
13. Nyman, E. M., Rho, M. 1976. *Nucl. Phys.* A268:408
14. Bethe, H. A. 1971. *Ann. Rev. Nucl. Sci.* 21:93
15. Day, B. D. 1978. *Rev. Mod. Phys.* 50:495
16. Kümmel, H., Lührmann, K. H., Zabolitzky, J. G. 1978. *Phys. Rep.* 36C:1
17. Feenberg, E. 1979. *Theory of Quantum Fields.* New York: Academic
18. Pandharipande, V. R., Wiringa, R. B. 1979. *Rev. Mod. Phys.* 51:821
19. Ripka, G. 1979. *Phys. Rep.* 56C:1
20. Ripka, G. 1979. *Nucl. Phys.* A314:115
21. Jackson, A. D., Lande, A., Smith, R. A. 1982. *Phys. Rep.* 86C:55
22. Goldstone, J. 1957. *Proc. R. Soc. London* A239:267
23. Jeukenne, J. P., Lejeune, A., Mahaux, C. 1976. *Phys. Rep.* 25C:84
24. Brandow, B. H. 1966. *Phys. Rev.* 152:863

25. Friman, B. L., Niskanen, J., Nyman, E. M. 1982. *Nucl. Phys.* A383:285
26. Bethe, H. A. 1965. *Phys. Rev.* 138:804B
27. Holinde, K., Machleidt, R. 1975. *Nucl. Phys.* A247:495
28. Day, B. D. 1969. *Phys. Rev.* 187:1269
29. Day, B. D. 1981. *Phys. Rev. Lett.* 47:226
30. Castillejo, L. C., Jackson, A. D., Jennings, B. K., Smith, R. A. 1979. *Phys. Rev.* B20:3631
31. Lantto, L. J., Jackson, A. D., Siemens, P. J. 1977. *Phys. Lett.* 68B:311
32. Diatlov, I. T., Sudakhov, V. V., Ter-Martirosian, K. A. 1957. *JETP* 5:631
33. Feynman, R. P., Cohen, M. 1956. *Phys. Rev.* 102:1189
34. Owen, J. C. 1979. *Ann. Phys.* 118:373
35. Lagaris, I. E., Pandharipande, V. R. 1981. *Nucl. Phys.* A359:331
36. Clark, J. W., Mead, L. R., Krotscheck, E., Kürten, K. E., Ristig, M. L. 1979. *Nucl. Phys.* A328:45
37. Krotscheck, E., Smith, R. A., 1983. *Nucl. Phys.* To be published
38. Jackson, A. D., Krotscheck, E., Meltzer, D. E., Smith, R. A. 1982. *Nucl. Phys.* A386:125
39. Krotscheck, E. 1982. *Preprint*. Univ. Ill. Urbana
40. Carlson, J., Pandharipande, V. R., Wiringa, R. B. 1982. *Preprint ILL-(NU)-82-10*. Univ. Ill. Urbana
41. Coon, S. A., Scadron, M. S., McNamee, P. C., Barrett, B. R., Blatt, D. W. E. 1979. *Nucl. Phys.* A317:242
42. Durso, J. W., Jackson, A. D., VerWest, B. J. 1980. *Nucl. Phys.* A345:471
43. Lee, T. D. 1975. *Rev. Mod. Phys.* 47:267
44. Barshay, S., Brown, G. E. 1975. *Phys. Rev. Lett.* 34:1106
45. Brown, G. E. 1979. In *Mesons in Nuclei*, ed. M. Rho, D. H. Wilkinson, p. 331. Amsterdam: North-Holland
46. Matsui, T., Serot, B. D. 1982. *Preprint ITP-708*. Stanford Univ.
47. Jackson, A. D., Krotscheck, E., Rho, M. 1983. *Nucl. Phys.* To be published
48. Jackson, A. D., Jennings, B. K. 1980. *Phys. Rep.* 66C:141
49. Blaizot, J. P., Gogny, D., Grammaticos, B. 1976. *Nucl. Phys.* A265:315

PHYSICS WITH THE CRYSTAL BALL DETECTOR

Elliott D. Bloom

Stanford Linear Accelerator Center, Stanford University, Stanford, California 94305

Charles W. Peck

High Energy Physics, California Institute of Technology, Pasadena, California 91125

CONTENTS

1. INTRODUCTION AND OUTLINE 144
2. DESCRIPTION OF THE APPARATUS AND ITS PERFORMANCE 149
3. THE CHARMONIUM ^3P_J STATES 153
   3.1 *Dominant Features in the Inclusive Photon Spectrum of the ψ'* 153
   3.2 *The Photon Cascade $\psi' \to \gamma\chi_J \to \gamma\gamma J/\psi$* 154
   3.3 *Results from the Full Analysis of the Inclusive Spectrum and Some Comparisons to Exclusive Results...* 157
4. HADRONIC TRANSITIONS FROM THE ψ' TO THE J/ψ 162
   4.1 *The Transitions $\psi' \to \eta(\pi^0)J/\psi$* 162
5. THE CHARMONIUM $^1S_0$ STATES 164
   5.1 *Evidence for the 1^1S_0 in Inclusive γ Spectra of the ψ' and the J/ψ* 164
   5.2 *Hadronic Decays of the 1^1S_0* 165
   5.3 *Evidence for the 2^1S_0 State in the Inclusive γ Spectrum of the ψ'* 166
   5.4 *Discussion* 167
6. RADIATIVE TRANSITIONS FROM THE J/ψ 168
   6.1 *The Fundamental Character of the Gluonic Mesons in QCD* 168
   6.2 *The "End Point" of the Inclusive γ Spectrum at the J/ψ* 169
   6.3 *Gamma Transitions to Well-Known Particles Using Exclusive Decays* 170
   6.4 *The Gluonium Candidates, $\iota(1440)$ and $\vartheta(1640)$* 174
   6.5 *Other Radiative Transitions* 179
7. SEARCHES FOR THE F MESON, THE AXION, AND THE $^1P_1$ STATE 181
   7.1 *The Inclusive η Cross Section* 181
   7.2 *The Search for $J/\psi \to \gamma$ Axion* 183
   7.3 *The Search for the Decays $\psi' \to \pi^0\,^1P_1$* 184
8. MEASUREMENTS OF R_h IN THE E_{CM} RANGE OF 5.0 TO 7.4 GeV 185

143

9. TWO-PHOTON PHYSICS	189
9.1 *Measurement of* $\Gamma(f \to \gamma\gamma)$ *from* $\sigma(\gamma\gamma \to f \to \pi^0\pi^0)$	189
9.2 *Measurement of* $\Gamma(A_2 \to \gamma\gamma)$ *from* $\sigma(\gamma\gamma \to A_2 \to \pi^0\eta)$	191
9.3 *Other States*	192
10. MEASUREMENTS IN THE REGION FROM CHARM THRESHOLD TO 4.5 GeV	192
11. SUMMARY AND FUTURE PROSPECTS	193

1. INTRODUCTION AND OUTLINE

At the 1974 PEP Summer Study (1), one of the projects was to explore the possibilities and limitations of detectors optimized to measure photons produced in high energy e^+e^- collisions. It was realized that a device that had high detection efficiency over a large solid angle and that could measure the energy of photons in the region above a few tens of MeV with high precision (in the range of a few percent) would provide a unique capability offered by no existing apparatus. Thus it could possibly yield important and otherwise unattainable information about these fundamental interactions. Furthermore, if it also measured the directions of both photons and charged particles well enough, even a nonmagnetic version of such a device would be able to compete with the large general-purpose magnetic spectrometers then in existence in the reconstruction of certain simple, few-particle final states. And finally, a device designed to absorb all the electromagnetic energy in an event would in fact quickly and directly measure a large fraction of its total energy. This prompt information could form the basis for an admirable trigger having very different biases from those used by the magnetic spectrometers. Thus such a device would be an interesting complementary technique for the investigation of e^+e^- physics. In particular an efficient "all-neutral" trigger would be possible.

Although the thrust of the summer's work had been directed toward instrumentation for PEP (an e^+e^- storage ring at the Stanford Linear Accelerator Center, allowing beam energies up to 15 GeV), which was still in the planning stages at that time, a keen interest in the idea developed among a group of people[1] from Caltech, Harvard, Stanford-HEPL, and SLAC and this led to serious work in the Fall of 1974 toward producing a formal proposal for the existing lower energy storage ring, SPEAR. The startling discoveries of the $J/\psi(3100)$ and $\psi'(3700)$ in November and December of that year spurred on these efforts, especially as people realized there was the likely possibility of a rich gamma ray spectroscopy in the range of a few hundred MeV. Eventually this work led the group to submit a proposal for a nonmagnetic, large solid angle detector whose principal component was a spherical shell of NaI(Tl) with a 10-inch inner radius and

---

[1] A group from Princeton joined the collaboration in 1977.

a 26-inch outer radius. The device was quickly dubbed the "Crystal Ball" and it has been universally called that ever since. The proposal was approved in the Spring of 1975 and the construction of the detector was completed three years later in the Spring of 1978. Section 2 describes the configuration and performance of the detector.

The Ball was installed at SPEAR in the Fall of 1978 and took data there on e^+e^- collisions in the energy region from 3.1 to 7.4 GeV during the 40 months of calendar time until December 1981. SPEAR actually supplied beam during about half of this time. We spent about five months collecting about 2×10^6 hadronic events at each of the two $^3S_1$ states, the $J/\psi(3100)$ and the $\psi'(3700)$. Typical luminosities at these energies were 0.5×10^{30} cm^{-2} s^{-1} and 1.8×10^{30} cm^{-2} s^{-1}, respectively. About one month was spent at the $\psi''(3770)$ collecting 4×10^4 hadronic events and the rest of the time was spent at energies in the continuum, almost all of which were above charm threshold. We obtained a cumulative exposure of 24.0 pb^{-1} in this region. At the highest energy at which we took data, 7.4 GeV, SPEAR provided a peak luminosity of about 2.0×10^{31} cm^{-2} s^{-1}.

In the Spring of 1982 the Ball was moved as an intact experiment to the Deutsches Elektronen Synchrotron (DESY), Hamburg, Germany to run on the DORIS II e^+e^- storage ring in order to make a parallel study of the Υ system. Data taking in the 10-GeV region, in which Doris II is optimized, had just begun as this paper was being prepared.

This brief review surveys all the Crystal Ball physics results that had been completed as of December 1982. The available space does not permit any detailed discussion of either the experimental details or the theoretical framework that provides the proper setting for the experimental findings described here. Also, the historical perspective is that of participants in the Crystal Ball experiment, one of many that has made major contributions to this field. However, the interested reader can find a discussion of many of the theoretical questions in (2) and (3) and in the literature cited therein; appropriate experimental references and only limited theoretical references are given in this review. A general survey of the physics of psionic matter up to 1977 can be found in (3a). Finally, in Section 11 we briefly discuss some of the analysis projects currently in progress as well as our expectations for results from the just begun exposure of the Crystal Ball at DORIS II.

The principal accomplishments of the Crystal Ball experiment have resulted from the study of radiative and certain hadronic transitions involving the charmonium states. Figure 1 shows the energy level diagram of this system and it also indicates the several radiative and hadronic transitions that have been the focus of the Crystal Ball efforts. The refreshing simplicity of this first-known heavy quark spectrum compared to the corresponding situation among the light quarks (u,d,s) has played an important role in the recent development of particle physics. This positronium-like structure gives strong qualitative evidence for the fundamental $c\bar{c}$ interpretation of charmonium and quantitative details about the energies and transition rates can be compared with phenomenological models motivated by quantum chromodynamics (QCD).

When the Crystal Ball experiment began, however, there were several outstanding difficulties with the then-favored, and now well-established, $c\bar{c}$ model. A total of five states had been reported in two-photon cascade transitions between ψ' and the J/ψ and one state had been reported below the J/ψ in the 3γ decay mode. Preferred quantum numbers for three of the intermediate mass states were indirectly inferred (4–6) from their hadronic decay patterns and mass ordering and these caused them to be identified with the three ^3P_J states. However, the experimental situation concerning the candidate $^1S_0$ states seemed to present an insurmountable challenge to the beautiful $c\bar{c}$ interpretation (7–9).

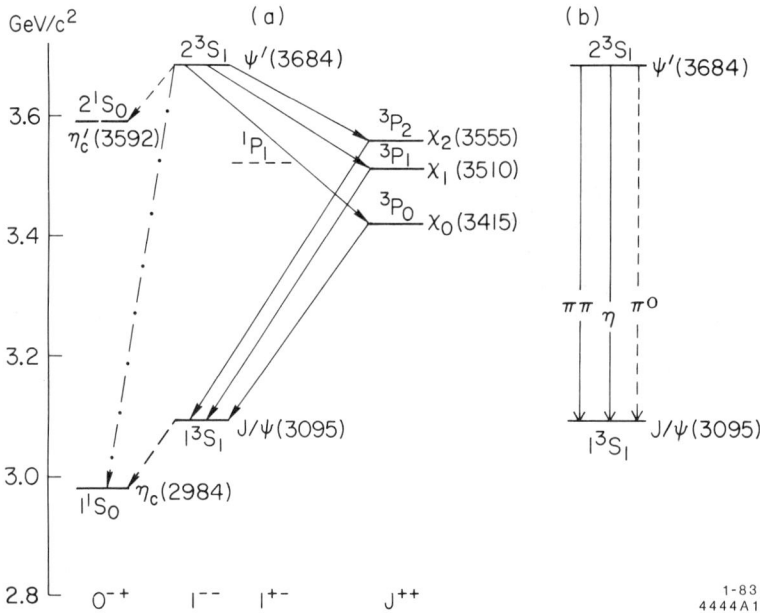

Figure 1 (a) The current status of the charmonium spectrum below charm threshold. All the observed photon transitions between these states are shown. Solid lines indicate electric dipole transitions; broken lines indicate allowed magnetic dipole transitions (between states with the same radial wave function); and broken-dotted lines show "hindered" magnetic dipole transitions (between states with different radial wave functions). The $^1P_1$ state is yet to be observed. (b) The observed hadronic transitions between the ψ' and the J/ψ. The $\pi\pi$ transitions are allowed, the η transition is $SU(3)_F$ forbidden, and the π^0 transition is $SU(2)_F$ forbidden.

In brief, the $^1S_0$ problem was as follows. In 1977, the DASP collaboration (10, 11) observed a significant signal at 2.83 ± 0.03 GeV/c^2 in the distribution of the highest $\gamma\gamma$ mass from the decay $J/\psi \to \gamma\gamma\gamma$. This state, the X(2830), immediately became a candidate for the 1^1S_0 state, the η_c. Confirming evidence for the state was reported by a Serpukov group using the reaction $\pi^- p \to \gamma\gamma n$ (11a). The measured product branching ratio from the DASP collaboration, $B[J/\psi \to \gamma X(2830)] \cdot B[X(2830) \to \gamma\gamma]$, was $(1.2 \pm 0.5) \times 10^{-4}$. However, no evidence for $J/\psi \to \gamma X(2830)$ was seen in the inclusive γ spectrum from the J/ψ by the SPEAR experiment SP-27 (12), which set an upper limit of 2% for $B[J/\psi \to \gamma X(2830)]$. These results were incompatible with any reasonable $c\bar{c}$ model since this interpretation predicts $B[J/\psi \to \gamma X(2830)]$ to be an order of magnitude larger than the limit set by SP-27. Furthermore, it predicts $B[X(2830) \to \gamma\gamma]$ to be about five times smaller than the lower limit inferred from the DASP and SP-27 results combined. Finally, a hyperfine splitting of 265 MeV is surprisingly large within the $c\bar{c}$ model.

The second serious problem concerned the 2^1S_0 state, the η'_c. Initially, some evidence for an η'_c candidate was reported (13) at a mass of 3455 MeV/c^2 in the cascade process $\psi' \to \gamma 2^1S_0 \to \gamma\gamma J/\psi$ by the Mark I experiment at SPEAR. This observation was not confirmed by a subsequent experiment (14), the DESY-Heidelberg collaboration at DORIS, which independently investigated the radiative cascade process. On the other hand, the DESY-Heidelberg experiment presented evidence for an alternative intermediate state at 3591 MeV/c^2 as a possible η'_c candidate. However, their reported branching ratio for $\psi' \to \gamma\chi(3591) \to \gamma\gamma J/\psi$ was orders of magnitude greater than predicted by the model, if this state were taken to be the η'_c.

One of the first processes measured with the newly commissioned Crystal Ball was $J/\psi \to 3\gamma$. These new observations provided both higher statistics and better resolution than the earlier ones, but they did not confirm the X(2830). A lower limit of 2.2×10^{-5} was initially (15) set for the product branching ratio [this limit was subsequently lowered to 1.6×10^{-5} (90% C.L.); see Section 6.3]. Nor did later data confirm either the $\chi(3455)$ or the $\chi(3591)$ in the radiative cascades from the ψ' to the J/ψ. Thus the experimental status of the two expected $^1S_0$ states was again open.

The first evidence for the η_c in the Crystal Ball came from the inclusive γ spectra observed from the ψ' and, shortly thereafter, that from the J/ψ. Somewhat later, with a doubling of the ψ' data sample, evidence for the η'_c was also found in the ψ' inclusive γ spectrum. The current status of these two charmonium states is discussed in Section 5 below. With these two contributions from the Crystal Ball, there is only one qualitative feature of the expected $c\bar{c}$ spectrum for which no experimental evidence has yet been found, namely, the $^1P_1$ state. Section 7 summarizes our current limits on certain decay modes involving this state.

The radiative transitions involving the three ^3P_J states give rise to the several prominent peaks in the ψ' inclusive γ spectrum shown in Figure 3. So characteristic, in fact, is this spectrum that it has become the logo of the Crystal Ball experiment. The careful study of all the systematics (efficiencies and resolutions) necessary to obtain the branching ratios and natural widths of these states from the inclusive γ spectrum has been recently completed and is discussed in Section 3. The radiative cascade exclusive channels $\psi' \to \gamma\, ^3P_J \to \gamma\gamma e^+e^-$ or $\gamma\gamma\mu^+\mu^-$ were susceptible to more rapid analysis and Section 3 also summarizes our results on product branching ratios, masses, and angular distributions (which strongly support the earlier spin assignments for these states). As a by-product of our study of these exclusive channels, we also made measurements on the three transitions $\psi' \to \pi^0\pi^0 J/\psi$, $\psi' \to \eta J/\psi$, and $\psi' \to \pi^0 J/\psi$. The first simply corroborated the much better results from $\psi' \to \pi^+\pi^- J/\psi$, but the other two yielded significant improvements over earlier work. These hadronic transitions are discussed in Section 4.

Radiative transitions from the J/ψ are especially interesting since their primary mechanism is expected (in the context of QCD) to be $J/\psi \to \gamma gg$ with the two gluons in a singlet state of both color and flavor. Thus any gg bound states that are even under charge conjugation and less massive than the J/ψ are likely to be excited in this decay. At least two candidates for such objects have been observed in the Crystal Ball data. One, with a mass of about 1440 MeV/c^2 was thought to be the 1^{++}, E(1420) meson when it was first found in J/ψ decays by the Mark II experiment. The existence of the state was quickly confirmed by the Crystal Ball. However, only after the partial-wave analysis of twice the initial data sample did the Crystal Ball collaboration find that the 0^{-+} assignment for the state was favored. This state was then named $\iota(1440)$. A second gluonium candidate, the $\vartheta(1640)$, was found by the Crystal Ball in the $\eta\eta$ decay mode and the preferred spin-parity assignment is 2^+. Finally, searching in the channel $J/\psi \to \gamma\eta\pi\pi$, we find no evidence that the 1440-MeV/c^2 state decays into $\eta\pi\pi$ but we do see both the expected signal of $\eta' \to \eta\pi\pi$ and an unexpected very broad enhancement at an $\eta\pi\pi$ mass of 1710 MeV/c^2. The present status of these several interesting possibilities as well as the Crystal Ball's observations on the modes $J/\psi \to \gamma X$, where X = π^0, η, η', f, and f', is discussed in Section 6.

In addition to the extensive search for the $^1P_1$ charmonium state mentioned earlier, two other searches with negative results have been carried out and are described in Section 7. The first was an attempt to corroborate a strong enhancement in the inclusive η cross section reported by the DASP experiment. This had been interpreted as evidence for $e^+e^- \to F + \cdots \to \eta + \cdots$ where the F is the charmed strange meson. However, no significant enhancement was observed by the Crystal Ball. The second search with negative results was for evidence of the axion. Since there are quite sharp theoretical predictions for radiative decays of the J/ψ and Υ into the axion, we made a detailed investigation of our J/ψ data looking for this decay; nothing was found.

Finally, in addition to the charmonium studies, comprising the bulk of the Crystal Ball results, this experiment has also collected a body of data in the energy region above charm threshold. To date, in addition to the inclusive η cross sections mentioned earlier, we have made total hadronic cross section measurements (R_h) up to the highest energies at SPEAR (Section 8), we have observed production of the f and A_2 by two-photon collisions from which we obtain the decay rates $\Gamma(f \to \gamma\gamma)$ and $\Gamma(A_2 \to \gamma\gamma)$ (Section 9), and, finally, we have made several measurements in the region from charm threshold to 4.5 GeV (Section 10).

2. DESCRIPTION OF THE APPARATUS AND ITS PERFORMANCE

Over the years, many methods have been developed and extensively used for measuring the energy of high energy photons. By the mid-1970s, however, the pioneering work of R. Hofstadter and his colleagues (16) had shown that the technique of total absorption shower counters made of thallium-doped sodium iodide, NaI(Tl), was unsurpassed in the combination of high detection efficiency and energy resolution. Consequently, in spite of the technical difficulties occasioned by the extremely hydroscopic nature of NaI(Tl), this technique, supplemented by fine segmentation of the material, was selected to form the basis of a detector covering nearly the full 4π-sr solid angle about the e^+e^- collision point, the Crystal Ball. The final result of the design was a detector consisting of four main parts. These were a central charged-particle detection system, two hemispherical shells of NaI(Tl), endcaps of tracking chambers followed by sodium iodide covering the beam entry holes into the spherical shell, and a small-angle luminosity monitor. Figure 2 shows the geometric arrangement of the two major components of the detector. Details about the apparatus can be found in (17).

The central tracking system consisted of three concentric cylindrical ionization detectors covering 71%, 83%, and 94% of 4π sr, respectively. The middle detector (18) was a proportional chamber with two gaps, and the other two detectors were magnetostrictive spark chambers. For particles that were detected in both spark chambers, both direction and origin along the beam line could be determined. Those that failed to be detected in both spark chambers were only "tagged," i.e. identified as being charged.

The heart of the detector, of course, was the spherical shell, 16 radiation lengths thick, made of sodium iodide. This thickness is sufficient to contain essentially the entire longitudinal development of electromagnetic showers in our energy range. As shown in Figure 2, the shell is actually a dense packing of truncated triangular pyramids of NaI(Tl). These are optically

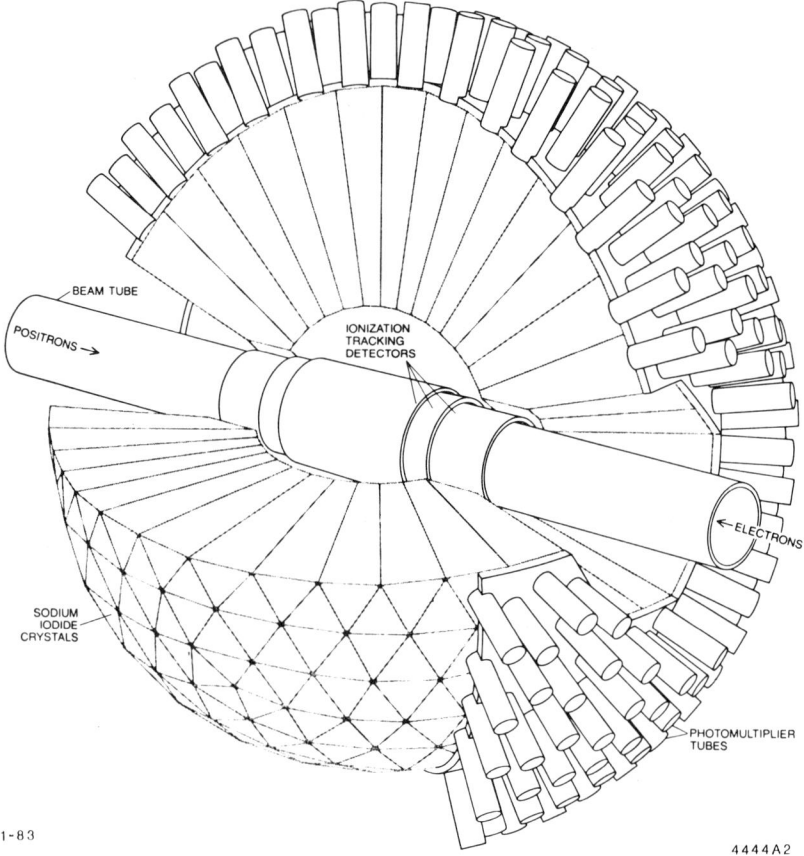

Figure 2 The two principal elements of the Crystal Ball detector, the charged-particle tracking chambers in the 25-cm diameter cavity of the shell, and the NaI(Tl) shell itself. The middle chamber is a continuously sensitive wire proportional chamber and the other two are magnetostrictive spark chambers. The shell itself is segmented into optically separated triangular pyramids in a solidly packed geometry based on an icosahedron. Each pyramid is viewed from the outside by a single photomultiplier. [From *Quarkonium*, by E. D. Bloom and G. J. Feldman. Copyright © 1982 by Scientific American, Inc. All rights reserved.]

isolated one from another, and each is viewed from the outside by a single photomultiplier tube. The only materials separating the individual crystals are thin layers of white paper and aluminum foil (except for the plane separating the two hemispheres). The shell consists of a total of 672 of these crystals and it covers 93% of 4π sr. The missing 7% is due to beam entry holes, but these are almost completely covered by the endcaps.

With this geometric arrangement, we not only measure the amount of energy deposited in the NaI with little loss, but we also obtain information about the transverse structure of this energy deposition. Being minimum ionizing and lacking strong interactions, high energy muons leave simple tracks, with a deposited energy of about 200 MeV distributed over no more than two or three crystals. Electrons and photons with energy greater than about 20 MeV produce electromagnetic showers and deposit all of their energy in a reasonably characteristic pattern covering about 13 crystals. Finally, most hadrons strongly interact in the Ball since it is about one absorption length thick. They thus give rise to somewhat more irregular patterns than electromagnetic showers and the total deposited energy bears little relation to the hadron's energy. This geometric arrangement provides no information about the longitudinal distribution of an energy deposition, but we have found that careful statistical analysis of a transverse pattern is a useful technique for resolving some particle identification ambiguities.

The parameter of particular interest in this detector is its energy resolution for electromagnetic showers. For the energy range of interest, the standard deviation σ_E of this resolution is well approximated by $(0.0255 \pm 0.0013)E^{3/4}$ where E and σ_E are in GeV. Thus, for example, we measure the energy of a 1.55-GeV Bhabha-scattered electron to an accuracy of 36 MeV and that of a 100-MeV photon to an accuracy of 4.6 MeV. An example of the utility of this relatively high resolution is that we can extract the natural widths of the charmonium $^1S_0$ and $^3P_{0,2}$ from our inclusive γ distributions. More generally, the goodness of our photon energy resolution has proved invaluable in allowing us to reliably identify certain reactions by the technique of kinematically constrained statistical fitting, which in turn leads to some of the physics results discussed below. It should be noted, however, that because of the size of electromagnetic showers and the edge effects of the beam holes, the good energy resolution is only available over 85% of 4π sr. A detailed description of tests made on a prototype of the detector and the signal processing method used can be found in (19).

A second parameter of considerable interest is the resolution with which the direction of a photon can be determined. By examining the profile of its shower's energy deposition we can determine the direction of a photon to much better than the size of one module. The limitation on the accuracy of angles determined in this manner is caused by shower fluctuations. The Crystal Ball has achieved a resolution with $\sigma_{\vartheta_\gamma} = 1.5$ to $2°$, where ϑ_γ is the polar angle from the photon's true direction. There is a slight energy dependence in this angular resolution.

An important design goal in this apparatus was to cover as much as possible of the solid angle around the collision point with high efficiency particle detectors. This was achieved by covering the necessary beam holes in the ball with endcaps, 20 radiation lengths thick of individually packaged NaI(Tl) hexagonal prisms covered by two gaps of spark chambers. These brought the total coverage to 98% of 4π sr. Primarily because of edge effects, the energy resolution for photons and electrons going outside the central 85% of 4π sr of the main ball was relatively poor and strongly direction dependent. Consequently, the endcaps were primarily used as veto counters. They allowed us to determine the topology of events with very high confidence, and this was of crucial importance for reducing backgrounds in some of the physics measurements given later.

Finally, for many of our measurements an absolute luminosity determination was necessary. This was provided by a small-angle Bhabha-scattering detector consisting of four counter elements, symmetrically disposed about the beam and centered at a $4°$ angle to the beam line. Each of the four elements was identical, consisting of three scintillators followed by a shower counter, and covered a solid angle of 4.2×10^{-4} sr. The system provided a counting rate of about 0.7 Hz at the ψ' with our typical luminosity. The accuracy of luminosity determination was better than 3% with this monitor, as checked by using large-angle Bhabha events observed in the full Ball.

The apparatus was triggered and events written on tape when at least one of several overlapping conditions was satisfied. Each of these triggers was based on a coincidence between a beam-crossing signal and the analog sum of signals from the Ball and each required that this sum, proportional to the total energy in the Ball, be greater than some threshold. Generally, a further requirement was also imposed and the more restrictive it was, the lower the total energy threshold. The simplest trigger involved no other requirements and its total energy threshold was normally about 1 GeV. More restrictive triggers involved such event features as charged particles being detected in the proportional chambers, or a requirement on the general pattern of energy deposition in the Ball. In general, the hardware trigger conditions were highly efficient for the classes of events that have been studied with the Crystal Ball, and the Monte Carlo simultations done to determine detection efficiencies have included these hardware trigger conditions.

Data acquisition and general system monitoring were performed by a PDP11/t55 computer. Chestnut et al [20] discuss in detail both the

3. THE CHARMONIUM ^3P_J STATES

3.1 *Dominant Features in the Inclusive Photon Spectrum of the ψ'*

After the discoveries of the J/ψ (21, 22) and ψ' (23) in 1974, four experiments measured the inclusive photon spectrum from the ψ' with increasing levels of sensitivity. The first experiment was a two-crystal NaI(Tl) detector (24); it could only place upper limits on radiative transitions to the $^3P_J(\chi_J)$ states. A magnetic detector, measuring converted photons, was able to measure the photon transition to the χ_0 state (13), but was not able to inclusively observe the other transitions. A moderately segmented NaI(Tl) detector (12) finally measured the photon transitions to each of the χ_J states and also inclusively observed the cascade transitions from the χ_2 and χ_1 to the J/ψ. Finally, Figure 3 shows the inclusive spectrum at the ψ' from the Crystal Ball detector, the most sensitive experiment so far. The main spectrum in the figure is from the analysis of approximately 0.9×10^6 ψ' events (the last half of the full sample) obtained at SPEAR. Severe cuts have been made in this spectrum to enhance structure. First, all photons are required to have $|\cos \vartheta_\gamma| < 0.85$, where ϑ_γ is the angle between the photon and the beam direction. The cosine of the angle between each photon and any charged particle is required to be less than 0.9. Pairs of γ's with invariant mass consistent with the mass of the π^0 have been eliminated. Finally, the lateral shower energy deposition in the NaI(Tl) crystals is required to be consistent with a single electromagnetic shower. This "pattern cut" removes most of those minimum ionizing charged particles that were not identified by the tracking-chamber system, many of the spurious energy deposits resulting from interacting charged particles, and some of the high energy π^0's in which the electromagnetic showers from the two photons from the π^0 decay overlap. The pattern cut used for the spectrum in Figure 3, one of many algorithms possible, was designed to optimize the efficiency for photons with energy, E_γ, less than or about 100 MeV.

As is seen in the figure, the photon transitions from the ψ' to the χ_J states and the cascade transitions from the χ_J states to the J/ψ stand out clearly in this inclusive spectrum. Indeed, the strength of these transitions in our detector has allowed frequent checks of the NaI(Tl) energy calibration and resolution over the course of our stay at SPEAR. Typically, two days of reasonable data taking at SPEAR, yielding approximately 2.5×10^4 ψ' decays, allowed an accurate determination of the transition energies to the χ_J states.

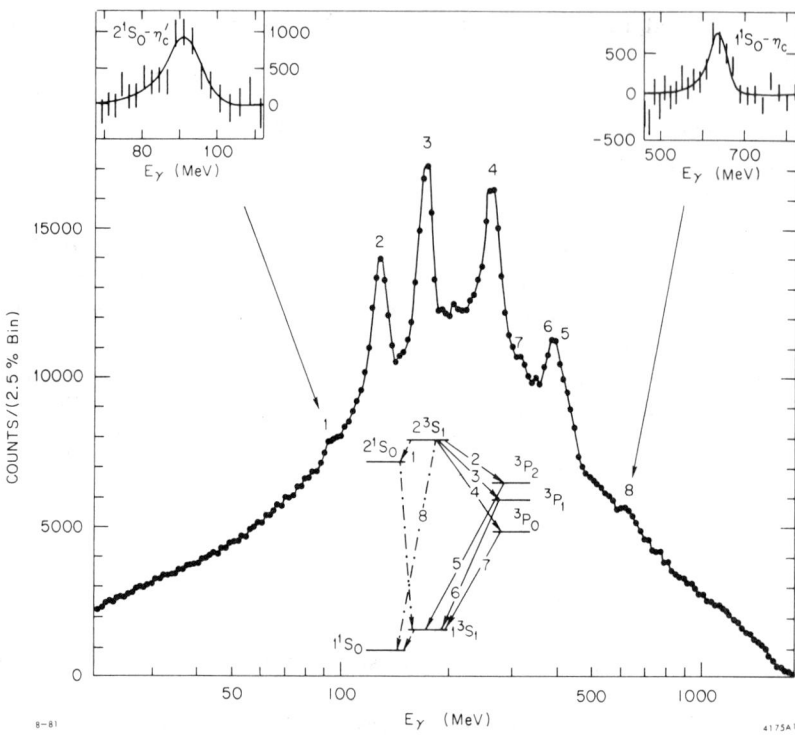

Figure 3 Inclusive γ spectrum at the ψ′. Note that the spectrum is $\Delta N/\Delta(\log E) \cong E\, dN/dE$. The upper inserts show the background-subtracted signals for the η_c and η_c' candidate states. The numbers over the spectrum key the observed spectral features with the expected radiative transitions in the charmonium spectrum inset.

3.2 *The Photon Cascade,* $\psi' \to \gamma\chi_J \to \gamma\gamma J/\psi$

A study of the radiative transitions from the ψ' to the χ_J states and the cascade radiative decays from the χ_J states by means of the sequence, $\psi' \to \gamma\chi_J$, $\chi_J \to \gamma J/\psi$, $J/\psi \to \ell^+\ell^-$, where $\ell^+\ell^-$ is e^+e^- or $\mu^+\mu^-$, provides a method for identifying the χ_J states (17) that is almost free of background. Indeed, it was in this reaction sequence that the $\chi_{1,2}$ were first observed (4, 25). Additionally, an analysis of the angular correlations in the cascade final state of Crystal Ball data (17) has permitted a direct measurement of the spin of the $\chi_{1,2}$ states and of the multipole coefficients describing the two individual radiative transitions for each of these states. The decays, $\psi' \to \eta(\pi^0)J/\psi \to \gamma\gamma\ell^+\ell^-$ exhibit the same topology as the cascade reactions; these processes have also been studied by the Crystal Ball collaboration (17) in order to separate them from χ_J events as well as for their own sake (cf Section 4.1). Note that the decay, $\psi' \to \pi^0 J/\psi$, which is forbidden by isospin symmetry, has been observed by the Crystal Ball (17) and the Mark II (26) detectors at SPEAR. The details of the Crystal Ball data analysis for cascade reactions are discussed in (17) and the references cited therein.

Figure 4a shows the Dalitz plot of the final event sample (from the first half of the full data set) containing 1206 $\gamma\gamma e^+ e^-$ and 1280 $\gamma\gamma\mu^+\mu^-$ decays prior to kinematic fitting. These same events are also shown on the Dalitz plot in Figure 4b after they have been kinematically fit to the hypothesis that they arise from $\psi' \to \gamma\gamma J/\psi \to \gamma\gamma \ell^+\ell^-$ (5-C for e^+e^-, 3-C for $\mu^+\mu^-$). The fit kinematics restricts all of the surviving 2234 $\gamma\gamma\ell^+\ell^-$ events to fall within the outer envelope illustrated in Figures 4a, b; the cuts to the data restrict the events to fall within the inner envelope.

The decay, $\psi' \to \pi^0\pi^0 J/\psi$, $\pi^0 \to \gamma\gamma$, $J/\psi \to \ell^+\ell^-$, in which two photons go undetected or have energies less than 20 MeV, is a background of $\sim 5\%$ to

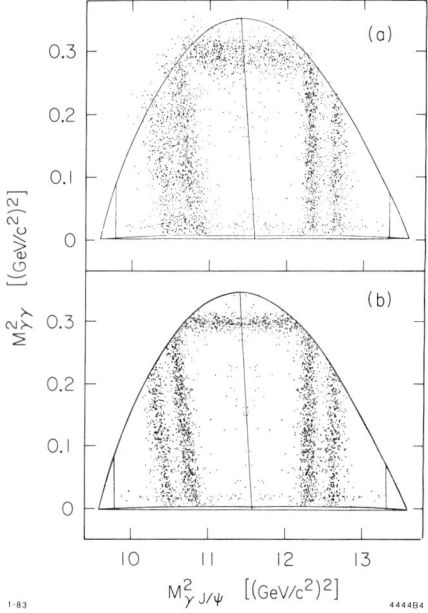

Figure 4 (a) Dalitz plot showing events from the two exclusive decays $\psi' \to \gamma\gamma e^+e^-$ and $\psi' \to \gamma\gamma\mu^+\mu^-$. The kinematic boundary is the outer one shown, and inscribed within it are the boundaries imposed by the event selection cuts. Each event appears twice in this plot, once to the right of the almost vertical central dividing line, once to the left. The combination with the lower energy photon is on the right and the clear verticality of the bands shows that the lower energy photon is the first emitted. Horizontal bands corresponding to the η and π^0 are also evident. (b) The same as (a) but after kinematic fitting. The $\psi' \to \gamma\gamma e^+e^-$ events are subjected to five constraints and $\psi' \to \gamma\gamma\mu^+\mu^-$ to three. The main effect of fitting is to remove background and to improve the energy resolution of the higher energy photon. The latter significantly sharpens the bands on the left and those for the η and the π^0.

the events of Figure 4a. This background, as well as all other backgrounds that have been considered (17), is totally negligible in the final fit event sample shown in Figure 4b.

In both Figures 4a and 4b the horizontal band at the top occurs at the η mass; that near the bottom occurs at the π^0 mass. Two strong signals for $\chi_1(3508)$ and $\chi_2(3554)$ appear as vertical bands to the right of the symmetry line shown, which has slope, $d(m_{\gamma\gamma}^2)/d(M_{\gamma J/\psi}^2) = -2$.

The Doppler-shifted bands on the left of the symmetry line (each event is plotted twice, once for each $\gamma J/\psi$ mass combination) are tilted with a slope of -1. The mass resolution for the π^0, η, and the low solution $\gamma J/\psi$ mass is better in Figure 4b than in the unfitted plot (Figure 4a). This is due to the fact that the kinematic fit reduces the absolute energy error of the higher energy photon to that of the lower energy one.

After separating the η and π^0 bands, the populous states at $\gamma J/\psi$ masses of 3554 and 3508 MeV/c^2 contain 479 and 943 events, respectively. Three of the 20 events associated with $\chi_0(3413)$ are expected to arise from the reaction $\psi' \to \pi^0\pi^0 J/\psi$. The cuts on the data restrict the $\gamma J/\psi$ mass to the range 3129 to 3644 MeV/c^2; in this region we find no evidence for a fourth χ state.

The branching ratio for a particular $\gamma\gamma\ell^+\ell^-$ decay channel is obtained by taking the number of events observed in the channel, correcting for detection efficiency (from 0.5 to 0.25 for various channels, and typically about 0.4), photon conversion and charged-particle identification efficiencies (0.95 and 0.96, respectively), and dividing by the total number of ψ' produced and the branching ratio for the decay of the J/ψ into dileptons. The J/ψ dilepton branching ratio (27) is the dominant systematic error (13%) in this measurement. The branching ratios obtained by the Crystal Ball are shown in Table 1 in comparison with those obtained from other experiments. There is good agreement for the χ_2 and χ_1 measurements; however, only the Crystal Ball measures a significant χ_0 branching ratio. Only upper limits are given for $\chi(3455)$ and $\chi(3591)$.

Additional information obtained from the Crystal Ball measurements of the photon cascade decays included the spins of the $\chi_{1,2}$ states and the multipolarity of the γ transitions. The particles participating in the cascade sequence $e^+e^- \to \psi'$, $\psi' \to \gamma'\chi_J$, $\chi_J \to \gamma J/\psi$, $J/\psi \to \ell^+\ell^-$ define the five angles, $\cos \vartheta' = \hat{e}^+ \cdot \hat{\gamma}'$, $\cos \vartheta_{\gamma\gamma} = \hat{\gamma}' \cdot \hat{\gamma}$, $\tan \varphi' = [\hat{e}^+ \cdot (\hat{\gamma}' \times \hat{\gamma})]/\{\hat{e}^+ \cdot [(\hat{\gamma}' \times \hat{\gamma}) \times \hat{\gamma}']\}$, $\cos \vartheta = \hat{\ell}^+ \cdot \hat{\gamma}$, $\tan \varphi = [\hat{\ell}^+ \cdot (\hat{\gamma}' \times \hat{\gamma})]/\{\hat{\ell}^+ \cdot [(\hat{\gamma}' \times \hat{\gamma}) \times \hat{\gamma}]\}$.

The angular distribution function $w(\cos \vartheta', \varphi', \cos \vartheta_{\gamma\gamma}, \cos \vartheta, \varphi, \mathbf{p})$, detailed in (28), which describes the above cascade sequence is a function of the five angles, and of the multipole parameters $\mathbf{p} = (J, a'_j, a_j)$, where a'_j and a_j describe the multipole structure for the two radiative transitions.

Table 1 Comparison of Crystal Ball results for $\psi' \to \gamma\gamma J/\psi$ with those from other experiments[a]

State (MeV/c^2)	Crystal Ball	Mark II (26)	Mark I (13)	DESY-Heidelberg (14)
		$B(\psi' \to \gamma\gamma J/\psi)(\%)$		
$\chi(3553.9 \pm 0.5)$*	1.26 ± 0.22	1.1 ± 0.3	1.0 ± 0.6	1.0 ± 0.2
$\chi(3508.4 \pm 0.4)$*	2.38 ± 0.40	2.4 ± 0.6	2.4 ± 0.8	2.5 ± 0.4
$\chi(3412.9 \pm 0.6)$**	0.06 ± 0.02	<0.56	0.2 ± 0.2	0.14 ± 0.09
$\chi(3455)$	<0.02	<0.13	0.8 ± 0.4	<0.25
$\chi(3591)$	<0.04	—	—	0.18 ± 0.06
		$B(\psi' \to m J/\psi)(\%)$		
η	2.18 ± 0.38	2.5 ± 0.6	4.3 ± 0.8	3.6 ± 0.5
π^0	0.09 ± 0.03	0.15 ± 0.06	—	—

[a] Limits are at the 90% confidence level. Masses as measured by the Crystal Ball are denoted by an asterisk, and those measured by Mark II by a double asterisk. There is an additional 4-MeV/c^2 systematic uncertainty on all the masses.

The multipole coefficients are $a_j(a'_j)$ and they satisfy the relation, $\Gamma(\chi_J \to \gamma J/\psi) \propto \sum_{j=1}^{J+1} |a_j|^2$, and similarly for a'_j. The explicit form of the multipole coefficients is given in (28). Given the standard charmonium model, one expects the electric dipole amplitudes to dominate the transitions. Thus, the coefficients a_3 and a'_3, which are possible in the spin-2 case, can be expected to be very small and they were set to zero.

The data were analyzed by means of a histogram over the five angles. A maximum-likelihood comparison was made to a binned Monte Carlo simulation that was acceptance corrected and constrained to have a total number of events equal to that in the experimental sample. Table 2 contains the results of the likelihood fit.

The multipolarities of the radiative transitions for the $\chi_{1,2}$ are thus found to be predominantly dipole. An earlier analysis (6) also found this to be the case for the χ_1, but only when its spin was assumed to be 1. The data from the Crystal Ball study yield high confidence levels for the spin and multipole values preferred in the standard charmonium models (2).

3.3 Results from the Full Analysis of the Inclusive Spectrum and Some Comparisons to Exclusive Results

In this section we focus on the results of a detailed study (29), using the Crystal Ball, of the radiative transitions from ψ' to the χ_J states. Measurements of the natural line widths of the χ_J states will also be discussed briefly. The results are derived from 1.8×10^6 ψ' hadronic decays

Table 2 Results of likelihood fit of data for $\psi' \to \gamma\gamma J/\psi \to \gamma\gamma \ell^+\ell^-$ to correlated angular distributions for various χ spin values[a]

Hypothesis	$-2\ln(L/L_{\max})$	a'_2	a_2
$\chi(3508)$ data:			
$J_\chi = 1$	0	$+(0.077^{+0.050}_{-0.045})$	$-(0.002^{+0.020}_{-0.008})$
$J_\chi = 2$	16		
$J_\chi = 0$	162		
$\chi(3554)$ data:			
$J_\chi = 2$	0	$+(0.132^{+0.098}_{-0.075})$	$-(0.333^{+0.292}_{-0.116})$
$J_\chi = 1$	20		
$J_\chi = 0$	40		

[a] The multipole amplitudes have been normalized so that $\sum_{j=1}^{J+1} |a_j|^2 = 1$, and, for spin 2, a_3 has been set to zero.

selected using criteria designed to reject cosmic rays, beam gas, and QED events. These criteria rejected all but a negligible part of the background while maintaining a 94% efficiency for the hadronic events.

The tracks from the hadronic events were selected for the inclusive photon analysis in four different ways. This was done to compare the effects of the different sets of cuts and the resulting different background shapes on the measured photon branching ratios and $\chi_{0,1,2}$ line widths. The following cumulative selection criteria were applied to the data to yield the four ψ' inclusive photon spectra shown in Figure 5:

(a) Removal of tracks with $|\cos \vartheta_j| > 0.85$ where ϑ_j is the angle of the track to the positron beam direction. This solid angle restriction ensures that each particle in the spectrum is a fiducial volume of the NaI(Tl) that has a uniform energy resolution and scale. Since both charged and neutral tracks are accepted into this spectrum, an enormous peak at about 200 MeV is observed corresponding to minimum ionizing charged particles passing through the detector. The peak presents a very large background, which dwarfs the χ_J lines. However, these lines are still highly significant and measurable.

(b) Removal of charged tracks using tracking chamber information. Most charged particles are removed by this cut, as is evidenced by the great reduction in the relative size of the peak at ~ 200 MeV; however, the persistence of a remnant bump at the minimum ionization energy indicates some small inefficiency in charged-particle identification.

(c) Removal of neutral tracks close to charged tracks, $\cos \vartheta_{ij} < 0.9$, and removal of neutral pairs that reconstruct to a π^0 mass. These last cuts

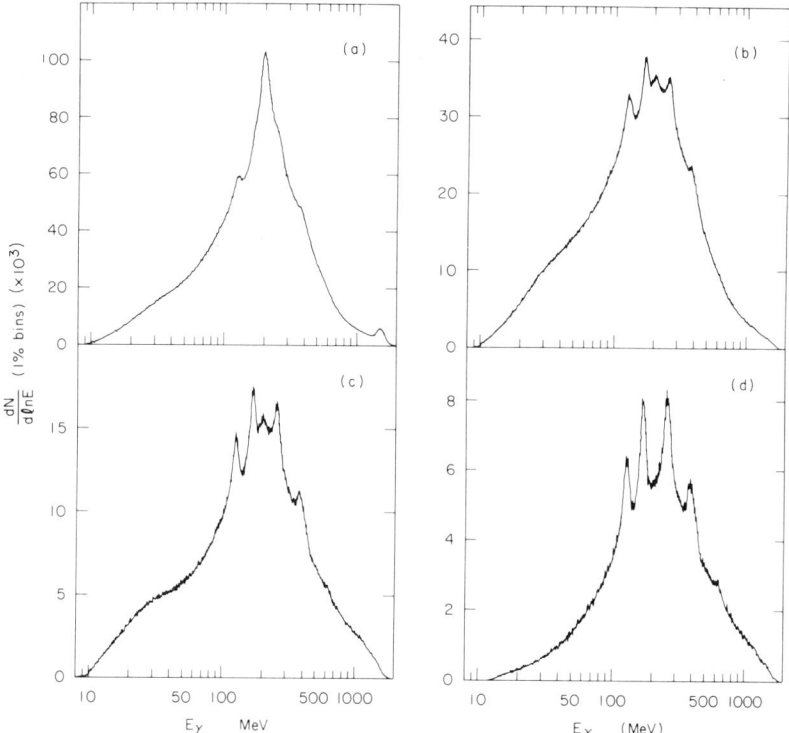

Figure 5 Inclusive γ spectra at the ψ' used in the measurement of $\psi' \to \gamma\chi_J$ and $\psi' \to \gamma\eta_c(2984)$. (a) All tracks neutral and charged with $|\cos \vartheta| < 0.85$. (b) Same as (a), except that tracks tagged as charged by the tracking chambers are removed. (c) Same as (b), except that photons resulting from reconstructed π^0 decays and those near interacting charged particles are removed. (d) Same as (c), except that each track is required to have a lateral energy deposition pattern consistent with that of an electromagnetically showering particle.

improve the signal-to-noise ratio by about a factor of two while reducing γ detection efficiency by about a factor of 0.7.

(d) Removal of tracks identified as minimum ionizing charged particles by their lateral energy deposition in the NaI(Tl) crystals. These charged particles were not rejected in (b) because of the charged-particle identification inefficiency of the tracking chambers. In this heavily cut spectrum, the minimum ionizing signal is negligible. The signal-to-noise ratio of the photon transitions has been maximized so that the $\psi' \to \gamma\eta_c$ transition is clearly visible at $E_\gamma \sim 640$ MeV (cf Section 5.1). Note that because of the fine (1%) binning of the data shown in histograms of Figure 5, the signal at $E_\gamma \sim 92$ MeV arising from the transitions $\psi' \to \gamma\eta_c'$ is not clearly visible (cf Section 5.3).

The signals corresponding to the χ_J radiative transitions were obtained from fits to the spectra of Figures 5 (29). The results from the fits are summarized in Figure 6 after corrections for photon detection efficiency, photon conversion probability, and the photon angular distributions arising from the different spins of the χ_J states have been made.

By comparing the branching ratios $B(\psi' \to \gamma\chi_J)$ extracted from the four spectra, one is able to assess the magnitude of the systematic errors contributing to the measurement. As is seen in Figure 6, the variation among the four branching ratio values for each line is consistent within the statistical errors of the measurements. The fact that consistent results are obtained with such widely different looking spectra gives one confidence in the finally extracted branching ratios.

A second check is the comparison of the cascade branching ratios $B(\psi' \to \gamma\chi_{1,2}) \cdot B(\chi_{1,2} \to \gamma J/\psi)$ as measured using the Doppler-broadened secondary transition lines seen in the inclusive photon spectra of Figure 5 with the values obtained from the exclusive events discussed in Section 3.2.

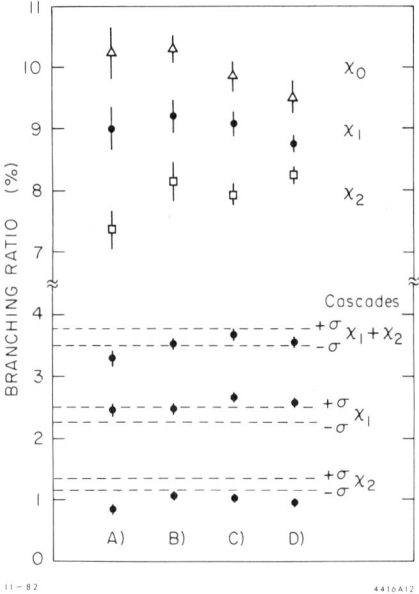

Figure 6 The upper part of the figure shows the observed values of $B(\psi' \to \gamma\chi_J)$ as obtained from independent analysis of each of the four spectra shown in Figure 5. The lower part compares the cascade product branching ratios $B(\psi' \to \gamma\chi_{1,2}) \cdot B(\chi_{1,2} \to \gamma J/\psi)$ from the four spectra (*dots*) with the direct measurements of these products from our analysis of the exclusive events $\psi' \to \gamma\gamma\ell^+\ell^-$ (*dashed bands*). Since separation of the overlapping lines from the two photons $\chi_1 \to \gamma J/\psi$ and $\chi_2 \to \gamma J/\psi$ in the inclusive spectra is difficult, the comparison with the sum is also shown.

The results of the cascade measurement are shown as the points on the bottom part of Figure 6. The dashed bands show the exclusive measurement of the same transitions given in Table 1. As is seen in this figure, the inclusive measurement for the χ_2 is somewhat lower than the exclusive measurement, while for χ_1 it is somewhat higher. However, the sum of the χ_1 and χ_2 branching ratios yields good agreement between the inclusive and exclusive measurements. This effect has also been reproduced in Monte Carlo calculations. It is due to the overlap of the two transitions in the inclusive spectra. That the sum of the inclusive lines is in good agreement with the sum of the exclusive measurements allows an uncertainty in the absolute normalization of the inclusive result of less than 16%, the absolute error in the exclusive measurement [remember that this error is dominated by the uncertainty in $B(J/\psi \to \ell^+\ell^-)$].

The final results of the analysis, including branching ratios and the values for the natural line widths of the χ_J states, are shown in Table 3. For the branching ratios the first error is dominated by the statistical uncertainty and point-to-point errors in the photon detection efficiency. The second error is an estimate of the overall normalization error due mainly to a $\pm 5\%$ uncertainty in both the hadronic event selection efficiency and the overall photon detection efficiency.

Agreement between these Crystal Ball branching ratio measurements and those of the lower statistics experiment of Biddick et al (12) is within the experimental errors. However, our branching ratios to the χ_J states are consistently higher, and within the point-to-point errors of our measurements there is an indication for an increase in rate from χ_2 to χ_0 transitions. In nonrelativistic models, $\Gamma(\psi' \to \chi_J) \propto (2J+1)E_\gamma^3$, and thus we expect $\Gamma'_0 : \Gamma'_1 : \Gamma'_2 = 1:1:1$ where $\Gamma'_J = \Gamma(\psi' \to \gamma\chi_J)/[(2J+1)E_\gamma^3(\chi_J)]$. As shown in Table 3, we obtain $1 : 1.07 \pm 0.08 : 1.39 \pm 0.11$, in reasonable agreement

Table 3 Results from the $\psi' \to \gamma\chi_J$ [a]

Datum	χ_0	χ_1	χ_2
E_γ (MeV)	$258.4 \pm 0.4 \pm 4$	$169.6 \pm 0.3 \pm 4$	$126.0 \pm 0.2 \pm 4$
$\Gamma(\chi_J)$ (MeV)	(13.5–20.4)	<3.8	(0.85–4.9)
$B(\psi' \to \gamma\chi_J)$ (%)	$9.9 \pm 0.5 \pm 0.8$	$9.0 \pm 0.5 \pm 0.7$	$8.0 \pm 0.5 \pm 0.7$
Ratio $\left\{\dfrac{B(\psi' \to \gamma\chi_J)}{E_\gamma^3(2J+1)}\right\}$	1:	1.07 ± 0.08 :	1.39 ± 0.11
$B(\chi_J \to \gamma J/\psi)$ (%)	0.60 ± 0.17	28.4 ± 2.1	12.4 ± 1.5

[a] When two errors are given, the first error is statistical and the second is systematic. Ranges and upper limits are at 90% confidence levels.

with the simple theory. However, our absolute branching ratios are lower by a factor of two to three than the predictions of the simple nonrelativistic charmonium models (29). Models that include relativistic corrections, variations of the 2S and ^3P_J wave function shapes resulting from higher order corrections, and coupled channels achieve better agreement with the data.

The measurement of the natural line widths of the χ_J states is a tricky one since the Crystal Ball's photon energy resolution is comparable to or greater than these widths. It does appear, however, that the χ_0 is much broader than predicted by QCD, while the χ_1 and χ_2 widths are in good agreement with QCD within errors (29, 30).

4. HADRONIC TRANSITIONS FROM THE ψ' TO THE J/ψ

Figure 1b shows the hadronic transitions that have been observed between the ψ' and J/ψ. All of these transitions were observed by at least two experiments and the $\pi\pi$ and η transitions have been observed by many experiments. As the $\pi\pi$ transitions can easily be observed in the charge mode $[B(\psi' \to \pi^+\pi^- J/\psi) = 33 \pm 2\%$ (27)], excellent measurements of this mode have been made by other detectors stressing charged-particle detection. The Crystal Ball has measured the neutral $\pi^0\pi^0$ mode (31), as a check on measurements of the η and π^0 transitions. Comparison of the neutral $\pi^0\pi^0$ to the charged (32) $\pi^+\pi^-$ mass distributions show the shapes of the two distributions to be the same within error, as is expected from isospin symmetry.

4.1 The Transitions $\psi' \to \eta(\pi^0)J/\psi$

The study of these processes is related to that of the ^3P_J state cascades and so is detailed in (17) (cf Section 3.2). The $m_{\gamma\gamma}$ distribution for all fitted events is shown in Figure 7a. Of the events in this figure, 412 candidates for the η events are separated from χ_J and π^0 events by using the cut $m_{\gamma\gamma} > 525$ MeV/c^2. This cut loses no η events, but does admit some χ_1 events into the η sample.

Monte Carlo calculations determined that 21 χ_1 events and 5 $\pi^0\pi^0$ events are expected in the η sample. The resulting η mode branching ratio is compared in Table 1 with other measurements. The Crystal Ball and Mark II results (26) are in good agreement, while the other measurements shown are larger than our measurement by about a factor of two.

Existence of the transition $\psi' \to \pi^0 J/\psi$ is apparent in the Dalitz plots of

Figures 4a,b. A π^0 signal is observed in the diphoton mass plot by removing the dominant background from cascade photons using a cut on the $\gamma J/\psi$ masses. A subtraction of events from the $m_{\gamma\gamma}$ plot of Figure 7a with $(M_{\gamma J/\psi})_{\text{high}}$ in the ranges 3410 ± 5 and 3530 ± 60 MeV/c^2, and $m_{\gamma\gamma} > 525$ MeV/c^2, results in the distribution shown in Figure 7b. These data have been fitted to a Gaussian peak with a quadratic background distribution. The fit yields 23 events above background having $m_{\gamma\gamma} < 200$ MeV/c^2. The resulting π^0 mode branching ratio is compared in Table 1 with another measurement from the Mark II (26). The two measurements are in good agreement. This decay violates isospin symmetry. A review of the theoretical literature relevant to our measurement can be found in (17).

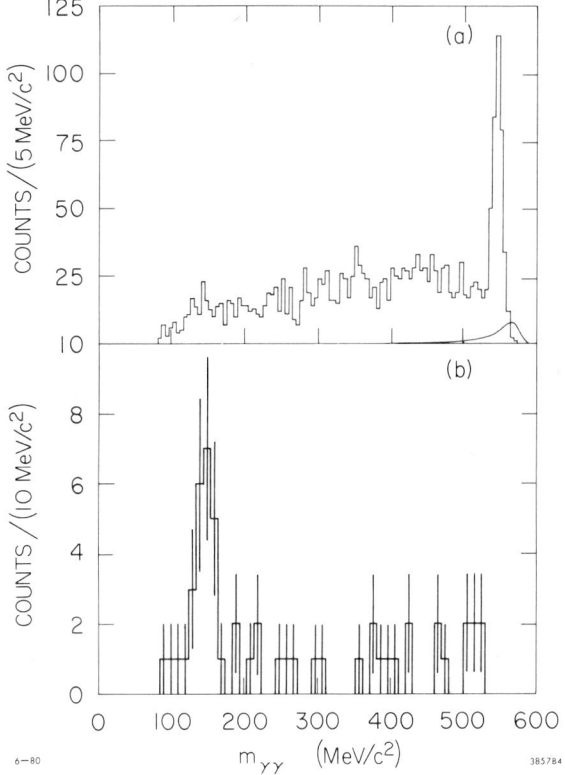

Figure 7 Diphoton masses of fitted events for $\psi' \to \gamma\gamma J/\psi \to \gamma\gamma \ell^+\ell^-$. (a) The peak due to the η; the smooth line is a 10-times-magnified calculated curve for the expected contamination from $\psi' \to \pi^0\pi^0 J/\psi$. (b) The same as (a) except that events consistent with $\psi' \to \eta J/\psi$, and $\psi' \to \gamma\chi_J$ have been removed. These cuts allow the π^0 peak to show clearly.

5. THE CHARMONIUM $^1S_0$ STATES

The Crystal Ball discoveries that created the most excitement were the lack of a signal in $J/\psi \to \gamma\gamma\gamma$ at $M_{\gamma\gamma} = M_{X(2830)}$ (15), which had been reported by the DASP collaboration (10, 11) (cf Section 6.4), and the discovery of an η_c candidates state at $M_{\eta_c} = 2984 \pm 4$ MeV/c^2 by means of the radiative transitions from the ψ' (33) and J/ψ (34). Since the original observations were made, the Crystal Ball has doubled both the ψ' and J/ψ data sets to about 2×10^6 hadronic decays each. This increase in data has allowed a more precise determination of the η_c parameters (29). Furthermore, it has also resulted in the discovery of an η_c' candidate at $M_{\eta_c'} = 3592 \pm 5$ MeV/c^2 via a radiative transition from the ψ'.

The two states at 2984 and 3592 MeV/c^2 can be naturally associated with the 1^1S_0 and 2^1S_0 charmonium states, the η_c and η_c'. As we see in this section, their properties fall within the range of theoretical expectations. Thus, with the work described in previous sections we have come from a state of relative confusion and uncertainty concerning the validity of charmonium as a model of the J/ψ system to one of good agreement between theory and experiment in most cases.

5.1 Evidence for the 1^1S_0 in Inclusive γ Spectra of the ψ' and the J/ψ

The analysis of the inclusive photon spectra from the 1.8×10^6 ψ' and 2.2×10^6 J/ψ decays when studying the $\eta_c(2984)$ is very similar to that described in Section 3.3 and is detailed in (29). However, not only were the four spectra from the ψ' shown in Figure 5 and the corresponding four from the J/ψ (not shown) used, but a fifth spectrum from both the J/ψ and ψ' was also included in the analysis. The pattern cuts for this fifth spectrum were designed to improve the efficiency for detection of low energy photons, at the expense of reduced efficiency for removing minimum ionizing charged particles. It is shown for the J/ψ in Figure 8. The inserts on the upper left of Figures 3 and 8 show the result of one of the simultaneous fits made to correspondingly cut J/ψ and ψ' inclusive photon spectra. For the radiative transition to the η_c, the η_c mass and width are constrained to be the same for both spectra. The results of the fits to each of the five pairs of spectra were compared as a consistency check. An additional check was made by measuring the mass and width in the γ spectrum coming from events containing exactly two observed charged particles.

The results of this analysis (29) are $M_{\eta_c} = 2984 \pm 5$ MeV/c^2, $\Gamma_{\eta_c} = 11.5^{+4.5}_{-4.0}$ MeV, $B(J/\psi \to \gamma\eta_c) = (1.27 \pm 0.36)\%$, and $B(\psi' \to \gamma\eta_c) = (0.28 \pm 0.06)\%$. The errors are dominated by the statistical uncertainties, except for the mass error, which is mainly due to the uncertainty in our absolute

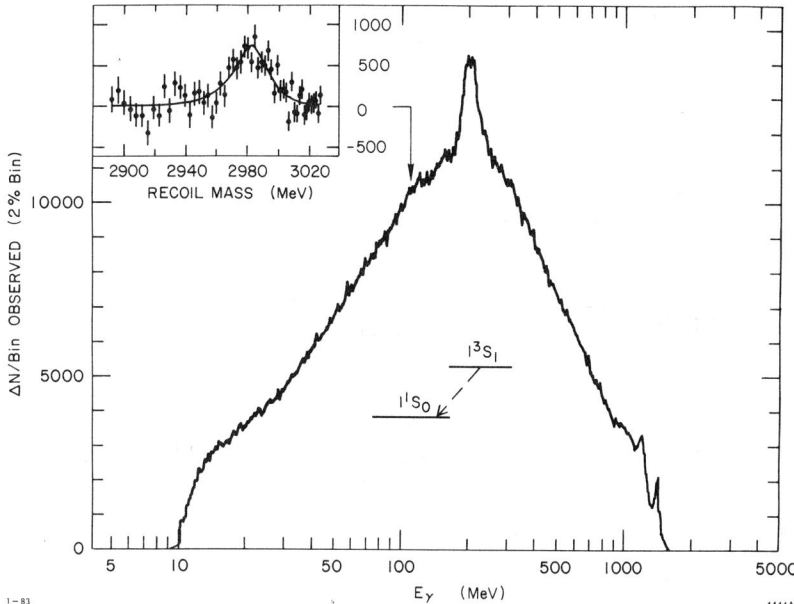

Figure 8 Inclusive γ spectrum at the J/ψ, $\Delta N/\Delta(\log E) \cong E \, dN/dE$. The strong peak at 200 MeV is due to charged particles that were not tagged by the tracking system. The inset shows the background-subtracted signal from the η_c candidate state. The two prominent peaks near the high energy end point of the spectrum are from the monochromatic photons in the reactions $\psi \to \gamma \eta'$ and $\psi \to \gamma \iota(1440)$. The photons from $\psi \to \gamma \eta \, (\pi^0)$ were eliminated from this spectrum by a hard QED cut.

energy calibration. These values are in good agreement, within errors, with previously reported Crystal Ball values (34).

5.2 *Hadronic Decays of the 1^1S_0*

Confirmation of the $\eta_c(2984)$ came soon after its discovery in the ψ' and J/ψ inclusive γ spectra with the first reports of the observation of its hadronic decays by the Mark II collaboration at SPEAR (35). A number of decay modes were seen, as is shown in Table 4.

We also have looked for exclusive decays of the $\eta_c(2984)$ into hadrons by performing kinematic fits to exclusive final states with multiple photons and two charged hadrons (34, 36). Remember that the Crystal Ball measures both the energy and angle of electromagnetically showering particles but for charged hadrons (π, K) it measures only the angles well. Secondary interactions of the charged hadrons in the sodium iodide complicate the fitting of some events, but special pattern recognition algorithms have been developed to deal with this effect.

Table 4 η_c branching ratio measurements

	Mark II (35)	
Decay mode	$B(\psi' \to \gamma\eta_c) \cdot B(\eta_c \to X)$	$B(\eta_c \to X)^a$
$p\bar{p}$	$(8^{+8}_{-4}) \times 10^{-6}$	$(2.9^{+3.0}_{-1.6}) \times 10^{-3}$
$\pi^+\pi^-\pi^+\pi^-$	$(5.7^{+3.9}_{-2.4}) \times 10^{-5}$	$(2.0^{+1.5}_{-0.9}) \times 10^{-2}$
$\pi^+\pi^-K^+K^-$	$(4.0^{+6.0}_{-2.5}) \times 10^{-5}$	$(1.4^{+2.1}_{-0.9}) \times 10^{-2}$
$\pi^+\pi^-p\bar{p}$	$<5 \times 10^{-5}$ (90% C.L.)	$<2.3 \times 10^{-2}$ ($\sim$90% C.L.)
$K^\pm\pi^\mp K^0_s$	$(1.5^{+0.8}_{-0.6}) \times 10^{-4}$	$(5.4^{+3.3}_{-2.4}) \times 10^{-2}$

	Crystal Ball	
Decay mode	$B(J/\psi \to \gamma\eta_c) \cdot B(\eta_c \to X)$	$B(\eta_c \to X)^b$
$\eta\pi^+\pi^-$	$(3.1 \pm 1.9) \times 10^{-4}$	$(2.6^{+1.8}_{-1.7}) \times 10^{-2}$
$\gamma\gamma^c$	$<1.6 \times 10^{-5}$ (90% C.L.)	$<1.8 \times 10^{-3}$ ($\sim$90% C.L.)
$K^+K^-\pi^0$	$<1.5 \times 10^{-4}$ (90% C.L.)	$<1.7 \times 10^{-2}$ ($\sim$90% C.L.)

[a] Uses Crystal Ball value for $B(\psi' \to \gamma\eta_c)$.
[b] Uses Crystal Ball value for $B(J/\psi \to \gamma\eta_c)$.
[c] See Section 6.4.

Events with a three-photon, two-charged-particle topology were selected from the sample of J/ψ hadronic decays and subjected to a 3-C kinematic fit to the hypotheses, $J/\psi \to \gamma\eta\pi^+\pi^-$ and $\gamma\eta K^+K^-$, $\eta \to \gamma\gamma$. The energy spectrum for the low energy radiated photon, arising from events that have a χ^2 greater than 0.1 for the $\eta\pi^+\pi^-$ hypothesis, showed a significant signal above background at the $\eta_c(2984)$ mass, within errors. No comparable signal was seen for the ηK^+K^- hypothesis.

The $\gamma\eta\pi^+\pi^-$ data, in principle, contain additional information on the width of the $\eta_c(2984)$. However, given the limited statistics of this measurement, which comes from only half the presently available J/ψ data, we believe the inclusive measurement of the width to be more reliable at this time. From the signal of 18 ± 6 events, we obtain the product branching ratio $B(J/\psi \to \gamma\eta_c) \cdot B(\eta_c \to \eta\pi\pi)$, and branching ratio $B(\eta_c \to \eta\pi\pi)$ given in Table 4. In addition, to compare directly with the Mark II observation of the $K^\pm K^0_s \pi^\mp$ final state of the $\eta_c(2984)$, our upper limit for the $K^+K^-\pi^0$ final state is also given. Note that the Crystal Ball value must be doubled before comparing with the Mark II result, owing to isospin; we assume $I = 0$ for $\eta_c(2984)$. Also, for completeness, the $\gamma\gamma$ final-state branching ratio is given here (cf Section 6.4).

5.3 Evidence for the 2^1S_0 State in the Inclusive γ Spectrum of the ψ'

As mentioned in Section 5, a candidate for the 2^1S_0 state or η'_c has been found by the Crystal Ball using inclusive photon decays of the ψ'. In this section we briefly describe our evidence for the state. A more complete description can be found in (37).

The event selection for the analysis as well as the photon selection criteria used are the same as those described in Section 3.3 with two minor changes. First, events with more than 10 charged or more than 10 neutral observed tracks are not considered. Second, a somewhat different lateral shower energy deposition pattern in the NaI(Tl) crystals is used to define photons than in the analysis described in Section 3.3 and 5.1. In this case, an extra premium was placed on good efficiency for $E_\gamma < 100$ MeV. The main spectrum of Figure 3 results from about half the 1.8×10^6 ψ' decays, cut as described in Section 3.1. A signal at 3592 ± 5 MeV is evident in this spectrum. The insert on the upper left of Figure 3 shows the result of performing a fit to the region containing the structure at $E_\gamma \sim 90$ MeV; this insert contains a spectrum obtained from all 1.8×10^6 ψ' decays. A clear signal is obtained with 4.4σ to 6σ significance, depending on how the fit is performed. The properties obtained from the fit for the η_c' candidate state are: $M_{\eta_c'} = 3592 \pm 5$ MeV/c^2, $\Gamma_{\eta_c'} < 8$ MeV (95% C.L.), and $B(\psi' \to \gamma\eta_c')$, in the range 0.2% to 1.3% with a confidence level of 95%. The confidence interval for the uncertainty in the branching ratio includes the correlation with $\Gamma_{\eta_c'}$.

It should be noted that the DESY-Heidelberg group reported evidence (14) for a state at a mass of 3592 ± 7 MeV/c^2 in the exclusive channel $\psi' \to \gamma\gamma J/\psi$, $J/\psi \to \mu^+\mu^-$. However, as reviewed in Section 3.2, we have looked for evidence of such a state in the cascade decays and found none. If we assume that the object we observe in the inclusive spectrum is the η_c', then it is expected (37) that $B(\psi' \to \gamma\eta_c') \cdot B(\eta_c' \to \gamma J/\psi) < 10^{-6}$. This estimate is based on our measured value of $B(\psi' \to \gamma\eta_c')$ and on theoretical calculations (7) for the η_c' total width and radiative transition rate. The estimate for the hindered magnetic dipole transition $\eta_c' \to \gamma J/\psi$ was based upon our measurement of the similar transition $\psi' \to \gamma\eta_c$, which reduces the estimate's sensitivity to the details of the wave functions. Such a small product of branching ratios has not been accessible to any experiment.

5.4 Discussion

Clear signals have been seen for states at $M = 2984 \pm 5$ MeV/c^2, and $M = 3592 \pm 5$ MeV/c^2 by the Crystal Ball detector; the Mark II has confirmed the state at 2984. These states are obvious candidates for the $1^1S_0(\eta_c)$ and $2^1S_0(\eta_c')$ states of charmonium. What evidence makes these tentative assignments plausible?

First, the η_c is seen to decay into three pseudoscalars and not two. This allows only $0^-, 1^+, \ldots$ assignments for the J^P of the state. As discussed in (3), the measured radiative transition branching ratio $B(J/\psi \to \gamma\eta_c)$ is in reasonable agreement with both the nonrelativistic and QCD sum rule calculations, which assume $J^{PC} = 0^{-+}$ for the observed state. In addition, the branching ratio $B(\psi' \to \gamma\eta_c)$ was also predicted by theory (38) and these predictions are in agreement with the observation. The mass splitting 1^3S_1–1^0S_1 is predicted by a number of theories, including nonrelativistic models and QCD sum rule calculations (38, 39) and these are again in good agreement with the data. Finally, the width of the η_c is predicted to be 8.3 ± 0.5 MeV using QCD with higher order corrections (40) and the experimental value of $\Gamma_{\eta_c} = 11.5^{+4.5}_{-4.0}$ MeV agrees within the error. The important partial width $\Gamma(\eta_c \to \gamma\gamma)$ which has been predicted to be 4.2 ± 0.4 keV using the QCD sum rules (41) is well below the Crystal Ball upper limit of $\Gamma(\eta_c \to \gamma\gamma) < 20$ keV (90% C.L.) (cf Section 6.4). One can thus conclude with some certainty, given the above evidence, that the Crystal Ball state at 2984 ± 4 MeV/c^2 is truly the 1^1S_0 of charmonium.

Unfortunately, the case is not so clear for the η_c' candidate at $M = 3592 \pm 5$ MeV/c^2. Relatively little is known about this state. No exclusive decays have been seen, and only an upper limit exists on its width. Within the limits of uncertainty concerning 2^3S_1–3^3D_1 mixing (42), the agreement between theory and experiment is reasonable for the ψ'–η_c' mass splitting. Also, the observed value for $B(\psi' \to \gamma\eta_c')$ agrees with nonrelativistic model calculations within the large range allowed by observations. However, confirmation and more information are needed on this state before a firm connection to the 2^1S_0 state of charmonium can be made.

6. RADIATIVE TRANSITIONS FROM THE J/ψ

Other than for J/$\psi \to \gamma\eta_c$, $\eta_c \to$ hadrons, interest in radiative decays of the J/ψ first centered on 3γ decays, e.g. $\gamma\eta(\gamma\gamma)$ or $\gamma\eta'(\gamma\gamma)$, and particularly on searches for $\gamma\eta_c(\gamma\gamma)$. However, in recent years the possibility of observing gluonic meson states, particles made up entirely of gluons, has also stimulated much interest in J/ψ radiative decays.

6.1 *The Fundamental Character of Gluonic Mesons in QCD*

The existence of an extensive spectrum of colorless, flavorless bound states of two or more gluons has been firmly predicted by QCD (3). These gluonic bound states were given the name "gluonic mesons" by their inventors, H. Fritzsch and M. Gell-Mann (43). It is expected that the lower mass gluonic meson states are bound states of mostly two gluons; in analogy to quarkonium, a bound state of a quark and antiquark, these systems are called gluonium. It is also expected that, because of their relatively lower masses predicted to lie in the range of 1 to 2 GeV, gluonium states should be by far easier to observe than the higher mass gluonic mesons. Although the existence of gluonium has not yet been experimentally

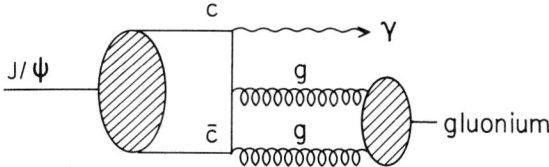

Figure 9 Lowest order QCD diagram for the radiative decay of the J/ψ into a gluonic meson.

established, the interest in this new form of matter has increased considerably since the observation of two new mesons, the $\iota(1440)$ (44, 45) and the $\vartheta(1640)$ (46). These are seen in a reaction thought to be a copious source of gluonic meson states (47), namely, $J/\psi \to \gamma x$. The mechanism is shown diagramatically in Figure 9. According to lowest order QCD calculations, the hadronic decays of quarkonium $^3S_1$ states, such as the J/ψ, proceed mainly via annihilation of the $q\bar{q}$ system into three gluons. Although this process might seem well suited to the production of gluonium states, it is not since each pair of the three final-state gluons must be in a color octet state. This follows from the fact that the overall state must be a color singlet and each pair recoils against one color octet gluon. However, if a photon is radiated with two gluons in the decay, as shown in Figure 9, the recoiling gluon pair must form a color singlet state, which is even under charge conjugation.

Perturbative QCD indicates (48, 49) a partial width for the process $J/\psi \to \gamma gg$ of about 8 keV, which is relatively large. Various authors (47) have used duality principles, and other ideas, together with the perturbative result to show that gluonium states should be copiously produced in this process. However, the experimental search for such states has proven to be a difficult and confusing one with a number of guiding theoretical principles losing credibility as the field has matured (3).

6.2 *The "End Point" of the Inclusive γ Spectrum at the J/ψ*

One can qualitatively appreciate the major features of the radiative decays of the J/ψ by viewing the "end point" of the inclusive γ spectrum as measured by the Crystal Ball detector (42), and shown in Figure 10.

Relatively narrow peaks at the $\iota(1440)$ and $\eta'(958)$ are evident, and there is also a broad structure centered at a recoil mass of about 1700 MeV/c^2 (the ϑ has a mass close to 1700 MeV/c^2 but it is not as broad as the structure seen in the inclusive spectrum). The tails of the $\iota(1440)$ structure include the regions where radiative transitions to the f(1270), D(1285), and f'(1515) would appear.

Transitions to the $\eta(549)$ should also be seen, but these are suppressed in this spectrum by the event selection cuts (42). Likewise, even if the $J/\psi \to \gamma\pi^0$

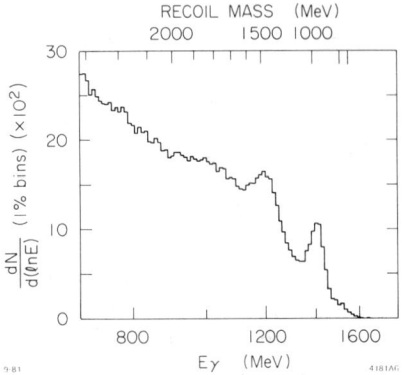

Figure 10 The same as Figure 8 showing details near the high energy end point of the spectrum.

rate were large, no signal would be seen because of these cuts. Up to the time of this writing, very little quantitative analysis of the spectrum in Figure 10 has been done, and so only the above qualitative information can be drawn from it. A strong analysis effort started recently in the Crystal Ball collaboration will, we hope, remedy this situation.

6.3 *Gamma Transitions to Well-Known Particles Using Exclusive Decays*

Experimental measurements have been reported by the Crystal Ball collaboration for the processes $J/\psi \to \gamma\pi^0, \gamma\eta, \gamma\eta', \gamma f$, and an estimate for $\gamma f'$, $f' \to \eta\eta$. Crystal Ball measurements for the η and η' have been published (15); however, new measurements by the Crystal Ball collaboration (50) derived from the full data sample of 2×10^6 J/ψ decays, about twice the data of the previously published results, disagree somewhat with the older measurements. The new measurements are in agreement within the errors with three measurements from other experiments (10, 50a, 51).

Table 5 shows the most recent Crystal Ball results for the various decays. Note that the new results on the η' use the $\eta\pi^+\pi^-$, $\eta\pi^0\pi^0$, and $\gamma\rho^0$ decay modes as well as the $\gamma\gamma$ decay mode, which was the only one used in the old result. The Dalitz plot for $J/\psi \to 3\gamma$ from all our data is shown in Figure 11. Prominent signals are seen for the η and η'. No signal is seen at $M_X = 2830$ MeV/c^2 and $M_{\eta_c} = 2984$ MeV/c^2. Upper limits for these processes are given in Table 5. The direct decay $J/\psi \to 3\gamma$ as well as the QED process $e^+e^- \to 3\gamma$ also contribute to the Dalitz plot.

As an example of the analysis of an hadronic final state of the η', Figure 12 shows the signal for $J/\psi \to \gamma\eta', \eta' \to \gamma\rho, \rho \to \pi^+\pi^-$. These events satisfy a 2-C

Table 5 Crystal Ball measurements (except as noted) of $J/\psi \to \gamma\pi^0, \gamma\eta, \gamma\eta', \gamma f, \gamma f'$ [a]

$B(J/\psi \to \gamma\pi^0) = (3.6 \pm 1.1 \pm 0.7) \times 10^{-5}$
$B(J/\psi \to \gamma\eta) = (0.88 \pm 0.08 \pm 0.11) \times 10^{-3}$

η' decay mode	$B(J/\psi \to \gamma\eta') \times 10^{-3}$
$\eta' \to \eta\pi^+\pi^-$	$3.9 \pm 1.0 \pm 1.1$
$\eta' \to \eta\pi^0\pi^0$	$4.2 \pm 0.6 \pm 0.6$
$\eta' \to \gamma\rho^0$	$4.1 \pm 0.4 \pm 0.6$
$\eta' \to \gamma\gamma$	$4.4 \pm 0.9 \pm 0.5$
Average	$4.1 \pm 0.3 \pm 0.6$

$B(J/\psi \to \gamma f) = (1.48 \pm 0.25 \pm 0.30) \times 10^{-3}$
$B(J/\psi \to \gamma f') \cdot B(f' \to \eta\eta) = (0.9 \pm 0.9) \times 10^{-4}$
$B(J/\psi \to \gamma f') \cdot B(f' \to K\bar{K})$[b] $= (1.8 \pm 1.0) \times 10^{-4}$
$B(J/\psi \to \gamma X) \cdot B(X \to 2\gamma) < 1.6 \times 10^{-5}$ (90% C.L.),
for $2600 < M_X < 3000$ MeV/c^2 and $\Gamma_X \lesssim 25$ MeV
$B[J/\psi \to 3\gamma \text{ (direct)}] < 5.5 \times 10^{-5}$ (90% C.L.)

---

[a] Where two errors are given, the first is statistical and the second systematic.
[b] Mark II (64).

fit to the hypothesis $\gamma\gamma\pi^+\pi^-$. They also were subjected to several more constraints:

1. The high energy neutral track was required to satisfy a lateral energy deposition in the NaI(Tl) crystals expected of a high energy photon, rather than two photons from a high energy π^0.
2. Photon pairs forming a π^0 or an η were excluded.
3. The energy of the charged particles had to be less than 1360 MeV.
4. The $\pi\pi$ mass was cut about the ρ mass.

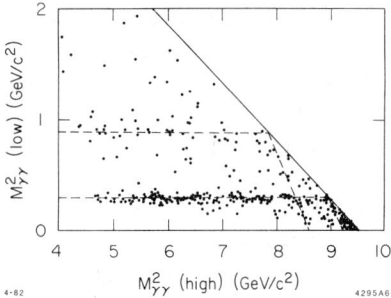

Figure 11 Dalitz plot for $J/\psi \to \gamma\gamma\gamma$. The two sets of dashed lines indicate where events from $J/\psi \to \gamma\eta$ and $J/\psi \to \gamma\eta'$ should be clustered. Except for a QED background, no other signals are seen.

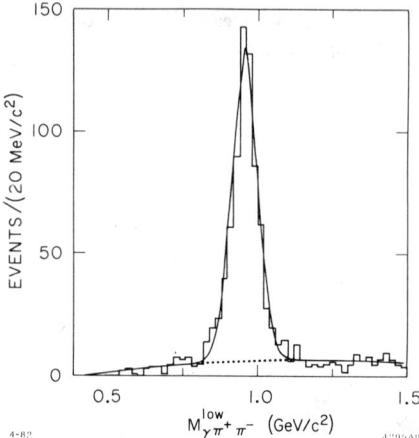

Figure 12 Distribution of the $\gamma_1\pi^+\pi^-$ mass from events satisfying the hypothesis $J/\psi \to \gamma_2\gamma_1\pi^+\pi^-$ where $E_{\gamma_1} < E_{\gamma_2}$ and $m_{\pi\pi}$ has been cut about the ρ^0 mass. The solid curve shows a fit to an η' peak plus a smooth background (*dotted*).

These requirements removed the strong $J/\psi \to \pi\rho$, and $J/\psi \to \gamma\eta'$, $\eta' \to \eta\pi^+\pi^-$ backgrounds. A Monte Carlo calculation gives an efficiency of 24% for the 666 events found and displayed in Figure 12. The ratio, $B(J/\psi \to \gamma\eta')/B(J/\psi \to \gamma\eta)$ has been of theoretical interest with the QCD sum rules (52), and other models (52), yielding values in the range from 3.7 to 4.0. In order to further compare data to theory we calculate from our data, $B(J/\psi \to \gamma\eta')/B(J/\psi \to \gamma\eta) = 4.7 \pm 0.6$.

Table 5 also shows a new Crystal Ball result for $J/\psi \to \gamma\pi^0$ (50). This result is in good agreement with the only other measurement of this quantity by DASP (10).

Although the process $J/\psi \to \gamma f(1270)$ has been well studied in other experiments (53), the analysis of this process in the Crystal Ball (54) provides a useful check on the analysis techniques employed in the ι and ϑ studies (cf Section 6.4). It also provides a check that the Crystal Ball efficiencies are well understood in this complex $\gamma\pi^0\pi^0$ (5γ) final state. In addition, our measurement provides confirmation of previous results, and has also yielded the most precise determination available of the helicity amplitudes for the process $J/\psi \to \gamma f$. Figure 13 shows the $\pi^0\pi^0$ invariant mass distribution; a prominent f(1270) signal is seen with 178 ± 30 events. Figure 14 shows contours of equal probability as a function of x and y, $x = A_1/A_0$ and $y = A_2/A_0$, where A_0, A_1, and A_2 are the f helicity amplitudes (55). The errors on our measurement are small enough that a quantitative comparison with theory can be made. Theoretical predictions for pure M2 and E3 transitions (E1 is offscale), QCD (56), and tensor meson dominance

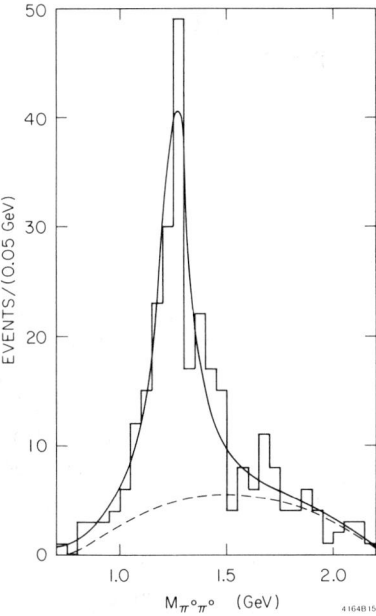

Figure 13 Distribution of the $\pi^0\pi^0$ mass from events that satisfy the 4-C fits to the hypothesis $J/\psi \to \gamma\pi^0\pi^0$. The solid curve shows a fit to an f peak plus a smooth background. The dashed curve represents the background contribution.

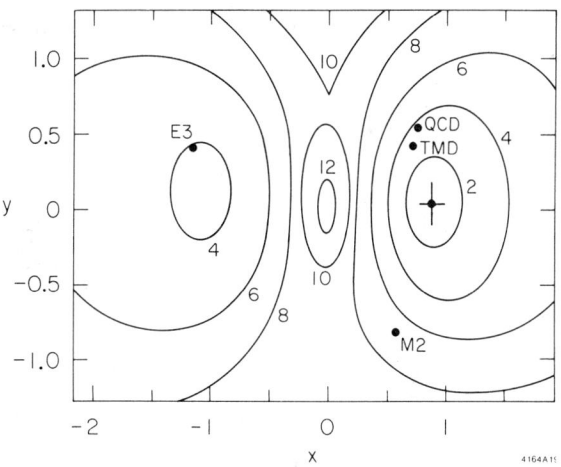

Figure 14 Contours of equal probability as a function of x and y, the ratios of helicity amplitudes in the decay of the f in $J/\psi \to \gamma f \to \gamma\pi^0\pi^0$. The data point with error bars represents the measurement and the others are theoretical predictions (see text). Numbers next to the curves are in units of standard deviations.

(TMD) (57) are also shown in Figure 14. All of these predictions are inconsistent with the experimental measurement. In particular, the QCD calculation based on two-gluon exchange (56) is more than three standard deviations from the experimental point.

As discussed in the next section, a by-product of the $\vartheta(1640)$ study has been a rough measurement of $B(J/\psi \to \gamma f') \cdot B(f' \to \eta\eta)$. This result is listed in Table 5, along with the Mark II measurement of $B(J/\psi \to \gamma f') \cdot B(f' \to K\bar{K})$.

6.4 The Gluonium Candidates, $\iota(1440)$ and $\vartheta(1640)$

A state at 1440 MeV/c^2 was first seen in the reaction, $J/\psi \to \gamma K^{\pm} K_s^0 \pi^{\mp}$, by the Mark II collaboration at SPEAR (44). They tentatively identified it as E(1420), a state with $J^{PC} = 1^{++}$ because their experiment was unable to determine the J^P value. The existence of this state was soon confirmed by the Crystal Ball collaboration at SPEAR (36) using the reaction, $J/\psi \to \gamma K^+ K^- \pi^0$. However, much more J/ψ data was needed (2.2×10^6 decays in total) before the Crystal Ball collaboration was able to measure the J^P of the state as 0^- (45).

This 0^{-+} state may have been previously observed in $p\bar{p}$ annihilations (58). The state seen in the $p\bar{p}$ case was named E(1420). However, the 0^{-+} assignment from that experiment was not considered conclusive (59) and so the name "E(1420)" was subsequently assigned to the $J^{PC} = 1^{++}$ state seen in $\pi^- p$ interactions (60). Thus, the Crystal Ball and Mark II experiments (in collaboration) have named the 0^{-+} state seen in J/ψ radiative decays the $\iota(1440)$ (44).

Figure 15a shows the $K^+ K^- \pi^0$ invariant mass distribution for events

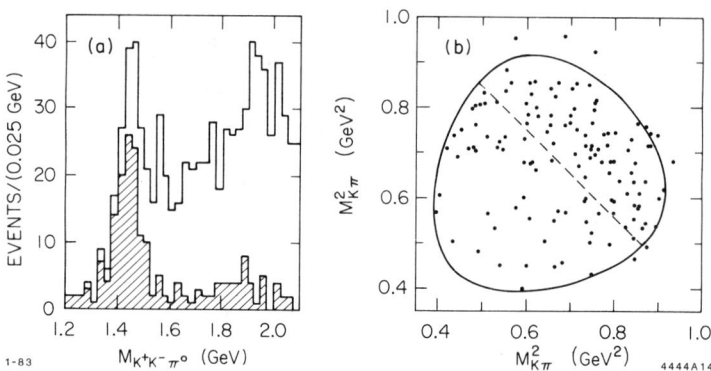

Figure 15 (a) Distribution of the $K^+ K^- \pi^0$ mass from events consistent with the hypothesis $J/\psi \to \gamma K^+ K^- \pi^0$. The events in the shaded region satisfy the further requirement $M_{K^+ K^-} < 1125$ MeV/c^2. (b) Dalitz plot for $K^+ K^- \pi^0$ events from $J/\psi \to \gamma K^+ K^- \pi^0$ with $1400 < M_{K\bar{K}\pi} < 1500$ MeV/c^2. The solid curve shows the boundary for $M_{K\bar{K}\pi} = 1450$ MeV/c^2 and the dashed line shows $M_{K\bar{K}} = 1125$ MeV/c^2.

that satisfy 3-C fits to the process $J/\psi \rightarrow \gamma K^+K^-\pi^0$. This analysis is based on 2.2×10^6 produced J/ψ events.

The $K\bar{K}\pi$ Dalitz plot from the Crystal Ball is shown in Figure 15b. A low $K\bar{K}$ mass enhancement (in the upper right corner of the plot) is evident. This enhancement has been associated with the $\delta(980)\pi$ decay of the resonance. No evidence for K^* bands, which would indicate a $K^*\bar{K}$+c.c. decay, is observed, although the situation is potentially confusing because of the limited phase space available for the decay and the fact that the K^* bands overlap in the region of the δ. The Mark II results are consistent with this. They find the ι to decay primarily into $\delta\pi$.

Before discussing the Crystal Ball spin analysis of the $\iota(1440)$, we review the status of the E(1420). The best estimate of the mass (27) is $M_E = 1418 \pm 10$ MeV/c^2. This is somewhat lower than, but not inconsistent with, the average of the Mark II and Crystal Ball measurements of the ι mass, $M_\iota = 1440 \pm 10$ MeV/c^2. The widths of the E($\Gamma_E = 50 \pm 10$ MeV) and the $\iota(\Gamma_\iota = 55 \pm 20$ MeV) are also consistent. Thus the mass and width measurements of the ι do not clearly identify it as a state different from the E.

As mentioned previously, the spin of the E was established in an experiment that analyzed the reaction $\pi^-p \rightarrow K_sK^\pm\pi^\mp n$ at 3.95 GeV/c (60). The results of a partial-wave analysis of the $K\bar{K}\pi$ system determined $J^{PC} = 1^{++}$ for the E, thus making it the SU(3) nonet partner of the D(1285) and the A_1. An additional result of the partial-wave analysis of Dionisi et al (60), is that the E decays primarily into $K^*\bar{K}$+c.c. with $B(E \rightarrow K^*\bar{K}$+c.c.$)/B[E \rightarrow (K^*\bar{K}$+c.c.) or $\delta\pi] = 0.86 \pm 0.12$.

The spin of the $\iota(1440)$ was determined from a partial-wave analysis of the Crystal Ball data (45). Contributions from five partial waves were included: 1. $K\bar{K}\pi$ flat (phase space); 2. $\delta^0\pi^0$-0^-; 3. $\delta^0\pi^0$-1^+; 4. $K^*\bar{K}$+c.c.-0^-; 5. $K^*\bar{K}$+c.c.-1^+. Note that $J^P = 0^+$ is not allowed for a state decaying into three pseudoscalars. $J^P = 1^-$, although allowed for $K^*\bar{K}$+c.c., would require the Dalitz plot to vanish at the boundaries, which is inconsistent with the data of Figure 15b. Amplitudes with $J \geq 2$ were not considered. Contributions from all partial waves except the $K\bar{K}\pi$ phase space contribution were allowed to interfere with arbitrary phase. The $K\bar{K}\pi$ contribution due to phase space was assumed to be incoherent. The full angular decay distributions in each case were included in the amplitudes. The ι and K^* helicities were allowed to vary in the fits. The δ and K^* parameters were taken to be the standard values (27). In other words, a standard isobar analysis (61) was done here.

The analysis was done for events with $K\bar{K}\pi$ masses between 1300 and 1800 MeV/c^2. The data were divided into five bins of 100 MeV/c^2 each. The standard procedure of eliminating those partial waves that do not contribute significantly to the likelihood was utilized (i.e. the number of events contributed by a given partial wave was required to be larger than the error on that number). The only significant contributions were from $K\bar{K}\pi$ flat, $\delta^0\pi^0$-0^-, and $K^*\bar{K}$+c.c.-1^+. These contributions, corrected for detection efficiency, are shown as a function of $K\bar{K}\pi$ mass in Figure 16. The $K^*\bar{K}$+c.c.-1^+ contribution is relatively small and independent of mass. On the other hand, the $\delta\pi$-0^- contribution shows clear evidence for the resonant structure in the ι signal region ($1400 \leq M_{K\bar{K}\pi} < 1500$ MeV/c^2). This establishes the spin-parity of the ι as 0^-. (The C-parity is required to be even because of the production mechanism.) In addition, contrary to the case of the E(1420), the principal decay of the ι is into $\delta\pi$ and $B(\iota \to K^*\bar{K}+\text{c.c.})/B[\iota \to (K^*\bar{K}+\text{c.c.})$ or $\delta\pi] < 0.25$ (90% C.L.).

Since a number of assumptions went into the partial-wave analysis, and, in particular, only a limited number of partial waves were considered, checks were made to show that the results of the analysis were valid. First, maximum likelihood fits were made to the restricted hypothesis that, in each mass interval, only one partial-wave contribution in addition to the flat contribution was allowed. The relative probabilities resulting from fits to the data in the signal region ($1400 \leq M_{K\bar{K}\pi} < 1500$ MeV/c^2) are given in

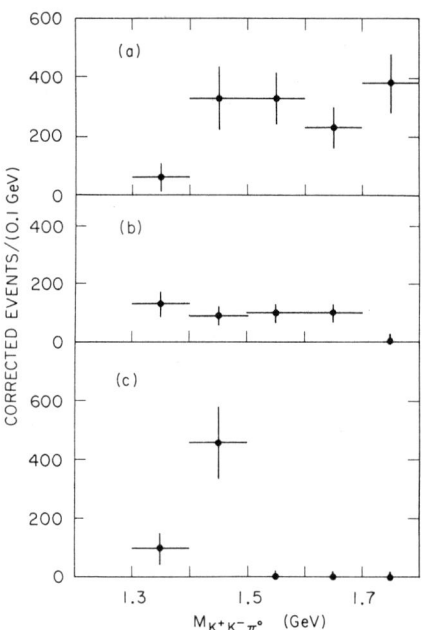

Figure 16 Partial-wave contributions to $J/\psi \to K^+K^-\pi^0$ as a function of $K\bar{K}\pi$ mass for (a) $K\bar{K}\pi$ flat, (b) $K^*\bar{K}$+c.c. with $J^P = 1^+$, and (c) $\delta\pi$ with $J^P = 0^-$.

Table 6 Relative partial-wave probabilities for various hypotheses for the structure of the $K\bar{K}\pi$ system in $J/\psi \to \gamma K^+ K^- \pi^0$ ($1400 \leq M_{K\bar{K}\pi} < 1500$ MeV/c^2)

Partial-wave contribution	Relative probability
flat + $\delta\pi$–0^-	1.0
flat + $\delta\pi$–1^+	0.006
flat + $K^*\bar{K}$ + c.c.–0^-	10^{-7}
flat + $K^*\bar{K}$ + c.c.–1^+	0.01

Table 6. Note that compared to the $\delta\pi$–0^- hypothesis, the next best hypothesis ($K^*\bar{K}$ + c.c.–1^+) has a relative probability of only 1%. This establishes that there is not a strong correlation between the $\delta\pi$ and $K^*\bar{K}$ + c.c. amplitudes. The properties of the ι as measured by the Mark II and Crystal Ball collaboration are shown in Table 7. Also shown is the Crystal Ball upper limit (cf Section 6.5) for $B(J/\psi \to \gamma\iota) \cdot B(\iota \to \eta\pi\pi)$. This upper limit is in mild conflict with the hypothesis that the $K\bar{K}\pi$ decay of the ι is dominated by $\delta\pi$ as hypothesized above (see Table 7), although some theoretical interpretations can avoid this conflict (62). Note that $\delta\pi$ dominance of the ι decay is an important element in our spin-parity analysis of the ι.

The $\vartheta(1640)$ was first observed in the process, $J/\psi \to \gamma\eta\eta$, $\eta \to \gamma\gamma$ by the Crystal Ball collaboration (46). The analysis was based on the full data sample. Figure 17a shows the $\eta\eta$ invariant mass distribution for events consistent with $J/\psi \to \gamma\eta\eta$ after a 5-C fit has been performed. Only events with $\chi^2 < 20$ are shown. Because of the limited statistics, it is not possible to establish whether the ϑ peak is one or two peaks (the ϑ and f'). However, it is

Table 7 Parameters for the $\iota(1440)$[a]

Parameter	Crystal Ball	Mark II (44)
M (MeV/c^2)	1440^{+20}_{-15}	1440^{+10}_{-15}
Γ (MeV)	55^{+20}_{-30}	50^{+30}_{-20}
$B(J/\psi \to \gamma\iota) \cdot B(\iota \to K\bar{K}\pi)$[b]	$(4.0 \pm 0.7 \pm 1.0) \times 10^{-3}$	$(4.3 \pm 1.7) \times 10^{-3}$ [c]
$B(J/\psi \to \gamma\iota) \cdot B(\iota \to \eta\pi\pi)$[d]	$<2 \times 10^{-3}$ (90% C.L.)	—
C	+	+
J^P	0^-	—

[a] Where two errors are given, the first is statistical and the second systematic.
[b] $I = 0$ is assumed in the isospin correction.
[c] This product branching ratio has been increased by 19% as compared to the value published in (44). This accounts for the differential efficiency correction from the spin-1 to spin-0 case, as discussed in the reference.
[d] Note that one experiment gives $B(\delta \to \eta\pi\pi)/B(\delta \to K\bar{K}) = 1.4 \pm 0.6$ (62a), while $\iota \to \delta\pi$ has been measured as the dominant decay for the $K\bar{K}\pi$ final state.

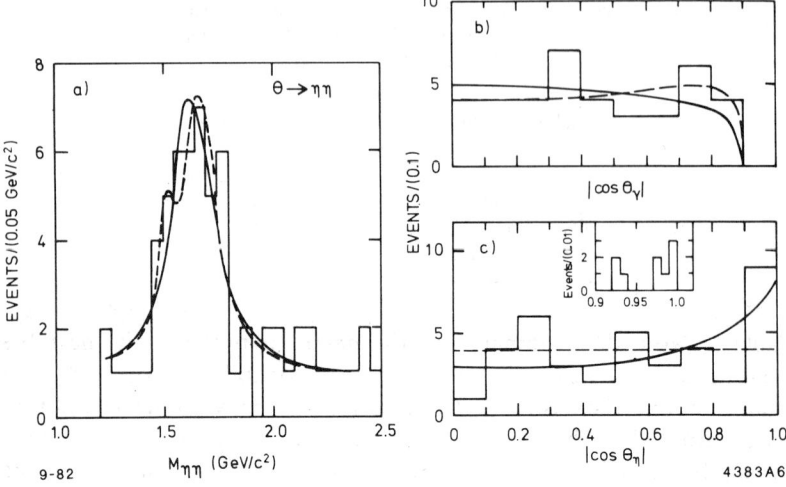

Figure 17 (a) The $\eta\eta$ mass distribution from the process $J/\psi \to \gamma\eta\eta$ for $M_{\eta\eta} < 2.5\,\text{GeV}/c^2$. The solid curve is a fit to a flat background plus one Breit-Wigner resonance. The dashed curve is a fit to a flat background plus two Breit-Wigner resonances, one with the mass and width of the f′ but fitted amplitude and the other with all three parameters fitted. A flat background is also included. (b) $|\cos\vartheta_\gamma|$ and (c) $|\cos\vartheta_\eta|$ distributions for $J/\psi \to \gamma\vartheta$, $\vartheta \to \eta\eta$. Solid curves are best-fit distributions for a ϑ spin of 2 and the dashed curves are expected distributions for spin 0. The inset shows the $|\cos\vartheta_\eta|$ distribution on an expanded scale.

probably most reasonable to assume that the f′ is present and fit for its amplitude. This was not done in (46); however, it was done in (63) and we also use the results from the fit including two resonances. The spin of the ϑ was determined from a maximum likelihood fit to the angular distribution $W(\vartheta_\gamma, \vartheta_\eta, \varphi_\eta)$ for the process $J/\psi \to \gamma\vartheta$, $\vartheta \to \eta\eta$. The parameter ϑ_γ is the polar angle of the γ with respect to the beam axis, and $(\vartheta_\eta, \varphi_\eta)$ are the polar and azimuthal angles of one of the η's with respect to the γ direction in the ϑ rest frame ($\varphi_\eta = 0$ is defined by the electron beam direction). The probability for the spin-0 hypothesis relative to the spin-2 hypothesis is 0.045. Spins greater than 2 were not considered. Note that the $\eta\eta$ decay mode establishes the parity of the state as even. Figures 17b and 17c show the $|\cos\vartheta_\gamma|$ and $|\cos\vartheta_\eta|$ distributions, respectively. Although the spin determination depends on information that cannot be displayed in these projections, it is clear that the $|\cos\vartheta_\eta|$ distribution plays the major role in the preference for spin 2. This is primarily due to the excess of events with $|\cos\vartheta_\eta| > 0.9$. There is no evidence that these events are anomalous.

The Crystal Ball and the Mark II (64) have searched for $J/\psi \to \gamma\vartheta$, $\vartheta \to \pi\pi$. Figure 13 shows the Crystal Ball results for the π^0's from 2.2×10^6 J/ψ decays. The binning in $M_{\pi\pi}$ is 50 (MeV/c^2) per bin. As summarized

Table 8 Parameters for the $\vartheta(1640)$[a]

Parameter	Crystal Ball	Mark II (64)
M (MeV/c^2)	1670 ± 50	1700 ± 30
Γ (MeV)	160 ± 80	156 ± 20
$B(J/\psi \to \gamma\vartheta) \cdot B(\vartheta \to \eta\eta)$	$(3.8 \pm 1.6) \times 10^{-4}$	—
$B(J/\psi \to \gamma\vartheta) \cdot B(\vartheta \to K\bar{K})$[b]	—	$(12.0 \pm 1.8 \pm 5.0) \times 10^{-4}$
$B(J/\psi \to \gamma\vartheta) \cdot B(\vartheta \to \pi\pi)$[b]	$<6 \times 10^{-4}$ (90% C.L.)	$<3.2 \times 10^{-4}$ (90% C.L.)
C	+	+
J^P	2^+ (95% C.L.)	2^+ (78% C.L.)

[a] Where two errors are given, the first is statistical and the second systematic.
[b] $I = 0$ is assumed in the isospin correction.

in Table 8, only upper limits were obtained from both the Crystal Ball and Mark II experiments.

The Mark II collaboration has obtained confirming evidence for the ϑ in the process $J/\psi \to \gamma\vartheta$, $\vartheta \to K^+K^-$ (64). They find the spin-parity assignment 2^+ to be favored at the 78% C.L. A summary of the Mark II results on the ϑ is also given in Table 7.

6.5 Other Radiative Transitions

The Mark II collaboration (65) reports a signal in the process $J/\psi \to \gamma\rho^0\rho^0$, $\rho^0 \to \pi^+\pi^-$. They interpret their $\rho^0\rho^0$ spectrum in this process as a combination of $\gamma\rho^0\rho^0$ phase space and a Breit-Wigner resonance. A maximum likelihood fit to this hypothesis yields, $M_{\text{res}} = 1650 \pm 50$ MeV/c^2, $\Gamma_{\text{res}} = 200 \pm 100$ MeV. These values are comparable to the mass and width of the ϑ shown in Table 7. Also, they obtain, $B(J/\psi \to \gamma\rho^0\rho^0, M_{\rho^0\rho^0} < 2 \text{ GeV}) = (1.25 \pm 0.35 \pm 0.4) \times 10^{-3}$. Assuming that the $\rho\rho$ is in the decay in an $I = 0$ state, we have $B(J/\psi \to \gamma\rho\rho, M_{\rho\rho} < 2 \text{ GeV}) = (3.75 \pm 1.05 \pm 1.3) \times 10^{-3}$. This branching ratio is approximately equal to the $\iota(1440)$ and η' branching ratios. As a strong note of caution, the Mark II collaboration states that much more data is needed to establish the connection, if any, between the $\rho\rho$ structure and the ϑ meson.

The Crystal Ball collaboration (66) has also found additional structure in the region of the ϑ by examining the process, $J/\psi \to \gamma\eta\pi^+\pi^-$, $\eta \to \gamma\gamma$. Figure 18 shows the $M_{\eta\pi^+\pi^-}$ and $M_{\eta\pi^0\pi^0}$ distributions obtained from the analysis of 2.2×10^6 J/ψ decays. A large signal at $M_{\eta\pi\pi} = M_{\eta'}$ is evident, and in addition, there is a broad enhancement centered at about 1710 MeV/c^2.

Examination of the Dalitz plots for the $\eta\pi^+\pi^-$ events (66) with $1600 < M_{\eta\pi\pi} < 1850$ MeV/c^2 shows no structure. Thus the broad enhancement is not strongly associated with a δ, or any other, resonance in either $\eta\pi^\pm$ or $\pi^+\pi^-$. Three possible interpretations are suggested for this new enhancement.

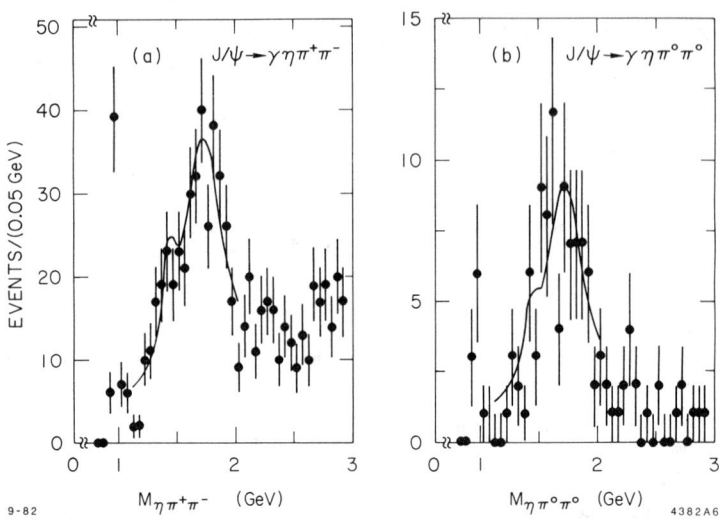

Figure 18 $\eta\pi\pi$ mass spectrum from (a) $J/\psi \to \gamma\eta\pi^+\pi^-$ and (b) $J/\psi \to \gamma\eta\pi^0\pi^0$. The curves are fits including contributions from the $\iota(1440)$ as described in the text.

First, the $\eta\pi\pi$ mass distribution for events with a prompt γ may be quite different from Lorentz invariant phase space. Then the enhancement could arise from the (nonresonant) decay of the J/ψ to a photon plus two gluons. Secondly, the enhancement could be a group of resonances. A third possibility is that it is a single resonance. The data may be fit with a single Breit-Wigner line shape. For the fit, the $\eta\pi^+\pi^-$ and $\eta\pi^0\pi^0$ mass spectra are fit simultaneously with the mass and width parameters constrained to be the same for both channels. A constant background was assumed for the $\eta\pi^0\pi^0$ channel. For $\eta\pi^+\pi^-$, the background was determined by fitting the $\gamma\gamma\pi^+\pi^-$ mass spectrum for events with a $\gamma\gamma$ mass combination in the η sidebands ($320 < M_{\gamma\gamma} \leq 470$ MeV/c^2 or $610 < M_{\gamma\gamma} < 760$ MeV/c^2). The fit has a χ^2 of 66 for 69 degrees of freedom and yields, $M = 1710 \pm 45$ MeV/c^2, $\Gamma = 530 \pm 110$ MeV, where the errors include estimates of the systematic uncertainty.

Using the number of events in the peak, as determined by the fit, and an efficiency obtained from Monte Carlo calculations of 18% (6.6%) for $J/\psi \to \gamma\eta\pi^+\pi^-(\gamma\eta\pi^0\pi^0)$, one obtains the branching ratios, $B(J/\psi \to \gamma\eta\pi^+\pi^-) = (3.5 \pm 0.3 \pm 0.7) \times 10^{-3}$, $B(J/\psi \to \gamma\eta\pi^0\pi^0) = (2.3 \pm 0.3 \pm 0.7) \times 10^{-3}$, where the first error is statistical and the second is systematic. These branching ratios, when added, are comparable to or larger than those for the ι and η'.

The fit shown in Figure 18 also includes a term for the ι, from which the upper limit in Table 7 was obtained. The implications of this low value for $\iota \to \eta\pi\pi$ are discussed in Section 6.4.

It is of interest to note that if the presently known contributions to radiative decays of the J/ψ in the ϑ region are added together, one obtains, $B[J/\psi \to \gamma\vartheta(\text{region})] \geq B(J/\psi \to \gamma\vartheta + \gamma\rho\rho + \gamma\eta\pi\pi) = (1.1 \pm 0.2) \times 10^{-2}$. This is the second largest branching ratio seen in the J/ψ radiative decays, being just less than that of the $\eta_c(2984)$.

The interpretation of the character of the ι and ϑ and the other new states seen for the first time in radiative transitions from the J/ψ is complex (3). In particular, the discussion of whether some of these states are gluonic mesons is beyond the scope of this review.

7. SEARCHES FOR THE F MESON, THE AXION, AND THE $^1P_1$ STATE

7.1 The Inclusive η Cross Section

In 1977, the DASP collaboration reported (67) a strong increase in the inclusive η production in e^+e^- collisions at $E_{CM} \cong 4.4$ GeV (and possibly at 4.17 GeV) relative to the production at 4.03 GeV. They interpreted this as evidence for production of the charmed strange F meson, which is expected to have a strong branching fraction into η's (68). Correlations with electrons and low energy γ's (expected from $F^* \to \gamma F$) strengthened this interpretation. Furthermore, they observed a cluster of events at $E_{CM} = 4.42$ GeV fitting the hypothesis $e^+e^- \to FF^* \to \gamma F\eta\pi$. Based on all of this, they reported that $R(e^+e^- \to F\bar{F}X) \cdot B(F \to \eta x) = 0.46 \pm 0.10$ in the E_{CM} range from 4.36 to 4.49 GeV, where $R(e^+e^- \to F\bar{F}X) = \sigma(e^+e^- \to F\bar{F}X)/\sigma(e^+e^- \to \mu^+\mu^-)$. The final state is written $F\bar{F}X$ to take into account the possibility that F-meson pairs may occur via production of excited F mesons, e.g. $e^+e^- \to F^*\bar{F} \to \gamma F\bar{F}$.

In order to study this interesting phenomenon, the Crystal Ball data were analyzed for inclusive η production. The data sample consisted of hadronic events from six fixed center-of-mass (c.m.) energies and seven c.m. energy bands. The fixed points were the J/ψ, ψ', ψ'', a point at 3.670 GeV in the continuum just below the J/ψ to act as a control from below charm threshold, and two energies above charm threshold (4.028 and 5.200 GeV). The seven energy bands covered a range in E_{CM} from 3.878 to 4.500 GeV. These data were taken in fine scans with steps in E_{CM} of between 2 and 12 MeV. For purposes of measuring $R_\eta = \sigma(e^+e^- \to \eta x)/\sigma(e^+e^- \to \mu^+\mu^-)$, the seven energy bins were chosen to correlate with observed structure in $R(e^+e^- \to \text{hadrons})$ (69).

The method for obtaining the number of produced η mesons at each energy was to study the inclusive $\gamma\gamma$ distribution in the vicinity of the η mass. In all cases, a clear enhancement at the η mass was visible to the naked eye. The number of observed η's was obtained by standard statistical fitting of

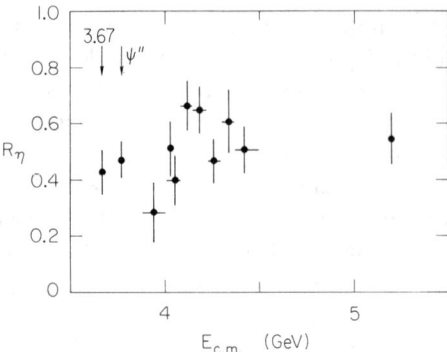

Figure 19 R_η as a function of E_{CM}. The first two points are for $E_{CM} = 3.67$ GeV and the ψ''. The ψ' point is offscale. The error bars include the point-to-point systematic uncertainty, but not the estimated 20% overall systematic uncertainty.

the observed distribution to a smooth background function plus a resolution function of adjustable size centered on the η mass. The observed number was then corrected for the branching ratio for $\eta \to \gamma\gamma$ and the η detection efficiency, which ranged from 38% at the J/ψ to 27% at 5.2 GeV. Uncertainties in the detection efficiencies due to our uncertain knowledge of the details of e^+e^- annihilation physics are included in the limits finally obtained.

Figure 19 shows our results for R_η. The offscale values at the J/ψ and ψ' are excluded[2] and the other points have been corrected for the radiative tails of these two resonances. Although there may be some correlation with the total hadronic cross section, we see that there is no dramatic difference in R_η below and above charm threshold. If we assume that the contribution to R_η due to non-charm physics is constant and that all excess in R_η is due to F decays, we can set limits on $R(e^+e^- \to F\bar{F}X) \cdot B(F \to \eta x)$ by comparing the values for R_η above charm threshold with that below it at 3.67 GeV. The 90% C.L. limits are all below 0.32 and, for the energy band from 4.365 to 4.500 GeV, it is 0.19. This disagrees with the DASP result (67). Most of the disagreement is due to the fact that the earlier experiment saw essentially no η signal at 4.03 GeV whereas the Crystal Ball observed almost the same strength at 4.03 GeV as at other energies, even below charm threshold. At energies above about 4.1 GeV, the cross sections reported by the two experiments are on the average compatible. Up to the time of this writing, the Crystal Ball has found no firm evidence for the elusive charmed strange F meson.

[2] When the results are expressed in terms of f_η, the average number of η's per hadronic event, the two resonances are not special; f_η has a value of about 0.13 and shows little variation over this energy range.

7.2 The Search for $J/\psi \to \gamma$ Axion

Because of its exceptionally large solid angle coverage by charged-particle and photon detectors with essentially 100% detection efficiency, and because of its moderately good time resolution (about 3 ns), the Crystal Ball is well adapted to search for certain exotic phenomena, especially those of the class $e^+e^- \to \gamma X$ where X escapes detection for some fundamental reason. An example of such a reaction involving known particles is that in which X is $\nu\bar{\nu}$ resulting from either direct production, or the decay of a light neutral spin-1 gauge boson as suggested by some supersymmetric theories (70), or, at higher energies, the decay of the Z^0. Another possibility, which has been searched for in the Crystal Ball and is reported in (71), is the radiative decay of the J/ψ into an axion. The axion (a) is the Goldstone boson appearing from the breaking of a chiral U(1) symmetry, which has been postulated to avoid large P- and CP-invariance violations in QCD (72). If the number of quark generations is assumed known, then this theory has only one free parameter, the ratio, x, of the vacuum expectation values of the two Higgs fields present in the theory. However, it does endow the axion with a sufficiently long life and weak enough interactions that it would escape detection in the Ball. The theory reliably predicts that $B(J/\psi \to \gamma a) = (5.7 \pm 1.4) \times 10^{-5} \, x^2$. Positive evidence for an axion or axion-like particle was reported by Faissner et al (73) with a mass $m_a = 250 \pm 25 \text{ keV}/c^2$ and $x = 3.0 \pm 0.3$ [but these values seem inconsistent with other experiments (74)]. They imply that there should be about 800 events in the Ball with $|\cos \vartheta| < 0.8$ and these events would have the distinctive signature of a single photon with beam energy. No significant numbers of such events are seen.

The dominant background in this search comes from cosmic rays and most of these can be eliminated by restricting attention to the bottom hemisphere of the Ball, which simply reduces the overall detection efficiency to 30%. After making a cosmic ray background subtraction using events out of time with the beam, we obtain a 90% C.L. upper limit of 6.2 events in an energy range from 1.3 GeV to 2 energy-resolution standard deviations above the beam energy. This implies an upper limit of 1.4×10^{-5} (90% C.L.) on the branching fraction. The corresponding upper limit on x is 0.6, in disagreement with the result in (73).

A definitive test of the standard axion model, which eliminates any dependence on x, has been proposed (75) by setting limits on both $J/\psi \to \gamma a$ and $\Upsilon \to \gamma a$. Recent results from the LENA collaboration at DORIS (76) and the CUSB collaboration at CESR (76a) have established that $B(\Upsilon \to \gamma a)$ is less than 9.1×10^{-4} (90% C.L.) and 3.5×10^{-4} (90% C.L.), respectively. These results together with the above Crystal Ball limit on $\psi \to \gamma a$ violate the results of the standard axion model and it now seems necessary to retreat to an even more elusive axion, such as has been proposed in grand unified theories (77).

7.3 The Search for Decays $\psi' \to \pi^0\,{}^1P_1$

The only predicted $c\bar{c}$ bound state for which no evidence exists is the ${}^1P_1$ with $J^{PC} = 1^{+-}$. Its mass is expected to be approximately equal to the center of gravity of the ${}^3P_J$ states, or about 3520 MeV (78). Experimental determination of its mass is important since any significant deviation from the above value would suggest a long-range spin-spin term in the quarkonium potential. We have searched extensively in our large ψ' data sample for evidence of this state and have not found it (79).

Single photon transitions between the ψ' and ${}^1P_1$ states are forbidden by C conservation and so one must investigate double photon transitions. Four possibilities are indicated in Figure 20. Estimates based on related measured rates indicate that only the $\psi' \to \pi^0\,{}^1P_1$ process can be reasonably expected to have a branching ratio in the percent range. This process would lead to a monochromatic π^0 in ψ' decay having an energy that is expected to be below about 200 MeV; 165 MeV is favored. Figure 21a shows the inclusive π^0 distribution observed in ψ' decays. The evident structures at about 200 MeV and just above 400 MeV are expected backgrounds. They are due to fake π^0's generated with the monochromatic photons from $\psi' \to {}^3P_J$ transitions, and the reactions $\psi' \to \pi^0\pi^0 J/\psi$ and $\psi' \to \pi^0 J/\psi$. No other structure is evident and 95% C.L. limits of less than 1.09% have been set for $B(\psi' \to \pi^0\,{}^1P_1)$ for any ${}^1P_1$ mass between 3440 and 3535 MeV/c^2. In particular, at the favored mass of 3520 MeV/c^2, the limit is 0.42%.

In an effort to reduce backgrounds and so increase sensitivity at the expense of a less general result, we have also searched for evidence of the cascade decay $\psi' \to \pi^0\,{}^1P_1 \to \gamma\gamma\gamma\,\eta_c$, where the η_c is constrained in mass but not decay mode. Study of this particular configuration is motivated

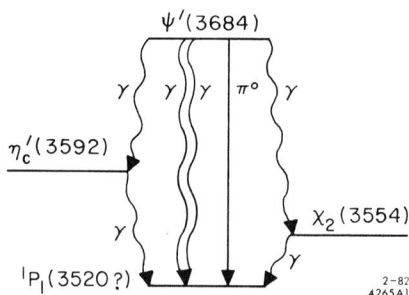

Figure 20   Possible mechanisms contributing to the decay $\psi' \to \gamma\gamma\,{}^1P_1$.

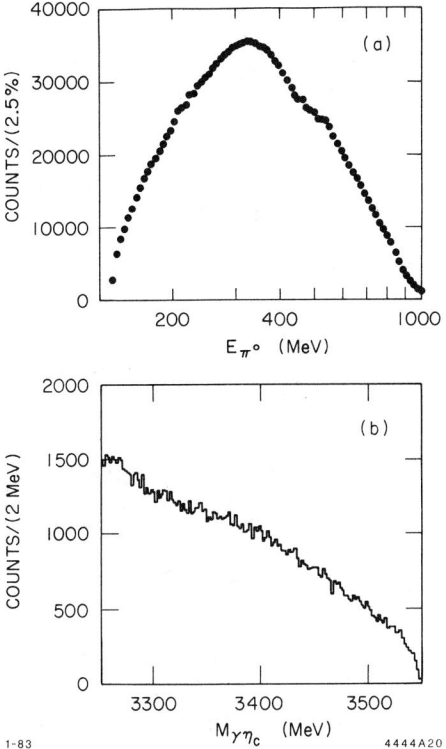

Figure 21 (a) The inclusive π^0 energy spectrum from ψ' decays. (b) Distribution of the $\gamma\eta_c$ mass for events satisfying the hypothesis $\psi' \to \pi^0\gamma\eta_c$.

by a reasonable expectation that $B(^1P_1 \to \gamma\eta_c)$ is in the vicinity of 50%. Figure 21b shows the γ(prompt)η_c mass distribution for events fitting the hypothesis $\psi' \to \pi^0\gamma\eta_c$ and again there is no evidence for the $^1P_1$ state. We have set 95% confidence limits of less than 0.35% for $B(\psi' \to \gamma\,^1P_1) \cdot B(^1P_1 \to \gamma\eta_c)$ for $^1P_1$ masses in the same range as above. At the preferred mass, the limit is 0.14%. Although the $^1P_1$ state has not been found, the limits set are sufficiently low to be theoretically interesting.

8. MEASUREMENTS OF R_h IN THE E_{CM} RANGE OF 5.0 TO 7.4 GeV

One of the most fundamental and difficult measurements that can be done at an e^+e^- storage ring is that of the total hadronic cross section, $\sigma_h = \sigma(e^+e^- \to \text{hadrons})$. In order to remove the straightforward effects of QED and so reveal the strong interaction effects more clearly, it is customary to normalize the hadronic cross section to the theoretical, lowest order, purely QED cross section $\sigma_{\mu\mu}$ for $e^+e^- \to \mu^+\mu^-$, which, in the energy range of interest here, is $(4\pi/3)\alpha^2(\hbar c)^2/E_{\rm CM}^2 = 86.8/E_{\rm CM}^2$ (nb, GeV). In the energy range well above charm threshold and below bottom threshold, the prediction of QCD for the normalized cross section, $R_{\rm h} = \sigma_{\rm h}/\sigma_{\mu\mu}$, is, to first order, $R_{\rm QCD} = 3\sum_i Q_i^2[1+\alpha_{\rm s}(s)/\pi + \cdots]$, where $s = E_{\rm CM}^2$, $i = ({\rm u,d,s,c})$ is the quark flavor index, Q_i is the charge of the ith quark, and $\alpha_{\rm s}(s)$ is the QCD running coupling constant (80). The sum over Q_i^2 yields a value of 10/3 and, at $E_{\rm CM} = 6$ GeV, the first order[3] QCD result increases this by about 6% for $\Lambda_{\overline{\rm MS}} = 100$ MeV.

In 1980, Barnett et al (80) made a careful comparison of all available measurements of $R_{\rm h}$ (81) to the predictions of QCD. They concluded that above 5.5 GeV a potentially serious discrepancy, in the range of 15 to 17%, existed. This was just outside the quoted systematic uncertainties of 10% and there was the exciting possibility that it represented new phenomena or difficulties with QCD. This prompted the Crystal Ball collaboration to undertake a series of new $R_{\rm h}$ measurements from 5.0 GeV to the top of the SPEAR range, 7.4 GeV (81a). To do this, a total exposure of 12.7 pb^{-1} was distributed over eleven energies in this range.

The two ingredients in the experimental determination of $R_{\rm h}$ are the integrated luminosity of the exposure and the number of hadronic events produced (and corrected for QED radiative effects). The luminosity was obtained both by the small-angle Bhabha-scattering monitor and by observation of large-angle QED events in the Ball itself. These independent determinations of the integrated luminosity agreed to about 2% and their average was used.

The determination of the radiatively corrected number of hadronic events produced by annihilations involved three steps. First, criteria were developed and applied to efficiently distinguish individual annihilation hadronic events from five classes of backgrounds: cosmic rays, beam-gas collision, QED events, two-photon collisions, and $\tau\bar{\tau}$ events. Next, properly normalized statistical subtractions were made to eliminate the residual backgrounds that slipped through the first step. And finally, the resulting count was corrected for the detection efficiency of both the triggering hardware and the event selection software of the first step. The purely QED radiative effects were also included in this step. Pure samples from each of the five event classes were obtained either experimentally (cosmic rays and beam-gas events) or by Monte Carlo simulation. These were examined and efficient criteria for event selection developed. These criteria were applied to

---

[3] The next higher order in the QCD calculation using the modified minimal subtraction renormalization scheme is known (80). It contributes only 0.7% to the prediction, well below present experimental precision.

CRYSTAL BALL DETECTOR PHYSICS 187

both the data and the five pure samples. The residuals from the backgrounds were then subtracted from the data, and the result was corrected by the detection efficiency determined from the Monte Carlo simulation of the annihilation events. The Monte Carlo calculation incorporated all radiative correction effects in the event generation algorithms and so this last step automatically includes these corrections.

The most important backgrounds were those due to cosmic ray events, beam-gas collisions, and τ decays. Criteria to identify the first two of these were based on the spatial distribution of energy in the Ball and were developed by studying events out of time with the beam and those obtained in runs with the beams separated. The cosmic ray background was reduced to negligible levels by these criteria and the residual beam-gas background, which had to be removed by a statistical subtraction, was typically at the 10 to 12% level. No attempt was made to identify $\tau\bar{\tau}$ events or those arising from $\gamma\gamma$ collisions and so these backgrounds were removed by subtraction only. The pure samples of these two background classes were obtained by Monte Carlo simulation. Together, these two background sources gave a subtraction of about 12%. Finally, the QED contamination was easily reduced to negligible levels by criteria involving leading-particle energy and number of observed particles.

Table 9 Measured values of $R_h = \sigma(e^+e^- \to \text{hadrons})/\sigma(e^+e^- \to \mu^+\mu^-)$[a]

E_{CM} (GeV)	R_h	$(\delta R_h/R_h)_{stat}$	$(\delta R_h/R_h)_{point\,sys}$ (%)
		1981 data	
5.00	3.46	3.3	3.4
5.25	3.60	2.8	1.2
5.50	3.33	2.9	2.4
5.75	3.40	3.1	1.4
6.00	3.25	2.8	2.3
6.25	3.31	2.8	1.4
6.50	3.33	2.8	2.2
6.75	3.38	2.3	1.5
7.00	3.34	2.9	1.5
7.25	3.56	3.0	2.2
7.40	3.32	4.0	2.9
		1980 data	
5.20	3.51	3.5	3.5
6.00	3.43	3.5	3.5
6.75	3.38	3.8	3.5
7.40	3.67	3.5	3.5

[a] The error on R_h consists of three parts, a statistical error $(\delta R_h/R_h)_{stat}$; a systematic error that depends upon E_{CM}, $(\delta R_h/R_h)_{point\,sys}$, and a systematic error that is uniformly applicable to all the 1981 data of 5.3% and to all of the 1980 data of 7.0%.

The important questions of detection efficiency and radiative corrections were answered by subjecting simulated annihilation events generated by the LUND81 (82) Monte Carlo program (including radiative corrections) to the same criteria as the data. The result of this was the product of detection efficiency and the radiative correction factor. This product had a typical value of 1.06.

Table 9 gives the results of the experiment. The point-to-point systematic errors are given in the table and include the effects of uncertainty in the normalization of the several background subtractions and the statistical errors in the several Monte Carlo simulations. In addition, there is an overall systematic scale uncertainty of 5.3% (for the 1981 data) arising from various effects: 3.0% from radiative corrections, 3.0% from the detection efficiency, 2.5% from the luminosity, 1.4% from $\tau\bar{\tau}$ events, 1.3% from beam-gas interactions, and 0.6% from two-photon collisions. Figure 22 compares these results with those of other experiments (82a) and with QCD predictions. Clearly, the trend of the data is completely consistent with that of QCD. The absolute value of the measurements is about 6% lower than

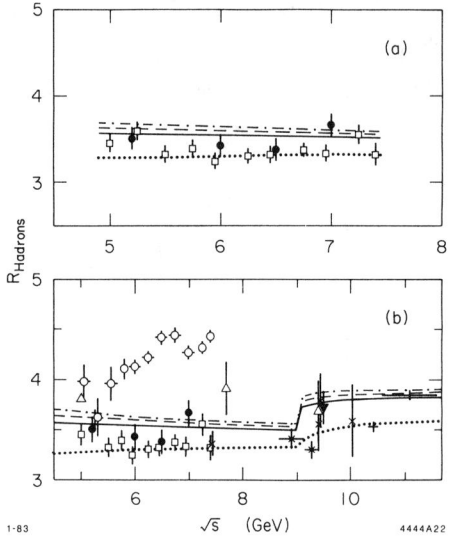

Figure 22 (*a*) Crystal Ball measurements of R_h compared to theoretical prediction. The solid points are from data taken in 1980 and the open squares are from a much larger data sample taken in 1981. The dotted curve is the simple quark-parton model prediction and the others are QCD predictions with $\Lambda_{\overline{MS}}$ = 100 MeV (*solid*), 200 MeV (*dashed*), and 300 MeV (*dash-dot*). The error bars do not include the 5.3% (1981 data) or 7.0% (1980 data) overall systematic error. (*b*) Comparison of Crystal Ball results (*solid circles* and *open squares*) with other measurements (82a) (Mark I, *open circles*; PLUTO, *open triangles*; LENA, *crossed dot*; DASP II, *solid triangle*; CUSB, *plus sign*; DESY-Heidelberg, *cross*). The curves are as in (*a*).

the QCD predictions for reasonable values of the QCD parameter $\Lambda_{\overline{MS}}$. This disagreement, of course, is easily accommodated by the systematic scale uncertainty of the data. We conclude that these results rule out the possibility of any new phenomena in this energy range, at least, at the level suggested by the Mark I data.

9. TWO-PHOTON PHYSICS

In addition to studying the physics of e^+e^- annihilations, the Crystal Ball experiment has been able to investigate certain two-photon reactions. More precisely, we can investigate the reaction $e^+e^- \rightarrow e^+e^- +$ hadrons in the configuration that each of the two leptons scatters through a very small angle. Because of the small scattering angle, the outgoing electrons are not detected. To lowest order in QED, then, the hadrons result from the collision of two photons, which though virtual, are very nearly on their mass shell. The Crystal Ball is particularly adapted to study the case that the hadrons decay into only photons and, to date, the work on the four-photon final state has been completed (83). The data for this analysis came from an integrated luminosity of 21 pb^{-1} distributed over an E_{CM} range from 3.9 to 7.0 GeV.

9.1 *Measurement of* $\Gamma(f \rightarrow \gamma\gamma)$ *from* $\sigma(\gamma\gamma \rightarrow f \rightarrow \pi^0\pi^0)$

All neutral events with exactly four energy clusters inside $|\cos \vartheta| < 0.9$, each with more than 20 MeV and with energy deposition patterns consistent with photons, were selected. Furthermore, it was required that the endcaps contain less than 40 MeV, and that the total invariant mass of the event be in the range from 720 MeV to E_{CM}. The resulting sample of events shows a strong peak at zero in the square of the total transverse momentum as expected for 2γ events. Only those in this peak [the cut was at 0.03 (GeV/c)^2] were subsequently used.

In essentially all of the final sample, photon pairings could be made that were consistent with either $\pi^0\pi^0$ or $\pi^0\eta$ being the primary hadrons. Figure 23 (*top*) gives the invariant mass distribution of the $\pi^0\pi^0$ sample, which clearly shows a strong signal near the f mass and no other significant structure. Of special note here is the smallness of the background. This is in contrast to earlier experiments that detected the charged-pion decay mode of the f and tended to be troubled with large nonresonant $\pi^+\pi^-$ and QED $\mu^+\mu^-$ backgrounds (84–87).

To obtain the cross section for $\gamma\gamma \rightarrow \pi^0\pi^0$, the $\pi^0\pi^0$ mass spectrum was corrected for the variation in the $\gamma\gamma$ flux over it (88) and detection efficiency. Figure 23 (*bottom*) gives the resulting cross section with the added restriction that $|\cos \vartheta^*| < 0.7$ where ϑ^* is the angle between the beam and the outoing π^0 direction in the $\pi^0\pi^0$ rest frame. The solid curve shows a fit with three contributions: a relativistic Breit-Wigner function (including slight spreading due to the experimental mass resolution) with mass and width parameters taken from the Particle Data Group compilation (27) for the f; the same for a possible S*(980); and a straight line to describe $\pi^0\pi^0$ nonresonant background. As is clear, the curve does not give a good fit since the data's mass peak is lower than that of the curve by about 40 MeV. The broken curve is the fit obtained when the f mass and width are allowed to be free, the best fit values being 1238 ± 14 MeV/c^2 and 248 ± 38 MeV/c^2, respectively. This mass shift could, in fact, be accommodated within the estimated systematic error of about 2% and the statistical error, but other $\gamma\gamma$ experiments have observed a very similar effect (85, 86), which suggests that

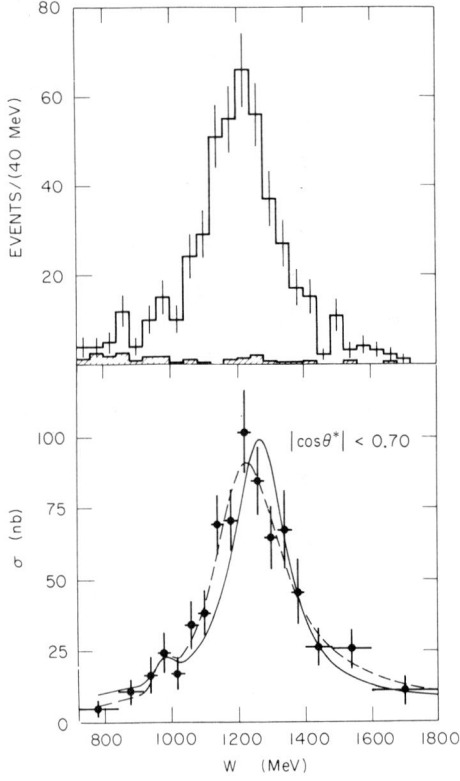

Figure 23 The top part of the figure shows the $\pi^0\pi^0$ mass distribution for 4γ events consistent with $e^+e^- \to e^+e^-\gamma\gamma \to e^+e^-\pi^0\pi^0$. The shaded histogram shows the non-$\pi^0\pi^0$ background. The bottom part of the figure gives $\sigma(\gamma\gamma \to \pi^0\pi^0)$ for |cos ϑ^*| < 0.7. The curves are described in the text.

the effect is due to some underlying physical mechanism rather than an instrumental artifact. Possible sources of this effect are interference with nonresonant background (89) or the $\varepsilon(1300)$. Finally, there is the interesting possibility that the f may be mixed with a predicted gluonic meson almost degenerate with it (90). In this last case, different production and/or decay channels would yield different resonance shapes, and so f production in the $\gamma\gamma$ channel could possibly give phenomenological resonance parameters different from those found in other hadronic interactions.

Previous determinations of $\Gamma(f \to \gamma\gamma)$ from two-photon collisions have assumed the theoretical prediction that the f is produced predominantly with helicity 2 (91, 92). Because of the negligible backgrounds in this experiment, it is possible to verify this theoretical expectation by observing the ϑ^* angular distribution. This is shown in Figure 24. It is clear that the spin-2 assumption with helicity-2 domination gives a good fit; the other helicity contributions are consistent with zero. If we assume that the mass peak is due to the f and that the decay is purely helicity 2, then we obtain $\Gamma_{f \to \gamma\gamma} = 2.7 \pm 0.2 \pm 0.6$ keV, the first error being statistical and the second systematic. This agrees well with the results from other experiments (84–87).

9.2 Measurement of $\Gamma(A_2 \to \gamma\gamma)$ from $\sigma(\gamma\gamma \to A_2 \to \pi^0\eta)$

After the $\pi^0\pi^0$ events were removed from the 4γ sample, the resulting events were essentially all $\pi^0\eta$ and their invariant mass peaks at around 1300 MeV/c^2. Identifying this peak with the $A_2(1320)$ and assuming pure helicity 2, we can extract the $\gamma\gamma$ partial width of the A_2. We obtain $\Gamma(A_2 \to \gamma\gamma) = 0.77 \pm 0.18 \pm 0.27$ keV. The naive quark model with ideal mixing predicts a ratio of 9/25 for $\Gamma(A_2 \to \gamma\gamma)/\Gamma(f \to \gamma\gamma)$, which is in agreement with the Crystal Ball observations of $0.29 \pm 0.07 \pm 0.07$ for this ratio.

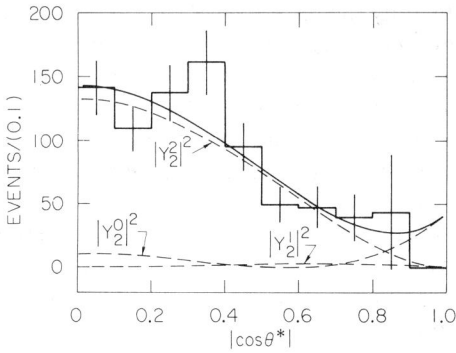

Figure 24 Acceptance-corrected distribution for $|\cos \vartheta^*|$ for $\pi^0\pi^0$ events in the f mass region (1040–1480 MeV/c^2). The solid curve is the best-fit spin-2 distribution and the dashed curves show the contributions from each of the three helicity amplitudes.

9.3 Other States

In addition to the two measurements discussed above, we can also set several limits based on the absence of signals. First, as is seen in Figure 23b, there is no evidence for $\gamma\gamma \to S^*(980) \to \pi^0\pi^0$. From this, we can set the limit $\Gamma(S^* \to \gamma\gamma) \cdot B(S^* \to \pi\pi) < 0.8$ keV. This limit is considerably smaller than the value of about 20 keV expected by most theoretical estimates (92), but consistent with a single-quark-exchange calculation (93), which predicts less than 0.4 keV for mesons in the 0^{++} nonet. Finally, no signal for $\gamma\gamma \to \eta\eta$ was observed. This allows two 95% C.L. limits to be set. First a limit of 0.05 can be put on $\Gamma(f \to \eta\eta)/\Gamma(f \to \pi\pi)$. This is consistent with the limit of 0.016 obtained in earlier work (94). And secondly, this absence implies that $\Gamma[\vartheta(1640) \to \gamma\gamma] \cdot B[\vartheta(1640) \to \eta\eta]$ is less than 5 keV.

10. MEASUREMENTS IN THE REGION FROM CHARM THRESHOLD TO 4.5 GeV

It has been known for a long time that the energy region above charm threshold, from the ψ'' at 3.77 GeV to about 4.5 GeV is rich in charmed physics phenomena. The ψ'' itself is known to be a "D factory," 4.03 GeV is a "D* factory," F and F* mesons are expected, and, more generally, there should be a rich spectrum of excited D and F mesons produced in e^+e^- collisions in this energy range (95). The strong structure in R_h in this region (81) is ample evidence for this but so far no details have been fully resolved above 4.03 GeV. In order to investigate this potentially interesting area of physics, the Crystal Ball accumulated an exposure at the $\psi''(3770)$ yielding about 1.3×10^4 produced ψ'', and an exposure of 11.3 pb^{-1} distributed over the range 3.8 to 4.5 GeV. Although a great deal of effort has gone into analyzing this data, only the results on R_η discussed above are considered complete. Preliminary reports on some of the other work have been given, however, and we give a short review of them here.

The global structure of the physics in this region is shown by the energy dependence of the normalized hadronic cross section itself, charged and neutral multiplicities, and charged and neutral energy fraction. Preliminary results on some of these from the Crystal Ball data are given in (96). The R_h measurements confirm the structure seen in other experiments (81): clear peaks at 3.77, 4.03, and 4.4 GeV, a broad peak with possibly some substructure around 4.16 GeV followed by a broad "valley" in the 4.2 to 4.3 GeV region. The statistical precision of this set of measurements is very high but much work remains to be done to reduce the point-to-point systematic uncertainties to fully exploit it. Analysis of R_h over the ψ'' excitation curve gives resonance parameters (97) in reasonable agreement with earlier work (98).

The observed neutral energy fraction is quite smooth through the whole region and so seems to be insensitive to the underlying physics. However, the observed neutral multiplicity does show large changes. Together, these two facts suggest that low energy γ's, π^0's, and η's could be useful indicators and these have been stressed in the data analysis. The resonance at E_{CM} = 4.03 GeV serves as a source of almost monochromatic low energy π^0's and photons. This results from the combination of a large D* cross section at this energy and low Q values for both the production channels, $D^*\bar{D}^*$ and $D^*\bar{D}+\bar{D}^*D$, and the decay channels $D^* \to \pi^0 D$ and γD. These circumstances give rise to an inclusive photon spectrum at this energy which is quite complex since there are significant contributions from all eight sources (π^0 and γ from both charged and neutral D* decays arising from both $D^*\bar{D}^*$ and $D^*\bar{D}$+c.c.). Qualitatively, the π^0 and γ contributions are resolvable since the former is peaked around 70 MeV and the latter around 135 MeV (the Doppler broadening gives rise to only a slight overlap). Further information is provided by the π^0 energy spectrum, which shows strong peaking at small kinetic energies.

By contrast to the strong structure in the π^0 and γ inclusive spectra at E_{CM} = 4.03 GeV, the corresponding ones below D* threshold at the $\psi''(3770)$ are smooth and featureless. We have used these as background functions in quantifying the effects due to the D* mesons. The spectra allow a new determination of the D^{*0}–D^0 mass difference of $142.2 \pm 0.5 \pm 1.5$ MeV/c^2 in agreement with earlier Mark I results (99). However, because of the relatively large number of cross sections (those for $e^+e^- \to D^{*0}\bar{D}^{*0}$, $D^{*+}D^{*-}$, $D^{*0}\bar{D}^0$+c.c., $D^{*+}D^-$+c.c.) and branching ratios (those for $\bar{D}^{*0} \to \pi^0 D^0$ and γD^0, $\bar{D}^{*+} \to \pi^0 D^+$ and γD^+) that are involved, the finite resolution of the apparatus, and the limited statistics, it is only possible to measure directly certain combinations of the physically interesting quantities. Guided by theoretical calculations (100) and Mark I measurements (99), we can make reasonable assumptions about some of these quantities and so obtain $\sigma(D^{*0}\bar{D}^0+\text{c.c.})/\sigma(D^{*0}\bar{D}^{*0}) \cong 1.6$ and $B(D^{*0} \to \gamma D^0) \cong 0.37$. These results are consistent with those of Mark I (99).

11. SUMMARY AND FUTURE PROSPECTS

Although the people who studied photon detectors for e^+e^- storage rings at the 1974 PEP Summer Study did not know about the soon-to-be-discovered charmed quarks, subsequent events have shown that the practical realization of their ideas has borne rich rewards in understanding this sector of nature. The Crystal Ball detector grew out of that work and, as this brief review has shown, it has proved to be an especially versatile instrument in spite of features (or lack of them) that at first sight would make it seem to be very specialized and restricted in application. Of course, the dominant strength of this detector has been, and always will be, the measurement of monochromatic photons, and it was this capability that allowed the Ball to resolve the old problems with the charmonium interpretation of psionic matter. However, the measurements of R_h, for example, demonstrate the instrument's ability to determine global properties of e^+e^- annihilations. At the other extreme, kinematically constrained fitting to very specific final states that include just two charged particles but are rich in photons has been successfully exploited.

Work is currently in progress on several projects involving our large SPEAR data sample. These include searches for the F meson by means of specific exclusive channels, study of D decays, further work on D* physics, completion of the work on R_h just above charm threshold, measurement of certain interesting exclusive hadronic final states in J/ψ and ψ' decays, and further work on radiative decays of the J/ψ and ψ' and in two-photon physics. However, an increasingly large fraction of the group's efforts is going into new ventures in Υ physics. Within the next few years, we expect to have sufficiently large data samples at the Υ, Υ', and other energies to be able to make contributions toward understanding upsilonic matter comparable to what we have done in the psionic sector. The difficulties are formidable since the rate of data accumulation at the higher energies is considerably smaller than in the J/ψ region and the events are more complex, but the work has begun. Finally, we expect to utilize the higher $\gamma\gamma$ flux in the 10-GeV energy range to explore further questions in two-photon physics.

ACKNOWLEDGMENTS

The work reported here would not have been possible without the dedicated, innovative, and, at times, brilliant efforts of the many members of the Crystal Ball collaboration and their supporting engineers and technicians, the staffs of the accelerators at SLAC, Harshaw Chemical Co., and other people who played important roles in building the detector. The members of the Crystal Ball collaboration were: C. Edwards, R. Partridge, C. Peck, F. Porter (Caltech); D. Antreasyan, Y. Gu, W. Kollmann, M. Richardson, K. Strauch, K. Wacker, A. Weinstein (Harvard); D. Aschman, T. Burnett, M. Cavalli-Sforza, D. Coyne, C. Newman, H. Sadrozinski (Princeton); D. Gelphman, R. Hofstadter, R. Horisberger, I. Kirkbride, H. Kolanoski, K. Königsmann, R. Lee, A. Liberman, J. O'Reilly, A. Osterheld, B. Pollock, J. Tompkins (Stanford-HEPL); E. Bloom, F. Bulos, R.

Chestnut, J. Gaiser, G. Godfrey, C. Keisling, W. Lockman, M. Oreglia, D. Scharre (SLAC). The work of the collaboration was supported in part by the Department of Energy under contracts DE-AC03-76SF00515 (SLAC), DE-AC02-76ER03064 (Harvard), DE-AC03-81ER40050 (Caltech), and DE-AC02-76ER03072 (Princeton); by the National Science Foundation contracts PHY81-07396 (HEPL), PHY79-16461 (Princeton), and PHY75-22980 (Caltech); and by fellowships from the NATO Fellowship, the Chaim Weizmann Fellowship, and the Sloan Foundation, which provided partial support to members of the collaboration.

Literature Cited

1. Bloom, E. D., et al. 1974. *1974 PEP Summer Study Rep.*, PEP Note 155; Mast, T., Nelson, J. 1974. *1974 PEP Summer Study Rep.* PEP Note 153. Stanford Univ. Press
2. Bloom, E. D. 1981. Proc. Summer Inst. Particle Phys., Jul. 27–Aug. 7, SLAC-245:1, and references therein
3. Bloom, E. D. 1982. 21st Int. Conf. on High Energy Physics, Paris, France, July 26–31, C3:407. J. Phys. C3, Suppl. 12. (This reference has an extensive review of gluonic mesons and contains references to the relevant theoretical literature.)
3a. Chinowsky, W. 1977. Ann. Rev. Nucl. Sci. 27:393
4. Feldman, G. J., et al. 1975. Phys. Rev. Lett. 35:821
5. Chanowitz, M. S., Gilman, F. J. 1976. Phys. Lett. B63:178
6. Tanenbaum, W., et al. 1978. Phys. Rev. D17:1731
7. Appelquist, T., Barnett, R. M., Lane, K. 1978. Ann. Rev. Nucl. Part. Sci. 28:387
8. Shifman, M. A. 1978. Phys. Lett. B77:80
9. Vainshtein, A., et al. 1978. Yad. Fiz. 28:465
10. Braunschweig, W., et al. 1977. Phys. Lett. B67:243
11. Wiik, B. H., Wolf, G. 1978. DESY 78/23. Hamburg: DESY
11a. Apel, W. D., et al. 1978. Phys. Lett. B72:500
12. Biddick, C. J. et al. 1977. Phys. Rev. Lett. 38:1324
13. Whitaker, J. S., et al. 1976. Phys. Rev. Lett. 37:1596
14. Bartel, W., et al. 1978. Phys. Lett. B79:492
15. Partridge, R., et al. 1980. Phys. Rev. Lett. 44:712
16. Hughes, E. B., et al. 1972. IEEE Trans. Nucl. Sci. 19:126
17. Oreglia, M., et al. 1982. Phys. Rev. D25:2259
18. Gaiser, J. E., et al. 1979. IEEE Trans. Nucl. Sci. 26:173
19. Chan, Y., et al. 1978. IEEE Trans. Nucl. Sci. 25:333
20. Chestnut, R., et al. 1979. IEEE Trans. Nucl. Sci. 26:4395
21. Aubert, J. J., et al. 1974. Phys. Rev. Lett. 33:1404
22. Augustin, J. E. 1974. Phys. Rev. Lett. 33:1406
23. Abrams, G. S., et al. 1974. Phys. Rev. Lett. 33:1453
24. Simpson, J. W., et al. 1975. Phys. Rev. Lett. 35:699
25. Braunschweig, W., et al. 1975. Phys. Lett. B57:407
26. Himel, T. M., et al. 1980. Phys. Rev. Lett. 44:920
27. Aguilar-Benitez, M., et al. 1982. *Particle Data Tables*; Particle Data Group, CERN and Univ. Calif. Berkeley, April 1982; Phys. Lett. B111:1
28. Karl, G., Meshkov, S., Rosner, J. 1976. Phys. Rev. D13:1203
29. Gaiser, J., et al. 1983. SLAC-PUB-2899. Stanford: SLAC. To be submitted to Phys. Rev. Lett. (A complete discussion of the analysis can be found in Gaiser, J. 1982. PhD thesis, Stanford Univ. SLAC-255. Unpublished.)
30. Bloom, E. D. 1983. In *Physics in Collision: High-Energy ee/ep/pp Interactions*, ed. P. Carlson, W. P. Trower. New York: Plenum
31. Oreglia, M., et al. 1980. Phys. Rev. Lett. 45:959
32. Himel, T. M. 1979. PhD thesis, Stanford Univ. SLAC-223 (unpublished)
33. Bloom, E. D. 1979. Proc. *1979 Int. Symp. on Lepton and Photon Interactions at High Energy*. Batavia, Illinois, August 23–29, p. 92. Batavia: Fermilab

34. Partridge, R., et al. 1980. *Phys. Rev. Lett.* 45:1150
35. Himel, T. M., et al. 1980. *Phys. Rev. Lett.* 45:1146
36. Aschman, D. 1980. *Proc. 15th Rencontre de Moriond*, Les Arcs, France, March 15–21, 2:83. Derux, France: Editions Frontieres
37. Edwards, C., et al. 1982. *Phys. Rev. Lett.* 48:70
38. Eichten, E., et al. 1980. *Phys. Rev.* D21:203
39. Novikov, V. A., et al. 1977. *Phys. Lett.* B67:409; Schifman, M., et al. 1979. *Nucl. Phys.* B147:448
40. Barbieri, R., et al. 1976. *Phys. Lett.* B61:465; Barbieri, R., et al. 1980. *Phys. Lett.* B95:93; Barbieri, R., et al. 1981. *Phys. Lett.* B106:497
41. Reinder, J. L., et al. 1982. *Rutherford Lab. Preprint RL-82-017*
42. Porter, F. C. 1981. *SLAC-PUB-2796*. Stanford: SLAC
43. Fritzsch, H., Gell-Mann, M. 1972. *Proc. 16th Int. Conf. on High Energy Physics*, Chicago, Ill. 2:135. Batavia: Fermilab; and personal cummunication
44. Scharre, D. L., et al. 1980. *Phys. Lett.* B97:329
45. Edwards, C., et al. 1982. *Phys. Rev. Lett.* 49:259
46. Edwards, C., et al. 1982. *Phys. Rev. Lett.* 48:458
47. Brodsky, S., et al. 1978. *Phys. Lett.* B73:203; Koller, K., Walsh, T. 1978. *Nucl. Phys.* B140:449; Bjorken, J. D. 1980. *Proc. Summer Inst. on Particle Physics, SLAC Report No. 224*. Stanford: SLAC
48. Appelquist, T., et al. 1975. *Phys. Rev. Lett.* 34:365
49. Chanowitz, M. S. 1975. *Phys. Rev.* D12:918; Okun, L. G., Voloshin, M. B. 1976. *ITEP Preprint No. ITEP-95-1976*
50. Königsmann, K. C. 1982. *17th Rencontre de Moriond; Workshop on New Spectroscopy*, Les Arcs, France, March 20–26; also *SLAC-PUB-2910 (1980)*. Stanford: SLAC
50a. Scharre, D. L. 1980. Presented at 6th Int. Conf. on Exp. Meson Spectrosc., Upton, N.Y., Apr. 25–26. *Meson Spectroscopy*, p. 329. Brookhaven Natl. Lab., N.Y.
51. Bartel, W., et al. 1977. *Phys. Lett.* B64:483; *Phys. Lett.* B66:489
52. Novikov, V. A., et al. 1979. *Phys. Lett.* B86:347; Novikov, V. A., et al. 1980. *Nucl. Phys.* B165:55; Karl, G. 1977. *Nuovo Cimento* 38:315; Fritzsch, H., Jackson, J. D. 1977. *Phys. Lett.* B66:365; Pham, T. N. 1979. *Phys. Lett.* B87:267
53. Alexander, G., et al. 1978. *Phys. Lett.* B72:493; Brandelik, R., et al. 1978. *Phys. Lett.* B74:292; Alexander, G., et al. 1978. *Phys. Lett.* B76:652
54. Edwards, C., et al. 1982. *Phys. Rev.* D25:3065
55. Kabir, P. K., Hey, A. J. G. 1976. *Phys. Rev.* D13:3161. Note that the first occurrence of $\sin^2 \vartheta_M$ in Eq. (6) of this reference should be replaced by $\sin 2\vartheta_M$.
56. Krammer, M. 1978. *Phys. Lett.* B74:361
57. Gampp, W., Genz, H. 1978. *Phys. Lett.* B76:319
58. Baillon, P., et al. 1967. *Nuovo Cimento* A50:393
59. Montanet, L. 1980. *Proc. 20th Conf. on High Energy Physics*, Madison, Wisconsin, 17–23 July, ed. L. Durand, L. G. Pondrom. New York: Am. Inst. Phys.
60. Dionisi, C., et al. 1980. *Nucl. Phys.* B169:1
61. Herndon, D. J., Söding, P., Cashmore, R. J. 1975. *Phys. Rev.* D11:3165
62. Palmer, W. F., Pinsky, S. S. 1982. *DOE/ER/01545-328, Ohio State Univ. Preprint*
62a. Stanton, R. N., et al. 1979. *Phys. Rev. Lett.* 42:346
63. Scharre, D. L. 1982. *Proc. Orbis Scientiae*, Coral Gables, Fla., ed. A. Perlmutler. New York: Plenum; also *SLAC-PUB-2880*. Stanford: SLAC
64. Franklin, M. E. B. 1982. PhD thesis, Stanford Univ., SLAC-254 (unpublished)
65. Burke, D. L., et al. 1982. *Phys. Rev. Lett.* 49:632
66. Newman-Holmes, C. 1982. *SLAC-PUB-2971*. Stanford: SLAC
67. Brandelik, R., et al. 1977. *Phys. Lett.* B70:132; Brandelik, R., et al. 1979. *Phys. Lett.* B80:412; Brandelik, R., et al. 1979. *Z. Phys.* C1:233
68. Einhorn, M. B., Quigg, C. 1975. *Phys. Rev.* D12:2015; Ellis, J., Gaillard, M. K., Nanopoulos, D. V. 1975. *Nucl. Phys.* B100:313; Quigg, C., Rosner, J. L. 1978. *Phys. Rev.* D17:239; Fakirov, D., Stech, B. 1978. *Nucl. Phys.* B133:315
69. Partridge, R., et al. 1981. *Phys. Rev. Lett.* 47:760
70. Fayet, P., Mezard, M. 1981. *Phys. Lett.* B104:226
71. Edwards, C., et al. 1982. *Phys. Rev. Lett.* 48:903
72. Pecci, R. D., Quinn, H. R. 1977. *Phys. Rev. Lett.* 38:1440, *Phys. Rev.* D16:1791; Weinberg, S. 1978. *Phys. Rev. Lett.* 40:223; Wilczek, F. 1978. *Phys. Rev. Lett.* 40:279

73. Faissner, H., et al. 1981. *Phys. Lett.* B103:234
74. Goldman, T., Hoffman, C. M. 1978. *Phys. Rev. Lett.* 40:220; Bechis, D. J., et al. 1979. *Phys. Rev. Lett.* 42:1511; Zehnder, A. 1981. *Phys. Lett.* B104:494
75. Porter, F. C., Königsmann, K. C. 1982. *Phys. Rev.* D25:1993, D26:716
76. Niczyporuk, B., et al. 1982. *DESY 82-068.* Hamburg: DESY
76a. Sivertz, M., et al. 1982. *Phys. Rev.* D26:717
77. Dine, M., Fischler, W., Srednicki, M. 1978. *Phys. Lett.* B104:199; Kim, J. 1979. *Phys. Rev. Lett.* 43:193; Wise, M. B., Georgi, H., Glashow, S. I. 1981. *Phys. Rev. Lett.* 47:402; Wilczek, F. 1982. *Phys. Rev. Lett.* 49:1549
78. Eichten, E., et al. 1975. *Phys. Rev. Lett.* 34:369
79. Porter, F. C., et al. 1982. *17th Rencontre de Moriond Workshop on New Flavors*, Les Arcs, France, Jan. 24–30, p. 27
80. Barnett, R. M., Dine, M., McLerran, L. 1980. *Phys. Rev.* D22:594
81. Siegrist, J. 1979. PhD thesis, Stanford Univ., *SLAC Rep. No. SLAC-225.* Stanford: SLAC
81a. Lockman, W., et al. 1983. *SLAC-PUB-3030.* Stanford: SLAC
82. Sjostrand, T. 1982. *LU-TP-82-3*
82a. Siegrist, J. L., et al. 1982. *Phys. Rev.* D26:991 (Mark I); Rice, E., et al. 1982. *Phys. Rev. Lett.* 48:906 (CUSB); Berger, C., et al. 1979. *Phys. Lett.* B81:410 (PLUTO); Bock, P., et al. 1980. *Z. Phys.* C6:125 (DESY-Heidelberg); Albrecht, H., et al. 1982. *Phys. Lett.* B116:383 (DASP II); Niczyporuk, B., et al. 1982. *Z. Phys.* C15:299 (LENA)
83. Edwards, C., et al. 1982. *Phys. Lett.* B110:82
84. Berger, C., et al. 1980. *Phys. Lett.* B94:254
85. Roussarie, A., et al. 1981. *Phys. Lett.* B105:304
86. Brandelik, R., et al. 1981. *Z. Phys.* C10:117
87. Biddick, C. J., et al. 1980. *Phys. Lett.* B97:320
88. Bonneau, G., Gourdin, M., Martin, F. 1973. *Nucl. Phys.* B54:573
89. Brodsky, S. L., Lepage, G. P. 1981. *Phys. Rev.* D24:1808
90. Rosner, J. L. 1981. *Phys. Rev.* D23:2625; Donoghue, J. F., Johnson, K., Li, B. A. 1981. *Phys. Lett.* B99:416
91. Gilman, F. J. 1979. *Int. Conf. on Two-Photon Interactions*, ed. J. F. Gunion, p. 215. Univ. Calif., Davis
92. Greco, M. 1980. *Proc. Int. Workshop on $\gamma\gamma$ Collisions*, Amiens, France, p. 311. *Lect. Notes Phys.*, Vol. 34. Berlin: Springer-Verlag
93. Babcock, J., Rosner, J. 1976. *Phys. Rev.* D14:1286
94. Emms, M. J., et al. 1975. *Nucl. Phys.* B96:155
95. De Rújula, A., Georgi, H., Glashow, S. L. 1976. *Phys. Rev. Lett.* 37:785
96. Tompkins, J. C. 1980. *Quantum Chromodynamics*, ed. A. Mosher, p. 556. Stanford: SLAC
97. Sadrozinski, H. F. W. 1980. *Proc. 20th Int. Conf. on High Energy Physics*, Madison, Wisconsin, July 17–23
98. Rapidis, P., et al. 1977. *Phys. Rev. Lett.* 39:5261; Bacino, W., et al. 1978. *Phys. Rev. Lett.* 40:671
99. Goldhaber, G., et al. 1977. *Phys. Rev. Lett.* B69:503
100. Ono, S. 1976. *Phys. Rev. Lett.* 37:655

SUM RULE APPROACH TO HEAVY QUARK SPECTROSCOPY

M. A. Shifman

Institute of Theoretical and Experimental Physics, Moscow 117259, USSR

CONTENTS

INTRODUCTION	199
Outlining the Problem	199
General Picture of QCD Vacuum	202
Short Summary of Results	205
METHOD OF QCD SUM RULES	208
Quantum Mechanical Example (Harmonic Oscillator)	208
Charmonium	210
Find the Gluon Condensate Yourself	214
Bottomonium Family—Peculiarities	217
Super-Heavy Quarks	219
OTHER ASPECTS OF QUARKONIUM THEORY	222
Photon Decays	223
Mesons with Open Flavor	226
Limits of Applicability of the Method	227
CONCLUSIONS AND PERSPECTIVES	230

INTRODUCTION

Outlining the Problem

The first new particle, J/ψ, which opened to us the world of heavy quarks, was discovered about nine years ago (4, 5), but today this event already seems to be ancient history. At that time, in 1974, the quark-gluon picture of hadronic matter was only one of a few competing alternatives—attractive but not proven—and quantum chromodynamics (QCD) was taking its first steps (29, 33, 56).

Now nobody questions the existence of quarks and gluons; they have become an indispensible part of our intuition. The triumph of QCD, the theory describing their interactions, is absolute. This quiet revolution might have happened much later if it were not for the impetus given by the discovery of heavy quarks and subsequent theoretical investigations.

Two heavy quarks, c and b, are known for certain and the third one, t, is expected to appear. They can combine with each other or with light quarks u, d, and s (the latter will be denoted generically by q) to form various hadronic families, $c\bar{q}$, bqq, $t\bar{t}$, etc. Members of the psi (charmonium) family headed by J/ψ have the structure $c\bar{c}$, while the upsilon (Υ) system consists of $b\bar{b}$, sometimes called bottomonium. (See the chapter by Franzini & Lee-Franzini in this volume for a review of the upsilon resonances.)

As early as 1975, soon after the pioneering paper of Appelquist & Politzer (2) revealing the nature of J/ψ, it was assumed that charmonium might well be the "hydrogen atom" of hadronic physics (25). In a sense, although not literally, this point of view has been confirmed. The heavy quark systems indeed have played a distinguished role in the development of the theory of strong interactions. The natural question is: Why did they happen to be more useful than dozens of light hadrons, good old acquaintances?

To give at least a partial answer one cannot help saying a few words about a special aspect of quantum chromodynamics (QCD). Of course, it would be impossible to characterize all its achievements and problems in this short survey intended for nonspecialists. Exhaustive reviews can be found elsewhere (e.g. 17, 23, 47, 52, 57).

At the fundamental level the dynamics of quarks and gluons is described by the well-known QCD Lagrangian

$$L_{QCD} = -\frac{1}{4} G^a_{\mu\nu} G^a_{\mu\nu} + \sum_\psi \bar{\psi}(i\slashed{D} - m_\psi)\psi, \qquad 1.$$

where $G^a_{\mu\nu}$ is the gluon field strength tensor, ψ is the quark field, m_ψ is the quark mass, and the summation runs over all quark flavors, ψ = u, d, s, c, ... Because of asymptotic freedom (33, 56) the effective coupling constant g_s (it enters via the covariant derivative $\slashed{D}$) is small at short distances; quarks and gluons are nearly free and their behavior is completely controlled by standard perturbation theory. However, the old traditional hadrons, charmonium, and even bottomonium are formed at larger distances, and the corresponding spectra are essentially determined by the notorious effects of color confinement, a mysterious phenomenon that so far is not understood completely by theorists.

The main difficulty blocking quantitative analysis of confining forces is the fact that they switch on at distances where $g_s \sim 1$. Thus, hadronic physics is deprived of a genuine small parameter like $\alpha = 1/137$. Still, we qualitatively understand that large-distance dynamics reflects nontrivial structure of the QCD vacuum, much more complicated than one can judge starting from perturbative expansions. Explicit examples, for instance, the famous instantons (8), indicate that the QCD vacuum is populated by violent nonperturbative fluctuating fields that are responsible for the formation of the spectrum and all its peculiarities. Thus, the vacuum resembles a complicated medium; valence quarks and gluons have to live in it, not in the empty space.

The vacuum medium consists of two phases. One is built from gluon fields, the other from the quark fields. It is important that the quark phase includes only light pairs, $\bar{u}u$, $\bar{d}d$, or $\bar{s}s$. Indeed, a virtual pair of heavy quarks, say $\bar{c}c$, from the uncertainty principle lives in the vacuum of order of $(2m_c)^{-1}$, a time much shorter than the characteristic scale inherent to the vacuum fields. As a result, the effect of heavy quark pairs is negligible. Moreover, such vacuum pairs can be consistently accounted for within the ordinary perturbation theory.

It is intuitively clear that the fact of existence of the "quark medium" affects drastically the spectrum of old light hadrons built from u, d, and s quarks: suffice it to mention the π meson, whose masslessness in the chiral limit is entirely due to collective phenomena. At the same time, charmonium and bottomonium levels below the open flavor threshold are practically insensitive to the vacuum fluctuations of u, d, and s fields. They depend on the structure of the gluon phase only.

Thus, the situation with heavy quarkonium is very favorable. Charmonium and bottomonium are excellent probes of large-distance dynamics because we can concentrate on one isolated aspect of the problem of hadrons, an obvious advantage.

There are a few quantitative approaches used in connection with heavy quark spectroscopy. The oldest one, dating back to the original paper (2), is based on the nonrelativistic model, which reduces the effect of the gluon medium to a phenomenological static potential acting between bound quarks, say $\bar{c}c$ or $\bar{b}b$. Such an approximation is justified theoretically if a characteristic frequency of the quark motion is much smaller than that inherent to the medium,

$$\omega_{\text{quark}} \ll \omega_{\text{glue}}. \qquad 2.$$

Details of the model are discussed in many nice reviews (see, for example, 3, 26, 31, 39, 40, 58). Here we only note that the condition written in Equation 2 is valid in the full sense of this word only for very large masses, $m_Q \gtrsim 10$ GeV. (The subscript Q marks heavy quarks; recall that $m_c \approx 1.35$ GeV, $m_b \approx 4.80$ GeV.) Besides that, the relation between the parameters of the model and the QCD Lagrangian is obscure.

This survey is devoted to another approach—the so-called QCD sum rules (68, 70, 81), which express the masses and coupling constants of low-lying quarkonium levels in terms of a few fundamental parameters. Two of them, g_s and m_Q enter the Lagrangian, and the third parameter measures an average gluon field in the QCD vacuum. Thus the sum rule approach presents a bridge between purely mathematical investigations of non-Abelian gauge theories, on the one hand, and hadronic physics on the other.

General Picture of QCD Vacuum

If in classical field theory the vacuum configuration is usually simple—the lowest energy is achieved for vanishing fields—the situation in quantum field theory is less trivial because the fields are permanently fluctuating. Even in electrodynamics one enounters certain difficulties, namely the energy of zero-point oscillations

$$E_{\text{vac}} = \sum_n \frac{1}{2} \omega_n$$

diverges at large frequencies. The problem is solved by virtue of a subtraction. Since only the difference between the energies of the excited and the lowest states is observable, the sum $\sum \omega_n$ can be treated as a constant; for instance, one can put it equal to zero without affecting any physical consequences. This does not mean, of course, that the presence of the zero-point oscillations is unimportant for measurable quantities. Introducing particles (electrons, positrons) perturbs the vacuum fields, and this perturbation—now the energy difference—results in the Lamb shift of atomic levels.

Naturally, in QED we are dealing with a very small effect, of order $\alpha^3 = (1/137)^3$. In QCD there is no small free parameter, and the most global characteristics of the spectrum are determined in this way. Are we able to account for the violent gluon field fluctuations that go beyond the standard perturbation theory?

The most straightforward strategy is adopted by those theorists who exploit the lattice version of QCD by performing the so-called Monte Carlo simulations. The discrete lattice means a finite number of degrees of freedom, and in Euclidean space the problem reduces to one in statistical mechanics. One simply generates in computers various field configurations randomly, each of them being taken with the weight $\exp(-S)$ where S is the corresponding (Euclidean) action (see 43, 87):

$$S = \frac{1}{4} \int d^4x \, G^a_{\mu\nu}(x) G^a_{\mu\nu}(x).$$

After a few successive iterations, an equilibrium distribution of fields is determined or, in other words, the class of configurations dominating the partition function. The equilibrium distribution serves as a numerical model of the QCD vacuum. Each particular value of the bare coupling constant $(g_s)_{\text{bare}}$ results in its own "vacuum." Then the quark propagators are calculated numerically in the given field, which allows one to fix correlation functions of colorless currents, say

$$\Pi_{\mu\nu}^{(c)}(x) = \langle \text{vac} | T\{\bar{c}(x)\gamma_\mu c(x), \bar{c}(0)\gamma_\nu c(0)\} | \text{vac} \rangle, \qquad 3.$$

where $\bar{c}\gamma_\mu c$ produces 1^{--} charmonium states from the vacuum. By studying the dependence of $\Pi_{\mu\nu}^{(c)}$ on $(g_s)_{\text{bare}}$ and x for large x, one can extract, in principle, the charmonium spectrum in the continuous limit.

This approach, very attractive from the pragmatic point of view, only strengthens the desire to go as far as possible with analytic methods in spite of the fact that they require some additional approximations.

If x is very small, the injected quarks have no time to interact with the vacuum medium and so propagate as free particles. However, in accordance with the uncertainty principle, small x values mean bad energy resolution, or, in more exact terms, for small x the correlation function in Equation 3 is saturated by a large number of charmonium levels, each contributing roughly equally.

With increasing x the (radially) excited states become less important and finally die away since their weight relative to the lowest-lying level is proportional to $\exp[-(M'-M)|x|]$ (here M' and M are the masses of the excited and ground levels with given quantum numbers, $J^{PC} = 1^{--}$ in the case at hand). Thus we are left with the ground state and can concentrate on its investigation.

The price we must pay is obvious—now it is necessary to account for the interaction of the injected quarks with the vacuum medium. If we are not very ambitious and do not try to put $|x| \to \infty$ in the mathematical sense, the price may be not very high, however. Indeed, for moderate values of $|x|$ the expansion of the quark propagator in $|x|^2 |\mathbf{E}_{\text{vac}}^a|$ or $|x|^2 |\mathbf{H}_{\text{vac}}^a|$ is meaningful, where $\mathbf{E}_{\text{vac}}^a$, $\mathbf{H}_{\text{vac}}^a$ are typical vacuum color fields. Notice that the gauge potential A_μ^a will not enter final formulas because the correlation function in Equation 3 is gauge invariant. Terms linear in $\mathbf{E}_{\text{vac}}^a$, $\mathbf{H}_{\text{vac}}^a$ will not enter either since there is no preferred orientation. The leading dynamical effect in $\Pi_{\mu\nu}^{(c)}$ will reduce evidently to

$$\langle \text{vac} | g_s^2 G_{\mu\nu}^a G_{\mu\nu}^a | \text{vac} \rangle \qquad 4.$$

multiplied by some known function of x.

Thus, we can fix $\Pi_{\mu\nu}^{(c)}(x)$ in the intermediate x domain without complete information on the vacuum structure. The matrix element in Equation 4 measures the mean field squared and was introduced first by Vainshtein et al (81). Its value should, of course, be calculable in the final theory of color confinement. At present we consider it as a free parameter, to be extracted from phenomenology.[1] It is important that one and the same parameter appears in the sum rules for charmonium and bottomonium for all quantum numbers, $J^{PC} = 1^{--}, 0^{-+}, 0^{++}, 2^{++}$, etc.

The central question is whether the expansion in the vacuum field is applicable for x large enough to guarantee the saturation of $\Pi^c_{\mu\nu}(x)$ and other similar correlation functions by the ground state alone. It is not evident a priori that the two requirements can be met simultaneously. If they are, we get a (quasi) theory of ground states based directly on first principles.

Since there are no small parameters like $\alpha = 1/137$, the answer depends crucially on numerics. As we see below, the balance between both requirements is really achieved in a certain domain of x and this, in turn, ensures the success of the sum rule method. The underlying physical reason seems to be the following. The characteristic frequencies of bound quarks, $\bar{c}c$ or $\bar{b}b$, are of order

$$\omega_{\text{quark}} \sim M_{\psi'} - M_{J/\psi} \sim M_{\Upsilon'} - M_{\Upsilon} \sim 0.6 \text{ GeV}.$$

At the same time the space-time scale of vacuum fields amounts to $R_{\text{vac}} \sim 0.7$–1 fermi or 3.5–5 GeV^{-1}. This fact—a relatively small value of the product

$$(\omega_{\text{quark}} R_{\text{vac}})^{-2} = \frac{1}{4} - \frac{1}{9}$$

—means that the quarks indeed perceive only coarse averaged vacuum characteristics and the expansion in $|x|^2 |\mathbf{E}^a_{\text{vac}}|$ should be convergent. As was already mentioned, in most practical applications it is sufficient to keep only the first nontrivial term proportional to the field squared (see Equation 4). In some cases further refinements may be useful. In particular, accounting for the cubic term helps to extend the procedure and somewhat increases the accuracy of predictions.

The numerical value of the gluon condensate was extracted from charmonium sum rules in the original paper (81):

$$\left\langle \text{vac} \left| \frac{g_s^2}{4\pi^2} G^a_{\mu\nu} G^a_{\mu\nu} \right| \text{vac} \right\rangle \approx 0.012 \text{ GeV}^4. \qquad 5.$$

Since then several similar analyses have been performed in a slightly different way, in particular for the Υ channel (10, 14, 59, 83). The estimate in Equation 5 is essentially confirmed, although there are some indications that the true value of the gluon condensate may be a bit larger, by 10–30%.

---

[1] In principle, some rough estimates within the instanton calculus (69, 78) or lattice Monte Carlo simulations (7, 21, 22, 44) are also available.

Below we also touch upon sum rules for mesons with open flavor, say $b\bar{q}$. The basic role here belongs to another vacuum characteristic, namely

$$\langle \text{vac} |\bar{q}q| \text{vac}\rangle \approx -(250 \text{ MeV})^3. \qquad 6.$$

This expectation value measures the mean vacuum density of u, d, and s quark pairs. The fact that the quark condensate (Equation 6) does exist has been known for a long time from low energy pion physics (30). Soft pion techniques allow one to establish the following relation:

$$(m_u + m_d)\langle \text{vac} |\bar{u}u + \bar{d}d| \text{vac}\rangle = -f_\pi^2 m_\pi^2, \qquad 7.$$

where m_u, m_d are the quark masses, m_π is the pion mass, and f_π is the $\pi \to \mu\nu$ decay constant, $f_\pi \approx 133$ MeV. All we need now in order to fix the quark condensate is an estimate of $m_u + m_d$. It is worth emphasizing that Equation 7 contains the so-called current quark masses that enter the QCD Lagrangian and determine chiral symmetry-breaking effects. They are surprisingly small; one can hardly doubt this fact today. Quantitatively, the analyses of Leutwyler (45) and Weinberg (86) imply

$$m_u + m_d = 11 \text{ MeV},$$

which leads, in turn, to Equation 6. This result, as any other assumption, has been checked within the sum rules themselves. Moreover, rough estimates of $\langle \text{vac} |\bar{q}q| \text{vac}\rangle$ were obtained recently in an instanton-based model (78) and in the lattice Monte Carlo simulations (34). Both approaches give numbers that do not contradict Equation 6.

Short Summary of Results

Members of the bottomonium and especially charmonium families are rather numerous now. For the latter, five ground levels of different quantum numbers are established experimentally (see Table 1). Historically, the first problems solved within the framework of the QCD sum rules referred to $J/\psi \to e^+e^-$ width (81) and J/ψ-η_c mass splitting (68). If the theoretical number for $\Gamma(J/\psi \to e^+e^-)$ brought no surprises, being in nice agreement with experiment, the issue of the η_c mass represented one of most dramatic pages, a culmination point in the development of QCD-based ideas. In accordance with the experimental data of those days, the ground-state paracharmonium was identified with an X(2.83)—a particle that has disappeared without a trace now. True, both quantities $\Gamma(J/\psi \to X\gamma)$ and $M_{J/\psi} - M_X$ seemed suspicious in any theoretical scheme. Still, the potential model was unable to state definitely: "This candiate is fictitious." Because of its phenomenological nature, it could, with some stretching, accom-

Table 1 The lowest-lying charmonium levels with different quantum numbers

Name	Nonrelativistic spectroscopic notation[a]	J^{PC}	Mass (MeV)	Currents producing the corresponding states from the vacuum
J/ψ	1^3S_1	1^{--}	3097	$\bar{Q}\gamma_\mu Q$
η_c	1^1S_0	0^{-+}	2980	$\bar{Q}i\gamma_5 Q$
χ_0	1^3P_0	0^{++}	3415	$\bar{Q}Q$
χ_1	1^3P_1	1^{++}	3510	$\bar{Q}\gamma_\mu\gamma_5 Q^b$
χ_2	1^3P_2	2^{++}	3556	$i\bar{Q}\gamma_\mu\overleftrightarrow{D}_\nu Q + (\mu \leftrightarrow \nu)^b$

[a] The convention used throughout the review is the following: $(n_r+1)^{2S+1}L_J$ where n_r is the radial quantum number (the number of nodes in the radial wave function).
[b] The current also produces lower spin states, which, however, do not show up in an appropriate invariant structure.

modate the X(2.83). In contrast, the sum rule prediction was unambiguous (68):

$$M_{\eta_c} = 3.00 \pm 0.03 \text{ GeV}.$$

It is perhaps worth nothing that this result was obtained two years before the SLAC discovery of the genuine η_c particle with mass 2.98 GeV (35, 55).

The range of applications of the sum rule method to charmonium became much wider after the work of Reinders et al (59) devoted to 1^3P_J levels. Using essentially the same values for the gluon condensate, $\alpha_s = g_s^2/4\pi$ and m_c, the authors found positions of the P levels with accuracy not worse than 20 MeV. For 1^3P_J levels agreement with experiment is excellent; the still-unobserved 1^1P_1 state is predicted to have a mass

$$M(1^1P_1) = 3.51 \pm 0.01 \text{ GeV}.$$

A new step is made in Refs. (42, 60), where three-point correlation functions of charmed quarks are considered and gluon condensate effects are calculated. This allows one to obtain $\Gamma(\eta_c \to 2\gamma)$ and work out other physical consequences.

The upsilon family is much younger, and at the time of writing the only observed ground level was $\Upsilon(1^3S_1)$.[2] The analysis of the upsilon spectrum has some peculiarities (we return to this point later). The complications are of a technical nature, not of principle, and the results for resonance masses

---

[2] Evidence for the 1^3P_J levels has now been found. See Section 4.2.2 of the chapter by P. Franzini and J. Lee-Franzini in this volume.

and coupling constants are as always expressed in terms of

$$\left\langle \text{vac} \left| \frac{\alpha_s}{\pi} G^a_{\mu\nu} G^a_{\mu\nu} \right| \text{vac} \right\rangle, \quad \alpha_s, \quad m_b$$

where m_b is the current b-quark mass. The latter was fixed from the upsilon sum rules (83) as

$$m_b \approx 4.80 \text{ GeV}.$$

The same work gives the position of the bottomonium P levels,

$$M(1^3P_J) = 9.83 \pm 0.03 \text{ GeV},$$

where the number on the right-hand side refers to the center of gravity.[3] First data obtained recently at CESR locate 1P states at 9.90 GeV, somewhat higher than the theoretical prediction. Keeping in mind that experimental studies of the upsilon P levels are only beginning, the discrepancy probably should not be taken too seriously.

Another interesting and unambiguous result is the ortho-paraupsilon splitting, which turns out to be rather small (85)

$$M_\Upsilon - M_{\eta_b} = 35\text{--}40 \text{ MeV}.$$

We return to a more detailed discussion of this point later.

A few words about spectroscopy of hadrons with open flavor. Systems with one heavy quark occupy, in a sense, an intermediate position between heavy quarkonium and old light hadrons. This makes them unique in many aspects, in particular, as was already mentioned, the corresponding sum rules are extremely sensitive to the quark condensate. Shuryak (77) analyzed the sum rules in the limit $m_Q \to \infty$; $b\bar{q}$ mesons are also discussed by Reinders et al (62). One of the striking findings is a surprisingly large mass difference for mesons with quantum numbers 0^-, 1^- on one hand and 0^+, 1^+ on the other. It amounts to 800 MeV, while more traditional estimates, based on, say, potential models, imply a considerably smaller splitting. The positive parity mesons with c or b quarks have not been observed so far. Their discovery at the proper place would be a crucial test of the predictive power of the approach.

Finally, one more issue that admits an elegant solution based on the notion of the gluon condensate is the so-called pre-Coulomb behavior of very heavy quarkonium (46, 84). If $m_Q \to \infty$ the quarks are bound essentially by the Coulomb force and the size of the $Q\bar{Q}$ system is much smaller than the confinement radius. The vacuum fields shift the $Q\bar{Q}$ levels

---

[3] P levels in QCD sum rules are discussed also in Refs. (14, 59).

and renormalize couplings. Deviations from Coulomb formulas are evidently determined in the leading approximation by the gluon condensate (Equation 5). Unlike the original sum rules where the expansion parameter is of a numerical nature, here the higher order corrections can be made arbitrarily small if the quark masses are sufficiently large. Based on the gluon condensate, we can give in this limit an exhaustive theory for any level we would like to study. An adequate formalism was developed in (46, 84) and results useful for practical applications in bottomonium are collected in the next section. They are expected to be even more important for toponium, the still-unobserved bound $\bar{t}t$ state of the third heavy quark t. At present, the experimental lower limit on the t mass is

$$m_t > 18.5 \text{ GeV}.$$

METHOD OF QCD SUM RULES

In the previous section we sketched the basic ideas underlying the QCD sum rule method and summarized the results, leaving aside all technical aspects. Now we show how the ideas are realized practically and explain the gross features of the corresponding formalism. Numerous computational details and subtle points are omitted, so that nonexperts can, we hope, follow the material without special efforts.

We start with a simple example from quantum mechanics familiar to everybody. It demonstrates nicely that a power series expansion for a Green's function, valid formally at short τ (τ stands for Euclidean time), being extrapolated to larger τ, fixes the position of the lowest-lying level to a good accuracy (9, 82).

Quantum Mechanical Example (Harmonic Oscillator)

Of course, in quantum mechanics the Schrödinger equation allows one to find the spectrum of any system, at least in principle. Here we determine the position of the ground level for the three-dimensional harmonic oscillator in another way that may seem extremely awkward. It has a unique advantage—it can be generalized to QCD, where there is no analogue of the Schrödinger equation.

Consider, a particle of mass m moving in the spherically symmetric potential

$$V(r) = \frac{1}{2}m\omega^2 \mathbf{r}^2.$$

For S-wave states the following sum plays the role of the correlation function in Equation 3:

$$S(\tau) = \sum_{n=0,2,4,\ldots} |R_n(0)|^2 \, e^{-E_n\tau}, \qquad 8.$$

where R_n is the S-wave radial wave function and E_n is the corresponding energy eigenvalue. Odd-n levels do not contribute because their wave functions vanish at the origin. One immediately recognizes in $S(\tau)$ the so-called time-dependent Green's function, namely,

$$S(\tau) = 4\pi G(\mathbf{x}_2 = 0, t_2 = -i\tau \,|\, \mathbf{x}_1 = 0, t_1 = 0),$$

and the exact expression for $G(2|1)$ taken from any textbook implies

$$S(\tau) = \frac{2}{\sqrt{2\pi}} \left(\frac{m\omega}{\sinh \omega\tau} \right)^{3/2}.$$

By analyzing the closed form for $\tau \to \infty$ and comparing it with the definition in Equation 8, one can readily obtain exhaustive information on the S-wave spectrum. For instance, the ground-state energy evidently reduces to the following simple expression:

$$E_0 = \lim_{\tau \to \infty} [-(d/d\tau) \ln S(\tau)] = \frac{3}{2}\omega \lim_{\tau \to \infty} (\coth \omega t) = \frac{3}{2}\omega.$$

We shall assume, however, that just as in QCD the large-τ asymptotics is unknown, and all we are able to calculate theoretically are the first few terms of a power expansion in the small-τ domain where formally $S(\tau)$ receives contributions from a large number of levels. The zeroth order term corresponds to free motion, the next one is the first order correction, and so on:

$$-(d/d\tau) \ln S(\tau) = \frac{3}{2}\omega \left[\frac{1}{\omega\tau} + \frac{\omega\tau}{3} - \frac{(\omega\tau)^3}{45} + \frac{2(\omega\tau)^5}{945} + \cdots \right].$$

Now, it is impossible to let τ tend to infinity in the mathematical sense. Figure 1 shows, however, that there exists an intermediate stability plateau where, on the one hand, the perturbation expansion still works and, at the same time, the ground level almost saturates the sum. For instance, with only the first two terms the prediction for E_0 is $\frac{3}{2}\omega(2/\sqrt{3})$, 15% higher than the actual position. In the middle of the stability plateau, the third (neglected) term of the expansion does not exceed 10%. If four terms are kept, $E_0 = \frac{3}{2}\omega \times 1.06$. Moreover, even a rough idea of the contribution of higher levels (they can be accounted for in the quasi-classical approximation, say) drastically improves the estimate of E_0 and makes the accuracy better than 1%.

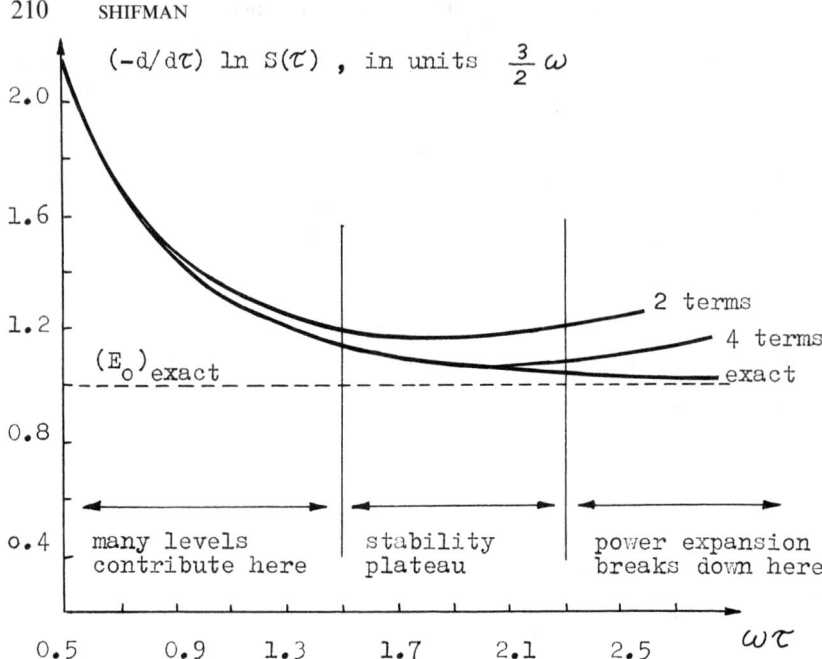

Figure 1 Harmonic oscillator: power expansion of $(-d/d\tau) \ln S(\tau)$ versus $\omega\tau$. The lower curve is the exact result and the other two curves correspond to two and four terms in the expansion.

An analogous procedure also yields $|R_0(0)|^2$. In the case at hand, the theoretical uncertainty is evaluated by direct comparison with the exact result. In QCD one has to be more sophisticated and estimate the uncertainty by analyzing the power series themselves. It is assumed, of course, that there are no irregularities in the series and high (unknown) power corrections are not abnormally large.

Charmonium

The first step is to select the correlation function containing a meson we would like to study. For instance, Equation 3 corresponds to 1^{--} charmonium states. Practically, instead of Equation 3, it is more convenient to deal with Fourier-transformed amplitudes of the type

$$\Pi(q) = i \int d^4x \, e^{iqx} \langle \text{vac} | T\{J_{\mu\ldots}(x), J_{\nu\ldots}(0)\} | \text{vac} \rangle, \qquad 9.$$

where $J_{\mu\ldots}$ is an appropriate current (see Table 1). Far from the physical cut beginning at $q^2 = 4m_c^2$ the two-point function in Equation 9 is calculable in the form of an expansion in powers of the QCD coupling constant and the properties of the vacuum color fields, as described in the introduction. On the other hand, the discontinuity along the cut is measurable; it represents a sum of the resonance and continuum contributions. By means of a dispersion relation, $\Pi(q)$ can thus be calculated from observed quantities, or alternatively parameterized in terms of the sought-for properties (masses, widths) of specific states.

Further maneuvres depend on the particular physical quantity we are interested in. Generally speaking, $\Pi(q)$ of Equation 9 is composed of a few invariant structures and they should be arranged in a form ensuring maximal sensitivity to the quantity of interest. The choice is not unique, though, and reflects the preference of investigator. Historically, the following ratio was introduced to fix the mass (68):

$$\frac{\frac{1}{(n+1)!}\frac{d^{n+1}}{(dq^2)^{n+1}}\Pi(q^2)}{\frac{1}{n!}\frac{d^n}{(dq^2)^n}\Pi(q^2)}, \quad q^2 \to 0. \qquad 10.$$

It plays a role analogous to $(d/d\tau) \ln S(\tau)$ of the previous section devoted to the harmonic oscillator. Moreover, n/m_c has the meaning of τ—the larger n, the larger distances are probed.

For the vector current the ratio in Equation 10 reduces, by means of the dispersion representation for $\Pi(q^2)$, to

$$r_n = \frac{\int \frac{R_c(s)\,ds}{s^{n+1}}}{\int \frac{R_c(s)\,ds}{s^n}},$$

where $R_c(s)$ is a normalized cross section of charm production in e^+e^- beams (it includes J/ψ, ψ', higher resonances and continuum):

$$R_c(s) = \frac{\sigma(e^+e^- \to \text{charm})}{\sigma_{\text{QED}}(e^+e^- \to \mu^+\mu^-)}.$$

At large n the contribution of the lowest-lying state dominates the dispersion integrals. Consequently

$$r_n \to M_{J/\psi}^{-2}, \quad n \to \infty.$$

We calculate $\Pi(q^2)$ theoretically and find r_n from Equation 10 for moderate n and extrapolate the result to larger n. Asymptotically large n are beyond reach since the expansion (recall that it contains nonperturbative terms as well as the standard perturbative corrections from gluon exchange) blows up. Indeed, as was first shown in (68),

$$r_n = \frac{n^2-1}{n^2+\frac{3}{2}n}\frac{1}{4m_c^2}\left\{1 + A_n\frac{\alpha_s(m_c)}{\pi} - \frac{(6n+14)n(n+1)(n+2)}{(2n+3)(2n+5)}\right.$$
$$\left. \times \frac{4}{9}\pi\frac{\langle\text{vac}|\alpha_s G^2|\text{vac}\rangle}{(4m_c^2)^2} + \cdots\right\},$$

where A_n is a relatively slowly varying coefficient function, less than or of order unity for $n \leq 10$.[4] The upper bound on n, impossible to pass starting perturbatively from short distances, is set by the last term whose n dependence is extremely steep. The limit lies at $n = 7$, and here the J/ψ contribution to r_n exceeds 95%. Thus, the J/ψ mass is fixed to 1% accuracy (Figure 2a).

A few comments are in order here. First of all, the "prediction" described above is not actually a prediction, since both the current quark mass m_c and the gluon condensate are not known accurately enough from independent sources. So, we have to sacrifice the sum rule in the vector channel in order to fit the input parameters. Figure 2a demonstrates then the existence of the stability domain with absolute J/ψ dominance at $n \approx 6$-7 and the quality of the fit. If $n \leq 5$ the higher states are non-negligible, for $n \geq 8$ higher power terms appear.

The value of the gluon condensate has already been quoted, and m_c, the so-called on-shell mass, is equal to 1.35 GeV. With these parameters in hand other channels are open for analysis. We return to that issue shortly.

Another remark concerns higher order nonperturbative terms, in particular, the triple condensate

$$\langle\text{vac}|g_s^3 f^{abc} G_{\mu\nu}^a G_{\nu\alpha}^b G_{\alpha\mu}^c|\text{vac}\rangle.$$

Its contribution to r_n has been found in a recent work (50). Having included G^3, the authors succeeded in making the stability plateau wider (Figure 2b).

Returning to physical consequences, we start the discussion with the initially controversial η_c particle. The corresponding sum rule was considered in the original paper (68) with a quite definite result (see Introduction). Later, a new analysis was undertaken in the η_c channel (59). It essentially confirmed the previous number for M_{η_c} with one new element, however. Having invoked a simple technical trick, the authors (59) managed to make the stability plateau broader. They observed that the point of differentiation in Equation 10 is a free parameter that can be optimized. True, if $q^2 \neq 0$ the expression for the G^2 coefficient becomes

---

[4] Notice, however, that $A_n \sim \sqrt{n}$ if $n \to \infty$; we discuss this point in the section devoted to bottomonium.

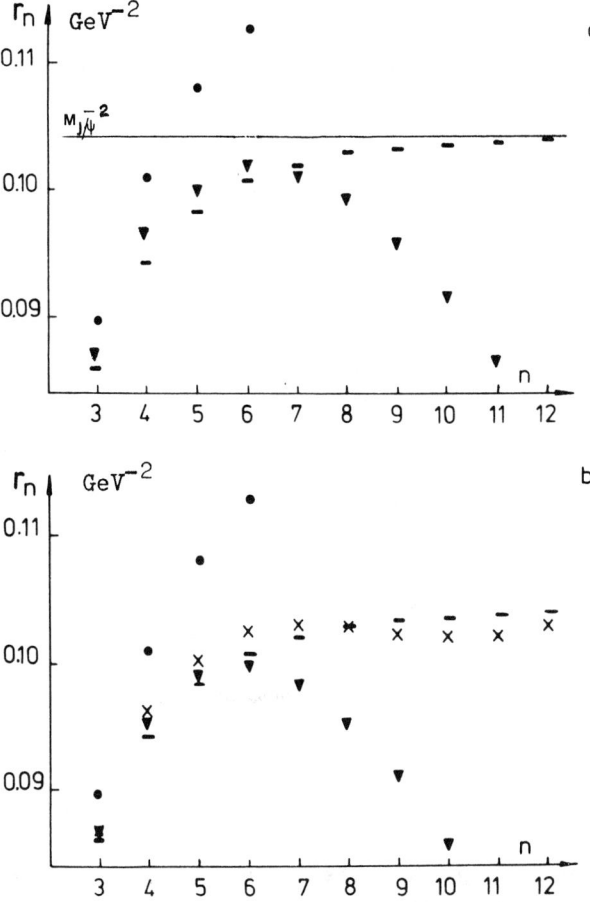

Figure 2 Ratio $r_n = \int ds\, s^{-(n+1)} R_c(s)/\int ds\, s^{-n} R_c(s)$ versus n. (*a*) bars = experiment; circles = no power corrections; nablas = G^2 term included ($\langle \text{vac}\,|(\alpha_s/\pi)G^2|\,\text{vac}\rangle = 0.012$ GeV^4). (*b*) bars = experiment; circles = no power terms; nablas = G^2 only; crosses = G^2 and G^3 terms included.

much more cumbersome. They made up for it by being able to fix the P-level positions. All the principal steps are the same as explained above, only the form of the current varies. It is instructive to reproduce here a few plots from Ref. (59) (Figure 3).

Not to make a false impression, it is worth mentioning other characteristics, namely, resonance coupling constants (residues). The most important of them, $\langle \text{vac}\,|\bar{c}\gamma_\mu c|J/\psi\rangle$, determines the J/ψ electronic width. Notice, that the residues as they appear in the sum rules include all relativistic factors, a valuable advantage over more traditional non-

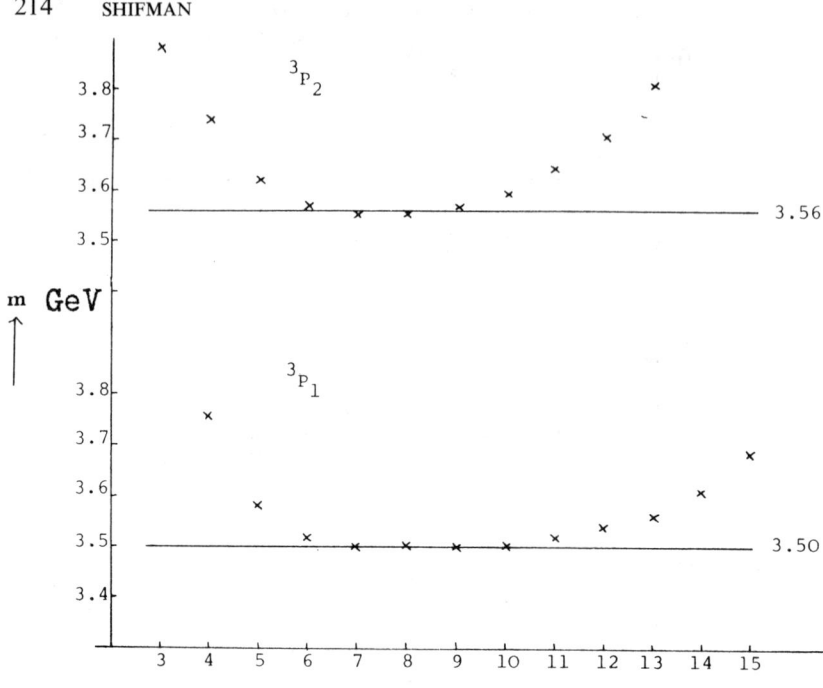

Figure 3 Masses of 1^3P_2 and 1^3P_1 charmonium levels from QCD sum rules (59).

relativistic approach. The ratio r_n is evidently insensitive to the residue since

$$\text{Im}\,\Pi \sim g_R^2 \delta(s - M_R^2),$$

and g_R drops out in the ratio. The coupling constant can be extracted, for instance, from $(d/dq^2)^n \Pi$. The J/ψ electronic width found in (59) is approximately 5 keV, which agrees very well with experiment (4.6 ± 0.4).

Find the Gluon Condensate Yourself

In principle, everyone familiar with the Feynman diagram technique is able to find the coefficient functions corresponding to various gluonic operators. There is nothing novel in the whole procedure. Practically, however, the question "How?" is asked rather often since the language of the Wilson operator expansion (53, 88)—the formal basis of the approach—is not quite conventional. If the reader is uninterested in technical aspects he/she may safely omit the remainder of this section.

We do not make historical excursions here; instead we turn directly to an elegant method invented during the last two or three years (24, 50, 67, 80). It is much simpler than that used originally and, what is also important, it is easily formulated for analytic computer operations. The method actually dates back to the classical Schwinger QED calculations in external fields (66). Its key element is the so-called Schwinger gauge for $(A^a)_{\text{vac}}$.

With respect to heavy quarks the vacuum gluon field can be considered as an external given field. Thus, any gauge condition is admissible. The correlation functions in Equation 9 are gauge independent, and the gauge is primarily the question of convenience. We impose the constraint

$$x_\mu A^a_\mu(x) = 0.$$

This is just the Schwinger gauge rediscovered in QCD in a series of independent papers (19, 24, 28); sometimes the name fixed-point gauge is also used in the literature. Of course, it breaks the Lorentz invariance, which is restored only in the final answer. The most essential point is that in the Schwinger gauge $A^a_\mu(x)$ can be expressed in terms of $G^a_{\mu\nu}(0)$ and *covariant* derivatives, namely (71),

$$A_\mu(x) = \frac{1}{2 \cdot 0!} x_\rho G_{\rho\mu}(0) + \frac{1}{3 \cdot 1!} x_\alpha x_\rho D_\alpha G_{\rho\mu}(0)$$

$$+ \frac{1}{4 \cdot 2!} x_\alpha x_\beta x_\rho D_\alpha D_\beta G_{\rho\mu}(0) + \cdots. \qquad 11.$$

The quark motion in the external field is described by a Green's function $S(x, y)$ whose expansion is quite standard

$$iS(x, y) = iS^{(0)}(x-y) + \int d^4 z\, iS^{(0)}(x-z) iA(z) iS^{(0)}(z-y) + \cdots.$$

Here $S^{(0)}$ is the free quark propagator and $A(z)$ is given in Equation 11. The second term represents quark scattering by the vacuum field. Higher order terms correspond to multiple scatterings.

In momentum space the correlation function in Equation 9 for the vector current, say, takes the form

$$\Pi_{\mu\nu}(q) = i \int \text{Tr}\, \{\gamma_\mu S(p) \gamma_\nu \tilde{S}(p-q)\} \left(\frac{d^4 p}{(2\pi)^4}\right) \qquad 12.$$

with

$$S(p) = \int S(x, 0)\, e^{ipx}\, d^4 x, \qquad \tilde{S}(p) = \int S(0, x)\, e^{-ipx}\, d^4 x.$$

Notice that $S(p)$ and $\tilde{S}(p)$ do not coincide with each other, generally speaking. If it were not for this subtlety the expression for $\Pi_{\mu\nu}(q)$ would look superficially identical to that for the free quark loop.

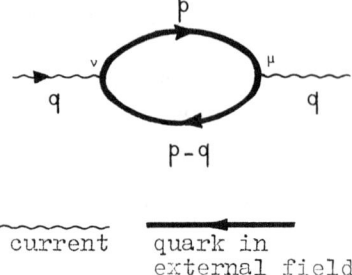

Figure 4 The quark current correlation function in the external gluon field.

The graphic representation of Equation 12 is given on Figure 4. More specifically, the representation of Equation 11 implies

$$A_\mu(k) = -\frac{i(2\pi)^4}{2} G_{\rho\mu}(0) \frac{\partial}{\partial k_\rho} \delta^{(4)}(k) + \frac{(-i)^2(2\pi)^4}{3}$$

$$\times [D_\alpha G_{\rho\mu}(0)] \frac{\partial^2}{\partial k_\rho \partial k_\alpha} \delta^{(4)}(k) + \cdots. \qquad 13.$$

Each gluon line is accompanied by a k integration. Since $A_\mu(k)$ reduces to derivatives of the delta function, the integration is trivial. Actually, it is equivalent to differentiation of all propagators either preceding or following the given vertex.

To illustrate the procedure let us consider a typical example: the G^2 contribution in the vector current correlation function (67). In this exercise it suffices to keep only the first term in the expansion of $A(k)$ (see Equation 13). Moreover, G^2 can evidently emerge only if $A(k)$ is inserted twice (see Figure 5, where the crosses indicate the vacuum gluon field).

Using the fact that

$$\langle \text{vac} | G^a_{\mu\nu}(0) G^b_{\alpha\beta}(0) | \text{vac} \rangle$$

$$= \left(\frac{1}{8}\delta^{ab}\right)\left(\frac{1}{12}(g_{\mu\alpha}g_{\nu\beta} - g_{\mu\beta}g_{\nu\alpha})\right)\langle \text{vac} | G^2 | \text{vac} \rangle$$

and taking the color trace, we get

$$[\Pi_{\mu\nu}(q)]_a = -\frac{i}{96}\langle g^2 G^a_{\mu\nu} G^a_{\mu\nu}\rangle (g_{\rho\sigma}g_{\alpha\beta} - g_{\rho\beta}g_{\alpha\sigma})\int\frac{d^4p}{(2\pi)^4}\frac{\partial}{\partial k_{1\rho}}\frac{\partial}{\partial k_{2\sigma}}$$

$$\times \text{Sp}\{\gamma_\mu(\slashed{p}-m)^{-1}\gamma_\alpha(\slashed{p}-\slashed{k}_1-m)^{-1}$$

$$\times \gamma_\nu(\slashed{p}'+\slashed{k}_2-m)^{-1}\gamma_\beta(\slashed{p}'-m)^{-1}\}_{k_1=k_2=0},$$

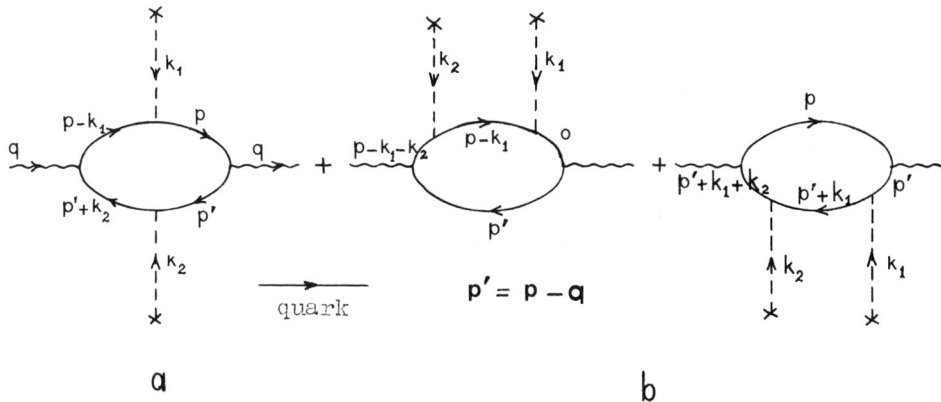

Figure 5 The G^2 correction in the correlation function of heavy quarks. The vacuum gluon field is indicated with dashed lines marked by crosses.

$$[\Pi_{\mu\nu}(q)]_b = (\ldots)2Sp\{\gamma_\mu(\not{p}-m)^{-1}\gamma_\alpha$$
$$\times (\not{p}-\not{k}_1-m)^{-1}\gamma_\beta(\not{p}-\not{k}_1-\not{k}_2-m)^{-1}\gamma_\nu(\not{p}'-m)^{-1}\}_{k_1=k_2=0},$$

where subscripts a, b refer to diagrams 5a,b and Sp implies taking the trace over the Lorentz indices only.

Contracting μ and ν and differentiating with respect to k_1 and k_2, we are left with the following remarkably simple expression:

$$[\Pi_{\mu\mu}]_a = -i\langle g^2 G^2\rangle \int \frac{d^4p}{(2\pi)^4} \frac{pp'}{(p^2-m^2)^2(p'^2-m^2)^2},$$

$$[\Pi_{\mu\mu}]_b = -i4m^2\langle g^2 G^2\rangle \int \frac{d^4p}{(2\pi)^4} \frac{pp'-2p^2}{(p^2-m^2)^4(p'^2-m^2)}.$$

Thus, the problem is reduced to standard Feynman integrals. As the reader can see the algorithm is universal and is trivially generalized to higher order gluonic operators.

Bottomonium Family — Peculiarities

The papers devoted to b quarks in QCD sum rules are now rather numerous (14, 59, 83, 85). Some of them, however, ignore the fact that approximations true in charmonium are not automatically valid for heavier quarks. In particular, perturbative gluon exchanges, almost inperceptible for $c\bar{c}$, grow and renormalize $b\bar{b}$ correlation functions by an order of magnitude.

To understand why this happens, consider the moments

$$M_n = \int s^{-(n+1)} R_b(s) \, ds,$$

where R_b is the cross section of bottom production in e^+e^- collisions measured in units of $\sigma_{QED}(e^+e^- \to \mu^+\mu^-)$. Chiefly because the more massive upsilon states are fractionally much closer together in mass, the position of the stability plateau shifts to much larger n values than in the J/ψ case. Indeed, the lower bound of the range of n we work in is determined by dominance of the ground state, Υ—the Υ' contributions should not exceed, say, 5% of that of Υ. In other words,

$$\frac{\Gamma(\Upsilon' \to e^+e^-)}{M_{\Upsilon'}^{2n+1}} \simeq \frac{1}{20} \frac{\Gamma(\Upsilon \to e^+e^-)}{M_{\Upsilon}^{2n+1}}. \qquad 14.$$

Numerically

$$\Gamma(\Upsilon \to e^+e^-) \simeq 1.2 \text{ keV}, \qquad \Gamma(\Upsilon' \to e^+e^-) \approx 0.55 \text{ keV},$$

$$M_\Upsilon \approx 9.46 \text{ GeV}, \qquad M_{\Upsilon'} \approx 10.02 \text{ GeV} \qquad (\Delta M^2/M^2 \approx 0.12).$$

With these numbers Equation 14 yields $n \gtrsim 20$, to be compared with $n = 6, 7$ in the charmonium sum rules where $\Delta M^2/M^2 \approx 0.4$ (Figure 2a). From the purely theoretical side, the shift in n is attributed to the fact that the nonperturbative term $n^2 \langle \text{vac} |G^2| \text{vac}\rangle/m_c^4$, is replaced now by $n^2 \langle \text{vac} |G^2| \text{vac}\rangle/m_b^4$.

On the other hand, for large n the ordinary α_s series blows up as $\alpha_s(n)^{1/2}$. The origin of a factor of $(n)^{1/2}$ is quite transparent. Introducing the quark velocity $v = (1 - 4m_b^2/s)^{1/2}$ and rewriting M_n in the form

$$M_n = (4m_b^2)^{-n} \int R_b (1-v^2)^{n-1} \, dv^2,$$

we see that for large n the dominant velocity becomes small as $v \sim (n)^{-1/2}$. The limit $n \to \infty$ is equivalent to extremely nonrelativistic propagation, and in this situation each Coulomb quantum yields the well-known α_s/v factor (Figure 6), which converts to $\alpha_s(n)^{1/2}$ in expressions for the moments. With $(n)^{1/2} \sim 5$ and $\alpha_s \approx 0.15$ the expansion parameter is of order unity and the whole infinite series $(\alpha_s n^{1/2})^k$ should be summed up. It would be completely wrong to keep (as is sometimes done) only the first one-gluon term.

Can one perform the summation? The answer is affirmative and conceptually simple. The only modification needed is the following. The free quark Green's function describing the motion between consecutive scatterings by the vacuum field should be replaced by the Coulomb one. Technical difficulties are rather significant, however, since integrations now

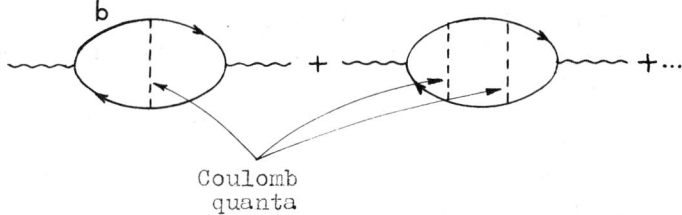

Figure 6 Coulomb corrections in the two-point function of b-quark currents.

involve hypergeometric functions. Still, they can be and have been overcome (83). The corresponding sum rules turn out to be very sensitive to α_s, the quark-gluon coupling constant, not by accident, of course, since the effects $(\alpha_s n^{1/2})^k$ sum to a sharp function of α_s whose absolute value is of order 10. Thus, one obtains one of the most precise methods for measuring α_s. Moreover, a stringent constraint emerges on $\Gamma(\Upsilon \to e^+e^-)$, which reads (83)

$$\Gamma(\Upsilon \to e^+e^-) \approx 1.15 \text{ keV}.$$

The experimental number has evolved somewhat with time and is now very close to the theoretical one. For skeptics the fact may serve as an additional proof of the reliability of results. The issue of reliability becomes especially acute in light of recent CESR results (15), which are interpreted as observation of bottomonium 1P levels with $M = 9.90$ GeV. The sum rules in their present form indicate that $M(1P) \leq 9.86$ GeV (83).

Notice that there are no free parameters in the theoretical analysis except the b-quark mass adjusted in such a way as to fit M_Υ.

Super-Heavy Quarks

With increasing m_Q the quarks become more and more nonrelativistic and finally find themselves in the domain where the interactions reduce to ordinary Coulomb forces due to one-gluon exchange. The latter bind the quarks at distances r of order $(m_Q \alpha_s)^{-1}$. Since the radius of the orbit is so small, confinement effects almost do not show up, and thus the super-heavy $Q\bar{Q}$ system is a mere copy of positronium. Everything is trivially predictable and, hence, uninteresting. The copy is not perfect, however, because even at short distances the quarks still experience the influence of the vacuum fields. The gluon condensate induces corrections to Coulomb formulas. The relative size of the corrections is governed by the quark mass. Of most interest, of course, is the intermediate domain where they are neither too large nor too small. There is a good chance of finding the t quark in just this domain.

Since the size r of the unperturbed system is small, there is no need to write current correlation functions and derive sum rules. The problem is formulated in the following way (46, 84): the Coulomb bound system is immersed in an external field, constant on the scale of r, randomly oriented, and weak. The interaction is of the dipole type, $\mathbf{d}^a \cdot \mathbf{E}^a_{\text{vac}}$, where $\mathbf{E}^a$ is the chromoelectric field, while the effects due to chromomagnetic field are suppressed by two powers of α_s (recall that the quarks are nonrelativistic). In this formulation corrections to the energy eigenvalues and wave function renormalization are calculable. In principle, all that is needed is the knowledge of quantum mechanics. Operationally the task is not as easy as it might at first seem, owing to the complexity of the Coulomb Green's functions. Details of the derivation are exhaustively discussed in the original papers (46, 48, 84); we concentrate on results.

Random orientation means that terms linear in $\mathbf{E}^a_{\text{vac}}$ drop out and deviations start with $\mathbf{E}^a_{\text{vac}} \cdot \mathbf{E}^a_{\text{vac}}$. The Lorentz invariance of the vacuum state allows one to rewrite this combination in the following way

$$\mathbf{E}^a_{\text{vac}} \cdot \mathbf{E}^a_{\text{vac}} = -\frac{1}{4} \langle \text{vac} | G^a_{\mu\nu} G^a_{\mu\nu} | \text{vac} \rangle.$$

As usual, we get the answer in terms of the gluon condensate.

The basic parameter that sets the scale of corrections is not the quark mass itself but rather the radius of the orbit or equivalently the bound quark momentum determined from the equation

$$k_n = \frac{m_Q}{n} \frac{2}{3} \alpha_s(k_n). \qquad 15.$$

Here n is the principle quantum number and $\alpha_s(k_n)$ is the running quark-gluon coupling constant. In particular, the cubic gluon condensate $\langle \text{vac} | G^3 | \text{vac} \rangle$ will enter with an extra k_n^{-2} factor as compared to the bilinear condensate. In order to give a feeling for how large this factor is, let us notice that for the upsilon ground state ($n = 1$, $m_b = 4.80$ GeV)

$$k_1 \approx 0.96 \text{ GeV}$$

if the value of Λ is chosen in the interval 100–150 MeV in accordance with modern data (17).[5] Equation 15 shows that k_n grows roughly linearly with m_Q and falls off with increasing n. The latter fact has far-reaching consequences discussed later.

With all these definitions we are finally able to quote some of the most interesting results. In particular, the energy levels are

$$M_{nl} = 2m_Q - \frac{k_n^2}{m_Q} \left\{ 1 - \frac{m_Q^2}{k_n^6} n^2 a_{nl} \left\langle \text{vac} \left| \frac{\pi \alpha_s}{18} G^a_{\mu\nu} G^a_{\mu\nu} \right| \text{vac} \right\rangle \right\}, \qquad 16.$$

[5] Λ parameterizes the running quark-gluon coupling constant. At the one-loop levels $\alpha_s(k) = 2\pi/[b \ln(k/\Lambda)]$, $b = (11/3)N_c - (2/3)N_f$, where N_c is the number of colors and N_f the number of quark flavors.

where l is the orbital momentum and a_{nl} is a known coefficient function of order unity (for example, $a_{10} \approx 1.65$, $a_{20} \approx 1.78$, etc). Amusingly, this formula works nicely for Υ (it gives $M_\Upsilon \approx 9.46$ GeV starting from $m_b = 4.8$ GeV), although the correction term is rather large here:

$$\frac{m_b^2}{k_1^6} a_{10} \left\langle \text{vac} \left| \frac{\pi \alpha_s}{18} G_{\mu\nu}^a G_{\mu\nu}^a \right| \text{vac} \right\rangle \sim 0.3.$$

Equation 16 is definitely useless for excited $b\bar{b}$ levels. (For $n = 2$ the lower boundary of the pre-Coulomb domain shifts to $m_Q \geq 20$ GeV.) It shows that toponium *will* form nearly Coulomb states at least for a few low-lying levels. In contrast, the role of the Coulomb force in charmonium is practically negligible.

If $m_Q \geq 20$ GeV, the narrow quarkonium levels below the open flavor threshold are numerous—their number is no less than seven—and, proceeding from the lowest level to higher ones, one should observe a peculiar evolution and change of dynamical regimes: from a trivial Coulomb attraction with small deviations to a well-developed picture of large-distance confinement.

For the wave functions, deviations from the asymptotic Coulomb form turn out to be numerically larger by a factor of 3 than for the energies. For instance, the renormalization of $|\psi(0)|^2$ for the ground S level amounts to

$$\frac{|\psi_1(0)|^2}{|\psi_1(0)_{\text{Coulomb}}|^2} = 1 + \frac{m_Q^2}{k_1^6} 4.93 \left\langle \text{vac} \left| \frac{\pi \alpha_s}{18} G_{\mu\nu}^a G_{\mu\nu}^a \right| \text{vac} \right\rangle. \qquad 17.$$

Thus, it would be impossible to estimate the Υ electronic width (proportional to $|\psi_1(0)|^2$) starting from the Coulomb asymptotics.

In the far asymptotic Coulomb regime $m_Q \to \infty$

$$\frac{\Gamma(1^3S_1 \to e^+e^-)}{(\text{quark charge})^2} = 4\pi \frac{\alpha^2}{m_Q^2} \frac{k_1^3}{\pi} \left(1 - \frac{16\alpha_s}{3\pi}(m_Q)\right) \sim m_Q,$$

where we have kept only the one-photon contribution; a trivial factor accounting for the Z boson can be easily inserted if necessary. With decreasing m_Q the electronic width decreases unless the *positive* renormalization, Equation 17, becomes appreciable at $m_Q \sim 20$ to 10 GeV. For such masses the gluon condensate stabilizes the curve at the level

$$\frac{\Gamma(1^3S_1 \to e^+e^-)}{(\text{quark charge})^2} \approx 9 \text{ keV}. \qquad 18.$$

Surprisingly, the rate in Equation 18 coincides within experimental uncertainties with corresponding empirical numbers for $\rho, \omega, \varphi, J/\psi, \Upsilon$. (The latter all coincide with each other, which may be even more surprising.)

The theory of the pre-Coulomb behavior allows one also to find spin splittings. For instance, for the ground state ($n = 1$) we have

$$M(1^3S_1) - M(1^1S_0) = \frac{32\pi}{9} \frac{\alpha_s(m_Q)}{m_Q^2} |\psi_1(0)|^2$$

$$+ \frac{688}{153} \left\langle \text{vac} \left| \frac{\pi\alpha_s}{18} G^a_{\mu\nu} G^a_{\mu\nu} \right| \text{vac} \right\rangle (4m_Q k_1^2)^{-1}.$$

The domain of applicability of this relation is determined primarily by the wave function renormalization (see Equation 17). If we stick to purely theoretical information, we have to choose, as was already mentioned, $m_Q > 10$ GeV. However, the bound can be significantly lowered if, instead of Equation 17, we use independent data on $|\psi_1(0)|^2$ from experiment (85). In this way we come to

$$M(1^3S_1) - M(1^1S_0) = 8 \frac{\Gamma(1^3S_1 \to e^+e^-)}{9 \, (\text{quark charge})^2} \frac{\alpha_s(m_Q)}{\alpha^2} \left[1 + 6 \frac{\alpha_s(m_Q)}{\pi} \right]$$

$$+ \frac{172}{153} \frac{1}{m_Q k_1^2} \left\langle \text{vac} \left| \frac{\pi\alpha_s}{18} G^a_{\mu\nu} G^a_{\mu\nu} \right| \text{vac} \right\rangle,$$

and this expression is valid for the upsilon system (where it gives $M_\Upsilon - M_{\eta_b} \approx 36$ MeV) and even, with lesser accuracy, for charmonium.

Morozov (48) studied spin splittings in P levels with the same approach. For $l = 1$ the domain of validity is strongly shifted to higher masses, perhaps, $m_Q \geq 40$–50 GeV. For such quarks it is predicted

$$M(1^1P_1) - M(1^3P_0) \approx 9 \text{ MeV},$$

$$M(1^1P_1) - M(1^3P_1) \approx 2 \text{ MeV},$$

$$M(1^3P_2) - M(1^1P_1) \approx 3 \text{ MeV}.$$

All relevant formulas, though rather cumbersome, are collected in (48).

OTHER ASPECTS OF QUARKONIUM THEORY

In previous sections we sketched the life of a heavy quark pair in the surrounding vacuum medium. This is not a precise photographic picture, of course—some details are not clearly seen, others are somewhat displaced—but it is hoped that the gross features are reproduced correctly. In any case, we have tried to follow as closely as possible the basic notions of QCD. Unfortunately, the quantitative scheme emerging in this way is not completely universal and leaves open important questions referring, say, to radial excitations. As far as lowest-lying levels are concerned, the methods are quite satisfactory.

Apart from spectroscopy, however, there are other topics deserving attention. In particular, decays of heavy quarkonium is a rich field of research promising interesting findings. QCD sum rules can say much on the subject. Here we discuss only a few typical examples.

Photon Decays

The first example is the decay $\eta_c \to 2\gamma$. It has not yet been seen experimentally, so that predictions of various theoretical schemes will compete for some time. In the purely nonrelativistic approximation, one readily finds that $\Gamma(\eta_c \to 2\gamma)/\Gamma(J/\psi \to e^+e^-) = 4/3$ so that

$$\Gamma(\eta_c \to 2\gamma) = 6.2 \pm 0.5 \text{ keV}, \qquad 19.$$

where the empirical number for the J/ψ electronic width is substituted. As was explained above, the accuracy of nonrelativistic approximation in charmonium should not be very high, and one may expect noticeable deviations from Equation 19. Are they really large? This is not merely an academic question since $\Gamma(\eta_c \to 2\gamma)$ plays a key role in at least two important relations: one of them fixes the ratio $\Gamma(\eta_c \to \text{hadrons})/\Gamma(\eta_c \to 2\gamma)$—the so-called Appelquist-Politzer recipe—the other one gives the ratio $\Gamma(J/\psi \to \eta_c\gamma)/\Gamma(\eta_c \to 2\gamma)$ (see below).

Generalization of the sum rule method referring to decay coupling constants was sketched in early works (52). It involves no conceptually new elements. The only difference is that, instead of two-point functions, one considers three-point functions induced by appropriate currents. For instance, the amplitude relevant to $\eta_c \to 2\gamma$ is the following

$$A_{\mu\nu}(q, q_1, q_2) = e^2 \int d^4x \, d^4y \, \exp[-i(qx + q_2 y)]$$

$$\times \langle \text{vac} \,|T\{\bar{c}(0)\gamma_\mu c(0), \bar{c}(x)i\gamma_5 c(x), \bar{c}(y)\gamma_\nu c(y)\}|\, \text{vac}\rangle.$$

It describes the conversion of the pseudoscalar current with momentum q into two photons with momenta q_1 and q_2 via a c-quark loop. Further manipulations are standard: calculate $A_{\mu\nu}$ taking account of the gluon condensate, differentiate with respect to q^2 at $q^2 = 0$ and find the stability plateau, saturate with η_c and get the product of the couplings $\langle \text{vac}\,|\bar{c}i\gamma_5 c|\eta_c\rangle\langle\eta_c|2\gamma\rangle$. The first one is assumed to be known from the corresponding sum rules.

First estimates along these lines were made by Novikov et al (52) and the program has been realized in full in recent works (42, 60). I discuss the work of Reinders et al (60) here since it is more clearly written. Figure 7 displays

$$M^{2n+1}_{\eta_c} \frac{1}{n!} (d/dq^2)^n A(q^2) \text{ versus } n.$$

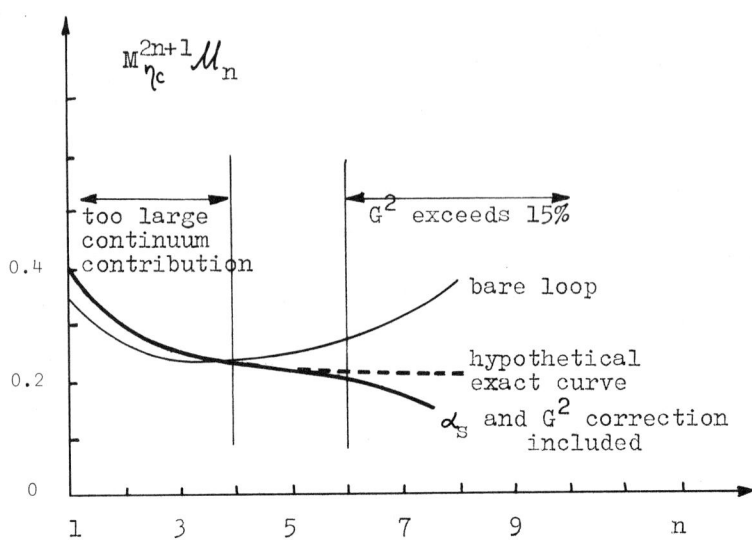

Figure 7 Decay $\eta_c \to 2\gamma$: the moments of the invariant amplitude versus n. The moments are defined in the following way:

$$\mu_n = \frac{1}{n!}(d/dq^2)^n A(q^2)|_{q^2=0}.$$

The dashed curve represents an extrapolation, which is our guess for the exact solution.

Here $A(q^2)$ is the invariant function, with kinematical factors eliminated, namely

$$A_{\mu\nu} = 3\alpha \varepsilon_{\mu\nu\alpha\beta}(q_1)_\alpha (q_2)_\beta A.$$

The exact curve should approach a constant asymptote proportional to $\langle \text{vac} | \bar{c}i\gamma_5 c|\eta_c\rangle \langle \eta_c|2\gamma\rangle$ as $n \to \infty$ since only the η_c intermediate state survives in this limit. Practically, one must stop at $n = 6$ where the G^2 correction amounts to 15%. On the other hand, for $n \geq 4$ the higher state contribution becomes less than 10%. Hence, the height of the (quasi)plateau at $n = 4$–6 allows one to judge the true asymptote with an accuracy no worse than 5%. As seen from the figure the asymptote is ~ 0.21, which implies in turn,

$$\Gamma(\eta_c \to 2\gamma) \approx 4.5 \text{ keV}. \qquad 20.$$

This result lies significantly below the naive estimate of Equation 19.

More generally, the same three-point function calculated at $q_1^2 \neq 0$ and saturated by resonances simultaneously in two channels, pseudoscalar and vector, also yields the $J/\psi \, \eta_c \gamma$ coupling constant (see 41).

However, once the $\eta_c \to 2\gamma$ width is known there is no need to redo tedious computations. A more economic alternative leading to a simple relation between $\Gamma(J/\psi \to \eta_c\gamma)$ and $\Gamma(\eta_c \to 2\gamma)$ is the following. One starts with the $\eta_c \to 2\gamma$ amplitude with both photons on the mass shell and writes it as a dispersion integral in one of the photons. The imaginary part in the intergrand actually represents a sum of three distinct contributions: (a) J/ψ; (b) ψ' and other radial excitations; (c) a continuum of charmed meson pairs $D\bar{D}$, $D\bar{D}\pi\pi$, etc. The J/ψ intermediate state dominates, however. As seen from Figure 8, the corresponding piece in the amplitude is determined by two constants fixing $\gamma \to J/\psi$ and $J/\psi \to \eta_c\gamma$ transitions. Higher radial excitations enter in similar manner.

The absolute dominance of J/ψ is easy to understand in view of the fact that the $\psi' \to \eta_c\gamma$ transition is forbidden in nonrelativistic limit by orthogonality of the radial wave functions. In the real world one can show that the ψ' contamination is less than 3%. Moreover, the continuum threshold lies rather high, at approximately $(s_0)^{1/2} = 4$ GeV. This means that the corresponding imaginary part originating from $D\bar{D}$, $D\bar{D}\pi\pi$, etc can be approximated in the dual sense by a perturbative curve determined by quark and gluon intermediate states, $c\bar{c}$ or $c\bar{c}$ + gluon. In the latter case an α_s factor is explicit while in the former it appears implicitly since kinematics requires hard gluon exchange between c and $\bar{c}$. The characteristic gluon offshellness is $(s_0 - 4m_c^2) \approx 10$ GeV^2 and, hence, it is $\alpha_s(10\,\text{GeV}^2)$ that figures in all expressions—the expansion is completely justified. With the perturbative estimate of the continuum, one obtains the following formula (72, 73):

$$\Gamma(J/\psi \to \eta_c\gamma) = \frac{2}{9} \frac{\Gamma(\eta_c \to 2\gamma)}{\Gamma(J/\psi \to e^+e^-)} \alpha \frac{M_{J/\psi}^4}{M_{\eta_c}^3} \left(1 - \frac{M_{\eta_c}^2}{M_{J/\psi}^2}\right)^3 (1 - 0.28\alpha_s).$$

If the estimate of Equation 20 is indeed correct, the $J/\psi \to \eta_c\gamma$ width turns out to be close to 1.8 keV, to be compared with the recent SLAC result (15) of $0.8^{+0.4}_{-0.2}$ keV. The experimental number is 2.5 standard deviations lower than that expected theoretically. Does this fact cast a shadow on the theory? Personally I believe in the theory, and it seems to me premature to draw any definite conclusions.

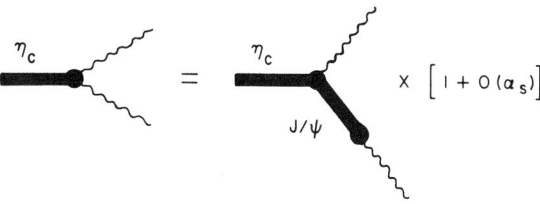

Figure 8 The J/ψ dominance in the transition $\eta_c \to 2\gamma$.

Mesons with Open Flavor

Some families, for instance $b\bar{c}$, resemble—in essence repeat—heavy quarkonium at least as far as their strong interactions are concerned. We undertake here a short excursion into the world of mesons built from one heavy and one light quark. The novel feature of such systems is a strong coupling to the quark condensate. Since our purpose is essentially pedagogical, we assume that $m_Q \to \infty$; in contrast, the light quark mass is set equal to zero. In this limit the heavy quark plays the role of a static force center; recoil effects can be neglected. Moreover, one can systematically expand in $1/m_Q$ and use nonrelativistic language, in particular, work with energy measured from the threshold (m_Q), not with q^2.

It is quite evident that states differing only by spin alignment are degenerate if $m_Q = \infty$. Thus, the spectra in 0^- and 1^- channels are the same. The degeneracy also takes place in 0^+ and 1^+ channels. The currents producing corresponding states from the vacuum are

$$j^{PS} = i\bar{q}\gamma_5 Q(0^-); \qquad j^V_\mu = \bar{q}\gamma_\mu Q(1^-);$$
$$j^S = \bar{q}Q(0^+); \qquad j^A_\mu = \bar{q}\gamma_\mu\gamma_5 Q(1^+).$$

In analyzing the current correlation functions, one should keep in mind that the physical threshold now lies numerically far from the point $q^2 = 0$ and, instead of differentiating at $q^2 = 0$, it is much more convenient to choose another reference point, separated from the physical threshold by a fixed energy interval. Thus, we choose $q^2 = (m_Q - \mu)^2$, where μ is a free parameter that will control the resolution ability of the sum rules. If μ is sufficiently large, we are in the asymptotic freedom domain and can rely on the external field expansion described above. External fields now include both gluon and light quark vacuum fluctuations.

In the limit $m_Q \to \infty$ with appropriate definitions of the amplitudes m_Q drops out of all relations. In particular, the sum rule for the pseudoscalar current takes the form (77)

$$\frac{1}{\pi}\int_0^\infty dE\, e^{-E/\mu}\, \text{Im}\,\Pi^{(0^-)}(E) = \frac{3\mu^3}{\pi^2} - \frac{1}{2}\langle \text{vac}\,|\bar{q}q|\,\text{vac}\rangle$$
$$+ \frac{1}{64\mu^2}\langle \text{vac}\,|ig_s\bar{q}\sigma_{\mu\nu}G^a_{\mu\nu}t^a q|\,\text{vac}\rangle$$
$$+ \frac{1}{2304\mu^3}\langle \text{vac}\,|(\bar{q}\gamma_\mu t^a q)^2|\,\text{vac}\rangle + \cdots,$$

where t^a is the Gell-Mann matrix acting in the color space. Here the first term corresponds to the bare loop (Figure 9a), the second and the fourth

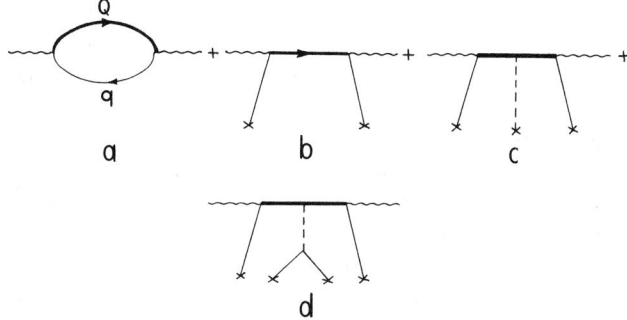

Figure 9 Graphic representation of the two-point function induced by the current $\bar{Q}i\gamma_5 q$. Diagrams (a)–(d) correspond to four different terms in the external field expansion. The vacuum fields are marked with crosses.

terms describe the coupling to the quark condensates (Figures 9b, d), and the third term represents the combined effect of quark and gluon vacuum fields (Figure 9c). Saturating the imaginary part by the pseudoscalar meson analogous to D (B), plus continuum starting at some energy E_c, we find both the resonance position and the continuum threshold. The resonance residue is also fixed (see Table 2, taken from Ref. 77). Applying these results, for instance, to mesons with the quark content $b\bar{u}$ (they are called B mesons) we get $M_B = (5.20 \pm 0.10)$ GeV and $f_B \approx 140$ MeV, while the experimental value of their mass is 5275 ± 10 MeV.

Limits of Applicability of the Method

The reader might already have noticed some similarity between the sum rule method and the old local duality concept (13, 16, 32, 65), which states that a physical cross section averaged over some energy interval coincides with that for bare quarks. Likewise, the sum rules teach us that the ground-state contribution to current correlation functions is imitated in certain conditions by the bare loop plus moderate power (nonperturbative) corrections due to gluon and quark vacuum condensates. A kinship is

Table 2 Parameters of lowest-lying mesons with open flavor ($Q\bar{q}$) in the limit $m_Q \to \infty$.

Quantum numbers	$M_R - m_Q$ (GeV)[a]	$E_c = \sqrt{s_c} - m_Q$ (GeV)[b]	$f_R^2 M_R$ (GeV^3)[c]
$0^-, 1^-$	0.4 ± 0.1	0.9 ± 0.1	$5 \cdot 10^{-3}$
$0^+, 1^+$	1.2 ± 0.2	1.8 ± 0.2	$5 \cdot 10^{-2}$

[a] M_R stands for the resonance mass.
[b] $\sqrt{s_c}$ denotes the continuum threshold.
[c] Definitions of f_R are as follows: $\langle 0^- | \bar{Q}\gamma_\mu\gamma_5 q | \mathrm{vac} \rangle = -if_R \bar{p}_\mu$; $\langle 1^- | \bar{Q}\gamma_\mu q | \mathrm{vac} \rangle = f_R M_R \varepsilon_\mu$.

undeniable, although differences between the two approaches are deep. The local duality principle is rather vague since the way of averaging is ambiguous—neither the weight function nor the interval of smearing are fixed theoretically. Actually, it is a semiquantitative approach. In contrast, within the sum rule method everything is specified, and once the quark masses and basic vacuum matrix elements are determined there is no freedom left. The accuracy of the results is under theoretical control. The price we must pay for this advantage is rather high. The sum rule method, in its present form, says next to nothing about radial excitations that are strongly suppressed and represent a weak background to the dominant ground-state contribution. They are simply not seen from the Euclidean domain.

Another limitation of principle prevents application to high orbital excitations, i.e. mesons with large spin. As was already mentioned, the QCD sum rules express a refined duality; they work successfully only if the ground state (in a given channel) is more or less dual to the bare loop. Only then there exists the stability plateau where, on the one hand, nonperturbative terms are moderate and the external field expansion is applicable, and, on the other hand, the ground-state contribution dominates over all others. Also conceivable is the opposite, unlucky, situation in which the resonance saturation is achieved in the domain where the power terms are large and summation of the infinite nonperturbative series is necessary, a task going far beyond our present abilities. Under these circumstances we should have made a step back and confessed that the lowest-lying resonance is not resolved and all we can predict is the continuum spectral density. Unfortunately, just such a situation is realized for mesons with large spin. The sum rules practically do not constrain the properties of resonances with $J \geq 4$.

In order to understand why, recall the simple dynamical nature of quark-hadron duality. It merely reflects the fact that a current produces the quark pair at short times, shorter than the characteristic interaction time. If so, confinement effects on a larger time scale play the role of a large box that makes the spectrum discrete. They cannot, however, renormalize the cross section; the sum over a few close discrete levels should still reproduce the bare quark curve. In other words, the necessary condition for the quark-hadron duality to be valid is

$$\tau_{\text{prod}} \ll \tau_{\text{int}}. \qquad 21.$$

In contrast, if $\tau_{\text{prod}} > \tau_{\text{int}}$ the confinement effects are essential in the process of quark production and there is no reason to think that the physical spectral density will coincide with the bare one even after averaging.

Let us see what the implications of Equation 21 are. For simplicity we consider massless quarks, although the final conclusion is independent of mass.

Thus, a $q\bar{q}$ pair with angular momentum J is injected in the vacuum with total energy E in the center-of-mass frame. It is evident that the size R of the region where the pair is produced is of order $J/p = J/E$. Then, since the quark's velocity is that of light,

$$\tau_{\text{prod}} \sim J/E. \qquad 22.$$

On the other hand, the characteristic interaction time

$$\tau_{\text{int}} \sim 1/V_{\text{int}}$$

where V_{int} describes the dipole interaction of the $q\bar{q}$ pair with the vacuum gluon field, $V_{\text{int}} \sim \mathbf{d}^a \cdot \mathbf{E}^a_{\text{vac}}$. This estimate immediately yields

$$V_{\text{int}} \sim gRE_{\text{vac}} \sim \mu^2 J/E.$$

The dimensional parameter μ is related to the gluon condensate. Roughly speaking, $\mu \sim \langle \text{vac} |\alpha_s G^a_{\mu\nu} G^a_{\mu\nu}| \text{vac}\rangle^{1/4}$. The interaction time is thus

$$\tau_{\text{int}} \sim E/\mu^2 J. \qquad 23.$$

Comparing now Equations 22 and 23, we see that the condition $\tau_{\text{prod}} \ll \tau_{\text{int}}$ actually means

$$E^2 \gg \mu^2 J^2.$$

In other words, the quark-hadron duality in current spectral densities takes place provided that $s \gg \mu^2 J^2$.

On the other hand, if the Regge trajectories are linear, the lowest resonances are situated at $s \sim [\alpha'(0)]^{-1} J$, where $\alpha'(0)$ is the trajectory slope. The linear dependence on J means that only states with small J fall in the region of quark-hadron duality. Resonances with spin of order unity turn out to be inside the duality domain. For $J = 3$ they are close to the boundary, while for larger J the lowest-lying state becomes separated from the duality domain by a rapidly growing gap. Within this gap the physical spectral density is nonvanishing, but not dual to the quark density.

This fact explains the success of the QCD sum rules in application to $\bar{Q}Q$ states with $J = 0, 1, 2$. The same arguments simultaneously tell us that the sum rules are useless for large-spin mesons, since the gluon condensate in the corresponding correlation functions increases with n too rapidly to permit saturation by the lowest state. A formal analysis of the issue, with all relevant mathematical details, is given in Ref. (75).

CONCLUSIONS AND PERSPECTIVES

Theorists working on a universal theory of hadrons are still exploring. So far, a few approximate approaches have been developed and successfully exploited in practical applications. One of them, based on the QCD sum rules, may be considered as a natural heir to such old concepts as local duality and dispersion sum rules (51). It has incorporated them, extended them, and grown to a powerful computational scheme. It is over its youthful period, so it is not premature to draw definite conclusions. In this short review we have tried to present all its strong and weak points.

We have shown that the QCD sum rules allow one to calculate masses and couplings for charmonium and the upsilon family to a rather high degree of accuracy. The procedure is perfectly reliable, as is repeatedly verified, sometimes in a most dramatic way. The basic notions are justified from the physical point of view and are quite transparent. One uses only the Lagrangian parameters (quark masses and α_s) and a limited number of vacuum expectation values parameterizing the nonperturbative QCD vacuum. The same basic ideas give rise to such elegant theories as that of the pre-Coulomb behavior.

The literature devoted to heavy quarkonium in the QCD sum rules is extensive. Some of the predictions are already confirmed by experiment, others are awaiting their turn. The field is far from being exhausted, however. The spectrum of charmed mesons and baryons is not yet investigated completely. One does not anticipate any unforeseen difficulties of principle in this problem. The analysis of various three-point functions also seems very interesting since it promises to give exhaustive and unambiguous information on transitions between quarkonium levels, for instance, $\chi_J \to J/\psi\gamma$, $\eta_b \to 2\gamma$, etc. This work is already in progress. Many other important questions referring to heavy quarks may find answers within the framework of the QCD sum rules.

The sphere of application of the approach is actually much wider. An impressive body of results has been accumulated during the last few years and touches upon all aspects of the hadronic phenomenology. Apart from the examples already quoted, suffice it to mention the analysis of classical vector (70) and tensor (1, 61) mesons (ρ and f nonets), traditional baryon spectroscopy (11, 12, 18, 36, 38, 64), and highly nontrivial predictions (6, 54, 79, 89) for exotic states built from light quarks and gluons. Related ideas inspired theoretical investigations of Wilson's operator expansion (e.g. 20, 24, 53) and QCD low energy theorems (54). Quite recently a rich and novel field was opened. It was shown (27, 37, 49, 63) that the method can be generalized to include the decay couplings of the old light hadrons and form factors, say, the pion electromagnetic form factor, for intermediate momentum transfers. As far as I know, in this respect the QCD sum rules have no competitors thus far.

What may be even more important, the approach can be reversed. The spectrum of hadrons and their properties yield direct information on the vacuum structure. The sum rules translate observable quantities into the language of vacuum matrix elements. Thus they present an indispensible tool for studying these most fundamental physical objects. The theorists who are building vacuum models use the results to test their constructions. Each particular problem solved adds something to our knowledge of the QCD vacuum, and this knowledge, perhaps, will evolve to give, some beautiful day, the long-sought complete theory.

I am grateful to my coauthors and teachers Drs. A. Vainshtein and V. Zakharov with whom I shared all the difficulties and surprises in the process of working on the sum rules. Practically all original results reported here were obtained in collaboration with them. The method would never have emerged without the preceding work done together with Dr. V. Novikov, Prof. L. Okun', and Dr. M. Voloshin. It is a pleasure to thank them for numerous discussions that formed my opinions on the subjects broached in this review. A similar treatment of some of the issues was given previously in (74, 76).

Literatured Cited

1. Aliev, T. M., Shifman, M. A. 1982. *Phys. Lett.* 112B:401–5
2. Appelquist, T., Politzer, H. D. 1975. *Phys. Rev. Lett.* 34:43–45
3. Appelquist, T., Barnett, R. M., Lane, K. 1978. *Ann. Rev. Nucl. Part. Sci.* 28:387–99
4. Aubert, J. J., et al. 1974. *Phys. Rev. Lett.* 33:1404–6
5. Augustin, J. E., et al. 1974. *Phys. Rev. Lett.* 33:1406–8
6. Balitsky, I. I., Dyakonov, D. I., Yung, A. V. 1982. *Phys. Lett.* 112B:71–74
7. Banks, T., et al. 1981. *Nucl. Phys.* B190:692–98
8. Belavin, A. A., Polyakov, A. M., Schwarz, A. S., Tyupkin, Y. S. 1975. *Phys. Lett.* 59B:85–87
9. Bell, J. S., Bertlmann, R. A. 1981. *Nucl. Phys.* B177:218–36
10. Bell, J. S., Bertlmann, R. A. 1981. *Nucl. Phys.* B187:285–92
11. Belyaev, V. M., Ioffe, B. L. 1982. *Zh. Eksp. Teor. Fiz.* 83:876–91
12. Belyaev, V. M., Ioffe, B. L. 1983. Determination of baryon masses from QCD sum rules. Strange baryons. *Zh. Eksp. Teor. Fiz.* In press
13. Bertlmann, R. A. 1981. *Acta Phys. Austriaca* 53:305–38
14. Bertlmann, R. A. 1982. *Nucl. Phys.* B204:387–412
15. Bloom, E. D. 1982. Presented at 21st Int. Conf. on High Energy Physics, Paris
16. Bramon, A., Etim, E., Greco, M. 1972. *Phys. Lett.* 41B:609–12
17. Buras, A. 1981. A tour of perturbative QCD. *Proc. Int. Symp. on Lepton and Photon Interactions at High Energies*, ed. W. Pfeil, pp. 636–88. Univ. Bonn
18. Chung, Y., et al. 1982. *Nucl. Phys.* B197:55–75
19. Cronström, C. 1980. *Phys. Lett.* 90B:267–69
20. David, F. 1982. *Nucl. Phys.* B209:433–60
21. Di Giacomo, A., Rossi, G. C. 1981. *Phys. Lett.* 100B:481–84
22. Di Giacomo, A., Pafutti, G. 1982. *Phys. Lett.* 108B:327–31
23. Dokshitzer, Y. L., Dyakonov, D. I., Troyan, S. I. 1980. *Phys. Rep.* 58:271–395
24. Dubovikov, M. S., Smilga, A. V. 1981. *Nucl. Phys.* B185:109–32
25. Eichten, E., et al. 1975. *Phys. Rev. Lett.* 34:369–72

26. Eichten, E., et al. 1980. *Phys. Rev.* D21:203–33
27. Eletsky, V. L., Ioffe, B. L., Kogan, Y. I. 1983. *Phys. Lett.* 122B:423–26
28. Fateev, V., Schwarz, A., Tyupkin, Y. 1976. Report Lebedev Physical Institute No. 155, Moscow (unpublished)
29. Fritzsch, H., Gell-Mann, M. 1972. Current algebra: quarks and what else? *Proc. Int. Conf. on High Energy Physics, 16th, Batavia*, ed. J. D. Jackson, A. Roberts, 2:135–68. Batavia, Ill: Fermi Natl. Accel. Lab.
30. Gell-Mann, M., Oakes, R. J., Renner, B. 1968. *Phys. Rev.* 175:2195–99
31. Gottfried, K. 1977. *Proc. Int. Symp. on Lepton and Photon Interactions at High Energies, Hamburg*, ed. F. Gutbrod, pp. 667–710. Hamburg: DESY
32. Greco, M. 1973. *Nucl. Phys.* B63:398–412
33. Gross, D. J., Wilczek, F. 1973. *Phys. Rev. Lett.* 30:1343–46
34. Hamber, H., et al. 1982. *Phys. Lett.* 108B:314–16
35. Himel, T. L., et al. 1980. *Phys. Rev. Lett.* 45:1146–49
36. Ioffe, B. L. 1981. *Nucl. Phys.* B188:317–41; B191:591–92
37. Ioffe, B. L., Smilga, A. V. 1982. *Phys. Lett.* B114:353–57
38. Ioffe, B. L. 1983. *Z. Phys.* In press
39. Jackson, J. D. 1977. *Proc. Eur. Conf. on Particle Physics, Budapest*, ed. L. Jenik, I. Montvay, 1:603–29. Budapest: Cent. Res. Inst. Phys.
40. Jackson, J. D., Quigg, C., Rosner, J. L. 1978. *Proc. Int. Conf. on High Energy Physics, 19th*, pp. 391–408. Tokyo: Phys. Soc. Jpn.
41. Khodjamirian, A. Y. 1980. *Phys. Lett.* 90B:460–64
42. Kirschner, R., Schiller, A. 1982. *Z. Phys.* C 16:141–43
43. Kogut, J. B. 1979. *Rev. Mod. Phys.* 51:659–714
44. Kripfganz, J. 1981. *Phys. Lett.* 101B:169–72
45. Leutwyler, H. 1974. *Phys. Lett.* 48B:431–34
46. Leutwyler, H. 1981. *Phys. Lett.* 98B:447–50
47. Marciano, W., Pagels, H. 1978. *Phys. Rep.* 36:139–276
48. Morozov, A. Y. 1982. *Yad. Fiz.* 36:1302–13 (Transl. *Sov. J. Nucl. Phys.* 36:755–61)
49. Nesterenko, A. V., Radyishkin, A. V. 1982. *Phys. Lett.* 115B:410–14
50. Nikolaev, S. N., Radyishkin, A. V. 1983. *Nucl. Phys.* B213:285–304
51. Novikov, V. A., Okun', L. B., Shifman, M. A., Voloshin, M. B., Vainshtein, A. I., Zakharov, V. I. 1977. *Phys. Rev. Lett.* 38:626–29
52. Novikov, V. A., et al. 1978. *Phys. Rep.* 41:3–133
53. Novikov, V. A., et al. 1980. *Nucl. Phys.* B174:378–96
54. Novikov, V. A., Shifman, M. A., Vainshtein, A. I., Zakharov, V. I. 1981. *Nucl. Phys.* B191:301–69
55. Partridge, R., et al. 1980. *Phys. Rev. Lett.* 45:1150–54
56. Politzer, H. D. 1973. *Phys. Rev. Lett.* 30:1346–49
57. Politzer, H. D. 1974. *Phys. Rep.* 14:129–209
58. Quigg, C., Rosner, J. L. 1979. *Phys. Rep.* 56:169–235
59. Reinders, L. J., Rubinstein, H. R., Yazaki, S. 1981. *Nucl. Phys.* B186:109–46
60. Reinders, L. J., Rubinstein, H. R., Yazaki, S. 1982. *Phys. Lett.* 113B:411–14
61. Reinders, L. J., Rubinstein, H. R., Yazaki, S. 1982. *Nucl. Phys.* B196:125–46
62. Reinders, L. J., Rubinstein, H. R., Yazaki, S. 1981. *Phys. Lett.* 104B:305–9
63. Reinders, L. J., Rubinstein, H. R., Yazaki, S. 1983. *Nucl. Phys.* B213:109–21
64. Reinders, L. J., Rubinstein, H. R., Yazaki, S. 1983. *Phys. Lett.* 120B:209–13 (Erratum 122B:487)
65. Sakurai, J. J. 1973. *Phys. Lett.* 46B:207–10
66. Schwinger, J. 1970. *Particles, Sources and Fields*, Vols. 1, 2. Reading, Mass: Addison-Wesley
67. Smilga, A. V. 1982. *Yad. Fiz.* 35:473–84 (Transl. *Sov. J. Nucl. Phys.* 35:271–77)
68. Shifman, M. A., Vainshtein, A. I., Zakharov, V. I., Voloshin, M. B. 1978. *Phys. Lett.* 77B:80–83
69. Shifman, M. A., Vainshtein, A. I., Zakharov, V. I. 1978. *Phys. Lett.* 76B:471–74
70. Shifman, M. A., Vainshtein, A. I., Zakharov, V. I. 1979. *Nucl. Phys.* B147:385–534
71. Shifman, M. A. 1980. *Nucl. Phys.* B173:13–31
72. Shifman, M. A. 1980. *Z. Phys.* C4:345–353, C6:282
73. Shifman, M. A., Vysotsky, M. I. 1981. *Z. Phys.* C10:131–38
74. Shifman, M. A. 1981. See Ref. 17, pp. 242–78
75. Shifman, M. A. 1982. *Yad. Fiz.* 36:1290–1301
76. Shifman, M. A., Khoze, V. A. 1983. *Usp. Fiz. Nauk* 140:3–74
77. Shuryak, E. V. 1982. *Nucl. Phys.* B198:83–101
78. Shuryak, E. V. 1982. *Nucl. Phys.* B203:93–115, 116–39, 140–56

79. Shuryak, E. V. 1982. Pseudoscalar mesons and instantons. *Preprint CERN-TH-3351.* Geneva: CERN
80. Shuryak, E. V., Vainshtein, A. I. 1982. *Nucl. Phys.* B199:451–81
81. Vainshtein, A. I., Zakharov, V. I., Shifman, M. A. 1978. *Zh. Eksp. Teor. Fiz. Pis'ma Red.* 27:60–64 (Engl. transl. *JETP Lett.* 27:55–59)
82. Vainshtein, A. I., et al. 1980. *Yad. Fiz.* 32:1622–35 (Transl. *Sov. J. Nucl. Phys.* 32:840–47)
83. Voloshin, M. B. 1980. Non-relativistic bottonium in the physical QCD vacuum. *Preprint ITEP-21.* Moscow: ITEP (unpublished)
84. Voloshin, M. B. 1981. Non-perturbative effects in pre-Coulomb levels of heavy quarkonium. *Preprint ITEP-30.* Moscow: ITEP (unpublished)
85. Voloshin, M. B. 1982. *Yad. Fiz.* 35:1016–26 (Transl. *Sov. J. Nucl. Phys.* 35:592–95)
86. Weinberg, S. 1978. In *Festschrift for I. I. Rabi,* ed. L. Motz. New York Acad. Sci.
87. Wilson, K. G. 1977. In *New Phenomenon in Subnuclear Physics (1975 Erice Lectures),* ed. A. Zichichi. New York: Plenum
88. Wilson, K. 1969. *Phys. Rev.* 179:1499–1511
89. Zhitnitsky, A. R., Zhitnitski, I. R. 1982. Are the scalar mesons $S^*(980)$, $\delta(980)$, $K(1500)$ gluekonia? *Preprint IYaF-82-65.* Novosibirsk: Inst. Nucl. Phys. (unpublished)

BAG MODELS OF HADRONS

Carleton E. DeTar

Department of Physics, University of Utah, Salt Lake City, Utah 84112

John F. Donoghue

Department of Physics and Astronomy, University of Massachusetts, Amherst, Massachusetts 01003

CONTENTS

1. INTRODUCTION 235
   1.1 *Quantum Chromodynamics and MIT Bag Model* 237
   1.2 *Other Bag Models* 238
   1.3 *Soliton Bag Model* 238
2. THE STATIC CAVITY APPROXIMATION 239
   2.1 *Doing Without the Approximation* 239
   2.2 *Quark Eigenmodes for the Sphere* 240
   2.3 *A Simple Calculation and the Zero-Point Energy* 242
   2.4 *Gluon Eigenmodes for the Sphere* 242
   2.5 *Interactions and Cavity Perturbation Theory* 243
   2.6 *Lowest Excitations of the Light Hadrons* 246
   2.7 *Refinements-Correcting for Center-of-Mass Motion* 246
3. A VARIETY OF BAG STATES 247
   3.1 *Strings* 247
   3.2 *Heavy Quark-Antiquark Bound States* 247
   3.3 *Glueballs* 248
   3.4 *Exotics and Cryptoexotics* 249
4. INTERACTIONS 251
   4.1 *Electroweak Interactions* 251
   4.2 *P Matrix and Multiquark Bag States* 253
   4.3 *The Nucleon-Nucleon Interaction* 254
   4.4 *Hybrid Chiral Models for Pion-Bag Interactions* 255
5. CRITIQUE AND OUTLOOK 259
   5.1 *Limitations of the Models* 259
   5.2 *Conclusions and Outlook* 260

1. INTRODUCTION

During the early development of the quark model the light hadrons were treated as bound states of quarks moving nonrelativistically in a confining potential. However, it became clear that the nonrelativistic treatment was wanting. Ordinarily, nonrelativistic systems have excitation energies that are small compared to the component masses. In mesons and baryons, however, these energies are comparable to the quark masses of these models. As quantum chromodynamics (QCD), the SU(3) gauge theory of interacting colored quarks and gluons, grew in acceptance, the need for a new approach became even more compelling. Although it was hoped that the hadron spectroscopy, structure, and interaction could ultimately be deduced from first principles, the complexities of QCD compelled us to resort to approximate models, such as the Massachusetts Institute of Technology (MIT) bag model (1), the Stanford Linear Accelerator Center (SLAC) bag model (2), and the soliton bag model (3). These bag models attempt to incorporate three desirable features of hadronic structure that were omitted from the earlier nonrelativistic QCD approach: (*a*) the QCD property of short-distance asymptotic freedom, which on the one hand permits the use of perturbation theory in describing the short-distance interaction of quarks and gluons, and on the other hand forbids the propagation of colored fields to large distances; (*b*) the introduction of gluons as hadronic constituents and the mediators of the short-distance interaction between quarks; and (*c*) a relativistic and gauge-invariant framework. A variant of the MIT bag model, the hybrid chiral bag model (4), attempts to incorporate chiral symmetry, a feature of QCD missing also in the other bag models.

In this brief, critical review we describe the good and bad features of bag models with principal emphasis upon the more developed MIT model. The MIT model has enjoyed more popularity because of its simplicity. However, most of its successes and its drawbacks are undoubtedly shared by the other models. The plan of the review is as follows. In the remainder of the introduction we discuss the relationship between the bag models and QCD, describing the various models briefly. In Section 2 we describe the method for obtaining the light hadron spectrum in the MIT model and discuss problems associated with the cavity approximation. A variety of bag states are described in Section 3. Bag interactions in Section 4 include the important subject of electromagnetic structure as well as bag-bag interactions and, in particular, pion-bag interactions. Finally, in Section 5, we review recent attempts at describing the vacuum structure of QCD and its relationship to future developments of bag models. For a quick introduction the casual reader may want to read the introductory and concluding sections and Sections 2.1 to 2.3.

We have not attempted to be complete and have omitted some significant developments in the interest of coherence and brevity. The purpose of this review is to provide an entrée to the literature and a critical discussion of key difficulties. There are several summary lectures and reviews (5–10).

1.1 Quantum Chromodynamics and MIT Bag Model

We begin by elaborating briefly upon the relationship between the MIT bag model and QCD. Quantum chromodynamics is defined by the Lagrangian density

$$\mathscr{L}_{\text{QCD}} = -\frac{1}{4}F^{a\mu\nu}F^a_{\mu\nu} + \frac{i}{2}\bar{q}\overleftrightarrow{\partial}_\mu\gamma^\mu q - \frac{1}{2}g\bar{q}A^a_\mu\gamma^\mu\lambda^a q - \bar{q}Mq, \qquad 1.$$

where

$$F^a_{\mu\nu} = \partial_\mu A^a_\nu - \partial_\nu A^a_\mu + gf_{abc}A^b_\mu A^c_\nu \qquad 2.$$

is the color electromagnetic field tensor with color index, $a = 1, 2, \ldots, 8$, A^a_μ is the color vector potential, g is the bare strong interaction coupling constant, q is the quark field spinor, f_{abc} are the structure constants for SU(3), λ^a are the matrix generators of SU(3), and $\overleftrightarrow{\partial}_\mu$ is the derivative on the right minus the derivative on the left. Quantum chromodynamics is renormalizable and asymptotically free at short distances ($r \ll 1$ fm) (11, 12), where perturbation theory is valid. The theory very likely confines quarks and gluons into color singlet bound states at large distances ($r \approx 1$ fm); the best evidence for this feature comes from numerical studies of lattice gauge theories (e.g. 13). Experimental support for the theory is mounting, but it has been difficult to find definitive tests at short distances (14). On the other hand, a wealth of experimental data is available at large distances where theoretical predictions from first principles are scarce.

In anticipation of further progress in understanding the large-distance behavior of QCD, various phenomenological models have been proposed that incorporate guesses at this behavior. The MIT bag model is an example of such a model. It is elegant and simple. It describes hadrons as extended objects of quarks and gluons immersed in a vacuum that exerts an inward pressure B. Long-distance effects of QCD are represented by this bag pressure. Short-distance effects are treated in QCD perturbation theory, confined to a finite region of space. The Lagrangian density is (apart from some technicalities concerning boundary conditions)

$$L_{\text{bag}} = (L_{\text{QCD}} - B)\theta_V \qquad 3.$$

where θ_V is zero outside the space-time volume occupied by the quark and gluon fields and one inside. The use of the step function θ_V to define the bag surface renders the theory difficult to quantize, but leads naturally to a simple semiclassical treatment, namely the cavity approximation; this makes possible an extensive phenomenological analysis, which we review in Sections 2 to 4. The boundary condition for the fermions in effect renders them infinitely massive outside the bag, at the same time breaking the chiral symmetry explicitly, an undesirable result. Attempts at remedies are discussed in Section 4.4. An alternative definition of the bag surface involving a step function of the fields was proposed by Johnson (15). This alternative model has fewer excited states.

Whether the model is a good approximation to QCD is an open question, although there have been attempts to derive it analytically (16–18). Even in the absence of a rigorous justification from QCD, the MIT bag model is important in its own right since it embodies some of the essential features widely believed to characterize the light hadrons—namely, that they are bound systems of interacting quarks and gluons moving at relativistic speeds, confined to a region of space of size about 1 fm. The theory also respects color gauge invariance. In short it is the prototypical relativistic shell model for hadronic structure. The widely used cavity approximation to the model has many of the good and bad features of the nuclear shell model.

1.2 Other Bag Models

Prior to the introduction of the MIT model, Bogoliubov (19) proposed a relativistic square well of fixed radius as a model for hadrons but did not include a stabilizing pressure. Bardeen et al (2) developed the SLAC model, in which a scalar field plays the role of the bag. In their model, quarks exist primarily in a thin spherical shell and are not spread throughout the hadron. A thin shell picture is in contradiction with experiment, and the model has been supplanted by the related soliton model (see Section 1.3). Hasenfratz & Kuti (6) reviewed the bag model and proposed a variant in which a surface tension is introduced to supplement the bag pressure B.

Among currently popular variants of the MIT model are the hybrid chiral models, which contain an elementary pion field that interacts with the quarks through a chirally invariant coupling. These models have come under extensive study recently, largely because of their utility in nuclear physics (10). We discuss them in Section 4.4.

1.3 Soliton Bag Model

A theory that permits interpolation between the MIT model and the SLAC version has been proposed by Friedberg & Lee (3). The model has an elementary scalar field that serves to define the bag geometry. The Lagrangian is

$$\mathcal{L} = -\frac{1}{4}\kappa(\sigma)F^{a\mu\nu}F_{a\mu\nu}$$
$$+ \bar{q}\left(\frac{i}{2}\overleftrightarrow{\partial}_\mu\gamma^\mu - \frac{1}{2}gA_\mu^a\lambda^a\gamma^\mu - M + f\sigma\right)q - \frac{1}{2}\overleftrightarrow{\partial}_\mu\sigma\partial^\mu\sigma - U(\sigma). \quad 4.$$

Compared with the fundamental QCD Lagrangian (Equation 1), this

Lagrangian has the following new features: a dynamical scalar field σ with self-interaction given by $U(\sigma)$, a dielectric constant κ that depends on σ, and a Yukawa coupling with strength f between the scalar field and the quark field. The scalar field self-interaction ordinarily causes it to develop a nonzero vacuum-expectation value. However, in the presence of quarks or gluons, a soliton-like or bag-like solution for the classical field equations emerges, i.e. the quark, gluon, and scalar field energy densities are nonzero over a localized region of space. Outside the bag the fermion has a large effective mass determined by the nonzero vacuum-expectation value of the scalar field. Inside the bag the fermion has essentially only its bare mass M. There is also an effective bag pressure coming from the self-interaction of the scalar field. The dielectric constant κ depends on σ in such a way that inside the bag $\kappa = 1$ and outside $\kappa = 0$, leading to confinement of color electric flux. The scalar field is to be regarded as an effective composite field and so is usually treated semiclassically. This approximation is analogous to the static cavity approximation of the MIT bag model. With the introduction of the scalar field and its attendant interactions, the model has many more phenomenological parameters and is more difficult to work with than the MIT model. Wanted or not, the scalar field plays a dynamical role in the theory (20). Like the MIT and SLAC models, the soliton model is not chirally symmetric, but presumably there are corresponding "hybrid" models with chiral symmetry (10). One advantage of the model is its greater flexibility. One expects that a soft bag surface will emerge as a consequence of QCD. Perhaps in this way the soliton model is more realistic. Interest in the model is reviving (21, 22).

2. THE STATIC CAVITY APPROXIMATION

2.1 Doing Without the Approximation

The Lagrangian density of the MIT model (Equation 3) defines a Poincaré-invariant theory of hadrons and their interactions. Because the action involves an integration over a finite region of space, the location of the bag surface is itself a dynamical field, related through the Euler-Lagrange equations to the quark and gluon fields by a complicated, nonlinear expression. This feature has so far frustrated attempts to formulate a full quantum theory except in one space and one time dimension (1). However, in the approximations of a large bag with small surface fluctuations (23) and of a bag with only long wavelength oscillations (24), a Poincaré-invariant quantum theory can be developed. For reasons of simplicity, however, most applications resort to the static cavity approximation, described below, in which the surface is frozen (25). Indeed, in popular usage the static cavity approximation has become nearly synonymous with the MIT bag model, although many of its shortcomings, e.g. the lack of Poincaré invariance, are not present in the Lagrangian of Equation 3. For a corresponding treatment of small oscillations of the soliton bag, see (22).

2.2 Quark Eigenmodes for the Sphere

Consider the bag with only quarks present. In detail the action is

$$W = \int dt \left[\int_V d^3x \left(\frac{i}{2} \bar{q} \overleftrightarrow{\partial}_\mu \gamma^\mu q - \bar{q} M q - B \right) - \frac{1}{2} \int_S d^2x \, \bar{q}q \right], \qquad 5.$$

where V is the volume occupied by the bag with surface S. The surface term is added so that the quarks move as if they had an infinite mass outside the bag (1, 4). The surface term is required for consistency, and it breaks the chiral symmetry of the massless theory. An extremum of the action under varations in q and V occurs if

$$(i\partial_\mu \gamma^\mu - M)q = 0 \quad \text{in } V, \qquad 6.$$

$$\left. \begin{array}{l} in^\mu \gamma_\mu q = q \\ \frac{1}{2} n_\mu \partial^\mu (\bar{q}q) = B \end{array} \right\} \quad \text{on } S, \qquad \begin{array}{l} 7. \\ 8. \end{array}$$

where n_μ is the covariant inward normal to the surface. The first two equations describe a free Dirac particle of mass M moving in a cavity with a boundary condition that is tantamount to requiring that the normal component of the vector current $J_\mu = \bar{q}\gamma^\mu q$ vanish at the surface. Therefore, the vector current is conserved. The third condition, the nonlinear boundary condition, requires that the outward pressure of the quark field balance the bag pressure B.

The cavity approximation freezes the bag volume V at a shape and location that agrees with the nonlinear boundary condition in a time-averaged sense. For static bag configurations this condition is equivalent to requiring that the energy be minimum as a function of shape and size.

The static cavity Hamiltonian is easily derived from Equation 5:

$$H = \int_V d^3x \, [q^\dagger(-i\boldsymbol{\alpha}\cdot\mathbf{V})q + q^\dagger \beta M q + B]. \qquad 9.$$

Thus the bag pressure B is equivalent to a constant energy density everywhere inside the bag. This feature and the confinement of color electric flux (Section 2.4) enforce quark confinement since an infinite quark separation requires an infinite bag volume, hence, an infinite energy (1).

The quark fields are usually expressed in terms of an expansion in cavity normal modes that satisfy Equations 6 and 7. For spherical cavities of radius R they are easily expressed in terms of spinor spherical harmonics. We display here the lowest zero-mass cavity eigenfunction, a $J^P = \frac{1}{2}^+$ state, sometimes designated $s_{1/2}$:

$$q_s = N \begin{pmatrix} j_0(\omega r) U \\ i\sigma \cdot \hat{r} j_1(\omega r) U \end{pmatrix} e^{-i\omega t}, \qquad \qquad 10.$$

where j_0 and j_1 are spherical Bessel functions, $\omega R = 2.043...$ is the lowest root of the linear boundary condition $j_0(\omega R) = j_1(\omega R)$, U is a two-component spinor, and N is fixed by the normalization condition $\int d^3x\, q_s^\dagger q_s = 1$. The next lowest orbitals are of odd parity: $p_{3/2}$ at $\omega = 3.204/R$ and $p_{1/2}$ at $\omega = 3.812/R$ for $M = 0$, respectively. For nonspherical cavities a variational approach can be used (26, 27). The linear boundary condition makes it awkward, if not impossible, to separate the equation in the usual coordinate systems.

The noninteracting cavity normal modes provide a basis for expanding the quark field and defining the canonical creation and annihilation operators for the quarks and antiquarks:

$$q(x,t) = \sum_{j,l,m,n} [q_{jlmn}(x) \exp(-i\omega_{jln} t) b_{jlmn} + q^c_{jlmn}(x) \exp(i\omega_{jln} t) d^\dagger_{jlmn}].$$

11.

The creation and annihilation operators of Equation 11 can then be used to define a Fock space for a cavity of a fixed geometry. These Fock states are the basis states for the bag model. This approach is different from the usual treatments of the Schrödinger equation, in which one deals directly with two- or three-particle wave functions. In particular there is no connection between the Fock spaces for inequivalent cavities. The static cavity approximation does not admit the "quantum superposition" of bags with different cavities or the "inner product" ("overlap") of such bag states; these concepts are meaningful only when the bag surface is treated quantum mechanically (23, 24). However, in practice, when calculating a transition amplitude between states of slightly different radii (e.g., the electroweak transitions. See Section 4.1), the small region of nonoverlapping bags is often not included. The nonlinear boundary condition is imposed upon the states as the expectation value $\langle :\frac{1}{2} n \cdot \partial \bar{q} q: \rangle = B$ or by minimizing the energy of the state with respect to shape and size, but it is not imposed upon the individual orbitals. Thus spherical bags may be formed from quarks in "nonspherical" orbitals, such as $p_{3/2}$. Because of difficulties in treating nonspherical cavities, slightly nonspherical states are often approximated as spheres, and the pressure balance is achieved on average by minimizing the energy with respect to radius alone.

2.3 A Simple Calculation and the Zero-Point Energy

Let us consider the construction of a spherical baryon made of n massless quarks or antiquarks in the $s_{1/2}$ orbital. The expectation value of the Hamiltonian of Equation 9 on such a state is

$$E = \langle H \rangle = 2.043n/R + \frac{4\pi}{3}BR^3 - \sum_{njlm} \omega_{njlm}. \qquad 12.$$

The first term is the kinetic energy of n quarks, the second is the bag volume energy, and the third is the energy of the Dirac sea, i.e. the zero-point energy of the fermion modes. Gluon modes described below also contribute to the zero-point energy.

In a field theory defined on an infinite volume, the last term is unimportant. In a finite system, however, its influence is related to the Casimir effect (28). It should appear in all bag models that treat the surface classically. In computing its contribution it is necessary to subtract an infinite part. The calculation of the finite part of the zero-point energy is delicate and controversial (29–32). In applications it has been treated phenomenologically by replacing it by a term $-Z_0/R$ where Z_0 is an adjustable parameter. The phenomenological applications favor $Z_0 \sim 1$ to 1.8, while the best theoretical estimate (32) of its value is small and negative. This discrepancy is not yet well understood. Quite possibly, a more realistic modeling of the bag boundary would change the result. The phenomenological value of Z_0 may contain other contributions to the energy as well [e.g. the center-of-mass correction (Section 2.7) and the quark-self energies (Section 2.5)].

With this phenomenological treatment of the zero-point energy it is straightforward to minimize E with respect to R in Equation 12, giving

$$R^4 = (2.043n - Z_0)/4\pi B, \qquad 13.$$

$$E = \frac{4}{3}(2.043n - Z_0)^{3/4}(4\pi B)^{1/4}. \qquad 14.$$

This example illustrates the basic features of spectroscopic calculations in the static cavity approximation (25, 33, 34). To complete the physical description of the spectrum, it remains to include the level splitting induced by the color interactions among the quarks and to remove from E the contribution from the kinetic energy of motion of the center of mass. These topics occupy the remainder of this section.

2.4 Gluon Eigenmodes for the Sphere

With only gluons present the action is

$$W = \int dt \int d^3x \left(-\frac{1}{4}F^a_{\mu\nu}F^{a\mu\nu} - B \right). \qquad 15.$$

The Euler-Lagrange equations are

$$\partial^\mu F^a_{\mu\nu} = 0 \quad \text{in } V,$$

$$\left.\begin{array}{r} n^\mu F^a_{\mu\nu} = 0 \\ -\tfrac{1}{4}F^a_{\mu\nu}F^{a\mu\nu} = B \end{array}\right\} \text{on } S. \qquad 16.$$

In the limit of zero coupling, $F^a_{\mu\nu} = \partial_\mu A^a_\nu - \partial_\nu A^a_\mu$, and the first two equations are sourceless Maxwell equations with the boundary conditions $\hat{\mathbf{r}} \cdot \mathbf{E}^a = 0$ and $\hat{\mathbf{r}} \times \mathbf{B}^a = 0$, where $\mathbf{E}^a$ and $\mathbf{B}^a$ are the color electric and magnetic fields. These boundary conditions are of great importance for quark and gluon confinement, since they require that the total color charge of the state vanish. It is elementary to construct the cavity normal modes for the gluon field (35, 36). The lowest eigenmodes are a $J^P = 1^+$ magnetic dipole mode with $\omega R = 2.74$, a $J^P = 2^-$ magnetic mode with $\omega R = 3.87$, and a $J^P = 1^-$ electric dipole mode with $\omega R = 4.49$.

2.5 Interactions and Cavity Perturbation Theory

The free quark and gluon normal modes provide a basis for a perturbative treatment of the interacting theory. The justification for an expansion of physical quantities in powers of the color coupling constant $\alpha_s = g^2/4\pi$ is made a posteriori from a phenomenological determination of α_s and a calculation of higher order terms. Little is known about higher order corrections, but with typical, rather large phenomenological values of $\alpha_s \approx 2$, there is just cause for concern. Nevertheless, the lowest order calculations may provide a foundation for future improvements.

The formulation of the cavity perturbation theory is straightforward. Feynman rules are modified only in that the propagators for quarks and gluons are not free propagators, but are cavity propagators. The method of (37, 38) involves calculating these propagators by a direct summation over cavity modes—a rather cumbersome procedure. A somewhat more practical representation of cavity propagators uses a multiple reflection expansion (39). See (40) for details and further references. Another method, which separates the free and confining parts of the propagator, has been implemented successfully (29, 41, 42).

Let us now consider the calculation of the effects of the color interaction upon the energies of bags containing quarks and antiquarks to second order in the color coupling g in Rayleigh-Schrödinger perturbation theory. In Coulomb gauge the full interacting Hamiltonian to $O(g^2)$ that contributes to these states is (43)

$$H = H_0 + H_I \qquad 17.$$

$$H_0 = \int d^3x \left[q^\dagger(-i\boldsymbol{\alpha} \cdot \nabla + \beta M)q + \frac{1}{2}(\mathbf{E}^a_\perp)^2 + \frac{1}{2}(\mathbf{B}^a)^2 + B \right] \qquad 18.$$

$$H_{\text{I}} = -\int d^3x\, j^a \cdot A^a + \frac{1}{2}\int d^3x \int d^3x'\, \rho^a(x) D(x,x') \rho^a(x'), \qquad 19.$$

where $j^{\mu a} = g\bar{q}\lambda^a \gamma^\mu q$, and $\rho^a = j^{0a}$, $E_\perp^a = \dot{A}^a$, $B^a = \nabla \times A^a$. $D(x,x')$ is the scalar Neumann-Green function for the spherical cavity satisfying $\nabla_x^2 D = -4\pi \delta^3(x-x')$. The first term in H_I is $O(g)$, couples the quark state to a state containing an additional gluon, and gives an energy shift in second-order Rayleigh-Schrödinger perturbation theory. The second term is $O(g^2)$ and gives the instantaneous Coulomb interaction between quarks and gluons. These contributions are depicted in Figure 1 for the three-quark baryon state. Contributions a, b, and d are called "self-energy" diagrams and c and e, "exchange" diagrams.

The self-energy diagrams are particularly difficult to calculate, because it is necessary to sum over all cavity modes for both the intermediate quark and gluon. The sum is divergent at short distances just as it is in unconfined perturbation theory. A straightforward summation of the series is therefore dangerous. Chin et al (44) find a rather large contribution from these terms, but in the mode sum method this may be due to the difficulties of handling the infinite self-energy. Recent work by Hansson & Jaffe (40) using the

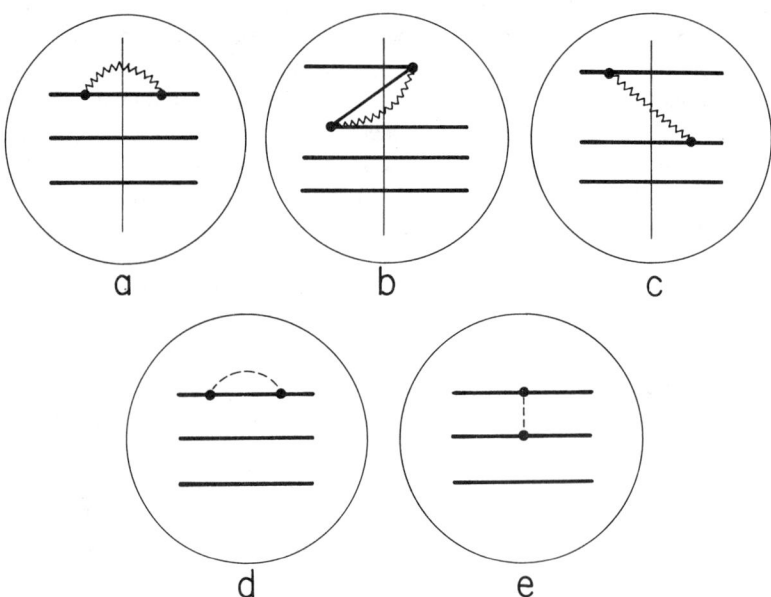

Figure 1 Coulomb gauge Rayleigh-Schrödinger perturbation theory diagrams for the energy level shift to second order in g^2 for a three-quark hadron. The wavy line is a physical gluon and the dotted line depicts the direct Coulomb interaction. The vertical line intersects lines for quarks and gluons occurring in the intermediate state.

multiple reflection expansion, in which the infinite renormalization is straightforward, find a much smaller self-energy. See also (42, 45).

In the earliest work on the light hadrons (43) the self-energy terms were handled in a rather ad hoc fashion as follows: the contributions a and b were omitted altogether and, for the contribution d, the intermediate quark line was kept in the same orbital as in the initial and final states. With all quarks in the $s_{1/2}$ orbital, the contribution c is purely magnetic and is simply of the form

$$\Delta E_c = -\sum_{i>j} \int \mathbf{B}_i^a \cdot \mathbf{B}_j^a \, d^3x, \qquad 20.$$

where $\mathbf{B}_i^a$ is the color magnetic field produced by the ith quark. The contributions d and e are simply

$$\Delta E_{de} = \sum_{i,j} \frac{1}{2} \int \mathbf{E}_i^a \cdot \mathbf{E}_j^a \, d^3x \qquad 21.$$

where $\mathbf{E}_i^a$ is the color electric field produced by the ith quark. The color magnetic splittings are of the form

$$\sum_{i>j} \lambda_i^a \sigma_i \lambda_j^a \sigma_j,$$

i.e. a color magnetic moment interaction. This form gives the correct sign for the N-Δ splitting as well as the π-ρ splitting. The color electric splittings are negligible for the light hadrons. Equations 20 and 21 were included in Equation 12, and the determination of the masses of the nucleon, Δ, ρ, and π as in Equation 14 was straightforward. There are four masses and three parameters, namely α_c, B, and Z_0. With massless down quarks DeGrand et al (43) adjusted the parameters to give the correct masses for the ρ, N, and Δ with the result that $\alpha_s = 2.2$, $B^{1/4} = 146$ MeV, and $Z_0 = 1.84$. Finally, they predicted $m_\pi = 0.28$ GeV. Compared with the experimental value of 0.14 GeV, the prediction is poor, but it may be improved with further refinements of the model described below. Moreover, with the inclusion of the strange quark and a fourth parameter, its mass ($m_s = 280$ MeV), the model predicts eight masses quite well and also gives a respectable accounting for various static properties, such as the axial vector coupling.

The η and the η' present special problems. A straightforward application of the static cavity approximation would lead to unacceptably low masses for these states. This typical quark model result is commonly known as the U(1) problem. It is assumed that it can be resolved within a quark model framework by taking into account the annihilation into two (or more) gluons (e.g. 46).

2.6 Lowest Excitations of the Light Hadrons

While the spectroscopy of the ground states is reasonably satisfactory, that of the excited states is less so. With naive Fock-space methods an extensive series of excited states can be constructed. However, there is a problem of interpretation. The rigid cavity approximation gives rise to spurious states; moreover, the problem of isolating the c.m. motion becomes both more difficult and more important. The methods of Rebbi (23) mentioned in Section 2.1 suggest a resolution to this problem. DeGrand and Jaffe adapted these methods to a study of the p-wave excitations in the cavity approximation, but further excited states are difficult to treat (47, 48). The Roper resonance $N(1470)$ has been studied as an analogous breathing excitation (49), although the static approach is apparently satisfactory (50).

The DeGrand-Jaffe results for the p-wave energy levels are disappointing, particularly when compared with impressive achievements in a nonrelativistic quark model (51). However, the difficulty may lie in too rigid an adherence to the various approximations used in the $s_{1/2}$ hadrons, particularly, in the treatment of self-energies and the center-of-mass kinetic energy correction. This chapter of bag model history is not yet closed (cf 52).

2.7 Refinements-Correcting for Center-of-Mass Motion

One of the awkward features of the cavity approximation that is shared by shell models in general is that the center of mass (c.m.) of the many-body state is in motion, and the kinetic energy of motion is unavoidably included in the total orbital energy. This contribution should be removed from the bag energy to obtain the mass. In the nuclear shell model it is possible to project out nonrelativistic states of definite c.m. momentum. The underlying translationally invariant quantum Hamiltonian is known, and the corrected energy of the state can be determined (53). In the cavity approximation, however, the quantum Hamiltonian is defined only with respect to a particular cavity and is not translationally invariant. No projection onto momentum eigenstates exists. A further complication exists in relativistic interacting theories in that Lorentz boosts change the Fock-space composition of a given state.

There is no unambiguous procedure for correcting for the c.m. motion. Nevertheless, several approaches have been tried (7, 31, 54–58). In the method of Donoghue & Johnson (54) the bag state with nucleon quantum numbers is not to be identified precisely with the nucleon; rather, it is considered more generally as a wave packet of nucleon momentum states with overall $\langle \mathbf{p} \rangle = 0$ but $\langle p^2 \rangle \neq 0$. The bag energy is then not precisely the particle mass, but

$$E_{\text{bag}} = \langle H \rangle = \langle \sqrt{p^2 + m^2} \rangle. \qquad 22.$$

For most states (those with $m \gg 1/R$), the nonrelativistic expansion is possible: $E_{bag} = m + \langle p^2 \rangle /2m$. The correction may be estimated in several ways. Johnson (31) gave the simplest estimate for n quarks in the $s_{1/2}$ orbital, namely, $\langle p^2 \rangle \approx n\Omega^2/R^2$ with $\Omega = 2.043$, the dimensionless wave number for a quark.

For heavy bound states the c.m. correction is equivalent to a shift in Z_0. With this refinement and further modification incorporating some effects of the renormalization scale dependence of the color coupling constant, Donoghue & Johnson were able to produce a fit to the hadron spectrum similar in quality to that of DeGrand et al (43), and they obtained a zero-mass pion as well in the limit of zero quark mass. Such a result is not natural in the bag model, but is desirable if the theory is to be derived from QCD. The parameters of this fit ($Z_0 = 1$, $m_u = m_d = 33$ MeV, $m_s = 330$ MeV) are close to those of DeGrand et al.

3. A VARIETY OF BAG STATES

3.1 *Strings*

The ground-state particles have been assumed to correspond to a spherical bag. However, the bag states of high angular momentum J are likely to deform into rotating tubes with quarks and/or antiquarks at the ends and a flux of color fields connecting them. Such a structure resembles a "string" with a constant energy per unit length or string tension T_0. Dynamical models of strings predict linear Regge trajectories with $J = \alpha' M^2 + \alpha_0$, where α', the Regge slope, is related to the string tension by $\alpha' = 1/(2\pi T_0)$ (59). Phenomenologically $\alpha' \approx 0.9$ GeV^{-2} so that $T_0 \simeq 900$ MeV fm^{-1}. Johnson & Thorn (60) showed that the bag model, like the string model, generates a linear string tension and Regge slope of the appropriate size. The bag string tension is $T_0 = (32\pi\alpha_s B/3)^{1/2}$. With the values of α_s and B required for the light hadron spectroscopy (43), one obtains $T_0 = 856$ MeV fm^{-1} in good agreement with the experimental value.

3.2 *Heavy Quark-Antiquark Bound States*

The spectrum of the J/ψ and Υ particles and their excitations have been successfully explained using an interquark potential that interpolates between a Coulombic form at short distances and a linear potential at long distances. Many parametrizations of such a potential are possible, although the resulting shape is very similar in all cases.

The bag model potential between heavy quarks and antiquarks has been investigated by two groups (61, 62). They both used the adiabatic Born-Oppenheimer approximation in which the static energy of the Q$\bar{\text{Q}}$ system is calculated for fixed quark sources. This energy is then used as the potential in the Schrödinger equation. At short distances a Coulombic form of the potential is essentially automatic. At long distances the bag forms a flux tube that generates a linear potential, as discussed in the last section. Therefore, it is not surprising that the bag model is able to reproduce the phenomenological form of the heavy quark potential. It is interesting that the linear behavior sets in before the sphere is much distorted.

The two groups fit the potential with different models. Both agree that the short-distance Coulombic term should have a small color coupling α_s. Haxton & Heller (61) use a running coupling that grows with increasing interquark separation, as suggested by asymptotic freedom. Hasenfratz et al (62) use a constant α_s. Phenomenology requires the correct string tension at large distances. Since the string tension depends on the product $\alpha_s B$, it is not surprising that Haxton and Heller get a smaller value of $B^{1/4}$ (145 MeV) than Hasenfratz et al (235 MeV). However, Haxton and Heller get a smaller value of α_s at a distance of 1 fm ($\alpha_s = 1.0$) than in the light hadrons (43) ($\alpha_s = 2.2$). The source of the discrepancy is not known. However, the asymptotic freedom scaling of α_s may be different than supposed at large distances. Also, there may be important nonlocal or velocity-dependent terms in the potential due to the gluon cloud surrounding the heavy quark. Models in the same spirit have been constructed for mesons and baryons with one heavy quark, and they offer the possibility of a unified treatment of the heavy-heavy and heavy-light mesons (63).

3.3 Glueballs

In addition to bound states made up of quarks, it is expected that in QCD bound states of gluons—"glueballs" or "gluonium"—should exist. As a model for these states, the bag has several advantages over other constituent models. In the bag model the gluon fields are massless gauge-invariant Lagrangian fields. Moreover, since the model already treats states with relativistic quarks successfully, it is expected to do as well for states with necessarily relativistic gluons. A common set of parameters suffices. However, the extension of the bag model to glueballs is not entirely straightforward; new subtleties appear. The full structure of these states is not yet clear.

As we mentioned in Section 2.4 the lowest-lying gluon modes in a spherical cavity are these: $J^P = 1^+$ at $\omega = 2.744/R$, $J^P = 2^-$ at $\omega = 3.87/R$, and $J^P = 1^-$ at $\omega = 4.49/R$. The glueballs that can be formed from these cavity modes were first given by Jaffe & Johnson (64) and were corrected and refined by Donoghue et al (65). The lowest energy two-gluon states have $J^{PC} = 0^{++}, 2^{++}$, while the first excited states have $J^{PC} = 0^{-+}, 2^{-+}$. The lowest three-gluon states have $J^{PC} = 0^{++}, 1^{+-}, 3^{+-}$. Contrary to some erroneous claims (e.g. 66), the scalar and pseudoscalar states satisfy the nonlinear boundary condition in a spherical cavity. States with other

J^{PC} do not have a uniform outward pressure, but have nevertheless been approximated as spheres.

When the masses of these glueballs were calculated with the same parameters as found from the $Q\bar{Q}$ and QQQ spectrum, but without the spin-spin interaction, the two-gluon ground states ($0^{++}, 2^{++}$) were found to have a mass of 1 GeV, with the excited states ($0^{-+}, 2^{-+}$) 300 MeV above that. The three-gluon ground states ($0^{++}, 1^{+-}, 3^{+-}$) are at 1.5 GeV.

The inclusion of interactions between the gluons to order α_s shifts this spectrum considerably and presents a problem for the model. The spin interactions to order α_s were first calculated for the 0^{++} and 2^{++} glueballs by Thorn (unpublished), and the results for 0^{-+} and 2^{-+} were subsequently given by Barnes et al (67) and by Carlson et al (68). The size of these splittings is larger than that of quark states, because the effective color charge of the gluon is 9/4 that of quarks. If one uses the parameters and methods of DeGrand et al (43), one obtains $m(0^{++}) \approx 0.1$ GeV, $m(0^{-+}) \approx 0.4$ GeV, $m(2^{++}) \approx 1.3$ GeV and $m(2^{-+}) \approx 1.6$ GeV. The two spin-zero states are much too light to be phenomenologically acceptable. However, it may be premature to reject the model entirely, because these particles have special quantum numbers. The 0^{++} glueball has the quantum numbers of the vacuum and is expected to mix with it. To describe this mixing requires going beyond the model (see Section 5.1). The 0^{-+} glueball has the same quantum numbers as the η and η' and it mixes strongly with them (69, 70). In the cavity approximation the η' also has too small a mass. Evidently, the model lacks the ingredient that resolves the U(1) problem and raises the mass of the η' (71). Perhaps the mass of the 0^{-+} glueball is raised by the same process (46). At least for these channels the unadorned model is probably deficient.

Various authors have suggested modifications in the model to repair the predicted spectrum of scalars and pseudoscalars. Barnes et al (67) use a different additive energy scale for quark and gluon states. Chanowitz & Sharpe (72) include a phenomenological gluon self-energy. Both groups fit the 0^{-+} mass to the $\iota(1440)$. Carlson et al (73) determine self-energy parameters from a specific model for the bag constant, but obtain the usual excessively light 0^{++} and 0^{-+} states. Clearly, we need a theoretical understanding of the gluon self-energies, a resolution to the U(1) problem, and a better understanding of the magnitude of lowest order, if not higher order, spin splitting (i.e. the value of α_s) before the glueball spectrum can be predicted with confidence.

3.4 *Exotics and Cryptoexotics*

In addition to glueballs there are a variety of bag states that do not occur in the simplest quark models. One can form color singlet combinations of constituents from $Q\bar{Q}Q\bar{Q}$, $Q\bar{Q}G$, $QQQQ\bar{Q}$, $QQQG$, and $QQQQQQ$ (64,

74–84). Some of them have quantum numbers that do not occur for the combinations $Q\bar{Q}$ or QQQ. They are called exotics. The others, the "cryptoexotics" (75), have quantum numbers that also occur for $Q\bar{Q}$ and QQQ. Those without gluons are generally above threshold for falling apart into two lower mass color singlet groups. The present understanding is that most or all of these are not resonances, i.e. poles in the S matrix; instead they are poles in the P matrix (see Section 4.2).

The $Q\bar{Q}G$ and QQQG states have been less extensively studied until recently, although they play an indirect role in the well-studied splittings of the $Q\bar{Q}$ and QQQ states. In their investigation of charmonium, Hasenfratz et al (62) studied the effect of adding a gluon to the heavy $Q\bar{Q}$ pair. They find two states of considerable interest, one of them exotic, both with a mass near 10.5 GeV for the case of $b\bar{b}$. One has $J^{PC} = 1^{--}$ and would be directly produced in e^+e^- reactions. No such state has been seen with an electronic width $\Gamma_{ee} > 0.07$ keV [for comparison $\Gamma_{ee}(\Upsilon) = 1.2$ keV] (85). However, the $b\bar{b}$-gluon wave function has a node at the origin, due to the repulsive coulombic force between a Q and $\bar{Q}$ in a color octet, which leads to a reduced electronic width. Until more work is done, the existence of this state cannot be ruled out. The other state has $J^{PC} = 1^{-+}$, quantum numbers that cannot be obtained with $Q\bar{Q}$ (i.e. it is exotic). The discovery of such a state would be strong evidence for gluonic excitations.

Barnes & Close (81) have studied the $Q\bar{Q}G$ $J^{PC} = 1^{-+}$ state that can be formed from light quarks. They locate its mass near 1.3 GeV. The spectrum of QQQG has been studied recently by Golowich et al (82) and by Barnes & Close (83). These authors predict many new states in the baryon spectrum above 1.5 GeV that mix strongly with QQQ baryons. The existence of these states may prove to be a problem, since the baryon spectrum has been extensively mapped in this region, and most known states are accounted for by the three-quark baryons. However, even in the QQQ sector more states are predicted than are seen (51). Perhaps this lack of observation can be attributed to small couplings to pions and photons. Until the couplings of the QQQG baryons have been estimated we do not know whether they should have been seen.

With the possibility of adding extra gluons and $Q\bar{Q}$ pairs to the low-lying quark and gluon states and the existence of radial and orbital excitations of all states, the predicted baryon and meson spectra become densely populated above 1.5 GeV. This result does not seem to be a peculiarity of the bag model alone, but appears to be a property of QCD, which should be manifest in all realistic models. However, the theoretical and phenomenological interpretation of this spectrum remains a problem.

4. INTERACTIONS

4.1 *Electroweak Interactions*

The bag model has been extensively used as a tool in the study of the electromagnetic and weak properties of hadrons. We are not able to describe all applications, but instead concentrate on the static properties of the baryons (25, 34, 43, 86). We describe first the uncorrected results of (43). Four types of corrections have been studied: (a) c.m. recoil, (b) gluon exchange, (c) pion cloud, and (d) bag deformation. No study includes them all. The first two are discussed here; the last two, in Section 4.4.

For the proton charge radius DeGrand et al (43) obtain

$$\langle r^2 \rangle_{EM} = 0.55R^2 = (0.73 \text{ fm})^2$$
$$[\text{experiment}: \quad \langle r^2 \rangle_{EM} = (0.90 \text{ fm})^2] \qquad 23.$$

for massless quarks. [For the values of this and other quantities at nonzero quark mass, see (34, 86).] This result is significant because it means that the bag model very nearly correctly predicts the size of the proton. Such a connection between the mass and radius is nontrivial. In fact, using the proper radius is often more important than having the correct mass, because it is the radius that determines the scale of most matrix elements.

The magnetic moment of a state can be calculated by the matrix element of the first moment of the electromagnetic current

$$\langle \mu \rangle = \left\langle \int d^3 x \frac{1}{2} \mathbf{x} \times \mathbf{J}(x) \right\rangle. \qquad 24.$$

This definition of the moment includes the contributions due to both the flow of charge and the spin of the quark. In the nonrelativistic limit, it reduces to the usual

$$\mu = \sum_i \frac{Q_i \sigma_i}{2m_i}, \qquad 25.$$

while for massless quarks in the proton, it is

$$|\mu_p| = 0.20Re. \qquad 26.$$

The parameters of DeGrand et al yield too small a gyromagnetic ratio ($g_p = 2m_p \mu_p/e = 1.9$ compared with the experimental value $g_p = 2.79$). However, we will see that c.m. recoil corrections improve this value. The magnetic moments of other baryons follow a typical quark model pattern. For example, $\mu_n/\mu_p = -2/3$ (experiment: -0.685). The moment of the s quark is 24% smaller than that of the d quark. For the hyperons this pattern gives a reasonable accounting of the magnetic moments, although there are some discrepancies (notably the Σ^+ moment).

For the weak interactions the most important static quantity is the axial vector coupling constant. For spin-up nucleons it is calculated from

$$g_A = \left\langle P \left| \int d^3x\, \bar{\psi}_u(x)\gamma_3\gamma_5\psi_d(x) \right| N \right\rangle = \frac{5}{3}\left(1 - \frac{4}{3}L\right), \qquad 27.$$

where L is determined from the lower component of the quark wave function. In the nonrelativistic limit $L = 0$, and the result is the familiar prediction of SU(6). For massless quarks the result is reduced to $g_A = 1.09$, and for reasonably small masses one can reproduce the experimental value $g_A = 1.254$. The SU(3) structure of g_A is described by D and F octet couplings, with $D/(D+F) = 0.60$, which is fairly close to the experimental ratio $D/(D+F) = 0.64 \pm 0.02$. Other weak interaction current matrix elements, such as weak magnetism and the second-class axial form factor, have been calculated (87). They may be measurable in new high statistics studies of semileptonic hyperon decay.

Recently Maxwell & Vento (88) calculated all the $O(\alpha_s)$ corrections to the static properties of the nucleons. Previous groups (35, 89–92) had studied various incomplete subsets of $O(\alpha_s)$ diagrams. Despite the large coupling constant ($\alpha_s = 2.2$), the changes in the observables are quite small—generally less than 10%. Particularly notable is their finding that the charge radius of the neutron is too small and of the wrong sign, despite the suggestion in a different context by Carlitz et al (93) that the order α_s corrections generate the correct value. However, a pion cloud gives the correct sign and magnitude (see Section 4.4). The smallness of the $O(\alpha_s)$ corrections is encouraging in that it implies that perturbation theory in a cavity may be well behaved. To our knowledge this is the first complete $O(\alpha_s)$ calculation.

In the results quoted above the static bag values were not corrected for c.m. motion or recoil. For example in the wave packet method for handling the c.m. motion (54) one finds that the static bag magnetic moment is related to that of the proton by

$$\mu_{\text{static bag}} = \mu_p\left(1 - \frac{1}{2}\frac{\langle p^2\rangle}{m^2}\right) \qquad 28.$$

where $\langle p^2 \rangle$ is the average momentum fluctuation of the proton. Including these corrections and using the parameters of (54), one obtains $g_p = 2.5$ and $g_A = 1.27$. Other groups obtained qualitatively similar corrections by attempting to construct moving bags and including the effects of recoil (94, 95).

The vector current matrix elements also determine the M1 radiative decays (i.e. $\Delta \to N\gamma$, $\Sigma^* \to \Sigma\gamma$ etc) but with less satisfactory results. These decays are due to magnetic-moment-like transitions, but the rates depend upon the squares of the moments. Even allowing the quark magnetic moment to increase, as was done by Hackman et al (96), the $\Delta \to N\gamma$ transition remains too low by a factor of about two. This failure can probably be attributed to the static no-recoil approximation and the sensitivity of the result to small changes in the wave function.

The bag model has also been useful in the study of the hadronic weak interactions. The hadronic wave functions have been used to calculate matrix elements for the weak, nonleptonic decays of baryons and mesons (97–99), to study parity violation in nuclear physics (100), and to estimate effects of CP violation (101–103). Recently, in grand unified theories the bag model has been used to complete the calculation of the proton and neutron lifetimes and branching ratios (104–106).

Jaffe and co-workers have applied the bag model to deep inelastic scattering (107–111). At its earliest stages the object was to verify that the bag model has the scaling properties expected from the parton model. This result was accomplished by use of the multiple reflection expansion, in which the short-distance singularities of the cavity propagators are explicitly related to those of the free field propagators. Recent attention has focused upon the normalization of the asymptotic behavior of the structure functions and higher twist terms, which are not calculable in QCD perturbation theory (110, 111).

4.2 P Matrix and Multiquark Bag States

Three methods of studying low energy hadron-hadron scattering in the bag model are described in this and the following two subsections: the P-matrix method, the deformation energy method, and the hybrid chiral bag method.

The P-matrix method of Jaffe & Low (112) is an adaptation of an old technique in nuclear physics, the R-matrix method, which relates properties of the scattering of a nucleon from a nucleus of atomic number A to the existence of virtual compound nuclear states with $A+1$. In the case of the scattering of two $Q\bar{Q}$ mesons, the P-matrix method relates certain features of the scattering phase shift to the existence of virtual $Q^2\bar{Q}^2$ bag states. As in the R-matrix method the P-matrix method divides the scattering process into two parts: a long-range part $r > b$ in which a simple two-body potential description is valid and a short-range part $r < b$ in which the bag model is used. The matching separation b is of the order of the bag radius.

To make the connection between the properties of the $Q^2\bar{Q}^2$ states and the two-body potential model, it is necessary to interpret these states in potential-model language. A number of such states can be constructed in the standard manner described in Section 3.4 that have both quarks and antiquarks in the $s_{1/2}$ orbital (64, 74–84). Many of these $Q^2\bar{Q}^2$ states may not appear in nature; e.g. the $I^G(J^P) = 0^+(0^+)$ $u\bar{u}d\bar{d}$ state with a theoretical mass of about 650 MeV may not correspond to any resonance in this channel. Such a state would readily fall apart into two pions. However, by requiring the quarks to remain in the $s_{1/2}$ orbital, any decay is precluded. Therefore, such a bag state is not very likely to be a legitimate hadronic bound state or resonance. Rather, Jaffe and Low argue that it should be regarded as an eigenstate of an artifically constrained Hamiltonian, to wit: it is assumed that restricting the quarks to the $s_{1/2}$ orbital corresponds to requiring the two-meson wave function to vanish at a relative separation $r = b$. Using rough estimates of wave packet size, they obtain $b \approx 1.4R$ for this case. The phenomenological analysis is then straightforward. The scattering wave function for $r > b$ can be reconstructed from the measured phase shift and the assumed long-range potential. At the c.m. energy corresponding to the energy of the $Q^2\bar{Q}^2$ bag state, the wave function should have a node at $r = b$. To formalize this procedure, Jaffe and Low introduce the "P matrix," which has poles whenever a node in the wave function crosses $r = b$. They also present a method for estimating residues of P-matrix poles in terms of fractional parentage coefficients relating the $Q^2\bar{Q}^2$ states to the open external meson-meson channels.

The P-matrix analysis has been applied with good results to a few other meson-meson channels (112), meson-nucleon channels (113), baryon-baryon channels (114), and charmonium (115). The successes of the approach are remarkable and encouraging, especially in light of the assumptions made. The interpretation of the restriction to the $s_{1/2}$ orbital in terms of a particular constraint on the relative wave function and the neglect of internal two-body interactions in determining the relationship between b and R might make one hope for improvements in the theoretical analysis. Nevertheless, its simplicity is appealing.

4.3 *The Nucleon-Nucleon Interaction*

A rather different bag model approach to low energy hadron-hadron interactions was adopted by DeTar (116, 117). This approach studies the dependence of the bag energy upon a large-scale deformation of the quark configuration, in particular, a deformation that results in the separation of the two clusters of quarks that form the individual hadrons. The deformation is defined by introducing a Lagrange constraint that fixes a parameter δ, which measures the separation of the clusters. At $\delta = 0$ the quarks are all in $s_{1/2}$ orbitals, and at finite δ configurations involving pairs of quarks, promoted to $p_{3/2}$ orbitals, are mixed in a manner leading to the disengagement of the hadron clusters. In the two-nucleon problem an appreciable separation of clusters occurs in a roughly spherical cavity before deformations of the spherical shape are required.

The calculation using the parameters of DeGrand et al (43) reveals a short-range repulsive core (116), in qualitative agreement with phenomenological potentials. The tensor component of the interaction can also be isolated and it is in similar qualitative agreement (117). However, a quantitative comparison is not warranted, because the deformation energy is not a potential and the separation of clusters is not exactly a two-body coordinate. The deformation energy is more properly regarded as the energy of a constrained, interacting two-baryon wave packet including the kinetic energy of relative motion of the two clusters. In this regard the strong attraction at intermediate range found in (116) and (117) is not in accordance with what is expected, since the deuteron is only weakly bound. This discrepancy can probably be cured by a small adjustment of the color coupling constant.

The qualitative features of the two-nucleon interaction found in the bag model, especially a repulsive core of short range, were also found in various earlier and subsequent analyses of nonrelativistic quark models (118–122). Harvey (122) expanded upon this analysis by including a richer set of configurations and found that the repulsive core "disappeared." However, Harvey's surprising result was due primarily to a definition of "separation" that did not force the clusters into a fully overlapped configuration at zero "separation." Indeed, the repulsive core was at work at Harvey's calculation. A repulsive core makes a [4,2] orbital symmetry in the even-parity NN channel energetically favorable in the absence of a constraint (123). This feature is common to the various calculations (118–122). Thus they appear to be in qualitative agreement despite differences in language.

With a nonrelativistic model one may use the deformation energy calculation as a basis for a resonating group or generator coordinate analysis (124), which leads to a determination of phase shifts (120, 121). However, the cavity approximation in the bag model precludes such an approach because a quantum mechanical superposition of different cavities is not meaningful. The Born-Oppenheimer approximation is also not well justified, because the kinetic energy of relative cluster motion is not negligible. How do we proceed? One suggestion (125) involves attempting to establish a parallel between a deformation energy calculation based on the bag model and one based on an effective two-body potential. The method has not been tested, however. See also (126).

4.4 *Hybrid Chiral Models for Pion-Bag Interactions*

4.4.1 INTRODUCING THE PION FIELD   With a few rather plausible assumptions based upon chiral symmetry it is possible to construct a simple hybrid model for pion-bag interactions. Because of its low mass and consequent long range the pion plays an important role in the structure of hadrons as revealed through magnetic moments, charge radii, and electromagnetic transition rates and, of course, in determining the lifetimes of states that decay through pion emission. Thus, since the earliest work of Chodos & Thorn (4) and Kikkawa et al (127, 128), the hybrid models have attracted considerable attention.

Recent interest was stimulated by a provocative suggestion of Brown et al (129, 130) that the pion cloud exerts a large inward pressure on the bag, reducing its radius by a factor of two or more. A drastic reduction in the bag radius would lead us back to the meson field theory of the 1950s, and downplay progress made in the quark model since then. Fortunately, modern work has shown that the model can accommodate a pion cloud without such drastic results. Thomas has recently reviewed this work thoroughly (10). We shall be brief, therefore.

The argument for embellishing the bag model goes as follows: with massless up and down quarks, QCD is chirally symmetric, i.e. the Lagrangian (Equation 1) is invariant under the infinitesimal transformation of the quark spinor

$$q(x) \to q(x) + \frac{1}{2}\tau \cdot \boldsymbol{\phi} \gamma_5 q(x), \qquad 29.$$

where τ are the Pauli matrices for isospin and $\boldsymbol{\phi}$ is a constant infinitesimal vector. Associated with this symmetry is the conserved axial vector-isovector current

$$\mathbf{A}_{\mu Q}(x) = \frac{1}{2} : \bar{q}(x) \gamma_5 \gamma_\mu \tau q(x) : . \qquad 30.$$

Now the bag model is supposed to follow in some sense from QCD. But the model as described by the Lagrangian (Equation 3) breaks chiral symmetry explicitly. The trouble is caused by the confinement of the quarks. The surface term in Equation 5 is essential for the conservation of the vector current. It has the effect of making the quarks seem to be infinitely massive outside the bag. But a mass term breaks the chiral symmetry explicitly. In effect confinement and the breaking of chiral symmetry are related in a profound way. It is widely believed that the breaking of chiral symmetry in QCD for massless up and down quarks is not explicit but dynamical, i.e. the symmetry manifests itself in the Goldstone manner (131) with the pion being the Goldstone boson. Because confinement and the appearance of the Goldstone mode are thought to be long-distance phenomena, it is plausible to attempt to repair the defects of the model of Equation 5 by introducing a phenomenological zero-mass-pion field outside the bag, coupling in a chirally invariant way to the quarks. (The gluons are irrelevant at this level of discussion and so we omit them here.) Thus the hybrid action becomes (8)

$$W = \int dt \left\{ \int_V d^3x \left[\frac{i}{2} \bar{q} \overleftrightarrow{\partial}_\mu \gamma^\mu q - B \right] - \frac{1}{2} \int_S d^2x \, \bar{q} \exp(i\tau \cdot \pi \gamma_5/f) q \right.$$
$$\left. + \frac{1}{2} \int_{\bar{V}} d^3x \, [\dot{\pi}^2 - (\nabla \pi)^2] \right\}, \qquad 31.$$

where $\bar{V}$ is the region outside the bag. The action is then invariant under the transformation (Equation 29) together with the appropriate "nonlinear" transformation of the pion field

$$\pi \to \pi - f[\boldsymbol{\phi} - (1 - \zeta \cot \zeta) \hat{n} \times (\boldsymbol{\phi} \times \pi)], \qquad 32.$$

where $\zeta = |\pi|/f$ and $\hat{n} = \pi/|\pi|$. This model features a nonlinear realization of the chiral symmetry (132). Other realizations are possible (4, 133–135).

To lowest order in ζ the conserved axial vector-isovector current is

$$\mathbf{A}_\mu(x) = \mathbf{A}_{\mu Q}(x) \theta_V + \partial_\mu \pi \theta_{\bar{V}}, \qquad 33.$$

where $\mathbf{A}_{\mu Q}(x)$ is given by Equation 30, $\theta_{\bar{V}}$ is one outside and zero inside the bag, and $\theta_V = 1 - \theta_{\bar{V}}$, and to the same lowest order $\partial^2 \pi = 0$. [Here $\pi(x)$ is defined only outside the bag.] Current conservation is assured, because the normal component of the axial vector current inside the bag (due to the quarks) is equal to the normal component outside (due to the pion):

$$n_\mu \mathbf{A}_Q^\mu(x) = n_\mu \partial^\mu \pi \quad \text{on } S. \qquad 34.$$

By insisting upon a chirally symmetric action we have arrived at a theory of pion-bag interactions. Is this theory uniquely required by chiral symmetry? Unfortunately, it is not. One popular alternative involves extending the pion field continuously inside the bag (4, 136, 137). Equation 34 is replaced by a discontinuity condition $n_\mu \partial^\mu \pi|_+ - n_\mu \partial^\mu \pi|_- = n_\mu \mathbf{A}_{\mu Q}(x)$. Although it may be plausible that the Goldstone modes should occur only in the external, nonperturbative vacuum (129), one can argue that such an inward continuation is useful in order to represent the important effects of additional pionic $Q\bar{Q}$ pairs that are generated by the external field (10, 137). Other alternative theories involve various canonical transformations of the action (134, 135). Although such a transformation does not result in a different theory, different results are obtained in second and higher order perturbation theory in the pion-quark coupling $1/f_\pi$ (10, 138). The problem here is to find the transformation that gives the least error when the perturbation series is truncated—a very difficult task in practice.

In addition to the ambiguities listed above, the hybrid chiral models are beset with problems of principle. They appear to describe two unrelated pions—one, the familiar Q$\bar{\text{Q}}$ bag state; the other, the elementary field. This difficulty presumably arises because in nature the pion plays a dual role as a near Goldstone boson and a Q$\bar{\text{Q}}$ bound state. The static cavity approximation localizes the bound states, thereby preventing the Q$\bar{\text{Q}}$ pion state from filling its other role.

4.4.2 APPLICATIONS OF HYBRID CHIRAL MODELS Most applications of the hybrid chiral model involve a perturbative expansion in powers of $1/f_\pi$, the pion-bag coupling constant. Since f_π, the pion decay constant, has dimensions of inverse length, the small dimensionless expansion parameter is $1/f_\pi R$, where R is the bag radius. Clearly, a perturbative expansion is likely to be in trouble for small bags. For special configurations of the internal symmetries, nonperturbative solutions to the classical static field equations are known (4, 133), but, owing to the complexity of the action, none has been found with the correct isospin and spin of the nucleon, nor any including quantum fluctuations in quark and pion fields. A perturbative treatment of the hadrons can be justified a posteriori by estimating the higher order terms, based on the corrected bag radius R. Generally satisfactory results have been obtained for masses of the light hadrons (137, 139, 140), their magnetic moments (137, 139–144), nucleon charge radii (137, 139, 145, 146), and the low energy pion-nucleon scattering phase shift (136, 147). For a detailed discussion see Thomas (10). A complication arises with the nucleon axial vector decay constant g_A. Models that continue the pion field into the bag obtain satisfactory results for this parameter, but models that exclude the pion from the bag interior obtain values 30–50% higher than the experimental constant (8, 133, 148). A similar difficulty plagues the pion-nucleon coupling constant g, since the model connects the two constants through the Goldberger-Treiman relation, $g = g_A M_N/f_\pi$. Proponents of the exclusionary model have suggested adopting a device of Glashow (149) involving a large admixture of d-wave-quark configurations to remedy the problem (150). These components result from deformations of the quark orbitals due to the pion and the gluon fields. These are higher order effects and are probably small (88).

The lowest order pionic corrections to the baryon masses and other parameters are generally of the same order of magnitude as the gluon corrections. Thus without more progress in understanding higher order gluon effects, further refinements of the structural and spectroscopic predictions are not warranted. Treating the pion as a point-like field coupling to a sharply defined and unyielding surface is likely to introduce unwanted artifacts at high pion momenta. Therefore, the model is limited to low energy pion-bag interactions. Even so, there are indications (151) that the model may overestimate pionic effects. These restrictions and the ambiguities in formulating the hybrid action will limit further progress in applications of the model. We must hope for more hints from QCD, the underlying theory, to constrain and refine the model.

5. CRITIQUE AND OUTLOOK

5.1 *Limitations of the Models*

There are two levels at which to discuss limitations of the bag models: (*a*) the most commonly used form of the models, the static cavity approximation of the MIT model and the corresponding treatment of the scalar field in the soliton model may have defects that are not shared by the full Lagrangian models; (*b*) it may be that the full models are themselves not a good approximation to QCD.

The classical approximation to the MIT and soliton models describes the surface incorrectly and introduces various shell model problems, namely, spurious excitations and the need for correcting for the c.m. motion. In both models the bag surface, treated correctly, has excitations that are tied to excitations of the internal fields. These excitations are important for obtaining the complete and correct spectrum and for restoring translational invariance (22, 23). Other problems of the classical surface approximation, shared by both models, are that the natural Fock space for the quarks and gluons is tied to a particular classical bag, and so is the quantum Hamiltonian, thereby preventing a meaningful superposition of states with different bag shapes or positions. Thus, standard shell model corrections for the c.m. motion are precluded. The cavity approximation of the MIT model also has some divergences that are at least partially due to the rigidity of the walls. Examples of this are the zero-point energy (32) and the vacuum expectation value of $\bar{q}q$. At a more practical level the various parameters of the model that are determined phenomenologically may include extraneous effects. For example, the zero-point energy constant Z_0 is known to be different depending on whether the c.m. corrections are included (54); Z_0 may also include some of the effects of the quark-self energies or energies associated with the boundary. The strong coupling constant, as determined from the first-order hyperfine splittings, is large, but it may include some of the effects of a pion cloud (8) or higher order gluon exchange. The Fock-space treatment neglects the effects of open channels for decay, which can modify the bound state properties. In short the classical surface approximation may be useful for a semiquantitative determination of static properties of isolated hadrons, but one should not insist upon precise agreement with experiment, nor should it be extended incautiously to processes involving large momentum transfers.

Both models have a rich structure that has not been fully explored. However, there are several important features of QCD that are apparently missing. Most obvious is the mistreatment of chiral symmetry breaking. The hybrid chiral models of Section 4.4 patch up this defect at the price of introducing an extra pion. Better remedies would be welcome. Higher order perturbative attributes of QCD, such as the trace anomaly (152) and asymptotic freedom may have already been incorporated into the bag constant and structure, but if so, there may be double counting problems in cavity perturbation theory. For similar reasons it is not clear whether instanton effects (17) should be considered within the model, or whether they have already been partially incorporated. Finally, the model may give an inaccurate description of the vacuum (see below). As we learn more about QCD it is quite likely that the MIT model will need to be modified in order to provide a closer description of nature. Perhaps the more flexible soliton model will fill this need.

It is important to keep these concerns in mind when interpreting existing results and applying the model to new physical problems. The cavity approximation, although an oversimplification, has still been of proven value when sensibly used.

5.2 Conclusions and Outlook

As we stated above, we believe further progress in the bag model must come from a better understanding of QCD. We conclude by discussing some promising recent developments that may affect the model. The bag theory is basically a simple model of the vacuum. A "perturbative" vacuum of finite extent is found inside the bag, while the "true" vacuum is found outside. The formation of the bag can be viewed as a phase change between the two types of vacuum. In what sense does QCD support this view?

There have been many recent attempts to characterize the QCD vacuum. Of particular relevance to the bag model is recent work by Hansson et al (153). They set out to determine the structure of the vacuum wave function by using a variational argument. Their "trial" wave function was inspired by the bag model, but their intention was to describe general features of QCD. Their work starts from the realization that with the usual perturbative model of the vacuum a $J^{PC} = 0^{++}$ glueball state can be made with $m^2 < 0$. This result implies that the simple vacuum is not the true vacuum. A negative squared mass occurs in the simple model, because the spin force between two gluons paired into a color singlet, spin scalar state is strongly attractive. The potential energy of attraction can overcome the kinetic energy needed to create the two gluons. Such a situation was encountered in the calculation of the lightest bag model 0^{++} glueball (C. B. Thorn, unpublished), which leads to a state with $m^2 < 0$ when c.m. corrections are included (65). Hansson et al show that the perturbative vacuum can lower its energy by filling up with scalar glueballs. They calculate the energy of glueballs containing two and four gluons and find that the energy of the four-gluon state is higher. Therefore the vacuum energy reaches a minimum when the glueballs start to overlap.

There are other attempted variational descriptions of the QCD vacuum not specifically based on the bag model. There is the work of Savvidy (154), Tomboulis & Pagels (155), Fukuda (156), Donoghue (157), Feynman (158), and, for lattice gauge theories, Patkös (159). There is still no completely satisfactory description, however. Having established an approximate picture of the vacuum, one may proceed to characterize the low-lying excitations. Here the lattice gauge theories have made some progress (13), but much remains to be done. The limitations of lattice models require that we must in the end develop a bag-like phenomenological model, if further progress is to be made in elucidating the structure and interactions of the hadrons. Herein lies the challenge for future work.

ACKNOWLEDGMENTS

Preparation of this review was supported in part by the US National Science Foundation under grant numbers PHY82-42390 and PHY80-16167 and by the CERN Theory Division.

Literature Cited

1. Chodos, A., Jaffe, R. L., Johnson, K., Thorn, C. B., Weisskopf, V. F. 1974. *Phys. Rev. D* 9:3471–95
2. Bardeen, W. A., Chanowitz, M. S., Drell, S. D., Weinstein, M., Yan, T.-M. 1975. *Phys. Rev. D* 11:1094–1136
3. Friedberg, R., Lee, T. D. 1978. *Phys. Rev. D* 18:2623–31
4. Chodos, A., Thorn, C. B. 1975. *Phys. Rev. D* 12:2733–43
5. Johnson, K. 1975. *Acta Phys. Pol. B* 6:865–92
6. Hasenfratz, P., Kuti, J. 1978. *Phys. Rep.* 40C:75–179
7. DeTar, C. E. 1980. In *Quantum Flavordynamics, Quantum Chromodynamics and Unified Theories*, ed. K. T. Mahanthappa, J. Randa, NATO Adv. Stud. Inst. Ser. B 54:393–439. New York: Plenum. 494 pp.
8. Jaffe, R. L. 1982. In *Pointlike Structures Inside and Outside Hadrons*, ed. A. Zichichi, p. 99. New York: Plenum
9. Jaffe, R. L. 1980. In *Field Theory and Strong Interactions*, Acta Phys. Austr. Suppl. XXII, pp. 269–339. Vienna: Springer-Verlag
10. Thomas, A. 1983. *Adv. Nucl. Phys.* In press
11. Gross, D. J., Wilczek, F. A. 1973. *Phys. Rev. Lett.* 30:1343–46
12. Politzer, H. D. 1973. *Phys. Rev. Lett.* 30:1346–49
13. Creutz, M. 1982. In *Particles and Fields—1981, Testing the Standard Model*, ed. C. A. Heusch, W. T. Kirk, AIP Conf. Proc. 81:238–44. New York: AIP. 599 pp.
14. Politzer, H. D. 1983. *Proc. Int. Conf. on High Energy Physics, 21st, Paris, 1982*, ed. P. Petiau, M. Porneuf. *J. Phys. Colloque C3*, Vol. 43, No. 12, pp. 659–69, Suppl.
15. Johnson, K. 1978. *Phys. Lett.* 78B:259–62
16. Shuryak, E. V. 1978. *Phys. Lett.* 79B:135–37
17. Callan, C. G., Dashen, R. F., Gross, D. J. 1979. *Phys. Rev. D* 19:1826–55
18. Adler, S. L. 1981. *Phys. Rev. D* 23:2905–15; 1982. *Phys. Lett.* 110B:302–6
19. Bogoliubov, P. N. 1967. *Ann. Inst. Henri Poincaré Sec. A* 8:163–90

20. Huang, K., Stump, D. R. 1976. *Phys. Rev. D* 14:223–45
21. Goldflam, R., Wilets, L. 1982. *Phys. Rev. D* 25:1951–63
22. Breit, J. D. 1982. *Nucl. Phys. B* 202:147–72
23. Rebbi, C. 1975. *Phys. Rev. D* 12:2407–21; 1976. *Phys. Rev. D* 14:2362–73
24. Johnson, K. 1977. *Proc. Scottish Univ. Summer Sch. Phys. 17th, St. Andrews, 1976*, ed. I. M. Barbour, A. C. Davis, pp. 245–96, Edinburgh: SUSSP Publ.
25. Chodos, A., Jaffe, R. L., Johnson, K., Thorn, C. B. 1974. *Phys. Rev. D* 10:2599–2604
26. DeTar, C. E. 1978. *Phys. Rev. D* 17:302–22
27. Hahn, K., Goldflam, R., Wilets, L. 1983. *Phys. Rev. D* 27:635–43
28. Casimir, H. B. G. 1948. *Proc. K. Ned. Akad. Wet. B* 51:793–96
29. Milton, K. A., de Raed, L. L., Schwinger, J. 1978. *Ann. Phys. (NY)* 115:388–403
30. Milton, K. A. 1980. *Phys. Rev. D* 22:1441–43
31. Johnson, K. 1980. *Particles and Fields—1979*, ed. B. Margolis, D. G. Stairs, AIP Conf. Proc. 59:353–71. New York, AIP
32. Milton, K. A. 1983. *Phys. Rev. D* 27:439–43
33. Golowich, E. 1975. *Phys. Rev. D* 12:2108–12
34. Barnes, T. 1975. *Nucl. Phys. B* 96:353–64
35. Golowich, E. 1978. *Phys. Rev. D* 18:927–34
36. Barnes, T. 1979. *Nucl. Phys. B* 158:171–88
37. Lee, T. D. 1979. *Phys. Rev. D* 19:1802–19
38. Close, F. E., Horgan, R. R. 1980. *Nucl. Phys. B* 164:413–426
39. Balian, R., Duplantier, B. 1978. *Ann. Phys. (NY)* 112:165–208
40. Hansson, T. H., Jaffe, R. L. 1982. *MIT-CTP 1026*. Cambridge, Mass: Mass. Inst. Technol.
41. Bender, C. M., Hays, P. 1976. *Phys. Rev. D* 14:2622–32
42. Baacke, J., Igarashi, Y., Kasperidus, G. 1983. *Z. Phys. C* 17:161–70
43. DeGrand, T. A., Jaffe, R. L., Johnson, K., Kiskis, J. 1975. *Phys. Rev. D* 12:2060–76
44. Chin, S. A., Kerman, A. K., Yang, X. H. 1981. *MIT-CTP 919*. Cambridge, Mass: MIT
45. Breit, J. D. 1982. *CU-TP 229*. New York, NY: Columbia Univ.
46. Donoghue, J. F. 1983. *UMHEP-178*. Amherst, Mass: Univ. Mass.
47. DeGrand, T. A., Jaffe, R. L. 1976. *Ann. Phys. (NY)* 100:425–56
48. DeGrand, T. A. 1976. *Ann. Phys. (NY)* 101:496–519
49. DeGrand, T. A., Rebbi, C. 1978. *Phys. Rev. D* 17:2358–63
50. Bowler, K. C., Walters, P. J. 1979. *Phys. Rev. D* 19:3330–35
51. Isgur, N., Karl, G. 1977. *Phys. Lett.* 72B:109–13; 1978. *Phys. Lett.* 74B:353–56; 1978. *Phys. Rev. D* 18:4187–4205; 1979. *Phys. Rev. D* 20:1191–94
52. Fiebig, H. R., Schwesinger, B. 1983. *Nucl. Phys. A* 393:349–71
53. Peierls, R. E., Yoccoz, J. 1957. *Proc. Phys. Soc. Sec. A* 70:381–87
54. Donoghue, J. F., Johnson, K. 1980. *Phys. Rev. D* 21:1975–85
55. Barnhill, M. V. 1979. *Phys. Rev. D* 20:723–26; 1982. *Phys. Rev. D* 25:860–66
56. Wong, C. W. 1981. *Phys. Rev. D* 24:1416–19
57. Tsai Hsu, Yang Zhung-le, Liu Lianson, 1980. *HZPP-80-2*, Wuhan, People's Republic of China: Hua-Zhong Teacher's College
58. Dethier, J. L., Goldflam, R., Henley, E. M., Wilets, L. 1983. *Phys. Rev. D* 27:2191–98
59. Jacob, M., ed. 1974. *Dual Theory*, Vol. 1. Amsterdam: North Holland. 398 pp.
60. Johnson, K., Thorn, C. B. 1976. *Phys. Rev. D* 13:1934–39
61. Haxton, W. C., Heller, L. 1980. *Phys. Rev. D* 22:1198–1208
62. Hasenfratz, P., Horgan, R. R., Kuti, J., Richard, J. M. 1980. *Phys. Lett.* 95B:299–305; 1981. *Phys. Scr.* 23:914–21
63. Izatt, D. L., DeTar, C. E., Stevenson, M. A. 1982. *Nucl. Phys. B* 199:269–89
64. Jaffe, R. L., Johnson, K. 1976. *Phys. Lett.* 60B:201–4
65. Donoghue, J. F., Johnson, K., Li, B. A. 1981. *Phys. Lett.* 99B:416–20
66. Robson, D. 1980. *Z. Phys. C* 3:199–207
67. Barnes, T., Close, F. E., Monaghan, S. 1982. *Nucl. Phys. B* 198:380–406
68. Carlson, C., Hansson, T. H., Peterson, C. 1983. *Phys. Rev. D* 27:2167–81
69. Carlson, C. E., Hansson, T. H. 1982. *Nucl. Phys. B* 199:441–50
70. Donoghue, J. F., Gomm, H. 1983. *Phys. Lett.* 121B:49–52
71. 't Hooft, G. 1975. *Phys. Rev. Lett.* 37:8–11
72. Chanowitz, M., Sharpe, S. 1982. *LBL-14865*. Berkeley, Calif: Lawrence Berkeley Lab.

73. Carlson, C. E., Hanson, T. H., Peterson, C. 1983. *Phys. Rev. D* 27:1556–64
74. Jaffe, R. L. 1981. In *Proc. Intl. Symp. on Lepton and Photon Interactions, 10th, Bonn*, ed. W. Pfeil, p. 395. Bonn Univ.
75. Jaffe, R. L. 1976. *Phys. Rev. D* 15:267–80, 281–89
76. Aerts, A. T. M., Mulders, P. J. G., deSwart, J. J. 1978. *Phys. Rev. D* 17:260–74; 1980. *Phys. Rev. D* 21:1370–87
77. Mulders, P. J. G., Aerts, A. T. M., deSwart, J. J. 1978. *Phys. Rev. Lett.* 40:1543–46
78. Högasen, H., Sorba, P. 1978. *Nucl. Phys. B* 145:119–40
79. Strottmann, D. 1979. *Phys. Rev. D* 20:748–67
80. Wong, C. W., Liu, K. F. 1980. *Phys. Rev. D* 21:2039–40
81. Barnes, T., Close, F. E. 1982. *Phys. Lett.* 116B:365–68
82. Golowich, E., Haqq, E., Karl, G. 1983. *Phys. Rev. D* 28:161–69
83. Barnes, T., Close, F. E. 1983. *Phys. Lett.* 123B:89–92
84. Högåsen, H., Wroldsen, J. 1982. *Phys. Lett.* 116B:369–72
85. CLEO Collaboration. 1983. *Phys. Rev. Lett.* 50:807–14
86. Donoghue, J. F., Golowich, E., Holstein, B. R. 1975. *Phys. Rev. D* 12:2875–85
87. Donoghue, J. F., Holstein, B. R. 1982. *Phys. Rev. D* 25:206–12
88. Maxwell, O. V., Vento, V. 1982. *IPNO/TH 82-11*. Orsay: Inst. Phys. Nucl.
89. Donoghue, J. F., Golowich, E. 1978. *Phys. Rev. D* 18:927–34
90. Halprin, A., Sorba, P. 1977. *Phys. Lett.* 66B:177–80
91. Close, F. E., Horgan, R. R. 1980. *Nucl. Phys. B* 164:413–26
92. Kobzarev, I., Martem'yanov, V., Shchepkin, M. 1979. *Yad. Fiz. (USSR)* 30:504–9 (transl. *Sov. J. Nucl. Phys.* 30:261–64)
93. Carlitz, R., Ellis, S. D., Savit, R. 1977. *Phys. Lett.* 68B:443–46
94. Hwang, W. Y. P. 1983. *Z. Phys. C* 16:331
95. Pićek, I., Tadić, D. 1982. *Univ. Zagreb Report*. Zagreb, Yugoslavia: Univ. Zagreb
96. Hackman, R., Deshpande, N., Dicus, D. A., Teplitz, V. L. 1978. *Phys. Rev. D* 18:2537–46
97. Donoghue, J. T., Golowich, E. 1976. *Phys. Rev. D* 14:1386–99; 1977. *Phys. Lett.* 69B:437
98. Donoghue, J. F., Golowich, E., Holstein, B. R., Ponce, W. A. 1981. *Phys. Rev. D* 23:1213–16
99. Tadić, D. Trampetić, J. 1981. *Phys. Rev. D* 23:144–54
100. Desplanques, B., Donoghue, J. F., Holstein, B. R. 1980. *Ann. Phys. (NY)* 124:449–95
101. Baluni, V. 1979. *Phys. Rev. D* 19:2227–30
102. Donoghue, J. F., Hagelin, J. S., Holstein, B. R. 1982. *Phys. Rev. D* 25:195–205
103. Trampetić, J. 1982. *Max Planck Institute Report*. Munich, W. Germany: Max Planck Inst.
104. Donoghue, J. F. 1980. *Phys. Lett.* 92B:99–102
105. Golowich, E. 1980. *Phys. Rev. D* 22:1148–55
106. Donoghue, J. F., Golowich, E. 1982. *Phys. Rev. D* 26:3092–99
107. Jaffe, R. L. 1975. *Phys. Rev. D* 11:1953–68
108. Jaffe, R. L., Patrascioiu, A. 1975. *Phys. Rev. D* 12:1314–18
109. Jaffe, R. L. 1981. *Ann. Phys. (NY)* 132:32–52
110. Jaffe, R. L., Ross, G. 1980. *Phys. Lett.* 93B:313–17
111. Jaffe, R. L., Soldate, M. 1981. *Phys. Lett.* 105B:467–72
112. Jaffe, R. L., Low, F. E. 1979. *Phys. Rev. D* 19:2105–18
113. Roiesnel, C. 1979. *Phys. Rev. D* 20:1646–55
114. Mulders, P. J. 1982. *Phys. Rev. D* 25:1269–79; 26:3039–56
115. Dosch, H. G. 1982. *Z. Phys. C* 13:329–31
116. DeTar, C. E. 1978. *Phys. Rev. D* 17:323–39; 1979. *Phys. Rev. D* 19:1028(E)
117. DeTar, C. E. 1979. *Phys. Rev. D* 19:1451–64
118. Liberman, D. A. 1977. *Phys. Rev. D* 16:1542–44
119. Warke, C. S., Shanker, R. 1980. *Phys. Rev. C* 21:2643–62
120. Ribeiro, J. E. F. T. 1980. *Z. Phys. C* 5:27–35
121. Oka, M., Yazaki, Y. 1981. *Phys. Lett.* 90B:41–44
122. Harvey, M. 1981. *Nucl. Phys. A* 352:301–25, 326–42
123. Obukhovsky, I. T., Neudatchin, V. G., Smirnov, Yu. F., Tchuvilsky, Yu. M. 1979. *Phys. Lett.* 88B:231–33
124. Hill, D. L., Wheeler, J. A. 1952. *Phys. Rev.* 89:1102–45
125. DeTar, C. E. 1983. In *Asymptotic Realms in Physics*, ed. A. H. Guth, K.

Huang, R. L. Jaffe, pp. 118–27. Cambridge: MIT
126. Henley, E., Kisslinger, L., Miller, G. 1983. *Carnegie-Mellon Rep.* Pittsburgh: Carnegie-Mellon Univ.
127. Kikkawa, K. 1978. *Phys. Rev. D* 18:2606–22
128. Kikkawa, K., Kotani, T., Sato, M., Kenmoku, M. 1979. *Phys. Rev. D* 19:1011–18
129. Brown, G. E., Rho, M. 1979. *Phys. Lett.* 82B:177–80
130. Brown, G. E., Rho, M., Vento, V. 1979. *Phys. Lett.* 84B:383–87
131. Goldstone, J. 1961. *Nuovo Cimento (Ser. 10)* 19:155–64
132. Gell-Mann, M., Lévy, M. 1960. *Nuovo Cimento (Ser. 10)* 16:705–26
133. Vento, V., Rho, M., Nyman, E. M., Jun, J. H., Brown, G. E. 1980. *Nucl. Phys. A* 345:413–34
134. Thomas, A. W. 1981. *J. Phys. G* 7:L283–86
135. Szymacha, A., Tatur, S. 1981. *Z. Phys. C* 7:311–16
136. Théberge, S., Thomas, A. W., Miller, G. A. 1980. *Phys. Rev. D* 22:2838–52; 1981. *Phys. Rev. D* 23:2106(E)
137. DeTar, C. E. 1981. *Phys. Rev. D* 24:752–61, 762–74
138. Kälbermann, G., Eisenberg, J. M. 1979. *J. Phys. G* 5:35–38
139. Myhrer, F., Brown, G. E., Xu, Z. 1981. *Nucl. Phys. A* 362:317–30
140. Myhrer, F., Xu, Z. 1982. *Phys. Lett.* 108B:372–76
141. Thomas, A. W., Théberge, S., Miller, G. A. 1981. *Phys. Rev. D* 24:216–29
142. Brown, G. E., Rho, M., Vento, V. 1980. *Phys. Lett.* 97B:423–26
143. Hulthage, I., Myhrer, F., Xu, Z. 1981. *Nucl. Phys. A* 364:322–32
144. Théberge, S., Thomas, A. W. 1982. *Phys. Rev. D* 25:284–86
145. Thomas, A. W., Théberge, S., Miller, G. A. 1981. *Phys. Rev. D* 24:216–29
146. Myhrer, F. 1982. *Phys. Lett.* 110B:353–58
147. Théberge, S., Miller, G. A., Thomas, A. W. 1982. *Can. J. Phys.* 60:59–72
148. Barnhill, M. V., Halprin, A. 1980. *Phys. Rev. D* 21:1916–18
149. Glashow, S. L. 1979. *Physica* 96A:27–30
150. Vento, V., Baym, G., Jackson, A. D. 1981. *Phys. Lett.* 102B:97–101
151. de Kam, J., Pirner, H. 1982. *Nucl. Phys. A* 389:640–54
152. Collins, J., Duncan, A., Joglekar, S. 1977. *Phys. Rev. D* 16:438–49
153. Hansson, T. H., Johnson, K., Peterson, C. 1982. *Phys. Rev. D* 26:2069–85
154. Savvidy, G. K. 1977. *Phys. Lett.* 71B:133–34
155. Tomboulis, E., Pagels, H. 1978. *Nucl. Phys. B* 143:485–502
156. Fukuda, R. 1980. *Phys. Rev. D* 21:485–503
157. Donoghue, J. F. 1982. In *Proc. French-American Semin. on Theoretical Aspects of Quantum Chromodynamics*, ed. J. Dash, p. 136. Marseille: Cent. Phys. Théor.
158. Feynman, R. P. 1981. *Nucl. Phys. B* 188:479–512
159. Patkös, A. 1982. *Phys. Lett.* 110B:391–94

FUSION REACTIONS BETWEEN HEAVY NUCLEI[1]

J. R. Birkelund and J. R. Huizenga

Departments of Chemistry and Physics and the Nuclear Structure Research Laboratory, University of Rochester, Rochester, New York 14627

CONTENTS

1. INTRODUCTION 265
2. EXPERIMENTAL RESULTS 270
   - 2.1 *Excitation Functions of Evaporation Residues* 288
   - 2.2 *Excitation Functions of Fission Fragments* 290
   - 2.3 *Sub-barrier Fusion* 294
   - 2.4 *Incomplete Fusion* 295
3. ONE-DIMENSIONAL ENTRANCE CHANNEL MODELS 297
   - 3.1 *Critical Distance Models* 298
   - 3.2 *One-Dimensional Potential Dynamical Models* 298
   - 3.3 *Fusion Barriers and Limiting Angular Momenta* 301
4. OTHER FUSION MODELS 302
   - 4.1 *Yrast or Compound Nucleus Limitation Models* 302
   - 4.2 *Empirical Models* 303
   - 4.3 *Time-Dependent Hartree-Fock Models* 303
5. A NEW ENTRANCE CHANNEL LIMITATION 304
   - 5.1 *The Conditional Saddle and the Extra Push Energy* 306
   - 5.2 *The Unconditional Saddle and the Extra Extra Push Energy* 308
   - 5.3 *Experimental Results for Systems with Large* $(Z^2/A)_{eff}$ 310
6. SUMMARY AND CONCLUSIONS 313

1. INTRODUCTION

The most general theoretical models (1–3) of nuclear fusion reactions conceive the process as a motion of the target and projectile across a potential energy surface created by the forces between the two nuclei. The internuclear potential energy is a function of the separation of the two nuclei, the angular momentum of the system, and the deformation of the

---

[1] Work supported by the US Department of Energy.

two nuclei, or of the composite system formed when the target and projectile matter distributions overlap. During the reaction, nucleons (protons and neutrons) are exchanged between the target and projectile and, as a consequence, energy and angular momentum are transferred from the relative motion to the intrinsic degrees of freedom of the target-like and projectile-like fragments or the composite system. In addition, the deformation of the system begins an evolution that may proceed from shapes describing two separate nuclei to a shape no longer showing evidence of separate target and projectile. The rates of evolution of the system are not identical in all its degrees of freedom. In fact, the mass asymmetry degree of freedom evolves much more slowly than the other degrees of freedom of the system. It is this slow evolution of the mass asymmetry degree of freedom of the system that allows the possibility of an experimentally observable distinction between fusion and other types of reactions between heavy nuclei. Models of the fusion reaction attempt to describe the essential features of this process with a few parameters.

The type of reaction that may occur between two heavy nuclei can be related to the entrance channel parameters of total mass, bombarding projectile energy, and classical impact parameter, or orbital angular momentum. These factors determine the degree of interpenetration of the target and projectile matter distributions occurring during the reaction.

An example of the classification of the possible reactions at projectile energies above the reaction barrier is shown in Figure 1, where the various possible reaction types are shown as a function of orbital angular momentum. Above the angular momentum l_{max}, no nuclear reactions occur. Between l_f and l_{max} various types of damped reactions may occur. In this range of impact parameters, the shape of the reacting system preserves a distinction between the target and projectile nuclei; such reactions were termed "dinucleus" by Swiatecki (3). In the dinucleus regime, there are no exit channel barriers or saddle points in the potential energy surface, and the reactions lead to fragments that may be identified as target-like or projectile-like, even though considerable mass and energy exchange may have occurred. The classification of these dinucleus or damped reactions has been discussed by Schröder & Huizenga (4).

At angular momenta below l_f, the trajectory of the system is influenced by a conditional saddle in the internuclear potential energy surface (2). This conditional saddle is a stationary point in all deformation parameters, except the mass asymmetry.

However, since the mass asymmetry has been found experimentally to be a slowly evolving parameter, the conditional saddle may have a considerable influence on the trajectory, especially causing a considerable increase in the reaction time, which allows both the evolution of the system to shapes that no longer describe separate nuclei and a relaxation of the mass asymmetry. Systems with shapes more compact than the conditional saddle shape were termed "mononucleus" by Swiatecki. In the mononucleus region the system closely approaches statistical equilibrium in its intrinsic degrees of freedom, and the exit channel shows few or none of the characteristics of the entrance channel, except those required by the conservation laws. However, for heavy systems, with the parameter $(Z^2/A)_{eff}$ sufficiently large as discussed below, mononucleus reactions may not be identical to compound nucleus reactions. This is because the unconditional, or fission, saddle is a more compact configuration than the conditional saddle. Hence trajectories passing inside the conditional saddle may not also lead behind the unconditional saddle. For this reason, we refer to compound nucleus reactions as those on trajectories that pass inside the unconditional saddle. For light systems with sufficiently small $(Z^2/A)_{eff}$, at low angular momentum, there is no conditional saddle affecting the trajectories, and all mononucleus reactions are compound nucleus reactions.

In terms of the entrance channel angular momentum, as shown in Figure 1, mononucleus reactions outside the fission saddle occur between angular momenta l_f and l_{crit} and are termed fusion-like in Figure 1, whereas

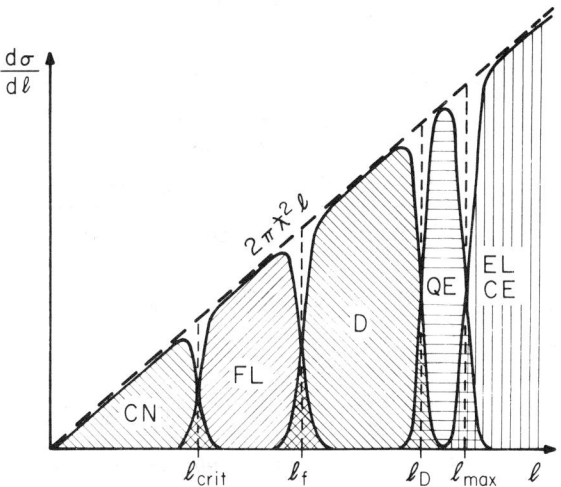

Figure 1 Schematic illustration of the l dependence of the partial cross section for compound nucleus (CN), fusion-like (FL), damped (D), quasi-elastic (QE), Coulomb excitation (CE) and elastic (EL) processes. The long-dashed line represents the geometrical partial cross section $d\sigma/dl = 2\pi\lambda^2 l$. Vertical dashed lines indicate the extensions of the various l windows in a sharp cutoff model with the characteristic l values noted at the abscissa. Hatched areas represent the diffuse l windows assumed in a smooth cutoff model. The division of the total cross section among the various processes is strongly dependent on the target and projectile nuclei and on the bombarding energy. (From 167.)

compound nucleus reactions occur at angular momenta less than l_{crit}. As discussed below, if $(Z^2/A)_{\text{eff}}$ is below a critical value, than the values of l_{crit} and l_f are the same. The fusion cross section is taken to be the sum of compound and fusion-like reactions, that is, all mononucleus reactions.

Experimentally, the angular momentum limits for the reaction components are determined from a sharp cutoff model in which the cross section is given by

$$\sigma_t = \pi \lambda^2 [l_i(l_i+1) - l_j(l_j+1)], \qquad 1.$$

where l_i and l_j are the upper and lower angular momentum bounds of the reaction type t. In reality, the angular momentum limits to the various reaction categories are not sharp (5–8), although most experimental analyses have not accounted for the diffuse transition regions.

Besides the angular momentum restrictions to the various reaction components, there are also projectile energy limits imposed by the shape of the internuclear potential energy surface, which exhibits barriers, or influences trajectories in a manner requiring an additional energy threshold before a reaction region can be reached. These energy requirements are illustrated in Figure 2, in which the projectile energy boundaries between various reaction categories are shown for a heavy system that may exhibit fusion-like reactions. This figure shows the additional energy above the

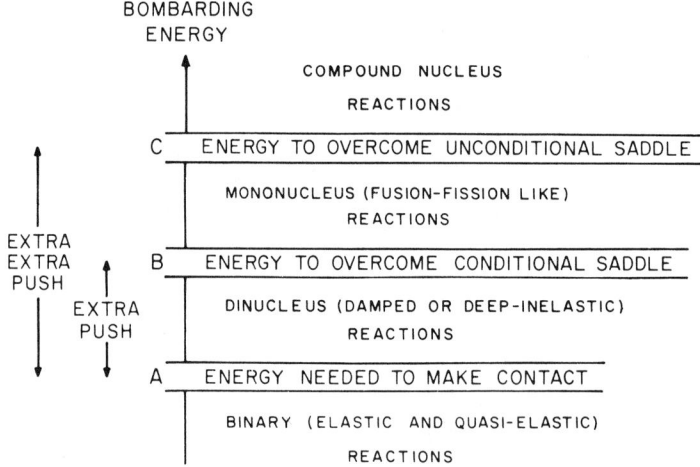

Figure 2 Schematic illustration of the relation between three critical energies, four types of nuclear reactions (see Figure 1), and two kinds of extra push. In some situations the critical energies *A*, *B*, *C* may merge (pairwise or all together), squeezing out the regimes corresponding to damped and/or fusion-fission-like reactions. In other situations one or both of the upper boundaries (*B* and *C*) may dissolve, making the adjoining regions merge into continuously graduated reaction types. (From 364.)

FUSION OF HEAVY NUCLEI 269

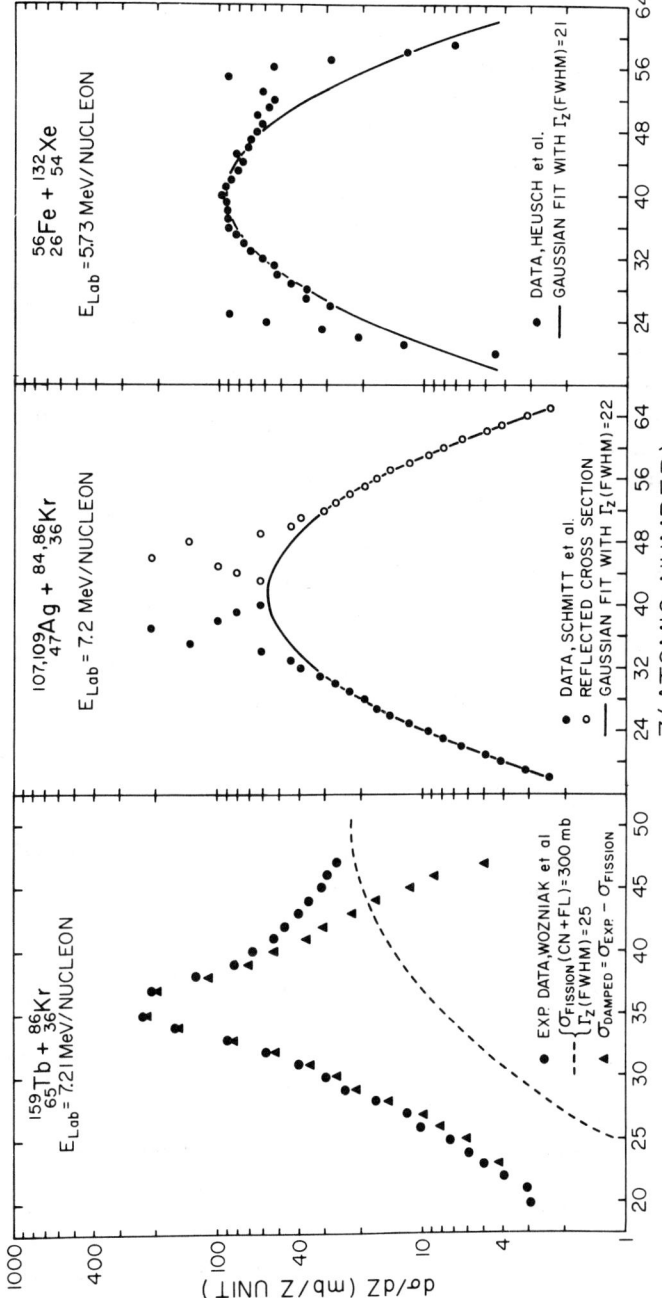

Figure 3 Element distributions (121, 370, 371). The Gaussian curves are interpreted as being due to CN and FL reactions. (From 213.)

spherical barrier required to go from dinucleus to mononucleus reactions as an "extra push" energy and the additional energy called the "extra extra push" to reach the compound nucleus reaction region. The calculation of these energy thresholds is discussed below.

Once a mononucleus has been formed, the excitation energy and spin, together with the potential energy surface, determine its further decay into the exit channel. If a compound nucleus is not formed, or if the spin or excitation energy is sufficiently high, fission fragments will be observed in the exit channel. For a compound nucleus, if the spin and excitation energy are sufficiently low, decay may be by emission of nucleons or charged particles and an evaporation residue results in the exit channel. This is illustrated by Figure 3 where the Z spectrum of observed fragments is shown for several heavy systems. Dinucleus reactions are manifested by the sharp peaks in the spectra near the Z of the target and projectile, and fission fragments from mononucleus reactions are seen as a broad distribution centered at half of the total Z of the system. For some systems, evaporation residues are also observable, with a distribution approximately centered at a Z value just below that of the total Z.

This review concentrates on models of the fusion process that have been used to explain the macroscopic features of fusion excitation functions at energies above the fusion reaction threshold. Structure observed in fusion excitation functions for light systems was reviewed recently by Cormier (9) and is not discussed here. Other recent reviews of aspects of fusion include (10–14).

2. EXPERIMENTAL RESULTS

The fusion cross section σ_f as a function of the projectile energy E generally rises rapidly as the projectile energy increases over the barrier until, at an available energy a few MeV per nucleon over the barrier, the cross section reaches a maximum and remains constant or begins a slow decline as the projectile energy increases further. At each projectile energy above the barrier there is a maximum angular momentum for fusing trajectories l_f. Near the energy at which the cross section reaches its maximum, this angular momentum is predicted by some models to reach its largest possible magnitude and remain approximately constant, or begin a slow decline as the projectile energy increases. However, the functional dependence of l_f on projectile energy and system mass is not well known experimentally. As a fraction of the reaction cross section, the fusion cross section becomes less significant as the projectile energy rises. This limitation of the fusion cross section at high energies has been clearly demonstrated to be dependent on the entrance channel in some cases. Such a demonstration can be made if the compound nucleus is formed under the same conditions by different entrance channels (15–24). However, in other cases, by examination of the angular momentum limits to fusion and by comparison to entrance channel models of fusion, some systems have been interpreted in terms of a compound nucleus (or exit channel) limit to the fusion cross section (23, 25–37).

Examples of the behavior of two fusion excitation functions are shown in Figure 4 for the $^{10}B + ^{16}O$ and $^{12}C + ^{14}N$ reactions (21), which produce the compound nucleus ^{26}Al. These two entrance channels lead to very different maximum cross sections for fusion; they are outside the differences expected from the different nuclear sizes and reduced masses of the entrance channels. These cases are not well described by the fusion models available, and are discussed further in later sections.

A compilation of experiments that report heavy ion fusion cross sections is given in Table 1. The table includes columns indicating which fusion reaction exit channel was observed, and the technique and laboratory projectile energy range over which measurements were made. The significance of the other columns is discussed below.

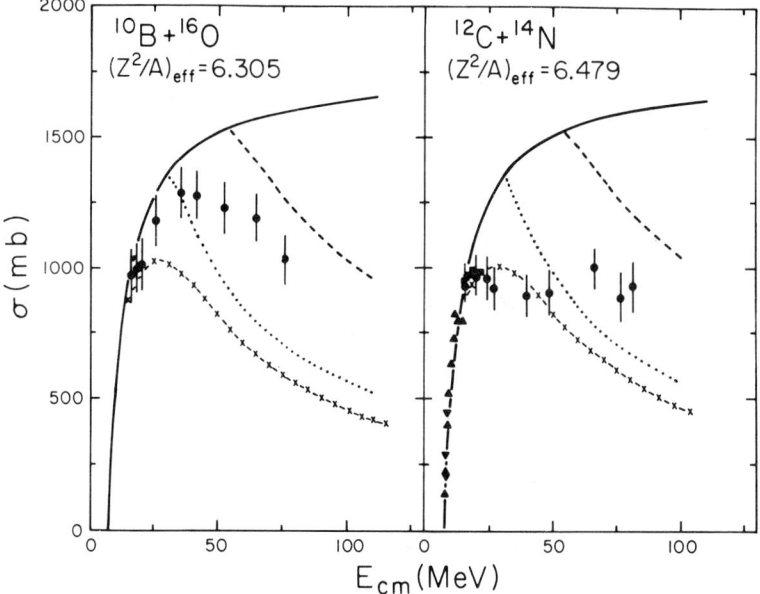

Figure 4 Comparison of experimental and theoretical fusion cross sections for the $^{10}B + ^{16}O$ and $^{12}C + ^{14}N$ reactions. The dotted and dashed curves are calculated with Equation 26 and correspond to sliding and rolling collisions respectively. The -x-x- curve is calculated with the one-dimensional classical dynamical model (Equations 8–11). The solid curve is calculated with Equation 7. References to these data are in Table 1.

Table 1 Reactions with measured cross sections

No.	Reaction	Energy range (MeV)	Z^2/A^b (eff)	Barrier[a] Radius (fm)	Barrier[a] Potential (MeV)	l (pocket)[c]	Detected fragments[d]	Method[e]	Ref.
1	$^9Be+^{16}O$	5– 14	5.30	7.54	5.53	11.7	e	g	38
2	$^9Be+^{18}O$	1– 6	4.99	7.70	5.42	12.5	e	g	39
3	$^{10}B+^{14}N$	7– 18	5.90	7.41	6.14	11.6	e	g	40
4	$^{10}B+^{14}N$	27– 75	5.90	7.41	6.14	11.6	e	z	22
5	$^{10}B+^{14}N$	86–180	5.91	7.41	6.15	11.6	e	z	41
6	$^{10}B+^{16}O$	24–130	6.30	7.49	6.95	12.6	e	z	21
7	$^{10}B+^{17}O$	32– 81	6.11	7.58	6.87	13.0	e	z	42
8	$^{10}B+^{17}O$	54–124	6.11	7.58	6.87	13.0	e	z	28
9	$^{12}C+^{6}Li$	10– 36	4.21	7.12	3.28	7.3	e	m	43
10	$^{12}C+^{7}Li$	10– 38	3.91	7.30	3.20	8.0	e	m	43
11	$^{12}C+^{11}B$	14– 54	5.22	7.45	5.24	11.2	e	m	44
12	$^{12}C+^{12}C$	8– 14	6.00	7.42	6.32	11.9	e	g	45
13	$^{12}C+^{12}C$	10– 40	6.00	7.41	6.31	11.9	e	g	45a
14	$^{12}C+^{12}C$	14– 62	6.00	7.41	6.31	11.9	e	z	47
15	$^{12}C+^{12}C$	14– 64	6.00	7.41	6.31	11.9	e	z	48
16	$^{12}C+^{12}C$	27– 60	6.00	7.41	6.31	11.9	e	z	30
17	$^{12}C+^{12}C$	28– 62	6.00	7.41	6.31	11.9	e	g	49
18	$^{12}C+^{12}C$	40– 40	6.00	7.41	6.31	11.9	e	z, m	31
19	$^{12}C+^{12}C$	45–197	6.00	7.41	6.31	11.9	e	z	50
20	$^{12}C+^{12}C$	52– 92	6.00	7.41	6.31	11.9	e	z	22
21	$^{12}C+^{13}C$	6– 23	5.76	7.53	6.23	12.5	e	g	51
22	$^{12}C+^{13}C$	15– 52	5.76	7.53	6.23	12.5	e	z	47
23	$^{12}C+^{14}N$	7– 20	6.47	7.50	7.28	13.1	e	g	52
24	$^{12}C+^{14}N$	13– 35	6.47	7.50	7.28	13.1	e	g	53

FUSION OF HEAVY NUCLEI   273

25	$^{12}C+^{14}N$	27– 60	6.47	7.50	7.28	13.1	e	z	30
26	$^{12}C+^{14}N$	32– 52	6.47	7.50	7.28	13.1	e	z	47
27	$^{12}C+^{14}N$	34–178	6.47	7.50	7.28	13.1	e	z	54
28	$^{12}C+^{14}N$	43–178	6.47	7.50	7.28	13.1	e	z	35
29	$^{12}C+^{15}N$	20– 60	6.25	7.60	7.20	13.7	e	z	47
30	$^{12}C+^{15}N$	27– 60	6.25	7.60	7.20	13.7	e	z	30
31	$^{12}C+^{16}O$	10– 24	6.92	7.59	8.24	14.2	e	g	55
32	$^{12}C+^{16}O$	15– 72	6.92	7.59	8.24	14.2	e	g	56
33	$^{12}C+^{16}O$	16– 33	6.92	7.59	8.24	14.2	e	z	57
33a	$^{12}C+^{16}O$	17– 26	6.92	7.59	8.24	14.2	e	z	75
34	$^{12}C+^{16}O$	30– 63	6.92	7.59	8.24	14.2	e	z	58
35	$^{12}C+^{16}O$	30– 63	6.92	7.59	8.24	14.2	e	z	47
36	$^{12}C+^{16}O$	35– 62	6.92	7.59	8.24	14.2	e	z	59
37	$^{12}C+^{16}O$	40– 65	6.92	7.59	8.24	14.2	e	g	60
38	$^{12}C+^{16}O$	45– 80	6.92	7.59	8.24	14.2	e	m	61
39	$^{12}C+^{16}O$	60– 60	6.92	7.59	8.24	14.2	e	z,m	62
40	$^{12}C+^{17}O$	16– 34	6.71	7.68	8.16	14.8	e	z	57
41	$^{12}C+^{17}O$	29– 73	6.71	7.68	8.16	14.8	e	z	42
42	$^{12}C+^{17}O$	34– 80	6.71	7.68	8.16	14.8	e	z	63
43	$^{12}C+^{18}O$	17– 35	6.51	7.76	8.08	15.3	e	z	57
44	$^{12}C+^{18}O$	29– 70	6.51	7.76	8.08	15.3	e	z	47
45	$^{12}C+^{18}O$	30– 70	6.51	7.76	8.08	15.3	e	z	48
46	$^{12}C+^{18}O$	32–140	6.52	7.77	8.08	15.3	e	z,m	24
47	$^{12}C+^{18}O$	100–100	6.51	7.76	8.08	15.3	e	m	64
48	$^{12}C+^{19}F$	30– 70	7.13	7.75	9.10	15.8	e	z	47
49	$^{12}C+^{19}F$	31– 77	7.13	7.75	9.10	15.8	e	z	48
50	$^{12}C+^{19}F$	50– 76	7.13	7.75	9.10	15.8	e	m	65
51	$^{12}C+^{19}F$	92– 92	7.13	7.75	9.10	15.8	e	m	66
52	$^{12}C+^{20}Ne$	40– 69	7.71	7.74	10.13	16.2	e	z	67
53	$^{12}C+^{20}Ne$	54– 80	7.72	7.75	10.13	16.3	e	z	68
54	$^{12}C+^{20}Ne$	66– 66	7.71	7.74	10.13	16.2	e	z,m	69

Table 1 continued

No.	Reaction	Energy range (MeV)	Z^2/A[b] (eff)	Barrier[a] Radius (fm)	Barrier[a] Potential (MeV)	l (pocket)[c]	Detected fragments[d]	Method[e]	Ref.
55	$^{12}C + ^{20}Ne$	67–160	7.71	7.74	10.13	16.2	e	z	34
56	$^{12}C + ^{28}Si$	61– 94	9.07	8.02	13.74	19.6	e	z,g	70
57	$^{12}C + ^{46}Ti$	266–266	10.90	8.54	20.10	25.1	e	z,m	71
58	$^{13}C + ^{6}Li$	9– 35	4.04	7.22	3.23	7.4	e	m	43
59	$^{13}C + ^{7}Li$	10– 34	3.75	7.41	3.16	8.4	e	m	43
60	$^{13}C + ^{10}B$	14– 54	5.26	7.44	5.25	11.1	e	m	44
61	$^{13}C + ^{12}C$	6– 13	5.76	7.53	6.23	12.5	e	g	45
62	$^{13}C + ^{12}C$	40– 40	5.76	7.53	6.23	12.5	e	z,m	31
63	$^{13}C + ^{13}C$	6– 14	5.54	7.64	6.15	13.2	e	g	45
64	$^{13}C + ^{13}C$	8– 32	5.54	7.64	6.15	13.2	e	g	72
65	$^{13}C + ^{13}C$	30– 60	5.54	7.64	6.15	13.2	e	z,m	31
66	$^{13}C + ^{14}C$	40– 40	5.33	7.74	6.08	13.8	e	z,m	31
67	$^{13}C + ^{14}N$	25– 62	6.23	7.62	7.19	13.9	e	z	42
68	$^{13}C + ^{14}N$	86–180	6.23	7.62	7.19	13.9	e	z	73
69	$^{13}C + ^{17}O$	54–140	6.45	7.80	8.06	15.7	e	z,m	24
70	$^{13}C + ^{18}O$	100–100	6.26	7.87	7.98	16.2	e	m	64
71	$^{14}N + ^{12}C$	248–248	6.47	7.50	7.28	13.1	e	z	36
72	$^{14}N + ^{14}N$	9– 22	7.00	7.59	8.41	14.5	e	g	55
73	$^{16}O + ^{14}N$	10– 24	7.48	7.68	9.52	15.8	e	g	52
74	$^{16}O + ^{16}O$	13– 23	8.00	7.76	10.77	17.2	e	g	74
75									
76	$^{16}O + ^{16}O$	27– 66	8.00	7.76	10.77	17.2	e	z	76
77	$^{16}O + ^{16}O$	29– 72	8.00	7.76	10.77	17.2	e	z	47
78	$^{16}O + ^{16}O$	35– 80	8.00	7.76	10.77	17.2	e	m	61

FUSION OF HEAVY NUCLEI

#	System	Range							Ref
79	$^{16}O+^{16}O$	55–55	8.00	7.76	10.77	17.2	e	z,m	69
80	$^{16}O+^{16}O$	60–60	8.00	7.76	10.77	17.1	e	z,m	62
81	$^{16}O+^{16}O$	60–80	8.00	7.76	10.77	17.2	e	g	77
82	$^{16}O+^{16}O$	119–141	8.00	7.76	10.77	17.2	e	z	34
83	$^{16}O+^{17}O$	25–62	7.76	7.85	10.66	17.9	e	z	42
84	$^{16}O+^{27}Al$	105–145	9.96	8.19	16.68	23.4	e	z	78
85	$^{23}Na+^{19}F$	45–128	9.46	8.23	15.81	23.9	e	z	79
86	$^{24}Mg+^{12}C$	20–63	8.43	7.88	11.95	17.3	e	z	80
87	$^{24}Mg+^{12}C$	49–89	8.43	7.89	11.95	17.3	e	z,m	81
88	$^{24}Mg+^{16}O$	30–81	9.77	8.06	15.62	21.0	e	z	82
89	$^{24}Mg+^{16}O$	40–51	9.77	8.06	15.62	21.0	e	g	46
90	$^{24}Mg+^{18}O$	32–72	9.22	8.24	15.31	23.1	e	z	82
91	$^{24}Mg+^{19}F$	67–87	10.11	8.23	17.25	24.4	e	z	83
92	$^{24}Mg+^{20}Ne$	45–105	10.95	8.22	18.90	25.1	e	g	84
93	$^{24}Mg+^{24}Mg$	42–83	12.00	8.37	22.45	27.9	e	z,m	81
94	$^{24}Mg+^{26}Mg$	42–69	11.53	8.50	22.05	29.5	e	z,m	81
95	$^{24}Mg+^{32}S$	67–132	13.84	8.62	29.23	32.7	e	z	85
96	$^{24}Mg+^{32}S$	160–160	13.84	8.62	29.23	32.7	e	z,m	86
97	$^{25}Mg+^{32}S$	160–160	13.56	8.69	28.99	33.7	e	z,m	86
98	$^{26}Mg+^{12}C$	20–56	8.08	8.01	11.78	18.3	e	z	80
99	$^{26}Mg+^{16}O$	29–81	9.38	8.19	15.39	22.2	e	z	82
100	$^{26}Mg+^{16}O$	40–114	9.38	8.19	15.39	22.2	e	z	79
101	$^{26}Mg+^{32}S$	160–160	13.30	8.75	28.75	34.6	e	z,m	86
102	$^{26}Mg+^{208}Pb$	124–208	25.23	11.60	113.70	64.9	f	k	87,88
103									
104	$^{27}Al+^{12}C$	44–180	8.58	8.01	12.76	18.6	e	t,z	33,89
105	$^{27}Al+^{12}C$	45–70	8.58	8.01	12.76	18.6	e	z	90
106	$^{27}Al+^{14}N$	157–262	9.30	8.10	14.73	20.7	e	z	91
107	$^{27}Al+^{15}N$	27–70	9.00	8.21	14.57	21.9	e	z	92
108	$^{27}Al+^{16}O$	25–50	9.96	8.19	16.68	22.6	e	g	93
109	$^{27}Al+^{16}O$	27–42	9.96	8.19	16.68	22.6	e	z	94

Table 1 *continued*

No.	Reaction	Energy range (MeV)	Z^2/A[b] (eff)	Barrier[a] Radius (fm)	Barrier[a] Potential (MeV)	l (pocket)[c]	Detected fragments[d]	Method[e]	Ref.
110	^{27}Al+^{16}O	34–81	9.96	8.19	16.68	22.6	e	m	95
111	^{27}Al+^{16}O	40–100	9.96	8.19	16.68	22.6	e	z	83
112	^{27}Al+^{16}O	50–80	9.96	8.19	16.68	22.6	e	m	96
113	^{27}Al+^{16}O	81–168	9.96	8.19	16.68	22.6	e	z	97
114	^{27}Al+^{16}O	87–161	9.96	8.19	16.68	22.6	e	t	98
115	^{27}Al+^{17}O	27–42	9.68	8.28	16.51	23.7	e	z	94
116	^{27}Al+^{18}O	27–42	9.41	8.37	16.36	24.8	e	z	94
117	^{27}Al+^{18}O	34–81	9.41	8.37	16.36	24.8	e	m	95
118	^{27}Al+^{20}Ne	32–151	11.17	8.36	20.22	26.0	e	z	99
119	^{27}Al+^{20}Ne	50–290	11.17	8.36	20.22	26.0	e	m	100
120	^{27}Al+^{20}Ne	87–198	11.17	8.35	20.22	26.0	e	t	98
121	^{27}Al+^{20}Ne	120–120	11.17	8.35	20.22	26.0	e	z	101
122	^{27}Al+^{20}Ne	138–210	11.17	8.35	20.22	26.0	e	z	97
123	^{27}Al+^{28}Si	65–81	13.24	8.63	27.63	32.7	e	g	102
124	^{27}Al+^{32}S	67–132	14.15	8.75	31.22	35.1	e	z	85
124a	^{27}Al+^{32}S	160–160	14.15	8.75	31.22	35.1	e	z,m	86
125	^{27}Al+^{32}S	336–336	14.15	8.75	31.22	35.1	e	z	97
126	^{27}Al+^{35}Cl	70–170	14.36	8.86	32.79	37.0	e	z	103
127	^{27}Al+^{84}Kr	490–490	19.31	10.00	62.26	54.1	e	z,m	104
128	^{27}Al+^{208}Pb	128–159	26.88	11.58	123.40	63.6	f	k	87,88
129	^{28}Si+^{9}Be	30–60	6.93	7.96	9.21	15.6	e	z	105
130	^{28}Si+^{12}C	20–49	9.07	8.01	13.74	18.8	e	z	106
131	^{28}Si+^{12}C	26–50	9.07	8.01	13.74	18.8	e	z,g	70
132	^{28}Si+^{12}C	50–82	9.08	8.02	13.74	18.8	e	z,m	81

FUSION OF HEAVY NUCLEI

	Reaction	Range							Ref.
133	$^{28}Si + ^{16}O$	21– 61	10.54	8.19	17.71	22.8	e	z	106
134	$^{28}Si + ^{16}O$	25– 50	10.54	8.19	17.71	22.8	e	g	93
135	$^{28}Si + ^{16}O$	34– 81	10.54	8.19	17.71	22.8	e	m	95
136	$^{28}Si + ^{16}O$	50– 65	10.54	8.19	17.71	22.8	e	g	107
137	$^{28}Si + ^{18}O$	34– 81	9.95	8.37	17.61	25.2	e	m	95
138	$^{28}Si + ^{24}Mg$	52– 80	13.96	8.50	25.87	30.4	e	z,m	81
139	$^{28}Si + ^{28}Si$	55– 83	14.00	8.63	29.83	33.2	e	z,m	81
140	$^{28}Si + ^{28}Si$	65– 90	14.00	8.63	29.82	33.2	e	g	108
141	$^{28}Si + ^{29}Si$	53– 83	13.52	8.75	29.38	34.8	e	z,m	81
142	$^{28}Si + ^{30}Si$	53– 83	13.52	8.75	29.38	34.8	e	z,m	81
143	$^{29}Si + ^{12}C$	20– 49	8.90	8.07	13.65	19.3	e	z	106
144	$^{29}Si + ^{16}O$	21– 61	10.35	8.25	17.84	23.4	e	z	106
145	$^{30}Si + ^{12}C$	20– 49	8.75	8.13	13.56	19.7	e	z	106
146	$^{30}Si + ^{16}O$	21– 61	10.17	8.31	17.73	24.0	e	z	106
147	$^{30}Si + ^{132}Xe$	712–1082	23.31	10.85	93.20	64.7	f	z,m	109
148	$^{40}Ca + ^{16}O$	40–214	12.50	8.53	24.51	27.3	e	z	110
149	$^{40}Ca + ^{16}O$	45– 63	12.50	8.53	24.51	27.3	e	z	111
150	$^{40}Ca + ^{20}Ne$	44– 70	14.05	8.69	30.21	31.5	e	z	112
151	$^{40}Ca + ^{20}Ne$	151–151	14.05	8.69	30.21	31.5	e	z	113
152	$^{40}Ca + ^{32}S$	67–132	17.88	9.09	46.55	41.0	e	z	85
153	$^{40}Ca + ^{40}Ca$	107–195	20.00	9.30	56.96	47.5	e	z	114
154	$^{40}Ca + ^{40}Ca$	112–165	20.00	9.31	57.00	47.5	e	z	115
155	$^{48}Ca + ^{36}S$	117–147	15.38	9.64	43.88	51.44	e	z	116
156	$^{48}Ca + ^{208}Pb$	229–384	31.87	12.21	180.00	72.70	f	k	87,88
157	$^{47}Ti + ^{16}O$	310–310	12.51	8.80	26.17	30.3	e	c,z	117
158	$^{48}Ti + ^{12}C$	81–180	10.71	8.62	20.21	24.9	e	t,z	89
159	$^{48}Ti + ^{35}Cl$	70–170	18.22	9.47	52.34	48.6	e		103
160	$^{48}Ti + ^{35}Cl$	104–104	18.22	9.47	52.34	48.6	e	m,r	118
161	$^{50}Ti + ^{32}S$	121–166	17.55	9.44	49.39	47.3	f	m	25
162	$^{50}Ti + ^{208}Pb$	238–400	34.40	12.17	198.10	63.4	f	k	87,88
163	$^{52}Cr + ^{14}N$	157–262	12.15	8.80	24.95	28.7	e	z	91

Table 1 continued

No.	Reaction	Energy range (MeV)	Z^2/A[b] (eff)	Barrier[a] Radius (fm)	Potential (MeV)	l (pocket)[c]	Detected fragments[d]	Method[e]	Ref.
164	$^{52}Cr + ^{208}Pb$	272–416	36.86	12.12	216.10	51.2	f	k	87, 88
165	$^{54}Fe + ^{6}Li$	10–100	8.11	8.37	12.26	15.6	e	r	119
166	$^{54}Fe + ^{35}Cl$	149–157	20.28	9.56	61.36	50.2	e	z	103
167	$^{56}Fe + ^{6}Li$	10–100	7.95	8.45	12.17	15.9	e	r	119
168	$^{56}Fe + ^{35}Cl$	70–170	19.91	9.64	60.91	51.7	e	z	103
169	$^{56}Fe + ^{52}Cr$	264–264	23.13	10.09	82.36	63.1	e,f	z	120
170	$^{56}Fe + ^{132}Xe$	756–756	32.33	11.42	164.00	73.3	f	z	121
171	$^{58}Fe + ^{208}Pb$	304–464	37.96	12.25	231.40	41.7	f	k	87, 88
172	$^{59}Co + ^{16}O$	119–161	13.73	9.05	31.38	33.2	e	t	98
173	$^{58}Ni + ^{12}C$	63–180	12.22	8.83	24.89	27.1	e	t,z	89
174	$^{58}Ni + ^{14}N$	126–126	13.38	8.90	28.92	29.9	f	k	122
175	$^{58}Ni + ^{14}N$	157–262	13.28	8.93	28.82	30.1	e	z	91
176	$^{58}Ni + ^{16}O$	38–46	14.37	8.99	32.81	32.7	e	g,r	123
177	$^{58}Ni + ^{16}O$	44–77	14.37	8.99	32.81	32.7	e	g	124
178	$^{58}Ni + ^{32}S$	67–132	20.70	9.53	62.37	48.7	e	z	85
179	$^{58}Ni + ^{35}Cl$	70–170	21.05	9.64	65.56	51.6	e,f	z	103, 125
180	$^{58}Ni + ^{35}Cl$	165–165	21.05	9.65	65.56	51.6	f	z	126
181	$^{58}Ni + ^{40}Ar$	167–227	20.76	9.87	67.92	57.4	e	z	127
182	$^{58}Ni + ^{40}Ca$	113–170	23.21	9.74	76.36	53.9	e	z	128
183	$^{58}Ni + ^{58}Ni$	187–220	27.03	10.17	102.50	64.2	e	z,s	129, 130
184	$^{60}Ni + ^{14}N$	126–126	13.14	8.97	28.68	30.5	f	k	122
185	$^{60}Ni + ^{16}O$	38–46	14.12	9.06	32.55	33.4	e	g,r	123
186	$^{60}Ni + ^{35}Cl$	70–170	20.69	9.71	65.11	53.0	e	z	103, 125
187	$^{60}Ni + ^{40}Ca$	113–170	22.81	9.81	75.84	55.6	e	z	128

FUSION OF HEAVY NUCLEI

188	$^{62}Ni+^{35}Cl$	70–170	20.34	9.79	64.68	54.4	e	z	103, 125
189	$^{62}Ni+^{35}Cl$	155–170	20.34	9.79	64.68	54.4	f	z	126
190	$^{62}Ni+^{35}Cl$	200–215	20.34	9.79	64.68	54.4	e,f	z	131
191	$^{62}Ni+^{40}Ca$	113–170	22.43	9.88	75.35	57.1	e	z	128
192	$^{62}Ni+^{58}Ni$	215–215	26.15	10.32	101.30	67.3	e	z	132
193	$^{64}Ni+^{35}Cl$	70–170	20.01	9.85	64.26	55.6	e,f	z	103, 125
194	$^{64}Ni+^{58}Ni$	187–220	25.73	10.38	100.70	69.3	e	z,s	129, 133
195	$^{64}Ni+^{64}Ni$	171–215	24.50	10.59	98.88	75.4	e	z,s	129, 133
196	$^{64}Ni+^{208}Pb$	335–512	39.04	12.36	246.70	22.2	f	k	87, 88
197	$^{63}Cu+^{12}C$	44– 97	12.12	8.96	25.40	28.2	e	t,z	89
198	$^{63}Cu+^{12}C$	126–126	12.12	8.96	25.40	28.2	e	t	134
199	$^{63}Cu+^{16}O$	38– 51	14.24	9.13	33.47	34.2	e	g	135
200	$^{63}Cu+^{16}O$	40– 85	14.24	9.13	33.47	34.2	e	g	136
201	$^{63}Cu+^{16}O$	97–168	14.17	9.15	33.39	34.3	e	t	134
202	$^{63}Cu+^{20}Ne$	169–210	15.96	9.31	41.17	39.6	e	t	134
203	$^{63}Cu+^{24}Mg$	161–341	17.59	9.45	48.80	44.1	e,f	z	137
204	$^{65}Cu+^{16}O$	38– 51	14.00	9.19	33.23	34.8	e	g	135
205	$^{65}Cu+^{63}Cu$	225–325	26.28	10.52	106.60	72.7	e	r	138
206	$^{65}Cu+^{84}Kr$	494–604	28.23	10.95	127.40	77.4	e,f	z	27
207	$^{65}Cu+^{86}Kr$	366–716	27.90	11.00	126.80	78.4	e	z,m	139
208	$^{70}Ge+^{86}Kr$	310–470	29.68	11.06	139.10	77.9	e	z,m	140
209	$^{72}Ge+^{40}Ar$	167–227	21.27	10.20	75.25	64.0	e	z	127
210	$^{72}Ge+^{84}Kr$	350–494	29.62	11.07	139.0	78.3	e	r	141
211	$^{74}Ge+^{58}Ni$	171–215	27.33	10.57	113.00	72.6	e	z,s	133
212	$^{74}Ge+^{64}Ni$	171–215	26.03	10.78	111.10	79.4	e	z,s	133
213	$^{74}Ge+^{86}Kr$	310–470	28.87	11.18	137.70	80.8	e	z,m	140
214	$^{76}Ge+^{32}S$	108–225	20.55	10.00	68.20	56.5	e,f	m	142
215	$^{76}Ge+^{86}Kr$	310–470	28.49	11.24	137.10	90.0	e	g	140
216	$^{78}Se+^{14}N$	126–126	13.74	9.39	33.39	34.5	f	k	122
217	$^{89}Y+^{12}C$	107–197	13.56	9.46	32.60	32.5	e	z	143
218	$^{90}Zr+^{35}Cl$	70–170	23.93	10.29	88.10	61.9	e	z	103

Table 1 continued

No.	Reaction	Energy range (MeV)	Z^2/A (eff)[b]	Barrier[a] Radius (fm)	Barrier[a] Potential (MeV)	l (pocket)[c]	Detected fragments[d]	Method[e]	Ref.
219	^{90}Zr+^{86}Kr	366–716	32.73	11.40	168.60	88.3	e	z,m	139
220	^{92}Mo+^{16}O	187–187	16.80	9.66	46.11	39.7	e,f	z	120
221	^{95}Mo+^{14}N	126–126	15.25	9.67	40.24	37.2	f	k	122
222	^{95}Mo+^{40}Ar	200–300	24.15	10.59	95.32	70.2	f	k	144
223	^{98}Mo+^{12}C	107–197	13.84	9.62	34.57	33.8	e	z	143
224	^{110}Pd+^{52}Cr	211–255	28.97	11.06	133.30	79.8	e	z	145
225	^{107}Ag+^{12}C	107–197	14.68	9.76	38.21	35.1	e	z	143
226	^{107}Ag+^{14}N	126–126	16.07	9.83	44.35	38.7	f	k	122
227	^{107}Ag+^{40}Ar	296–296	25.41	10.7	105.00	72.5	e	z	127
228	^{107}Ag+^{16}O	97–168	17.21	9.94	50.23	42.4	e	t	134
229	^{109}Ag+^{12}C	126–126	14.60	9.78	38.12	35.2	e	t	134
230	^{109}Ag+^{14}N	126–126	15.90	9.88	44.16	39.0	f	k	122
231	^{109}Ag+^{20}Ne	173–210	19.35	10.12	61.84	48.8	e	t	134
232	^{109}Ag+^{40}Ar	169–337	25.27	10.80	104.70	73.0	e,f	z	27
233	^{109}Ag+^{86}Kr	716–716	34.92	11.68	193.20	90.1	e	z,m	139
234	^{116}Cd+^{84}Kr	350–494	34.96	11.75	196.10	91.0	e	r	141
235	^{114}In+^{32}S	336–336	25.29	10.58	98.93	63.2	e	z	146
236	^{112}Sn+^{35}Cl	135–177	26.65	10.61	106.90	65.1	e	z	147
237	^{112}Sn+^{35}Cl	165–165	26.65	10.61	106.90	65.1	f	z	126
238	^{116}Sn+^{35}Cl	70–170	26.16	10.69	106.10	66.7	e,f	z	103
239	^{116}Sn+^{35}Cl	155–170	26.16	10.69	106.10	66.7	f	z	126
240	^{116}Sn+^{35}Cl	200–215	26.16	10.69	106.10	66.7	e,f	z	131
241	^{116}Sn+^{40}Ar	187–340	26.01	10.89	110.50	74.0	e,f	z	18
242	^{120}Sn+^{35}Cl	135–177	25.69	10.77	105.40	68.1	e	z	147

FUSION OF HEAVY NUCLEI

243	^{120}Sn + ^{35}Cl	165–165	25.69	10.77	105.40	68.1	f	z	126
244	^{120}Sn + ^{56}Fe	330–465	31.46	11.26	154.00	81.4	e,f	z,m	148
245	^{122}Sn + ^{40}Ar	145–200	25.33	11.01	109.40	76.2	e	g	140
246	^{122}Sn + ^{56}Fe	330–465	31.19	11.30	153.60	82.4	e,f	z,m	148
247	^{124}Sn + ^{35}Cl	70–170	25.24	10.85	104.70	69.4	e,f	z	103
248	^{124}Sn + ^{35}Cl	165–170	25.24	10.85	104.70	69.4	f	z	126
249	^{124}Sn + ^{40}Ar	145–200	25.11	11.05	109.00	76.8	e	g	140
250	^{124}Sn + ^{94}Zr	360–430	37.01	12.00	221.90	90.4	e	s	149a
251	^{121}Sb + ^{40}Ar	162–300	25.86	10.99	111.70	75.6	f	k	144
252	^{121}Sb + ^{40}Ar	167–296	25.86	10.99	111.70	75.6	e	z	127
253	^{121}Sb + ^{40}Ar	282–340	25.95	10.98	111.90	75.4	e,f	z	27
254	^{123}Sb + ^{86}Kr	335–388	35.64	11.88	206.00	92.1	e	s	149a
255	^{133}Cs + ^{16}O	113–166	17.95	10.32	56.77	45.6	f	k	150
256	^{133}Cs + ^{20}Ne	103–128	20.31	10.48	70.07	52.1	e,f	c,m	151
257	^{139}La + ^{86}Kr	505–710	37.42	12.06	226.50	89.0	f	z	152
258	^{141}Pr + ^{12}C	92–137	15.86	10.24	45.91	38.7	e,f	c,m	151
259	^{141}Pr + ^{16}O	108–166	18.64	10.42	60.39	46.5	f	k	150
260	^{141}Pr + ^{35}Cl	135–170	27.80	11.03	121.50	69.8	e	z	147
261	^{141}Pr + ^{35}Cl	155–170	27.80	11.03	121.50	69.8	f	z	126
262	^{141}Pr + ^{35}Cl	200–215	27.80	11.03	121.50	69.8	e,f	z	131
263	^{148}Nd + ^{16}O	50–81	18.45	10.52	60.83	47.2	e	g	17
264	^{148}Nd + ^{18}O	50–81	17.51	10.71	59.72	53.1	e	g	17
265	^{150}Nd + ^{16}O	50–85	18.31	10.55	60.66	47.4	e	g	17
266	^{150}Nd + ^{20}Ne	127–175	20.73	10.71	74.89	54.0	e	g	153
267	^{150}Nd + ^{20}Ne	175–175	20.73	10.71	74.89	54.0	e,f	z	154
268	^{144}Sm + ^{32}S	138–231	28.33	10.93	121.10	65.3	f	c	155
269	^{144}Sm + ^{40}Ar	144–180	28.75	11.24	132.80	76.8	e,f	g,z	156
270	^{144}Sm + ^{40}Ar	190–190	28.75	11.24	132.80	76.8	e,f	m,s	157
271	^{148}Sm + ^{16}O	60–75	19.06	10.51	62.98	47.1	e	r	158,159
272	^{148}Sm + ^{40}Ar	144–180	28.33	11.31	132.10	78.1	e,f	g,z	156
273	^{150}Sm + ^{16}O	60–75	18.92	10.54	62.80	47.4	e	r	158,159

Table 1 continued

No.	Reaction	Energy range (MeV)	Z^2/A[b] (eff)	Barrier[a] Radius (fm)	Barrier[a] Potential (MeV)	l (pocket)[c]	Detected fragments[d]	Method[e]	Ref.
274	^{152}Sm + ^{12}C	40– 63	15.97	10.39	47.59	39.7	e	g	17
275	^{152}Sm + ^{16}O	60– 75	18.78	10.57	62.61	47.6	e	r	158, 159
276	^{154}Sm + ^{16}O	60– 75	18.64	10.60	62.44	47.8	e	r	158, 159
277	^{154}Sm + ^{16}O	137–137	18.64	10.60	62.44	47.8	e	t	15
278	^{154}Sm + ^{16}O	137–137	18.64	10.60	62.44	47.8	f	t	160
279	^{154}Sm + ^{32}S	138–231	27.32	11.10	119.50	67.8	f	c	155
280	^{154}Sm + ^{40}Ar	144–180	27.74	11.41	131.00	79.7	e,f	g,z	156
281	^{154}Sm + ^{40}Ar	222–340	27.74	11.41	131.00	79.7	e,f	z	18
282	^{158}Gd + ^{12}C	126–126	16.13	10.47	48.79	40.2	f	t	160
283	^{158}Gd + ^{12}C	126–126	16.13	10.47	48.79	40.2	e	t	15
284	^{159}Tb + ^{7}Li	17– 36	10.27	10.33	25.31	28.7	e	g	17
285	^{159}Tb + ^{11}B	115–115	14.12	10.51	41.47	39.2	e	t,z	15, 161
286	^{159}Tb + ^{11}B	115–115	14.12	10.51	41.47	39.2	e	z	16
287	^{159}Tb + ^{16}O	93–166	19.20	10.65	65.18	48.2	f	k	150
288	^{159}Tb + ^{19}F	87–193	20.02	10.82	72.27	54.7	f	c	162
289	^{159}Tb + ^{22}Ne	101–229	20.84	10.96	79.32	60.5	f	k	150
290	^{162}Dy + ^{4}He	12– 27	8.66	10.01	17.62	18.3	e	g	17
291	^{164}Dy + ^{40}Ar	222–340	28.54	11.52	138.20	80.2	e,f	z	18
292	^{165}Ho + ^{14}N	83–145	17.98	10.64	58.79	45.0	f	c	162
293	^{165}Ho + ^{14}N	126–126	17.98	10.64	58.79	45.0	f	k	122
294	^{165}Ho + ^{16}O	93–166	19.38	10.72	66.76	48.8	f	k	150
295	^{165}Ho + ^{16}O	94–167	19.38	10.72	66.76	48.8	f	t	163
296	^{165}Ho + ^{20}Ne	116–204	21.95	10.87	82.43	55.3	f	t	163
297	^{165}Ho + ^{20}Ne	220–220	21.95	10.87	82.43	55.3	f	m,c	164

FUSION OF HEAVY NUCLEI

#	Reaction	Range							Ref
298	^{165}Ho + ^{40}Ar	190–190	28.88	11.52	140.20	80.0	e,f	m,s	157
299	^{165}Ho + ^{40}Ar	194–194	28.88	11.52	140.20	80.0	f	m	165
300	^{165}Ho + ^{40}Ar	226–300	28.88	11.52	140.20	80.0	f	k	144
301	^{165}Ho + ^{56}Fe	351–510	35.66	11.77	197.10	79.3	f	z	166,167
302	^{165}Ho + ^{84}Kr	450–492	40.72	12.24	261.10	74.2	f	k	168
303	^{165}Ho + ^{84}Kr	600–714	40.72	12.24	261.10	74.2	f	k	169
304	^{170}Er + ^{16}O	93–166	19.34	10.79	67.37	49.2	f	k	150
305	^{170}Er + ^{30}Si	125–168	25.58	11.30	112.80	70.7	e,f	z,m	170
306	^{169}Tm + ^{11}B	71–114	14.48	10.63	43.57	40.1	f	c	162
307	^{169}Tm + ^{12}C	74–124	16.73	10.59	52.05	41.1	f	k	150
308	^{169}Tm + ^{12}C	89–124	16.73	10.59	52.05	41.1	f	t	163
309	^{169}Tm + ^{16}O	87–166	19.69	10.77	68.50	49.1	f	k	150
310	^{169}Tm + ^{16}O	94–167	19.69	10.77	68.50	49.1	f	t	163
311	^{169}Tm + ^{37}Cl	148–186	28.74	11.47	136.90	76.7	e	s	149
312	^{174}Yb + ^{12}C	70–124	16.70	10.66	52.50	41.5	f	c	162
313	^{174}Yb + ^{16}O	81–166	19.65	10.83	69.10	49.5	f	k	150
314	^{174}Lu + ^{16}O	81–166	19.87	10.84	70.06	49.6	f	k	150
315	^{175}Lu + ^{11}B	69–114	14.60	10.70	44.55	40.5	f	c	162
316	^{175}Lu + ^{12}C	74–124	16.88	10.66	53.23	41.6	f	c	162
317	^{175}Lu + ^{19}F	98–142	20.73	11.00	77.70	56.1	f	z	171
318	^{175}Lu + ^{40}Ar	190–190	29.65	11.62	147.20	80.3	e,f	m,s	157
319	^{176}Hf + ^{40}Ar	190–190	29.97	11.63	149.20	80.0	e,f	m,s	157
320	^{181}Ta + ^{16}O	77–102	20.04	10.90	71.60	50.1	f	z	172
321	^{181}Ta + ^{16}O	88–167	20.04	10.90	71.60	50.1	f	t	163
322	^{181}Ta + ^{19}F	80–126	20.91	11.07	79.42	56.7	e,f	z,m	170
323	^{181}Ta + ^{20}Ne	99–211	22.72	11.05	88.43	56.6	f	t	163
324	^{181}Ta + ^{40}Ar	194–194	29.93	11.69	150.50	80.6	e,f	m,s	157
325	^{182}W + ^{12}C	70–124	17.20	10.74	55.12	42.1	f	c	162
326	^{182}W + ^{12}C	77–167	17.20	10.74	55.12	42.1	e,f	z	18
327	^{182}W + ^{12}C	121–167	17.20	10.74	55.12	42.1	e,f	z	171
328	^{182}W + ^{16}O	81–166	20.25	10.91	72.55	50.1	f	k	150

Table 1 continued

No.	Reaction	Energy range (MeV)	Z^2/A^b (eff)	Barrier[a] Radius (fm)	Potential (MeV)	l (pocket)[c]	Detected fragments[d]	Method[e]	Ref.
329	^{186}W + ^{84}Kr	492–492	42.25	12.41	283.40	62.2	f	c	168
330	^{185}Re + ^{12}C	79–124	17.27	10.77	55.70	42.3	f	t	163
331	^{185}Re + ^{16}O	94–167	20.34	10.94	73.31	50.4	f	t	163
332	^{186}Re + ^{20}Ne	124–124	22.97	11.10	90.42	56.9	f	m	165
333	^{187}Re + ^{12}C	99–124	17.17	10.80	55.56	42.4	f	t	163
334	^{187}Re + ^{16}O	167–167	20.22	10.97	73.14	50.5	f	t	163
335	^{197}Au + ^{4}He	16–103	9.23	10.41	20.35	19.8	e	r	173
336	^{197}Au + ^{12}C	68–124	17.55	10.91	58.00	43.2	f	e	174, 175
337	^{197}Au + ^{12}C	126–126	17.55	10.91	58.00	43.2	e	t	143
338	^{197}Au + ^{12}C	126–126	17.55	10.91	58.00	43.2	f	e	176
339	^{197}Au + ^{14}N	126–126	19.17	10.99	67.23	47.4	f	k	122
340	^{197}Au + ^{16}O	77–166	20.68	11.08	76.36	51.2	f	k	150
341	^{197}Au + ^{16}O	97–168	20.68	11.08	76.36	51.2	e	t	134
342	^{197}Au + ^{16}O	160–160	20.68	11.08	76.36	51.2	f	e	177
343	^{197}Au + ^{16}O	168–168	20.68	11.08	76.36	51.2	f	e	176
344	^{197}Au + ^{20}Ne	210–210	23.45	11.22	94.33	57.6	e	t	134
345	^{197}Au + ^{20}Ne	220–220	23.50	11.22	94.33	57.6	f	e, m	164
346	^{197}Au + ^{20}Ne	260–260	23.45	11.22	94.33	57.6	f	e	178
347	^{197}Au + ^{40}Ar	183–248	30.94	11.85	160.70	80.7	f	m	179
348	^{197}Au + ^{40}Ar	217–217	30.94	11.85	160.70	80.7	f	z	180
349	^{197}Au + ^{40}Ar	222–340	30.94	11.85	160.70	80.7	e, f	z	18
350	^{197}Au + ^{63}Cu	365–443	40.40	12.15	249.60	66.5	f	m	181
351	^{197}Au + ^{63}Cu	605–605	40.40	12.15	249.60	66.5	f	k	182
352	^{208}Pb + ^{16}O	80–102	20.81	11.19	78.46	52.0	f	z	172

FUSION OF HEAVY NUCLEI

#	Reaction	Range							Ref
353	^{208}Pb+^{16}O	90– 99	20.81	11.19	78.46	52.0	f	z,m	183
354	^{208}Pb+^{32}S	180–266	30.66	11.66	150.50	68.9	f	m	184
355	^{208}Pb+^{40}Ar	208–208	31.18	11.96	165.20	81.1	e,f	m,s	157
356	^{209}Bi+^{4}He	69–140	9.37	10.54	21.13	20.3	f	k	185
357	^{209}Bi+^{12}C	61– 73	17.83	11.03	60.28	44.0	e,f	r,e,t	186
358	^{209}Bi+^{12}C	126–126	17.83	11.03	60.28	44.0	f	e	176
359	^{209}Bi+^{12}C	126–126	17.83	11.03	60.28	44.0	e	t	134
360	^{209}Bi+^{16}O	77–166	21.01	11.20	79.38	52.0	f	k	150
361	^{209}Bi+^{16}O	117–168	21.01	11.20	79.38	52.0	e	t	134
362	^{209}Bi+^{16}O	168–168	21.01	11.20	79.38	52.0	f	e	176
363	^{209}Bi+^{20}Ne	210–210	23.83	11.34	98.07	58.3	e	t	134
364	^{209}Bi+^{20}Ne	220–220	23.83	11.34	98.07	58.4	f	e,m	164
365	^{209}Bi+^{84}Kr	500–525	44.59	12.51	312.60	29.0	f	k	168
366	^{209}Bi+^{84}Kr	600–712	44.59	12.51	312.60	29.0	f	k	169
367	^{232}Th+^{6}Li	24– 38	12.15	10.93	33.22	28.4	f	e	187
368	^{232}Th+^{7}Li	24– 38	11.40	11.13	32.68	33.1	f	e	187
369	^{233}U+^{4}He	15– 27	9.73	10.78	22.94	21.2	f	e	188
370	^{233}U+^{4}He	27–140	9.73	10.78	22.94	21.2	f	c	185
371	^{235}U+^{20}Ne	175–252	24.73	11.59	106.50	59.6	f	k	189
372	^{238}U+^{4}He	15– 27	9.61	10.84	22.83	21.2	f	e	188
373	^{238}U+^{4}He	21– 41	9.61	10.84	22.83	21.2	f	c	190
374	^{238}U+^{4}He	140–140	9.61	10.84	22.83	21.2	f	k	191
375	^{238}U+^{6}Li	24– 38	12.23	10.99	33.79	28.7	f	e	187
376	^{238}U+^{7}Li	24– 38	11.48	11.19	33.25	33.4	f	e	187
377	^{238}U+^{11}B	54–114	15.85	11.36	54.55	44.8	f	k	190
378	^{238}U+^{12}C	65–124	18.34	11.32	65.21	45.5	f	k	190
379	^{238}U+^{14}N	77–145	20.04	11.41	75.61	49.9	f	k	190
380	^{238}U+^{16}O	86–166	21.63	11.49	85.90	53.6	f	k	190
381	^{238}U+^{20}Ne	103–208	24.55	11.62	106.20	59.8	f	k	190
382	^{238}U+^{40}Ar	250–300	32.50	12.23	181.20	80.4	f	k	144
383	^{238}U+^{56}Fe	538–538	40.26	12.39	255.50	61.6	f	k	192

Table 1 continued

| No. | Reaction | Energy range (MeV) | Barrier[a] | | | | Detected fragments[d] | Method[e] | Ref. |
			Z^2/A^b (eff)	Radius (fm)	Potential (MeV)	l (pocket)[c]			
384	$^{238}U + ^{84}Kr$	500–500	46.15	11.37	341.50		f	c	168
385	$^{237}Np + ^4He$	17– 23	9.74	10.82	23.11	21.3	e,f	t,r	193

Cross Reference to Projectiles by Line Numbers[f]

^{4}He: 290, 335, 356, 369, 370, 372–374, 385
^{6}Li: 9, 58, 165, 167, 367, 375
^{7}Li: 10, 59, 284, 316, 368
^{9}Be: 129
^{10}B: 60
^{11}B: 11, 285, 286, 306, 315, 377
^{12}C: 12–20, 61, 62, 71, 86, 87, 98, 104, 105, 130–132, 143, 145, 158, 173, 197, 198, 217, 223, 225, 229, 258, 274, 282, 283, 307, 308, 312, 316, 325–327, 330, 333, 336–338, 357–359, 378
^{13}C: 21, 22, 63–65
^{14}C: 66
^{14}N: 3–5, 23–38, 67, 68, 72, 73, 106, 163, 174, 175, 184, 216, 221, 226, 230, 292, 293, 339, 379
^{15}N: 29, 30, 107
^{16}O: 1, 6, 31–39, 74–82, 88, 89, 99, 100, 108–114, 133–136, 144, 146, 148, 149, 157, 172, 176, 177, 185, 199–201, 204, 220, 228, 255, 259, 263, 265, 271, 273, 275–278, 287, 294, 295, 304, 309, 310, 313, 314, 320, 321, 328, 331, 334, 340, 341–343, 352, 353, 360–362, 380
^{17}O: 7, 8, 40–42, 69, 83, 115
^{18}O: 2, 43–47, 70, 90, 116, 117, 137, 264
^{19}F: 48–51, 85, 91, 288, 317, 322
^{20}Ne: 52–55, 92, 118–122, 150, 151, 202, 231, 256, 266, 267, 296, 297, 323, 332, 344–346, 363, 364, 371, 381
^{22}Ne: 289
^{24}Mg: 93, 138, 203
^{26}Mg: 94
^{27}Al: 84
^{28}Si: 56, 123, 139, 140

^{29}Si: 141
^{30}Si: 142, 305
^{32}S: 95–97, 101, 124, 125, 152, 161, 178, 214, 235, 268, 279, 354
^{36}S: 155
^{35}Cl: 126, 159, 160, 166, 168, 179, 180, 186, 188, 189, 190, 193, 218, 236, 237–240, 242, 243, 247, 248, 260–262
^{37}Cl: 311
^{40}Ar: 181, 209, 222, 227, 232, 241, 245, 249, 251–253, 269, 270, 272, 280, 281, 291, 298–300, 318, 319, 324, 347–349, 355, 382
^{40}Ca: 153, 154, 182, 187, 191
^{46}Ti: 57
^{52}Cr: 169, 224
^{56}Fe: 244, 246, 301, 383
^{58}Ni: 183, 192, 194, 211
^{64}Ni: 195, 212
^{63}Cu: 205, 350, 351
^{84}Kr: 127, 206, 210, 234, 302, 303, 329, 365, 366, 384
^{86}Kr: 207, 208, 213, 215, 219, 233, 254, 257
^{94}Zr: 250
^{132}Xe: 147, 170
^{208}Pb: 102, 128, 156, 162, 164, 171, 196

[a] Barriers calculated using the modified proximity potential (304, 305) and the Coulomb potential of Bondorf et al (309).
[b] $(Z^2/A)_{eff}$ calculated according to the definition of Swiatecki (3).
[c] Largest value of angular momentum that shows a pocket in the effective potential calculated with the modified proximity potential and the Bondorf Coulomb potential.
[d] Detected fragments in the measurement are denoted by:
  e — Evaporation residues
  f — Fission fragments.
[e] Methods of cross-section measurement are denoted by:
  c — Coincident fission fragment measurements
  e — Fission fragments distinguished by energy measurements
  g — Characteristic gamma-ray emission identifying evaporation residues
  k — Kinematic correlation of fission fragments
  m — Mass identification of fission fragments or residues by time of flight
  r — Recoil residue detection by radiochemical techniques
  s — Separation of residues in a recoil mass spectrometer or velocity filter
  t — Track measurement in mica or emulsions
  z — Z identification by delta E-E techniques.
[f] Line numbers in the cross reference are in order of increasing target mass.

2.1 Excitation Functions of Evaporation Residues

The factor inhibiting the formation of evaporation residues for light systems is the limitation of the fusion cross section. For heavy systems, there is an additional limitation to evaporation residue formation due to competition from the fusion-fission process. There is an intermediate range of system masses where the fusion cross-section limitation on residue formation is effective at low projectile energies, and the fusion-fission limitation becomes effective at higher projectile energies.

The de-excitation of the composite nucleus, including competition from the fission process were necessary, has usually been modeled by statistical evaporation calculations (65, 69, 86, 136, 194–198). These calculations are applied to the evaporation residue mass and charge distributions. By use of Monte Carlo techniques the angular distributions of evaporation residues may also be obtained (199–201).

In some asymmetric systems, the presence of evaporation residues cannot be determined by measurement of the fragment mass and atomic number alone, since products of damped or transfer reactions may have similar mass and charge characteristics. Evaporation calculations and kinematic characteristics have been used in some cases to assist in the separation of evaporation residues from other reaction components (41, 54, 202).

Fission competition is usually included in evaporation calculations with an angular-momentum-dependent, fission barrier height given by the rotating liquid drop model (RLDM) (203). Although the magnitude of the fission barrier height predicted by the RLDM may be too high (194, 204), the application of the statistical model, including fission competition, to evaporation residue data (103, 122, 127, 139, 151, 170, 197, 205–207) and direct examination of angular momentum limits to evaporation residue production (27, 33, 35, 36, 54, 140) confirm that the fission barrier height declines with increasing angular momentum and reaches zero close to the angular momentum limit predicted by the RLDM (208, 209). The RLDM predicts that, as the mass of the nucleus increases, the angular momentum that the nucleus may stably sustain in the absence of fission rises to a peak value of about 90 at a mass of about 120 and then begins to decline to zero at a mass of about 310. Experimental observations (210, 211) in the mass range $128 < A < 160$ have confirmed the existence of evaporation residues with spin $I \approx 70$, in agreement with the rotating liquid drop limit. However, for systems that may form evaporation residues, in addition to the angular momentum limit to the survival of the residue, the fission competition is enhanced by an increase in the excitation energy of the system at angular momenta below the rotating liquid drop limit. An approximate calculation of the excitation energy at which fission begins to be more likely than evaporation may be obtained from the systematics of Wilcke et al (212).

Thus, although the RLDM predicts that stable rotating nuclei can survive if the mass is less than about 310, this high mass limit will only be correct at zero excitation energy, which cannot be achieved in a reaction between heavy nuclei. Thus for heavy systems, evaporation residues have very small or negligible cross sections. Even so, there is an intermediate mass region where evaporation residues dominate the fusion cross section in the near barrier region of projectile energies. The ^{62}Ni+^{35}Cl and ^{116}Sn +^{35}Cl systems, with excitation functions shown in Figure 5, are examples of intermediate mass systems. For such systems, as the projectile energy rises, the rising angular momentum will limit formation of evaporation residues, and fission will become an important exit channel for fusion. For systems at the heavy end of this intermediate mass range, fission competition may become important before the RLDM angular momentum limit is reached, because of the increasing excitation. For light systems no fission is observed, and the angular momentum stability limit of the nucleus limits the fusion cross section directly; higher angular momentum trajectories lead to damped or transfer reactions.

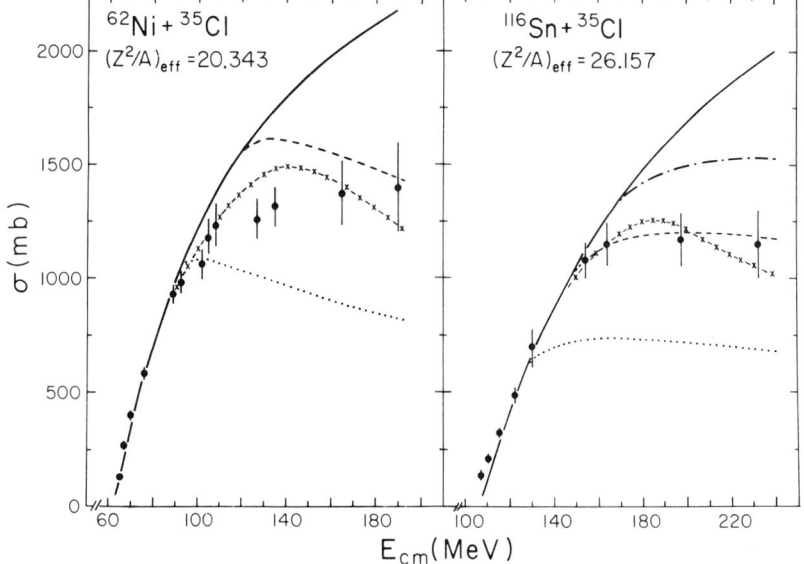

Figure 5 Comparison of experimental and theoretical fusion cross sections for the ^{62}Ni +^{35}Cl reactions. See caption of Figure 4. The dot-dashed curve corresponds to sticking collisions. References to data are in Table 1.

2.2 Excitation Functions of Fission Fragments

For intermediate mass systems at sufficiently high excitation energy and/or angular momentum and for heavy systems, fission fragments are produced in fusion reactions. If the angular momentum of the system is below the rotating liquid drop angular momentum limit l_{RLDM}, the system will form a compound nucleus before fission by following a trajectory that passes inside the unconditional saddle on the potential energy surface. However, for heavy systems, fission fragments have been observed from trajectories with angular momenta well above the liquid drop limit (27, 88, 109, 121, 127, 144, 166, 179, 211) and such trajectories pass only inside the conditional saddle defined earlier. Fission from trajectories with angular momentum above the liquid drop angular momentum limit has been termed "fast fission," and various signatures for its occurrence have been sought (214). It has been suggested that the mass distribution widths of fragments from systems with $l > l_{RLDM}$ are abnormally broad (165, 215). Although an increase in the width of the mass distribution is observed as the angular momentum of the system increases, this effect is expected to be due, at least in part, to the rise in the temperature of the system and the shape of the potential energy surface (211, 216, 217), which makes a definitive signature for the so-called fast fission process difficult to extract from the mass distribution.

Fission fragment angular distributions provide information on the shape of the fissioning system (218–220). Such measurements may also lead to a signature for the fast fission process. However, the interpretation of fission fragment angular distributions in terms of nuclear saddle shapes is complex (164) and subject to several uncertainties when a fission barrier does not exist.

In conventional transition state theory (220–222) the density of levels in the transition state nucleus with quantum numbers I and K is given approximately by

$$\rho(I, K) \propto \exp\{[E - E(I, K)]/T\}. \qquad 2.$$

In this expression, I is the total angular momentum and K is its projection on the body-fixed symmetry axis of the transition nucleus. Furthermore, the total energy E, the rotational energy $E(I, K)$ of state (I, K), and the nuclear temperature T are all values at the well-defined deformation of the transition state. For constant energy E, and upon substitution of the rotational energy of a nucleus in its saddle point deformation into Equation 2, one obtains

$$\begin{aligned}&\rho(K) \propto \exp[-K^2/2K_0^2(I)] \qquad K \leq I \\ &\rho(K) = 0 \qquad\qquad\qquad\qquad\quad K > I\end{aligned} \qquad 3.$$

where the values of K have a Gaussian distribution with a centroid at $K = 0$. The variance $K_0^2(I)$ is given by (220, 222)

$$K_0^2(I) = \frac{\mathscr{I}_{\text{sph}}}{\hbar^2} \frac{T}{[\mathscr{I}_{\text{sph}}/\mathscr{I}_{\text{eff}}(I)]}, \qquad 4.$$

where T is the nuclear temperature at the saddle-point deformation and $\mathscr{I}_{\text{sph}}$ is the rigid-body moment of inertia of a sphere. The effective moment of inertia, defined by

$$\mathscr{I}_{\text{eff}}(I) = \mathscr{I}_{\parallel}(I)\mathscr{I}_{\perp}(I)/[\mathscr{I}_{\perp}(I) - \mathscr{I}_{\parallel}(I)], \qquad 5.$$

determines the magnitude of $K_0^2(I)$, where $\mathscr{I}_{\parallel}(I)$ and $\mathscr{I}_{\perp}(I)$ are the moments of inertia parallel and perpendicular to the nuclear symmetry axis, respectively. For a Gaussian K distribution and a reaction in which the projection M of the total angular momentum on a space-fixed axis is zero, the fission fragment angular distribution is given by (220, 223)

$$W(\theta) \propto \sum (2I+1)T_I \frac{\sum\limits_{K=-I}^{I} \frac{1}{2}(2I+1)|D_{M=0,K}^{I}(\theta)|^2 \exp\left[-\frac{K^2}{2K_0^2(I)}\right]}{\sum\limits_{K=-I}^{I} \exp\left[-\frac{K^2}{2K_0^2(I)}\right]}, \qquad 6.$$

where the summation extends over all spin values I contributing to fission. In Equation 6 the transmission coefficients are written as T_I, since $l = I$ when $M = 0$. The $D_{M,K}^I(\theta)$ functions are defined and their use described elsewhere (220).

The relationship between the experimental fission fragment angular distribution and the deformation of the transition state is contained in Equation 6, where the K distribution is related to the angular momentum shape-dependent parameter $\mathscr{I}_{\text{sph}}/\mathscr{I}_{\text{eff}}(I)$ by Equation 4. If no fission barrier exists, or if the system trajectory remains outside the fission barrier, it is not clear where on the trajectory the K distribution is frozen in.

Back et al (224) have measured fission fragment anisotropies from ^{32}S-induced fission on targets of ^{197}Au, ^{232}Th, ^{238}U, and ^{248}Cm; they observed anisotropies much larger than those predicted by the RLDM. Rossner et al (164) have measured angular distributions from fission reactions induced by ^{20}Ne projectiles at 11 MeV per nucleon on several targets, and have discussed in detail the difficulties of deducing a nuclear saddle shape from fission fragment angular distribution data. These difficulties are illustrated in Figures 6 and 7, where fission fragment angular anisotropies are shown for ^{20}Ne-induced reactions on ^{165}Ho, ^{197}Au, and ^{209}Bi. The RLDM leads to the expectation that $\mathscr{I}_{\text{sph}}/\mathscr{I}_{\text{eff}}(I)$ should be a decreasing function of I, reaching toward zero when the entrance channel angular momentum $l = l_{\text{RLDM}}$. The dash-dot curve in Figure 6 is the result of calculating the fission fragment anisotropy using the angular-momentum-dependent RLDM prescription for $\mathscr{I}_{sph}/\mathscr{I}_{eff}(I)$, and it can be seen that the theory greatly underestimates the anisotropy. If the $I = 0$ value of $\mathscr{I}_{sph}/\mathscr{I}_{eff}$ is taken from the rotating liquid drop model, then the anisotropy is greatly overestimated, as shown by the curve label LDM in Figure 6. A fixed, I-independent, value of $\mathscr{I}_{sph}/\mathscr{I}_{eff}$ is chosen to give the values of K_0^2 and the anisotropies shown in Figure 6 by the thin solid lines. If the same value of K_0^2 is used, and account is also taken of the dealignment effect of neutron or charged-particle emission, before scission of the mononucleus, then the

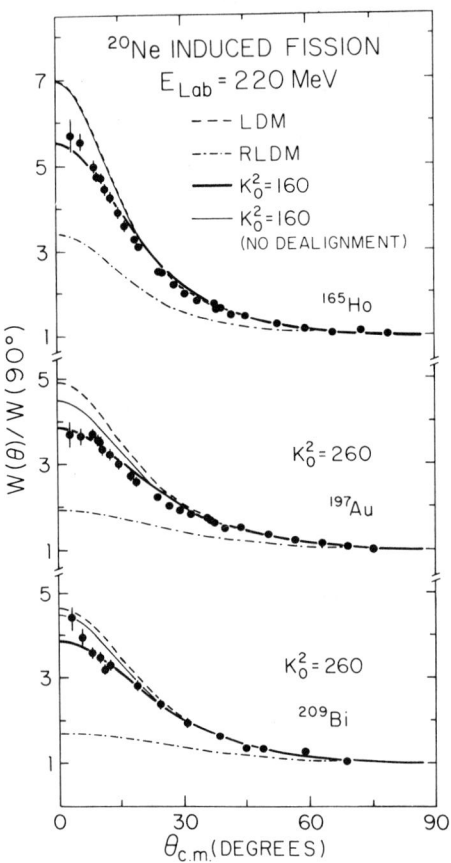

Figure 6 Effect of dealignment of pre-fission recoil nuclei (due to six pre-fission neutrons) on the experimental fission fragment angular distribution. Four curves are included for each reaction, three of which take account of the dealignment of the pre-fission recoil nuclei. The heavy solid line leads to the "best-fit" value of K_0^2. The thin solid line utilizes this K_0^2 value but neglects the dealignment effect. The dashed and dash-dotted lines represent calculations with the nonrotating $[\mathscr{I}_{sph}/\mathscr{I}_{eff}(I = 0)]$ LDM and RLDM calculations, respectively. (From 164.)

thick solid lines of Figure 6 are obtained as the best fit to the data. However, the best-fit curve of Figure 6 is obtained for a constant value of K_0^2 or $\mathscr{I}_{sph}/\mathscr{I}_{eff}$, and good agreement between theory and data may also be obtained by I-dependent functional forms of $\mathscr{I}_{sph}/\mathscr{I}_{eff}$, as shown in Figure 7. Here the functional form of $\mathscr{I}_{sph}/\mathscr{I}_{eff}(I)$ is shown on the right side of Figure 7, and shown on the left is the anisotropy corresponding to fission of the composite nucleus (solid lines), a composite nucleus with six pre-scission-evaporated neutrons (dash-dot lines), and a composite system in

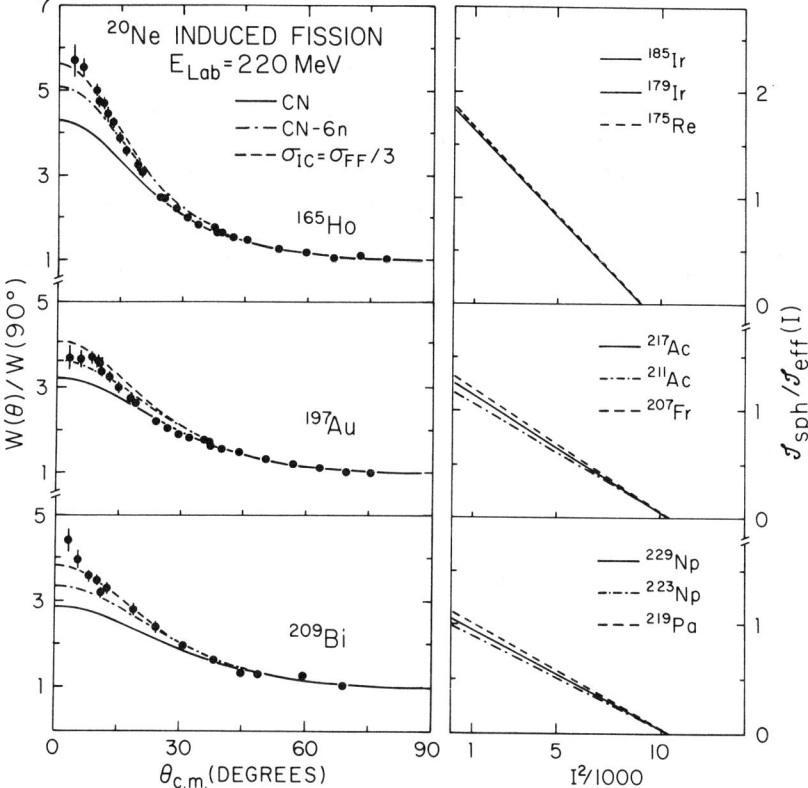

Figure 7 Comparison of experimental fission fragment angular distributions with theory based on spin-dependent values of $\mathscr{I}_{sph}/\mathscr{I}_{eff}(I)$. The functional forms of $\mathscr{I}_{sph}/\mathscr{I}_{eff}(I)$ are shown on the right side of this figure. The three theoretical curves, shown on the left along with the data (*solid points*), are for no pre-fission particle emission (*solid line*), emission of six pre-fission neutrons (*dash-dotted line*), and one third of the fission cross section due to incomplete fusion as well as the emission of six pre-fission neutrons (*dashed line*).

The functional forms of $\mathscr{I}_{sph}/\mathscr{I}_{eff}(I)$ used in the curves shown are not unique, and by altering these, it is possible to obtain even better agreement between the angular distribution data and the calculations. (From 164.)

which one third of the cross section is due to incomplete fusion reactions in which one alpha particle from the projectile is not captured by the target. Thus, it is difficult to make an unambiguous interpretation of fission fragment angular distributions in terms of nuclear saddle shapes during a heavy ion reaction.

2.3 Sub-barrier Fusion

Resonances in the fusion excitation functions of some systems have been sought in the sub-barrier region (225–231). In addition, the average behavior of the fusion excitation functions in the sub-barrier region has shown interesting effects not yet fully explained. The effects are illustrated in Figure 8 where reduced fusion excitation functions are shown in the sub-barrier region for several reactions between Ni isotopes and between Ni and Ge isotopes (133). The reduced excitation function is obtained by dividing the cross section by a barrier radius parameter R_0^2 and the energy by a barrier height V_0; this eliminates variations in the cross section between different systems that may be attributed to size effects alone. As can be seen from the figure there are large additional differences between the excitation functions in the sub-barrier region. Other measurements have

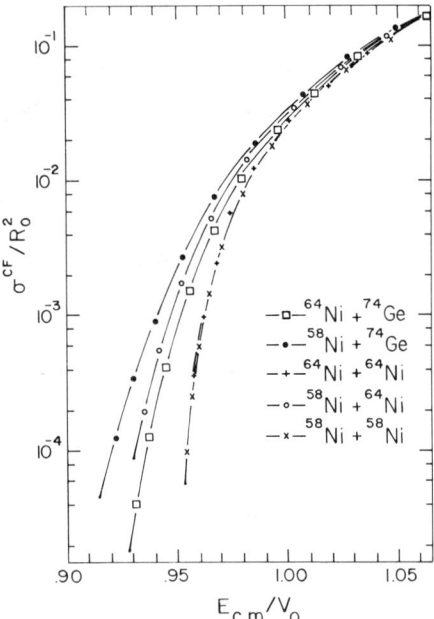

Figure 8 Reduced excitation functions for evaporation residues. The experimental excitation functions are scaled by using fitted barrier heights and positions (From 133.)

shown similar effects (129, 130, 232–234) that seem to be related to valence nucleon effects and to deformation effects (156, 158, 159, 235).

Theoretical considerations of sub-barrier fusion have included barrier penetration models (236–242), the incoming wave boundary condition model (40, 52, 243), the optical model (244), the WKB model (245), and a deformed nucleus, coupled channel model (246). For some systems barrier penetration models have been fitted to the data (55, 125, 128, 247–250) and the de-excitation of fusion residues in the sub-barrier region has been considered using the statistical model (51, 94). Specific models to explain the variation in the average behavior of the excitation function include a deformation model (251), a dynamic single-particle process (133), and a barrier penetration model including zero-point oscillations (252, 253).

2.4 Incomplete Fusion

The most detailed examinations of incomplete fusion reactions have come from measurements of characteristic gamma radiation from reaction residues in coincidence with light charged particles (254–258). On the basis of the excitation functions for ^{160}Gd (^{12}C,α) and ^{160}Gd (^{12}C,2α) reactions, Siwek-Wilczynska et al (257, 259) proposed that incomplete fusion reactions can be understood on the basis of the disappearance of a pocket in the one-dimensional internuclear potential energy as the angular momentum increases. In order to reduce the effective angular momentum of the composite nucleus and to restore a pocket into the internuclear potential energy as the entrance channel angular momentum is increased, an increasing fraction of the projectile escapes and carries away some of the angular momentum. This simple model has been extended to take into account phase space considerations in the formation of a compound nucleus (260, 261). The model of Siwek-Wilczynska et al implies that incomplete fusion processes are confined to the high angular momentum components of the fusion and fusion-like reactions. This hypothesis is supported by measurements of γ-ray multiplicities and yrast line γ-ray transitions in incomplete fusion reactions (254, 262–264). Other models proposed to explain incomplete fusion include projectile breakup followed by capture of one fragment (265), a classical trajectory model with random nucleon transfer included (266), and a fragmentation model (267, 268).

In an incomplete fusion reaction, since a large portion of the projectile is not captured by the target, there is a deficit in the linear momentum of the composite nucleus when compared to the projectile momentum. This incomplete momentum transfer may be observed directly for evaporation residues by direct measurement of their velocity (269, 269a, 270) and for fission fragments by the measurement of the correlation angle between the emitted fission fragments (187, 189, 191, 271–278). In such a fission fragment angular correlation measurement, the incomplete momentum transfer reactions show a larger correlation angle than do full momentum fission reactions. This is illustrated in Figure 9 for the ^{197}Au + ^{20}Ne reaction at a bombarding energy of 292 MeV (277). In this figure, the correlation angle spectrum for fission fragments is shown. The arrows indicate the expected fission fragment correlation angles for full fusion-fission at 135.5° and incomplete fusion at successively larger angles corresponding to the escape of larger fractions of the projectile mass composed of increasing numbers of alpha particles.

As can be seen from Figure 9 a significant fraction of the fusion-fission cross section is composed of incomplete fusion processes. If the lost linear momentum in an incomplete fusion process is confined to the reaction plane defined by the correlated fission fragments, then the observed out-of-plane width in the fission fragment angular correlation is due to the subsequent de-excitation of the fission fragments by particle evaporation. If this observed out-of-plane width is used as a measure of the in-plane width of the angular correlation, and Gaussian curves are fitted to the in-plane correlation spectrum at the positions indicated by the arrows, the dashed curves shown in Figure 9 are obtained. These Gaussian curves provide a good fit to the experimental data and indicate the possible presence of incomplete momentum transfer processes in which fragments as large as

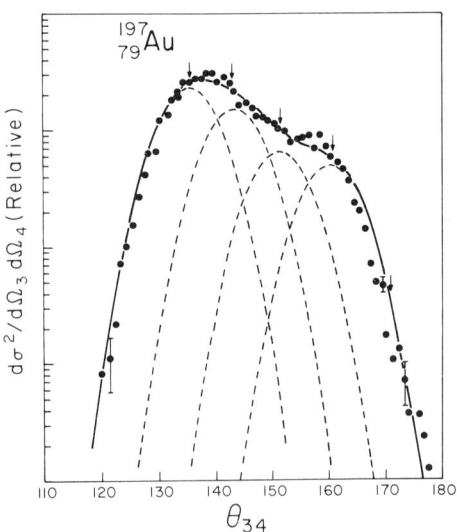

Figure 9 In-plane fission fragment correlation data for 292-MeV ^{20}Ne projectiles on ^{197}Au. The arrows from left to right indicate the predicted maximum for ^{20}Ne, ^{16}O, ^{12}C, ^{8}Be, and ^{4}He fusion at the same projectile energy per nucleon with the target. (From 277.)

^{12}C are not captured by the target. Measurements have also been made of charged particles in coincidence with correlated fission fragments (274–276). These measurements show that protons, deuterons, tritons, and alpha particles as well as alpha-cluster structured nuclei may be the uncaptured fragments in incomplete fusion reactions.

It has been suggested (279, 280) that the projectile velocity above the barrier v', is the important factor determining the threshold for incomplete fusion. However, the model of Siwek-Wylczynska et al (257) suggests that the total angular momentum is the fundamental factor determining the onset of incomplete fusion.

3. ONE-DIMENSIONAL ENTRANCE CHANNEL MODELS

One-dimensional entrance channel models of fusion all assume a spherically symmetric interaction potential between two spherical interacting nuclei. Thus, the potential energy, which contains nuclear, Coulomb, and centrifugal energy terms, may be expressed as a function of the nuclear radial separation only. Since the Coulomb and centrifugal energy terms are long range and arise from repulsive forces, the potential energy of the system initially rises as the nuclei approach each other. However, the short-range nuclear attractive forces may produce a local decrease or "pocket" in the potential energy as the nuclear matter density distributions of the target and projectile begin to overlap. At smaller interaction distances, the Coulomb and centrifugal potentials may again cause the total potential energy to rise if the total nuclear charge or angular momentum is sufficiently high. For high angular momentum, the pocket in the potential may be completely eliminated by the centrifugal force. In addition, for very high Z systems, no pocket will exist in the interaction potential at any value of the angular momentum.

The presence of the interaction potential energy pocket has an important influence on the trajectory of the interacting nuclei, and is crucial for fusion in the one-dimensional potential models. The one-dimensional potential also gives rise to the concept of a fusion potential energy barrier, which leads to the simplest expression for the fusion cross section, based on the conservation of energy and angular momentum at the barrier (281):

$$\sigma_f(E) = \pi R_B^2 (1 - V_B/E), \qquad 7.$$

where R_B and V_B are the radius and potential at the barrier top, and E is the center-of-mass kinetic energy of the system. This expression has often been used to determine barrier parameters from fusion cross sections (85, 282).

3.1 Critical Distance Models

Simple models based on energy and angular momentum conservation at the one-dimensional potential barrier predict the fusion cross section quite well for heavy ion reactions at energies not too far above the barrier, at least for systems that are not too heavy. However, as the projectile energy rises, the fusion cross section falls below the predictions of such a simple model. A model proposed to explain the fusion cross section magnitudes at high energies is the critical distance model (10, 283–286) in which it is assumed that all relative motion of the two nuclei is completely damped once the trajectories reach a critical separation distance. This critical distance is usually inside the fusion barrier. At near barrier projectile energies, the angular momentum is so low that once the system has crossed the barrier, it is not prevented by the potential energy from reaching the critical distance. Thus, at low energies the fusion cross section is still determined by the barrier radius and potential energy. At higher energies, because of the higher angular momentum reached by some trajectories, there will be trajectories that cross the barrier, but are prevented by the centrifugal potential from reaching the critical distance. At these high energies the radius and potential energy at the critical distance are the determining factors for the magnitude of the fusion cross section, which is lower than would be predicted by the barrier parameters.

A variation of the critical distance model assumes that fusion will occur only if the one-dimensional effective internuclear potential has a pocket that can trap the interacting nuclei (287). A refinement of this "pocket" model includes the effect of angular momentum transfer from orbital to intrinsic spin motion, and generally assumes that the target and projectile reach the rolling condition at the barrier (288–291).

These models predict a nonzero fusion cross section at very high energies and require adjustment of critical distance parameters for each system. Apart from an attempt to calculate the critical distance on the basis of the two-centered shell model (286), no comprehensive microscopic description of the critical distance exists.

3.2 One-Dimensional Potential Dynamical Models

Several dynamical calculations have been proposed for the fusion reaction. Generally these models are based on a one-dimensional effective conservative potential combined with dissipative forces and utilizing classical equations of motion (11, 266, 292–301). A one-dimensional potential quantum mechanical model has also been considered (302).

The differences between the classical models lie mainly in the choice of conservative and dissipative forces. Choices available for the nuclear potentials include the proximity potential (303–305), the Krappe-Nix folding potential (293, 306), the energy density potential (307), and the optical potential (308). Coulomb potentials that have been used include the point charge potential, the potential of Bondorf et al (309), and point plus distributed charge potentials (11). Dissipative forces include the proximity one-body friction (310) and surface friction (293).

Generally the equations of motion of these models take the form

$$\mu\ddot{r} = -\frac{\partial V_C}{\partial r} - \frac{\partial V_N}{\partial r} + \mu r\dot{\theta}^2 - f_r(r)\dot{r} \qquad 8.$$

$$\mu r^2 \dot{\theta} = L = L_0 - L_T - L_P \qquad 9.$$

$$\frac{dL_T}{dt} = -C_T \left[\frac{r}{C_T + C_P}\right]^2 f_\theta(r) [C_T(\dot{\theta}_T - \dot{\theta}) + C_P(\dot{\theta}_P - \dot{\theta})] \qquad 10.$$

$$\frac{dL_P}{dt} = -C_P \left[\frac{r}{C_T + C_P}\right]^2 f_\theta(r) [C_T(\dot{\theta}_T - \dot{\theta}) + C_P(\dot{\theta}_P - \dot{\theta})]. \qquad 11.$$

The degrees of freedom in this model are r, the radial separation of the nuclear centers; θ the angular orientation of the vector between the nuclear centers; and θ_P and θ_T, the angular orientation of the target and projectile, respectively. The conservative potentials are $V_C(r)$ for the Coulomb force, and $V_N(r)$ for the nuclear force. The first three terms of Equation 8 define the effective potential. The dissipative forces are expressed as radially dependent tangential and radial form factors $f_\theta(r)$ and $f_r(r)$, respectively, in a velocity-dependent force determined by the radial velocity (for the radial force) and the relative surface velocity (for the tangential force). Such a dissipative force is a sliding friction. The dimensions of the nuclei are given by their respective matter density radii (311) and the angular momentum in the relative and spin motion is represented by L, L_P, and L_T, while L_0 is the asymptotic entrance channel orbital angular momentum. The reduced mass of the system is μ. Models with minor differences in the choice of nuclear size parameters (301) and the inclusion of rolling friction (312) have also been proposed. The constants required in the equations can be determined from the liquid drop model. No further adjustable parameters are needed to obtain broad agreement between the model and excitation function data collected over a wide mass range (11).

Calculations have been performed with the proximity nuclear potential given by (303–305)

$$V_N(\zeta) = 4\pi\gamma b \bar{R} \Phi(\zeta) \qquad \zeta > 0 \qquad 12.$$

$$V_N(\zeta) = 4\pi\gamma b \bar{R} \Phi(\zeta = 0) + \gamma(4\pi\bar{R}b\zeta) \qquad \zeta \leq 0 \qquad 13.$$

where $\bar{R} = C_T C_P/(C_T + C_P)$. The surface energy coefficient $\gamma = 0.9517 \times$

$(1-1.7826I^2)$, in units of MeV fm^{-2}, where $I = (N-Z)/A$ and N, Z, and A refer to the combined system of the two interacting nuclei. The values of C_i (C_P and C_T) are calculated from the equivalent sharp surface radii R_i by the relations

$$R_i = 1.28A_i^{1/3} - 0.76 + 0.8A_i^{-1/3} \qquad 14.$$

and

$$C_i = R_i[1 - (b/R_i)^2 \ldots], \qquad 15.$$

where b is a constant related to the surface diffuseness with a value of approximately 1 fm. The indefinite integral of the universal interaction potential $\Phi(\zeta)$ (303) depends on the separation distance ζ of the two surfaces, where ζ is given by

$$\zeta = (r - C_T - C_P)/b. \qquad 16.$$

The frictional form factors are given by

$$f_r(r) = 4\pi n_0 \bar{R} b \Psi(\zeta) \qquad 17.$$

$$f_\theta(r) = 2\pi n_0 \bar{R} b \Psi(\zeta). \qquad 18.$$

In the present calculations we use Randrup's tabulated values (310) of the universal proximity function $\Psi(\zeta)$, which represents the form factor for energy dissipation. This model assumes that the interacting nuclei are spherical and that the internuclear potential is spherically symmetric, which allows the equations of motion to be expressed in the form of Equations 8–11. Fusion of the target and projectile occurs because the nuclei are trapped in the pocket existing in the effective potential if the angular momentum is not too high. It should be noted, however, that angular momentum transfer from relative to spin motion is an important factor in determining if the system will fuse. Whether or not fusion will occur cannot be determined by inspection of the conservative potential alone (296).

A comparison to data is shown in Figure 10 of a calculated fusion excitation function using the one-dimensional proximity potential and proximity friction. The data are for the ^{27}Al + ^{20}Ne reaction (100) over a wide energy range, with corrections made at the higher energies for incomplete momentum transfer. These data yield the best determined excitation function to date in this mass region. As can be seen from the figure, the calculations are in very good agreement with the data (313).

The parameters of the liquid drop model applied to classical trajectory calculations give good agreement with the macroscopic behavior of fusion excitation functions for a broad range of masses at energies above the fusion

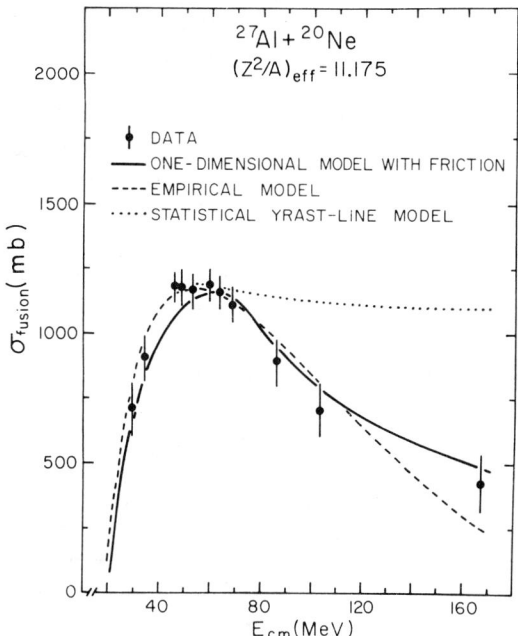

Figure 10 Experimental and theoretical fusion excitation functions for the ^{27}Al+^{20}Ne reaction. Fusion data (see Table 1) include corrections for incomplete momentum transfer. (From 313.)

barrier. However, for some light systems, there may be effects unaccounted for by the liquid drop model (21, 47, 80, 106). Examples of this are shown in Figure 4 where the excitation functions for the formation of ^{26}Al by two entrance channels are shown. The excitation function calculated by the classical trajectory model is shown in each case by the crosses. As can be seen, the model fails in the high energy region. The explanation for this failure is at present unknown.

In addition, the classical model fails for very heavy systems where it greatly overestimates the cross section. This has been attributed to effects of neck formation between the two interacting nuclei during the reaction, and is discussed further below.

3.3 *Fusion Barriers and Limiting Angular Momenta*

The fusion barrier radius and potential energy in the one-dimensional potential energy model may be measured experimentally by application of Equation 7 to fusion excitation functions near but still above the barrier (85, 103, 282). Such measurements, in conjunction with assumptions about the Coulomb potential, have also been used to extract information concerning the value of the nuclear potential in the barrier region (103, 295, 314). These procedures confirm that the proximity potential is a reasonable description of the nuclear potential at radii equal to or larger than the barrier radius; but there are considerable difficulties in using such a simple analysis if precise potential measurements are required (295). The values of the fusion barrier parameters may be approximately calculated from the properties of the liquid drop model (11, 287, 289), or they may be calculated from assumed nuclear and Coulomb potentials. Barrier parameters are shown in Table 1, calculated from the proximity potential combined with the Bondorf Coulomb potential. The systematics of barrier parameters have recently been discussed (12, 315).

As the angular momentum of the system increases, the centrifugal potential begins to reduce the depth of the pocket in the one-dimensional effective potential until, at some value of the angular momentum, the pocket disappears. This value of the angular momentum is also tabulated in Table 1. The maximum entrance channel angular momentum for fusion reactions l_f^{max} is larger than the highest value of angular momentum for which there is a pocket in the one-dimensional potential. This arises because of the effects of tangential friction between the two nuclei, which reduces the angular momentum in relative motion by transfer to intrinsic spin. If the rolling condition is reached then $l_f^{max} = (7/5)l_{pocket}$, which gives a good estimate of l_f^{max}. Wilczynski (287) has also proposed expressions for l_f^{max} based on the liquid drop model. The maximum cross section for fusion and the energy at which it is reached, may be obtained from an empirical expression (11) for cases in which $Z_p Z_T < 1700$.

4. OTHER FUSION MODELS

A compound nucleus limitation model, and an empirical model based on a power series in the projectile energy, are discussed below, together with a dynamical model based on time-dependent Hartree-Fock (TDHF) methods, which attempt to describe heavy nucleus interactions with the time-dependent mean field calculated from a knowledge of internucleon forces.

4.1 *Yrast or Compound Nucleus Limitation Models*

As has been discussed by Lefort (316) and Harar (317), the absence of nuclear levels in the compound nucleus at spins below the Yrast line may limit the fusion cross section at high projectile energies. Glas & Mosel (318–320) have shown, in a calculation of the Yrast line for s-d shell nuclei, that the Yrast line angular momenta were not limiting experimentally measured fusion cross sections. However, Lee et al (321) suggested that a critical level density was required before fusion could occur. Vandenbosch & Lazzarini (23) suggested that this critical level density is reached when the width and separation of the levels are equal. This idea has led to the statistical Yrast line model, which postulates a limiting fusion angular momentum line parallel to, but at lower angular momentum than, the Yrast line. In this model, the near barrier region of the excitation function is given by the simple barrier Equation 7, but the higher energy cross sections are limited by the density of available levels. The fusion cross section at high energy is given by

$$\sigma_f = \frac{\pi \mathscr{I}}{\mu}[1+(Q-\Delta Q)/E], \qquad 19.$$

where $\mathscr{I}$ is the moment of the inertia of the composite system, μ is the reduced mass of the target-projectile system, E is the center-of-mass energy, Q is the fusion Q value, and ΔQ is the energy shift between the Yrast and statistical Yrast lines, taken to be 10 ± 2.5 MeV by Lee et al (83).

Some data have been compared to the model by Lee et al (83). Although the data and the model agree near the peak of some fusion excitation functions, at higher energies the model clearly overestimates the fusion cross section. A comparison of the statistical Yrast model to the ^{27}Al $+^{20}$Ne fusion excitation function is shown in Figure 10, where the disagreement with the data is clearly shown at high energies. Recently, Matsuse et al (322) proposed that the high energy region of the excitation function be treated as if a critical distance was also required for fusion, thereby introducing additional parameters to preserve agreement between model and data.

4.2 Empirical Models

Horn et al (79, 323) have suggested that the fusion cross section may be expressed as a power series in the energy of the projectile in the center-of-mass reference frame. The expression for the cross section is

$$\sigma_f = 10\pi[-1.438Z_\mathrm{P}Z_\mathrm{T}bE^{-1}+(b^2-1.438Z_\mathrm{P}Z_\mathrm{T})+2bmE+m^2E^2], \qquad 20.$$

where Z_P and Z_T refer to the projectile and target proton numbers, and E is the projectile energy in MeV. Lozano & Madurga (324) have suggested slightly different parameters for the equation. This model, adjusted to fit a variety of fusion excitation functions, agrees well with a large amount of data, an example of which is shown in Figure 10 (dashed curve).

4.3 Time-Dependent Hartree-Fock Models

Time-dependent Hartree-Fock (TDHF) models have been extensively used for the theoretical investigation of the reactions between heavy nuclei, and have also been specifically applied to the fusion reaction in a number of cases (325–344). Early calculations were restricted to the interaction of infinite slabs of nuclear matter, but recent calculations have been conducted with a fully three-dimensional geometry, including deformations of the nuclear fragments. This model has been mainly used as a tool for theoretical investigations of heavy nucleus interactions, although good agreement with fusion cross sections has been achieved in a number of cases. Several interesting effects, which have been investigated experimentally, are predicted by TDHF calculations. These include a lower angular momentum cutoff in the fusion cross section, which has been called an "l window" (325, 326, 333, 334, 345), and emission of alpha particles at the point of scission (329). Experimental evidence and theoretical justifications for the "l window" have been sought (20, 61, 168, 346–353a) but the existence of such an effect is not established. There is, however, some evidence for the emission of charged particles from the neck joining the two fragments in fusion-fission reactions (354).

The application of TDHF models requires considerable computer resources, which has restricted the application of the model in fusion investigations. However, Bertsch (355) has suggested a simpler model, including the major features of TDHF models, that may allow wider application of TDHF-like models to fusion reactions.

5. A NEW ENTRANCE CHANNEL LIMITATION

For some time calculations have been performed of fusion cross sections expected from models that include a multidimensional potential energy surface (356–361). This allows for the deformation of projectile, target, and composite system. Recently, Swiatecki (2, 3) introduced a simple model based on such a multidimensional dynamic calculation with one-body dissipation and the effects of neck formation between the interacting nuclei. This theory has been successful in reproducing a wide variety of measured fusion excitation functions for very heavy systems where the one-dimensional potential models fail. The potential energy used in the Swiatecki model is based on the nuclear proximity potential between two deformable nuclear liquid drops. An example of the total potential energy surface, including Coulomb and nuclear potential, is shown in Figure 11 for the interaction of two nuclei of mass $A = 104$ and $Z = 41$. The potential is shown by dotted contour lines in the figure, where the contours are defined in terms of potential energy units of magnitude $4\pi R^2 \gamma$, where R is the projectile radius and γ the nuclear surface tension. The potential is a function of the separation of the nuclear centers ρ, the radius of the neck α, and the mass asymmetry of the projectile- and target-like fragments. In

Figure 11, however, the potential is shown for a mass asymmetry fixed at the entrance channel mass asymmetry, with ρ and α in units of the projectile radius R. The potential surface shows a saddle point at a potential value of 0.033.

The effects of the potential saddle on the trajectory of the system are also illustrated in Figure 11 for three dynamically calculated trajectories with entrance channel angular momentum $l = 0$ and for different projectile energies above the spherical touching point barrier. The two lower energy trajectories move along the abscissa until the touching point is reached at $\rho = 1$, when the rapid growth of the neck forces them away from the spherical touching point. Because of the dynamical influence of the potential energy surface, neither of the lower energy trajectories pass over the saddle point, although simple energy considerations would suggest that they should; rather they experience an increase in the neck size to a maximum value, and then a decrease as the nuclei move apart. For the third trajectory with an energy higher above the spherical barrier than the other two, the growth of the neck does not force the system away from the saddle point, but rather the system moves behind the saddle and experiences a large increase in reaction time. The times are indicated on the trajectories by filled dots, each separated in time by 1.2×10^{-22} s. The reaction time can

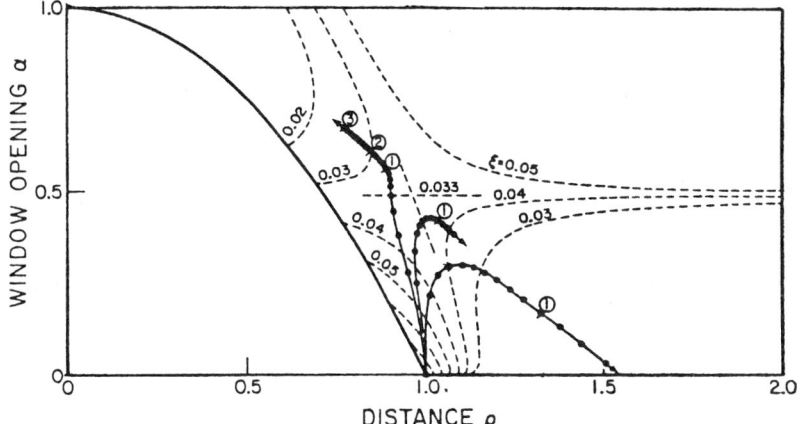

Figure 11 Example of a dynamical study of symmetric collisions between two medium-weight nuclei with $A_1 = A_2 \approx 104$ and $Z_1 = Z_2 \approx 41$. The dashed lines are equipotentials, with energy expressed in units of the surface energy of a single sphere (about 583 MeV) and with reference to the energy of the spherical configurations. The location of the saddle point is indicated by the crossing equipotentials in the α-ρ plane. Two of the trajectories, with energies 0 and 3.6 MeV above the interaction barrier, fail to fuse and reseparate after dissipating some energy. The third trajectory, with an energy of 10 MeV above the barrier, fuses. The dots are spaced by 1.2×10^{-22} s. The circled numbers represent time intervals of 1.2×10^{-21} s. (From 2.)

be inferred from the numbered circles, which give the total time from the spherical touching point in units of 1.2×10^{-21} s.

Thus, the saddle point, because of its dynamical influence on the system trajectories, forces the requirement of an additional projectile energy, above the spherical touching point barrier, before fusion can occur. For a configuration that always remains outside the saddle point, the neck size remains relatively small and a dinucleus results, producing target- and projectile-like fragments. This is the dinucleus shape region. For trajectories passing behind the saddle point, the neck may grow in size until the distinction between the two nuclei is eliminated. Such trajectories lead to mononucleus reactions.

For asymmetric systems there are two saddle points in the potential surface. One of these, the conditional saddle point, is not a true saddle, since the potential energy there is not stationary in the mass asymmetry coordinate. The other saddle, a stationary point for all coordinates in the potential surface, is a true saddle point called the unconditional, or fission, saddle point. For heavy asymmetric systems, the conditional saddle is more extended than the fission saddle. In such a case, if a trajectory passes behind the conditional saddle but does not pass inside the fission saddle, the system cannot be trapped, even if energy is lost from the relative motion, because a change of mass asymmetry leads to trajectories in which the nuclei separate. Behind the fission saddle, however, if sufficient relative energy is lost, the system will be trapped.

The dynamical influence of the two saddle points leads to the requirement that the projectile energy be above that necessary to reach the spherical touching point if the system is to move behind the saddle points. These additional energies are called the "extra push" to move the system behind the conditional saddle, and the "extra extra push" to move the system behind the unconditional saddle. The calculation of these energies and the conditions when they are necessary, according to the Swiatecki model, are discussed below.

It has also been suggested (362, 363) that fusion excitation functions for all values of $(Z^2/A)_{\text{eff}}$ can be reproduced by a new version of the surface friction model that includes statistical fluctuations and dynamical deformations of the interacting nuclei. However, the remainder of this section is limited to the Swiatecki model.

5.1 *The Conditional Saddle and the Extra Push Energy*

The important parameter in the determination of the additional energy above the spherical barrier required to cross the conditional saddle is $(Z^2/A)_{\text{eff}}$, defined by:

$$(Z^2/A)_{\text{eff}} = 4Z_1 Z_2 / A_2^{1/3} A_1^{1/3} (A_1^{1/3} + A_2^{1/3}) \qquad 21.$$

where Z_1, A_1 and Z_2, A_2 refer to the atomic number and mass of the projectile and target. The angular-momentum-dependent extra push is then given by

$$E_x(l) \begin{cases} = 0 \quad \text{for } (Z^2/A)_{\text{eff}} + (fl/l_{\text{ch}})^2 \leq (Z^2/A)_{\text{eff thr}} \\ = K[(Z^2/A)_{\text{eff}} + (fl/l_{\text{ch}})^2 - (Z^2/A)_{\text{eff thr}}]^2 + \cdots \\ \quad \text{for } (Z^2/A)_{\text{eff}} + (fl/l_{\text{ch}})^2 > (Z^2/A)_{\text{eff thr}} \end{cases} \qquad 22.$$

where

$$l_{\text{ch}}^2 \begin{cases} = (e^2 m r_0/4\hbar^2)[A_1^{4/3} A_2^{4/3}(A_1^{1/3} + A_2^{1/3})^2/(A_1 + A_2)] \\ = 0.01055 A_1^{4/3} A_2^{4/3}(A_1^{1/3} + A_2^{1/3})^2/(A_1 + A_2) \end{cases} \qquad 23.$$

and

$$K \begin{cases} = (32/2025)(3/\pi)^{2/3}(e^4 m/\hbar^2) a^2 [A_1^{1/3} A_2^{1/3}(A_1^{1/3} + A_2^{1/3})^2 \\ \quad \div (A_1 + A_2)] \text{ MeV} \\ = 7.601 \times 10^{-4} a^2 [A_1^{1/3} A_2^{1/3}(A_1^{1/3} + A_2^{1/3})^2/(A_1 + A_2)] \text{ MeV}. \end{cases} \qquad 24.$$

Here, m is the nuclear mass unit 931 MeV/c^2, and r_0 is the nuclear radius constant 1.225 fm. The angular momentum l is measured in units of $\hbar$.

There is a threshold below which the extra push energy is zero. This is given by

$$(Z^2/A)_{\text{eff thr}} = b\{1 - 1.7826[(N_1 - Z_1 + N_2 - Z_2)/(A_1 + A_2)]^2\}. \qquad 25.$$

Thus, as can be seen from Equation 22, if the angular momentum is sufficiently high, then all systems require a nonzero extra push. For systems in which $(Z^2/A)_{\text{eff}}$ is greater than $(Z^2/A)_{\text{eff thr}}$, an extra push is already required for $l = 0$.

The adjustable parameters in the expressions for the extra push are a, b, and f. The coefficients a and b have been empirically determined (364) to be $a = 12 \pm 2$ and $b = 35 \pm 1$. The quantity f represents the fraction of the entrance channel angular momentum remaining in the orbital motion after contact. We consider three possible values for f, corresponding to sliding, rolling, and sticking collisions. For spherical nuclei these, respectively, correspond to f values of 1, 5/7, and

$$(A_1^{1/3} + A_2^{1/3})^2 / \{(A_1^{1/3} + A_2^{1/3})^2 + \frac{2}{5}(A_1 + A_2)[A_1^{2/3} A_2^{-1} + A_2^{2/3} A_1^{-1}]\}.$$

The fusion cross section, at energies where the extra push is nonzero, is then given by (3)

$$\sigma_{\text{fus}} = \frac{\pi R_B^2}{E} \left[-\frac{\alpha_x \beta_x + 1/2}{\beta_x^2} + \left\{ \left(\frac{\alpha_x \beta_x + 1/2}{\beta_x^2} \right)^2 - \frac{\alpha_x^2 + V_B - E}{\beta_x^2} \right\}^{1/2} \right] \qquad 26.$$

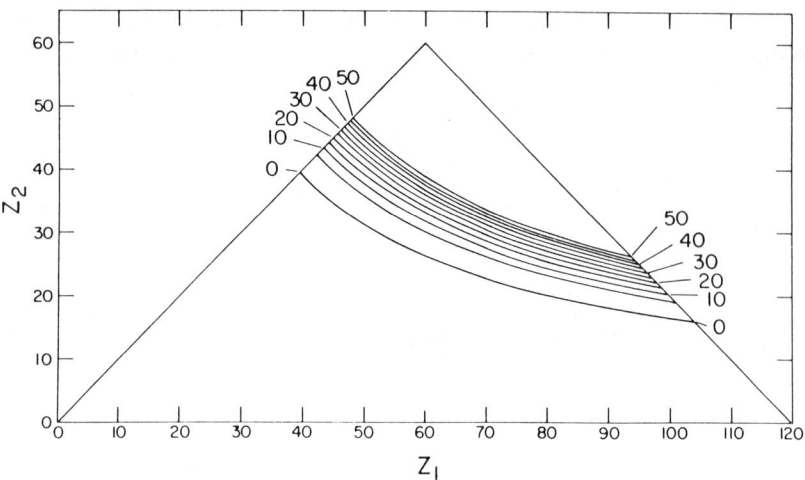

Figure 12 Contour lines of the extra push energy in MeV. (From 364.)

where

$$\alpha_x = K^{1/2}\left[\left(\frac{Z^2}{A}\right)_{\text{eff}} - \left(\frac{Z^2}{A}\right)_{\text{eff thr}}\right] \text{MeV}^{1/2} \qquad 27.$$

and

$$\beta_x = 8K^{1/2}(r_0/e^2)f^2/(A_1^{1/3}A_2^{1/3}). \qquad 28.$$

The quantities K, $(Z^2/A)_{\text{eff}}$ and $(Z^2/A)_{\text{eff thr}}$ are defined by Equations 24, 21, and 25, respectively, and $(e^2/r_0) = 1.1755$ MeV. At energies where the extra push is zero, the cross section in this model is calculated by the simple barrier formula given by Equation 7. The contour lines of extra push energy (364) for $l = 0$, are shown in Figure 12 as a function of the target and projectile Z. The nuclei represented in Figure 12 are assumed to be on the valley of stability where $(N-Z) = 0.4A^2/(200+A)$. This figure shows the rapid increase of the extra push as Z_1Z_2 increases.

5.2 The Unconditional Saddle and the Extra Extra Push Energy

The energy above the spherical touching barrier required to reach the unconditional saddle has been parameterized in a manner similar to the extra push energy (364). The main parameter determining whether the extra extra push energy is required is $(Z^2/A)_m$ defined by

$$(Z^2/A)_m = \left(\frac{Z^2}{A}\right)\left[\left(\frac{Z^2}{A}\right)_{\text{eff}}\right]^{1/2} - k\left(\frac{Z^2}{A}\right)\left\{1 - \left[\left(\frac{Z^2}{A}\right)_{\text{eff}} \bigg/ \left(\frac{Z^2}{A}\right)\right]^{1/2}\right\}^2. \qquad 29.$$

The "extra extra push" energy for central collisions ($l = 0$) is now given by

$$E_{xx} \begin{cases} = E_x & \text{for } (Z^2/A)_m \leq 0.84(Z^2/A)_{crit} \\ = K[(Z^2/A)_m - (Z^2/A)_{eff\ thr}]^2 & \text{for } (Z^2/A)_m > 0.84(Z^2/A)_{crit}, \end{cases} \quad 30.$$

where

$$(Z^2/A)_{crit} = 50.833\{1 - 1.7826[(N_1 - Z_1 + N_2 - Z_2)/(A_1 + A_2)]^2\}. \quad 31.$$

Contour lines are shown in Figure 13 for the extra extra push energy as a function of Z_1 and Z_2 for nuclei lying on the valley of beta stability. In the region below the cliff shown in Figure 13, the contour lines are identical with the extra push shown in Figure 12. Above the cliff the contours are given by Equation 30 with $k = 1$. However, neither $(Z^2/A)_m$ nor k have as yet been experimentally determined. When an extra extra push is required, the compound nucleus cross section is given by Equation 26, with α_x and β_x replaced by α_{xx} and β_{xx} defined by (364)

$$\alpha_{xx} = K^{1/2}\left[\left(\frac{Z^2}{A}\right)_m - \left(\frac{Z^2}{A}\right)_{eff\ thr}\right] \text{MeV}^{1/2} \quad 32.$$

$$\beta_{xx} = 8K^{1/2}(r_0/e^2)f_m^2/(A_1^{1/3} + A_2^{1/3}) \text{ MeV}^{-1/2}. \quad 33.$$

The additional parameter f_m is given by

$$f_m = [\eta(x)]^{1/4}f^{1/2}(25/24)A_1^{1/3}A_2^{1/3}(A_1^{1/3} + A_2^{1/3})^{1/2}(A_1 + A_2)^{-5/6}. \quad 34.$$

The values $\eta(x)$ as a function of the parameter $x = (Z^2/A)/(Z^2/A)_{crit}$ are tabulated in Table 2 of Reference (364). The relationships between the extra push and extra extra push energies and the reaction regions defined by the

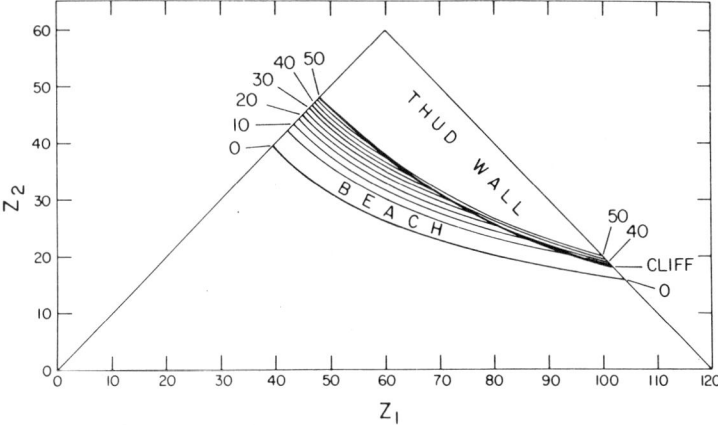

Figure 13 Contour lines of equal extra extra push energies in MeV. In the region below the cliff the contour lines are identical to those in Figure 12. (From 364.)

conditional and unconditional saddle points are shown in Figure 2. Bjornholm & Swiatecki (364) have considered the effects of the extra extra push on the production of superheavy nuclei.

5.3 Experimental Results for Systems with Large $(Z^2/A)_{eff}$

Considerable experimental data currently exist for systems with $(Z^2/A)_{eff}$ above the threshold for the extra push defined by Equation 25. These systems are predicted to show a fusion threshold higher than that expected from the spherical potential barrier, and cross sections below the expectations of the simple barrier Equation 7, and below the predictions of trajectory calculations with the one-dimensional potential. For lighter systems, when the angular momentum is high, the Swiatecki model also predicts that an extra push energy will be required, as can be seen from Equation 22.

Comparison of the model with data for systems with large $(Z^2/A)_{eff}$ is shown in Figures 14 and 15, with the model parameters discussed in Section 5.1, for sliding (dotted lines), rolling (dashed lines), and sticking (dash-dotted lines) collisions. The solid lines in the figures show the predictions of the simple barrier formula, Equation 7. For the systems with $(Z^2/A)_{eff} > 35$ (see Equation 25), shown in Figures 14 and 15, the model predicts that the excitation functions will show evidence of the extra push energy at the fusion threshold. This effect is confirmed for the ^{50}Ti + ^{208}Pb and ^{64}Ni + ^{208}Pb reactions, and may be true for the ^{165}Ho + ^{84}Kr reaction, where the cross sections were measured in very early experiments. The presence of an extra push at high angular momentum for lower values of $(Z^2/A)_{eff}$ is indicated for the ^{165}Ho + ^{40}Ar reaction in Figure 14 and also, although less

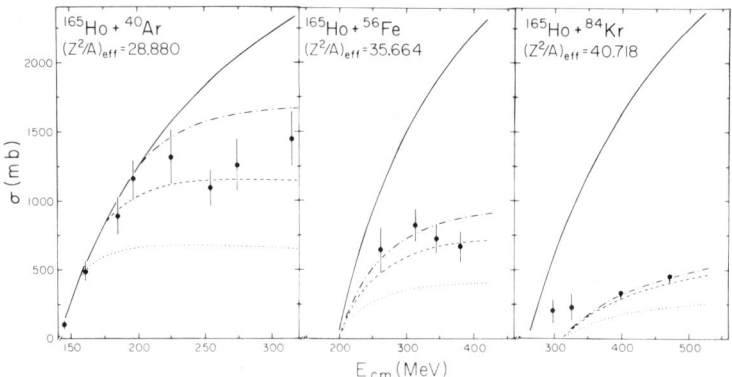

Figure 14 Comparison of experimental and theoretical fusion cross sections for three projectiles on a ^{165}Ho target. See captions of Figures 4 and 10 for identification of the curves. References to the data are in Table 1.

certainly, by the ^{26}Mg + ^{208}Pb reaction shown in Figure 15. In both of the latter cases, there is no evidence for, and the model does not predict, an extra push energy near the fusion threshold. In all of the cases shown in Figures 14 and 15 the rolling collision calculation seems to give the best agreement with the data.

For the systems in Figures 14 and 15 with $(Z^2/A)_{\text{eff}} > 30$, the one-dimensional potential trajectory calculations cannot reproduce the data. However, as shown in Figure 5 by the crosses, the one-dimensional potential calculations do agree well with the ^{116}Sn + ^{35}Cl fusion reaction excitation function even though $(Z^2/A)_{\text{eff}}$ is as high as 26.157. Experimental data for this reaction at higher energies would be useful in the comparison of the two models. For the ^{116}Sn + ^{35}Cl reaction shown in Figure 5, the Swiatecki model for rolling collisions is also in good agreement with the data. This is not true for the ^{62}Ni + ^{35}Cl reaction with $(Z^2/A)_{\text{eff}} = 20.343$, where the Swiatecki model for rolling collisions predicts too large a cross section. Also shown in Figure 5 are the Swiatecki model predictions for sliding collisions and, in the case of ^{116}Sn + ^{35}Cl, for sticking collisions. For the ^{62}Ni + ^{35}Cl reaction an adjustment of the angular momentum transfer parameter f will bring the model into better agreement with the data, but the agreement will not be superior to the classical trajectory calculation.

The question of whether the extra push is required for systems with small values of $(Z^2/A)_{\text{eff}}$ is considered in Figures 4 and 16 where the data are compared for sliding (dotted lines) and rolling (dashed lines) collisions in the Swiatecki model. The sliding collision calculation for ^{27}Al + ^{20}Ne is close to, but overestimates, the data at high energy and is not as good as the agreement between the data and the trajectory model shown in Figure 10. For the data shown in Figure 4, none of the model curves shown reproduce

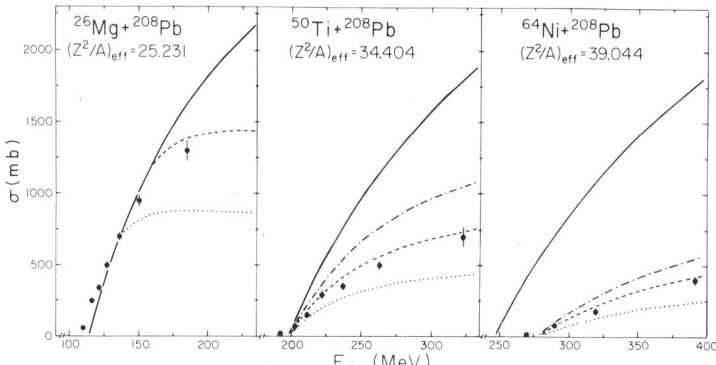

Figure 15 Comparison of experimental and theoretical fusion cross sections for ^{208}Pb-induced fission of three targets. See caption of Figure 14.

the data well. Bertsch & Mundinger (364a) suggested on theoretical grounds that above a critical velocity nuclear matter is insufficiently cohesive for fusion to occur. It is observed for lighter heavy ion systems with $(Z^2/A)_{\text{eff}} < 30$, the fusion cross section begins to deviate from Equation 7 at values of $(E - V_B)/\mu \geq 2.0 \pm 0.3$ MeV per nucleon mass (313). With this information it is possible to adjust the parameter l_{ch} (Equation 23), which controls the effect of angular momentum on the extra push energy, by normalizing the calculated cross section of the Swiatecki model at the point where the excitation function departs from the simple barrier prediction. Such a calculation, shown for the $^{27}\text{Al} + ^{20}\text{Ne}$ reaction by the crosses in Figure 16, does not substantially improve the agreement with the data for this reaction. Additional comparisons of the extra push model to data have also been made (365–367).

The experimental evidence on the extra extra push energy for heavy systems is essentially nonexistent. However, a very small yield of evaporation residue nuclei was observed in ^{54}Cr and ^{58}Fe bombardments (368, 369) of ^{209}Bi that led to the discovery of elements 107 and 109. The predictions concerning the formation of superheavy elements remains very uncertain. Not only is it very problematic that a compound nucleus can be

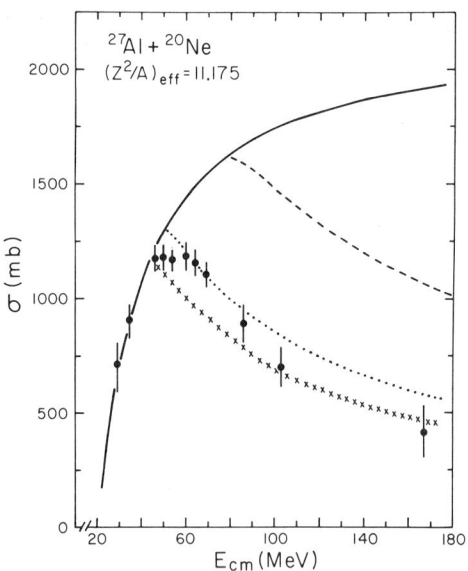

Figure 16 Same as Figure 9 except theoretical fusion excitation functions are based on the Swiatecki "extra push" model. The dashed and dotted curves correspond to rolling and sliding collisions, respectively. The crosses utilize a value of β_x normalized to experimental data. (From 313.)

formed in this mass region, but the high probability of fission of such a compound nucleus makes the likelihood of producing superheavy nuclei by heavy ion reactions very small.

6. SUMMARY AND CONCLUSIONS

The classification of the reactions between heavy nuclei and the definition of fusion are discussed in Section 1, where it is pointed out that fission fragments or evaporation residues, or both, may result in the exit channel of a fusion reaction. The measurement of fusion cross sections by observation of these exit channels is complicated by the presence at high energies of the incomplete fusion process discussed in Section 2.4. Although the threshold for incomplete fusion is not well established experimentally, the most developed model of the process relates its occurrence to high angular momentum in the entrance channel. More experimental measurements of the evaporation residue velocities and fission fragment angular correlations are needed to correct existing fusion excitation functions for the effects of incomplete fusion.

Several models for the calculation of the fusion excitation function are discussed in Sections 3 and 4. It is possible, however, to reproduce almost all of the available above barrier excitation functions with a model based on the systematic behavior of the liquid drop model, together with a one-body dissipation mechanism. In its one-dimensional potential form for spherical nuclei, as discussed in Section 3.2, this model reproduces the excitation functions for fusion of systems with $(Z^2/A)_{\text{eff}} \lesssim 27$ (see, for example, Figures 5 and 10). As $(Z^2/A)_{\text{eff}}$ increases above 27 the one-dimensional potential model increasingly overestimates the fusion cross section. When the effects of neck deformations are included in the liquid drop one-body dissipation model, as discussed in Section 5, then the model reproduces available data for systems with $(Z^2/A)_{\text{eff}} > 27$, as can be seen from Figures 14 and 15.

Further experimental evidence for the effect of neck formation has been sought in the mass and angular distributions of fission fragments, as discussed in Section 2.2. A clear signature for such effects is difficult to extract from mass and angular distributions, in part because of the uncertain theoretical basis for their interpretation. In the case of fission fragment angular distributions, the rotating liquid drop model is not in agreement with the data, but unambiguous interpretation of the anomaly in terms of nuclear shapes is not yet possible.

Additional features of fusion reactions not well understood include the phenomenon of sub-barrier fusion, which shows systematic behavior in disagreement with the liquid drop model and is discussed in Section 2.3, and the systematic behavior of the excitation function of some light systems, an example of which is shown in Figure 4.

Literatured Cited

1. Swiatecki, W. J. 1979. *Int. Sch. Nucl. Phys., Sicily, LBL-8950.* Berkeley: LBL
2. Swiatecki, W. J. 1981. *Phys. Scr.* 24:113
3. Swiatecki, W. J. 1982. *Nucl. Phys.* A376:275
4. Schröder, W. U., Huizenga, J. R. 1983. *Heavy Ion Science,* ed. D. A. Bromley. New York: Plenum
5. Adler, L., Gonthier, P., Ho, J. H. K., Khodai, A., Namboodiri, M. N., Natowitz, J. B., Simon, S. 1980. *Phys. Rev. Lett.* 45:696
6. Hofmann, H., Siemens, P. J. 1975. *Phys. Lett.* 58B:417
7. Seglie, E., Sperber, D. 1975. *Phys. Rev.* C12:1236
8. Natowitz, J. B., Namboodiri, M. N., Kasiraj, P., Eggers, R., Adler, L., Gonthier, P., Cerruti, C., Alleman, T. 1978. *Phys. Rev. Lett.* 40:751
9. Cormier, T. M. 1982. *Ann. Rev. Nucl. Sci.* 32:271
10. Lefort, M. 1980. *Heavy Ion Collisions,* ed. R. Bock, 2:46. Amsterdam: North-Holland
11. Birkelund, J. R., Tubbs, L. E., Huizenga, J. R., De, J. N., Sperber, D. 1979. *Phys. Rep.* 56:107
12. Vaz, L. C., Alexander, J. M., Satchler, G. R. 1981. *Phys. Rep.* 69:373
13. Krappe, H. J. 1979. *Proc. Symp. Deep-Inelastic Fusion Reactions With Heavy Ions, Lect. Notes Phys.,* 117:312. Berlin: Springer-Verlag
14. Bass, R. 1979. See Ref. 13, 117:281
15. Zebelman, A. M., Miller, J. M. 1973. *Phys. Rev. Lett.* 30:29
16. Kozub, R. L., Logan, D., Miller, J. M., Zebelman, A. M. 1974. *Phys. Rev.* C10:214
17. Broda, R., Ishihara, M., Herskind, B., Oeschler, H., Ogaza, S., Ryde, H. 1975. *Nucl. Phys.* A248:356
18. Della Negra, S., Gauvin, H., Jungclas, H., Le Beyec, Y., Lefort, M. 1977. *Z. Phys.* A282:65
19. Galin, J., Gatty, B., Guerreau, D., Rousset, C., Schlotthauer-Voos, U. C., Tarrago, X. 1974. *Phys. Rev.* C9:1126
20. Gauvin, H., Lebeyec, Y., Lefort, M., Hahn, R. L. 1974. *Phys. Rev.* C10:722
21. Gomez del Campo, J., Dayras, R. A., Biggerstaff, J. A., Shapira, D., Snell, A. H., Stelson, P. H., Stokstad, R. G. 1979. *Phys. Rev. Lett.* 43:26
22. Parks, R. L., Thornton, S. T., Dennis, L. C., Cordell, K. R. 1980. *Nucl. Phys.* A348:350
23. Vandenbosch, R., Lazzarini, A. J. 1981. *Phys. Rev.* C23:1074
24. Heusch, B., Beck, C., Coffin, J. P., Engelstein, P., Freeman, R. M., Guillaume, G., Haas, F., Wagner, P. 1982. *Phys. Rev.* C26:542
25. Barrette, J., Braun-Munzinger, P., Gelbke, C. K., Wegner, H. E., Zeidman, B., Gamp, A., Harney, H. L., Walcher, Th. 1977. *Nucl. Phys.* A279:125
26. Blann, M. 1972. *Phys. Rev. Lett.* 29:303
27. Britt, H. C., Erkkila, B. H., Stokes, R. H., Gutbrod, H. H., Plasil, F., Ferguson, R. L., Blann, M. 1976. *Phys. Rev.* C13:1483
28. Chan, Y. D., Digregorio, D. E., Ford, J. L. C. Jr., Gomez del Campo, J., Oritz, M. E., Shapira, D. 1982. *Phys. Rev.* C25:1410
29. Chulick, E. T., Namboodiri, M. N., Natowitz, J. B. 1973. *Physics Chemistry of Fission,* Vol. II. Vienna: IAEA
30. Conjeaud, M., Gary, S., Harar, S., Wielczko, J. P. 1978. *Nucl. Phys.* A309:515
31. Heusch, B., Beck, C., Coffin, J. P., Freeman, R. M., Gallmann, A., Haas, F., Rami, F., Wagner, P., Alburger, D. E. 1981. *Phys. Rev.* C23:1527
32. Kolata, J. J. 1980. *Phys. Lett.* 95B:215
33. Natowitz, J. B., Chulick, E. T., Namboodiri, M. N. 1973. *Phys. Rev. Lett.* 31:643
34. Saint-Laurent, F., Conjeaud, M., Harar, S., Loiseaux, J. M., Menet, J., Viano, J. B. 1979. *Nucl. Phys.* A327:517
35. Stokstad, R. G., Gomez del Campo, J., Biggerstaff, J. A., Snell, A. H., Stelson, P. H. 1976. *Phys. Rev. Lett.* 36:1529
36. Stokstad, R. G., Dayras, R. A., Gomez del Campo, J., Stelson, P. H., Olmer, C., Zisman, M. S. 1977. *Phys. Lett.* 70B:289
37. Volant, C., Conjeaud, M., Harar, S., Lee, S. M., Lepine, A., Da Silveira, E. F. 1975. *Nucl. Phys.* A238:120
38. Switkowski, Z. E., Wu, S.-C., Overley, J. C., Barnes, C. A. 1977. *Nucl. Phys.* A289:236
39. Roth, H. A., Christiansson, J. E., Dubois, J. 1980. *Nucl. Phys.* A343:148
40. Wu, S.-C., Overley, J. C., Barnes, C. A., Switkowski, Z. E. 1978. *Nucl. Phys.* A312:177
41. Oritz, M. E., Gomez del Campo, J., Chan, Y. D., Digregorio, D. E., Ford, J. L. C. Jr., Shapira, D., Stokstad, R. G., Sellschop, J. P. F., Parks, R. L., Weiser, D. 1982. *Phys. Rev.* C25:1436
42. Wieleczko, J. P., Harar, S., Conjeaud, M., Saint-Laurent, F. 1980. *Phys. Lett.* 93B:35

43. Denis, L. C., Abdo, K. M., Frawley, A. D., Kemper, K. W. 1982. *Phys. Rev.* C26:981
44. Mateja, J. F., Frawley, A. D., Dennis, L. C., Abdo, K., Kemper, K. W. 1982. *Phys. Rev.* C25:2963
45. Dasmahapatra, B., Cujec, B., Lahlou, F. 1982. *Nucl. Phys.* A384:257
45a. Satkowiak, L. J., De Young, P. A., Kolata, J. J., Xapsos, M. A. 1982. *Phys. Rev.* C26:2027
46. Racca, R. A., Prosser, F. W., Davids, C. N., Kovar, D. G. 1982. *Phys. Rev.* C26:2022
47. Kovar, D. G., Geesaman, D. F., Braid, T. H., Eisen, Y., Henning, W., Ophel, T. R., Paul, M., Rehm, K. E., Sanders, S. J., Sperr, P., Schiffer, J. P., Tabor, S. L., Vigdor, S., Zeidman, B., Prosser, F. W. Jr. 1979. *Phys. Rev.* C20:1305
48. Sperr, P., Braid, T. H., Eisen, Y., Kovar, D. G., Prosser, F. W. Jr., Schiffer, J. P., Tabor, S. L., Vigdor, S. 1976. *Phys. Rev. Lett.* 37:321
49. Kolata, J. J., Freeman, R. M., Haas, F., Heusch, B., Gallmann, A. 1980. *Phys. Rev.* C21:579
50. Namboodiri, M. N., Chulick, E. T., Natowitz, J. B. 1976. *Nucl. Phys.* A263:491
51. Dayras, R. A., Stokstad, R. G., Switkowski, Z. E., Wieland, R. M. 1976. *Nucl. Phys.* A265:153
52. Switkowski, Z. E., Stokstad, R. G., Wieland, R. M. 1977. *Nucl. Phys.* A279:502
53. Kuehner, J. A., Almqvist, E. 1964. *Phys. Rev.* 134:1229
54. Gomez del Campo, J., Stokstad, R. G., Biggerstaff, J. A., Dayras, R. A., Snell, A. H., Stelson, P. H. 1979. *Phys. Rev.* C19:2170
55. Switkowski, Z. E., Stokstad, R. G., Wieland, R. M. 1976. *Nucl. Phys.* A274:202
56. Chan, Y. D., Bohn, H., Vandenbosch, R., Bernhardt, K. G., Cramer, J. G., Sieleman, R. N., Green, L. 1978. *Nucl. Phys.* A303:500
57. Eyal, Y., Beckerman, M., Chechik, R., Fraenkel, Z., Stocker, H. 1976. *Phys. Rev.* C13:1527
58. Sperr, P., Vigdor, S., Eisen, Y., Henning, W., Kovar, D. G., Ophel, T. R., Zeidman, B. 1976. *Phys. Rev. Lett.* 36:405
59. Tabor, S. L., Eisen, Y., Kovar, D. G., Vager, Z. 1977. *Phys. Rev.* C16:673
60. Kolata, J. J., Freeman, R. M., Haas, F., Heusch, B., Gallmann, A. 1979. *Phys. Rev.* C19:408
61. Fernandez, B., Gaarde, C., Larsen, J. S., Pontoppidan, S., Videbaek, F. 1978. *Nucl. Phys.* A306:259
62. Weidinger, A., Busch, F., Gaul, G., Trautmann, W., Zipper, W. 1976. *Nucl. Phys.* A263:511
63. Hertz, A., Essel, H., Korner, H. J., Rehm, K. E., Sperr, P. 1978. *Phys. Rev.* C18:2780
64. Coffin, J. P., Engelstein, P., Gallmann, A., Heusch, B., Wagner, P., Wegner, H. E. 1978. *Phys. Rev.* C17:1607
65. Puhlhofer, F., Pfeffer, W., Kohlmeyer, B., Schneider, W. F. W. 1975. *Nucl. Phys.* A244:329
66. Kohlmeyer, B., Pfeffer, W., Puhlhofer, F. 1977. *Nucl. Phys.* A292:288
67. Tserruya, I., Barrette, J., Kubono, S., Braun-Munzinger, P., Gai, M., Uhlhorn, C. D. 1980. *Phys. Rev.* C21:1864
68. Shapira, D., Ford, J. L. C. Jr., Gomez del Campo, J. 1982. *Phys. Rev.* C26:2470
69. Cole, A. J., Longequeue, N., Menet, J., Lucas, J. J., Ost, R., Viano, J. B. 1980. *Nucl. Phys.* A341:284
70. Lesko, K. T., Lock, D. K., Lazzarini, A., Vandenbosch, R., Metag, V., Doubre, H. 1982. *Phys. Rev.* C25:872
71. Busch, F., Canty, M. J., Pfeffer, W., Kohlmeyer, B., Schafer, W., Schneider, W. F. W., Freiesleben, H., Puhlhofer, F. 1979. *Z. Phys.* A290:167
72. Charvet, J. L., Dayras, R., Fieni, J. M., Joly, S., Uzureau, J. L. 1982. *Nucl. Phys.* A376:292
73. Digregorio, D. E., Gomez del Campo, J., Chan, Y. D., Ford, J. L. C. Jr., Shapira, D., Ortiz, M. E. 1982. *Phys. Rev.* C26:1490
74. Spinka, H., Winkler, H. 1974. *Nucl. Phys.* A233:456
75. Frolich, H., Duck, P., Galster, W., Treu, W., Voit, H., Witt, H., Kuhn, W., Lee, S. M. 1976. *Phys. Lett.* 64B:408
76. Tserruya, I., Eisen, Y., Pelte, D., Gavron, A., Oeschler, H., Berndt, D., Harney, H. L. 1978. *Phys. Rev.* C18:1688
77. Kolata, J. J., Freeman, R. M., Haas, F., Heusch, B., Gallmann, A. 1979. *Phys. Rev.* C19:2237
78. Shapira, D., Ford, J. L. C. Jr., Gomez del Campo, J., Stelson, P. H. 1980. *Phys. Rev.* C21:1824
79. Horn, D., Ferguson, A. J., Hausser, O. 1978. *Nucl. Phys.* A311:238
80. Daneshvar, K., Kovar, D. G., Krieger, S. J., Davies, K. T. R. 1982. *Phys. Rev.* C25:1342
81. Gary, S., Volant, C. 1982. *Phys. Rev.* C25:1877

82. Tabor, S. L., Geesaman, D. F., Henning, W., Kovar, D. G., Rehm, K. E., Prosser, F. W. Jr. 1978. *Phys. Rev.* C17:2136
83. Lee, S. M., Higashi, Y., Nagashima, Y., Hanashima, S., Sato, M., Yamaguchi, H., Yamanouchi, M., Matsuse, T. 1981. *Phys. Lett.* 98B:418
84. Albinska, M., Belery, P., Delbar, Th., El Masri, Y., Gregoire, Gh., Michel, C., Vervier, J., Albinski, J., Grotowski, K., Kopta, S., Kozik, T., Planeta, R., Paic, G. 1983. *Phys. Rev.* C27:207
85. Gutbrod, H. H., Winn, W. G., Blann, M. 1973. *Nucl. Phys.* A213:267
86. Puhlhofer, F., Schneider, W. F. W., Busch, F., Barrette, J., Braun-Munzinger, P., Gelbke, C. K., Wegner, H. E. 1977. *Phys. Rev.* C16:1010
87. Bock, R., Chu, Y. T., Dakowski, M., Gobbi, A., Grosse, E., Olmi, A., Sann, H., Schwalm, D., Lynen, U., Muller, W., Bjornholm, S., Esbensen, H., Wolfli, W., Morenzoni, E. 1982. *Nucl. Phys.* A388:334
88. Sann, H., Bock, R., Chu, Y. T., Gobbi, A., Olmi, A., Lynen, U., Muller, W., Bjornholm, S., Esbensen, H. 1981. *Phys. Rev. Lett.* 47:1248
89. Natowitz, J. B., Chulick, E. T., Namboodiri, M. N. 1972. *Phys. Rev.* C6:2133
90. Betts, R. R., Lanford, W. A., Mortensen, M. H., White, R. L. 1976. *Proc. Symp. Macroscopic Features of Heavy Ion Collisions,* ANL/PHY-76-2. Argonne Natl. Lab., Ill.
91. Namboodiri, M. N., Chulick, E. T., Natowitz, J. B., Kenefick, R. A. 1975. *Phys. Rev.* C11:401
92. Prosser, F. W. Jr., Racca, R. A., Daneshvar, K., Geesaman, D. F., Henning, W., Kovar, D. G., Rehm, K. E., Tabor, S. L. 1980. *Phys. Rev.* C21:1819
93. Dauk, J., Lieb, K. P., Kleinfeld, A. M. 1975. *Nucl. Phys.* A241:170
94. Eisen, Y., Tserruya, I., Eyal, Y., Fraenkel, Z., Hillman, M. 1977. *Nucl. Phys.* A291:459
95. Rascher, R., Muller, W. F. J., Lieb, K. P. 1979. *Phys. Rev.* C20:1028
96. Back, B. B., Betts, R. R., Gaarde, C., Larsen, J. S., Michelsen, E., Tai, K.-H. 1977. *Nucl. Phys.* A285:317
97. Kozub, R. L., Lu, N. H., Miller, J. M., Logan, D., Debiak, T. W., Kowalski, L. 1975. *Phys. Rev.* C11:1497
98. Kowalski, L., Jodogne, J. C., Miller, J. M. 1968. *Phys. Rev.* 169:894
99. Van Sen, N., Darves-Blanc, R., Gondrad, J. C., Merchez, F. 1983. *Phys. Rev.* C27:194
100. Morgenstern, H., Bohne, W., Grabisch, K., Lehr, H., Stoffler, W. 1983. *Z. Phys.* In press
101. Natowitz, J. B., Namboodiri, M. N., Eggers, R., Gonthier, P., Geoffroy, K., Hanus, R., Towsley, C., Das, K. 1977. *Nucl. Phys.* A277:477
102. Medsker, L. R., Wilson, D. C., Fry, L. H. Jr. 1978. *Phys. Lett.* 74B:39
103. Scobel, W., Gutbrod, H. H., Blann, M., Mignerey, A. C. 1976. *Phys. Rev.* C14:1808
104. Schneider, W. F. W., Puhlhofer, F., Chestnut, R. P., Volant, C., Freiesleben, H., Pfeffer, W., Kohlmeyer, B. 1981. *Nucl. Phys.* A371:493
105. Eck, J. S., Leigh, J. R., Ophel, T. R., Clark, P. D. 1980. *Phys. Rev.* C21:2352
106. Jordan, W. J., Maher, J. V., Peng, J. C. 1979. *Phys. Lett.* 87B:38
107. Betts, R. R., Hindi, M. M., Dicenzo, S. B., Parker, P. D. 1980. *Phys. Rev.* C21:175
108. Medsker, L. R., Theisen, L. V., Fry, L. H. Jr., Clements, J. S. 1979. *Phys. Rev.* C19:790
109. Oeschler, H., Freiesleben, H., Hildenbrand, K. D., Engelstein, P., Coffin, J. P., Heusch, B., Wagner, P. 1980. *Phys. Rev.* C22:546
110. Vigdor, S. E., Kovar, D. G., Sperr, P., Mahoney, J., Menchaca-Rocha, A., Olmer, C., Zisman, M. S. 1979. *Phys. Rev.* C20:2147
111. Geesaman, D. F., Davids, C. N., Henning, W., Kovar, D. G., Rehm, K. E., Schiffer, J. P., Tabor, S. L., Prosser, F. W. Jr. 1978. *Phys. Rev.* C18:284
112. Van Sen, N., Darves-Blanc, R., Gondrad, J. C., Merchez, F. 1979. *Phys. Rev.* C20:969
113. Van Sen, N., Gondrad, J. C., Merchez, F., Darves-Blanc, R. 1980. *Phys. Rev.* C22:2424
114. Doubre, H., Gamp, A., Jacmart, J. C., Poffe, N., Roynette, J. C., Wilczynski, J. 1978. *Phys. Lett.* 73B:135
115. Tomasi, E., Ardouin, B., Barreto, J., Bernard, V., Gauvin, B., Magnago, C., Mazur, C., Ngo, C., Pasecki, E., Ribrag, M. 1982. *Nucl. Phys.* A373:341
116. Sperr, P., Korner, H. J., Mayer, W. A., Wagner, W. 1980. *Z. Phys.* A297:355
117. Gonthier, P., Ho, H., Namboodiri, M. N., Adler, L., Natowitz, J. B., Simon, S., Hagel, K., Terry, R., Khodai, A. 1980. *Phys. Rev. Lett.* 44:1387
118. Hille, P., Rudolph, K., Hille, M., Sperr, P., Mannhart, W., Munzer, H., Gaul, G. 1976. *Nucl. Phys.* A266:253
119. Jastrzebski, J., Karwowski, H., Sadler,

M., Singh, P. P. 1979. *Phys. Rev.* C19:724
120. Agarwal, S., Galin, J., Gatty, B., Guerreau, D., Lefort, M., Tarrago, X., Babinet, R., Girard, J. 1980. *Z. Phys.* A296:287
121. Heusch, B., Volant, C., Freiesleben, H., Chestnut, R. P., Hildenbrand, K. D., Puhlhofer, F., Schneider, W. F. W. 1978. *Z. Phys.* A288:391
122. Cabot, C., Ngo, C., Peter, J., Tamain, B. 1975. *Nucl. Phys.* A244:134
123. Robinson, R. L., Kim, H. J., Ford, J. L. C. Jr. 1974. *Phys. Rev. Lett.* C9:1402
124. Simpson, J. J., Tjom, P. O., Espe, I., Hagemann, G. B., Herskind, B., Neiman, M. 1977. *Nucl. Phys.* A287:362
125. Scobel, W., Mignerey, A. C., Blann, M., Gutbrod, H. H. 1975. *Phys. Rev.* C11:1701
126. Bisplinghof, J. F., David, P., Blann, M., Scobel, W., Mayer-Kuckuk, T., Ernst, J., Mignerey, A. C. 1978. *Phys. Rev.* C17:177
127. Gauvin, H., Guerreau, D., Lebeyec, Y., Lefort, M., Plasil, F., Tarrago, X. 1975. *Phys. Lett.* 58B:163
128. Sikora, B., Bisplinghof, J. F., Scobel, W., Beckerman, M., Blann, M. 1979. *Phys. Rev.* C20:2219
129. Beckerman, M., Salomaa, M., Sperduto, A., Enge, H., Ball, J., Di Rienzo, A., Gazes, S., Chen, Y., Molitoris, J. D., Mao, N.-F. 1980. *Phys. Rev. Lett.* 45:1472
130. Beckerman, M., Ball, J., Enge, H., Salomaa, M., Sperduto, A., Gazes, S., Di Rienzo, A., Molitoris, J. D. 1981. *Phys. Rev.* C23:1581
131. Sikora, B., Scobel, W., Beckerman, M., Bisplinghof, J., Blann, M. 1982. *Phys. Rev.* C25:1446
132. Sikora, B., Blann, M., Scobel, W., Bisplinghof, J., Beckerman, M. 1982. *Phys. Rev.* C25:885
133. Beckerman, M., Salomaa, M., Sperduto, A., Molitoris, J. D., Di Rienzo, A. 1982. *Phys. Rev.* C25:837
134. Natowitz, J. B. 1970. *Phys. Rev.* C1:623
135. Wells, J. C., Jr., Robinson, R. L., Kim, H. J., Ford, J. L. C. Jr. 1975. *Phys. Rev.* C12:1529
136. Langevin, M., Barreto, J., Detraz, C. 1976. *Phys. Rev.* C14:152
137. Borderie, B., Bimbot, R., Cabot, C., Gardes, D., Nowicki, L., Tamain, B., Gregoire, C., Ngo, C., Berlanger, M., Hanappe, F. 1980. *Z. Phys.* A298:235
138. Lucas, R., Poitou, J., Girard, J. 1979. *Nucl. Phys.* A321:501
139. Plasil, F., Ferguson, R. L., Britt, H. C., Erkkila, B. H., Goldstone, P. D., Stokes, R. H., Gutbrod, H. H. 1978. *Phys. Rev.* C18:2603
140. Britt, H. C., Erkkila, B. H., Goldstone, P. D., Stokes, R. H., Back, B. B., Folkmann, F., Christensen, O., Fernandez, B., Garrett, J. D., Hagemann, G. B., Herskind, B., Hillis, D. L., Plasil, F., Ferguson, R. L., Blann, M., Gutbrod, H. H. 1977. *Phys. Rev. Lett.* 39:1458
141. Gauvin, H., Lebeyec, Y., Lefort, M., Deprun, C. 1972. *Phys. Rev. Lett.* 28:697
142. Guillaume, G., Coffin, J. P., Rami, F., Engelstein, P., Heusch, B., Wagner, P., Fintz, P., Barrette, J., Wegner, H. E. 1982. *Phys. Rev.* C26:2458
143. Natowitz, J. B., Namboodiri, M. N., Chulick, E. T. 1976. *Phys. Rev.* C13:171
144. Tamain, B., Ngo, C., Peter, J., Hanappe, F. 1975. *Nucl. Phys.* A252:187
145. Von Oertzen, W., Fuchs, H., Gamp, A., Homeyer, H., Jahnke, U., Jacmart, J. C. 1980. *Z. Phys.* A298:207
146. Lu, N. H., Logan, D., Miller, J. M., Debiak, T. W., Kowalski, L. 1976. *Phys. Rev.* C13:1496
147. David, P., Bisplinghof, J. F., Blann, M., Mayer-Kuckuk, T., Mignerey, A. C. 1977. *Nucl. Phys.* A287:179
148. Blann, M., Akers, D., Komoto, T. A., Dietrich, F. S., Hansen, L. F., Woodworth, J. G., Scobel, W., Bisplinghof, J., Sikora, B., Plasil, F., Ferguson, R. L. 1982. *Phys. Rev.* C26:1471
149. Schier, W., Cheervenak, J., Di Rienzo, A. C., Enge, H., Grogan, D., Molitoris, J., Salomaa, M., Sperduto, S. 1981. *Phys. Rev.* C23:261
149a. Schmidt, K. A., Armbruster, P., Hessberger, F. P., Munzenberg, G., Reisdorf, W., Sahm, C. C., Vermuelen, D., Clerc, H. G., Keller, J., Schulte, H. 1981. *Z. Phys.* A301:21
150. Sikkeland, T. 1964. *Phys. Rev.* 135:B669
151. Plasil, F., Ferguson, R. L., Hahn, R. L., Obenshain, F. E., Pleasonton, F., Young, G. R. 1980. *Phys. Rev. Lett.* 45:333
152. Dyer, P., Webb, M. P., Puigh, R. J., Vandenbosch, R., Thomas, T. D., Zisman, M. S. 1980. *Phys. Rev.* C22:1509
153. Sarantites, D. G., Barker, J. H., Halbert, M. L., Hensley, D. C., Dayras, R. A., Eichler, E., Johnson, N. R., Gronemeyer, S. A. 1976. *Phys. Rev.* C14:2138
154. Halbert, M. L., Dayras, R. A.,

Ferguson, R. L., Plasil, F., Sarantites, D. G. 1978. *Phys. Rev.* C17:155
155. Back, B. B., Betts, R. R., Henning, W., Wolf, K. L., Mignerey, A. C., Lebowitz, J. M. 1980. *Phys. Rev. Lett.* 45:1230
156. Stokstad, R. G., Reisdorf, W., Hildenbrand, K. D., Kratz, J. V., Wirth, G., Lucas, R., Poitou, J. 1980. *Z. Phys.* A295:269
157. Sahm, C.-C., Schulte, H., Vermeulen, D., Keller, J., Clerc, H.-G., Schmidt, K.-H., Hebberger, F., Munzenberg, G. 1980. *Z. Phys.* A297:241
158. Stokstad, R. G., Eisen, Y., Kaplanis, S., Pelte, D., Smilansky, U., Tserruya, I. 1978. *Phys. Rev. Lett.* 41:465
159. Stokstad, R. G., Eisen, Y., Kaplanis, S., Pelte, D., Smilansky, U., Tserruya, I. 1980. *Phys. Rev.* C21:2427
160. Kowalski, L., Zebelman, A. M., Kandil, A., Miller, J. M. 1971. *Phys. Rev.* C3:1370
161. Zebelman, A. M., Kowalski, L., Miller, J. M., Beg, K., Eyal, Y., Jaffe, G., Kandil, A., Logan, D. 1974. *Phys. Rev.* C10:200
162. Sikkeland, T., Clarkson, J. E., Steiger-Shafrir, N. H., Viola, V. E. 1971. *Phys. Rev.* C3:329
163. Gilmore, J., Thompson, S. G., Perlman, I. 1962. *Phys. Rev.* 128:2276
164. Rossner, H. H., Hilscher, D., Holub, E., Ingold, G., Jahnke, U., Orf, H., Huizenga, J. R., Birkelund, J. R., Schröder, W. U., Wilcke, W. W. 1983. *Phys. Rev.* C27:2666
165. Lebrun, C., Hanappe, F., Lecolley, J. F., Lefebvres, F., Ngo, C., Peter, J., Tamain, B. 1979. *Nucl. Phys.* A321:207
166. Hoover, A. D., Birkelund, J. R., Hilscher, D., Schröder, W. U., Wilcke, W. W., Huizenga, J. R., Viola, V. E. Jr., Wolf, K. L., Breuer, H., Mignerey, A. C. 1982. *Phys. Rev.* C25:256
167. Birkelund, J. R., Hoover, A. D., Huizenga, J. R., Schröder, W. U., Wilcke, W. W. 1983. *Phys. Rev.* C27:882
168. Peter, J., Ngo, C., Tamain, B. 1975. *Nucl. Phys.* A250:351
169. Wolf, K. L., Roche, C. T. 1976. See Ref. 90, 1:295
170. Leigh, J. R., Hinde, D. J., Newton, J. O., Galster, W., Sie, S. H. 1982. *Phys. Rev. Lett.* 48:527
171. Miller, J. M., Logan, D., Catchen, G. L., Rajagopalan, M., Alexander, J. M., Kaplan, M., Ball, J. W., Zisman, M. S., Kowalski, L. 1978. *Phys. Rev. Lett.* 40:1074
172. Videbaek, F., Goldstein, R. B., Grodzins, L., Steadman, S. G., Belote, T. A., Garrett, J. D. 1977. *Phys. Rev.* C15:954
173. Kurz, H. E., Jasper, E. W., Fischer, K., Hermes, F. 1971. *Nucl. Phys.* A168:129
174. Gordon, G. E., Larsh, A. E., Sikkeland, T., Seaborg, G. T. 1960. *Phys. Rev.* 120:1341
175. Gordon, G. E., Larsh, A. E., Sikkeland, T. 1960. *Phys. Rev.* 118:1610
176. Britt, H. C., Quinton, A. R. 1960. *Phys. Rev.* 120:1768
177. Quinton, A. R., Britt, H. C., Knox, W. J., Anderson, C. E. 1960. *Nucl. Phys.* 17:74
178. Rivet, M. F., Gatty, B., Guillemont, H., Borderie, B., Bimbot, R., Forest, I., Galin, J., Gardes, D., Guerreau, D., Tamain, B., Novicki, L. 1982. *Z. Phys.* A307:365
179. Ngo, C., Peter, J., Tamain, B., Berlanger, M., Hanappe, F. 1977. *Z. Phys.* A283:161
180. Galin, J., Gatty, B., Guerreau, D., Lefort, M., Tarrago, X., Agarwal, S., Babinet, R., Gauvin, B., Girard, J., Nifenecker, H. 1977. *Z. Phys.* A283:173
181. Peter, J., Ngo, C., Plasil, F., Tamain, B., Berlanger, M., Hanappe, F. 1977. *Nucl. Phys.* A279:110
182. Shea, J. H., Durell, J. L., Foote, G. S., Grant, I. S. 1982. *Nucl. Phys.* A390:351
183. Sperr, P., Essel, H., Korner, H. J., Rehm, K. E. 1979. *Z. Phys.* A291:179
184. Tsang, M. B., Ardouin, D., Gelbke, C. K., Lynch, W. G., Xu, Z., Back, B. B., Betts, R., Saini, S., Baisden, P. A., McMahan, M. A. 1983. *Phys. Rev.* C28:747
185. Meyer, W. G., Viola, V. E., Jr., Clark, R. G., Read, S. M., Theus, R. B. 1979. *Phys. Rev.* C20:1716
186. Jin, G.-M., Xie, Y.-X., Zhu, Y.-T., Shen, W.-G., Sun, X.-J., Guo, J.-S., Liu, G.-X., Yu, J.-S., Sun, C.-C., Garrett, J. D. 1980. *Nucl. Phys.* A349:285
187. Freiesleben, H., Rizzo, G. T., Huizenga, J. R. 1975. *Phys. Rev.* C12:42
188. Freiesleben, H., Huizenga, J. R. 1974. *Nucl. Phys.* A224:503
189. Viola, V. E. Jr., Clark, R. G., Meyer, W. G., Zebelman, A. M., Sextro, R. G. 1976. *Nucl. Phys.* A261:174
190. Viola, V. E. Jr., Sikkeland, T. 1962. *Phys. Rev.* 128:767
191. Viola, V. E. Jr., Roche, C. T., Meyer, W. G., Clark, R. G. 1974. *Phys. Rev.* C10:2416
192. Shea, J. H., Durell, J. L., Foote, G. S., Grant, I. S. 1979. *Nucl. Phys.* A327:207
193. Lin, S. Y., Alexander, J. M. 1977. *Phys. Rev.* C16:688
194. Beckerman, M., Blann, M. 1977. *Phys. Rev. Lett.* 38:272

195. Beckerman, M., Blann, M. 1978. *Phys. Rev.* C17:1615
196. Stokstad, R. G. 1974. *Proc. Int. Conf. React. Complex Nuclei*, p. 327. Amsterdam: North-Holland
197. Blann, M., Plasil, F. 1972. *Phys. Rev. Lett.* 29:303
198. Puhlhofer, F. 1977. *Nucl. Phys.* A280:267
199. Horn, D., Enge, H. A., Sperduto, A., Graue, A. 1978. *Phys. Rev.* C17:118
200. Eyal, Y., Hillman, M. Program *JULIAN* (unpublished)
201. Gomez del Campo, J., Stokstad, R. G. 1981. Program *LILITA*, ORNL TM-7295 (unpublished)
202. Stokstad, R. G., Namboodiri, M. N., Chulick, E. T., Natowitz, J. B., Hanson, D. L. 1977. *Phys. Rev.* C16:2249
203. Cohen, S., Plasil, F., Swiatecki, W. J. 1974. *Ann. Phys. (NY)* 82:557
204. Beckerman, M., Blann, M. 1977. *Phys. Lett.* 68B:31
205. Cormier, T. M., Cosman, E. R., Lazzarini, A. J., Garrett, J. D., Wegner, H. E. 1976. *Phys. Rev.* C14:334
206. Cormier, T. M., Cosman, E. R., Lazzarini, A. J., Wegner, H. E., Garrett, J. D., Puhlhofer, F. 1977. *Phys. Rev.* C15:654
207. Hille, M., Hille, P., Gutbrod, H. H., Blann, M. 1975. *Nucl. Phys.* A252:496
208. Plasil, F. 1980. *Int. Conf. Nucl. Behaviour At High Angular Momentum.* Strasbourg, France. *J. Phys.* 41:C10–183
209. Blann, M., Komoto, T. T. 1982. *Phys. Rev.* C26:472
210. See Ref. 27
211. Huizenga, J. R., Birkelund, J. R., Schröder, W. U., Wilcke, W. W., Wollersheim, H. J. 1981. *Dynamics of Heavy-Ion Collisions*, p. 15. Amsterdam: North-Holland
212. Wilcke, W. W., Birkelund, J. R., Hoover, A. D., Huizenga, J. R., Schröder, W. U., Viola, V. E. Jr., Wolf, K. L., Mignerey, A. C. 1980. *Phys. Rev.* C22:128
213. Huizenga, J. R., Schröder, W. U., Birkelund, J. R., Wilcke, W. W. 1982. *Nucl. Phys.* A387:257c
214. Gregoire, C., Ngo, C., Tomasi, E., Remaud, B., Scheuter, F. 1982. *Nucl. Phys.* A387:37c
215. Borderie, B., Berlanger, M., Gardes, D., Hanappe, F., Nowicki, L., Peter, J., Tamain, B., Agarwal, S., Girard, J., Gregoire, C., Matuszek, J., Ngo, C. 1981. *Z. Phys.* A299:263
216. Faber, M. E. 1980. *Z. Phys.* A297:277
217. Faber, M. E. 1981. *Phys. Rev.* C24:1047
218. Chaudhry, R., Vandenbosch, R., Huizenga, J. R. 1962. *Phys. Rev.* 126:220
219. Reising, R. F., Bate, G. L., Huizenga, J. R. 1966. *Phys. Rev.* 141:1161
220. Vandenbosch, R., Huizenga, J. R. 1973. *Nuclear Fission*. New York: Academic
221. Bohr, A. 1956. *Proc. UN Int. Conf. Peaceful Uses of Atomic Energy*, Paper P/911, 2:151. New York: United Nations
222. Halpern, I., Strutinski, V. M. 1958. *Proc. 2nd UN Conf. Peaceful Uses of Atomic Energy*, Paper D/1513, 15:408. Geneva: United Nations
223. Huizenga, J. R., Behkami, A. N., Moretto, L. G. 1969. *Phys. Rev.* 177:1826
224. Back, B. B., Clerc, H.-G., Betts, R. R., Glagola, B. G., Wilkins, B. D. 1981. *Phys. Rev. Lett.* 46:1068
225. Hanson, D. L., Stokstad, R. G., Erb, K. A., Olmer, C., Sachs, M. W., Bromley, D. A. 1974. *Phys. Rev.* C9:1760
226. High, M. D., Cujec, B. 1976. *Nucl. Phys.* A259:513
227. Christensen, P. R., Switkowski, Z. E., Dayras, R. A. 1977. *Nucl. Phys.* A280:189
228. High, M. D., Cujec, B. 1977. *Nucl. Phys.* A278:149
229. High, M. D., Cujec, B. 1977. *Nucl. Phys.* A282:181
230. Hulke, G., Rolfs, C., Trautvetter, H. P. 1980. *Z. Phys.* A297:161
231. Kettner, K. U., Lorentz-Wirzba, H., Rolfs, C., Winkler, H. 1977. *Phys. Rev. Lett.* 38:337
232. Stokstad, R. G., Switkowski, Z. E., Dayras, R. A., Wieland, R. M. 1976. *Phys. Rev. Lett.* 37:888
233. Chatterjee, M. L., Potvin, L., Cujec, B. 1980. *Nucl. Phys.* A333:273
234. Cheung, H. C., High, M. D., Cujec, B. 1978. *Nucl. Phys.* A296:333
235. Jahnke, U., Rossner, H. H., Hilscher, D., Holub, E. 1982. *Phys. Rev. Lett.* 48:17
236. Avishai, Y. 1978. *Z. Phys.* A286:285
237. Avishai, Y. 1978. *Phys.* A285:333
238. Dethier, J. L., Stancu, F. 1981. *Phys. Rev.* C23:1503
239. Kodama, T., Nazareth, R. A. M. S., Moller, P., Nix, J. R. 1978. *Phys. Rev.* C17:111
240. Mathews, G. J., Moretto, L. G. 1979. *Phys. Lett.* 87B:331
241. Sherman, A., Sperber, D., Seglie, E. 1975. *Phys. Lett.* 59B:205
242. Descouvemont, P., Baye, D., Heenen, P. H. 1982. *Z. Phys.* A306:79
243. Christensen, P. R., Switkowski, Z. E. 1977. *Nucl. Phys.* A280:205

244. Dayras, R. A., Stokstad, R. G., Switkowski, Z. E., Wieland, R. M. 1976. *Nucl. Phys.* A261:478
245. Hussein, M. S. 1977. *Phys. Lett.* 71B:249
246. Hussein, M. S., Canto, L. F., Donangelo, R. 1980. *Phys. Rev.* C21:772
247. Lin, S. Y., Alexander, J. M. 1977. *Phys. Rev.* C16:688
248. Robinson, R. L., Bair, J. K., Johnson, C. H., Stelson, P. H., Dress, W. B., Jones, C. M. 1976. *Phys. Rev.* C14:2126
249. Haider, Q., Malik, F. B. 1982. *Phys. Rev.* C26:162
250. Haider, Q., Malik, F. B. 1982. *Phys. Rev.* C26:989
251. Stokstad, R. G., Gross, E. E., 1981. *Phys. Rev.* C23:281
252. Esbensen, H. 1981. *Nucl. Phys.* A352:147
253. Landowne, S., Nix, J. R. 1981. *Nucl. Phys.* A368:352
254. Inamura, T., Kojima, T., Nomura, T., Sugitate, T., Utsunomiya, H. 1979. *Phys. Lett.* 84B:71
255. Zolnowski, D. R., Yamada, H., Cala, S. E., Kahler, A. C., Sugihara, T. T. 1978. *Phys. Rev. Lett.* 41:92
256. Yamada, H., Zolnowski, D. R., Cala, S. E., Kahler, A. C., Pierce, J., Sugihara, T. T. 1979. *Phys. Rev. Lett.* 43:605
257. Siwek-Wilczynska, K., Du Marchie Van Voorthuysen, E. H., Van Popta, J., Siemssen, R. H., Wilczynski, J. 1979. *Phys. Rev. Lett.* 42:1599
258. Siwek-Wilczynska, K., Du Marchie Van Voorthuysen, E. H., Van Popta, J., Siemssen, R. H., Wilczynski, J. 1979. *Nucl. Phys.* A330:150
259. Wilczynski, J. 1979. See Ref. 13, 117:254
260. Wilczynski, J., Siwek-Wilczynska, K., Van Driel, J., Gonggrijp, S., Hageman, D. C. J. M., Janssens, R. V. F., Lukasiak, J., Siemssen, R. H. 1980. *Phys. Rev. Lett.* 45:606
261. Wilczynski, J., Siwek-Wilczynska, K., Van Driel, J., Gonggrijp, S., Hageman, D. C. J. M., Janssens, R. V. F., Lukasiak, J., Siemssen, R. H., Van Der Werf, S. Y. 1982. *Nucl. Phys.* A373:109
262. Inamura, T., Ishihara, M., Fukuda, T., Shimoda, T., Hiruta, H. 1977. *Phys. Lett.* 68B:51
263. Geoffroy, K. A., Sarantites, D. G., Halbert, M. L., Hensley, D. C., Dayras, R. A., Barker, J. H. 1979. *Phys. Rev. Lett.* 43:1303
264. Barker, J. H., Beene, J. R., Halbert, M. L., Hensley, D. C., Jaaskelainen, M., Sarantites, D. G., Woodward, R. 1980. *Phys. Rev. Lett.* 45:424
265. Udagawa, T., Tamura, T. 1980. *Phys. Rev. Lett.* 45:1311
266. Morison, W. W., Samaddar, S. K., Sperber, D., Zielinska-Pfabe, M. 1981. *Phys. Lett.* 99B:205
267. Wu, J. R., Lee, I. Y. 1980. *Phys. Rev. Lett.* 45:8
268. Wu, J. R., Lee, I. Y. 1980. *Proc. Int. Conf., Nucl. Phys. Berkeley*, Vol. 1, Abstracts, p. 634. US Govt. Print. Off. 695–972 (LBL-11118)
269. Birkelund, J. R., Wilcke, W. W., Tubbs, L. E., Huizenga, J. R., Hoover, A. D., Wollersheim, H. J., Schröder, W. U., Hilscher, D., Jahnke, U., Holub, E., Bohne, W., Morgenstern, H., Orf, H. 1981. *Bull. Am. Phys. Soc.* 26:539
269a. Chan, Y., Murphy, M., Stokstad, R. G., Tserruya, I., Wald, S., Budzanowski, A. 1983. *Phys. Rev.* C27:447
270. Morgenstern, H., Bohne, W., Grabisch, K., Kovar, D. G., Lehr, H. 1982. *Phys. Lett.* 113B:463
271. Nicholson, W. J., Halpern, I. 1959. *Phys. Rev.* 116:175
272. Sikkeland, T., Haines, E. L., Viola, V. E. Jr. 1962. *Phys. Rev.* 125:1350
273. Sikkeland, T., Viola, V. E. Jr. 1963. *3rd Conf. Reactions Between Complex Nuclei, Asilomar*, ed. A. Ghiorso, R. M. Diamond, H. E. Conzett, p. 232. Berkeley: Univ. Calif. Press
274. Awes, T. C., Gelbke, C. K., Back, B. B., Mignerey, A. C., Wolf, K. L., Dyer, P., Breuer, H., Viola, V. E. Jr. 1979. *Phys. Lett.* 87B:43
275. Back, B. B., Wolf, K. L., Mignerey, A. C., Gelbke, C. K., Awes, T. C., Breuer, H., Viola, V. E. Jr., Dyer, P. 1980. *Phys. Rev.* C22:1927
276. Awes, T. C., Poggi, G., Gelbke, C. K., Back, B. B., Glagola, B. G., Breuer, H., Viola, V. E. 1981. *Phys. Rev.* C24:89
277. Tubbs, L. E. 1982. Thesis, Univ. Rochester (Unpublished)
278. Viola, V. E., Back, B. B., Wolf, K. L., Awes, T. C., Gelbke, C. K., Breuer, H. 1982. *Phys. Rev.* C26:178
279. Fonte, R., Oeschler, H. 1980. *Phys. Lett.* 96B:265
280. Gerschel, C. 1982. *Nucl. Phys.* A387:297c
281. Blatt, J. M., Weiskopf, V. F. 1952. *Theoretical Nuclear Physics.* New York: Wiley
282. Gutbrod, H. H., Winn, W. G., Blann, M. 1973. *Phys. Rev. Lett.* 30:1259
283. Galin, J., Guerreau, D., Lefort, M., Tarrago, X. 1974. *Phys. Rev.* C9:1018
284. Glas, D., Mosel, U. 1974. *Phys. Rev.* C10:2620

285. Glas, D., Mosel, U. 1975. *Nucl. Phys.* A237:429
286. Glas, D., Mosel, U. 1976. *Nucl. Phys.* A264:268
287. Wilczynski, J. 1973. *Nucl. Phys.* A216:386
288. Bass, R. 1973. *Phys. Lett.* 47B:139
289. Bass, R. 1974. *Nucl. Phys.* A231:45
290. Birkelund, J. R., Huizenga, J. R., De, J. N., Sperber, D. 1978. *Phys. Rev. Lett.* 40:1123
291. Vax, L. C., Alexander, J. M. 1978. *Phys. Rev.* C18:2152
292. Gross, D. H. E., Kalinowski, H. 1974. *Phys. Lett.* 48B:302
293. Gross, D. H. E., Kalinowski, H. 1978. *Phys. Rep.* 45:175
294. De, J. N., Gross, D. H. E., Kalinowski, H. 1976. *Z. Phys.* A277:385
295. Birkelund, J. R., Huizenga, J. R. 1977. In *Proc. Symp. Heavy Ion Elastic Scattering, Rochester, October,* ed. R. DeVries. Rochester, NY: Univ. Rochester
296. Birkelund, J. R., Huizenga, J. R. 1978. *Phys. Rev.* C17:126
297. Mosel, U. 1979. *Phys. Rev.* C19:2167
298. Samaddar, S. K., Sperber, D., Zielinska-Pfabe, M. 1981. *Phys. Lett.* 98B:340
299. Seglie, E., Sperber, D., Sherman, A. 1975. *Phys. Rev.* C11:1227
300. Sperber, D. 1974. *Phys. Scr.* 10A:115
301. Vandenbosch, R. 1980. *Nucl. Phys.* A339:167
302. Hasse, R. W., Hahn, K. 1980. *Nukleonika* 25:133
303. Blocki, J., Randrup, J., Swiatecki, W. J., Tsang, C. F. 1977. *Ann. Phys. (NY)* 105:427
304. Blocki, J., Beck, F., Dworzecka, M., Feldmeier, H. 1980. *Proc. Int. Workshop Gross Properties of Nuclei and Nuclear Excitations,* Vol. VII, p. 157. Darmstadt: Inst. Kernphys. Tech. Hochsch.
305. Randrup, J. 1978. *Nucl. Phys.* A307:319
306. Krappe, H. J., Nix, J. R., Sierk, A. J. 1979. *Phys. Rev.* C20:992
307. Ngo, C., Tamain, B., Beiner, M., Lombard, R. J., Mas, D., Deubler, H. H. 1975. *Nucl. Phys.* A252:237
308. Zisman, M. S. 1980. See Ref. 268, p. 612
309. Bondorf, J. P., Sobel, M. I., Sperber, D. 1974. *Phys. Rep.* 15:83
310. Randrup, J. 1978. *Ann. Phys. (NY)* 112:356
311. Myers, W. D. 1973. *Nucl. Phys.* A204:465
312. Bangert, D., Freiesleben, H. 1980. *Nucl. Phys.* A340:205
313. Huizenga, J. R., Birkelund, J. R. 1982. *UR-NSRL 261,* December. Univ. Rochester, NY
314. Bass, R. 1977. *Phys. Rev. Lett.* 39:265
315. Gupta, S. K., Kailas, S. 1982. *Phys. Rev.* C26:747
316. Lefort, M. 1976. *Nukleonika* 21:111
317. Harar, S. 1979. See Ref. 13, 117:367
318. Glas, D., Mosel, U. 1978. *Phys. Lett.* 78B:9
319. Mosel, U. 1978. *Proc. Int. Conf. Nucl. Interactions, Canberra Lect. Notes Phys.* 92:185. New York: Springer-Verlag
320. Mosel, U., Diebel, M. 1979. See Ref. 13, 117:375
321. Lee, S. M., Matsuse, T., Arima, A. 1980. *Phys. Rev. Lett.* 45:165
322. Matsuse, T., Arima, A., Lee, S. M. 1982. *Phys. Rev.* C26:2338
323. Horn, D., Ferguson, A. J. 1978. *Phys. Rev. Lett.* 41:1529
324. Lozano, M., Madurga, G. 1980. *Phys. Lett.* 90B:50
325. Bonche, P., Grammaticos, B., Koonin, S. 1978. *Phys. Rev.* C17:1700
326. Bonche, P., Davies, K. T. R., Flanders, B., Flocard, H., Grammaticos, B., Koonin, S. E., Krieger, S. J., Weiss, M. S. 1979. *Phys. Rev.* C20:641
327. Cusson, R. Y., Meldner, H. W. 1979. *Phys. Rev. Lett.* 42:694
328. Davies, K. T. R., Feldmeier, H. T., Flocard, H., Weiss, M. S. 1978. *Phys. Rev.* C18:2631
329. Davies, K. T. R., Devi, K. R. S., Strayer, M. R. 1979. *Phys. Rev.* C20:1372
330. Davies, K. T. R., Devi, K. R. S., Strayer, M. R. 1980. *Phys. Rev. Lett.* 44:23
331. Davies, K. T. R., Devi, K. R. S., Strayer, M. R. 1980. See Ref. 268, p. 422
332. Davies, K. T. R., Devi, K. R. S., Strayer, M. R. 1980. *Phys. Rev. Lett.* 44:23
333. Devi, K. R. S., Dhar, A. K., Owen, J., Strayer, M. R. 1980. See Ref. 268, p. 415
334. Devi, K. R. S., Dhar, A. K., Strayer, M. R. 1981. *Phys. Rev.* C23:2062
335. Dhar, A. K., Nilsson, B. S. 1979. *Nucl. Phys.* A315:445
336. Flocard, H., Koonin, S. E., Weiss, M. S. 1978. *Phys. Rev.* C17:1682
337. Flocard, H., Weiss, M. S. 1978. *Phys. Rev.* C18:573
338. Koonin, S. E. 1976. *Phys. Lett.* 61B:227
339. Koonin, S. E., Davies, K. T. R., Maruhn-Rezwani, V., Feldmeier, H., Krieger, S. J., Negele, J. W. 1977. *Phys. Rev.* C15:1359
340. Krieger, S. J., Davies, K. T. R. 1978. *Phys. Rev.* C18:2567
341. Krieger, S. J., Davies, K. T. R. 1979. *Phys. Rev.* C20:167
342. Maruhn-Rezwani, V., Davies, K. T. R.,

Koonin, S. E. 1977. *Phys. Lett.* 67B: 134
343. Wong, C. Y., Maruhn, J. A., Welton, T. A. 1977. *Phys. Lett.* 66B: 19
344. Wong, C. Y., Davies, K. T. R. 1980. *Phys. Lett.* 96B: 258
345. Cusson, R. Y., Smith, R. K., Maruhn, J. A. 1976. *Phys. Rev. Lett.* 36: 116
346. Baldo, M., Broglia, R. A., Winther, A. 1980. *Phys. Lett.* 94B: 473
347. Broglia, R. A., Dasso, C. H., Pollarolo, G., Winther, A. 1978. *Phys. Rev. Lett.* 40: 707
348. Cabot, C., Gauvin, H., Lebeyec, Y., Lefort, M. 1975. *J. Phys.* 36: L289
349. Doukellis, G., Hlawatsch, G., Kolb, B., Natowitz, J. B., Rosner, G., Sedelmeyer, B., Walcher, T. 1978. *Proc. Hirschegg Workshop*, Vol. VII, p. 185. Darmstadt: Inst. Kernphys. Tech. Hochsch.
350. Hahn, R. L., Toth, K. S., Cabot, C., Gauvin, H., Lebeyec, Y. 1979. *Phys. Rev. Lett.* 42: 218
351. Lefort, M. 1975. *Phys. Rev.* C12: 686
352. Lefort, M. 1974. *Phys. Scr.* 10A: 101
353. Szanto De Toledo, A., Hussein, M. S. 1981. *Phys. Rev. Lett.* 46: 985
353a. Lazzarini, A., Doubre, H., Lesko. K. T., Metag, V., Seamster, A., Vandenbosch, R., Merryfield, W. 1981. *Phys. Rev.* C24: 309
354. Wilcke, W. W., Kosky, J. P., Birkelund, J. R., Butler, M. A., Dougan, A. D., Huizenga, J. R., Schröder, W. U., Wollersheim, H. J., Hilscher, D. 1983. *Phys. Rev. Lett.* 51: 99
355. Bertsch, G. F. 1982. MSUCL-385 Preprint, October. Mich. State Univ., Ann Arbor
356. Krappe, H. J. 1976. *Nucl. Phys.* A269: 493
357. Nix, J. R., Sierk, A. J. 1974. *Phys. Scr.* 10A: 94
358. Nix, J. R., Sierk, A. J. 1977. *Phys. Rev.* C15: 2072
359. Sierk, A. J., Nix, J. R. 1974. *Proc. 3rd Int. Atomic Energy Agency Symp. Physics Chemistry of Fission.* Vienna: IAEA
360. Sierk, A. J., Nix, J. R. 1976. See Ref. 90, 1: 407
361. Sierk, A. J., Nix, J. R. 1977. *Phys. Rev.* C16: 1048
362. Gross, D. H. E., Satpathy, L. 1982. *Phys. Lett.* 110B: 31
363. Frobrich, P. 1983. *Phys. Lett.* 122B: 338
364. Bjornholm, S., Swiatecki, W. J. 1982. *Nucl. Phys.* A391: 471
364a. Bertsch, G., Mundinger, D. 1978. *Phys. Rev.* C17: 1646
365. Blann, M., Akers, D. 1982. *Phys. Rev.* C26: 465
366. Sikora, B., Bisplinghof, J., Blann, M., Scobel, W., Beckerman, M., Plasil, F., Ferguson, R. L., Birkelund, J. R., Wilcke, W. W. 1982. *Phys. Rev.* C25: 686
367. Westmeier, W., Esterlund, R. A., Rox, A., Pazelt, P. 1982. *Phys. Lett.* 117B: 163
368. Munzenberg, G., Hofmann, S., Hessberger, F. P., Reisdorf, W., Schmidt, K. H., Schneider, J. R. H., Armbruster, P., Sahm, C. C., Thuma, B. 1981. *Z. Phys.* A300: 107
369. Munzenberg, G., Armbruster, P., Hessberger, F. P., Hofmann, S., Poppensieker, K., Reisdorf, W., Schneider, J. R. H., Schneider, W. F. W., Schmidt, K. H., Sahm, C. C., Vermeulen, D. 1982. *Z. Phys.* A309: 89
370. Wozniak, G. J., Schmitt, R. P., Glassel, P., Moretto, L. G. 1978. *Nucl. Phys.* A298: 169
371. Schmitt, R. P., Russo, P., Babinet, R., Jared, R., Moretto, L. G. 1977. *Nucl. Phys.* A279: 141

ELEMENTAL AND ISOTOPIC COMPOSITION OF THE GALACTIC COSMIC RAYS

J. A. Simpson

Enrico Fermi Institute and Department of Physics, University of Chicago, Chicago, Illinois 60637

CONTENTS

1. INTRODUCTION 324
2. OVERVIEW OF THE GALACTIC COSMIC RAY NUCLEAR COMPONENT 329
   2.1 *Energy Spectra* 330
   2.2 *Evidence for Galactic Origin* 331
   2.3 *Sources and Acceleration* 332
3. PROPAGATION OF COSMIC RAY NUCLEI IN THE GALAXY 333
   3.1 *Confinement in Magnetic Fields* 334
   3.2 *Changes in Energy and Composition During Propagation* 334
   3.3 *Propagation Pathlength Distributions* 335
   3.4 *Secondary Nuclei Production* 335
   3.5 *Models of Cosmic Ray Confinement and Propagation* 336
   3.6 *Nuclear Interaction Cross Sections* 337
   3.7 *Energy Dependence of the Elemental Composition* 338
4. PROPAGATION IN THE HELIOSPHERE 338
   4.1 *Extrapolation of Solar Modulated Spectra* 339
   4.2 *Effect of Solar Modulation on Isotopic Abundances* 342
5. EXAMPLES OF PROPAGATION FROM SOURCES 342
   5.1 *Elemental Composition* 342
   5.2 *Isotopic Composition* 344
   5.3 *Isotopic Monitors of Propagation* 346
   5.4 *Relative Flux Levels for Isotopic Measurements* 346
6. THE RADIOACTIVE NUCLIDES 347
   6.1 *Time Between Nucleosynthesis and Detection* 347
   6.2 *Time Between Nucleosynthesis and Acceleration* 347
   6.3 *Propagation Time in Galaxy* 347
7. EXPERIMENTAL METHODS 348
   7.1 *Resolution and Exposure Factor* 348
   7.2 *Nuclear Species $1 \leq Z \leq 28$* 348
   7.3 *Elemental Composition Above Nickel* 352

8.	SUMMARY OF THE ELEMENTAL ABUNDANCES AT EARTH AND AT THE COSMIC RAY SOURCE	353
	8.1 *Hydrogen and Helium*	353
	8.2 *Average Values of Source Abundances: Helium to Zinc*	353
	8.3 *Zinc to Molybdenum*	354
	8.4 *Molybdenum to Uranium and Beyond*	354
	8.5 *The Antinuclides of Hydrogen ($\bar{H}$) and Helium ($\bar{He}$)*	356
9.	SUMMARY OF OBSERVED AND SOURCE ISOTOPIC ABUNDANCES	357
	9.1 *General Considerations*	357
	9.2 *Hydrogen and Helium Isotopes*	357
	9.3 *Lithium, Beryllium, and Boron Isotopes*	358
	9.4 *Isotopes of Carbon to Iron*	359
10.	EFFECTS OF INJECTION/ACCELERATION IN SOLAR FLARES AND GALACTIC COSMIC RAYS	360
	10.1 *Elemental Abundances of Cosmic Ray Nuclei*	364
	10.2 *Solar Flare Accelerated Nuclei*	365
	10.3 *Dependence on Atomic Number and First Ionization Potential*	365
	10.4 *Solar Flare Isotope Tests for Mass Fractionation*	367
11.	THE SOLAR SYSTEM AND GALACTIC COSMIC RAY SOURCE ABUNDANCES	369
	11.1 *Hydrogen and Helium*	369
	11.2 *Carbon and Oxygen*	370
	11.3 *Nitrogen*	370
	11.4 *Neon, Magnesium, and Silicon*	370
	11.5 *Iron, Cobalt, and Nickel*	372
	11.6 *The Elements Zinc to Ruthenium*	372
	11.7 *Elements Palladium to Cerium*	373
	11.8 *Platinum-Lead and the Actinides*	373
12.	SUMMARY	375

1. INTRODUCTION

The determination of the elemental and isotopic abundances in matter from different astrophysical sites with different physical conditions—both in our solar system and in the galaxy—is important for critically testing current ideas and models for the nucleosynthesis of elements in stars and for the evolution of stars, the interstellar medium, and the solar system (e.g. 1). It is also essential for establishing the origins of the galactic cosmic rays. Historically, solar system material has provided the most comprehensive and accessible information on elemental abundances. Terrestrial, lunar, and meteoritic samples and solar spectroscopic measurements, supplemented by spectroscopic observations of other stars and interstellar matter—along with knowledge of the nuclear and chemical evolutionary history of matter—led to the compilations that are widely accepted as representative of the elemental abundances in the solar system (e.g. 2–6). The uncertainty in elemental abundance determinations has ranged generally from 10% to a factor of two or more for some elements (e.g. B, Kr, Ar, Xe, Hg). These abundances have constituted the reference or "bench mark" set of abundances with which the composition of matter outside the solar system is compared. Until recently it was assumed that the solar system abundances were a homogeneous sample of the solar nebula, since much of the information was derived from meteorites whose ages were $\geq 10^9$ years, and it is believed that since that time the solar surface and atmosphere had not mixed with the core where nucleosynthesis is occurring. However, the discoveries of isotopic anomalies in meteorites (7, 8) revealed that the early solar nebula contained an admixture of matter whose composition was consistent with extra-nebular matter, possibly of supernova origin (9). Since the contribution of anomalous material to the solar system is small ($<1\%$), it is widely believed that the general solar system composition is representative of the interstellar medium as it existed 4.5×10^9 years ago.

Well-established atomic or chemical fractionation effects may introduce biases into the observed elemental abundances without, in general, modifying the isotopic abundances of an element. [However, in dense matter at high temperatures in the solar system there exist isotopic fractionation effects in some meteorites due to thermodynamic processes (10).] Therefore, since isotopic abundances of an element reflect mainly the history of nuclear processes leading to their formation, isotopic measurements are fundamental for guiding the direction of both nuclear and cosmic ray astrophysics. Most of our knowledge of isotopic composition of elements in the solar system comes from meteorites and terrestrial and lunar samples; because of spectroscopic observational difficulties, only a few isotopes could be identified in the solar photosphere and none in the chromosphere or corona. Recently, measurements from satellites are extending the mass range of solar isotopic measurements to accelerated nuclei in the solar wind (e.g. 11) and in solar flares (12–14).

Within the past decade there have been dramatic advances in the measurement of isotopic compositions of extra-solar matter derived from electromagnetic spectra extending from radio emission to the extreme UV and from diverse astrophysical sites such as the interstellar medium (e.g. 15), giant molecular clouds, stars, supernova remnants, and so forth. There are both remarkable similarities and differences of relative abundances over wide astrophysical scales and they provide important clues for models of nucleosynthesis. However, it is not yet certain how important isotopic fraction effects are in some regions (e.g. 16). Since the observational techniques and observational "windows" in the electromagnetic spectrum (e.g. 17) cover selective and generally different narrow ranges of nuclear masses and elements, one must compile and attempt to normalize the diverse measurements in order to obtain a comprehensive view of the abundances in any given astrophysical setting to compare, for example, with solar system abundances.

The galactic cosmic ray nuclei are an important independent channel of information since cosmic ray nuclei are the only accessible sample of matter

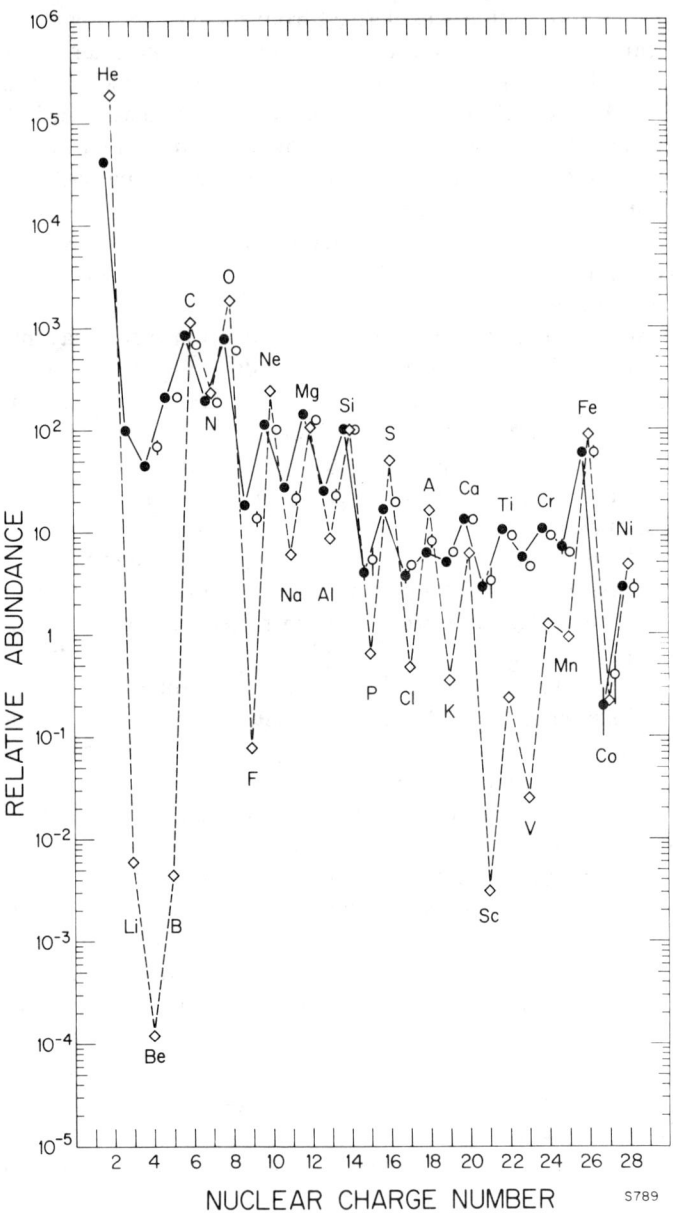

Figure 1 The cosmic ray element abundances (He-Ni) measured at Earth compared to the solar system abundances, all relative to silicon: (*solid circles*) low energy data, 70–280 MeV n^{-1} (20); (*open circles*) compilation of high energy measurements, 1000–2000 MeV n^{-1} (see Table 2); (*diamonds*) solar system (5).

from outside the solar system. This extra-solar-system sample is relatively modern since the measured lifetime for confinement in magnetic fields during propagation from source to observer is $\sim 10^7$ y (18, 19, 19a). Furthermore, it is a unique nuclear sample since it includes all the elements from hydrogen to the actinides. The relative elemental abundances of cosmic ray nuclei from hydrogen to nickel, as measured at the orbit of Earth, are plotted in Figure 1 and compared with the solar system reference abundances compiled by Cameron (5). Figures 2 and 3 illustrate recent measurements of the very much less abundant nuclei occurring in the mass range beyond nickel (21–25).

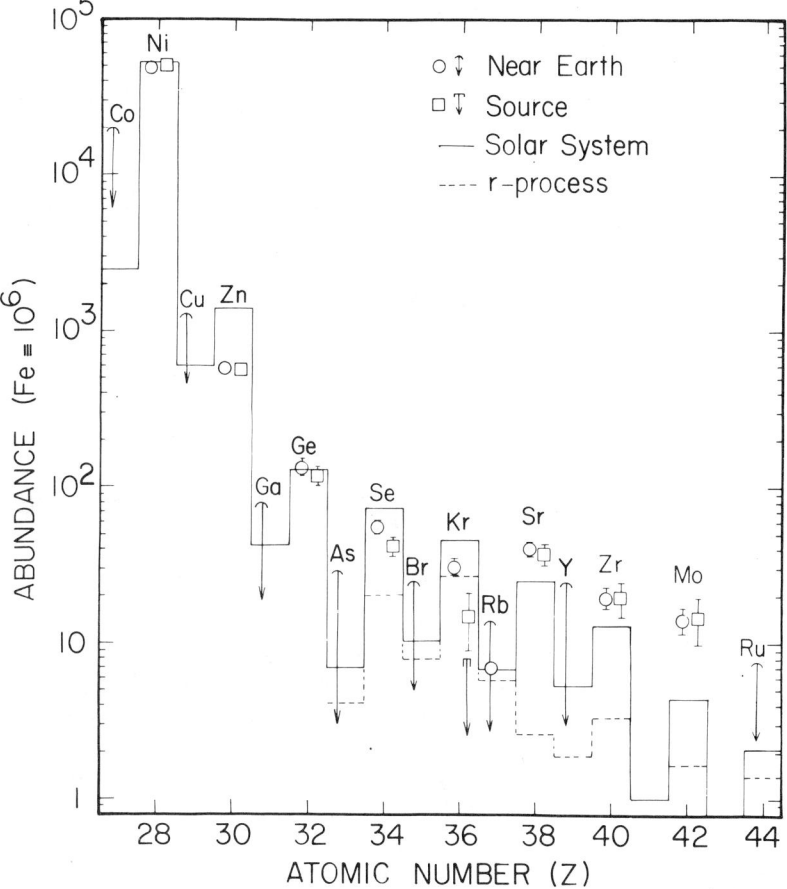

Figure 2 Comparison of cosmic ray element abundances (Co-Ru) measured at Earth and calculated at the cosmic ray source with solar system material (*solid histogram*) and the results of rapid neutron capture nucleosynthesis (*dashed histogram*) (21, 25).

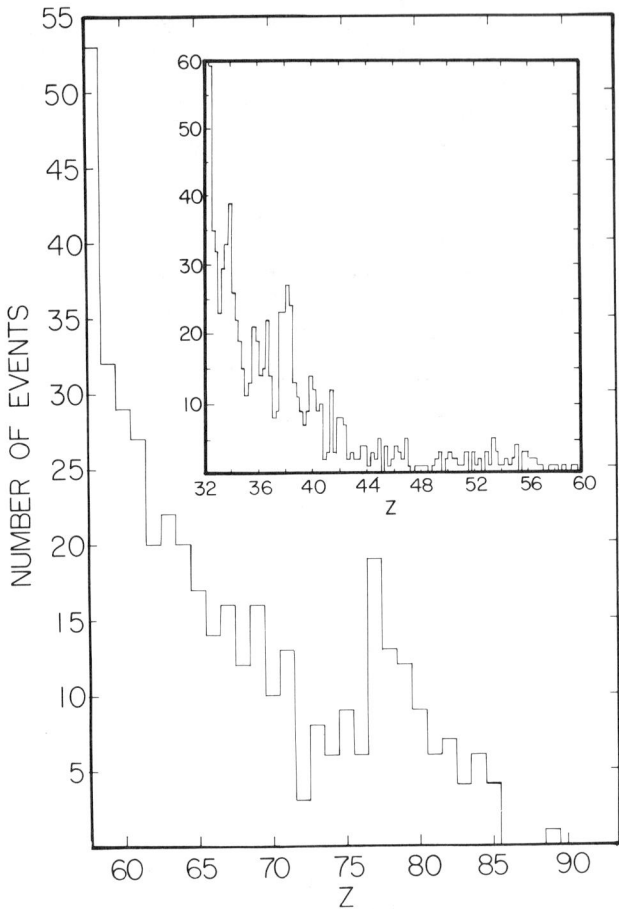

Figure 3 The charge spectrum measured at Earth for cosmic rays with $Z \geq 32$ (22).

The isotopic abundances of cosmic rays convey not only the "imprint" of their nucleosynthesis origins at astrophysical sites in the galaxy, but also their radioactive species reveal the time required for their acceleration or the history of their propagation in interstellar magnetic fields. The rapid development of instrumental methods in satellites, space probes, and balloons during the past 15–20 years has increased dramatically the capabilities for resolving the elements and isotopes in the cosmic rays. A recent illustration of isotopic resolution is shown in Figure 4 (26).

It is the purpose of this review to record the remarkable advances in recent years in our knowledge of cosmic ray nuclear composition (including anti-nuclei), which led to discovering significant differences in the com-

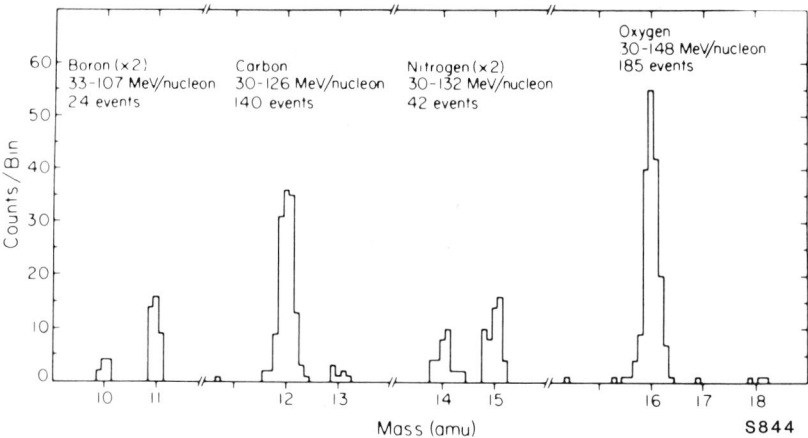

Figure 4 The mass spectrum measured at Earth for the cosmic ray elements boron, carbon, nitrogen, and oxygen (26).

position between solar system matter and the matter at the source(s) of the galactic cosmic rays—results having important astrophysical implications. The cosmic ray source abundances are of paramount importance for determining their origin, which is one of the major open questions in astrophysics. For example, is this matter a sample of interstellar or typical stellar matter, or is it a sample of matter blown off in a supernova explosion, or is it a mixture of sources? Thus, although the elemental and isotopic abundances of galactic cosmic rays are the best-established abundances in astrophysics, their origin is at present not uniquely identified with a galactic source. It is our purpose in this review to discuss the underlying assumptions and biases in determining source abundances and to summarize some of the consequent astrophysical implications. This review is intended for nonspecialists and, therefore, includes a brief summary of cosmic ray physics in addition to tables of measured and source abundances that can be applied to a wide array of astrophysical investigations. However, there are also some new results that may interest investigators in cosmic ray and nuclear astrophysics.

2. OVERVIEW OF THE GALACTIC COSMIC RAY NUCLEAR COMPONENT

The composition of the cosmic rays must be understood within the framework of cosmic ray physics and its astrophysical setting. For the reader not acquainted with the field, the following review and Sections 3

2.1 Energy Spectra

Following the discoveries that cosmic rays were mainly composed of protons (e.g. 27) and heavier nuclei (28) stripped of all their orbital electrons, extensive experimental investigations using balloons and satellites showed that the arriving flux is composed of ∼98% nuclei and ∼2% electrons and positrons [for a summary of the electron/positron component, see (19a, 29, 30)]. The nuclear component consists of ∼87% hydrogen, ∼12% helium, and ∼1% for all heavier nuclei in the general energy range where they have the highest intensity—namely, 10^8–10^{10} eV per nucleon. [In this review the unit of energy is the electron-volt (eV) = 1.6×10^{-12} erg. The further

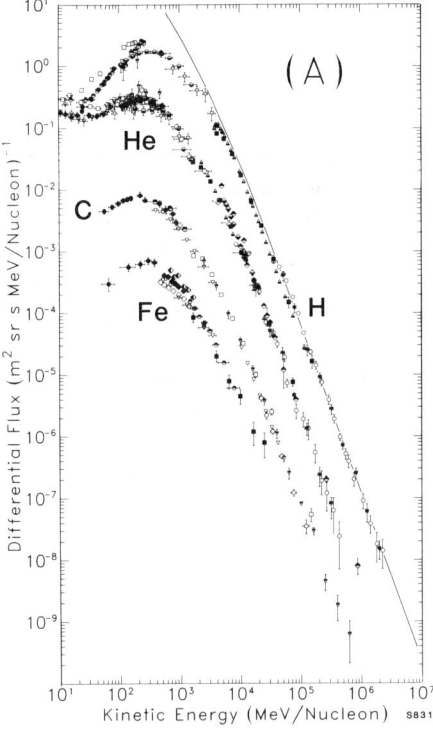

Figure 5a The energy spectra of the cosmic rays measured at Earth. Differential energy spectra for the elements (from top) hydrogen, helium, carbon, and iron. The solid curve shows the hydrogen spectrum extrapolated to interstellar space by unfolding the effects of solar modulation. The turn-up of the helium flux below ∼60 MeV n^{-1} is due to the additional flux of the anomalous ^{4}He component (31, 32).

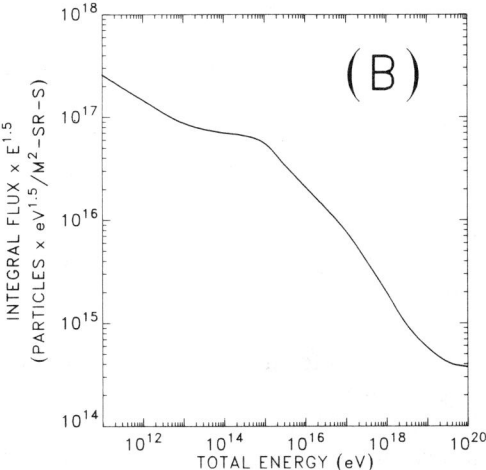

Figure 5b The energy spectra of the cosmic rays measured at Earth. The integral energy spectrum multiplied by $E^{1.5}$ for the nuclear component showing the spectrum at the highest energies (33). The composition above $\sim 10^{13}$ eV has not been established.

abbreviations MeV = 10^6 eV and GeV = 10^9 eV are also used.] The general shape of the observed differential energy spectra for H, He, C, and Fe in Figure 5a is typical of measurements at the orbit of Earth (1 AU $\equiv$ astronomical unit $\equiv$ mean distance between Sun and Earth). At high energy these spectra are well represented by $J_i(E) \propto E^{-\gamma}$ where $J_i(E)$ is the differential flux of nuclear species i at kinetic energy per nucleon E and $\gamma_i = 2.5$–2.7. The spectra of nuclei extend to energies in excess of $\sim 10^{20}$ eV, as shown in Figure 5b, where the integral flux multiplied by $E^{+1.5}$ is plotted. Changes in slope have been verified near $\sim 10^{15}$ to 10^{16} eV and above 10^{19}. Above $\sim 10^{15}$ eV the nuclear composition is still uncertain, with estimates ranging from pure H to pure Fe (e.g. 33). At the highest energies, $\gtrsim 10^{17}$ eV, the gyroradius for either H or Fe in interstellar magnetic fields (~ 1–2×10^{-6} gauss) becomes much larger than the scale size for containment in the disk of our galaxy, whose approximate thickness is shown in Figure 6. Thus, it appears that these ultra-high-energy nuclei probably are extra-galactic in origin (34).

2.2 Evidence for Galactic Origin

There is convincing evidence that the bulk of nuclear radiation below $\sim 10^{15}$ eV has its origin in, and is confined to propagate throughout, our galaxy. For example, nuclei of 10^{10} eV have gyroradii less than 10^{-5} parsec in the interstellar magnetic fields or $< 10^{-7}$ the thickness of the galactic disk (Figure 6). Since their motion is determined by the irregular interstellar

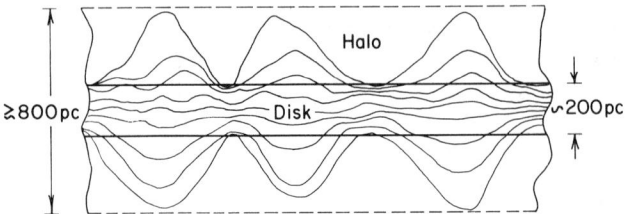

Figure 6 A schematic cross-sectional view of the galaxy indicating the disk and a possible halo and showing the approximate configuration of the galactic magnetic field [1 parsec (pc) $\approx 3 \times 10^{18}$ cm]. The disk thickness is in the range 200 to 400 pc.

fields, the cosmic rays appear to an observer as highly isotropic in the solar neighborhood. Measurements of the flux and distribution of gamma rays with energies $\gtrsim 70$ MeV arising from the reaction of cosmic ray protons with interstellar hydrogen, through neutral pion (π^0) decay, $P_{\text{cosmic ray}} + H \rightarrow \pi^0 + \text{nucleons} \rightarrow \gamma + \gamma + \text{nucleons}$, reveal a γ-ray intensity distribution roughly proportional to the distribution and density of matter and cosmic rays in the galaxy, thus indicating that cosmic rays are present throughout the galaxy. In addition, the γ-ray results suggest that the cosmic ray intensity shows a galactocentric gradient (35–38). Furthermore, the lifetime for containment in interstellar magnetic fields is $\sim 10^7$ y, as measured by the surviving fraction of the radioactive secondary nuclei ^{10}Be ($\tau_{1/2} \sim 1.5 \times 10^6$ y) (see Section 9.3). Therefore, the bulk of the cosmic ray nuclear flux at energies $\leq 10^{15}$ eV must be continually renewed in the galaxy since the record of cosmic ray bombardment in meteorites reveals that the average cosmic ray flux entering the solar system has been constant within approximately a factor of 2 over the past $\sim 4 \times 10^9$ y (e.g. 39) and, from ^{10}Be in deep sea sediments, has been constant to within $\pm 30\%$ over the past $\sim 10^6$ y (40). These observations lead to the assumption of a steady-state condition on a galactic scale, either with the cosmic rays confined to the disk region or extending into a magnetic field halo. This renewal of the cosmic rays requires an average energy input of $\sim 10^{40}$ erg s^{-1} to compensate for the loss of nuclei by nuclear and electronic collisions or escape from the galaxy.

2.3 Sources and Acceleration

There is, as yet, no conclusive proof of the mechanism accelerating the nuclei to cosmic ray energies, nor an identification of the astrophysical sites of the matter that becomes the cosmic radiation. For example, the nuclei could gain their principal energy from stochastic processes in the interstellar magnetic fields during propagation (e.g. 41) after injection from either the ambient interstellar medium, stellar flares, or stellar explosions, such as supernovae (however, see 41a,b,c and Section 3.7). Alternatively, shocks propagating in the interstellar medium or through massive clouds have recently been revived as a mechanism for acceleration (e.g. 42–47). Finally, acceleration in discrete sources, e.g. supernova, either at the time of their explosion, or by shocks within the supernova remnants also may provide the required energy (48–51) [for a recent review of acceleration mechanisms see (52, 53)]. Several astrophysical sources of energy sufficient to accelerate and sustain the bulk of the cosmic rays have been identified in recent years. Thus, the rate of energy input to the galactic cosmic rays is a necessary but not sufficient condition for identifying their origins. The determination of relative isotopic and elemental abundances of the source matter will be crucial for deciding between these and other alternatives, as discussed later in this review.

3. PROPAGATION OF COSMIC RAY NUCLEI IN THE GALAXY

Figure 7 is a schematic representation of the cosmic ray propagation between the source and the observer. An observer located at [a] in the interstellar medium would not only measure relative abundances different from the abundances of the primary nuclei accelerated in the source, but would also observe secondary nuclear species—both stable and radioactive—whose origins are the nuclear interactions occurring between cosmic ray nuclei and interstellar matter during propagation through the interstellar medium. Furthermore, as the nuclei propagate from interstellar magnetic fields into the interplanetary fields of solar origin in the heliosphere (Figure 7), their spectra at low energies undergo changes in

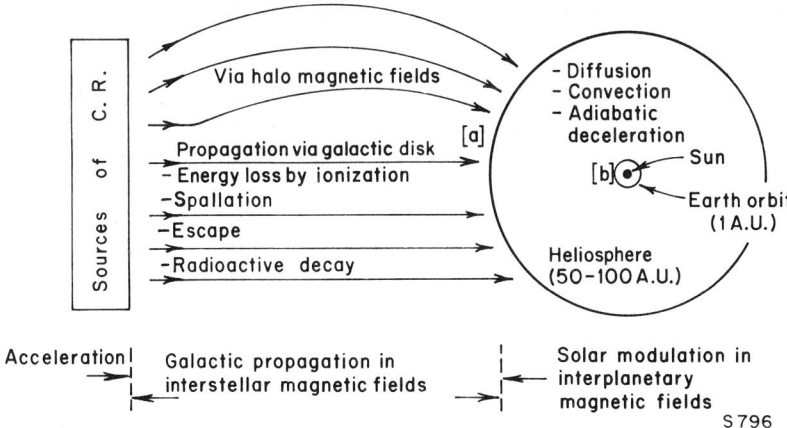

Figure 7 A schematic representation of the life history of a cosmic ray from acceleration in the source region to observation at Earth [b].

energy and intensity. These changes are called solar modulation. Thus, an observer within the heliosphere at [b] must extrapolate measured spectra using a solar modulation model to obtain the local interstellar spectrum at [a], in addition to determining the effects of propagation between the source and point [a]. Solar modulation effects are discussed in Section 4.

3.1 Confinement in Magnetic Fields

The transport mechanisms for cosmic rays in the galaxy are dominated by particle motion in the galactic magnetic fields. The cosmic rays, then, move in spiral trajectories around the magnetic field with gyroradii appropriate to the particle energy. Figure 6 shows a possible magnetic field configuration of our galaxy. Even the field lying approximately in the galactic disk is somewhat disordered, and field "loops" extend out of the disk into a galactic halo. A cosmic ray particle gyrating around such a field will interact with these and smaller irregularities and can be scattered by them so that the cosmic rays may become diffuse in the magnetic field. Therefore, diffusion theory is often employed to describe the large-scale propagation of the cosmic rays (50, 54, 55). However, Figure 6 also indicates that the magnetic field configurations permit access of the cosmic rays to the galactic halo where the propagation conditions are different (55a). Thus, a complete treatment of cosmic ray transport, in the diffusion approximation, must consider diffusion both in the galactic disk and also within the halo region. The particles observed at Earth may have traveled directly through the galactic disk or through the intermediary of a galactic halo, and current data are not yet sufficient to separate these different modes. A recent review of cosmic ray confinement was presented by Cesarsky (56).

3.2 Changes in Energy and Composition During Propagation

For discussion of cosmic ray composition, we are interested less in the details of particle transport and more in the changes in energy and composition experienced by the cosmic ray nuclei during interstellar propagation. For this purpose, cosmic ray propagation can be described in one dimension, x, the amount of matter (in g cm^{-2}) traversed by the particles. The exact spatial distribution of this matter becomes important only when considering radioactive isotopes, since the matter distance x_0, traversed in a time T is

$$x_0 = \rho \beta c T, \qquad 1.$$

where ρ is the density of the matter and β is the particle velocity in units of the speed of light, c. In this approach the propagation of the cosmic rays can be described by the equation

$$\frac{\partial J_i}{\partial x} = \underbrace{\frac{\partial}{\partial E}\left(\frac{dE}{dx} J_i\right)}_{\text{energy loss}} - \underbrace{\frac{N\sigma_i^t J_i}{\bar{A}}}_{\text{nuclear interactions}} - \underbrace{\frac{J_i}{\gamma \beta \rho c T_i}}_{\text{radioactive decay}}$$

$$+ \underbrace{\sum_{j \neq i} \frac{N\sigma_{ij} J_j}{\bar{A}}}_{\text{spallation production}} + \underbrace{\sum_{j \neq i} \frac{J_j}{\gamma \beta c T_{ij}}}_{\text{production by radioactive decay}}, \qquad 2.$$

where $J_i(x)$ is the flux of species i after propagating through an amount of matter x g cm^{-2} subject to the condition that $J_i(0)$ represents the source term. The other parameters in Equation 2 are: dE/dx = the rate of ionization energy loss; σ_i^t = the total inelastic cross section for species i; T_i = the mean lifetime at rest for radioactive decay of species i; σ_{ij} = the cross section for production of species i from fragmentation of species j; T_{ij} = the mean lifetime at rest for decay of species j into species i; N = Avogadro's number; $\bar{A}$ = the mean atomic weight of the interstellar gas ($\sim 94\%$ hydrogen, 6% helium); and γ is the Lorentz factor.

3.3 Propagation Pathlength Distributions

The cosmic rays arriving outside the heliosphere, [a] in Figure 7, may have arrived from a large variety of different paths through the interstellar medium implying a distribution of pathlengths (see below). Thus, the local interstellar flux of species i, F_i, is obtained by averaging the results of Equation 2 at different values of x over the cosmic ray pathlength distribution $P(x)$ as

$$F_i = \int_0^\infty J_i(x) P(x) \, dx, \qquad 3.$$

where $P(x)$ is the normalized probability distribution $[1 = \int_0^\infty P(x) \, dx]$ for the case where the pathlength distributions are energy dependent. Thus, the effects of cosmic ray diffusion and any conditions at the boundary of the confinement region are all contained in the pathlength distribution $P(x)$ (cf 57–59), which must be determined from the cosmic ray data.

3.4 Secondary Nuclei Production

In Figure 1 the large abundances of Li, Be, and B are examples of the generation of secondary nuclei during cosmic ray propagation, since during nucleosynthesis these light elements would have been destroyed in the sources. These secondary nuclei are used as probes of the amount of interstellar matter through which the primary nuclei pass between source and observer. For the light elements, the abundance ratio $(Li+Be+B)/(C+N+O) \approx 0.25$ at intermediate energies implies a pathlength of ~ 8 g cm^{-2} of interstellar matter.

The earliest analyses of secondary element production assumed a uniform "slab" of matter [$P(x)$ is a delta function] through which all of the cosmic rays pass. As the experiments improved and abundances became available in the region of iron, it was found that the uniform slab model predicted larger iron secondary-to-primary ratios than were observed [see, for example, review (60)]. This is due to the larger total inelastic cross section for iron compared for example, to carbon. For the (Sc-Mn)/Fe ratio, a slab of only a few g cm^{-2} was required to explain the observations. The resolution of this discrepancy forced the introduction of a more complicated but physically reasonable pathlength distribution of an exponential form $P(x) = (1/x_0) \cdot \exp(-x/x_0)$ where x_0 is the mean of the distribution. This was found to give a reasonably good fit to the data. Recent experimental evidence has suggested that there may be a deficiency of very short pathlengths in the distribution (61–64, 64a), but this result is still controversial (4, 65, 130).

3.5 *Models of Cosmic Ray Confinement and Propagation*

The measured ratio of secondary to primary nuclei in the cosmic radiation indicates that only a fraction of the primary cosmic rays have interacted with interstellar matter and suggests that part of the cosmic radiation escapes from the confinement region. This fact, and the measured constancy with time of the cosmic ray level during the extended past suggest that the cosmic rays are in a steady state in which the source input is balanced by the escape from the galaxy, nuclear spallation, and energy loss by ionization at low energy.

The interaction of cosmic rays with the galactic magnetic field suggests a diffusive propagation, as described above, in which the escape of the particles from the confinement region is specified by boundary conditions to the diffusion equation (e.g. 50). In a simplified alternative to such heterogeneous diffusion models we can choose not to specify distribution of sources and matter and substitute the diffusion term by a simple escape term. This becomes a homogeneous model called the "leaky box" model because the representative equation can also be derived from confinement in a homogeneous volume whose surface is frequently encountered by the cosmic ray particles having a constant but small probability of leaking out of the volume (66). In this case a simplified model for cosmic ray propagation can be constructed in which the loss by escape and nuclear spallation are balanced, for each species, by fragmentation production and the source term (66, 67). If ionization energy loss and radioactive decay are neglected, the equation for propagation in this model takes a particularly simple form:

$$\frac{dJ_i}{dx} = Q_i - \frac{j_i}{\lambda_e} - \frac{j_i \sigma_i^t N}{\bar{A}} + \sum_{j \neq i} \frac{j_i \sigma_{ij} N}{\bar{A}} = 0, \qquad 4.$$

where Q_i is the source term and λ_e is the mean-free path for escape from the confinement region. This set of equations, one for each species of interest, can be solved analytically to relate the source abundances to the measured cosmic ray composition (e.g. 68). It is only slightly more complicated to include decay and ionization energy loss in the solution. Inherent in this steady-state model is an exponential pathlength distribution, and the results of solving the propagation problem by Equations 2 and 3, with $P(x) = (1/x_0) \exp(-x/x_0)$ without energy loss, or via Equation 4 are identical for $x_0 = \lambda_e$, providing neither x_0 nor λ_e is a function of energy. Steady-state or "leaky box" models, of which this is an example, have been employed for over a decade to interpret most of the cosmic ray data.

Whether Equations 2 and 3 or Equation 4 are employed to determine the source abundances, the calculations require accurate values (and energy dependences) for the following parameters:

1. total inelastic cross sections, σ_i^t,
2. spallation cross sections, σ_{ij},
3. energy loss by ionization, dE/dx,
4. radioactive decay and the branching ratios for each species, and
5. the pathlength distribution (69, 70).

Recent comparisons of interstellar propagation calculations by six independent investigative groups have shown (71) that by using the same input parameters for different propagation computer codes used by these investigators yielded equivalent results to within $\sim 10\%$. The principal differences in published source abundances arise from the use of different values for the input parameters listed above. For example, differences in the value of total inelastic cross sections and their energy dependence used by different investigators can lead to significant differences in the calculated source abundances, especially for heavy nuclei such as iron. In computing source abundances it is important not to neglect energy loss by ionization, even for nuclei up to $\sim 10^9$ eV per nucleon (eV n^{-1}).

3.6 Nuclear Interaction Cross Sections

At present the principal uncertainties are associated with the total inelastic and spallation cross sections and the pathlength distribution. Some important cross sections are by now relatively well measured, such as for the reaction $^{12}C + ^{1}H \to ^{10}B + $ nucleons, while others must be calculated from nuclear physics models using semi-empirical methods, such as proposed by Rudstam (72, 73) or as extensively developed by Silberberg & Tsao (74–76). Fragmentation parameters have also been derived from the break-up of cosmic ray nuclei in the atmosphere or nuclear emulsion (77), or from accelerator beam measurements (cf 78, 80). Recently, extensive tables of measured cross sections were compiled (79, 80) and for many of the nuclear species the previously derived empirical values compare well. Overall, it appears that the uncertainties in cross sections and spallation data—and their energy dependences—introduce errors of $\sim \pm 20\%$ in the determination of source abundances for the abundant elements and isotopes, but substantially larger uncertainties for less abundant nuclear species where there is appreciable secondary production during propagation.

3.7 Energy Dependence of the Elemental Composition

An energy dependence in the ratios of secondary nuclei to primary nuclei was discovered in 1972 (81, 82) that becomes important above $\sim 2 \times 10^9$ eV per nucleon. The origin of this dependence is still not settled, but this effect can be interpreted as an energy (or magnetic rigidity) dependence at high energy of the leakage or escape mean-free path, λ_e. For $\lambda_e \propto E^{-\alpha}$, where $\alpha \approx 0.3$–0.6 (cf 69, 82a, 83) and E is the total energy of the cosmic rays, the calculations reproduce the data for $E \gtrsim 2 \times 10^9$ eV per nucleon. This energy dependence has recently been invoked to argue against cosmic ray acceleration in the diffuse interstellar medium, but in favor of models where acceleration occurs prior to extensive interstellar propagation (e.g. 41a,b,c).

At lower energies the experimental data again show an energy dependence with secondary-to-primary ratios decreasing with decreasing energy (63). For a dynamical halo model (cf 84) in which the cosmic rays stream out of the galaxy in a diffusion-convection model, an energy dependence arises from the velocity of outflow to a galactic halo (e.g. 56, 85, 86). This energy dependence is important only at low energies $< 1 \times 10^9$ ev per nucleon, and can be observed in the data only after correcting for the effects of solar modulation and ionization energy loss. Analysis of low energy cosmic ray data (63, 70) confirmed such an effect, but the results depend significantly on the assumed modulation level.

4. PROPAGATION IN THE HELIOSPHERE

Measurements of galactic cosmic ray composition are still limited to positions within the heliosphere (Figure 7), a region whose electrodynamical properties are established by the expansion of the solar corona forming an outward flowing solar wind (87, 88) carrying magnetic fields to a distant interface with the heliopause, which lies between the heliospheric boundary and the local interstellar medium. The boundary of the heliosphere with the heliopause is determined by both charge exchange between the solar wind and neutral interstellar gas and by the pressure balance of the interstellar magnetic field; it is estimated to lie in the radial range ~ 50–100 AU from the Sun (89, 90). A high energy charged particle propagating from [a] to [b] in Figure 7 will scatter and diffuse inward in the irregular magnetic fields, will tend to be convected outward as a result of the outward momentum flow of the solar wind, and will undergo adiabatic deceleration owing to the outward expansion of the magnetic fields (91). A particle that has a kinetic energy of 500 MeV per nucleon in the local interstellar medium may lose 1/3 or more of its kinetic energy before arriving at 1 AU. The resulting change in the spectra of galactic cosmic rays between [a] and [b] is a heliocentric phenomenon (92) that has so far been investigated to beyond $\simeq 25$ AU (32, 92a,b).

4.1 Extrapolation of Solar-Modulated Spectra

Solar modulation is dependent upon the mass-to-charge (A/Z) ratios of the nuclear species, with the effect on spectra being significant for nuclei of energies up to at least $\sim 10^4$ MeV n^{-1}. This includes the energy range over which the relative abundances of nuclear species are well measured and reported in this review. Although solar modulation is not well understood in detail, we can determine the level of modulation by comparing the electron spectrum measured at Earth with the local interstellar electron spectrum inferred from the measured synchrotron radiation at radio frequencies and from bremsstrahlung γ rays. By using a model of solar modulation we can derive local interstellar spectra for the nucleonic component of the cosmic rays. The electrodynamical properties of the heliosphere change over the 11-year solar activity cycle, so the modulation level changes throughout the solar cycle as shown in Figure 8b for the cosmic ray nuclear component (see reviews 93, 94). In Figure 5a we show an example of the extrapolation of the hydrogen spectrum measured at 1 AU to the nearby interstellar medium, based on current solar modulation theory. The fluxes for H and He at solar minimum are given in Table 1. For the relative abundances of elements shown in Table 2 with $A/Z \simeq 2$, the corrections for modulation effects are negligible over the relatively wide spectral range $\sim 10^8$–10^{10} eV n^{-1}. Measurements of the elemental abundances of primary nuclei, He through Ni, at both solar maximum and solar minimum show close agreement as illustrated in Figures 8a and 8b, which show measurements on satellites carried out at very different modulation levels. This is in contrast to the large change in the abundance

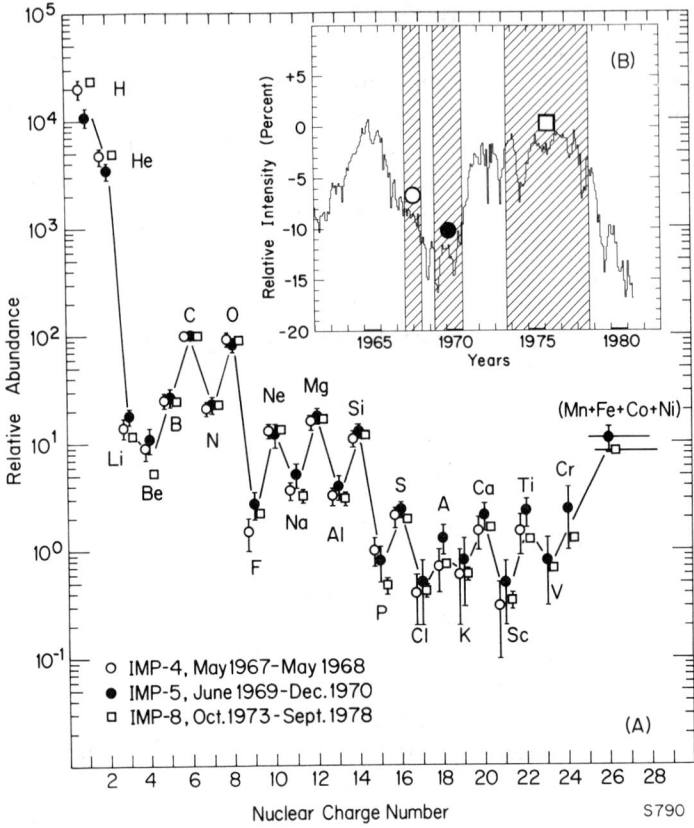

Figure 8 (a) The relative abundances of H to Ni (C ≡ 100) measured by the University of Chicago in satellites during different phases of the solar cycle: (*open and closed circles*) solar maximum modulation; (*squares*) solar minimum modulation. (b) The Climax Neutron monitor intensity for 1962–1981 with the periods of measurement for (a) indicated. (This neutron monitor records the relative intensity changes of the nuclear component of the cosmic rays $\geq 3 \times 10^3$ MeV per nucleon.)

Table 1 Cosmic ray hydrogen and helium

Element	100 MeV per nucleon[a]	50,000 MeV per nucleon[b]
H	1.54 ± 0.12[c]	$(4.0 \pm 0.4) \times 10^{-4}$[c]
He	0.33 ± 0.02[c]	$(1.9 \pm 0.2) \times 10^{-5}$[c]
H/He	4.7 ± 0.5	21 ± 3

[a] Values resulting from the best fit modulation curve to experimental points measured during the 1977 solar minimum at 1 AU (124, 125; M. Garcia-Munoz, private communication).

[b] From a power-law spectral fit at higher energies in Figure 5.

[c] Differential flux (nuclei/m^2-sr-s-MeV-n^{-1}).

ratios for secondary/primary nuclei of up to ~50% due to the changing level of adiabatic deceleration over a solar cycle.

The propagation calculations include propagation through both the interstellar medium and the heliosphere. They start from the measured abundances of the isotopes of Li, Be, and B, which are assumed to be zero at the sources. Using Li, Be, and B abundances measured at different levels of solar modulation, Garcia-Munoz, Mason & Simpson (95) and Hinshaw, Wiedenbeck & Greiner (96) placed important constraints on the values of the modulation level and the shape of the source spectra.

Table 2 Galactic cosmic ray elemental abundances at 1 AU (normalized to Si ≡ 100) (energy intervals given in MeV per nucleon)

Element	70–280[a]	600–1000[b,c]	Average at 1000–2000[c,d,e]
He	41700 ± 3000	27030 ± 580	
Li	100 ± 6	136 ± 3	
Be	45 ± 5	67 ± 2	69.4 ± 10
B	210 ± 9	233 ± 4	212 ± 10
C	851 ± 29	760 ± 16	684 ± 27
N	194 ± 8	208 ± 5	188 ± 6
O	777 ± 28	707 ± 15	607 ± 28
F	18.3 ± 1.3	17.0 ± 1.1	13.6 ± 2.3
Ne	112 ± 6	113 ± 3	100 ± 3
Na	27.3 ± 3.4	25.8 ± 1.1	21.3 ± 3.2
Mg	143 ± 6	142 ± 4	125 ± 12
Al	25.2 ± 3.0	28.2 ± 1.2	22.2 ± 3.2
Si	100	100	100
P	4.0 ± 0.7	5.3 ± 0.5	5.3 ± 1.6
S	16.4 ± 1.2	23.1 ± 1.1[b]	19.6 ± 0.9
Cl	3.6 ± 0.5	6.4 ± 0.5[c]	4.7 ± 0.4
A	6.3 ± 0.6	10.2 ± 0.7[c]	8.2 ± 1.2
K	5.1 ± 0.6	7.2 ± 0.5[c]	6.3 ± 0.4
Ca	13.5 ± 1.0	16.1 ± 0.9	13.1 ± 1.2
Sc	2.9 ± 0.5	4.5 ± 0.5	3.3 ± 1.1
Ti	10.7 ± 0.9	10.2 ± 0.7	9.1 ± 0.9
V	5.7 ± 0.6	6.7 ± 0.5	4.6 ± 0.3
Cr	10.9 ± 1.0	11.8 ± 0.8	9.1 ± 0.8
Mn	7.2 ± 1.2	8.2 ± 0.7	6.3 ± 0.4
Fe	60.2 ± 3.2	69.8 ± 2.0	60.5 ± 7.6
Co	0.2 ± 0.1	—	0.4 ± 0.2
Ni	2.9 ± 0.4	3.7 ± 0.5	2.8 ± 0.6
Cu			0.038 ± 0.006[e]
Zn			0.035 ± 0.005[e]

[a] (△) (20).
[b] (○) (128).
[c] (○) (129).
[d] (○) (130, 132).
[e] (△) (21, 131, 131a).

(△) ≡ Satellite   (○) ≡ Balloon

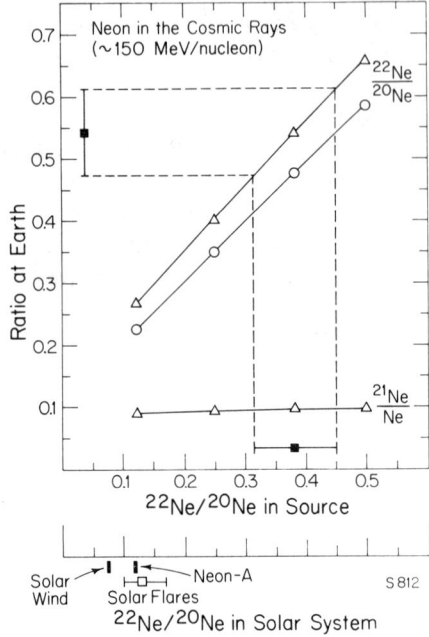

Figure 9 The transformation of the measured ^{22}Ne/^{20}Ne ratio at Earth (180) into the value at the cosmic ray source. The triangles show the result including solar modulation and the circles indicate the line to be used if modulation is neglected. Values of ^{22}Ne/^{20}Ne measured in solar system materials are indicated at the bottom (see text for details).

4.2 *Effect of Solar Modulation on Isotopic Abundances*

As an example of corrections required to determine isotopic source ratios of an element from 1 AU measurements (A/Z is different for different isotopes), we show in Figure 9 the transformation of the observed isotopic abundance ratio ^{22}Ne/^{20}Ne at ~ 150 MeV n^{-1} to the ratio at the cosmic ray source (cf 97, 180). Two curves are shown in Figure 9, one for no solar modulation and the other for a modulation level characteristic of the solar minimum period of the observations. The effect of solar modulation is 15–20% for the cosmic ray ratio at the source inferred from the measurements at Earth.

5. EXAMPLES OF PROPAGATION FROM SOURCES

5.1 *Elemental Composition*

By applying the foregoing principles of cosmic ray propagation through the interstellar medium, beginning with an assumed composition at the source, we can calculate the relative elemental abundances expected in the local interstellar medium and, after solar modulation, the relative abundances at

1 AU. For the simple example in Figures 10a and 10b, we have assumed $x_0 = 6$ g cm^2, $\bar{\rho} = 0.3$ hydrogen cm^{-3}, a modulation parameter $\phi = 500$ MV (near solar minimum), and no energy loss by ionization. All values are for a kinetic energy of $\sim 10^3$ MeV n^{-1}. By comparing Figure 1, Tables 2 and 3, and these illustrative histograms, the principal effects of propagation on the relative abundances become evident—including the large secondary production of Li, Be, and B and the sub-iron group of elements Sc to Mn. Figure 2 is another example wherein the observed abundances of very high

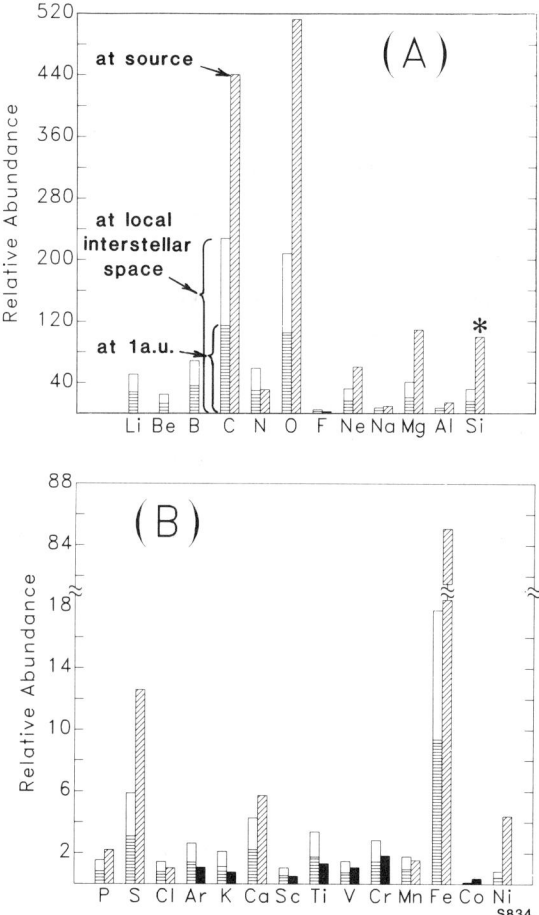

Figure 10 Illustrative comparison of elemental abundances at the galactic cosmic ray source (*slashed shaded bar*), with the calculated abundances for the local interstellar medium (*open* plus *horizontal shaded bar*), and for 1 AU after modulation (*horizontal shaded bar*). All values are for a kinetic energy of $\sim 10^3$ MeV n^{-1}. Source abundances are those from Table 3, normalized to Si $\equiv 100$. Solid bars indicate species whose abundance is very uncertain.

Table 3 "Average" cosmic ray galactic source abundances compared to "solar system" and "local galactic" abundances (normalized to Si ≡ 100)

Element	Cosmic ray source from (127)	Cosmic ray source from (21, 141)	"solar system"[a]	"local galactic"[b]
He				$(0.27 \pm 0.06) \times 10^6$
C	444 ± 57	420 ± 32	1110	1300 ± 300
N	31 ± 7	45 ± 18	231	230 ± 100
O	513 ± 66	505 ± 20	1840	2300 ± 500
F	$0 \to 5$	<2.5	0.078	$0.093 \,(1.6)^c$
Ne	60.5 ± 10	63 ± 8	240	$270 \,(1.7)^d$
Na	9.7 ± 2.0	5.6 ± 3.2	6	5.6 ± 0.9
Mg	109 ± 15.1	105 ± 6	106	105 ± 3
Al	14.4 ± 2.4	10.9 ± 3.9	8.5	8.4 ± 0.4
Si	100	100 ± 6	100	100 ± 3
P	2.2 ± 0.5	<2.5	0.65	0.96 ± 0.20
S	12.6 ± 2.0	14.3 ± 2.5	50	45 ± 13
Cl	1.0 ± 0.5	<1.6	0.47	$0.47 \,(1.6)^c$
Ar	$0 \to 2.1$	3.6 ± 0.8	10.6	$9.0 \,(1.7)^d$
K	$0 \to 1.5$	<1.9	0.35	0.36 ± 0.12
Ca	5.6 ± 2.2	7.1 ± 2.0	6.25	6.2 ± 0.8
Sc	$0 \to 1.0$	—	0.003	0.0035 ± 0.0005
Ti	$0 \to 2.6$	<2.4	0.24	0.27 ± 0.04
V	$0 \to 2.1$	—	0.025	0.026 ± 0.005
Cr	$0 \to 3.6$	<2.9	1.27	1.30 ± 0.12
Mn	$1.5^{+2.1}_{-0.6}$	<3.7	0.93	0.79 ± 0.17
Fe	85.1 ± 17.6	91 ± 5	90.0	88 ± 6
Co	$0 \to 0.7$	0.31 ± 0.13	0.22	0.21 ± 0.03
Ni	4.3 ± 0.8	4.8 ± 0.7	4.78	4.8 ± 0.6
Cu		0.061 ± 0.011		$0.052 \,(1.6)^c$
Zn		0.059 ± 0.011		$0.135 \,(1.6)^c$

[a] From (5).
[b] From (4).
[c] (1.6) indicates values within factor 1.6
[d] (1.7) indicates values within factor 1.7.

Z nuclei are "propagated back to the source" and then compared with solar system abundances (21).

5.2 Isotopic Composition

If instead of elemental abundances the individual isotopes of each element are followed in the calculation, the isotopic abundances that would be observed can be determined including secondary and tertiary production and radioactive decay of nuclear species during interstellar propagation. Using the relative isotopic abundances compiled by Cameron (5) for Ca and Fe and the recently determined neon, magnesium, and silicon isotopic ratios for cosmic ray source matter, we obtain the histograms in Figures 11a and 11b, showing the isotopic abundances of Ne, Mg, Si, Ca, and Fe, respectively, arriving at Earth. This example illustrates one of the important reasons for undertaking isotopic measurements, namely, determining more precisely the physics of the interstellar propagation.

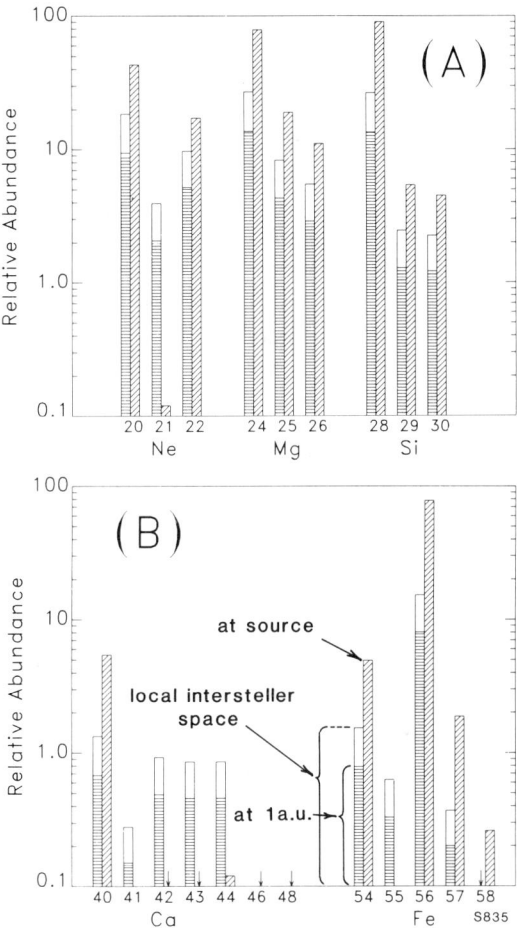

Figure 11 Illustrative comparison of isotopic abundances of Ne, Mg, and Si (Figure 11a) and Ca and Fe (Figure 11b). The galactic cosmic ray source abundances are assumed to be the solar system values given by Cameron (5) (*slashed shaded bar*), compared after propagation to the local interstellar medium (*open* plus *horizontal shaded bar*), and at 1 AU after solar modulation (*horizontal shaded bar*). All values are for $\sim 10^3$ MeV n^{-1}. Arrows indicate relative abundances <0.1.

5.3 Isotopic Monitors of Propagation

By selecting a source element that includes an isotope that is either rare or absent, the build-up of that isotope relative to the neighboring isotopes of the element during interstellar propagation provides a monitor of secondary production, i.e. an independent cross-check for the propagation calculations. For example, the secondary isotope ^{2}H in the quartet ^{1}H, ^{2}H, ^{3}He, and ^{4}He is a monitor of both secondary production and energy changes during propagation (see review, 98). A recent example is ^{21}Ne, which is negligible in source material so that any ^{21}Ne observed is secondary. Therefore, if the ratio of production cross sections $\sigma(^{21}\text{Ne})/\sigma(^{22}\text{Ne})$ is known, the observed ^{22}Ne abundance can be corrected for secondary production independent of the interstellar propagation model or pathlength distribution (for example, see 99). Other examples of missing or rare isotopes in elements include ^{17}O, ^{41}Ca, ^{55}Fe, or ^{33}S. Thus, by obtaining in the laboratory the ratios of production cross sections for nuclear species, it becomes possible with sufficient instrumental resolution and number of analyzed events to determine isotopic source abundances from measured abundances, even of relatively rare isotopes, with an error $\sim \pm 20\%$ for the relative abundances.

5.4 Relative Flux Levels for Isotopic Measurements

Figure 12 displays the relative fluxes of nuclear species arriving at 1 AU during a period of solar minimum modulation and after interstellar

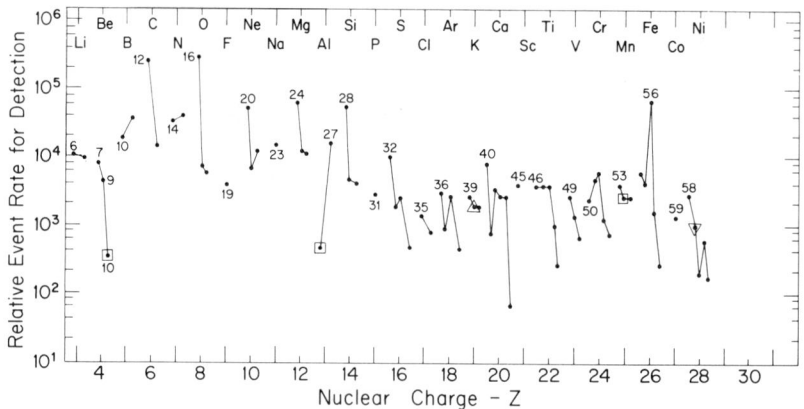

Figure 12 The relative event detection rate for galactic cosmic ray isotopes at Earth during a solar minimum modulation period. All radioactive decays have been included, and isotopes with calculated intensities below the scale have been omitted. Radioactive isotope "clocks" are indicated as measuring: △—time between nucleosynthesis and observation; ▽—time between nucleosynthesis and acceleration; □—time of confinement in the galaxy (see text for details).

propagation of the isotopes of the elements from sources whose isotopic composition was assumed to be like the solar system. Above Ni the elemental abundances decrease rapidly from Ni to U so that there are no contributions of secondary nuclei from the trans-nickel elements to the observed abundances below nickel.

6. THE RADIOACTIVE NUCLIDES

The cosmic ray radionuclides that originate during nucleosynthesis of the cosmic ray matter are called primary. Secondary radionuclides are produced as a result of nuclear collisions during propagation through the interstellar medium. Radionuclides, either primary or secondary, decay with different lifetimes and thus measurements of the surviving fraction of radionuclides relative to stable isotopes can yield the elapsed time for events in the life of cosmic ray matter (see review 100).

6.1 Time Between Nucleosynthesis and Detection

In the mass ranges where they can be measured at present are ^{40}K($\tau_{1/2}$ = 1.3 × 10^9 y) and ^{60}Fe($\tau_{1/2}$ = 3 × 10^5 y) (if measured at very high energies), although nucleosynthesis theories predict a very low intensity for the ^{60}Fe isotope. Trans-bismuth radionuclides cannot yet be measured but will be important for determining the origin of cosmic rays (see also Section 11). The primary heavy radionuclides include ^{247}Pu$_{94}$ (8 × 10^7 y), ^{237}Np$_{93}$ (2.1 × 10^6 y), ^{248}Cm$_{96}$ (4.7 × 10^5 y), and possibly ^{247}Cm$_{96}$ (1.5 × 10^7 y).

6.2 Time Between Nucleosynthesis and Acceleration

This time is determined by abundance measurements of radionuclides synthesized at the source that decay by electron capture, but that can survive in the cosmic rays if they are accelerated at the source before decaying. These radionuclides include ^{56}Ni ($\tau_{1/2}$ = 6.10 d), ^{57}Co ($\tau_{1/2}$ = 270 d), and ^{59}Ni [$\tau_{1/2}$ = 8 × 10^4 y (electron capture); $\tau_{1/2}$ > 8 × 10^6 y (β^+)]. They will be observed only if the time for electron stripping during acceleration is less than the electron capture half-life (cf 101, 233).

6.3 Propagation Time in Galaxy

Radionuclides useful for the determination of propagation time in the Galaxy include ^{10}Be ($\tau_{1/2}$ = 1.5 × 10^6 y), ^{26}Al ($\tau_{1/2}$ = 7.4 × 10^5 y), ^{36}Cl ($\tau_{1/2}$ = 3 × 10^5 y), and ^{54}Mn ($\tau_{1/2}$ ≈ 2 × 10^6 y). These nuclides are assumed to be formed in collisions of the primary cosmic rays with interstellar matter, and therefore their measurement gives the average density of interstellar matter traversed. ^{10}Be decay is discussed in Section 9.3 and the determination of lifetime from the manganese isotopes is discussed in Koch et al (102).

The abundances of the arriving primary and secondary radionuclides are very low relative to the dominant isotopes in the element range Li-Ni. The surviving abundances of some of the radionuclides are shown in Figure 12 for a period of minimum solar modulation.

7. EXPERIMENTAL METHODS[1]

7.1 Resolution and Exposure Factor

Although it is not the purpose of this review to include detailed information on the wide variety of measurement techniques used to obtain cosmic ray energy spectra or elemental and isotopic composition results, it is worthwhile to summarize some of the principal concepts, using as examples instruments that have provided data on isotopic or elemental abundances. For any experiment, the total number of measured nuclei of a given species is proportional to the instrument detection area, A, the solid angle for incident nuclei, Ω (giving the geometrical factor $A \times \Omega$ in cm^2 sr, for example), and to the measurement time, T, giving a total exposure factor $F = A\Omega T$. Electronic instrumentation (active detectors) with $A\Omega$ typically 1 to 10 cm^2 sr is carried on spacecraft free from the Earth's atmosphere for long times (1–5 years) or on balloons with $A\Omega \sim 10^2$–10^4 cm^2 sr for times of the order of 1 day. Passive detectors include nuclear emulsions, which must be recovered and developed in the laboratory. They record not only the incoming charged particles, but also the secondary particles that may result from nuclear interactions in the emulsions. Thus, they make possible the study of kinematics of a nuclear interaction. Meteorites and dielectric sheets (e.g. Lexan plastic) are also passive detectors that, upon chemical etching, reveal particle trajectory and charge information (104a, Section 7.3). The capability of a given instrument to measure adjacent nuclear charges Z (or masses A) depends upon the instrument's charge (or mass) resolution as given by the full width at half maximum (FWHM) of the peak distribution as a function of Z (or A). (For a Gaussian distribution FWHM = 2.36 σ, where σ is the characteristic width.) The σ required to resolve adjacent nuclear species varies with the relative abundance ratio of the two species in question.

7.2 Nuclear Species $1 \leq Z \leq 28$

Simultaneous measurements of energy loss and residual energy (i.e. bringing the nucleus to rest in the detector), as illustrated in Figures 13a and 13b, provide the highest mass and charge resolution for nuclei with energies

---

[1] For detailed reviews of experimental methods and instrumentation, see (100, 102, 103, 104, 104a).

≲ 500 MeV n⁻¹, where the nuclei have ranges less than the mean nuclear interaction length in most detector materials. This energy range includes the peak intensity of the cosmic rays at 1 AU (e.g. Figure 5a). The introduction of the semiconductor detector telescope on the IMP-1 satellite (Figure 13a), which resolved ^{1}H, ^{2}H, ^{3}He, ^{4}He, and the elements $1 \leq Z \leq 9$, marked the beginning of a rapid evolution that is producing semiconductor instruments of ever-increasing resolution (105). A major factor limiting the resolution is the inherent spread in pathlength in the energy loss detector due to the isotropic incidence of the cosmic rays. A three- to four-fold reduction in σ was obtained in the 1970s by reducing the width of the pathlength distribution using curved semiconductor detectors (Figure 13c, 106).

The importance of defining the particle trajectory is illustrated on Figure 14 where ^{40}Ar accelerator data are shown for three cases of trajectory definition in a solid-state telescope. The decrease in σ as the trajectory determination improves is striking. The ISEE-3 satellite included instruments (shown schematically in Figure 15a) that measured the trajectory of each particle, using in one case semiconductor detectors (107) and in the other instrument electron drift chambers (108). A mass resolution of $\sigma \lesssim$ 0.2 amu was demonstrated by these two ISEE-3 instruments for elements ranging from boron to iron (cf Figure 4). Position-sensing semiconductor detectors have been developed that can provide even higher resolution and

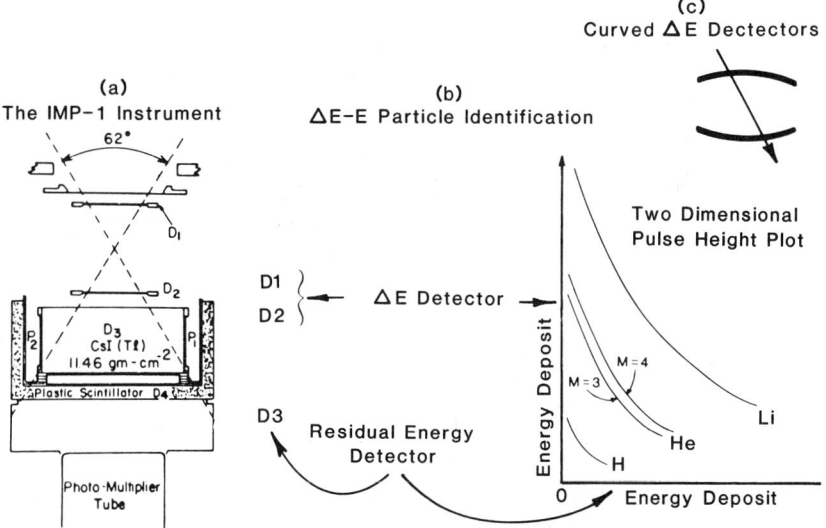

Figure 13 (a) Cross-sectional view of the IMP-1 telescope (105). (b) The $\triangle E - E$ particle identification technique. (c) Curved $\triangle E$ detectors (106).

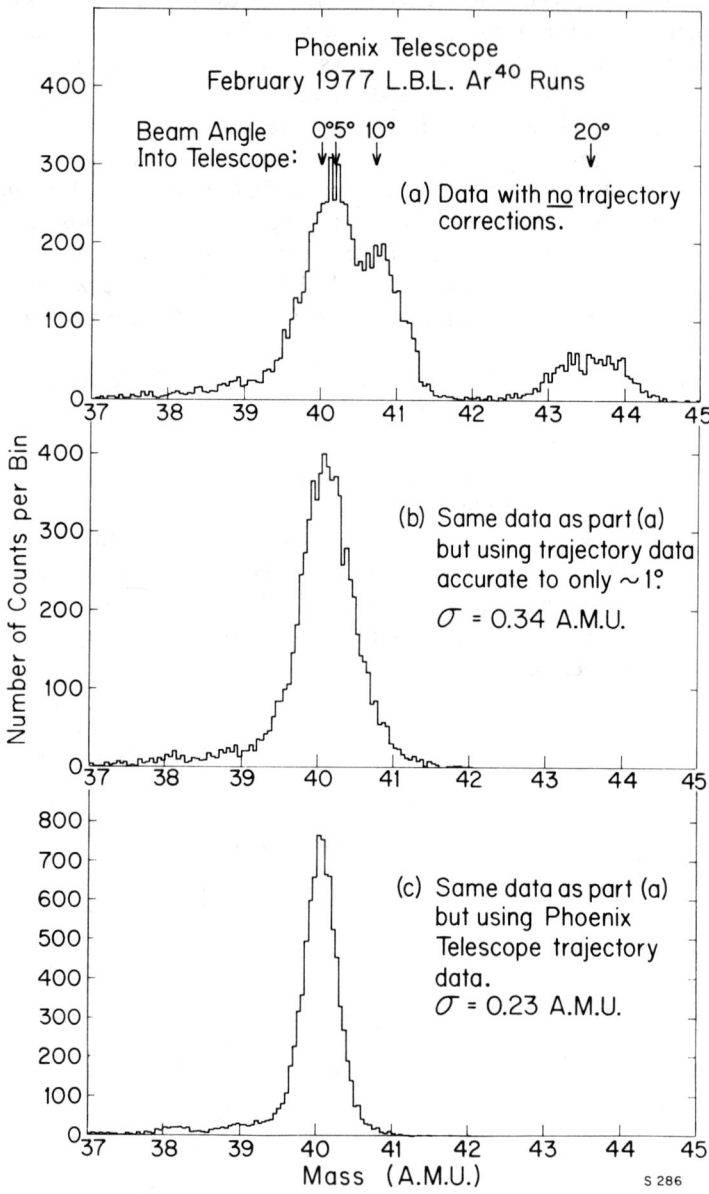

Figure 14 The mass resolution for an accelerator beam of ^{40}Ar nuclei at different angles showing the improvement obtained by the use of position-sensing detectors (Section 10.6) in a multirange solid-state telescope (e.g. 109a). (AMU ≡ atomic mass unit.)

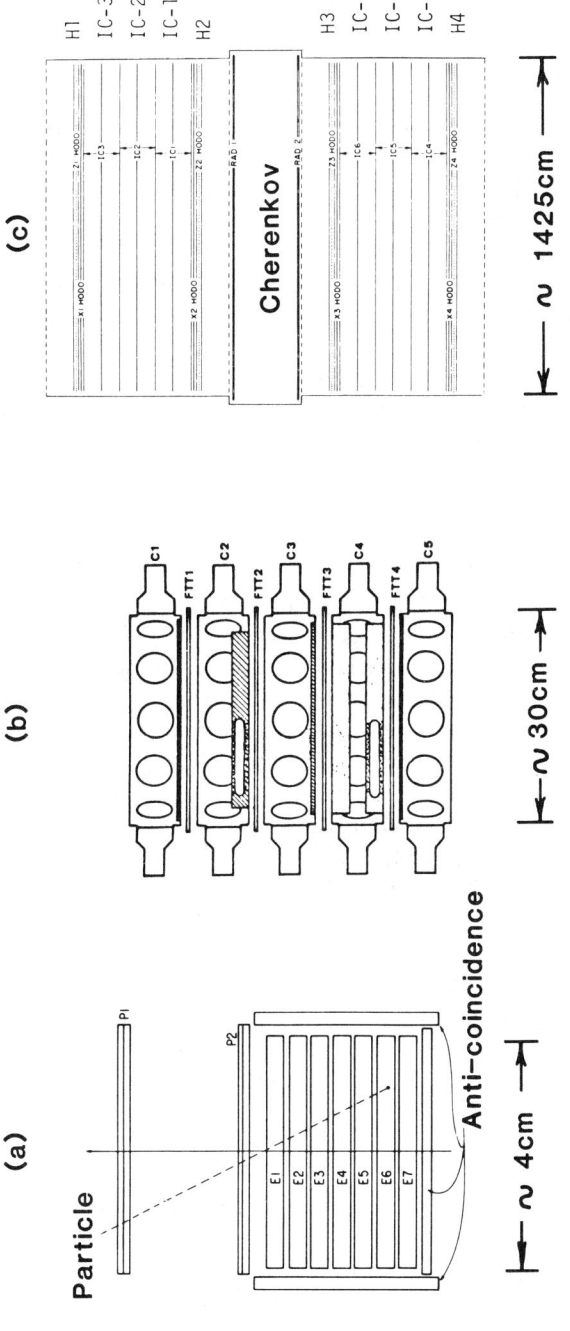

Figure 15 Comparison of recent cosmic ray satellite instruments. (*a*) Schematic diagram of an ISEE class telescope for the measurement of cosmic rays isotopes (107, 108). (*b*) The HEAO-C2 experiment of the Danish-French collaboration (116). (*c*) The ultraheavy nuclei instrument (HEAO-C3) for the study of cosmic rays with $Z \geq 30$ (117).

geometrical factors for future missions such as the International Solar Polar Mission (109, 109a).

The overall requirements for satellite measurements of the abundant isotopes (in Figure 12) are $\sigma \lesssim 0.25$, a precision of energy measurement $\sim \pm 0.3\%$, and observation times of the order 1–3 years, with typical $A\Omega = 5$–10 cm^2 sr. In order to measure the rare isotopes, $A\Omega$ must be increased by more than a factor of 10, σ must be $\lesssim 0.2$ with energy measurements made to $\pm 0.1\%$ for an observing time of ~ 1–3 years.

Isotopic measurements from balloon-borne experiments have employed a different technique—Čerenkov detectors combined with total energy or range measurements. The Čerenkov detector, a long-used technique in balloon experiments (see, for example, 110), provides an accurate measurement of the particle velocity, βc, above a threshold set by the index of refraction, n, of the detector material—i.e. $\beta n \gtrsim 1$. For a detector of thickness x, the light signal, I, is

$$dI/dx \propto \left[1 - \frac{1}{(\beta n)^2} \right]$$

(see, for example, 111, 111a). A recent instrument using curved scintillators, a Čerenkov detector, and a total energy scintillator, with a stated resolution of $\sigma = 0.4$ for the iron isotopes, has been developed by Webber (112).

At high energies, where the particle range exceeds the detector thickness, it is extremely difficult to measure isotopes, and the successful experiments, to date, have used the geomagnetic field to obtain mean masses using the method proposed by Peters and co-workers (113, 114). In this scheme a Čerenkov counter is used to accurately measure the velocity distribution of the incident nuclei (of a given charge Z) which is then compared to the distribution allowed by the Earth's field at the location of the experiment. In a balloon experiment, Dwyer (115) and Dwyer & Meyer (115a) demonstrated that this technique yielded mean masses in the region of oxygen consistent with solar system abundances but with substantially larger errors than for the low energy measurements. Another recent approach, also incorporating trajectory determination, is the HEAO-3 satellite experiment of the Copenhagen-Saclay collaboration (116, 176, 228), shown schematically in Figure 15b, with a reported charge resolution $\sigma \lesssim 0.15$ charge unit.

7.3 Elemental Composition Above Nickel

Above the Fe peak the relative abundances decrease by factors of $\sim 10^2$–10^6 and only elemental composition has so far been determined using both active and passive detector systems flown on balloons and in spacecraft. Recent satellite instruments that have achieved large exposure factors include the HEAO-3 experiment by Binns et al (117, see Figure 15c) and the

Ariel-VI experiment (23, 118), with charge resolution in the region of iron of $\sigma \sim 0.35$ and $\sigma \sim 0.55$, respectively. Historically, passive detectors with large F values were the first to observe the elements above nickel, starting with the meteoritic data of Fleischer et al (119) and followed by the nuclear emulsion studies of Fowler et al (120) and work with plastic track detectors (121, 122). In the latter technique, which has higher systematic errors, plastic sheets are chemically etched to reveal high Z particles with a resolution $\Delta Z \approx 1-2$ (104a, 123).

8. SUMMARY OF ELEMENTAL ABUNDANCES AT EARTH AND AT THE COSMIC RAY SOURCE

8.1 *Hydrogen and Helium*

Since the A/Z ratios for H and He nuclei differ by a factor of 2, the extrapolations of measured abundances to obtain the relative source abundances for these species are especially sensitive to the level of solar modulation, interstellar propagation, and the assumed form of the source spectra. Table 1 summarizes the fluxes and ratios of H and He at low and high energy per nucleon. Their fluxes can be normalized to Si = 100 through the He abundance at low energy in Table 2. Figure 5a shows the experimentally determined differential spectra for H and He, which, when fitted by a power-law spectral form $\propto E^{-\gamma}$ ($\gamma = 2.75 \pm 0.05$) gives at 50,000 MeV n^{-1} the high energy values in Table 1. If instead, a magnetic rigidity source spectrum is assumed for H, Webber & Lezniak (126) showed that H/He ≈ 7, which is comparable to the solar system abundance ratio. Since H and He fluxes are 10^3 and 10^2 times higher, respectively, than fluxes of the heavier nuclei, contributions of secondaries from heavier nuclei to the measured H and He abundances are negligible.

8.2 *Average Values of Source Abundances: Helium to Zinc*

The calculated elemental abundances of cosmic ray nuclei after acceleration at their sources in the galaxy are based on the 1 AU measurements over the energy range 70–2400 MeV per nucleon, which are summarized in Table 2. Although there is general agreement among different investigators on the propagation code (e.g. 71), as discussed in Section 3, there are substantial differences in the choice of input-parameters. Dwyer et al (127) and Garcia-Munoz et al (64, 70) have summarized the different assumptions, especially in the choice of pathlength distributions, which affect the calculated source abundances and their errors. To take account of these and other possible uncertainties, and the known errors in the calculations by different investigators, we have applied conservative overall errors to the source abundances presented in Table 3. For example, these large error limits include presently unresolved questions such as whether some of the differences in source abundances derived by different investigators in different energy ranges arise from an energy dependence in the source abundance or unknown energy dependences of the cross sections. In comparison with Table 3, the errors assigned to source composition by individual investigators for specific propagation models are generally smaller (20, 62, 127–129, 132–141).

Shapiro et al (135), using a compilation of measurements from several investigations from the 1960s and early 1970s, obtained elemental source abundances that are in satisfactory agreement with the summary in Table 3, with the outstanding exceptions of Ne, Ca, and Fe (cf 142, Figure 3; Ref. 127). There is now general agreement among cosmic ray investigators that the abundances of the principal elements (i.e. the even-Z nuclei) after acceleration are known to $\pm 20\%$. Substantially larger errors are to be assigned to most of the less abundant odd-Z nuclei, especially in the sub-iron group, as well as argon and nitrogen.

8.3 Zinc to Molybdenum

In this range the abundances are normalized to $Fe \equiv 10^6$ for convenience. The differences in measured values reported by various investigators appear to be due mainly to statistics and instrument charge resolution. The two satellite experiments, HEAO-3 (21, 25) and Ariel-VI (23, 24), have provided the most recent data, as shown in Table 4A. Results are presented only for even-Z elements, since the measurements give only upper limits for the abundances of the odd-Z species (e.g. 25). These investigators have calculated the source abundances shown in Table 4A by propagation back to the source assuming steady-state models and neglecting energy loss by ionization with a leakage or escape pathlength of 5 g cm^{-2}. Both groups used the fragmentation cross sections of Silberberg & Tsao (74, 75). Blake & Margolis (59) have studied this charge region in a calculation including the effects of energy loss by ionization and an energy-dependent propagation equation (e.g. Section 10). They discussed the differences that arise from choosing different pathlength distributions and from including the effects of an injection mechanism dependent on first ionization potential (Section 10). Additional recent calculations over portions of this nuclear charge range include those of Adams et al (143), Protheroe & Ormes (143a), and Brewster et al (144). Early work, including nuclear emulsions and plastic detectors, has been summarized by Fowler (145) and Wefel et al (146).

8.4 Molybdenum to Uranium and Beyond

Binns et al (22) (see Figure 3) and Fowler et al (23, 24) in their respective satellite experiments have extended the measurement of relative abun-

Table 4 Ultra-heavy ($Z \geq 30$) element abundances in the cosmic rays

		Earth[d]		Source	
	Element	HEAO-3[a,b]	Ariel-VI[c]	HEAO-3[a,b]	Ariel-VI[c]
A. Even Z ($30 \leq Z \leq 42$)					
26	Fe	$\equiv 10^6$	$\equiv 10^6$	$\equiv 10^6$	$\equiv 10^6$
30	Zn	584 ± 34	—	586 ± 36	—
32	Ge	138 ± 8	—	120 ± 10	—
34	Se	57 ± 5	83 ± 6	42 ± 6	59 ± 9
36	Kr	31 ± 4	46 ± 4	16 ± 5	29 ± 7
38	Sr	41 ± 4	48 ± 5	38 ± 6	41 ± 7
40	Zr	20 ± 3	28 ± 3	21 ± 4	24 ± 6
42	Mo	15 ± 3	13 ± 4	16 ± 4	5 ± 5
B. Charge Groups ($34 \leq Z \leq 100$)					
34–42		210 ± 30	216 ± 10		
44–48		20 ± 3	24 ± 3		
50–58		27 ± 6	23 ± 3		
60–74		12 ± 2	15 ± 2		
76–84		6 ± 2	7 ± 2		
86–100		≤ 1	0.5 ± 0.3		
$\dfrac{88 \leq Z \leq 100}{74 \leq Z \leq 87}$		≤ 0.03	0.06 ± 0.04		

[a] ($\triangle$) M. H. Israel (private communication, 1982) (see also 22).
[b] ($\triangle$) (21).
[c] ($\triangle$) (23, 24).
[d] Uncertainties are statistical only and do not include possible systematic uncertainties in the iron normalization ($\lesssim 20\%$).

($\triangle$) ≡ Satellite

dances to include the actinides. The observed number of events in their experiments are given in Table 4B, normalized to Fe $\equiv 10^6$. The measurements in both these experiments must still be considered preliminary until the final analysis of the experiments are complete. Since there are substantial differences in the results between these two experiments, both sets of data are presented. In this charge range, especially beyond $Z = 60$, propagation effects can be extremely important, and it is not possible to quote a set of reliable source abundances for these elements. Propagation studies for cosmic rays in this charge interval have been performed (147–149; see also the references given in Section 8.3). A brief summary of earlier work has been given by Fowler (118). There has been no confirmed report of the detection of nuclei with $Z \geq 96$.

8.5 The Antinuclides of Hydrogen ($\bar{H}$) and Helium ($\overline{He}$)

There are two possible sources for antinuclei in the cosmic radiation. First, of astrophysical interest, is the possibility of observing antinuclei of cosmological (primary) origin, several models for which were reviewed by Steigman (150), who showed that the observed gamma-ray fluxes place severe limits on the magnitude of anti-matter/matter mixing allowed in our galaxy. Experiments to search for antinuclei of charge $Z \geq 2$ have so far failed to detect any flux at a level $> 8 \times 10^{-5}$ of the nuclear component $Z \geq 2$ (e.g. 30, 151, 151a).

The second possibility is a secondary component of antinuclei—namely, antihydrogen and positrons produced from nuclear interactions of cosmic rays with the interstellar medium. The magnitude of secondary antiparticle fluxes is sensitive to the galactic propagation model parameters, especially the slope of the energy spectrum, containment time in the galaxy, and the galactic distribution of cosmic ray sources. The measurements of the $\bar{H}/H$ ratio are shown in Figure 16 and compared with predictions based on a "leaky box" model with $\lambda = 5$ g cm^{-2}, assuming a hydrogen differential energy spectrum $\propto E^{-2.6}$. The predictions are well below the measurements leading to proposals for alternate models (152) and suggesting a lifetime greater than that determined by ^{10}Be decay (Section 9.3). However, more recent work indicates that $\lambda \approx 8$–9 g cm^{-2} (Section 3.4) and, therefore, that the discrepancy is no longer as large. On the other hand, Stecker et al (153) have reopened the question of whether the antiproton fluxes are primary or secondary since the Buffington et al (154) measure-

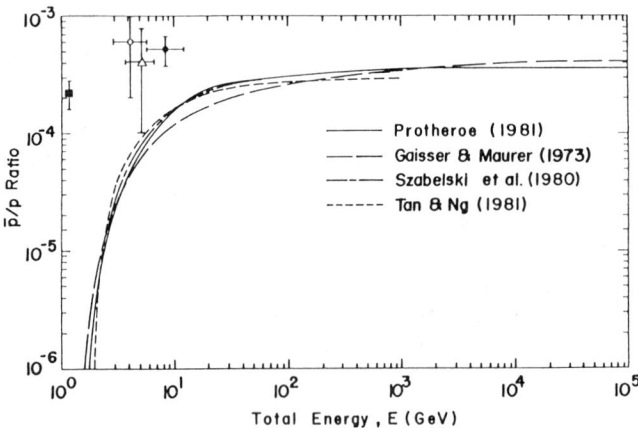

Figure 16 Comparison between experimental measurements and theoretical predictions (152, 189, 190a, 191) for the $\bar{p}/p$ ratio in the cosmic rays. The data points are: ●—(155); ○—(156); ■—(154). Adapted from (152), with addition of △—(157).

ments are well below the cutoff energy for secondary production of $\bar{H}$. Stephens (30) has reviewed recent $\bar{H}$ and e^+ measurements relative to the origin of the antiproton component.

9. SUMMARY OF OBSERVED AND SOURCE ISOTOPIC ABUNDANCES

9.1 General Considerations

Tables 5, 6, 7, 8, and 9 are summaries of recent isotopic abundance measurements, their ratios, and isotopic ratios in the source. Satellite experiments (with references denoted by $\triangle$) have been given highest priority in the selection of values since they have so far yielded measurements with the highest mass resolution and are not subject to the substantial atmospheric corrections for nuclear interactions required for balloon observations (with references denoted $\bigcirc$). Furthermore, measurements of Z and A, which identify individual nuclear species, have been assigned a higher priority than isotopic abundances derived from mean-mass measurements (references denoted by m). Since abundance ratios of stable primary nuclei measured in the energy range ~ 30 to $\sim 10^3$ MeV n^{-1} reveal no significant energy dependences (e.g. 142, 158), data in this energy interval have been included in the determination of average values. In a few cases where the choice is critical for theoretical models, and where the results are still controversial, alternate values of isotopic ratios are shown.

9.2 Hydrogen and Helium Isotopes

The first isotopes to be identified among the arriving galactic cosmic rays were ^{3}He and ^{4}He (160, 161) and ^{2}H (162). Deuterium is readily destroyed in high temperature stellar environments, and ^{3}He is rare in nature. Thus, these cosmic ray isotopes are produced in nuclear interactions during propagation of ^{1}H and ^{4}He in the interstellar medium [for the modes of ^{2}H

Table 5 Isotopic ratios of hydrogen and helium[a]

Isotope ratio	60 MeV n^{-1} [b]	80 MeV n^{-1} [c]	200 MeV n^{-1} [d]
^{2}H/^{1}H	—	$(4.4 \pm 0.5) \times 10^{-2}$	$(5.7 \pm 0.5) \times 10^{-2}$
^{3}He/^{4}He	—	$(9.5 \pm 1.5) \times 10^{-2}$	$(11.8 \pm 0.7) \times 10^{-2}$
^{2}H/^{4}He	0.21 ± 0.09[b]	—	0.31 ± 0.03

[a] Note: All measurements except (b) near solar minimum 1973–1977.
[b] ($\triangle$) Average of several measurements during the 1965–1969 period of increasing solar modulation (see 159 and references therein).
[c] ($\triangle$) (163–166).
[d] ($\bigcirc$) (125).

($\triangle$) ≡ Satellite   ($\bigcirc$) ≡ Balloon

and ^{3}He production see (98, 167, 168)]. The abundances of ^{1}H and ^{4}He are much greater than those of all other nuclei in the cosmic rays, and the cross sections for production of ^{2}H and ^{3}He are large so that, except for a small contribution from CNO nuclei, a close generic relationship exists within this group of four isotopes and provides unique information on the source spectra of ^{1}H and ^{4}He and the propagation modes of the galactic cosmic rays (98, 168). Both the ratios ^{2}H/^{4}He and ^{3}He/^{4}He have been used as monitors to determine the propagation parameters for the light isotopes (cf 98). Because both ^{2}H and ^{4}He have the same A/Z ratio, the ratio ^{2}H/^{4}He is expected to be little modified by solar modulation. However, the ratios ^{2}H/^{1}H and ^{3}He/^{4}He are more affected by solar modulation and, for this reason, only data near and in the solar minimum period 1973–1977 at 1 AU have been included in Table 5 for these ratios. From 1973 to 1978 the ^{4}He below ~ 60 MeV n^{-1} contains the anomalous ^{4}He component, which was first detected in the cosmic ray spectrum in 1972 (169). Thus, the data selected for Table 5 are in the energy range where there is practically no anomalous He.

9.3 Lithium, Beryllium, and Boron Isotopes

The observed fluxes of Li, Be, and B are all secondary nuclei produced in nuclear interactions during the acceleration of the primary nuclei before and/or during interstellar propagation of the primary nuclear species (see Section 3.7), with isotopic composition as shown in Table 6. Of special interest are the abundances of the Be isotopes, which include radioactive

Table 6   Isotopic composition of Li, Be, and B (energy intervals given in MeV per nucleon)

Isotope	175–318[a]	161–416[b]	175–510[c]	31–151[d]	60–185[e]
^{6}Li	0.53 ±0.06	—	0.52±0.06	0.51 ±0.04	—
^{7}Li	0.47 ±0.05	—	0.48±0.06	0.49 ±0.04	—
^{7}Be	0.63 ±0.07	0.63±0.07	0.63±0.06	0.585±0.026	0.546±0.029
^{9}Be	0.34 ±0.04	0.27±0.05	0.25±0.05	0.376±0.025	0.390±0.029
^{10}Be	$0.026^{+0.021}_{-0.024}$	0.10±0.04	0.12±0.05	0.039±0.014	0.064±0.015
^{10}B	0.24 ±0.04	0.33±0.04	0.41±0.04	0.29 ±0.02	—
^{11}B	0.76 ±0.07	0.67±0.05	0.59±0.04	0.71 ±0.02	—

[a] (○)(170). [d] (△)(18, 95, 173).
[b] (○)(171). [e] (△)(19).
[c] (○)(172; these investigators find a large energy dependence in the Be isotopic composition from 175 to 1860 MeV per nucleon).

(△) ≡ Satellite   (○) ≡ Balloon

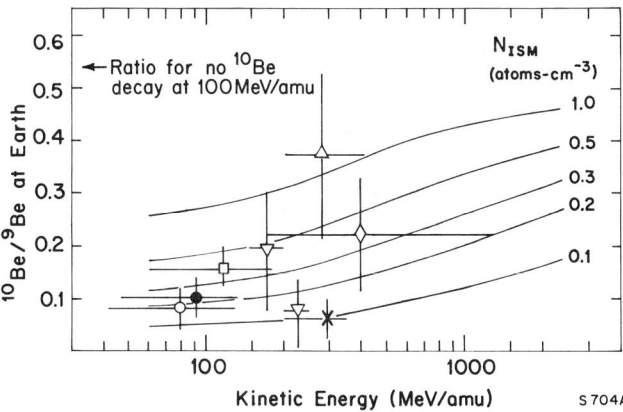

Figure 17 Comparison of the ^{10}Be/^{9}Be ratio measured at Earth with propagation calculations assuming different densities for the interstellar gas, N_{ism} [adapted from Garcia-Munoz et al (173)]. Data are from: ○—(18); ●—(173); □—(19); ▽—(170); △—(171); X—(190); ◇—(172).

^{10}Be ($\tau_{1/2} = 1.5 \times 10^6$ y). The decay of ^{10}Be relative to ^{9}Be yields the "clock" containment lifetime of the low and medium energy primary cosmic nuclei in interstellar and/or galactic halo magnetic fields (cf Section 6).

Two recent satellite measurements are shown in Table 6 to indicate the precision of the Be isotope measurements. Figure 17 includes all measurements of ^{10}Be/^{9}Be over the energy range ~ 30–1000 MeV n^{-1}, taking into account solar modulation. From the measured propagation time and the average amount of matter penetrated, we obtain from Equation 1 the *average* interstellar density of matter through which the cosmic ray nuclei propagate. The range of values for the interstellar density averaged over the propagation paths of primary nuclei is ~ 0.1 to 0.3 atom cm^{-3}. The energy dependence for the abundance of Be isotopes reported by Buffington et al (172) needs confirmation.

9.4 *Isotopes of Carbon to Iron*

Tables 7, 8, and 9 summarize both the current status of isotopic measurements of abundances and ratios at 1 AU and the extrapolation of these measurements to the cosmic ray source. Propagation mean-free paths and escape pathlengths are all consistent with values derived from the elemental abundance propagation calculations. The source abundances in Tables 7, 8, and 9 include the range of values reflecting differences in results obtained by different investigators.

Table 7a The isotopic composition of C, N, and O observed at 1 AU

Isotope	Low energy (30–400 MeV n^{-1})^j	High energy (400–3000 MeV n^{-1})^j
^{12}C	$0.940 \pm 0.006^{a,b,e,f,h}$	$0.933 \pm 0.013^{c,d,i}$
^{13}C	$0.060 \pm 0.006^{a,b,e,f,h}$	$0.067 \pm 0.013^{c,d,i}$
^{14}N	$0.452 \pm 0.026^{b,f,g,h}$	$0.470 \pm 0.024^{c,d,i}$
^{15}N	$0.548 \pm 0.025^{b,f,g,h}$	$0.530 \pm 0.024^{c,d,i}$
^{16}O	$0.966 \pm 0.005^{a,b}$	0.941 ± 0.008^{c}
^{17}O	$0.015 \pm 0.004^{a,b}$	0.016 ± 0.006^{c}
^{18}O	$0.018 \pm 0.002^{a,b}$	0.043 ± 0.005^{c}

^{a}(△) (97).　^{f}(△) (178).
^{b}(△) (26, 174).　^{g}(△) (179).
^{c}(○) (175).　^{h}(○) (171).
^{d}(△,m) (176).　^{i}(○,m) (115).
^{e}(△) (177).　^{j} Weighted average with respect to reported errors.
(△) ≡ Satellite　(○) ≡ Balloon　(m) ≡ Mean mass measurements

10. EFFECTS OF INJECTION/ACCELERATION IN SOLAR FLARES AND GALACTIC COSMIC RAYS

The galactic source abundances of the elements and their isotopic composition are derived from measurements that are representative of the composition *after* the nuclear matter has been injected and accelerated (Section 2.3) to cosmic ray energies. Therefore, it is essential to identify whether or not these electrodynamical processes, or the initial state of the ionized matter, may introduce atomic or mass biasing effects especially during injection that could modify significantly the relative "source" abundances of the elements or the isotopes of an element. These effects must be added to those known to be due to propagation in the interstellar and interplanetary media (see Sections 4 and 5). Evidence for biases in the elemental composition due to atomic effects have been found in both the composition of galactic and solar flare nuclei, whereas for the isotope ratios of elements $Z \geq 6$ no biasing effects have been found or predicted.

Aside from theoretical investigations, or correlations with atomic parameters, which are strongly dependent upon assumptions regarding the origin and acceleration mechanisms (Section 2), the only direct experimental searches for atomic and mass effects in accelerated matter are based on matter at the Sun accelerated in the solar wind or solar flares, which can accelerate nuclei to cosmic ray energies. Hence, in this section we introduce some solar flare nuclear composition measurements for comparison with the galactic cosmic rays to illustrate the magnitude of bias effects on elemental abundances and the absence of such effects in the isotopic composition above $Z \sim 4$.

Table 7b The isotopic composition of C, N, and O at the cosmic ray source (energy intervals given in MeV per nucleon)

Ratio	30–126[b]	45–94[f]	80–300[a,l]	410–486[c]	1200[j]	Solar system[k]
$^{13}C/C$	≤0.044		0.021±0.006 (0.012–0.017)	0.031±0.01		0.011
$^{14}N/O$	≤0.04	0.07±0.03		0.031±0.008	0.06±0.02	~0.12
$^{15}N/O$	≤0.09	0.0[m]		0.0[m]	0.0[m]	0.0005
$^{17}O/^{16}O$	≤0.012		$0.004^{+0.004}_{-0.003}$ (0.002–0.005)	0.0		0.0004
$^{18}O/^{16}O$	≤0.020		0.002±0.002 (0.002–0.005)	$0.003^{+0.006}_{-0.003}$		0.002

Footnotes a–j are defined in Table 7a.
[k] Cameron (5).
[l] Errors in parenthesis are propagation errors.
[m] Estimated value for ratio.

Table 8a Isotopic composition of Ne through Si at 1 AU (energy intervals given in MeV per nucleon)

Isotope	~60–230[a]	~100–300[b]	~30–180[c]	410–686[d]	~390–570[e]
^{20}Ne	0.58±0.03	$0.52^{+0.03}_{-0.02}$	$0.62^{+0.06}_{-0.11}$	0.619	0.564±0.071
^{21}Ne	0.10	$0.13^{+0.03}_{-0.02}$	$0.07^{+0.07}_{-0.03}$	0.087±0.033	0.174±0.046
^{22}Ne	0.32±0.03	$0.35^{+0.04}_{-0.03}$	0.31±0.08	0.293±0.027	0.262±0.047
^{24}Mg	0.71±0.04	$0.63^{+0.03}_{-0.02}$	$0.60^{+0.04}_{-0.07}$	0.689	0.590±0.051
^{25}Mg	0.13	0.18±0.02	$0.19^{+0.06}_{-0.04}$	0.145±0.034	0.212±0.044
^{26}Mg	0.16±0.05	0.19±0.02	$0.21^{+0.06}_{-0.04}$	0.167±0.016	0.198±0.033
^{26}Al				0.046±0.037	
^{27}Al				0.954	0.977±0.107
^{28}Si		$0.837^{+0.018}_{-0.023}$	$0.86^{+0.03}_{-0.11}$	0.942	0.819±0.067
^{29}Si		$0.092^{+0.018}_{-0.014}$	$0.07^{+0.07}_{-0.03}$		0.153±0.040
^{30}Si		$0.071^{+0.015}_{-0.011}$	$0.07^{+0.08}_{-0.03}$	0.058±0.023	0.028±0.025

[a] (△) (180, 181).
[b] (△) (182, 183).
[c] (△) (184).
[d] (○) (175).
[e] (○) (132, 185).

(△) ≡ Satellite   (○) ≡ Balloon

Table 8b Isotopic composition of the medium heavy nuclei Ne, Mg, and Si at the cosmic ray source compared with the solar system abundances (energy intervals given in MeV per nucleon)

Ratio	60–230[a]	100–300[b]	30–180[c]	410–686[d]	390–570[e]	Solar system[f]
^{21}Ne/^{20}Ne	—	$0.08 ^{+0.05}_{-0.04}$	<0.08	—	0.15 ± 0.04	0.003
^{22}Ne/^{20}Ne	0.38 ± 0.07	$0.50 ^{+0.10}_{-0.07}$	$0.33 ^{+0.20}_{-0.18}$	0.43 ± 0.07	0.30 ± 0.06	0.122
^{25}Mg/^{24}Mg	—	$0.21 ^{+0.04}_{-0.03}$	$0.24 ^{+0.11}_{-0.09}$	0.13 ± 0.04	0.27 ± 0.06	0.128
^{26}Mg/^{24}Mg	0.14 ± 0.05	$0.23 ^{+0.04}_{-0.03}$	$0.25 ^{+0.12}_{-0.09}$	0.20 ± 0.04	0.27 ± 0.05	0.141
(^{25}Mg + ^{26}Mg)/^{24}Mg			$0.49 ^{+0.23}_{-0.14}$			0.27
^{29}Si/^{28}Si	—	$0.085 ^{+0.023}_{-0.013}$	$0.04 ^{+0.09}_{-0.02}$		0.16	0.051
^{30}Si/^{28}Si	—	$0.056 ^{+0.018}_{-0.013}$	$0.05 ^{+0.08}_{-0.03}$			0.034
^{30}Si/(^{28}Si + ^{29}Si)	—			0.028 ± 0.02		0.032

Footnotes a–e are defined in Table 8a.
[f] Cameron (5).

($\triangle$) ≡ Satellite   (○) ≡ Balloon

Table 9 Isotopic composition of Fe (fraction of the element) (energy intervals given in MeV per nucleon)

	Abundances at 1 AU			Source abundances		
Isotope	320–500[a]	83–284[b]	646–900[c]	from (187)	from (186)	Solar system[d]
^{54}Fe		$0.10^{+0.08}_{-0.04}$	0.07 ± 0.03	$0.09^{+0.08}_{-0.05}$	0.048 ± 0.030	0.058
^{55}Fe		≤0.10		≤0.07[e]		0
^{56}Fe		$0.90^{+0}_{-0.15}$	0.90 ± 0.06	$0.91^{+0.05}_{-0.11}$	0.952	0.917
^{57}Fe		≤0.08		≤0.08		0.022
^{58}Fe		≤0.06	≤0.03	≤0.06	≤0.030	0.0033
^{54}Fe/Fe	≤0.10				0.041 ± 0.028	
^{58}Fe/Fe	≤0.10				≤0.028	

[a] (○) (186). [d] Cameron (5).
[b] (△) (187). [e] Assuming that ^{55}Fe does not decay.
[c] (○) (188).

Notes: 1. Young et al (132) report isotopic composition for Si, S, Ca, Fe, and Ni; Webber (188) reports isotopic composition for Cl through Fe.
2. Tarlé et al (186) report ^{40}Ca/Fe = 0.059 ± 0.019 at the source.

(△) ≡ Satellite   (○) ≡ Balloon

10.1 Elemental Abundances of Cosmic Ray Nuclei

As pointed out in the Introduction, the solar system abundances (e.g. 5) or the "local galactic abundances" (4) are widely used as the reference set for comparison with abundances derived from other astrophysical sites. In general, elemental abundance differences are either to be explained by biasing effects, as discussed above, or by physically significant differences in the composition relative to the solar system abundances. In Figure 18 we have plotted as a ratio the galactic cosmic ray source abundances of Tables 3 and 4 to the solar system abundances [including the addition of estimated errors by Meyer (4)] in the atomic number range 1 to 40. This ratio is normalized to unity at silicon ($Z = 14$). There are considerable uncertainties, especially in some of the solar system abundances and the odd-Z nuclei in the galactic cosmic rays. An error band of 1.0 ± 0.3 is shown, within which range more than 65% of all the element ratios above Ne are within 1σ of their individual estimated errors and, therefore, consistent with solar system abundances. Notable exceptions include S and Ar. For Ne and lighter nuclei, the galactic cosmic ray source relative abundances in Figure 18 appear significantly depleted. Alternatively, if the normalization had been taken at oxygen, as is often done, then (aside from H and He,

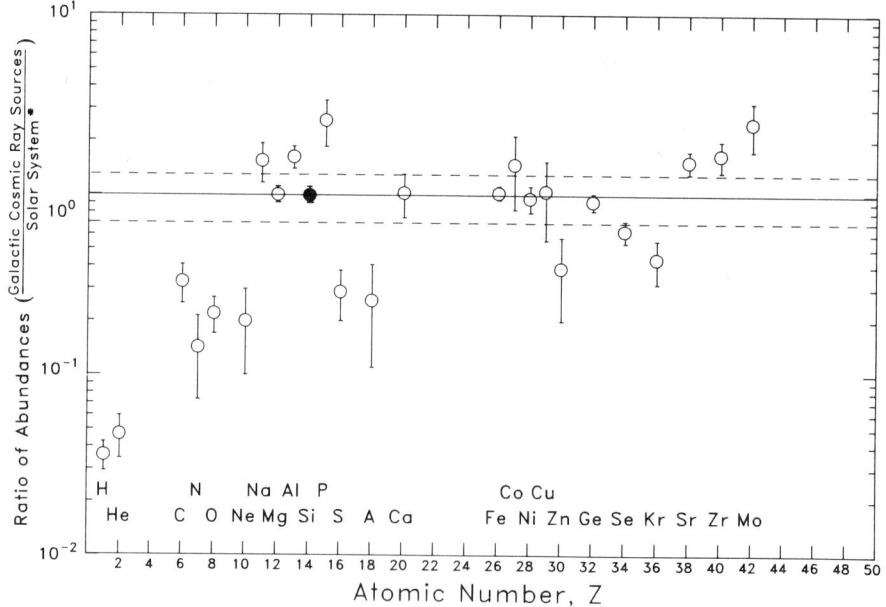

Figure 18 Cosmic ray sources and solar system abundances are based on Tables 1 and 3. Solar system values are from Cameron (5), with estimated errors by Meyer (4).

which are discussed in Section 11) it would appear that above Ne more than 65% of the abundance ratios would be enhanced above the solar system values. This question of normalization illustrates the conflict and difficulties associated with interpretations of elemental abundances that have appeared in the literature for over two decades.

10.2 Solar Flare Accelerated Nuclei

Solar flare accelerated nuclei provide direct evidence that the injection/acceleration effects are the origin of this enhancement (depletion) effect and the large abundance deviations of some of the elements.

First, at extremely low energies where solar flare nuclei arrive at 1 AU only partially stripped of their orbital electrons, the relative abundances of heavy nuclei were observed to display strong enhancements (e.g. 192, 193). Furthermore, recent work by Gloeckler and co-workers (194) has shown remarkably anomalous distributions of the charge states for nuclei at extremely low energies.

Second, the relative abundance of solar flare nuclei at much higher energies where the nuclei must be fully stripped were discovered not only to be preferentially enhanced relative to oxygen, but also to vary over more than an order of magnitude in the enhancements observed from flare to flare (195, 196). For example, the ratio Fe/O has a range extending from below the solar ratio of 0.04 to ~ 1 (197, 198).

This variability in elemental enhancements among solar flares makes it difficult to decide on the elemental composition of solar flare nuclei that most closely represents a sample of chromospheric or coronal matter accelerated at the flare site. This problem generally is attacked by averaging high energy data from a substantial number of flares for which there is essentially no enhancement of abundances relative to solar system abundances. With these constraints on solar flare selection, the solar flare abundances bear a close resemblance to coronal abundances (e.g. 198 and references therein; 198a).

10.3 Dependence on Atomic Number and First Ionization Potential

Of special interest was the discovery that this averaged subset of solar flare elemental abundances displayed essentially the same relative abundances as the galactic source abundances as a function of atomic number (196, their Figure 2) or as a function of first ionization potential (199, 200, 201). The possibility that first ionization potentials could account for the elemental cosmic ray abundances was first noted by Havnes (202) and Kristiansson (203).

Figure 19 is a recent plot of the ratio of galactic cosmic ray source

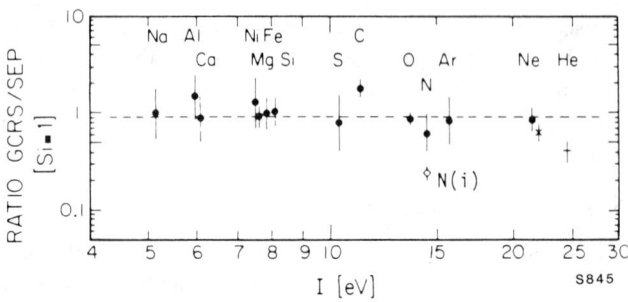

Figure 19 Comparison between the relative abundances in the galactic cosmic ray source (GCRS) and the abundances measured for solar (flare) energetic particle (SEP) events. The ratio (GCRS/SEP) is plotted as a function of the first ionization potential, I (eV), of the element. Two values are shown for GCRS nitrogen, inferred from elemental measurements (●) and from isotopic data ($\diamond, i$). (From 199.)

abundances to solar (flare) energetic particle abundances as a function of their first ionization potentials. This remarkable correlation between galactic cosmic rays and solar flare abundances constitutes the principal evidence at present that the acceleration mechanism for both the galactic cosmic rays (Figure 18) and solar flare nuclei introduce remarkably similar biasing effects on the elemental abundances relative to solar system composition.

Cassé & Goret (204) showed how the first ionization potential for elements organized the cosmic ray source abundances. This correlation of cosmic ray matter was extended by Binns et al (21) and Israel (25) to include nuclei up to $Z = 40$, as shown in Figure 20. These latter investigators and Blake & Margolis (59) have shown that, after correcting for first ionization potential, the abundances of nuclei above iron are remarkably like solar system abundances. Although the correlation with first ionization potential is impressive in organizing the observations, it does not fully explain the cosmic ray elemental source abundance enhancement nor the extreme enhancement variations of solar flare nuclei in some solar flares. For example, difficulties arise with galactic cosmic ray abundances of H, He, Ne, Ar, C, and N, so that a correlation in first ionization potential does not appear to be the whole physical explanation. In some cases the discrepancy may be due to uncertainties in solar system abundances. Furthermore, this correlation implies that the injection site must be at a temperature $T < 10^4$ K, but a unique temperature cannot simultaneously account for the ionization of the metals and rare gases. The dependence of the enhancement on Z or A/Z^* (where Z^* is the effective charge) has been explored (195, 196, 205) but is inconclusive. The volatility of elements as a factor has also been ruled out by recent evidence (e.g. 205). Although no detailed mechanism

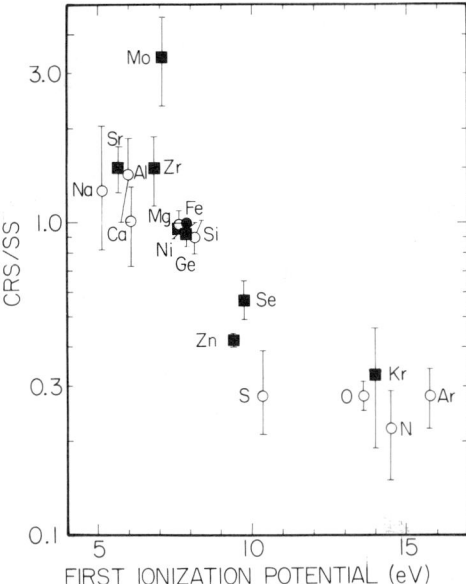

Figure 20 Comparison between the galactic cosmic ray source (CRS) abundance and solar system (SS) abundances (5). The ratio (CRS/SS) is plotted as a function of the first ionization potential of the element (from 25). Errors in (5) are not included.

has been proven to account for the observed galactic cosmic ray source composition, it is clear that atomic—rather than nuclear—effects are important during the injection and/or acceleration phase for some elemental abundances.

10.4 *Solar Flare Isotope Tests for Mass Fractionation*

Solar flare accelerated nuclei also provide tests to search for possible isotopic fractionation or nuclear interaction effects during injection and acceleration. The very light isotopes do not provide a valid test since there are many competing effects in the mass range 1 to 4. For example, partly because of spectral shape, the ratio H/He at a given energy per nucleon varies over several orders of magnitude from flare to flare and with time during a flare. Furthermore, a class of flares has been discovered where the abundance ratio ^{3}He/^{4}He (normally $\sim 2 \times 10^{-4}$ in the Sun) is enhanced by factors of $\sim 10^4$ to 10^5—the so-called ^{3}He-rich flares [206, 207; review by Klecker (208)]. Nuclear interactions do not account for this enhancement (209). This conclusion is further supported by the observed near absence of ^{2}H relative to ^{3}He (207, 210). Current theoretical work points to either a plasma resonance effect (211, 212), which could only be applicable to ^{3}He and a few other isotopes, or, as pointed out by Hayakawa (212a), to an atomic resonance line effect important only for ^{3}He relative to ^{4}He.

Tests to search for significant mass fractionation effects during solar flare acceleration are based on measurements at higher masses—namely, the ratios ^{22}Ne/^{20}Ne (12, 14), ^{26}Mg/^{24}Mg (213, 214), and, most recently, the isotopic ratios for C, N, and O (214a). These ratios are in close agreement with solar system ratios. Figure 21 is a plot that includes the ratio of the solar flare ratios to the solar system ratios of ^{22}Ne/^{20}Ne and ^{26}Mg/^{24}Mg. This ratio of ratios for the isotopes of an element is called the enhancement factor. However, there exists the outstanding problem of explaining the well-measured ^{22}Ne/^{20}Ne ratio in the solar wind (11), which is approximately a factor of two lower than that for the solar flare nuclei. Since, except for masses 1–4, the isotopic abundances of high energy nuclei are not found to be biased, it is possible that the difference may be explained by fractionation effects in the slow acceleration of solar wind (215). At present the results derived from solar flare studies lead to the tentative conclusion

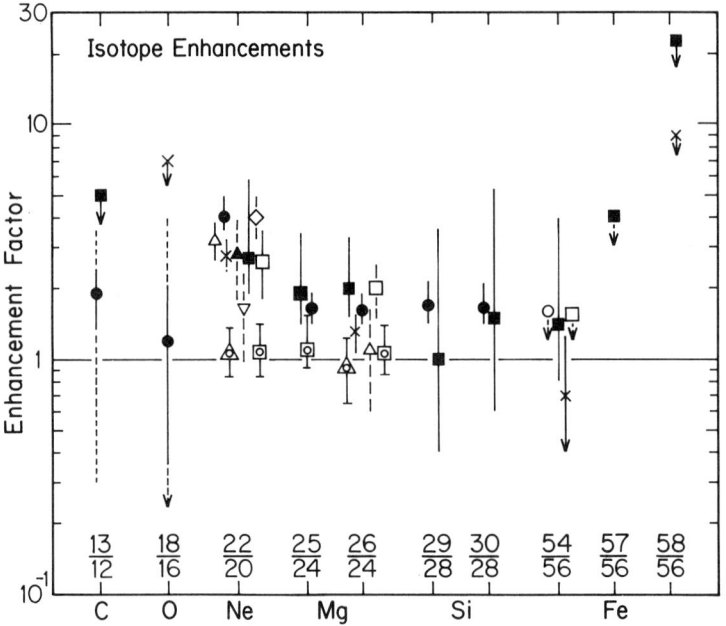

Figure 21 Enhancement (depletion) factors for galactic cosmic ray source and solar flare isotopic abundances relative to the solar system abundance compilation (5). Plotted is the measured ratio divided by the solar system ratio for galactic cosmic ray source isotope data: ●—(97, 182, 183); ■—(26, 174, 184, 187); △—(130, 181); ▲—(176); ○—(186); ▽—(115, 216); ◇—(217); □—(132, 185); X—(188, 218). Adapted from (142); with addition of solar flare isotopic abundances: ⊡—(14, 214); ⊿—(12, 13, 213); ⍁ indicates upper limit; dashed line indicates unresolved isotopes.

that the isotopic composition of galactic cosmic ray source matter carries the imprint of matter from the astrophysical site *before* injection/acceleration. Thus, the galactic cosmic ray source enhancement factor of ~ 3.5 for ^{22}Ne/^{20}Ne over the solar system ratios in Figure 21 is of special significance for identifying the astrophysical site(s) of cosmic ray sources in the galaxy, as discussed in Section 12.

11. THE SOLAR SYSTEM AND GALACTIC COSMIC RAY SOURCE ABUNDANCES

In this review we have placed considerable emphasis on understanding the physical phenomena and the modeling that underlie, and place limits on, the determination and meaning of source abundances derived from observations. Taking into account these factors, and recognizing the uncertainties in the values for the elemental abundances identified with the Sun and the solar system, there emerges a remarkable overall correspondence of these two sets of elemental abundances—especially for the "α-particle" stable nuclei for the elements Mg and beyond, as revealed in Figures 2 and 18. Equally remarkable and most significant are the differences for some elemental and several isotopic abundances, which are not accounted for by injection/acceleration, or which arise from propagation effects and uncertainties in nuclear interaction cross sections. These similarities and differences provide the clues and constraints for both the identification of astrophysical sites for the cosmic ray source matter, and for testing the validity of current models for nucleosynthesis of matter in the galaxy. Although it is beyond the scope of this review to examine in depth the astrophysical implications of the galactic cosmic ray composition, some examples of important differences and similarities between the galactic cosmic ray source and the solar system and solar abundances—and their significance—are discussed below.

11.1 *Hydrogen and Helium*

Since secondary production during interstellar propagation accounts for the observed abundances of ^{2}H and ^{3}He, it is assumed that the cosmic ray source isotopes of hydrogen and helium are ^{1}H and ^{4}He (Sections 8.1 and 9.2 and Tables 1 and 5). Because, as noted in Section 9.2, their ratios of A/Z differ by a factor of 2, the determination of the source abundance ratio ^{1}H/^{4}He is especially sensitive to the effects of source spectral differences, acceleration, propagation, and solar modulation. The range of values in Table 1 for the ratio ^{1}H/^{4}He reflects these difficulties. Although a correction for first ionization potential (Section 10) may be invoked to account for the ^{4}He abundance, this correction does not explain the low abundance of ^{1}H.

11.2 Carbon and Oxygen

The cosmic ray source abundance ratio for C/O (0.85 ± 0.02—Table 3) is a factor of ~ 1.5 larger than the ratio of 0.58 (± 0.03) derived from solar spectroscopy or solar flare accelerated particles (Figure 19). Since both nuclei are $A/Z = 2$ particles and there are only small corrections to the ratio to take account of their first ionization potentials, this high ratio may indicate that the source matter is closely connected with matter from supernovae (219–221). Recent theoretical advances have not changed this conclusion.

Figure 21 shows that the source compositions of rare C and O isotopes are not yet well enough determined, owing to propagation parameter uncertainties, to provide definitive tests of or models of composition.

11.3 Nitrogen

Since the flux of secondary N nuclei exceeds that from the sources and there are uncertainties in the production cross sections (e.g. 178), the determination of source abundances is subject to substantial errors that result in conflicting values reported by various investigators. Analyses based on the measured elemental ratio N/O lead to ratios in the range 0.05 to 0.08 that may be sensitive to the different ionization potentials of N and O (Tables 2 and 3). Alternatively, the isotopic measurements of ^{15}N and ^{14}N (Table 7a) rely upon the assumption that there is no ^{15}N in the source and, therefore, that it can be used as a monitor of secondary production of nitrogen (see Section 5.3). These isotope analyses at low energy yield N/O source ratios of 0.03–0.05 (cf 178). Although there are substantial differences in the calculated source abundance of N relative to O, both the elemental and isotopic measurements at low energies lead to the conclusion that the N/O abundance ratio (~ 0.02–0.08) is significantly lower in the cosmic radiation than for the Sun's photosphere, 0.12 ± 0.04 [(e.g. 222)], or corona, [0.14 ± 0.01 (223)], or probably the interstellar medium [~ 0.10–0.12]. If future work confirms that the N/O ratio from source matter is lower than for the interstellar medium by the factor ~ 1.5 to 2, the result would favor a supernova origin (e.g. 224) instead of the interstellar medium for the source matter (225).

11.4 Neon, Magnesium, and Silicon

11.4.1 NEON Neon was the first element in cosmic ray source matter found to have a ratio of isotopic abundances (e.g. ^{22}Ne/^{20}Ne) significantly different from the range of values for this ratio derived from all solar system or solar measurements. The first indications of an enhancement came from measurements in balloons (217, 226). Conclusive evidence came from a series of satellite experiments with higher mass resolution in the energy range 30–300 MeV n^{-1} by Garcia-Munoz et al (180, 181), Mewaldt et al (184), and Wiedenbeck & Greiner (182). These investigations showed that the enhancement factor for the source composition of ^{22}Ne/^{20}Ne was greater by the factor 3.5 ($\pm$0.7) (Table 8b, Figure 21) than for either solar system Neon-A in meteorites (e.g. 176, 227) or solar flares (12–14) and greater by a factor of 5.5 ($\pm$1) than the solar wind ratio of $\sim$0.07 (11). Mean-mass measurements (Section 7.2) at 1000–3000 MeV n^{-1} (216, 228) indicate that this enhancement is not energy dependent.

11.4.2 MAGNESIUM AND SILICON The high resolution satellite measurements summarized in Figure 21 show that the source abundance ratios ^{26}Mg/^{24}Mg and ^{25}Mg/^{24}Mg are enhanced by the factor 1.7 ($\pm$0.4) over the solar system and solar flare ratios (213, 214). This enhancement factor of 1.7 ($\pm$0.4) is reported also by Wiedenbeck & Greiner (183) for the ratios ^{29}Si/^{28}Si and ^{30}Si/^{28}Si.

The establishment of the above so-called anomalous enhancements of the neutron-rich isotopes of Ne, Mg, and Si, and the recognition that at least some of the cosmic ray source matter has evolved differently from solar system matter, have initiated theoretical investigations on scenarios for nucleosynthesis of the elements that could account for the source composition. It should be noted that already Arnett (229) had predicted enhanced isotopic ratios derived from nucleosynthesis models. Recently, for example, Woosley & Weaver (230) have examined hypotheses based on current models of stellar nucleosynthesis to explain the cosmic ray isotope enhancements. Of special interest is their proposal that massive stars with a higher fraction of metals than in the solar neighborhood at the time of formation of the Sun ("super metallicity model") will yield neutron-rich isotope enhancements (e.g. ^{22}Ne) under conditions of nucleosynthesis with an enhanced metallicity factor. However, since the predicted enrichment factors are approximately the same for Ne, Mg, and Si, their model could account for either Mg and Si (factor of 1.7) or ^{22}Ne (factor of 3.5) but not both unless the source material is a mixture of freshly synthesized and interstellar matter. A critical test of the high metallicity hypothesis would be the determination of the cosmic ray source ratio ^{34}S/^{32}S, since this ratio is not strongly dependent on the s-process. Cassé (201) reviewed additional alternatives to account for the enhanced neutron-rich isotopes, including a suggestion by Olive & Schramm (231) [see also Reeves (232)] that the solar system abundances could be different from the present interstellar medium if a series of supernova explosions injected matter into the solar nebula prior to solar system formation in an OB star association. Thus, if the cosmic ray source matter were interstellar in origin it would be different from the solar system—a scenario that might also explain the isotopic anomalies in meteorites (232, Section 1). However, the composition of the interstellar medium is still an unresolved problem.

11.5 *Iron, Cobalt, and Nickel*

11.5.1 IRON Both the elemental and ^{56}Fe abundances of iron in the cosmic ray source (Table 9) are within the limits of solar system abundances. However, the neutron-rich ^{57}Fe (from ^{57}Co by electron capture) and ^{58}Fe abundances in the cosmic rays still have only upper limits. The upper limit for the ratio ^{54}Fe/^{56}Fe includes both the solar system values and the enhanced values predicted for some nucleosynthesis models invoked (e.g. 230) to account for enhancements of ^{22}Ne etc.

11.5.2 COBALT AND NICKEL Although the isotopic abundances of Ni and Co have not been determined for cosmic rays sources, the elemental abundance ratios Fe/Co/Ni can provide evidence on the elapsed time, Δt, between nucleosynthesis and acceleration (see Section 6.2; 233). With the assumption that the observed species, after correction for spallation, are ^{58}Ni, ^{60}Ni, and ^{59}Co, the recent work of Koch-Miramond and collaborators (228) implies $\Delta t > 10^5$ y, rather than $\Delta t > 10^2$ y as reported earlier (234).

It is important to decide this question since $\Delta t > 10^5$ y implies that the source composition would not be fresh ejecta from explosive nucleosynthesis.

11.6 *The Elements Zinc to Ruthenium*

Since the nucleosynthesis of elements $Z \geq 30$ is by neutron capture (e.g. 235, 236), the determination of the source abundances can provide essential information on the relative importance of the two principal neutron capture processes—namely, the slow s-process and the rapid r-process. For example, the r-process produces much more Se and Kr than Sr, and more Te and Xe than Ba, relative to the s-process, which produces more Sr and less Ba than the r-process. Both processes lead to higher even-Z than odd-Z abundances. In a review Israel (25) discussed the history of the measurements for $Z \geq 30$ and summarized the most recent results obtained from the satellites Ariel-6 (23, 24) and HEAO-3, as discussed below. For the elements in the charge range $30 \leq Z \leq 44$ (21, 25; see Figure 2, Sections 7.3 and 8.3; Table 4), the overall decreasing abundances of even-Z nuclei with increasing charge assures that in first approximation these nuclei are mainly surviving source nuclei with relatively small contributions of secondary nuclei resulting from fragmentation of heavier nuclei during propagation. However, as shown in Figure 20 (25), their source abundances also display preferential enhancements (Section 10) so that corrections for this effect tend to increase the uncertainties of the source abundances of nuclei that could distinguish between r- and s-process (e.g. Sr, Zr, and Mo in Figure 2; Table 4). At this stage in the analysis of the HEAO-3 experiment the source abundances $30 \leq Z \leq 44$ show that pure r-process material is not dominant in the source composition but appears instead to be similar to the solar system and, therefore, a mixture of nucleosynthesis processes in the region beyond iron (21, 221). Ariel-6 source abundances appear to be consistent with this conclusion. It is well known that nucleosynthesis models for the r-process yield relatively higher abundances of odd-Z elements relative to the s-process in this mass range (Br, Rb, Y, etc) and would provide a sensitive test of the above conclusions. However, even though the HEAO-3 experiment combines a large geometrical factor, good statistics, and well-defined peaks for the even-Z nuclei, it does not appear possible yet to determine the abundances of these odd-Z nuclei.

11.7 Elements Palladium to Cerium

After the publication of the abundances in Table 4B, Binns et al (236a) analyzed further the abundant even-Z nuclei in the element range Pd to Ce. They report abundance ratios Sn/Te and Ba/Te that—with or without corrections for preferential enhancements (Section 10)—have ratios too high by a significant factor to be accounted for by pure r-processed material (see Section 11.6), but that are consistent with solar ratios when first ionization potentials are taken into account (see 59). The Ce may be consistent with this conclusion but is contaminated by fragmentation from heavier nuclei.

11.8 Platinum-Lead and the Actinides

Since the actinides (Th, U, Np, Pu, and Cm) are synthesized only by the r-process (see, however, 237), their abundances in the cosmic radiation are decisive indicators for determining the origin of cosmic ray matter. Even the shorter-lived species Np, Pu, and Cm can survive for the observed cosmic ray lifetime of $\sim 10^7$ y if their acceleration occurs immediately after being synthesized (Section 6). The most recent reports on the actinide abundances are based on the Ariel-6 (23, 24) and the HEAO-3 (22) measurements. The results from the HEAO-3 experiment, which appear to have the better charge resolution, are shown in Figures 3 and 22 and Table 4B. Because the propagation lifetime is $\sim 10^7$ y, the short-lived nuclear species in the charge range $84 \leq Z \leq 89$ have disappeared, leaving a gap above the peak distribution in the Pt-Pb element range (e.g. Figure 3). The Pt-Pb peak includes the range of elements $79 \leq Z \leq 87$ due to the lack of both instrumental charge resolution and the Z-dependence of the signals from the detectors (e.g. 238). Similarly, the actinides are defined by $88 \leq Z \leq 100$ and normalized to this "Pt-Pb" abundance as shown in Table 4B and Figure 22. Since this ratio is not very dependent upon parameters of interstellar propagation due to the small mass range and low actinide abundance, the ratio reflects the assumed initial conditions in the source as shown in Figure 22. As Figure 3 shows, Binns et al (22) find that there is one actinide event and 101 events assigned to the Pt-Pb region, leading to an actinide ratio of ≤ 0.01, or an upper limit of 0.03 (84% confidence level). This result is consistent with solar-system-like source abundances and not freshly synthesized r-processed material. However, these measurements do not distinguish between solar system and aged r-process abundances. These results reopen the question of the source of cosmic ray matter since, as discussed by Israel (25), the work of Shirk & Price (239) and Fowler et al (23, 24 and references therein) led to much higher actinide abundances (Figure 22) probably because of experimental uncertainties, and, therefore, to the conclusion that cosmic ray sources were dominated by r-processed material. Further experiments will require orders of magnitude increases in collecting power and higher charge resolution. No nuclei with charge $> Z = 96$ have been confirmed in the literature.

Although the ratio of the elements Pt/Pb also would be important for testing source composition models, no conclusive measurements of the ratio have yet been reported.

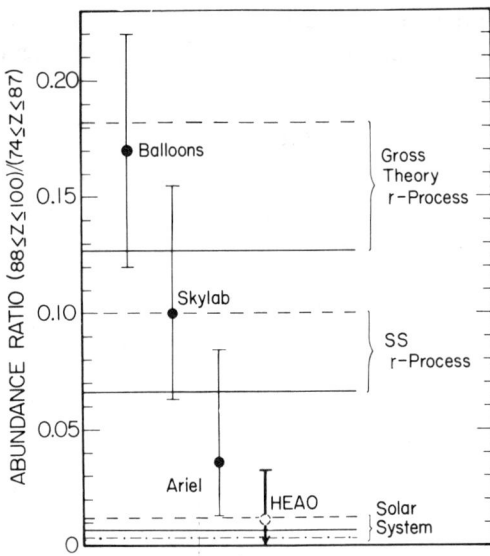

Figure 22 The measured abundance ratio "actinide"/"Pt+Pb" ($88 \leq Z \leq 100/74 \leq Z \leq 87$) from balloon and satellite experiments compared to theoretical predictions (from 25). See Figure 3, which shows the number of events assigned to the determination of this ratio.

12. SUMMARY

It is now established that there are important differences in elemental and isotopic abundance between the source matter of galactic cosmic rays and the solar system matter. These large differences (more than a factor of 2 in some cases, in contrast to $\leq 1\%$ for isotopic anomalies in solar system matter) may become essential clues to the identification of astrophysical sites for cosmic ray matter and to the nucleosynthesis processes that preceded the acceleration of the cosmic rays or galactic chemical evolution. There are also significant similarities demonstrating that the cosmic ray source material is not predominantly freshly made, r-processed material. With our present limited knowledge of abundances—especially isotopic— for different astrophysical sites, it appears that cosmic ray source matter may be a mixture of aged r-process and interstellar medium material. Clearly, the abundances in the cosmic ray sources, the interstellar medium, supernova remnants, molecular clouds, and so forth are all essential inputs for unraveling their respective nucleosynthesis histories. For example, the recent measurements by Jura & Smith (240) of rubidium in the interstellar medium indicate for the first time that r-process elements can be measured in the interstellar gas.

In this review we have placed considerable emphasis on how the cosmic ray sample at the source is transformed before reaching the observer, since the values of the resulting abundances depend upon this knowledge. For elements there are injection/acceleration effects, as well as propagation and nuclear interaction effects. The elemental abundances are biased by injection/acceleration; exactly how is still unknown but it seems clear that the first ionization potentials are to be taken into account. Investigations of solar flare acceleration mechanisms may be a key to resolving this problem since isotopic abundances of elements are not expected and, so far, have not been found to be biased by injection/acceleration (except for the very light ^{1}H, ^{3}He, ^{4}He isotopes—e.g. ^{3}He-rich flares). Thus, in general, isotopic abundance ratios (as well as some of the elemental abundances) are thought to be representative of cosmic ray composition prior to acceleration. Consequently, a major goal of cosmic ray research is the determination of the isotopic abundances, including the rare isotopes, and those in the region beyond iron. Unfortunately, the opportunities for carrying out these measurements in space vehicles in the future have practically vanished. This occurs at a time when the emerging field of gamma-ray line astronomy could strongly interact with cosmic ray physics.

It is important to be aware of the possibility that the observed galactic cosmic ray composition may be from a superposition of different kinds of sources and/or different spectra at very low and very high energies. Thus at the extremes of very low and very high energies, the elemental and isotopic compositions may be different from the broad, spectral region where the bulk of the cosmic ray nuclei have been measured, as reported in this review. It is clear that the composition revealed in the cosmic ray investigations—especially the evolution of isotopic measurements over the past 20 years—is one of the exciting areas of research in high energy astrophysics.

ACKNOWLEDGMENTS

The author wishes to extend his appreciation to M. Garcia-Munoz, T. G. Guzik and J. P. Wefel for their assistance of many kinds in the preparation of materials for this review. He is also indebted to J.-P. Meyer for generously sending in advance of publication the reports from the 17th International Conference on Cosmic Rays (Paris), and to many investigators—especially M. Israel and R. H. Mewaldt—for sending their papers in advance of publication. The author has attempted to include the most recent preprints with their publication references through the Spring of 1982. This review was supported in part by the Compton Fund at the University of Chicago and by NASA Grant NGL 14-001-006.

Literature Cited

1. Audouze, J., Tinsley, B. M., 1976. *Ann. Rev. Astron. Astrophys* 14:43
2. Suess, H., Urey, H. C. 1956. *Rev. Mod. Phys.* 28:53
3. Ross, J. E., Aller, L. H. 1976. *Science* 191:1223
4. Meyer, J. P. 1979. *Proc. 16th Int. Conf. Cosmic Rays (Kyoto)* 2:115
5. Cameron, A. G. W. 1982. *Essays in Nuclear Astrophysics*, ed. C. Barnes, R. N. Clayton, D. N. Schramm, p. 23. Cambridge Univ. Press
6. Anders, E., Ebihara, M. 1982. *Geochim. Cosmochim. Acta* 46:2363
7. Clayton, R. N., Grossman, L., Mayeda, T. K. 1973. *Science* 182:485
8. Lee, T. 1979. *Rev. Geophys. Space Sci.* 17:1591
9. Lattimer, J. M., Schramm, D. N., Grossman, L. 1978. *Astrophys. J.* 219:230
10. Clayton, R. N. 1981. *Adv. Phys. Geochem.* 1:85
11. Geiss, J., Buehler, F., Cerutti, H., Eberhardt, P., Filleux, Ch. 1972. In *Apollo 16 Preliminary Science Report, NASA Publ. SP-315*, p. 14. Washington: GPO
12. Dietrich, W. F., Simpson, J. A. 1979. *Astrophys. J. Lett.* 231:L91
13. Dietrich, W. F., Simpson, J. A. 1979. *Proc. 16th Int. Conf. Cosmic Rays (Kyoto)* 5:85
14. Mewaldt, R. A., Spalding, J. D., Stone, E. C., Vogt, R. E. 1979. *Astrophys. J. Lett.* 231:L97
15. Wannier, P. G. 1980. *Ann. Rev. Astron. Astrophys.* 18:399
16. Watson, W. D. 1974. *Astrophys. J.* 188:35
17. Penzias, A. 1981. *Astrophys. J.* 249:513
18. Garcia-Munoz, M., Mason, G. M., Simpson, J. A. 1977. *Astrophys. J.* 217:859
19. Wiedenbeck, M. E., Greiner, D. E. 1980. *Astrophys. J. Lett.* 239:L139
19a. Hartman, G., Müller, D., Prince, T. 1977. *Phys. Rev. Lett.* 38:1368
20. Garcia-Munoz, M., Simpson, J. A. 1979. *Proc. 16th Int. Conf. Cosmic Rays (Kyoto)* 1:270
21. Binns, W. R., Fickle, R. K., Garrard, T. L., Israel, M. H., Klarmann, J., Stone, E. C., Waddington, C. J. 1981. *Astrophys. J. Lett.* 247:L115
22. Binns, W. R., Fickle, R. K., Garrard, T. L., Israel, M. H., Klarmann, J., Stone, E. C., Waddington, C. J. 1982. *Astrophys. J. Lett.* 261:L117
23. Fowler, P. H., Masheder, M. R. W.,

Moses, R. T., Walker, R. N. F., Worley, A. 1981. In *Origin of Cosmic Rays*, ed. G. Setti, G. Spada, A. W. Wolfendale, p. 77. Dordrecht, Holland: Reidel
24. Fowler, P. H., Walker, R. N. F., Masheder, M. R. W., Moses, R. T., Worley, A. 1981. *Nature* 291:45
25. Israel, M. H. 1981. *Proc. 17th Int. Conf. Cosmic Rays (Paris)* 12:53
26. Mewaldt, R. A., Spalding, J. D., Stone, E. C., Vogt, R. E. 1981. *Astrophys. J. Lett.* 251:L27
27. Schein, M., Jesse, W. P., Wollan, E. O. 1941. *Phys. Rev.* 59:615
28. Freier, P. S., Lofgren, E. J., Oppenheimer, F., Bradt, H. L., Peters, B. 1948. *Phys. Rev.* 74:213
29. Prince, T. 1979. *Astrophys. J.* 227:676 (and references therein)
30. Stephens, S. A. 1981. *Proc. 17th Int. Conf. Cosmic Rays (Paris)* 13:89
31. Gloeckler, G. 1979. *Rev. Geophys. Space Phys.* 17:569
32. McKibben, R. B., Pyle, K. R., Simpson, J. A. 1982. *Astrophys. J. Lett.* 257:L41 (and references therein)
33. Rochester, G. D., Turver, K. E. 1981. *Contemp. Phys.* 22:425
34. Burbidge, G. R. 1974. *Philos. Trans. R. Soc. London* A277:481
35. Kniffen, D. A., Fichtel, C. E., Thompson, D. J. 1977. *Astrophys. J.* 215:765
36. Strong, A. W., Bignami, G. F., Caraveo, P. A., Lebrun, F., Paul, J. A., Mayer-Hasselwander, H. A., Kanbach, G., Hermsen, W., Willis, R. D. 1981. *Proc. 17th Int. Conf. Cosmic Rays (Paris)* 1:146
37. Paul, J. A. 1981. *Proc. 17th Int. Conf. Cosmic Rays (Paris)* 12:79
38. Mayer-Hasselwander, H. A., et al. 1982. *Astron. Astrophys.* 105:164
39. Forman, M. A., Schaeffer, O. A. 1979. *Rev. Geophys. Space Phys.* 17:552
40. Tanaka, S., Inoue, T. 1979. *Proc. 16th Int. Conf. Cosmic Rays (Kyoto)* 2:277
41. Fermi, E. 1949. *Phys. Rev.* 75:1169
41a. Eichler, D. 1980. *Astrophys. J.* 237:809
41b. Cowsik, R. 1980. *Astrophys. J.* 241:1195
41c. Fransson, C., Epstein, R. I. 1980. *Astrophys. J.* 242:411
42. Krimsky, G. F. 1977. *Dok. Acad. Nauk. SSSR* 234:1306
43. Axford, W. I., Leer, E., Skadron, G. 1977. *Proc. 15th Int. Conf. Cosmic Rays (Plovdiv)* 11:132
44. Blandford, R. D., Ostriker, J. P. 1978. *Astrophys. J. Lett.* 221:L29
45. Blandford, R. D., Ostriker, J. P. 1980. *Astrophys. J.* 237:793
46. Eichler, D. 1979. *Astrophys. J.* 229:419
47. Völk, H. J. 1981. *Proc. 17th Int. Conf. Cosmic Rays (Paris)* 13:131
48. Baade, W., Zwicky, F. 1934. *Proc. Natl. Acad. Sci. Am.* 20:259
49. Colgate, S. A., Johnson, M. H. 1960. *Phys. Rev. Lett.* 5:235
50. Ginzburg, V. L., Syrovatskii, S. I. 1964. *The Origin of Cosmic Rays*, transl. by H. S. H. Massey, ed. by D. ter Haar. New York: Macmillan. 426 pp.
51. Scott, J. S., Chevalier, R. A. 1975. *Astrophys. J. Lett.* 197:L5
52. Axford, W. I. 1981. See Ref. 23, p. 339
53. Axford, W. I. 1981. *Proc. 17th Int. Conf. Cosmic Rays (Paris)* 12:155
54. Ginzburg, V. L., Khazan, Ya. M., Ptuskin, V. S. 1980. *Astrophys. Space Sci.* 68:295
55. Owens, A. J. 1976. *Astrophys. Space Sci.* 40:357
55a. Simpson, J. A. 1983. In *Composition and Origin of Cosmic Rays*, ed. M. M. Shapiro, Ch. 1. Dordrecht, Holland: Reidel
56. Cesarsky, C. J. 1980. *Ann. Rev. Astron. Astrophys.* 18:289
57. Fichtel, C. E., Reames, D. V. 1968. *Phys. Rev.* 175:1564
58. Lezniak, J. A. 1979. *Astrophys. Space Sci.* 63:279
59. Blake, J. B., Margolis, S. H. 1981. *Astrophys. J.* 251:402
60. Shapiro, M. M., Silberberg, R. 1970. *Ann. Rev. Nucl. Sci.* 20:323
61. Shapiro, M. M., Silberberg, R. 1974. *Philos. Trans. R. Soc. London* A277:319
62. Garcia-Munoz, M., Mason, G. M., Simpson, J. A. 1977. *Proc. 15th Int. Conf. Cosmic Rays (Plovdiv)* 1:224
63. Garcia-Munoz, M., Margolis, S. H., Simpson, J. A., Wefel, J. P. 1979. *Proc. 16th Int. Conf. Cosmic Rays (Kyoto)* 1:310
64. Garcia-Munoz, M., Guzik, T. G., Simpson, J. A., Wefel, J. P. 1981. *Proc. 17th Int. Conf. Cosmic Rays (Paris)* 2:192
64a. Lezniak, J. A., Webber, W. R. 1979. *Astrophys. Space Sci.* 63:35
65. Protheroe, R. J., Ormes, J. F., Comstock, G. M. 1981. *Astrophys. J.* 247:362
66. Gloeckler, G., Jokipii, J. R. 1969. *Phys. Rev. Lett.* 22:1448
67. Cowsik, R., Pal, Y., Tandon, S. N., Verma, R. P. 1967. *Phys. Rev.* 158:1238
68. Cowsik, R. 1981. See Ref. 23, p. 93
69. Ormes, J. F., Freier, P. S. 1978. *Astrophys. J.* 222:471
70. Garcia-Munoz, M., Guzik, T. G., Margolis, S. H., Simpson, J. A., Wefel, J. P. 1981. *Proc. 17th Int. Conf. Cosmic Rays (Paris)* 9:195

71. Freier, P. S. 1981. *Proc. 17th Int. Conf. Cosmic Rays (Paris)* 2:182
72. Rudstam, G. 1955. *Philos. Mag.* 46:344
73. Rudstam, G. 1966. *Z. Naturforsch.* 21A:1027
74. Silberberg, R., Tsao, C. H. 1973. *NRL Rep. No. 7593* (unpublished)
75. Silberberg, R., Tsao, C. H. 1973. *Astrophys. J. (Suppl.)* 25:315
76. Silberberg, R., Tsao, C. H. 1977. *Proc. 15th Int. Conf. Cosmic Rays (Plovdiv)* 2:84
77. Waddington, C. J. 1960. *Philos. Mag.* 5:311
78. Westfall, G. D., Wilson, L. W., Lindstrom, P. J., Crawford, H. J., Greiner, D. E., Heckman, H. H. 1979. *Phys. Rev. C* 19:1309
79. Silberberg, R., Tsao, C. H., Shapiro, M. M. 1976. In *Spallation Nuclear Reactions and Their Applications*, ed. B. S. P. Shen, M. Merker, p. 49. Dordrecht, Holland: Reidel
80. Raisbeck, G. M., Yiou, F. 1976. See Ref. 79, p. 83
81. Juliusson, E., Meyer, P., Müller, D. 1972. *Phys. Rev. Lett.* 29:445
82. Smith, L. H., Buffington, A., Smoot, G. F., Alvarez, L. W., Wahlig, M. A. 1973. *Astrophys. J.* 180:987
82a. Caldwell, J. H. 1977. *Astrophys. J.* 218:269
83. Fontes, P., Meyer, J. P., Perron, C. 1977. *Proc. 15th Int. Conf. Cosmic Rays (Plovdiv)* 2:234
84. Jokipii, J. R. 1976. *Astrophys. J.* 208:900
85. Jones, F. C. 1979. *Astrophys. J.* 229:747
86. Margolis, S. H. 1981. *Proc. 17th Int. Conf. Cosmic Rays (Paris)* 9:215
87. Parker, E. N. 1958. *Phys. Rev.* 109:1874
88. Parker, E. N. 1958. *Astrophys. J.* 123:664
89. McKibben, R. B., Pyle, K. R., Simpson, J. A. 1982. *Astrophys. J. Lett.* 254:L23
90. Axford, W. I. 1972. In *Solar Wind*, ed. P. J. Coleman, C. P. Sonett, J. M. Wilcox, p. 609, NASA Publication SP-308. Washington, DC: GPO
91. Parker, E. N. 1966. *Planet. Space Sci.* 14:371
92. Fan, C. Y., Meyer, P., Simpson, J. A. 1960. *Phys. Rev. Lett.* 5:272
92a. McDonald, F. B., Lal, N., Trainor, J. H., VanHollebeke, M. A. I., Webber, W. R. 1981. *Astrophys J.* 249:L71
92b. McKibben, R. B., Pyle, K. R., Simpson, J. A. 1982. *Astrophys. J.* 254:L23
93. Jokipii, J. R. 1971. *Rev. Geophys. Space Phys.* 9:27
94. Fisk, L. A. 1979. In *Solar System Plasma Physics*, ed. E. N. Parker, C. F. Kennel, L. J. Lanzerotti, p. 174. New York: North-Holland
95. Garcia-Munoz, M., Mason, G. M., Simpson, J. A. 1977. *Proc. 15th Int. Conf. Cosmic Rays (Plovdiv)* 1:301
96. Hinshaw, G. F., Wiedenbeck, M. E., Greiner, D. E. 1981. *Proc. 17th Int. Conf. Cosmic Rays (Paris)* 9:191
97. Wiedenbeck, M. E., Greiner, D. E. 1981. *Proc. 17th Int. Conf. Cosmic Rays (Paris)* 2:76
98. Simpson, J. A. 1971. *Proc. 12th Int. Conf. Cosmic Rays (Hobart)* 8:324
99. Stone, E. C., Wiedenbeck, M. E. 1979. *Astrophys. J.* 231:606
100. Waddington, C. J. 1977. *Fundam. Cosmic Phys.* 3:1
101. Adams, J. H. Jr., Shapiro, M. M., Silberberg, R., Tsao, C. H. 1981. *Proc. 17th Int. Conf. Cosmic Rays (Paris)* 2:256
102. Koch, L., Engelmann, J. J., Goret, P., Juliusson, E., Petrou, N., Rio, Y., Soutoul, A., Byrnak, B., Lund, N., Peters, B., Rasmussen, I. L., Rotenberg, M., Westergaard, N. 1981. *Astron. Astrophys. Lett.* 102:L9
102a. Gloeckler, G. 1970. In *Introduction to Experimental Techniques of High-Energy Astrophysics*, ed. H. Ogelman, J. R. Wayland, pp. 1, NASA Publication SP-243 (document N71-11415 from the National Technical Information Service, Springfield, VA 22151)
103. Webber, W. R., Kish, J. 1972. *Nucl. Instrum. Methods* 99:237
104. Wefel, J. P. 1979. In *Workshop on the Radiation Environment of the Satellite Power System (SPS)*, ed. W. Schimmerling, S. B. Curtis, p. 117, Dept. Energy Publ. CONF-7809164 (UC-34b, 41, 48, 97c). Washington, DC: GPO
104a. Fleischer, R. L., Price, P. B., Walker, R. M. 1975. *Nuclear Tracks in Solids*. Berkeley: Univ. Calif. Press
105. Fan, C. Y., Gloeckler, G., Simpson, J. A. 1965. *J. Geophys. Res.* 70:3515
106. Perkins, M. A., Kristoff, J. J., Mason, G. M., Sullivan, J. D. 1969. *Nucl. Instrum. Methods* 68:149
107. Althouse, W. E., Cummings, A. C., Garrard, T. L., Mewaldt, R. A., Stone, E. C., Vogt, R. E. 1978. *IEEE Trans. Geosci. Electron.* GE15:204
108. Greiner, D. E., Bieser, F. S., Heckman, H. H. 1978. *IEEE Trans. Geosci. Electron.* GE16:163
109. Lamport, J. E., Mason, G. M., Perkins, M. A., Tuzzolino, A. J. 1976. *Nucl. Instrum. Methods* 134:71
109a. Simpson, J. A., et al. 1983. The High Energy Isotope Telescope: *Charge Particle Instruments for the International Solar Polar Mission*. In *Scientific and Experimental Aspects of*

the ISPM Mission, ed. K. P. Wenzel, R. Marsden. Eur. Space Agency
110. McDonald, F. B., Webber, W. R. 1959. *Phys. Rev.* 115:194
111. Jelley, J. V. 1958. *Čerenkov Radiation and its Applications*. New York: Pergamon
111a. Litt, J., Meunier, R. 1973. *Ann. Rev. Nucl. Sci.* 23:1
112. Webber, W. R. 1979. *Proc. 16th Int. Conf. Cosmic Ray (Kyoto)* 11:55
113. Lund, N., Rasmussen, I. L., Peters, B. 1971. *Proc. 12th Int. Conf. Cosmic Rays (Hobart)* 1:130
114. Peters, B. 1974. *Nucl. Instrum. Methods* 121:205
115. Dwyer, R., Meyer, P. 1979. In *Proc. 16th Int. Cosmic Ray Conf. (Kyoto)* 12:97
116. Copenhagen-Saclay Collaboration. 1981. *Adv. Space Res.* 1:173
117. Binns, W. R., Israel, M. H., Klarmann, J., Scarlett, W. R., Stone, E. C., Waddington, C. J. 1981. *Nucl. Instrum. Methods* 185:415
118. Fowler, P. H. 1977. *Nucl. Instrum. Methods* 147:183
119. Fleischer, R. L., Price, P. B., Walker, R. M., Maurette, M., Morgan, G. 1967. *J. Geophys. Res.* 72:355
120. Fowler, P. H., Adams, R. A., Cowen, V. G., Kidd, J. M. 1967. *Proc. R. Soc. London* A301:39
121. Fleischer, R. L., Price, P. B., Walker, R. M., Filz, R. C., Fukui, K., Friedlander, M. W., Holeman, E., Raajan, R. S., Tamhane, A. S. 1967. *Science* 155:187
122. Blandford, G. E., Friedlander, M. W., Klarmann, J., Walker, R. M., Wefel, J. P., Wells, W. C., Fleischer, R. L., Nichols, G. E., Price, P. B. 1969. *Phys. Rev. Lett.* 23:338
123. Price, P. B., Fleischer, R. L., Peterson, D. D., O'Cellaigh, C., O'Sullivan, D., Thompson, A. 1968. *Phys. Rev. Lett.* 21:630
124. von Rosenvinge, T. T., McDonald, F. B., Trainor, J. H., Webber, W. R. 1979. *Proc. 16th Int. Conf. Cosmic Rays (Kyoto)* 12:170
125. Webber, W. R., Yushak, S. M. 1979. *Proc. 16th Int. Conf. Cosmic Rays (Kyoto)* 1:383
126. Webber, W. R., Lezniak, J. A. 1974. *Astrophys. Space Sci.* 30:361
127. Dwyer, R. D., Garcia-Munoz, M., Guzik, T. G., Meyer, P., Simpson, J. A., Wefel, J. P. 1981. *Proc. 17th Int. Conf. Cosmic Rays (Paris)* 9:222
128. Webber, W. R. 1982. *Astrophys. J.* 255:329
129. Lezniak, J. A., Webber, W. R. 1978. *Astrophys. J.* 223:676
130. Dwyer, R., Meyer, P. 1981. *Proc. 17th Int. Conf. Cosmic Rays (Paris)* 2:54
131. Engelmann, J. J., Goret, P., Juliusson, E., Koch-Miramond, L., Masse, P., Petrou, N., Rio, Y., Soutoul, A., Byrnak, B., Jakobsen, H., Lund, N., Peters, B., Rasmussen, I. L., Rotenberg, M., Westergaard, N. 1981. *Proc. 17th Int. Conf. Cosmic Rays (Paris)* 9:97
131a. Tueller, J., Love, P. L., Israel, M. H., Klarmann, J. 1979. *Astrophys. J.* 228:582
132. Young, J. S., Freier, P. S., Waddington, C. J., Brewster, N. R., Fickle, R. K. 1981. *Astrophys. J.* 246:1014
133. Cartwright, B. G., Garcia-Munoz, M., Simpson, J. A. 1971. *Proc. 12th Int. Conf. Cosmic Rays (Hobart)* 1:215
134. Cassé, M., Goret, P., Cesarsky, C. J. 1975. *Proc. 14th Int. Conf. Cosmic Rays (Munich)* 2:646
135. Shapiro, M. M., Silberberg, R., Tsao, C. H. 1975. *Proc. 14th Int. Conf. Cosmic Rays (Munich)* 2:532
136. Webber, W. R., Damle, S. V., Kish, J. 1972. *Astrophys. Space Sci.* 15:245
137. Mewaldt, R. A., Fernandez, J. I., Israel, M. H., Klarmann, J., Binns, W. R. 1973. *Astrophys. Space Sci.* 22:45
138. Fisher, A. J., Hagen, F. A., Maehl, R. C., Ormes, J. F., Arens, J. F. 1976. *Astrophys. J.* 205:938
139. Orth, C. D., Buffington, A., Smoot, G. F., Mast, T. S. 1978. *Astrophys. J.* 226:1147
140. Scarlett, W. R., Freier, P. S., Waddington, C. J. 1978. *Astrophys. Space Sci.* 59:301
141. Perron, C., Engelmann, J. J., Goret, P., Juliusson, E., Koch-Miramond, L., Meyer, J. P., Soutoul, A., Lund, N., Rasmussen, I. L., Westergaard, N. 1981. *Proc. 17th Int. Conf. Cosmic Rays (Paris)* 9:118
142. Mewaldt, R. A. 1981. *Proc. 17th Int. Conf. Cosmic Rays (Paris)* 13:49
143. Adams, J. H. Jr., Shapiro, M. M., Silberberg, R., Tsao, C. H. 1979. *Proc. 16th Int. Conf. Cosmic Rays (Kyoto)* 2:163
143a. Protheroe, R. J., Ormes, J. F. 1981. *Proc. 17th Int. Conf. Cosmic Rays (Paris)* 9:114
144. Brewster, N. R., Freier, P. S., Waddington, C. J. 1981. *Proc. 17th Int. Conf. Cosmic Rays (Paris)* 9:126
145. Fowler, P. H. 1973. *Proc. 13th Int. Conf. Cosmic Rays (Denver)* 5:3627
146. Wefel, J. P., Schramm, D. N., Blake, J. B. 1977. *Astrophys. Space Sci.* 49:47
147. Margolis, S. H., Blake, J. B. 1981. *Proc. 17th Int. Conf. Cosmic Rays (Paris)* 2:44
148. Blake, J. B., Margolis, S. H. 1981. *Proc. 17th Int. Conf. Cosmic Rays (Paris)* 2:41
148a. Blake, J. B., Hainebach, K. L.,

Schramm, D. N., Anglin, J. D. 1978. *Astrophys. J.* 221:694
149. Tsao, C. H., Silberberg, R., Shapiro, M. M., Adams, J. H. 1981. *Proc. 17th Int. Conf. Cosmic Rays (Paris)* 9:130
150. Steigman, G. 1976. *Ann. Rev. Astron. Astrophys.* 14:339
151. Evenson, P. 1972. *Astrophys. J.* 176:797
151a. Smoot, G. F., Buffington, A., Orth, C. D. 1975. *Phys. Rev. Lett.* 35:258
152. Protheroe, R. J. 1981. *Astrophys. J.* 251:387
153. Stecker, F. W., Protheroe, R. J., Kazanas, D. 1981. *Proc. 17th Int. Conf. Cosmic Rays (Paris)* 9:211
154. Buffington, A., Schindler, S. M., Pennypacker, C. R. 1981. *Astrophys. J.* 248:1179
155. Golden, R. L., Horan, S., Mauger, B. G., Badhwar, D., Lacy, J. L., Stephens, S. A., Daniel, R. R., Zipse, J. E. 1979. *Phys. Rev. Lett.* 43:1196
156. Bogomolov, E. A., Lubyanaya, N. D., Romanov, V. A., Stepanov, S. V., Shulakova, M. S. 1979. *Proc. 16th Int. Conf. Cosmic Rays (Kyoto)* 1:330
157. Bogomolov, E. A., Lubyanaya, N. D., Ramonov, V. A., Stepanov, S. V., Shulakova, M. S. 1981. *Proc. 17th Int. Conf. Cosmic Rays (Paris)* 9:146
158. Garcia-Munoz, M., Mason, G. M., Simpson, J. A., Wefel, J. P. 1977. *Proc. 15th Int. Conf. Cosmic Rays (Plovdiv)* 1:230
159. Hsieh, K. C., Mason, G. M., Simpson, J. A. 1971. *Astrophys J.* 166:221
160. Apparao, M. V. K. 1961. *Phys. Rev.* 123:295
161. Shapiro, M. M., Hildebrand, B., O'Dell, F. W., Silberberg, R., Stiller, B. 1963. In *3rd Int. Space Sci. Symp.*, p. 1097. Amsterdam: North Holland
162. Fan, C. Y., Gloeckler, G., Simpson, J. A. 1966. *Phys. Rev. Lett.* 17:329
163. Mewaldt, R. A., Stone, E. C., Vogt, R. E. 1976. *Astrophys. J.* 206:616
164. Garcia-Munoz, M., Mason, G. M., Simpson, J. A. 1975. *Astrophys. J.* 202:265
165. Garcia-Munoz, M., Mason, G. M., Simpson, J. A. 1975. *Proc. 14th Int. Conf. Cosmic Rays (Munich)* 1:319
166. Teegarden, B. J., von Rosenvinge, T. T., McDonald, F. B., Trainor, J. H. 1975. *Astrophys. J.* 202:815
167. Ramaty, R., Lingenfelter, R. E. 1969. *Astrophys. J.* 155:587
168. Meyer, J. P. 1974. Thesis, Université Paris-Sud, Orsay (unpublished)
169. Garcia-Munoz, M., Mason, G. M., Simpson, J. A. 1973. *Astrophys. J. Lett.* 182:L81
170. Webber, W. R., Lezniak, J. A., Kish, J. C., Simpson, G. A. 1977. *Astrophys. Lett.* 18:125
171. Hagen, F. A., Fisher, A. J., Ormes, J. F. 1977. *Astrophys. J.* 212:262
172. Buffington, A., Orth, C. D., Mast, T. S. 1978. *Astrophys. J.* 226:355
173. Garcia-Munoz, M., Simpson, J. A., Wefel, J. P. 1981. *Proc. 17th Int. Conf. Cosmic Rays (Paris)* 2:72
174. Mewaldt, R. A., Spalding, J. D., Stone, E. C., Vogt, R. E. 1981. *Proc. 17th Int. Conf. Cosmic Rays (Paris)* 2:68
175. Webber, W. R. 1982. *Astrophys. J.* 252:386
176. Juliusson, E., Engelmann, J. J., Goret, P., Koch, L., Masse, P., Petrou, N., Rio, Y., Byrnak, B., Lund, N., Peters, B., Rasmussen, I. L., Rotenberg, M., Westergaard, N. 1981. *Proc. 17th Int. Conf. Cosmic Rays (Paris)* 9:101
177. Smith, B., McDonald, F. B. 1981. *Proc. 17th Int. Conf. Cosmic Rays (Paris)* 9:138
178. Guzik, T. G. 1981. *Astrophys. J.* 244:695
179. Wiedenbeck, M. E., Greiner, D. E., Bieser, F. S., Crawford, H. J., Heckman, H. H., Lindstrom, P. J. 1979. *Proc. 16th Int. Conf. Cosmic Rays (Kyoto)* 1:412
180. Garcia-Munoz, M., Simpson, J. A., Wefel, J. P. 1979. *Astrophys. J. Lett.* 232:L95
181. Garcia-Munoz, M., Simpson, J. A., Wefel, J. P. 1979. *Proc. 16th Int. Conf. Cosmic Rays (Kyoto)* 1:436
182. Wiedenbeck, M. E., Greiner, D. E. 1981. *Phys. Rev. Lett.* 46:682
183. Wiedenbeck, M. E., Greiner, D. E. 1981. *Astrophys. J. Lett.* 247:L119
184. Mewaldt, R. A., Spalding, J. D., Stone, E. C., Vogt, R. E. 1980. *Astrophys. J. Lett.* 235:L95
185. Freier, P. S., Young, J. S., Waddington, C. J. 1980. *Astrophys. J. Lett.* 240:L53
186. Tarlé, G., Ahlen, S. P., Cartwright, B. G. 1979. *Astrophys. J.* 230:607
187. Mewaldt, R. A., Spalding, J. D., Stone, E. C., Vogt, R. E. 1980. *Astrophys. J. Lett.* 236:L121
188. Webber, W. R. 1981. *Proc. 17th Int. Conf. Cosmic Rays (Paris)* 2:80
189. Gaisser, T. K., Maurer, R. H. 1973. *Phys. Rev. Lett.* 30:1264
190. Webber, W. R., Kish, J. C. 1979. *Proc. 16th Int. Conf. Cosmic Rays (Kyoto)* 1:389
190a. Szabelski, J., Wdowczyk, J., Wolfendale, A. W. 1980. *Nature* 285:386
191. Tan, L. C., Ng, L. K. 1981. *J. Phys. G* 7:123
192. Fleischer, R. L., Hart, H. R., Comstock, G. M. 1971. *Science* 171:1240
193. Price, P. B., Hutcheon, I., Cowsik, R.,

Barber, D. J. 1971. *Phys. Rev. Lett.* 26:916
194. Gloeckler, G., Weiss, H., Hovestadt, D., Ipavich, F. M., Klecker, B., Fisk, L. A., Scholer, M., Fan, C. Y., O'Gallagher, J. J. 1981. *Proc. 17th Int. Conf. Cosmic Rays (Paris)* 3:136 (and references therein)
195. Mogro-Campero, A., Simpson, J. A. 1972. *Astrophys. J. Lett.* 171:L5
196. Mogro-Campero, A., Simpson, J. A. 1972. *Astrophys. J. Lett.* 177:L37
197. Dietrich, W. F., Simpson, J. A. 1978. *Astrophys. J. Lett.* 225:L41
198. Cook, W. R., Stone, E. C., Vogt, R. E. 1980. *Astrophys. J. Lett.* 238:L97
198a. Meyer, J. P. 1981. *Proc. 17th Int. Conf. Cosmic Rays (Paris)* 3:149
199. Meyer, J. P. 1981. *Proc. 17th Int. Conf. Cosmic Rays (Paris)* 2:265
200. Webber, W. R. 1975. *Proc. 14th Int. Conf. Cosmic Rays (Munich)* 5:1597
201. Cassé, M. 1981. *Proc. 17th Int. Conf. Cosmic Rays (Paris)* 13:111
202. Havnes, O. 1971. *Nature* 229:548
203. Kristiansson, K. 1971. *Astrophys. Space Sci.* 14:485
204. Cassé, M., Goret, P. 1978. *Astrophys. J.* 221:703
205. Cesarsky, C. J., Rothenfluf, R., Cassé, M. 1981. *Proc. 17th Int. Conf. Cosmic Rays (Paris)* 2:269
206. Hsieh, K. C., Simpson, J. A. 1970. *Astrophys. J. Lett.* 162:L191
207. Garrard, T. L., Stone, E. C., Vogt, R. E. 1973. In *Proc. Symp. High Energy Phenomena on the Sun*, p. 341 (NASA SP-342). Washington, DC: GPO
208. Klecker, B. 1981. *Proc. 17th Int. Conf. Cosmic Rays (Paris)* 13:143
209. Anglin, J. D. 1975. *Astrophys. J.* 198:733
210. Anglin, J. D., Dietrich, W. F., Simpson, J. A. 1973. *Astrophys. J.* 19:457
211. Fisk, L. A. 1978. *Astrophys. J.* 224:1048
212. Kocharov, G. E., Kocharov, L. G., Charikov, Yu. E. 1980. *Preprint No. 68C.* Ioffe Physical Tech. Inst., Leningrad, USSR
212a. Hayakawa, S. 1983. *Astrophys. J.* 266:370
213. Dietrich, W. F., Simpson, J. A. 1981. *Astrophys. J. Lett.* 245:L41
214. Mewaldt, R. A., Spaulding, J. D., Stone, E. C., Vogt, R. E. 1981. *Astrophys. J. Lett.* 243:L163
214a. Mewaldt, R. A., Spaulding, J. D., Stone, E. C., Vogt, R. E. 1981. *Proc. 17th Int. Conf. Cosmic Rays (Paris)* 3:131
215. Geiss, J. 1982. *Space Sci. Rev.* 33:201
216. Dwyer, R., Meyer, P. 1979. *Proc. 16th Int. Conf. Cosmic Rays (Kyoto)* 12:97
217. Maehl, R. C., Fisher, A. J., Hagen, F. A., Ormes, J. F. 1975. *Astrophys. J. Lett.* 202:L119
218. Webber, W. R. 1981. *Proc. 17th Int. Conf. Cosmic Rays (Paris)* 2:261
219. Arnett, W. D. 1978. *Astrophys. J.* 219:1008
220. Arnett, W. D., Schramm, D. N. 1973. *Astrophys. J.* 182:L47
221. Wefel, J. P., Schramm, D. N., Blake, J. B., Pridmore-Brown, D. 1981. *Astrophys. J. (Suppl.)* 45:565
222. Lambert, D. L. 1978. *Mon. Not. R. Astron. Soc.* 182:249
223. McKenzie, D. L., Rugge, H. R., Underwood, J. H., Young, R. M. 1978. *Astrophys. J.* 221:342
224. Hainebach, K. L., Norman, E. B., Schramm, D. N. 1976. *Astrophys. J.* 203:245
225. Silberberg, R., Shapiro, M., Tsao, C. H. 1975. *Proc. 14th Int. Conf. Cosmic Rays (Munich)* 2:451
226. Prezler, A. M., Kish, J. C., Lezniak, J. A., Simpson, G., Webber, W. R. 1975. *Proc. 14th Int. Conf. Cosmic Rays (Munich)* 12:4096
227. Podosek, F. A. 1978. *Ann. Rev. Astron. Astrophys.* 16:293
228. Koch-Miramond, L. 1981. *Proc. 17th Int. Conf. Cosmic Rays (Paris)* 12:21
229. Arnett, W. D. 1971. *Astrophys. J.* 166:153
230. Woosley, S. E., Weaver, T. A. 1981. *Astrophys. J.* 243:651
231. Olive, K. A., Schramm, D. N. 1982. *Astrophys. J.* 257:276
232. Reeves, H. 1979. In *Proto-Stars and Planets*, ed. T. Gehrels, p. 399. Phoenix: Univ. Arizona Press
233. Soutoul, A., Cassé, M., Juliusson, E. 1978. *Astrophys. J.* 219:753
234. Tueller, J., Love, P. L., Israel, M. H., Klarmann, J. 1979. *Astrophys. J.* 228:582
235. Burbidge, E. M., Burbidge, G. R., Fowler, W., Hoyle, F. 1957. *Rev. Mod. Phys.* 29:547
236. Truran, J. W. 1981. *Prog. Part. Nucl. Phys.* 6:161
236a. Binns, W. R., Fickle, R. K., Garrard, T. L., Israel, M. H., Klarmann, J., Krombel, K. E., Stone, E. C., Waddington, C. J. 1983. *Astrophys. J.* 267:L96
237. Blake, J. B., Woosley, S. E., Weaver, T. A., Schramm, D. N. 1981. *Astrophys. J.* 248:315
238. Ahlen, S. P. 1980. *Rev. Mod. Phys.* 52:121
239. Shirk, E. K., Price, P. B. 1978. *Astrophys. J.* 220:719
240. Jura, M., Smith, W. H. 1981. *Astrophys. J. Lett.* 251:L43

MUON SCATTERING

J. Drees

University of Wuppertal, Department of Physics, 5600 Wuppertal 1, West Germany

H. E. Montgomery

EP Division, CERN, 1211 Geneva 23, Switzerland

CONTENTS

1. INTRODUCTION 384
   1.1 *Historical Notes* 384
   1.2 *Lepton Scattering as a General Technique* 385
   1.3 *General Situation with Deep Inelastic Scattering* 386
   1.4 *Summary of Muon Scattering Results* 387
2. FUNDAMENTALS 391
   2.1 *Kinematics* 391
   2.2 *Higher Order QED Corrections* 392
3. MUON EXPERIMENTS 394
   3.1 *Muon Beams and Their Use* 394
   3.2 *The Berkeley-Fermilab-Princeton (BFP) Experiment* 397
   3.3 *The Bologna-CERN-Dubna-München-Saclay (BCDMS) Experiment* 399
   3.4 *The European Muon Collaboration (EMC) Experiment* 401
   3.5 *Beam Measurement, Systematics* 403
4. GENERAL FEATURES OF THE DATA, THE PARTON MODEL 404
   4.1 *The Parton Model* 404
   4.2 *Determination of σ_L/σ_T* 407
   4.3 *Comparison of Experiments* 409
   4.4 *Neutron vs Proton* 413
   4.5 *Comparison of Iron and Deuterium Structure Functions* 416
5. HEAVY QUARK PRODUCTION 419
   5.1 *Theoretical Models* 419
   5.2 *J/ψ and ψ' Production Cross Sections* 420
   5.3 *Open Charm Production* 423
   5.4 *Upsilon Production* 427
   5.5 *Rare Multimuon Events* 428
6. SCALING VIOLATION OF STRUCTURE FUNCTIONS 429
   6.1 *QCD and Its Predictions* 429
   6.2 *Other Sources of Scaling Violations* 432
   6.3 *Moment Analysis of Muon Data* 433
   6.4 *Direct x, Q^2 Analysis of the Hydrogen Data* 434

6.5 Separation of Higher Twist Contributions from Leading Twist Effects............... 437
6.6 Including the Soft Gluon Limit x → 1 438
6.7 QCD Fits to F_2^N from Iron and Carbon Targets 439
6.8 Conclusions on QCD Analyses 440
7. WEAK ELECTROMAGNETIC INTERFERENCE... 441
7.1 Theoretical and Experimental Background 441
7.2 The Serpukhov Experiment 443
7.3 The SLAC ed Experiment 444
7.4 The BCDMS Measurement 444
8. CONCLUSIONS.. 447

1. INTRODUCTION

1.1 Historical Notes

The existence of the muon has been known for slightly more than forty years (1); it is one of the more venerable of our elementary particles. Its place in physics has, however, never been certain and despite its advancing years there is still no obvious lodging place. What was once the question of why a heavy partner for the electron should exist has now been embellished into the general question of why there should exist several generations (2) of fundamental fermions. The properties of the muon have, on the other hand, been extensively studied (3). We know rather well that it behaves (4) as QED would have it and as such it has gained a place among those particles well enough understood to be used to probe the behavior of others. It is with the more recent results of just this property of the muon, "muon scattering," that this article is most concerned.

The choice of contents of the article was controlled to some extent by the existence of two recent review articles. One (5) deals with muon scattering at the Fermilab accelerator and further contains an excellent review of preceding experiments. The second (6) deals with the hadron final state as seen in neutrino and muon deep inelastic scattering and does not as yet warrant major revision. Both of these articles are strongly recommended to the reader. It is further hoped that this rather long introductory section will permit the casual reader to appreciate the essential message given in more detail in the subsequent sections.

Section 1 therefore constitutes a self-contained summary of the review. The subsequent parts discuss in more detail and with more precision the points made. Section 2 covers fundamental definitions, kinematics, and expressions for cross sections, which are used extensively in later sections. Section 3 provides a description of various muon experiments, briefly mentioning those performed before 1978 but covering in more detail those experiments BFP, BCDMS, and EMC postdating the review of Francis & Kirk (5). Section 4 deals with those general features of the data that give a large measure of confidence that in the main the data are well described by the basic quark parton model. Section 5 deals with the production of heavy flavors, the J/ψ, open charm, upsilon, and the search for more exotic multimuon final states. Section 6 concerns the extremely important confrontation of the data with the theory (QCD) of strong interactions and Section 7 covers the study of weak interaction effects; finally, Section 8 summarizes and attempts to draw some conclusions.

1.2 *Lepton Scattering as a General Technique*

In general the most obvious way to study an object is to look at it, in technical terms to examine the interaction of photons with that object. The scale on which structure can be observed is inversely related to the energy of photons used. A most natural extension of this procedure is to study the internal structure of atoms, nuclei and nucleons (7), by scattering electrons or muons on them. Given that the interaction of the electrons or muons with the object can be described in QED by the exchange of a virtual photon, the experimenter is directly looking at the object under study. As the energy of leptons available to the experimental physicist has increased from a few MeV to many GeV, so the momentum transfer squared (Q^2) characterizing the lepton scatter has increased from a few MeV^2 to many GeV^2 and the relevant distance scale through the uncertainty principle has reduced to $\sim 10^{-15}$ cm. The classical work on nuclear form factors was performed by Hofstadter and collaborators (8). Later the quasi-elastic scattering of electrons from nucleons within the nucleus gave information on the motion of those nucleons within the nucleus (9). Later, still using liquid hydrogen and deuterium targets, researchers studied the elastic form factors (i.e. charge distributions) of the nucleons themselves (10). It was inelastic scattering from these nucleon targets that revealed the presence of yet smaller constituents within the nucleon by the apparent quasi-elastic scattering from these entities; partons, quarks. The detailed study of the motion and distributions of these partons continues to occupy a significant fraction of the total physics research effort.

In all the stages outlined above the lepton used was the electron. The reasons for this are clear; the electrons were easier to produce (an advanced electric lamp) and were easy to accelerate. However, these advantages become less significant at high energies because the electrons radiate large amounts of their energy when bent in a magnetic field (synchrotron radiation). Hence the largest electron synchrotron used for scattering experiments reached an energy of ~ 12 GeV (Cornell). Somewhat larger beam energies of ~ 26 GeV were obtained at the Stanford Linear Accelerator Center (SLAC).

An alternative source of well-behaved leptons is provided by the muon. Protons can be accelerated to very high energies more economically than electrons and they produce pions, which subsequently decay to muons. At high energies the effective lifetime of the muon is infinite in comparison to its transport time from source to experiment and with sufficiently sophisticated beam optics a very satisfactory beam can be produced despite the several stages in its production. These facts, in conjunction with the technical point that for muons with their higher mass the corrections for higher order QED effects (radiative corrections) are more manageable, have transferred the emphasis during the past ten years from electron scattering to muon scattering.

The standard set by the electron scattering experiments was nevertheless very high and, as described fully by Francis & Kirk, the first generation of high energy (>100 GeV) muon experiments sacrificed a certain amount of their advantage in energy by inadequate design and care over the muon beam and consequently over the experiments themselves. Only in the last three years or so have muon scattering results extended the kinetic range of electron scattering results with a comparable precision. It is with this phase of the history of lepton scattering that this review is concerned.

1.3 General Situation with Deep Inelastic Scattering

The goal of current deep inelastic lepton scattering experiments (both with muon and neutrinos) is a controlled study of the behavior and interactions of quarks and gluons. The last phase of experimental work with electrons indicated that the parton constituents (quarks) (11–13) of the nucleons had a comparatively weak interaction among themselves, manifested experimentally as the approximate "scaling" behavior. This observation, on the one hand, made experiments at high Q^2 feasible (because of the lack of Q^2 dependence) and, at the same time, stimulated significant development in the theoretical treatment of strong interactions.

Specifically it led to the presently conventional description of strong interactions, quantum chromodynamics (14–16), which is so promising that it is tentatively described as a theory, i.e. has no proscribing boundaries in its ambitions to describe all of strong interaction physics. The dynamics of the theory are especially well founded in the treatment of deep inelastic lepton scattering. Consequently the most critical experimental tests of the theory have been performed by these experiments and, conversely, a large fraction of lepton scattering experimental effort has been devoted to testing QCD.

Since the discovery of approximate scaling at SLAC (17–19), the "approximate" aspect has been studied extensively. That "approximate" it is was clearly established by one of the first high energy muon scattering experiments at Fermilab and that the "approximate" was approximately logarithmic was further established by neutrino and more precise muon scattering (20–23) experiments. The precision achieved in this phase of experiments was, however, such as to satisfy a test of the theory only in the early years of euphoria following many lifetimes without a strong interaction theory.

Both theory (24) and experiment (25, 26) are now much less naive than they were five years ago and measurements from muon and neutrino experiments now exist of the quark distributions in nucleons both free and embedded in nuclei, and of the quark-antiquark sea distributions, both u(ū) and d(d̄) as well as the more exotic s(s̄) and c(c̄) components. In addition, since the Q^2 dependence of these distributions is related to the partons' interactions via the exchange of virtual gluons, its measurement has permitted the indirect determination (27, 28) of the momentum distribution of these gluons. This, at any given time, totals approximately half of the momentum of the nucleon.

The theory (and this is its basic property) is hoped to be calculable using perturbation theory and this new generation of experimental precision has necessarily generated a second approximation in the calculations. These are inevitably more difficult to test; however, the fact that they were necessary at all should generate a feeling of hope for the theory. It should not generate a feeling of despair among the experimentalists. With luck it is trivial to disprove an incorrect theory, it is a long job to fully test a correct theory.

The currently active groups in this work not covered in the review by Francis & Kirk (5) are the Berkeley-Fermilab-Princeton experiment, which was the last of the line of FNAL "300-GeV" generation muon experiments, the Bologna-CERN-Dubna-München-Saclay Collaboration experiment, and the European Muon Collaboration experiment, the latter two experiments working on the SPS machine at CERN. The important point is that, by comparison to FNAL, the CERN 300-GeV muon beam is second generation in both quality and intensity.

1.4 *Summary of Muon Scattering Results*

Detailed kinematic description of deep inelastic muon scattering is given in Section 2 of this paper, but some definitions are needed here in this section.

The quark distributions and the structure functions of the nucleons are given as functions of the variable x. This variable is completely specified by measuring the lepton scattering parameters and, in the parton model, is the fraction of the target momentum carried by the constituents participating in the lepton scattering. The four-momentum transfer squared $(-Q^2)$ of the collision (also specified completely by the lepton kinematics) is a measure of the profundity of the collision with the nucleon. The muon scattering results are measurements of the two structure functions $F_1(x, Q^2)$ and $F_2(x, Q^2)$, or equivalent linear combinations such as $F_2(x, Q^2)$, $R(x, Q^2)$, required to describe the scattering under assumptions of Lorentz invariance and parity conservation. $R(x, Q^2)$ is related to the ratio of absorption cross sections for longitudinal and transverse virtual photons and its small value lends support to the quark parton model, which requires $R = 0$ as occurs for on-shell, massless, spin-1/2 scattering centers.

The measured structure function $F_2(x, Q^2)$ is illustrated in Figure 1 as a function of x for different values of Q^2. The qualitative tendency for F_2 to evolve with Q^2, in such a way that at higher and higher Q^2 the momentum is spread among more and more constituents giving distributions in momentum (x) peaked more and more to small x, is evident but, in this Q^2 range, not dramatic. The data shown are for a hydrogen target (29) but this same qualitative behavior is also shown by data on targets with higher A (30, 31).

QCD does not predict the quark distributions, merely their evolution with Q^2 so that the data are required to determine both the quark distributions and the scale parameter Λ of the strong interactions. Its discriminatory power derives from the requirement that such fits be

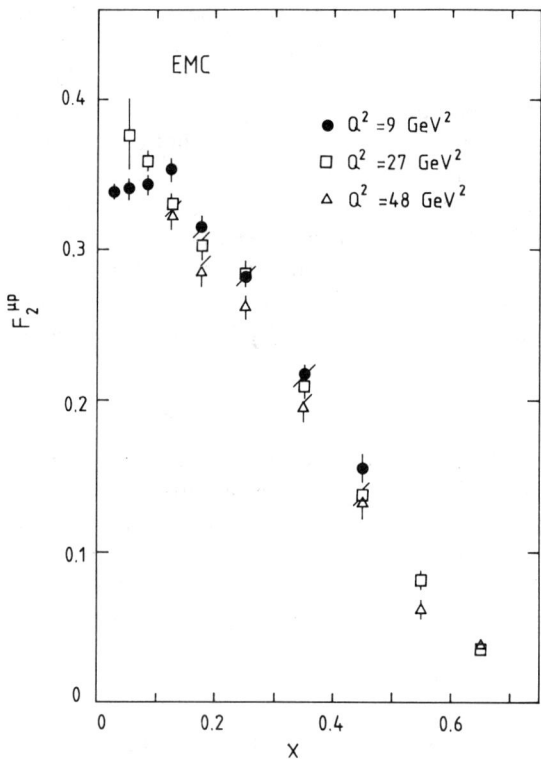

Figure 1 Illustration of the dependence of measurements of F_2 for different Q^2 (29).

acceptable. The currently derived values for Λ of the order of ~ 100 MeV (24, 25) are small enough to suggest that a convergence of the theory is possible. The change of this value, largely as a result of the muon experiments at high Q^2, from previously accepted values of 750 MeV in the space of three years does not, however, lead to a feeling that the value is stable. Indeed the difference between the two values is attributed to the presence of a multiparton scattering contribution to the theory, and a measure of the effective size and x dependence of such effects comes from an analysis (32) of the combined μp (EMC) and ep (SLAC) data sets and exploits the maximum Q^2 range available today. There is, however, no guarantee that another generation of experiments will not yield a value of $\Lambda \sim 10$ MeV with today's observed logarithmic nonscaling being interpreted in turn as a further higher twist.

Analyses of proton (or neutron) structure functions alone require a treatment that takes into account the transition of gluons to sea quark-antiquark pairs and vice versa. A simpler system is obtained if the difference $F_2^p - F_2^n$ is extracted from measurements on hydrogen and deuterium targets. The shape of this flavor nonsinglet combination (33) is shown in Figure 2 and shows the characteristic "quasi-elastic" peak in x expected of a valence quark distribution. The evolution is hard to measure with statistical accuracy and so far does not permit useful QCD analysis. The ratio F_2^n/F_2^p is, however, rather informative as a function of x. Its significance with respect to the quark parton content of the nucleon is discussed in Section 5.4.

It was remarked above that some of the structure function measurements are made on nuclear targets (carbon and iron) and that these latter

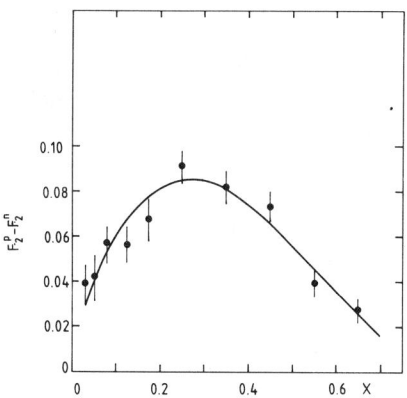

Figure 2 $F_2^{\mu p} - F_2^{\mu n}$ versus x, averaged over Q^2 (33). Only the statistical errors are shown. The curve represents a fit of the form $Ax^{0.5}(1-x)^{\alpha}(1+\beta x)$.

exhibited essentially the same behavior as the lighter (freer) nucleons. This can only be true to some approximation since nucleons in a nucleus do, after all, interact with each other and the importance of such effects is an experimental question (theory has as yet no quantitative calculations). Recent data (34) from EMC indeed indicate a difference in the x dependence of F_2(Fe) and F_2(D) at rather high Q^2 ($10 \rightarrow 170$ GeV^2). What is not yet clear is the Q^2 behavior of this effect and consequently its effect on QCD analyses of heavy target/high A nuclear data. If Λ is to be independent of target then there has to be some Q^2 dependence to offset the difference in the quark distribution within the Altarelli-Parisi evolution equations (35). On the other hand too strong a Q^2 dependence might indicate a higher twist effect, which could render obsolete QCD checks with iron, marble, and carbon.

Apart from light hadron production, treated by Renton & Williams (6), muon scattering experiments have been active in the study of the production of heavier hadrons (J/ψ, charm, beauty, Υ), which have significant branching ratios into channels producing one or more extra muons. Muon scattering experiments with their potentially very massive targets can attain high luminosity; they have not wasted this opportunity and provide a list of interesting results.

Through its dimuon decay channel, J/ψ production has been studied (36–41). Approximately half the production cross section at $\nu \sim 100$–200 GeV is inelastic; that is, the final state contains produced hadrons other than the ψ. The virtual photon cross sections is of the order of 25 nb compared to a total cross section of ~ 120 μb.

A much larger fraction of the total cross section is in the form of the open charm production (42–47), which is of the order of $\frac{1}{2}$–1 μb. Detection here is via the semileptonic decay of one or both of the produced charmed particles resulting in a di- or trimuon final state with missing undetected energy corresponding to the neutrino(s) energy.

Both the production of J/ψ and of open charm can be readily understood with the photon gluon fusion model (48, 49). This is a model of a first-order QCD process in which the virtual pair of $c\bar{c}$ quarks from a gluon in the target are put on mass shell by the interaction of the photon and result in a pair of charmed hadrons. If the total mass of this system is less than that necessary for charmed hadrons, then the $c\bar{c}$ pair is considered to fuse and produce some of the time a charmonium state. The observed ratio of (J/ψ)/ψ' production of ~ 5 is, however, difficult to accommodate and is an example of some detailed problems of which our understanding is not yet complete.

In a manner analogous to the J/ψ, the production of the highest mass state Υ of the vector mesons is expected, but with a cross section down by a factor 4×10^{-3}–4×10^{-5} at current energies. So far the experimental results are limits on the production cross section. These limits do not rule out the γGF mechanism, however, and a second round of experiments, both at present energies and at higher energies, is awaited with interest.

More exotic final states of multimuons with various charge combinations have been sought (50–52). So far no extremely surprising result has been obtained although the interest in this experimental technique has been justified by the observation of a few exotic trimuon events that could be ascribed to either open $B\bar{B}$ production or to a large CP-violating $D^0\overline{D^0}$ mixing probability.

One measurement, covered in more detail in Section 7, is that of the $\mu^+\mu^-$ asymmetry for polarized muons undergoing deep inelastic scattering (53). This measurement is sensitive to the interference between the single-photon-exchange amplitude and the single-Z^0-exchange amplitude of the weak interaction. An asymmetry is observed, with unarguable significance, which is as expected within the Glashow-Salam-Weinberg standard electroweak model. There is, as a result, a measurement of $\sin^2 \theta_W$ complementing that of the neutrino experiments and a confirmation that the coupling of the right-handed muons is zero with a significance similar to the current combined PETRA experiment results (54).

2. FUNDAMENTALS

2.1 Kinematics

A detailed discussion of the kinematics of deep inelastic lepton scattering can be found in (12, 55, 56). Here, for reasons of completeness, we summarize briefly the main formulae for the case in which only the scattered muon (electron) is detected.

In the lowest order, the process $\mu + N \to \mu +$ hadron is mediated by the exchange of one virtual photon. Denoting the four-momenta of the nucleon and the exchanged photon by p and q one can form two Lorentz scalar variables

$$Q^2 = -q^2 = Q^2_{\min} + 4kk' \sin^2 \frac{\theta}{4}$$

$$v = \frac{p \cdot q}{M} = E - E'$$

where E, E' (k, k') are the laboratory energies (absolute values of the laboratory momenta) of the incident and the scattered muon and θ is the scattering angle. M is the mass of the nucleon. In the following we concentrate on the region $Q^2 \gg Q^2_{\min}$ and neglect terms of order $Q^2_{\min}/Q^2$.

The mass of the hadronic final state is given by

$$W^2 = M^2 - Q^2 + 2Mv.$$

Frequently we use the ratios

$$x = \frac{Q^2}{2Mv} \quad \text{and} \quad y = \frac{v}{E}.$$

Summing over all undetected final states (and assuming Lorentz invariance, current conservation, and parity conservation) one derives the double differential cross section

$$\frac{d^2\sigma}{dQ^2\,dv} = \frac{4\pi\alpha^2}{Q^4} \frac{E'}{Ev}\left[2\frac{v}{M} F_1(Q^2, v)\sin^2\frac{\theta}{2} + F_2(Q^2, v)\cos^2\frac{\theta}{2}\right]. \qquad 1.$$

The information on the structure of the target is contained in the two functions F_1, F_2.

One can consider inelastic muon-nucleon scattering also as a collision between the virtual photon and the nucleon. Defining virtual photon-nucleon cross sections σ_T and σ_L for the absorption of transversely and longitudinally polarized photons, one can write

$$\frac{d^2\sigma}{dQ^2\,dv} = \frac{\alpha}{4\pi}\frac{1-x}{xME^2}\frac{1}{1-\varepsilon}[\sigma_T(Q^2, v) + \varepsilon\sigma_L(Q^2, v)], \qquad 2.$$

where

$$\varepsilon = \left[1 + 2\left(1 + \frac{v^2}{Q^2}\right)\tan^2\frac{\theta}{2}\right]^{-1}$$

is the polarization parameter, $0 \leq \varepsilon \leq 1$.

The structure functions and the virtual photon absorption cross sections are connected by

$$F_1 = \frac{2Mv - Q^2}{8\pi^2\alpha}\sigma_T$$

$$F_2 = \frac{2Mv - Q^2}{8\pi^2\alpha}\frac{v}{M}\frac{Q^2}{Q^2 + v^2}(\sigma_T + \sigma_L). \qquad 3.$$

In the limit $Q^2 \to 0$, σ_T approaches the total photoabsorption cross section for real photons while $\sigma_L \to 0$.

2.2 Higher Order QED Corrections

The purpose of deep inelastic scattering experiments is the evaluation of the structure functions. One therefore has to extract the cross section due to single-photon exchange from the directly measured cross section by calculating and correcting for higher order electromagnetic distributions

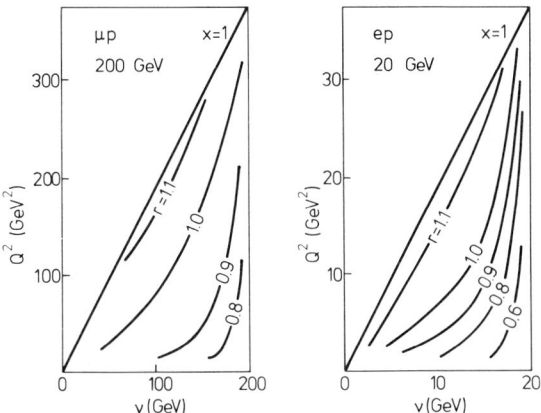

Figure 3 Size of radiative corrections for deep inelastic muon and electron scattering.

(57, 58). The typical size of radiative corrections is shown in Figure 3 for μp scattering at 200 GeV and for ep scattering at 20 GeV. Plotted are contour lines of constant values of the ratio

$$r = \frac{\left[\dfrac{d^2\sigma}{dQ^2\,dv}\right]_{\text{Born}}}{\left[\dfrac{d^2\sigma}{dQ^2\,dv}\right]_{\text{measured}}}$$

in the Q^2-v plane.

Because of the emission of hard real photons from the incident or scattered lepton, large corrections occur at v close to E, i.e. at large y. The calculation of this internal bremsstrahlung has been verified by measuring, together with the incident and scattered muon, a single hard photon in the expected angular range. A comparison of measurements and computations is shown in Figure 4 (59) where the ratio N_γ/N_μ is plotted versus y in various x intervals. N_γ denotes the number of events with a photon of energy $E_\gamma > 0.8v$ and without additional hadron tracks; N_μ is the total number of apparently inelastic events in the corresponding x, y bin. The curves present the calculation in lowest order of α folded with the geometrical acceptance of the apparatus. No obvious deviation from the calculation is observed. Nevertheless, the uncertainty of the calculation of radiative corrections is estimated to be 10–20%, therefore a lower cut $r \gtrsim 0.75$ is usually applied to the data.

A special case is the contribution due to the exchange of two photons between lepton and hadrons, which cannot be calculated in a model-independent way. It is, however, possible to determine the ratio of the real part of the two-photon-exchange amplitude to the single-photon-exchange

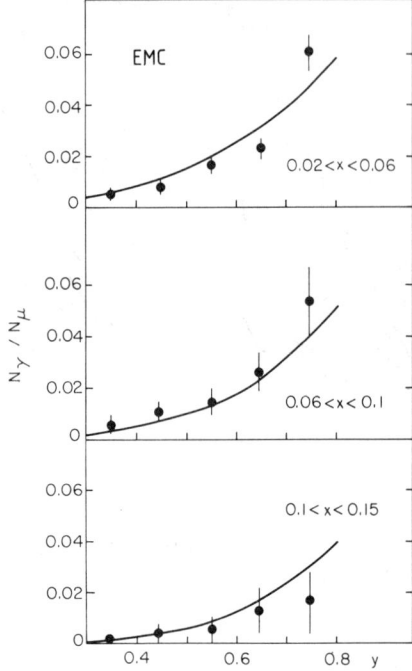

Figure 4 Ratio of events with bremsstrahlung photon(s) of energy $E_\gamma > 0.8yE$ to all events as a function of y for various x bins. The curves show calculations in lowest order of α folded with the experimental acceptance.

amplitude by measuring e^+/e^- or μ^+/μ^- cross-section ratios. Measurements performed at SLAC (60, 61) in the Q^2 region up to 15 GeV^2 and at Fermilab (62) for $15 \leq Q^2 \leq 50$ GeV^2 show that the cross-section ratios are consistent with unity and one can conclude that the two-photon-exchange part of the measured double differential cross section is small (less than 2%).

At large Q^2 the μ^+/μ^- ratios are expected to deviate from unity within a few percent owing to the interference of the electromagnetic current with the weak neutral current. We discuss results on recent precision measurements of this effect in Section 7.

3. MUON EXPERIMENTS

3.1 *Muon Beams and Their Use*

As mentioned earlier in the article, a rather complete historical survey of muon scattering experiments to 1978 was given by Francis & Kirk (5) and only a few brief comments are necessary here.

In the introduction it was remarked that lepton scattering was dominated by electrons until ~1970. The ease of production and utilization of electrons reflects very strongly in the character and ambition of the muon scattering experiments. In general they were performed in order to compare the various electron scattering experiments and thereby examine possible differences between muon and electron. This attitude seems to have persisted from the time of the experiments at Berkeley in 1961 to those in Brookhaven in ~1970. The only experiments that challenged the leading role of the electron were those in which a muon beam at SLAC was used for bubble (63) and streamer (64) chamber measurements of the hadronic final state. This competitiveness arose from the difficulty of using the high SLAC electron intensity for coincidence experiments given its extremely low duty cycle.

The real shift of emphasis came only when dedicated electron accelerators were no longer available at the highest energy and the primary source had to be the proton synchrotrons (FNAL and SPS). In this case both electrons and muons are at best tertiary beams and an objective comparison of the possibilities and capabilities could be made. The muon was chosen on three counts. Since the electron beam requires one further stage in production, its energy and also its intensity are lower. Since the muon has higher mass its electromagnetic interactions are less violent, that is to say its effective radiation length is a factor m_μ^2/m_e^2 longer and therefore long targets can give muon experiments a further advantage in luminosity. Finally, this property of the muon also implies that corrections for higher order QED effects, two-photon exchange, etc are correspondingly reduced. Everything else being equal, corrections for muon scattering at 200 GeV are similar to those for electron scattering at 20 GeV.

The penetrative quality of the muon, however, creates some difficulties in the production of a beam (65) of high quality. The muons are produced as follows. The extracted proton beam from the synchrotron is used to produce a beam of charged pions, which are allowed to decay over rather a long distance until they contain ~5% muons. The pions are then absorbed in several meters of comparatively low Z material (to reduce multiple scattering of the muons). The penetrating muons are finally transported to the experimental zone. With a normal hadron beam the beam shape/phase space can be trimmed efficiently by the use of simple mechanical collimators. For muon beams, magnetic bending of the unwanted muons must also be provided since the introduction of material only increases the scattering but does not stop the muons. Without the provision of magnetic collimation, the muon beam at the experiment is characterized by a halo/beam ratio of about 1/1. Halo has no unique definition but is typically those muons that traverse the apparatus further than 2–3 cm from the beam axis but within the 4 × 4 m² sensitive area of the apparatus. The use of magnetic scrapers at the CERN SPS was the main qualitative improvement over the FNAL 300-GeV muon beam. In the former the 1/1 halo/beam ratio pertained, in the latter the figure was typically $\leq 10\%$. The importance of this can be illustrated by a simple example. The conventional method of preventing halo particles from contributing to, even dominating, the experimental trigger is to use veto scintillator counters, which inhibit the trigger, and to suffer the consequent dead time. With a beam of 10^8 muons per second, a 1/1 halo, and resolving times of 10 ns, the dead time is 100%. With $\leq 10\%$ halo the dead time is correspondingly $\leq 10\%$.

The actual beam yields per extracted proton are given in Table 1 for a few combinations of energy and sign for the SPS muon beam. The FNAL values are typically a factor of 10 less because of a less well-designed (cheaper) pion decay channel. The actual fluxes used by the experiments at the SPS have been as high as $\sim 3 \times 10^7$ muons per spill in the years 1978–1980 and more recently fluxes of $\sim 5 \times 10^7$ have been used successfully. At the higher energies it is necessary to convince the scheduling committees to devote the accelerator to the muon beam running to actually achieve such fluxes.

The high energy muon experiments performed since 1970 fall into two categories based on the design of the apparatus. On the one hand there are those experiments aiming for maximum possible luminosity and using repetitive structures with iron core magnets to provide the necessary momentum analysis. These experiments have used several tens of meters of heavy material as the target and their configurations are such as to limit the final particle detection to muons.

On the other hand, the detection of hadrons is also interesting and two experiments, the CHIO (Chicago-Harvard-Illinois-Oxford) experiment at Fermilab and the EMC experiment at CERN, have therefore designed and

Table 1 Muon yields for SPS beam in different configurations

Momentum			Yield[a] μ^+ per incident proton ~ 1 interaction length Be target	
Proton	Pion	Muon	Monte Carlo (65)	Measured
400	115	100		1.9×10^{-5}
400	140	120	1×10^{-4}	1.4×10^{-5}
400	220	200	4×10^{-5}	1.0×10^{-5}
400	300	280	3×10^{-6}	1.6×10^{-6}
450	220	200		1.6×10^{-5}

[a] The yields are sensitive at the level of 30% to beam tuning parameters.

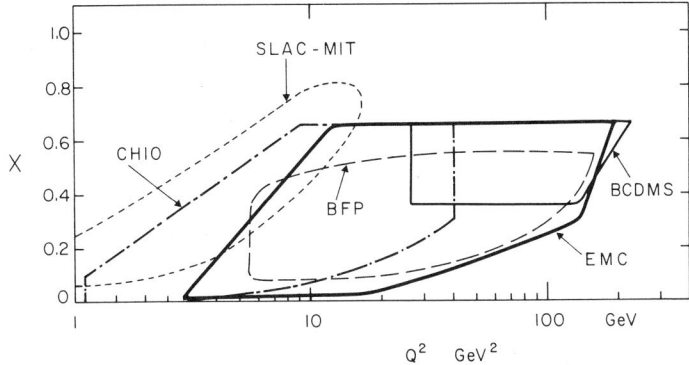

Figure 5 The acceptance (26) of major experiments in Q^2-x plane.

used large solid angle spectrometers with air gap magnets, in which the target used was typically a few meters of liquid hydrogen or deuterium. In the case of EMC they also participated in the nuclear target/high luminosity field with a calorimeter target but of limited length (~ 3 m). The experimental beam conditions outlined above are reflected in the luminosities used for physics and therefore in the statistical precision discussed in Section 4. The acceptance in the Q^2-ν plane of the major charged lepton experiments is illustrated in Figure 5.

3.2 The Berkeley-Fermilab-Princeton (BFP) Experiment

This experiment was the final one in the series of FNAL 300-GeV muon experiments. The experimental apparatus is also known as the Multi-Muon Spectrometer (MMS) since its primary aim was the detection and measurement of multimuon final states.

Measurement of multimuon final states requires, in order of importance, high luminosity, efficient muon detection, and calorimetric capability in the target. The MMS attempted to achieve these aims with a repeated structure of iron plates that were at one and the same time the target modules and the detector modules of the iron core magnet. The apparatus (38, 43, 51, 66) is illustrated in Figure 6.

The magnet consisted of 18 modules containing a total of 91 ten-centimeter plates of steel. The plates each had slits to permit the insertion of the current-carrying coils that provided a vertical dipole field of 19.7 kG in the central region of the detector. The muon beam was incident along the axis of the apparatus and during the data taking its effective size was ~ 20 cm vertical by 35 cm horizontal.

Associated with each of the 18 steel modules was a double package of multiwire proportional and drift chambers 178 cm high by 100 cm wide.

Given the size of the beam and the typical operating intensity of $\sim 2 \times 10^6$ muons per spill (1 s) it was possible to operate the chambers with no explicit dead region. Nevertheless the proportional chambers suffered a loss of efficiency to about 60–80% in the central region, different for sense and induced plane readout and improved along the length of the apparatus. Away from the beam the efficiencies were $\sim 95\%$.

Calorimeter scintillation counters were placed after each steel plate in the first 15 modules. Measurements of the pulse height after each of two stages of amplification provided the necessary dynamic range. The sampling thickness of this calorimeter was therefore 10-cm steel and led to a resolution in hadron energy measurement of $\Delta E/E = 1.5/[E \text{ (GeV)}]^{1/2}$. Calibration was provided by single-arm scattering events and the zero was also calibrated using elastic J/ψ production.

Triggering for the apparatus was provided by a series of 8 scintillation hodoscopes placed after each of the modules from number 4 onward. All triggers were inhibited by a signal from an upstream halo wall. The single-muon, deep inelastic trigger required a vertically scattered muon to appear in a sequence of three of the trigger hodoscopes and for there to be no small or zero angle muons in those same modules. That is, essentially a beam veto was applied and such a trigger has to be corrected for false vetoing, which can occur for a multimuon final state. The multimuon triggers, of which there were two, had no such requirement. The dimuon trigger required both

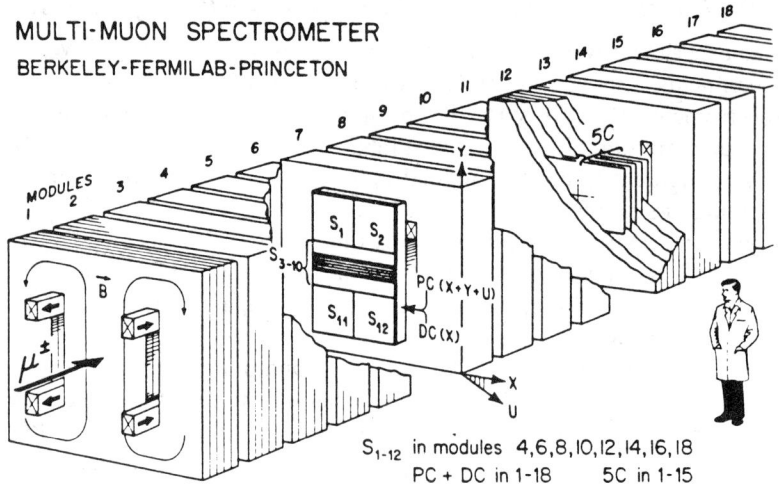

Figure 6 The BFP experimental apparatus. The modular structure of the apparatus is indicated. S_{1-12} are one set of trigger hodoscopes of which there are eight. PC $(x+y+u)$ are proportional chambers and DC (x) are drift chambers. Calorimeter counters (5C) are repeated in each of the 15 modules.

of (a) an acceptable trigger hodoscope pattern in three consecutive hodoscope modules and, (b) energy deposited in a combination of the calorimeter counters upstream of the struck trigger counters. The trimuon trigger did not make any requirement on the calorimeter counters; it merely demanded a pattern in three consecutive trigger hodoscopes that was compatible with a 3-muon (or greater) final state. For calibration and for use as input to the Monte Carlo simulation of the experiment, a beam trigger was also used (this trigger, of course, was heavily prescaled). The physics trigger rates were of the order of 10^{-5} per beam muon and data taking proceeded at a rate of ~ 60 events per spill.

The experiment first reported a preliminary analysis of its data in 1979 (66). Subsequently the analysis of its primary physics subject, multimuon final states, was completed and furnished a series of publications and theses on J/ψ, charm production, and rare multimuon processes (36–38, 42, 43, 50, 51). There is as yet no report on the final analysis of the structure function data.

3.3 The Bologna-CERN-Dubna-München-Saclay (BCDMS) Experiment

One of the two muon scattering experiments at the CERN SPS, the BCDMS experiment, is a huge, 50-meter long, 2.75-meter diameter iron toroid spectrometer (67) (Figure 7). The apparatus is subdivided into 10 supermodules. The 32 individual iron plates in each supermodule are 11 cm thick and there is a hole of radius 25 cm. The central hole provides space for the target and mounting. The target in each of the supermodules is independent and either hydrogen, deuterium, or carbon targets can be used. The data discussed in this paper concern only the latter.

Each of the supermodules is equipped with 8 planes, alternately vertical and horizontal, of multiwire proportional chamber that have "cut outs" in the region of the beam. The targets in each supermodule are continuous so

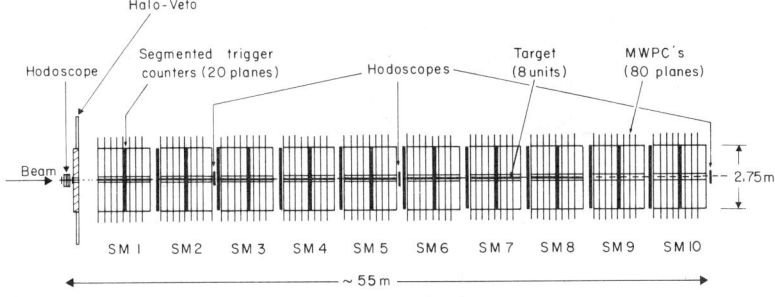

Figure 7 The BCDMS experimental apparatus. SM1–10 are the ten Super Modules, the repetitive units into which the apparatus is divided.

that there is no calorometric capability, neither in the region of the target nor in the outer part of the detector.

The trigger counters, two per supermodule, are pairs of effectively concentric half-ring liquid scintillator hodoscopes. This simplicity nevertheless allows a rather sophisticated trigger system due to the special properties of the toroidal system. A muon incident along the axis of the system, which then is scattered, executes a trajectory in its scattering plane (Figure 8) and is independent of azimuth. The trajectory and hence kinematics of the events are completely specified (up to corrections for multiple scattering and energy loss) by its maximum excursion from the axis and by the wavelength of the trajectory. In particular the sagitta Δ is directly related to the ratio Q^2/Q^2_{max}. Given the strong $1/Q^4$ dependence of muon scattering, a trigger directly sensitive to Q^2 is highly desirable and is easily achieved by demanding a series of trigger counters to fire with a ring number greater than a specified value. The trigger used is therefore the analogue of a time-over-threshold discriminator. A typical trigger requirement used was rings 1–7 or 2–7 firing in a sequence for four trigger counters. The latter yielded, on carbon, a trigger rate with respect to unvetoed beam of 4–20 × 10^{-6} depending on energy. In all cases a beam requirement is included in the trigger. The disadvantage of the apparatus triggering system is that it has little intrinsic discrimination against halo. Any halo particle incident on the apparatus executes a perfectly acceptable trajectory within the trigger definition; therefore heavy reliance is placed on the beam requirement and a bank of halo-veto scintillation counters.

The resultant trigger contained about 6% residual halo, independent of beam energy, and from 14% at 280 GeV to 70% at 120 GeV good deep inelastic events. The balance was dominated by multiple hits from the punch-through of hadron showers.

The experiment started effective data taking in 1979 and has published

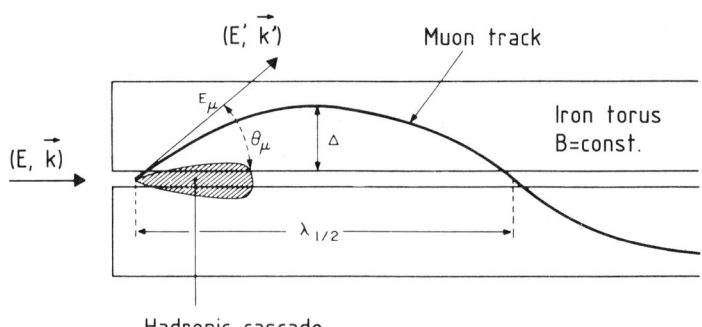

Figure 8 Trajectory of muon through BCDMS apparatus. The basic parameters of the trajectory are $\lambda_{1/2}$ and Δ, the latter is directly related to Q^2/Q^2_{max}.

results on both structure functions (30) and multimuon measurements (68). In addition a large data sample with alternatively positive and negative incident muons at 200 and 120 GeV was taken in 1980. These data have been analyzed to obtain a measurement of the weak electromagnetic interference between the photon and Z^0 exchanges (53) as described in Section 7.

3.4 The European Muon Collaboration (EMC) Experiment

In contrast to the experiments discussed in the previous two sections the EMC experiment (69) does not make use of a distributed target. The experiment is illustrated in Figure 9 and consists, for the data discussed in this paper, of a single 4-Tm air core dipole magnet. The total spectrometer length is ~ 30 meters and on this scale the targets, typically 3–6 m in length, are conceptually point-like. The scattered muon is detected in a series of multiwire proportional and drift chambers before, inside, and after the volume of the magnet. Because these detectors are also sensitive to the hadronic products of the muon interaction, the density of planes involved is rather higher (a total of 99) than in the previous two experiments discussed. Although the larger chambers are deadened in the region of the beam the effect on the acceptance is minimized by using a set of small proportional chambers in the beam region between the target and the magnet. These chambers will operate efficiently in fluxes of up to 10^8 particles per second on their 7-cm radius surface. Because of the open geometry of the basic spectrometer, a steel wall is introduced explicitly to stop all hadrons such that the particles detected after this absorber are only muons. The absorber

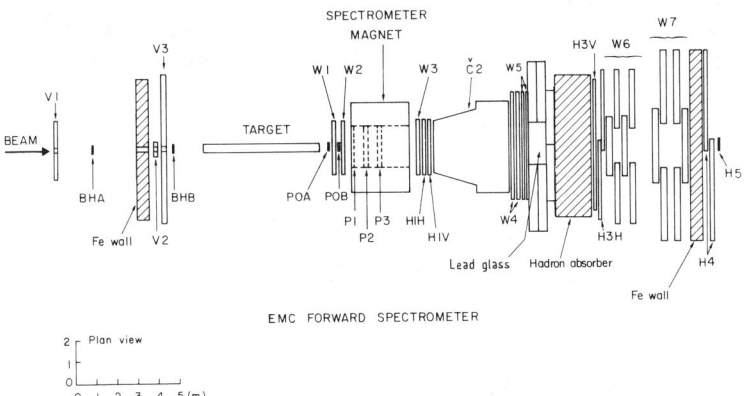

Figure 9 The EMC experimental apparatus. V_{1-3} are veto scintillator counters, H_1–H_4 are trigger hodoscopes, W_1–W_7 are large drift chambers while P_0–P_3 are proportional chambers; the hadron absorber is of iron.

has a hole to allow undisturbed passage of the beam, which is then transmitted to, and used by, the BCDMS experiment concurrently.

The target was variously 6 meters of liquid hydrogen or deuterium in a conventional thin steel vessel or an iron scintillator STAC target. This latter was in fact a very high resolution calorimeter, which, with its fine sampling, gave a resolution for hadronic energy deposition of $\Delta E/E = 0.56/[E(\text{GeV})]^{0.4}$. On each of the 36 phototubes, three digitizations of the analogue signal were made at different levels of amplification to provide the necessary dynamic range.

It is worth noting at this point that the main background, for the study of heavy quark production and exotic final states via their semileptonic decays to muons, is that produced by the decay of conventional hadrons, pions, and kaons. The quality of the experiment in this respect rests heavily on the effective density of the calorimeter/target and in this respect the EMC STAC proved to be somewhat superior to the BFP experiment.

The basic trigger system was provided by a series of five hodoscope planes, three of which were located after the hadron absorber. A serial system of matrix coincidences between individual elements of these hodoscopes attempted to establish two-dimensional target pointing and a minimum angle for a scattered muon. This system proved to be extremely effective and in particular the basic halo could not easily satisfy the trigger. This meant that, although the halo-veto wall had been designed to achieve inefficiencies of 10^{-6}, it was possible to operate most counters with an inefficiency $> 10^{-4}$. The trigger rate for a nominal $\frac{1}{2}°$ angle cut with the 6-m hydrogen target was typically 3×10^{-6} per incident muon. A large fraction of this turned out to be either a series of electromagnetic showers that confused the hodoscope logic both before and after the absorber or small angle muon scatters that satisfied the trigger by an additional photon or knock-on electron. The fraction of the trigger comprising genuine deep inelastic scatters was $\sim 10–20\%$.

With the iron scintillator STAC target it was also necessary to demand in the single-muon trigger the deposition of several GeV of hadronic energy. This proved to be an effective combatant for electromagnetic events and the resultant trigger contained $\geq 90\%$ genuine deep inelastic scattered muons.

The STAC target was used with specific multimuon triggers; these relaxed the requirement on minimum scattering angle and on deposition of energy, but required a number of separated clusters in the hodoscopes after the muon absorber. A minimum vertical angle separation was also required to suppress the copious electromagnetic tridents that appear as a fan in the bending plane of the magnet. The EMC experiment started taking data in August 1978 with the STAC target and to date has a series of published results on hadron production (70), on structure functions (29, 31–34, 71), and on multimuon production (39–41, 44–47).

The experiment has now been extended (72) to provide complete detection capability and particle identification for the hadronic final states. Results from this phase of the experiment have not yet been published and are not discussed here.

3.5 Beam Measurement, Systematics

In comparison with neutrino experiments, the measurement of the absolute beam flux is comparatively straightforward. The conventional method is to count with a scintillator hodoscope the total flux. Given the typical beam rates it is necessary to make dead-time corrections of 10–20% and also to determine the fraction of this counting rate corresponding to reconstructable muons. This latter requires that a sufficient number of "beam triggers" be accepted for offline analysis.

An alternative method (73), used so far only by EMC, is to open a gate at random times and to record the beam hodoscope and beam momentum spectrometer time digitizers. These "triggers" are then processed through the same offline beam reconstruction as the real data and a beam yield proportional to the total flux is obtained. Although care must be taken in the treatment of veto signals, the resultant yield is simply related to the total by the known random trigger rate. This method has the advantage that no dead-time corrections are necessary. The precision obtained is typically 1–3% depending on the data sample, although this is not an intrinsic limitation from the method.

In the foregoing discussions of the individual experiments, one common aspect was omitted. All experiments measure each individual incident muon energy in the case in which that muon scatters and satisfies the trigger. This is achieved by using a set of either multiwire proportional chambers or scintillator hodoscopes suitably arranged about bending elements in the muon beam. The resolution of this measurement is typically 0.3% and is usually matched by a direction measurement immediately before the experiment.

The measurement of structure functions, as is discussed below, is a precision game and is, given the adequate statistics of the muon experiments, dominated by systematic instrumental effects. This can be illustrated by Figure 10, which shows the relative error generated in the measurement of the structure function F_2 by a systematic mismeasurement of the absolute incident beam energy of 1%.

Effects such as these place a premium on the knowledge of and intercalibration of beam and scattered muon spectrometers. It is now standard practice to direct a low intensity beam directly into the apparatus to do just this job and typical levels of precision for the data discussed later are $\sim 0.3\%$.

Although this is not the place for an exhaustive discussion of all the

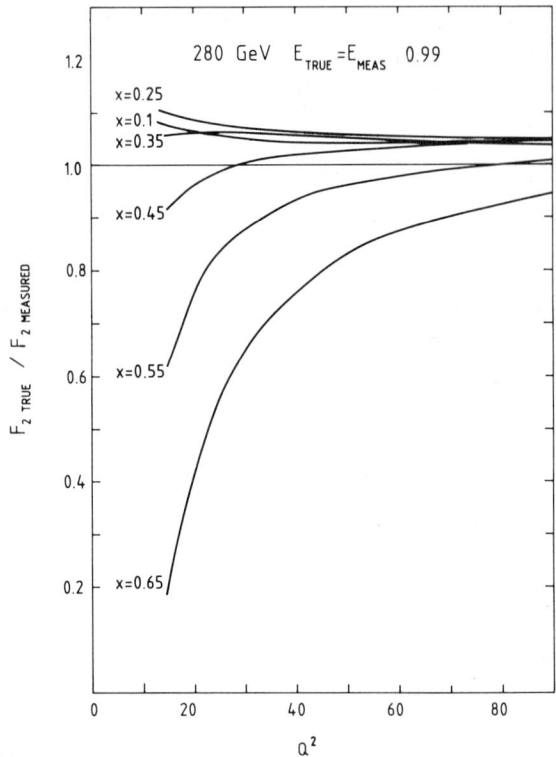

Figure 10 Effect of systematic error of 1% in beam energy on measurement of F_2, the ratio $F_{2\text{ true}}/F_{2\text{ measured}}$ is shown as a function of Q^2 for different x values.

possible systematic effects, their importance cannot be over emphasized. Normally such things are quasi-hidden in PhD theses and it is gratifying to see that the repeated public discussions have stimulated a subgroup of active experimentalists to publish (74) a study of these effects, including a comparison with comparable effects in competitive neutrino experiments.

4. GENERAL FEATURES OF THE DATA, THE PARTON MODEL

4.1 *The Parton Model*

The first experiments at energies high enough to observe the characteristic features of deep inelastic scattering and approximate scaling of the structure functions were performed at the Stanford Linear Accelerator Center in the years from 1967 onward. A review of the results can be found, for example, in (19, 75). The important information was that the structure functions are within 30% functions of the one variable x only. Such a property was anticipated by Bjorken (76), who had found that, in the limit of v and Q^2 large but x finite, the structure functions $F_1(Q^2, v)$, $F_2(Q^2, v)$ should not depend on any dimensional scale such as mass or length but should be a function of the ratio x only. This property of $F_{1,2}$ is therefore called Bjorken scaling.

Bjorken scaling is easily understood in the parton model where the nucleons are composed of point-like constituents or partons (11). At high energies the interaction time $1/v$ of the virtual photon is expected to be short enough so that the scattering can take place incoherently off a beam of quasi-free partons. If the initial four-momentum of the scattered parton is simply taken to be ξp (i.e. the parton has no momentum component transverse to the nucleon momentum), one obtains from energy momentum conservation for the elastic parton scattering process

$$(\xi p + q)^2 = m^2,$$

where m is the parton mass. For $Q^2 \gg M^2$ and m^2, we find

$$\xi = \frac{-q^2}{2pq} = \frac{Q^2}{2Mv} = x.$$

The only variable describing the scattering process is therefore x, which is found to be the fraction of the nucleons momentum carried by the struck quark before the interaction.

Including terms with M^2, but still neglecting m^2, yields

$$\xi = \frac{2x}{1 + (1 + 4M^2 x^2/Q^2)^{1/2}}, \qquad 4.$$

a quantity, often called the Georgi-Politzer variable (77), which includes target mass effects at finite Q^2 and is frequently used in the analysis of the data.

For F_2 one obtains for a general distribution of partons

$$F_2(x) = \sum_i e_i^2 x q_i(x),$$

where $q_i(x)$ is the probability of finding a parton of type i with momentum fraction x and charge e_i (in units of electron charge).

Energy momentum conservation requires

$$\int_0^1 dx \sum_i x q_i(x) = 1 - g,$$

where g is the fraction of the nucleon momentum carried by neutral partons, i.e. gluons. Assuming that the charged partons are quarks, it was found from the earlier experiments, by using the above equations, that about half of the nucleon momentum is carried by gluons.

It is common practice to denote the quark distribution functions for the individual quark flavors by $u(x)$, $d(x)$, $s(x)$, $c(x)$ and the corresponding antiquark distribution functions by $\bar{u}(x)$ etc. The functions u and d can be divided into a valence quark part u_v, d_v responsible for the proton quantum numbers and a sea quark part u_s, d_s for which one assumes $u_s = \bar{u}$, $d_s = \bar{d}$. In order to reproduce the proton quantum numbers, the following normalization conditions have to be satisfied:

$$\int_0^1 (u-\bar{u})\,dx = \int_0^1 u_v\,dx = 2$$

$$\int_0^1 (d-\bar{d})\,dx = \int_0^1 d_v\,dx = 1$$

$$\int_0^1 (s-\bar{s})\,dx = 0 \quad \text{etc.}$$

Because of isospin invariance, the distribution functions for the neutron are $u^n = d$, $d^n = u$, $s^n = s$.

In connection with the new theory of the strong interactions, quantum chromodynamics, it is useful to distinguish between quark distributions behaving like a nonsinglet under flavor SU(4), e.g.

$$V = u_v + d_v$$

or

$$\Delta = u - d + \bar{u} - \bar{d} + c - s + \bar{c} - \bar{s},$$

and flavor-singlet distributions, e.g.

$$\Sigma = u + \bar{u} + d + \bar{d} + s + \bar{s} + c + \bar{c}.$$

For the proton, we find

$$F_2^{\mu p} = \frac{1}{6}x\Delta + \frac{5}{18}x\Sigma.$$

Within QCD the parton distributions acquire a dependence on the momentum transfer implying that scaling violations, logarithmic in Q^2, must be expected. As we discuss in some detail in Section 6, it is indeed possible to determine the effective strong interaction coupling α_s from a study of the subtle scaling violations of the structure functions. Consequently, one of the most important aspects of the recent muon scattering experiments at CERN and at Fermilab is the accurate measurement of the deviation from Bjorken scaling.

4.2 Determination of σ_L/σ_T

A measurement of the ratio of longitudinal to transverse absorption cross sections

$$R = \sigma_L/\sigma_T = \frac{(F_L - F_2 Q^2/v^2)}{2xF_1} \qquad 5.$$

with

$$F_L = F_2 - 2xF_1$$

is important for several reasons. First, R contains information about the spin of the charged constituents; the parton model predicts $R = \infty$ for spin-0 partons, and $R = 4(k_T^2 + m^2)/Q^2$ for spin-1/2 partons allowing for an initial parton transverse momentum k_T due to the Fermi motion of the partons inside the nucleon and including parton mass effects. In QCD one expects $R \propto \alpha_s$ or more specifically in leading-order QCD (e.g. 16):

$$F_L(x, Q^2) = \frac{\alpha_s}{2\pi} x^2 \int_x^1 \frac{dz}{z^3} \left[\frac{8}{3} F_2(z, Q^2) + \frac{40}{9}\left(1 - \frac{x}{z}\right) z G(z, Q^2) \right]. \qquad 6.$$

The first term contributes mainly at large x ($x > 0.2$), the second term dominates at small x ($x < 0.1$). Good data on R would therefore either allow a direct determination of α_s or of the gluon distribution function $G(x, Q^2)$. A knowledge of R is also essential for the extraction of the structure function F_2 from the measured double differential cross section:

$$F_2 = \frac{Q^4}{4\pi\alpha^2} x \frac{d^2\sigma}{dQ^2\,dx} \left[1 - y - \frac{Mxy}{2E} + \frac{y^2(1 + 4M^2x^2/Q^2)}{2(1+R)} \right]^{-1}. \qquad 7.$$

Though a separation of σ_L and σ_T is simple in principle, in practice it is very hard since one has to compare absolutely normalized cross sections measured at fixed Q^2 and x but at different beam energies so that ε varies between values close to 0 and close to 1. This means measurements at small, but also at large, y where radiative corrections and various background contributions (e.g. muons from π, K decay) are large. Also different geometrical parts of the large solid angle muon detector are involved.

For these reasons the available information on R is still sparse. Figure 11 shows some example of straight-line fits of $\sigma_T + \varepsilon\sigma_L$ versus ε obtained by the EMC from measurements at $E = 120$, 200, and 280 GeV (71). The total error includes systematic and statistical uncertainties, the statistical error is indicated by the inner bars.

A summary of R data from charged lepton-proton scattering is collected in Figure 12. The errors include systematic and statistical uncertainties. The typical Q^2 values for the EMC data are $\langle Q^2 \rangle = 12$ GeV^2 at $x = 0.08$ and $\langle Q^2 \rangle = 30$ GeV^2 at $x = 0.2$. The data point from the CHIO collaboration

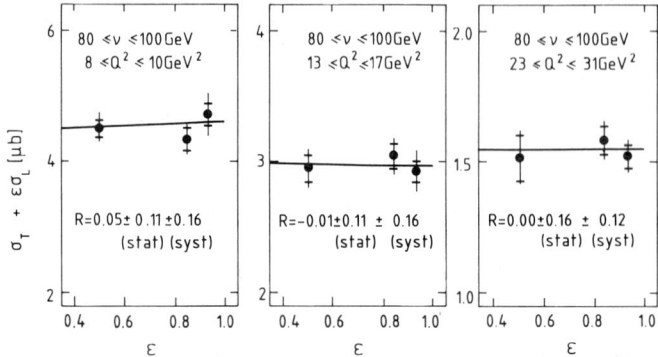

Figure 11 Straight-line fits to the EMC (71) measurements of $(\sigma_T + \varepsilon\sigma_L)$ versus ε for various v, Q^2 bins. The statistical error is indicated by the inner error bar.

(23) represents their average value for $x > 0.01$. For very small $x < 0.01$ the same group measures $R = 1.22^{+0.61}_{-0.67}$. The Q^2 range of the SLAC data (78, 79) is $1 \leq Q^2 \leq 4$ GeV^2 at $x = 0.2$ and $12 \leq Q^2 \leq 18$ GeV^2 at $x = 0.8$.

The dashed curve in Figure 12 indicates a typical QCD leading-order prediction calculated for a fixed value $v = 100$ GeV (assuming for the mass scale parameter a value of $\Lambda = 160$ MeV). In the kinematic region covered by the EMC data only small values of R are expected, consistent with the experimental result.

In the region of lower Q^2, covered by SLAC, larger values of R could be caused by contributions of integer-spin objects to the scattering. Abbott et al (80) estimated the effect of a significant diquark substructure in the

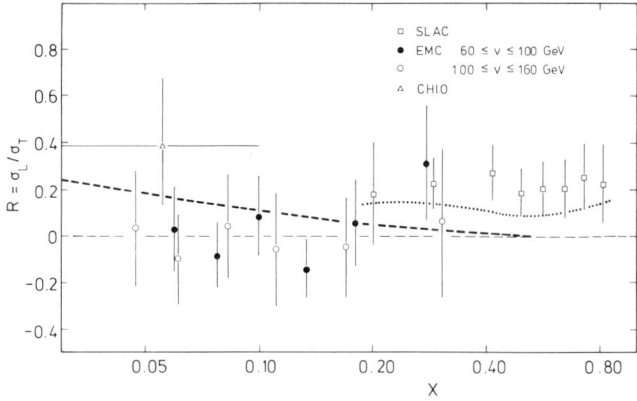

Figure 12 R data from charged lepton-proton scattering experiments. The dashed curve represents a typical QCD prediction, the dotted curve shows a prediction for diquark scattering plus QCD effects for the Q^2 range of the SLAC data.

Table 2 Average values of R

Experiment	Reaction	Q^2 (GeV^2)	R
SLAC-MIT (78)	ep	1–16	0.14 ± 0.07
	ed	1–16	0.17 ± 0.07
SLAC (79)	ep	3–18	0.22 ± 0.10
	ed	3–18	0.24 ± 0.10
CHIO ($x > 0.01$) (23)	μp	0.9–30	$0.38^{+0.29}_{-0.25}$
($0.01 < x < 0.1$)		0.4–30	$0.52^{+0.29}_{-0.25}$
EMC (71)	μp	$\langle Q^2 \rangle = 22.5$	0.00 ± 0.10
CDHS (81)	νFe	$\langle Q^2 \rangle = 20$	0.10 ± 0.07
($0.4 < x < 0.7$) (82)		$\langle Q^2 \rangle = 38$	$\leq 0.039 \pm 0.039$

nucleon leading to additional terms behaving like $1/Q^2$ and $1/Q^4$. The dotted curve in Figure 12 shows their analysis including diquark scattering plus QCD calculated for the average Q^2 values of the SLAC data in each bin. The EMC data are at higher Q^2 where diquark scattering effects are expected to be negligible.

Table 2 summarizes the average values of R obtained from electron or muon scattering by various groups. Also shown is a result from the neutrino scattering experiment of the CDHS collaboration (81, 82). The results measured at high Q^2 indicate that R is indeed very small. It therefore seems justified to use the assumption $R = 0$ for the evaluation of the structure function F_2 from the measured double differential cross section.

4.3 Comparison of Experiments

Data sets on electromagnetic structure functions are available from electron scattering experiments performed at SLAC and from muon experiments performed at Fermilab and at CERN. Table 3 summarizes some properties of the experiments, for instance the beam energies and the kinematic range explored. It should be noted that in each case the highest Q^2 values are reached only for high x values. The table also contains information on the typical statistical and systematic errors of the various experiments. The quoted relative errors are for F_2 data in the region near $x \sim 0.25$ ($x \sim 0.4$ for SLAC and $x \sim 0.35$ for BCDMS) and for Q^2 values close to the center of the measured Q^2 range. Overall normalization errors, which are the same for all data points, are not fully included in the quoted systematic errors. High precision is achieved by both the electron and the later muon experiments with systematic errors of 10% or less and statistical errors of 2–3%.

Table 3 Experiments on electromagnetic structure functions

Experiment, year	Reaction	Beam energies (GeV)	Structure function	x range	Q^2 range (GeV)^2	Typical error $\Delta F_2/F_2$ stat. (%)	syst. (%)
SLAC-MIT (78) 1979	ep	4.5–20	F_2^p, F_1^p	0.1–0.8	1–16	3	3
	ed		F_2^d, F_2^d			2	4
SLAC (79) 1982	ep	6.5–19.5	F_2^p, F_1^p	0.2–0.8	3–18	~6 total	
	ed		F_2^d, F_1^d				
CHIO (23) 1979	μp	96, 147, 219	F_2^p	0.0008–0.67	0.25–65	10	5–10
	μd	147	F_2^d	0.0013–0.67	0.25–40	20	—
MSU-Fermilab (83) (1979)	μFe	270	F_2^N	0.06–0.55	6–110	7	10
EMC (29) 1981–1982	μp	120, 280	F_2^p	0.03–0.65	2.5–170	3	5
(33)	μd	280	F_2^d	0.03–0.65	7–170	3	5
(31)	μFe	120, 250, 280	F_2^N	0.05–0.65	3–200	2	5
BCDMS (30) 1981–1982	μC	120, 200, 280	F_2^N	0.35–0.65	27–260	3	4

High statistics measurements of the structure function F_2 have also been obtained in neutrino and antineutrino scattering off heavy nuclei; Eisele (84) has reviewed recent results. Considering the fact that the experimental techniques and types of systematical errors are completely different for muon and neutrino experiments (74), it is very instructive to compare the results on F_2. The data depend sensitively on a precise determination of the momenta of the incident and the scattered muon. Therefore both trajectories of the muon before and after the interaction are accurately measured with beam hodoscopes and wire chambers in the CERN experiments. Because of the strong Q^2 dependence and in order not to fill the data tapes with low Q^2 events only, all muon experiments cut out some regions of small angle scattering. This, in turn, requires a thorough understanding of the acceptance of the detector. At large values of y and low values of x radiative corrections become large. The data therefore have to be restricted to a kinematical region where the corrections are less than 25%.

In neutrino experiments the results on F_2 depend on calorimetry, on scattering angle determination, and on a precise calculation of the incident neutrino flux, which in practice is the main experimental problem.

In Figure 13 fully analyzed F_2 data from the three high statistics

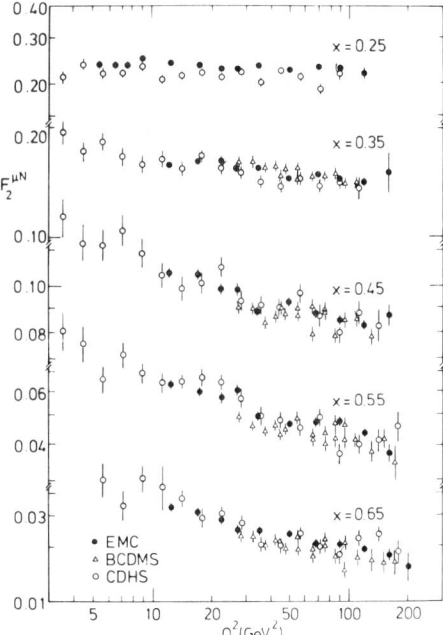

Figure 13 F_2 per nucleon as measured with heavy nuclear targets by the BCDMS, EMC, and CDHS collaborations as a function of Q^2 for various fixed $x \geq 0.25$ assuming $R = 0$.

experiments of the EMC, the BCDMS, and the CDHS collaborations are compared, all measured with heavy nuclear targets. The CDHS neutrino data are multiplied by the quark parton model factor of 5/18 representing the average charge squared of the quarks inside the nucleon. At large x the data extend to Q^2 values of 200 GeV^2. In none of the cases are corrections for the Fermi motion of the nucleons inside the heavy nuclei applied. There

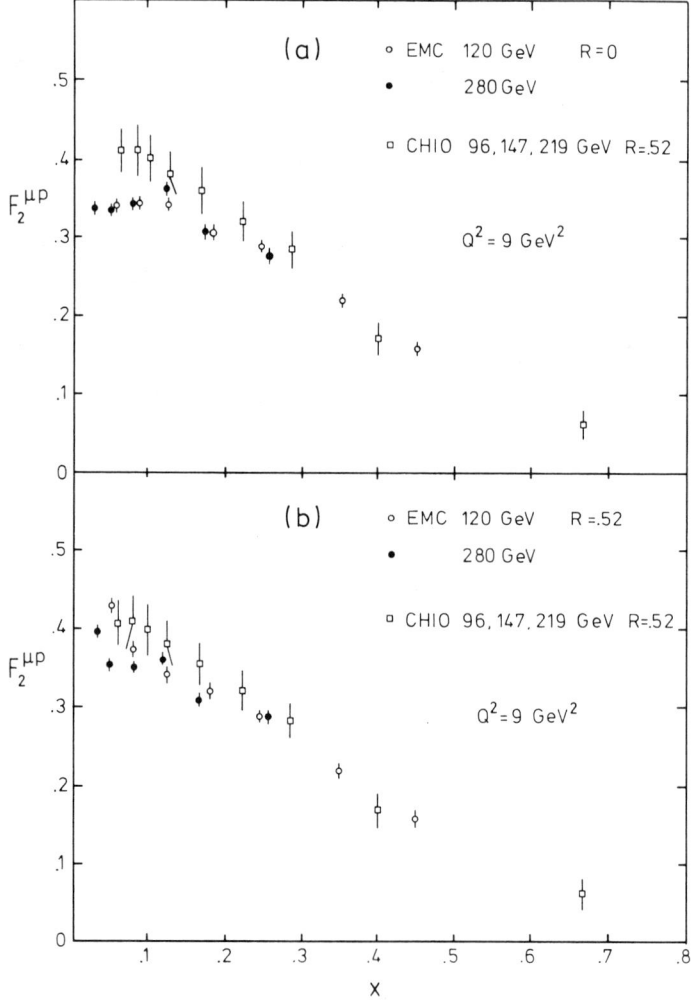

Figure 14 Comparison of μp data from EMC and CHIO. For the evaluation of F_2 from the measured double differential cross section, the assumptions made on R are: (a) $R = 0$ for the EMC and $R = 0.52$ for the CHIO data; (b) $R = 0.52$ in both cases.

is full agreement between the three experiments within their statistical and systematic errors.

All experiments verify on a highly significant level the pattern of scaling violation predicted by QCD; a rise of F_2 with Q^2 at small fixed x, a flat Q^2 dependence at x near 0.2, and a decrease with Q^2 at large x. There is some tendency toward the logarithmic Q^2 variation of F_2 becoming smaller with increasing Q^2 at large x. In principle such an effect is predicted by perturbative QCD, but the data indicate logarithmic derivatives $\partial F_2/\partial \ln Q^2$ at high Q^2 even smaller than expected on the basis of the extrapolation of QCD fits to the low Q^2 data. We return to this aspect in Section 6.

A somewhat troublesome point arises from a comparison of the μp data as published by the CHIO collaboration with the μp data of the EMC. In Figure 14 data of the two groups are plotted as a function x at $Q^2 = 9 \text{ GeV}^2$. In the analyses of the two groups there is one substantial difference in the evaluation of F_2 from the measured double differential cross sections; the CHIO group assumes a value of $R = 0.52$ at all x, Q^2, the EMC uses $R = 0$ at all x, Q^2. Clearly one would expect differences in F_2 due to these two assumptions even if the double differential cross sections are correctly measured by both groups. As shown in Figure 14a the F_2 values at small x differ by up to 20% for the two different assumptions on R. However, as indicated in Figure 14b the discrepancy partially disappears if a value of $R = 0.52$ is also used for the extraction of F_2 from the EMC measurements. Though the knowledge of R is rather poor, the summary of measured R values in Table 2 favors values closer to zero than to 0.52 in the x range measured at $Q^2 = 9 \text{ GeV}^2$. Note that in Figure 14b there is no longer agreement between the two EMC data sets measured with 120 and 280 GeV, which emphasizes the dislike of the data for a large R value.

4.4 Neutron vs Proton

A measurement of the neutron structure function $F_2^{\mu n}$, using data from both deuterium and hydrogen targets, allows the evaluation of the flavor nonsinglet function

$$F_2^{\mu p} - F_2^{\mu n} = \frac{1}{3}x(u-d+\bar{u}-\bar{d})$$

and of the ratio of d over u quark distribution functions since

$$\frac{F_2^{\mu n}}{F_2^{\mu p}} \simeq \frac{1+4d/u}{4+d/u} \quad \text{for} \quad x \gtrsim 0.3,$$

where sea quarks can be neglected. The first information on these functions was obtained from the ep and ed data of the SLAC-MIT group. New results from μd scattering at 280-GeV beam energy, combined with μp results, were published by the EMC (33). In Figure 2, $F_2^{\mu p} - F_2^{\mu n}$ is shown as evaluated after correcting the deuterium data for nuclear binding effects. The data represent an average over Q^2 with the average value of Q^2 increasing from 10 to 80 GeV^2 for x increasing from 0.03 to 0.65.

From these data one can try to evaluate the sum rule

$$3 \int_0^1 \frac{dx}{x} (F_2^{\mu p} - F_2^{\mu n}) = 1 + 2 \int_0^1 dx \, (\bar{u} - \bar{d}),$$

which counts the difference between the number of valence u quarks minus the number of valence d quarks in the proton if the sea of u and d quarks and antiquarks is equal, i.e. $\bar{u} = \bar{d}$. In an experiment one can measure only over a limited x range, and, because of the $1/x$ factor, the evaluation of the full integral is affected by substantial uncertainties. The method used here involves a fit to the data with a function of the form

$$F_2^{\mu p} - F_2^{\mu n} = A x^{0.5} (1-x)^{\alpha} (1 + \beta x),$$

which is also plotted in Figure 2. Performing the integration with the fitted function yields

$$3 \int_{0.03}^{0.65} \frac{dx}{x} (F_2^{\mu p} - F_2^{\mu n}) = 0.54 \pm 0.03 \text{ (stat.)} \pm 0.21 \text{ (syst.)}.$$

The dangerous attempt to extrapolate to $x = 0$ using the same fitted functions gives

$$3 \int_0^1 \frac{dx}{x} (F_2^{\mu p} - F_2^{\mu n}) = 0.72 \pm 0.06 \text{ (stat.)}^{+?}_{-0.39} \text{ (syst.)},$$

which can be considered as consistent with one but leaves room for $\bar{d} > \bar{u}$.

From an experimental point of view $F_2^{\mu n}/F_2^{\mu p}$ is better determined than the proton-neutron difference since some sources of errors cancel when the ratio of cross sections is determined. Figure 15 shows the EMC data points and the range of SLAC-MIT measurements (78), for $Q^2 > 2$ GeV^2 and $W > 2$ GeV, indicated by the shaded area plotted versus x. The two data sets agree within their statistical and systematic errors. Clearly, at high x scattering takes place mainly off u quarks. Both experiments indicate a tendency to reach the quark parton model bound $F_2^n/F_2^p = 1/4$ (i.e. $d = 0$) at high x. The naive assumption that u and d valence quarks have the same x dependence with $d/u = 1/2$ is excluded. Also, a model yielding $d/u = 1/5$, using the SU(6) spin-unitary spin wave function of the nucleons, and assuming that high x scattering takes place only off valence quarks carrying the nucleons helicity (85), is disfavored by the data.

The ratio u/d may also be obtained by comparing ep or μp with νp cross sections. This method has the advantage of using only hydrogen data and thus avoiding possible problems arising from corrections due to the Fermi motion of the nucleons in the deuteron. It suffers from lack of control of systematic errors. At large x where valence quarks dominate

$$\frac{F_2^{\nu p}}{F_2^{\mu p}} = \frac{18 d/u}{4 + d/u}.$$

Results using νp data obtained with the Big European Bubble Chamber (BEBC) (86) together with SLAC ep and EMC μp data are also shown in Figure 15. In order to allow a direct comparison with the $\mu n/\mu p$ ratios, we have chosen to plot the quantity $1/4 + 5F_2^{\nu p}/24F_2^{\mu p}$, which should be equal to $F_2^{\mu n}/F_2^{\mu p}$ in the region where scattering off sea quarks can be neglected. The data are averaged over Q^2 ranging from 2 to 70 GeV^2 for each bin.

By comparing the $\mu n/\mu p$ and $\nu p/\mu p$ data one finds, within errors, agreement in the region dominated by valence quark scattering, i.e. $x > 0.3$. On the other hand, there is a tendency for the $\mu n/\mu p$ ratios to be above the corresponding $\nu p/\mu p$ ratios at small x. In the quark parton model such an effect would be expected if there are more anti-d quarks than anti-u quarks in the proton at small x, precisely $\bar{d} > \bar{u} + 3/5 s$. This possibility has been discussed by Field & Feynman (87) and may be understood as a consequence of the Pauli exclusion principle. Since more valence u quarks

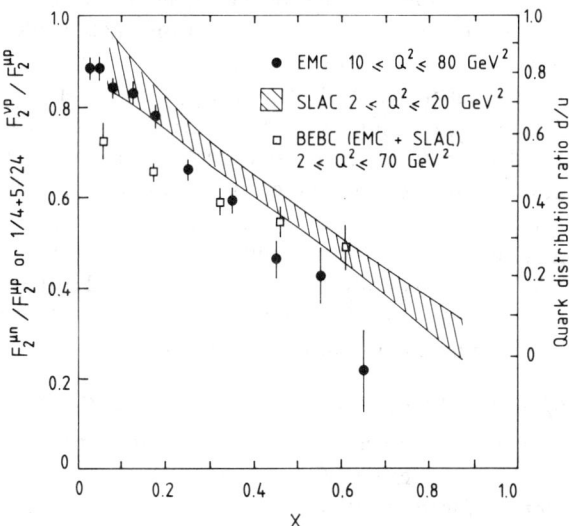

Figure 15 $F_2^{\mu n}/F_2^{\mu p}$ from EMC and in addition the νp data (see text). The shaded band indicates the F_2^{en}/F_2^{ep} data.

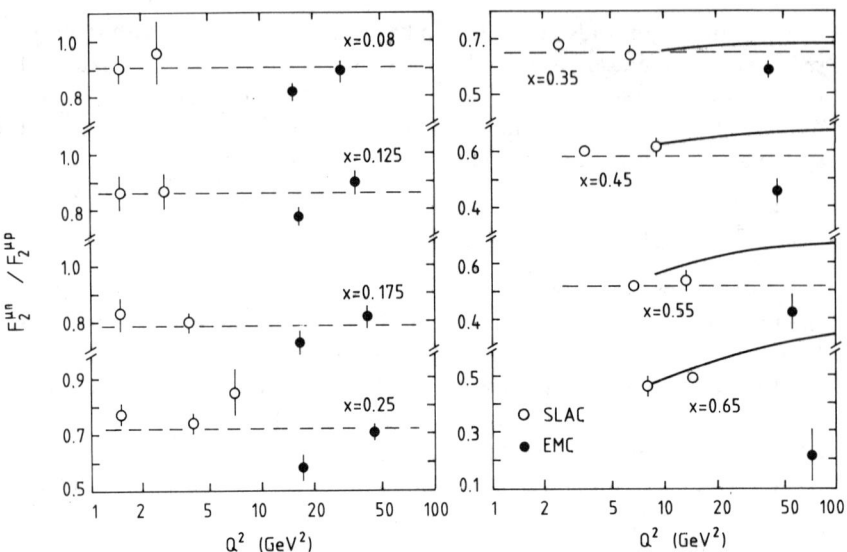

Figure 16 $F_2^{\mu n}/F_2^{\mu p}$ and F_2^{en}/F_2^{ep} data plotted as a function of Q^2 for fixed x bins. The dashed lines indicate the weighted mean at each x. The solid curve shows the expectations of a diquark model (89).

than valence d quarks are present even at small x, the number of $u\bar{u}$ pairs is suppressed compared to the number of $d\bar{d}$ pairs.

From QCD calculations one expects nearly no variation of the $F_2^{\mu n}/F_2^{\mu p}$ ratio in the Q^2 range covered by the data. For instance, a calculation with the Q^2 dependent quark distribution functions as parameterized by Glück et al (88) leads to an increase of the n/p ratio of 1% at $x = 0.08$ and a decrease of 4% at $x = 0.65$ if Q^2 is increased from 5 to 90 GeV^2.

Diquark models, on the other hand, expect a sizable variation with Q^2. The solid lines in Figure 16, where en/ep and μn/μp ratios are plotted versus Q^2 at fixed x, represent a prediction of Donnachie & Landshoff (89). In their model it is assumed that the entire scaling violation observed in the SLAC hydrogen and deuterium data with $x > 0.3$ is explained by a Q^2 dependent diquark term behaving like $1/(1+Q^2/M^2)$. Obviously, this assumption leads to an overestimate of diquark contributions.

4.5 Comparison of Iron and Deuterium Structure Functions

It has been generally assumed that from a measurement of deep inelastic muon or neutrino scattering off a heavy nucleus one can deduce the structure functions for free nucleons, after having performed corrections for the Fermi motion of the nucleons inside the nucleus. As a recent comparison of data measured by the EMC with iron and deuterium targets indicates, this assumption is not justified (34).

In order to emphasize the size of the observed effect we first discuss the expected behavior of the ratio of the structure function of a nucleus, with mass number A, to the sum of the structure functions for free protons and neutrons, weighted with their nucleon numbers $F_2^A/[ZF_2^p+(A-Z)F_2^n]$, when the nucleus is considered as a collection of slowly moving nucleons, each with an internal parton structure unchanged compared to the free nucleon case. The results of various calculations (90–92), differing in detail, for instance, by the ansatz for the wave functions of the two nuclei or on assumptions about the high momentum tail or the momentum balance, are collected in Figure 17. Despite some significant differences all curves show a strong increase of the ratio above unity for $x > 0.5$.

The experimental results are summarized in Figure 18 where the ratio of $F_2^N(\text{Fe})/F_2^N(\text{D})$ is plotted versus x. Here $F_2^N(\text{Fe})$ represents the iron data per nucleon after performing a correction for the non-isoscalarity of the Fe^{56} nucleus, which is less than 2.3% for $x < 0.7$. The data are averaged over Q^2 with a range of $9 \leq Q^2 \leq 27$ GeV^2 for $x = 0.05$ and $36 \leq Q^2 \leq 170$ GeV^2 for $x = 0.65$. No Fermi motion corrections have been applied. The plotted error bars indicate the statistical errors, the shaded band shows the systematic uncertainty of the x dependence of the structure function ratio.

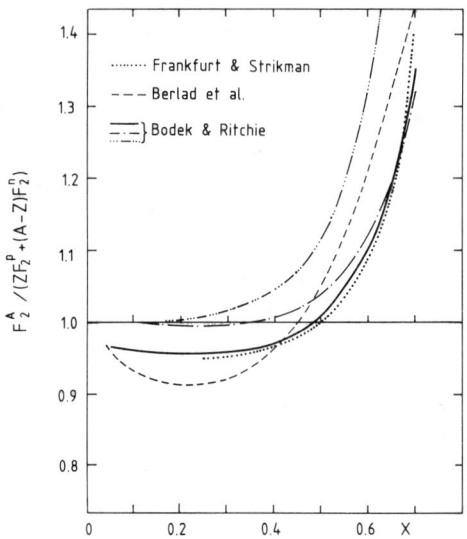

Figure 17 Fermi motion effects on ratio $F_2^{\mu A}/[ZF_2^{\mu p}+(A-Z)F_2^{\mu n}]$ with different assumptions on how to treat Fermi motion.

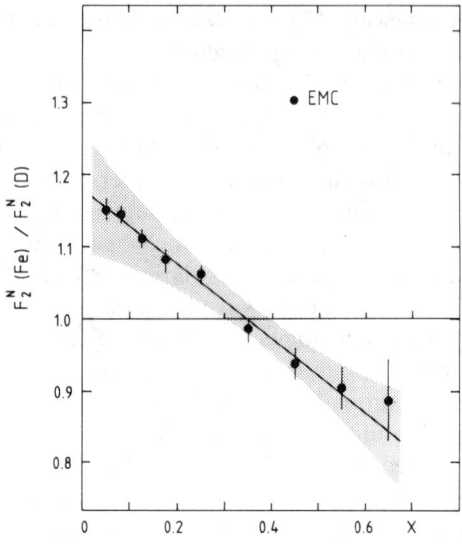

Figure 18 $2F_2^{\mu Fe}/56F_2^{\mu D}$ as function of x (34).

The x dependence of the ratio can be parameterized by

$$F_2^N(\text{Fe})/F_2^N(\text{D}) = a + bx \quad \text{for} \quad 0.05 \leq x \leq 0.65$$

with $b = -0.52 \pm 0.04$ (stat.) ± 0.21 (syst.). Normalization errors that do not change the x dependence of the ratio may change the absolute value by up to 7%.

At small x one observes a significant enhancement of the iron over deuterium ratio above unity. At large x, where one would expect a ratio larger than unity, owing to effects of nuclear binding, the measured value of the ratio is below one. For example at $x = 0.65$ the experimental value of about 0.9 might be compared with the value of 1.2 from the calculation of Bodek & Ritchie (solid curve in Figure 17). The iron/deuterium ratio is, within experimental errors, independent of Q^2 in the range covered by the EMC measurements. One should notice, however, that at small Q^2 a shadowing effect is observed for $x < 0.1$ (93, 94), which implies that the iron over deuterium ratio cannot be completely Q^2 independent.

At the time when these data were first presented the effect appeared to be quite surprising. We are not aware of any published prediction of such a behavior, though several effects are known that can modify the x dependence of quark distribution functions in a nucleus with high mass compared to free nucleon. For example,

1. Using a unified description of nuclear properties and of meson-nucleon interactions, one finds interaction-dependent internal properties of

the nucleon, in particular a substantially smaller mass. A fit to inelastic electron iron scattering data [taken at beam energies between 150 and 370 MeV (95)] results, for example, in a value of the nucleon physical mass in the nucleus of 700 MeV (96). In this connection it should be recalled that parton distributions are expressed in terms of the x variable defined as $x = Q^2/2Mv$ where M is the free proton mass. If the mass of the target quark cluster changes, x has to be redefined.

2. From electron scattering off light nuclei there is evidence for non-nucleonic components in the nucleus, i.e. for sizable probabilities of six- and nine-quark clusters in addition to the usual three-quark nucleonic cluster (97–100). Because of the existence of such clusters, $F_2^A(x)$ will remain finite at $x > 1$ above the kinematical limit for free nucleon scattering.

3. The quark-antiquark sea in a heavy nucleus may be composed of two contributions: the sum of the intrinsic quark seas associated with the individual nucleons plus an additional nuclear sea arising from the interactions between the nucleons in the nucleus, i.e. the exchange of mesons. The additional sea, counted per nucleon, is expected to increase with the mass number like $A - 1$ (101).

For a discussion of the influence of few-nucleon, short-range correlations in nuclei on the behavior of nuclear structure functions, we refer to the review of Frankfurt & Strikman (102).

Taking such effects into account, it seems dangerous to equate parton distributions extracted from lepton scattering off heavy nuclei to parton distributions for free nucleons since at present it is not possible to perform all corrections. It must be realized, however, that so far the only available information on the small strange and charmed quark distributions comes from high luminosity experiments using heavy targets.

Recently, attempts have been made to understand the observed iron versus deuterium effect (103, 104). Phenomenologically the explanations imply a large increase of the quark sea per nucleon in the iron nucleus compared to deuterium, a depletion of the valence quark distribution functions in iron at large x, and a reduction of the fraction of the nucleon momentum carried by gluons in iron.

At the time of submission we have received a communication (105) that a reanalysis of some SLAC data shows similar effects at lower Q^2.

5. HEAVY QUARK PRODUCTION

5.1 *Theoretical Models*

The quantum numbers of the photon $J^{PC} = 1^{--}$ led Sakurai (106) to treat photon (real and virtual) interactions in terms of the vector meson dominance model (VDM). In this model the importance of ρ, ω, and ϕ mesons in electro- and muoproduction is explicit and a straightforward extension would anticipate a similar role for J/ψ and Υ mesons. Alternatively, based on the quark content of the hadronic component of the photon, one can expect that above the charm threshold the photon can be considered to be, a significant fraction of the time, in a $c\bar{c}$ state. In the VDM the Q^2 dependence of any single-state M is expected to be $(Q^2+M^2)^{-2}$ so that with $M_{J/\psi}^2 \simeq 10$ GeV^2 the initial Q^2 dependence of J/ψ production would be expected to be less steep than that of the lower vector mesons and than that of the total cross sections giving a relative enhancement.

The complexity in detail of vector meson dominance when extended beyond simple ρ, ω, and ϕ dominance has led to its fall from favor although little definitive experimental data actually contradicts the model. In its place, heavy quarkonium and open charm muoproduction have been described in terms of the photon-gluon fusion (γGF) model (48, 107). This model is QCD motivated in that its simplest form employs the single Feynman diagram, corresponding to the Bethe-Heitler process of QED, which in order α_s QCD readily leads to the production of new quark states. The theoretical justification for its usage even for $Q^2 = 0$ is that the high mass of the produced quarks $M_{c\bar{c}}^2 \geq 10$ GeV^2 guarantees to some extent a high enough momentum transfer and hence the applicability of the QCD perturbation expansion. In its simplest form it is unappealing, since the process explicitly violates color conservation and the presence of innocuous extra soft fluons (108) has to be imagined. On the positive side the model is sensitive to many (perhaps too many) of the interesting quantities of strong interaction physics; the gluon distribution, the strong coupling constant α_s, and the mass M_c of the charmed quark.

It is, however, clear that within any theoretical model, either ancient or modern, production of charmonium, charm, bottomium, and bottom are expected to be important aspects of muoproduction and it is in this context that the data, the facts, are discussed.

5.2 J/ψ and ψ' Production Cross Sections

The production of J/ψ by muons proved to be very easy to observe in the spectrum of dimuons produced (Figure 19) (38). However, detailed understanding of the mechanism is complicated by several points. First, in order to achieve high luminosity, the experiments were performed with iron targets and this led to significant resolution and smearing problems in extracting the cross-section behavior. Second, if a diffractive mechanism is operative, "elastic" production will be enhanced by a coherent component from the complete nucleus, the strong t dependence associated with this being unmeasurable because of the poor resolution. Third, inelastic production is also present (it turns out to be as big as the elastic cross section) and neither EMC nor BFP experiments can define the inelastic and

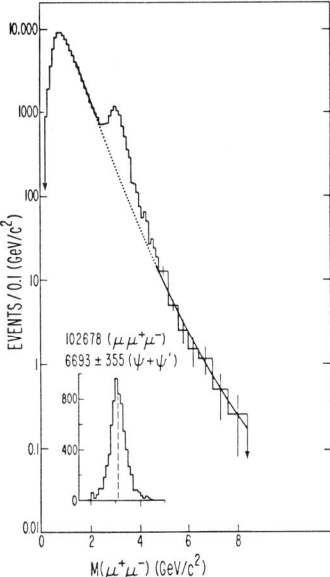

Figure 19 Dimuon mass spectrum from iron target data from BFP (38).

elastic production better than by observing, or not, a few GeV of energy in the calorimeter. Fourth, although less important, the acceptance of both experiments is restricted by a lower limit on detected muon energy.

These problems have led to detailed differences in the analysis of data that often prohibit direct comparison. For this reason we confine graphical comparison to the total photoproduction cross section (elastic plus inelastic), which is plotted as a function of photon energy in Figure 20. The data in this plot and the comments in this section are based on (38, 41, 49). Agreement is excellent and shows that the cross section is still rising slowly well above threshold.

The fraction that is inelastic has a similar behavior although this depends slightly on the definition of inelastic. For BFP 47% of the cross section has $z < 0.9$ and for EMC 65% with $z < 0.95$.

The extraction of the above cross sections demands an extrapolation to $Q^2 = 0$. The experiments indeed see a comparatively weak Q^2 dependence as expected from VDM arguments; however, the BFP fit to the form $(Q^2 + M_v^2)^{-2}$ yields $M_v = 2.2 \pm 0.2$ GeV somewhat far from the J/ψ mass. The EMC does not see a similar deviation except at low values of z.

The z dependence of J/ψ production, although peaked toward $z = 1$, extends below $z = 0.5$. The p_T dependence can be described by the form $(p_T)^{-n}$ with n varying from 2 at low z to 4 at high z.

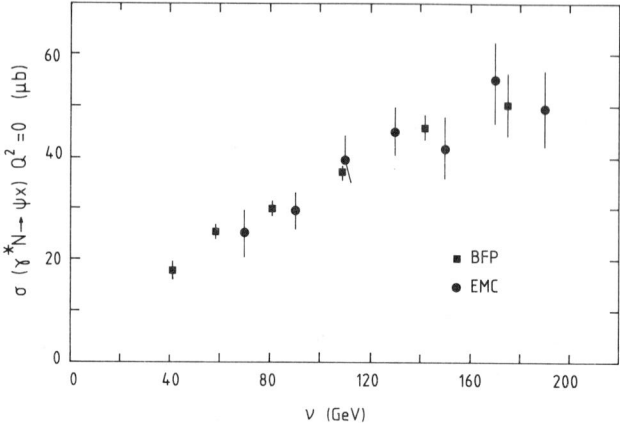

Figure 20 Comparison of $\sigma_T(\gamma^*N \to \psi X)$ from BFP and EMC as function of photon energy.

The discussion of which higher order form of the model is appropriate centers on whether all order α_s^2 diagrams should be included (109) or only those yielding directly a color singlet $c\bar{c}$ pair (110, 111). In the former case it is necessary to invoke the semi-local duality argument (108) to obtain the J/ψ yield; in the latter case a nonrelativistic J/ψ wave function may be used and the normalization is fixed. Baier & Rückl (112) find that the color singlet model agrees with the data at low z and high p_T^2, the region of validity of the model, but there is then $\sim 80\%$ of the total yield left unexplained. Halzen (113) arrives at a similar conclusion but then suggests that the remainder may be explained by nonperturbative effects.

The data are sufficiently extensive for attempts to be made to analyze the production and decay distributions of the J/ψ. This analysis is analogous to that which established natural parity exchange and s-channel helicity conservation (SCHC) of ρ production at lower energies. The angular distribution is given for SCHC by

$$W(\theta, \phi) = \frac{3}{16\pi}(1 + \cos^2\theta + 2\varepsilon R \sin^2\theta - \varepsilon \sin^2\theta \cos 2\phi),$$

where θ is the polar angle of the decay μ^+ measured in the helicity frame and ϕ measures the azimuthal decay angle with respect to the lepton scattering plane. Assuming $R \sim 0$, analysis of the data reduces to fitting for the $\cos^2\theta$ and the $\cos 2\phi$ terms. The EMC data (41) show no evidence for the presence of either term at least in the high z (elastic) region. The BFP experiment claims (37) evidence for SCHC on the basis of the ϕ distribution; however, there is no evidence for the corresponding $\cos^2\theta$ term. This issue cannot be considered resolved and the data also unfortunately turn out to be too weak statistically to help distinguish the different models (111, 114–116).

In the iron target data the experimental resolution is not good enough to resolve the ψ' (3.7 GeV) from the J/ψ (3.1 GeV) although a fit including both to the EMC data yields, after allowing for branching ratio, a production ratio of 0.22 ± 0.10. More direct measurement is obtained from the liquid target (D_2 and H_2) data. In these data there is even the hint of a positive ψ' signal and a fit to the data gives a production ratio 0.22 ± 0.12. These data are in good agreement with the photoproduction measurement of 0.20 ± 0.05 (117). With such yields it is difficult to see how the duality arguments, that all of below-threshold γGF production should appear in charmonia states (108), can be valid.

5.3 Open Charm Production

A high yield of dimuon production was first observed (118, 119) in the early experiments at FNAL and was attributed to charm production. The much higher luminosity of the current experiments has confirmed this. The signature of charm dimuons is the presence of an average of ~ 20 GeV of missing energy, corresponding to the neutrino in the semileptonic charm decay. In the EMC data there is also a trimuon (45, 47) sample that exhibits almost twice the missing energy now corresponding to two neutrinos missing. In addition, the acceptance-corrected ratio of trimuon to dimuon events is consistent with the known semileptonic branching ratio of D mesons.

The description of the data by the γGF model and the determination of the critical parameters of the model are a result of the analysis; however, the most convincing aspect of the model was that a description, good to 30% in all variables, was obtained using the parameters of the original predictions. The γGF model has therefore been used by both EMC and BFP as the basic Monte Carlo simulation for acceptance corrections. A possible alternative mechanism is the interaction of the virtual photon with a charm (anti-charm) quark in the nucleon sea. In this case the quark and antiquark are essentially uncorrelated and only the struck quark should appear in the forward direction. The model then has difficulties in describing the ratio of trimuons to dimuons.

The fragmentation of the charm quark is not specified in the γGF model; it must be determined from the data. The critical distributions are the $z_\mu = E_\mu/\nu$ distributions of the observed muons. The fragmentation variable Z_D of the charmed meson is defined as the ratio between its momentum and that of the charmed quark. This results in a Lorentz noninvariance and the EMC data favor a mechanism in which this definition is applied in the laboratory frame rather than say the photon gluon center-of-mass frame. This seems to indicate that the fragmentation chains of each member of the quark-antiquark pair are to be considered independently. The resultant function favored by the data is $D(Z_D) = \exp(1.6 \pm 1.6)Z_D$, which, despite its uncertainty, is much harder than the fragmentation functions of lighter quarks. There are theoretical predictions of a hard fragmentation function (120) although it is not yet clear whether a simple threshold behavior does not also lead, at current energies, to the same type of behavior. Since the acceptance is zero for low energy muons, this determination of the fragmentation function is important for the determination of any total cross sections. Even so the residual uncertainty is of the order of 20–30% since the function is not constrained by data as $Z_D \to 0$.

As mentioned above, a struck quark model is disfavored as the dominant production mechanism, but there have been predictions (121, 122) of a small 1% intrinsic charm presence in the nucleon to account for the ISR diffractive charm production cross sections. The data (46, 47) plotted as a function of $x_{Bj} = (Q^2 + m_c^2)/2M\nu$ are shown in comparison with this model in Figure 21. There is some argument (123) as to a threshold dependence, and also (coupled) the use of the variable $x_{Bj} = (Q^2 + 4m_c^2)/2M\nu$, in the computation of the model, but it is probably clear that anything greater than about 1/2% intrinsic $|uudc\bar{c}\rangle$ is strongly disfavored by the data.

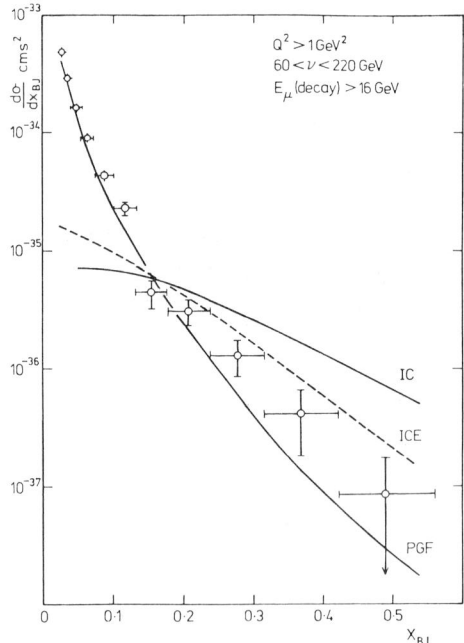

Figure 21 The x distribution of charm production compared to intrinsic charm and photon gluon fusion model predictions. The curve labelled IC is the standard intrinsic charm prediction, the ICE curve is similar with additional threshold suppression; PGF indicates the photon gluon fusion model.

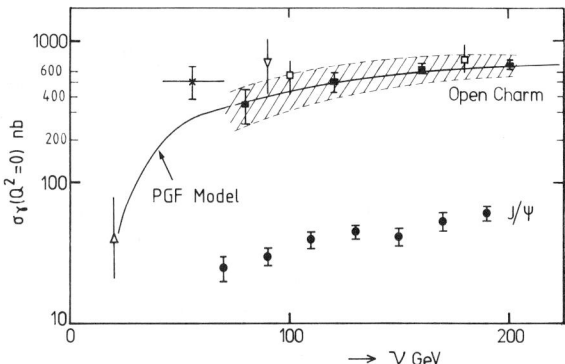

Figure 22 $\sigma_T(\gamma^*N \to c\bar{c}X)$ as function of v showing two sets of data: BFP (*open squares*) and EMC (*solid squares*). Also shown are photoproduction measurements SLAC (*upward pointing triangles*), WA4 (*crosses*), and CIF (*downward pointing triangles*) (124).

The Q^2 dependence for open charm production, similar to J/ψ production, is rather weak and characterized again by a $(Q^2 + M_v^2)^{-2}$ dependence with $M_v \simeq 3.9 \pm 0.1$ GeV (47) for the EMC data. As for the J/ψ data the BFP (43) values are systematically lower. Using these dependences, cross sections for $Q^2 = 0$ can be extracted and these are shown for the data from the two groups in Figure 22. Also shown in this plot are charm production measurements from photoproduction groups (124). The agreement between BFP and EMC data is excellent although the smoothness of the continuation to lower energies through the photoproduction experiments is not assured. In fact this may not be only a technical question since the experiments each have different, incomplete, acceptances and it is possible that an experiment with complete acceptance could yield a larger value than the present measurements. The total photoproduction cross section rises by ~ 3 μb from 50 to 200 GeV so the measured charm cross section clearly does not saturate this rise.

The charm production cross sections as a function of Q^2 and v can be formulated in terms of the structure function F_2 following

$$\frac{d^2\sigma}{dQ^2\,dv}(\mu N \to \mu c\bar{c}X) = \frac{4\pi\alpha^2}{Q^4 v}\left(1 - y + \frac{y^2}{2}\right) F_2(c\bar{c}).$$

The results of the two sets of data with a 10% adjustment (well within the scale errors of either data set) are compared in Figure 23 with the predictions of the γGF model with a particular set of parameters. The immediate feature observed is the strong Q^2 dependence, i.e. scale breaking. This strong dependence is to be contrasted with the gentle logarithmic behavior of the total F_2. This is emphasized in Figure 24 where the logarithmic slopes of the two quantities are plotted. At low x some 30% of the scale-breaking slope observed in the total F_2 can be attributed to the charm production. This behavior, due to the threshold behavior for the production of two heavy mesons, has no relationship to QCD scale breaking. In fact QCD is well formulated only for massless partons and the only clear way to treat charm production is to remove it from the observed F_2 values before fitting, although even this procedure leaves unanswered the questions of how many flavors should be included in the expression for α_s and also in the Altarelli-Parisi equations.

It was stated earlier that the γGF model with canonical parameters

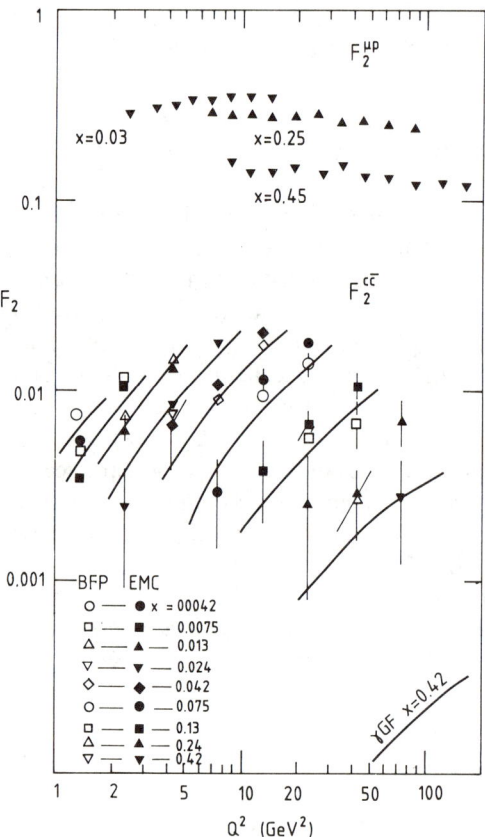

Figure 23 $F_2^{c\bar{c}}$ as function of x and Q^2. The two sets of data and example γGF predictions are shown; the BFP data are arbitrarily renormalized by 10%. The total $F_2^{\mu p}$ values are from EMC (29).

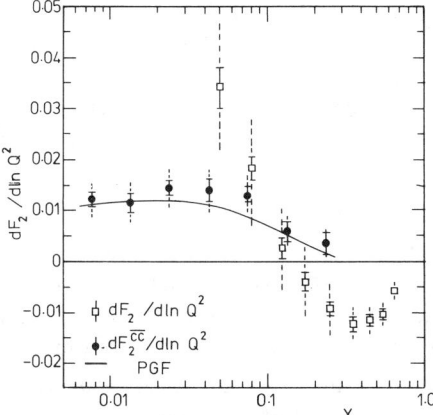

Figure 24 Comparison of $\partial \ln F_2 / \partial \ln Q^2$ for $F_2^{c\bar{c}}$ and total F_2 indicating the contribution of charm production to the observed F_2 scale breaking. The data are from EMC (47).

provides a good qualitative description of the data. Nevertheless it is interesting to perform a fit that attempts to determine all the parameters. The main effect of changes of either of $\alpha_s(\Lambda)$ or of M_c is to change the normalization of the data and hence independent determination of these two parameters is not possible. On the other hand, the predicted distribution shapes are sensitive to the form of the gluon distribution, particularly the ν distribution. EMC found (47) that a better fit than most was obtained with the simple counting-rule glue $xG(x) = 3(1-x)^5$. The data in different Q^2 ranges were fit to $xG(x) \simeq 3(1-x)^m$, where m was determined at each Q^2 and little variation was found. The data, which also include a point obtained by fitting the J/ψ production data, are compatible with the constant exponent $m = 5$ and hence favor either a CDHS (27) or Glück et al (88) glue; the Buras & Gaemers glue (125) among others (126) is ruled out. BFP (43) similarly found their data favored a conventional $(1-x)^5$ dependence for the glue distribution.

The z, p_T^2 distributions, as for J/ψ production, are potentially discriminative between the various models which go beyond the simple γGF model. The situation with respect to these models is not clear despite their confrontation with data (114) and it would be premature for an experimentalist to judge the theory at this time.

5.4 Upsilon Production

Whatever the production mechanism for J/ψ production, it is expected that the $b\bar{b}$ bound system of Υ particles can be produced in analogous fashion.

Three experiments—BFP (127), BCDMS (68), and EMC (41)—have searched for this production in their dimuon spectra with no success. The 90% confidence level (C.L.) upper limits obtained for $B \cdot \sigma(\mu N \to \mu \Upsilon x)$ are 22, 13, and 52×10^{-39} cm^2 per nucleon respectively. The BCDMS limit rules out some of the vector dominance and γGF predictions.

5.5 Rare Multimuon Events

The efficient detection of rare multimuon events was one of the design parameters of the BFP MMS spectrometer. With their luminosity and acceptance they have found a significant number of odd-sign trimuons and 4- and 5-muon events (51). From an analysis of dimuon events they have been able to place a lower limit on the mass $\overline{M_0}$ (50) of a higher mass neutral muon. Such a particle is desirable in certain extensions to the standard electroweak theory and, if it couples with Fermi coupling strength to a right-handed current and decays with branching ratio 10% to $\mu\mu\nu$, then the 90% C.L. lower limit on its mass is 9 GeV.

The exotic trimuon events, e.g. $\mu^+ \mu^\pm \mu^\pm$ from an incident μ^+ beam, can, in the BFP experiment, be interpreted as charm plus π, K decay background. In the EMC experiment there are many fewer events initially; however, here also the total is not significantly different from background, though one of the events has the appropriate kinematic characteristics for bottom hadron production. They use these events to place a limit (52) of 12×10^{-36} cm^2 on the production cross section for bottom hadrons. An alternative limit of 20% on the $D^0 - \overline{D^0}$, CP violating, mixing probability is obtained. BFP deduce a similar limit (17×10^{-36} cm^2) on $b\bar{b}$ production by examining in detail the kinematic behavior of their dimuon signal and finding no deviation from the charm production.

Of their 4-muon events, BFP have one inelastic event with kinematics suggesting that it could be $b\bar{b}$ production via the chain

$$\mu^+ N \to \mu^+ \bar{b} b X$$
$$\hookrightarrow \mu^- \bar{\nu}_\mu X$$
$$\hookrightarrow \psi X$$
$$\hookrightarrow \mu^+ \mu^-.$$

Their 5-muon events are all consistent with second-order radiative corrections to QED muon tridents.

6. SCALING VIOLATION OF STRUCTURE FUNCTIONS

6.1 QCD and Its Predictions

The main motivation for measuring structure functions with high precision, over a wide range of Q^2 values at a given x, arises from the fundamental connection to the new theory of strong interactions, quantum chromodynamics. QCD has the property of asymptotic freedom, i.e. at high momentum transfer Q^2 the effective coupling g_{eff} between color charged quarks and color charged gluons becomes small (14, 15, 128, 129). For recent reviews see (130, 131), which include a list of relevant references. Defining the effective strong interaction coupling $\alpha_s = g_{\text{eff}}^2/4\pi$ (similar to $\alpha = e^2/4\pi$ in QED), one finds at very high Q^2:

$$\alpha_s(t) = \frac{4\pi}{\beta_0 t}\left(1 - \frac{\beta_1}{\beta_0^2}\frac{\ln t}{t}\right), \qquad 8.$$

where the argument of α_s is

$$t = \ln(Q^2/\Lambda^2)$$

and $\beta_0 = (33-2f)/3$, $\beta_1 = (306-38f)/3$. The term f counts the number of flavors, e.g. u, d, s, c, etc. The mass scale parameter Λ arises from the interaction between quarks and gluons and is a fundamental parameter of QCD. The second term in the brackets of Equation 8 represents the second-order asymptotic freedom correction, or next-to-leading-order corrections, to α_s. Clearly for $\ln Q^2/\Lambda^2 \gg 1$ the correction is small (with the present values of Λ the correction term is of the order of 0.2 at Q^2 of about 100 GeV^2).

Since α_s vanishes for $Q^2 \to \infty$, one can use perturbation theory for a calculation of the Q^2 evolution of the structure functions and for the development of techniques to extract Λ from the data.

Before describing the methods used so far to confront the observed scaling violations of the structure functions with QCD predictions, let us briefly recall what can be intuitively expected. Increasing Q^2 of the virtual photon microscope means a probing of the internal structure of the target at smaller and smaller distances. One resolves a more complex internal structure since one can see partons created by the QCD splitting processes such as (a) a quark splits into a quark and gluon, (b) a gluon splits into a quark-antiquark pair, and (c) a gluon splits into two gluons (Figure 25). In consequence, considering the situation relative to a reference Q_0^2, the virtual photon finds, at $Q^2 > Q_0^2$, parton distributions created by a complicated

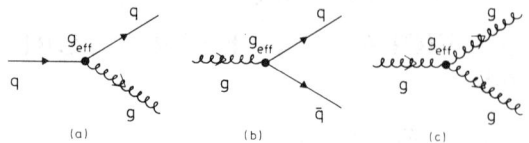

Figure 25 Basic splitting processes of QCD.

hierarchy of splitting processes. All parton distributions are then functions of x and t.

The Q^2 evolution is different for the different parton distributions. Valence quarks are only affected by bremsstrahlung process (a). They lose momentum, which is transferred to gluons, and, since gluons can split into $q\bar{q}$ pairs, also to sea quarks. While valence quark scattering at large x is reduced, sea quark scattering at small x is enhanced. The splitting process (c) contributes in an indirect way; the created gluons can again create a $q\bar{q}$ pair. Thus qualitatively we expect for F_2 the behavior sketched in Figure 26.

On the quantitative level two theoretically equivalent methods have been employed for the QCD analysis of the data:

1. Earlier analyses have been based on the evaluation of the moment integrals

$$M_n(Q^2) = \int_0^1 dx\, x^{n-2} \tilde{F}_i(x, Q^2), \qquad 9.$$

where $\tilde{F}_1 = xF_1$, $\tilde{F}_2 = F_2$ (and $\tilde{F}_3 = xF_3$ in neutrino scattering). The Q^2 development for the moments of flavor nonsinglet structure functions like $F_2^{p-n} \equiv F_2^p - F_2^n$ (or xF_3 in neutrino scattering) obeys a particularly simple law (130):

$$M_n^{NS}(Q^2) = M_n^{NS}(Q_0^2) \left[\frac{\alpha_s(t)}{\alpha_s(t_0)}\right]^{d_n^{NS}} \left[1 + R_n^{NS} \frac{\alpha_s(t) - \alpha_s(t_0)}{\pi}\right]. \qquad 10.$$

Flavor nonsinglet quantities are denoted by NS.

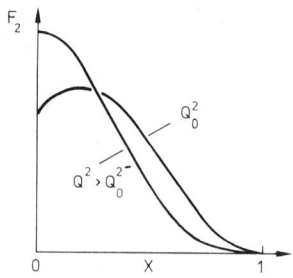

Figure 26 Evolution of the structure function F_2 with Q^2, as expected from QCD.

The powers d_n^{NS}, called anomalous dimensions, as well as the coefficients R_n^{NS}, have been explicitly calculated. The term R_n^{NS} is due to second-order QCD corrections. At this point we recall that the coefficients R_n^{NS} depend on the renormalization scheme used for calculation. Thus each second-order QCD analysis has to be performed in a specific renormalization scheme whose selection affects the outcome of the values of Λ and α_s. The question of selection of a renormalization scheme is discussed in some detail in (131). In the following all results are evaluated in the modified minimal scheme, the so-called $\overline{MS}$ scheme of Bardeen et al (132), which has the property that the numerical values $\Lambda_{\overline{MS}}$ resulting from QCD fits to the deep inelastic data are nearly equal to the numerical values resulting from a leading-order QCD analysis. The moment equations for the singlet contributions of structure functions like F_2 are more complicated. The corresponding expressions can be found in (132, 133), for example.

The analysis of moments has the advantage of clear theoretical predictions in next-to-leading-order QCD even for the more complicated singlet case. However, there are also serious disadvantages. One cannot use the high Q^2 data since, for kinematic reasons, they are measured only at a few x values. Also strong correlations exist between various orders n since the same data are used over and over again to evaluate the different order moments. Finally, at lower Q^2 the moments contain large contributions originating from elastic scattering and from resonance formation.

2. The method more recently applied by the experimental groups directly uses the full x, Q^2 variation of the structure functions by numerically solving the Altarelli-Parisi evolution equations (35). For nonsinglet structure functions or the valence part of F_2 (i.e. the region $x \gtrsim 0.3$) the evolution equation reads (24, 134)

$$Q^2 \frac{\partial}{\partial Q^2} F^{NS}(x, Q^2) = \int_x^1 dz\, F^{NS}\left(\frac{x}{z}, Q^2\right)$$
$$\times \left[\frac{\alpha_s(t)}{\pi} P^{(1)}(z) + \frac{\alpha_s^2(t)}{\pi^2} P^{(2)}(z)\right], \qquad 11.$$

where again second-order QCD corrections are included, the magnitude of which depends on the renormalization scheme used. The splitting functions $P^{(1)}$ and $P^{(2)}$ have been calculated, they contain the same information as d_n^{NS} and R_n^{NS}. The solutions of Equation 11 can, for instance, be of the form

$$F^{NS}(x, Q^2) = \int_x^1 dy\, b(x, y; Q^2, Q_0^2) F^{NS}(y, Q_0^2), \qquad 12.$$

where the kernel b can be expanded in a power series of α_s and is known up to the order α_s^2.

The application of this direct method is usually based on the numerical procedures developed by Gonzales-Arroyo et al (135, 136), by Abbott et al (80, 137), and by Bialas & Buras (138). Fits (including second-order QCD corrections) to the muon data can so far only be performed on the nonsinglet part of F_2. Only in leading-order QCD is the inclusion of the singlet part of F_2, the low x region, at present possible.

The main advantage of the method is that all data points can be incorporated. One can also easily study the effect of various cuts, for example in $W^2 \simeq Q^2(1-x)/x$ on the results of the fits. A minor disadvantage is the necessity to choose an explicit input parameterization of the x dependence at some reference Q_0^2, usually

$$F^{NS}(x, Q_0^2) = Cx^\alpha(1-x)^\beta(1+\gamma x)$$

where C, α, β, γ are free parameters to be adjusted to the data.

6.2 Other Sources of Scaling Violations

Unfortunately, there are other effects that can be held responsible for small scaling violations of the structure functions besides the Q^2 evolution of the parton distributions as described by perturbative QCD. Such effects have caused a considerable debate during the last few years and it has even been questioned whether meaningful results can be obtained from QCD fits to data (e.g. 139). We discuss the consequences of some aspects of these problems in Section 6.6.

In listing other possible sources of scaling violations, let us start with the ones that are understood. First, from Equation 3, one observes $F_2 \to 0$ for $Q^2 \to 0$. Though there exists no a priori argument at what Q^2 the threshold turn-on of F_2 is terminated, one knows from the SLAC data that the turn-on is complete for $Q^2 > 2$ GeV^2 in the region $x > 0.1$.

Second, new contributions to F_2 due to heavy quark production can rise from their threshold. As discussed in Section 5.3, the charmed piece of F_2 is now measured and the corresponding effects can be corrected.

Third, the Bjorken scaling variable x is the relevant scaling variable in the parton model only for very large Q^2 where target and quark mass effects can be neglected. Fortunately, a well-defined formalism exists for including target mass effects (140, 141) by using the variable ξ instead of x. The modified structure functions $F(\xi, Q^2)$, to be inserted in the Altarelli-Parisi or moment equations, can be calculated from the measured structure functions $F(x, Q^2)$.

While the effects mentioned so far are well understood, there are other effects (e.g. due to coherent scattering off diquarks, resonance production, quark masses, primordial transverse momentum) that cannot be controlled. These effects are summarized by the term "higher twist" effects.

They contribute terms, proportional to powers of $1/Q^2$, to the structure functions (e.g. 131). It has been attempted to include such terms by writing (80)

$$F_i(x, Q^2) = F_i^{QCD}(x, Q^2)[1 + h_4(x)/Q^2 + h_6(x)/Q^4 + \cdots], \qquad 13.$$

where F_i^{QCD} obeys the laws of perturbative QCD as outlined in the previous section. Despite some recent attempts (142) the functions h_4, h_6 are not fully calculable, though it is known that the influence of higher twist effects increases with x. Some information can be directly obtained from fits to the data, which are accurate enough to allow the evaluation of one of $h_4(x)$ or $h_6(x)$ neglecting the other. We return to this aspect in Section 6.5.

The credibility of the perturbative QCD approach depends strongly on the x range considered (e.g. 24). The region $0.02 \lesssim x \lesssim 0.7$, where most of the data at high Q^2 have been taken, is fortunately the least problematic. At large x higher twist effects become increasingly important and also large corrections due to higher order effects in α_s might be expected.

Considering the extremely complex situation that has evolved, one has to remark that much of the original excitement about scaling violations and perturbative QCD has been lost. In the following we concentrate on QCD analyses performed by the experimental groups. In most of this work uncertainties due to statistical and systematic errors are included. For a recent review of the work performed by theoretical groups see (24).

6.3 Moment Analysis of Muon Data

Published results of a moment analysis in lowest as well as in second-order QCD are available from the CHIO collaboration who used their μp and μd data together with ep and ed data from SLAC and SLAC-MIT. To take account of target mass corrections, the Nachtmann moments for $n = 2$ to 10 are evaluated; the Q^2 range extends from 3.25 to 40 GeV^2 with the main information in the region $Q^2 < 20$ GeV^2. From the present point of view it is interesting to recall their results obtained from fits to the $n = 2$ moments [i.e. $\int_0^1 dx\, F_2(x, Q^2)$] only, which are least affected by uncertainties due to higher twist effects.

As an elementary result one observes that, in the Q^2 range covered, 54% of the energy momentum of the proton is carried by quarks, the remaining fraction by gluons. The Q^2 variation of the $n = 2$ moment is small and the mass scale parameter Λ can only be determined within rather large errors. The result is $\Lambda_{LO} = 183 \pm 282$ MeV for the lowest order fit, a value consistent with the results from the later muon experiments, which obtain values in the range 100–200 MeV. However, a simultaneous fit to the $n = 2$, 4, 6 moments gives $\Lambda_{LO} = 637 \pm 153$ MeV, while a fit to the $n = 10$ moments gives $\Lambda_{LO} = 971 \pm 88$ MeV. The results indicate the problems involved in higher moment analyses and suggest the necessity to include higher twist effects and possibly higher order QCD corrections.

6.4 Direct x, Q^2 Analysis of the Hydrogen Data

As a consequence of the problems involved in the moment analysis, the more recent data of the EMC and the BCDMS groups have been analyzed by a numerical solution of the Altarelli-Parisi equations. In the following we discuss in some detail the QCD fits to the EMC μp data since these are the only available high statistics, high Q^2 data measured with a nucleon rather than with a nuclear target. For the extraction of $F_2^{\mu p}$ from the measured double differential cross section a value of $R = 0$ was used.

In a first step it was assumed that the F_2 data at large $x \geq 0.25$ can be compared to QCD predictions for a pure nonsinglet structure function (i.e. to a structure function containing only contributions from valence quark scattering) with the boundary condition

$$F_2^{NS}(x, Q_0^2) = Ax^\alpha(1-x)^\beta(1-\gamma x)$$

at $Q_0^2 = 5$ GeV^2. The parameters A, α, β, γ were simultaneously fitted together with the mass scale parameter Λ. The results of some representative fits assuming $f = 4$ flavors are summarized in Table 4 (32), where the first error indicates the statistical, the second the systematic, uncertainty.

The numerical value of Λ is found to be of the order of 100 MeV, changing only slightly when going from leading order to next-to-leading order in the $\overline{\text{MS}}$ scheme. Furthermore, the inclusion of the target mass formalism (third row of Table 4) causes only a small increase of Λ. The results on Λ are unaffected by changing the parameterization at the reference Q_0^2 or by changing the value of Q_0^2.

Table 4 Best-fit values from an analysis of the EMC hydrogen data

Type of analysis	Λ (MeV)
Nonsinglet $x \geq 0.25$	
leading order	$\Lambda_{\text{LO}} = 110^{+58+124}_{-46-69}$
next-to-leading order	$\Lambda_{\overline{\text{MS}}} = 139^{+68+156}_{-56-87}$
next-to-leading order with target mass corrections	$\Lambda_{\overline{\text{MS}}} = 154^{+70+173}_{-60+97}$
Singlet plus nonsinglet	
leading order	$\Lambda_{\text{LO}} = 81^{+36+44}_{-30-32}$

A problematic aspect of this procedure is that $F_2^{\mu p}$ is assumed to be a pure nonsinglet structure function at large x while in principle there might be singlet (sea quark) contributions to $F_2^{\mu p}$ even at $x \geq 0.25$. To investigate the influence on Λ, singlet fits in leading order have been performed including the terms depending on the gluon distribution. Here the gluon distribution is parameterized as $xG(x, Q_0^2) \propto (1-x)^n$ for $x > 0.25$. From the analysis of the neutrino and antineutrino data of the CDHS collaboration (27), one knows that the gluon distribution measured with an iron target can be parameterized as $xG(x, Q_0^2) = D(1-x)^{5.9}(1+3.5x)$ at $Q_0^2 = 5$ GeV^2. If the power n of the gluon distribution inside the proton is changed between 5 and 20, the values of Λ_{LO} vary between 123 and 115 MeV.

In a second step the analysis is extended into the region $x < 0.25$, where sea quark and gluon distributions are important, by using the singlet evolution equations. So far, this procedure could be applied only in leading-order QCD. The singlet term of $F_2^{\mu p}$ is parameterized as $F_2^S(x, Q_0^2) = Ax^\alpha(1-x)^\beta + B(1-x)^\gamma$; the two terms describe the valence and sea quark contribution. The ansatz used for the nonsinglet term is

$$F_2^{NS}(x, Q_0^2) = (F_2^p - F_2^n)/2 = C\sqrt{x}(1-x)^\delta(1-\varepsilon x).$$

The parameters C, δ, ε were determined from a fit to the EMC and SLAC $F_2^p - F_2^n$ data. Finally the gluon distribution was assumed to be of the form deduced from the CDHS data, with its normalization D being a parameter to be determined by the momentum sum rule of the fit. To avoid effects due to the charm threshold, which are important in the region of very low x, for $x < 0.08$ only the points at the lowest Q^2 were used in the fit in order to constrain F_2 for $x \to 0$. The result of the fit for $f = 4$ flavors is presented in the fourth line of Table 4. Assuming a much softer gluon distribution like $(1-x)^7$ or a much harder glue as predicted by Glück et al (88) causes Λ to vary from 70 to 110 MeV.

Figure 27 compares the EMC $F_2^{\mu p}$ data with leading-order QCD fits. The solid line indicates the nonsinglet fit for $x \geq 0.25$, the broken curve a singlet fit. The logarithmic slope $\partial F_2/\partial \ln Q^2$ of the data is shown in Figure 28 together with curves extracted from the nonsinglet (*curve A*) and singlet (*curve B*) fits. The trend of the logarithmic slope is well reproduced by the fits.

Finally, it should be mentioned that the influence of charm production on the outcome of QCD fits was also investigated. Subtracting the charmed piece of F_2 as measured by the EMC with an iron target from the hydrogen data before performing a leading-order fit with $f = 3$ flavors yields $\Lambda_{LO} = 139$ MeV.

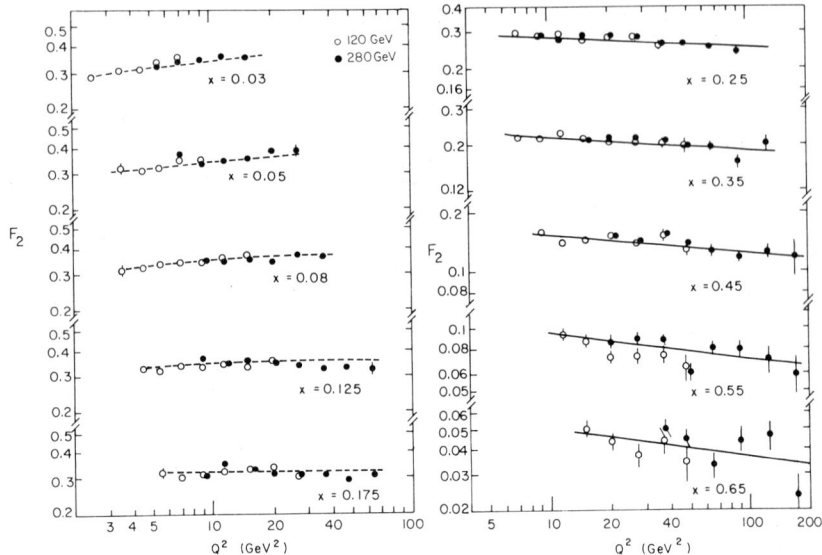

Figure 27 $F_2^{\mu p}$ from EMC with statistical errors compared to two different leading-order QCD fits.

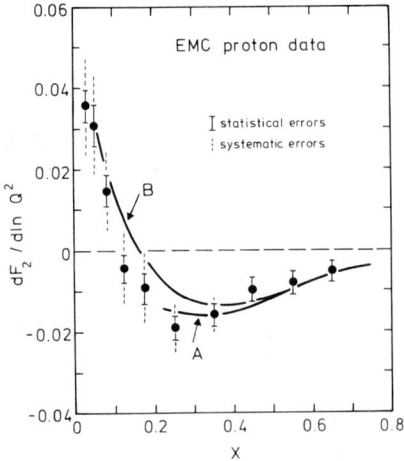

Figure 28 Logarithmic slope $\partial F_2 / \partial \ln Q^2$ of $F_2^{\mu p}$ as a function of x. Curve A represents the result of a nonsinglet analysis, curve B of a singlet plus nonsinglet analysis.

6.5 Separation of Higher Twist Contributions from Leading Twist Effects

A QCD analysis of the combined SLAC and EMC data clearly shows that higher twist terms have to be included in order to improve the quality of the fit. One can even try to separate the $1/Q^2$ or $1/Q^4$ power behavior due to higher twist contributions from the logarithmic Q^2 variation of the leading twist-2 part. Such studies are feasible because of the enormous range in Q^2 covered by the SLAC plus EMC proton data.

Generally, one can separate F_2 into a leading twist-2 and a higher twist part

$$F_2(x,Q^2) = F_2^{QCD}(x,Q^2) + F_2^{HT}(x,Q^2).$$

$F_2^{HT}(x,Q^2)$ contains terms proportional to $h_4(x)/Q^2$, $h_6(x)/Q^4$,..., where $h_4(x)$, $h_6(x)$ are unknown functions; F_2^{QCD} obeys the Q^2 evolution equations. In order to reduce somewhat the number of free parameters involved, fits to the hydrogen data have been performed in the region $x \geq 0.25$ using the simplified ansatz (32)

$$F_2(x,Q^2) = F_2^{QCD}(x,Q^2)[1+h_4(x)/Q^2].$$

F_2^{QCD} is considered as a nonsinglet structure function parameterized in next-to-leading order. Since the Q^2 dependence is studied for a finite number of x bins only, it is possible to determine the value of h_4 for each x together with the parameters of F_2^{QCD}. The fit value of $\Lambda_{\overline{MS}}$ is 124^{+66}_{-51} MeV. The resulting values of h_4 are shown in Figure 29. In this analysis, target mass corrections

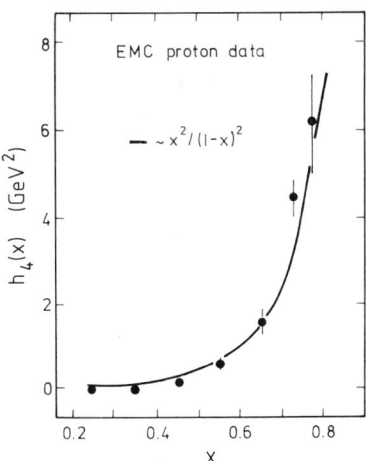

Figure 29 Higher twist coefficient h_4 as function of x. The solid line represents a function of the form $x^2/(1-x)^2$.

are not included. If target mass effects are taken into account, h_4 is reduced by about a factor of 1.5 while $\Lambda_{\overline{MS}}$ is reduced by only 24 MeV.

The function $h_4(x)$ is found to be consistent with zero for $x < 0.4$ rising sharply for $x > 0.5$. An adequate parameterization is $h_4(x) \propto x^2/(1-x)^2$ rather than the previously used forms $h_4(x) \propto x/(1-x)$ or $1/(1-x)$. The results do not significantly change if the sea contribution of F_2 as evaluated from the CDHS measurements is subtracted before fitting. It should be mentioned that the x dependence found for h_4 resembles the predictions of diquark models (143).

What is still missing is a combined analysis of the twist-2, -4, and -6 parts of F_2. This should be attempted in the future with the most extensive data possible.

6.6 Including the Soft Gluon Limit $x \to 1$

Despite the complexity of the QCD analysis described so far, still some aspects are missing. All perturbative tests of QCD necessarily rest on the hope that the perturbative series converges. To some extent this is supported by the small value of Λ or, equivalently, the small value of α_s/π (typically $\alpha_s/\pi = 0.05$), which causes the second-order corrections in Equations 10 and 11 to be small compared to the leading-order terms. One can show, however, that perturbation theory breaks down in the region of large x where $\alpha_s(t) \cdot \ln^2[1/(1-x)]$ is of order one and where a resummation of all orders of perturbation theory has to be performed (e.g. 24). In this region x is close to the kinematical limit and therefore the emission of real soft gluons is kinematically restricted and cannot fully compensate the effects due to virtual gluon contributions.

It has been argued (144) that the effect of the resummation can be incorporated in an improved evolution equation by replacing the argument t of α_s in Equation 11 by

$$t = \ln\left(\frac{Q^2}{\Lambda^2}\frac{1-z}{z}\right),$$

which corresponds to a rescaling from Q^2 to $Q^2(1-z)/z$. Recently Pennington et al (145) performed an analysis of deep inelastic data where they included in the argument of α_s two additional terms,

$$t = \left[\ln^2\left(\frac{Q^2}{\Lambda^2}\frac{1-z}{z} + C\right) + \pi^2\right]^{1/2}, \qquad 14.$$

so that $\alpha_s = 4\pi/\beta_0 t$ freezes to a constant value as $Q^2(1-z)/z \to 0$. This is due to the π^2 term, representing higher order perturbative corrections, and the C term, which is equivalent to the inclusion of higher twist effects.

In this context we are mainly interested in the change of the best fit value of the mass scale parameters Λ when the various forms of the argument of α_s are used. We therefore quote results for fits to the EMC hydrogen data treating $F_2^{\mu p}$ as a nonsinglet function for $x > 0.1$ with the following choices of the argument t of α_s:

1. $\ln(Q^2/\Lambda^2)$ of normal perturbative QCD,

2. $\ln\left(\dfrac{1-z}{z}Q^2/\Lambda^2 + C\right)$,

3. $\left[\ln^2\left(\dfrac{1-z}{z}Q^2/\Lambda^2 + C\right) + \pi^2\right]^{1/2}$.

The best-fit values obtained in next-to-leading order in the $\overline{\text{MS}}$ scheme (including second-order corrections to α_s) for the three cases are (145):

1. $\Lambda_{\overline{\text{MS}}} = 105 \pm 80$ MeV,
2. $\Lambda_{\overline{\text{MS}}} = 102 \pm 74$ MeV,
3. $\Lambda_{\overline{\text{MS}}} = 128 \pm 108$ MeV.

Systematic uncertainties of the data are not included in this analysis.

The values of Λ_{MS} change little for the various choices of the argument of α_s and are thus insensitive to the inclusion of higher order corrections or higher twist contributions. The numerical result is close to the one obtained by the EMC.

6.7 QCD Fits to $F_2^{\mu N}$ *from Iron and Carbon Targets*

The F_2 data with the highest statistical accuracy, extending into the Q^2 region of 200 GeV^2 and beyond, come from measurements of the BCDMS group with a carbon target (30) and from measurements of the EMC with an iron target (31). The analysis of both groups includes radiative corrections but no corrections for Fermi motion and nuclear binding effects. For the extraction of F_2 (quoted as F_2 per nucleon) a value of $R = 0$ is assumed.

Though one has to question whether parton distributions extracted from measurements with heavy nuclear targets are identical to those of the nucleon, it is worthwhile to perform QCD fits, since the mathematical structure of the QCD evolution equations remains unchanged. On the other hand, there is at present no possibility of excluding contributions to muon scattering from heavy nuclei arising from higher twist effects. Also other nuclear effects, e.g. the possible existence of non-nucleonic matter inside the nucleus, are not considered. Thus, in our opinion, the main interest in the QCD analysis of heavy target data is in the comparison with the results obtained for the hydrogen (or deuterium) case.

Table 5 Fits to iron and carbon data

Group	Type of fit	Λ_{LO} or $\Lambda_{\overline{MS}}$ (MeV)
EMC	NLO, nonsinglet	$173^{+29+158}_{-27-97}$
	LO, singlet	163^{+22+99}_{-22-64}
BCDMS	LO, nonsinglet	85^{+60+90}_{-40-70}
	NLO, nonsinglet	85^{+53+80}_{-40-67}

The EMC has performed a variety of QCD fits to their iron data similar to those described for the hydrogen data. The results for the next-to-leading-order nonsinglet fit for $x \geq 0.25$ and the leading-order singlet fit for all x are given in the first two lines of Table 5. Again the first error indicates the statistical and the second error the systematic uncertainty.

The BCDMS data extend over the region $0.3 < x < 0.7$. It therefore seems justified to consider F_2 as a nonsinglet structure function with the boundary condition $F_2(x, Q^2) \propto x^\alpha(1-x)^\beta$. The values of Λ resulting from fits to the 120- and 200-GeV data are collected in the last two lines of Table 5.

Recently a QCD moment analysis based on the BCDMS data, including the yet unpublished 280-GeV data, was performed by Fadeev et al (146). This method has the advantage that the full singlet formulae in next-to-leading order are available. However, the large number of free parameters involved in the fit (which are also strongly correlated) make it necessary to impose additional restrictions. In this case the ratios of the nth gluon moment to the nth quark singlet moment for $n = 4, 6$ are fixed to values between 0 and 1. The best-fit values of $\Lambda_{\overline{MS}}$ are in the range from 41 to 85 MeV (with a preferred value of about 70 MeV) and thus similar to the result of the nonsinglet fit of the BCDMS group (fourth line in Table 5).

The results for $\Lambda_{\overline{MS}}$ obtained from fits to the heavy nuclei data do not differ significantly from the results of the analysis of the hydrogen data. This might indicate that the Q^2 variation of F_2 as determined from the heavy target measurements is well represented by the leading twist Q^2 evolution equations and not by higher twist effects.

6.8 Conclusions on QCD Analyses

It has clearly been shown that the scaling violations observed in deep inelastic muon or neutrino scattering can be well understood in terms of perturbative QCD if the statistical and systematical uncertainties of the data are properly taken into account. However, the numerical value of the mass scale parameter Λ has changed since the first quantitative confron-

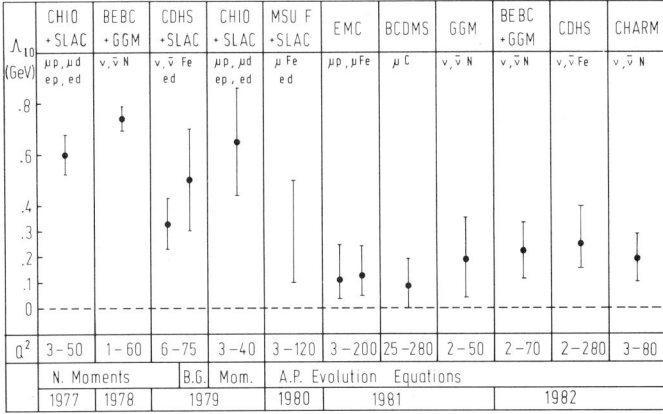

Figure 30 Development of Λ_{LO} values. The entries are for (147–149, 23, 150, 29, 31, 30, 151, 152, 82, 28). Also indicated is the Q^2 range of the different analyses, the method used, and the year of publication.

tation in 1977. The development of Λ values as obtained by various muon or neutrino groups from leading-order QCD fits to their data is demonstrated in Figure 30. With increasing range of Q^2 covered by the experiments, there is a clear trend toward smaller values of Λ, a direct consequence of the small scaling violations observed at high Q^2. The scaling violations measured by all experiments presenting results after 1980 can be parameterized by $50 < \Lambda_{LO} < 300$ MeV.

Considering the deviations of the F_2 data measured with a heavy nuclear target from those measured with a deuterium target and the presently not yet fully investigated consequences for the outcome of a QCD fit, we would, if asked to choose a single value of the mass scale parameter, select the result of a fit to the EMC hydrogen data

$$\Lambda_{\overline{MS}} = 139^{+170}_{-103} \text{ MeV}.$$

The corresponding value of the effective strong interaction coupling is

$$\alpha_s^{\overline{MS}}(Q^2 = 100 \text{ GeV}^2) = 0.144^{+0.028}_{-0.031}.$$

7. WEAK ELECTROMAGNETIC INTERFERENCE

7.1 *Theoretical and Experimental Background*

The now conventional unified theory of electroweak interactions (153) contains four intermediate bosons: two charged $W^\pm$; the neutral partner in the weak interaction, Z^0; and the mediator of electromagnetic interactions, the photon γ. Since the latter two are neutral, charged lepton scattering may proceed by the exchange of either, albeit with differing probability, and interference between the two amplitudes is also to be expected. Since the photon is massless but the Z^0 is not, the propagator of the Z^0 in the interference term causes a slower fall off as a function of Q^2. Nevertheless the smallness of the weak coupling constant results in an expected interference term of order $10^{-4}Q^2$ (GeV^2) and hence it is difficult to detect.

The technique used is to perform measurements in two different conditions to which the interference contributes with opposite sign. This is accomplished by making measurements in which the electric charge and/or helicity of the incident charged lepton is different. The characteristic asymmetries are then defined by:

$$B = \frac{d\sigma^+(\lambda) - d\sigma^-(-\lambda)}{d\sigma^+(\lambda) + d\sigma^-(-\lambda)}$$

$$A^+ = \frac{d\sigma^+(\lambda) - d\sigma^+(-\lambda)}{d\sigma^+(\lambda) + d\sigma^+(-\lambda)} \qquad 15.$$

$$A^- = \frac{d\sigma^-(\lambda) - d\sigma^-(-\lambda)}{d\sigma^-(\lambda) + d\sigma^-(-\lambda)}$$

where superscripts $\pm$ indicate electric charge and λ is the helicity of the incident muon. Note that these asymmetries occur for an unpolarized target in contrast to the case for pure single-photon exchange.

Taking B as an example, the electroweak theory with one photon and Z^0 exchange gives an expression containing the electromagnetic and weak coupling constants, the weak couplings, the axial vector a_μ and vector v_μ of the muon, and the axial vector couplings of the valence quarks a_d and a_u, to the Z^0 (see 154):

$$B = -K(a_\mu - \lambda v_\mu) \cdot \frac{6}{5}(a_d - 2a_u)g(y)Q^2,$$

$$K = \frac{G_F}{\sqrt{2}} \frac{1}{2\pi\alpha} = 1.8 \times 10^{-4} \text{ GeV}^{-2}, \qquad 16.$$

where G_F is the Fermi constant and α the fine structure constant,

$$g(y) = \frac{1-(1-y)^2}{1+(1-y)^2}$$

is a kinematic term.

In a general gauge theory a_μ and v_μ can be rewritten in terms of the couplings I_3^L and I_3^R for left- and right-handed isospin multiplets

$$v_\mu = I_3^L + I_3^R + 2 \sin^2 \theta_W$$

$$a_\mu = I_3^L - I_3^R.$$

The conventional assignments $I_3^L = -\frac{1}{2}$ and $I_3^R = 0$ and the currently accepted value $\sin^2 \theta \sim \frac{1}{4}$ result in $v_\mu \sim 0$ and the asymmetry B is axial-axial and hence parity conserving. This contrasts with the $A^\pm$, which contain only vector-axial vector (V-A) terms.

The possibility of measuring these effects has been discussed since values of $Q^2 \sim 10$ GeV^2 became reality and $Q^2 \sim 100$ GeV^2 seemed imminent. Three serious attempts at measurement have been made by "deep inelastic" experiments and they are discussed below. It is perhaps ironic that the first positive measurement (156) had a Q^2 in the range 1–2 GeV^2 at the lower limit of the deep inelastic regime.

Electron scattering experiments require explicit efforts to polarize the electron beams (157). Muons are, however, produced by the weak decay of the pion and the definite helicity of the neutrino forces a well-defined spin and helicity on the muon in the decay center of mass. On transformation to the laboratory system a strict relationship then exists between the ratio of muon p_μ and pion p_π momenta and the helicity of the muon. High energy muon beams that select a narrow band in p_μ/p_π are therefore "naturally" polarized (158). In particular $p_\mu/P_\pi \sim 1$ gives muons in their unnatural helicity state, i.e. right-handed μ^-, left-handed μ^+.

7.2 The Serpukhov Experiment

The first high energy muon experiment (159) to embark on a program to detect the weak electromagnetic interference was performed at the 70-GeV proton synchrotron at Serpukhov. The measurement attempted was that of A^- with muons of 21 GeV resulting from pions with 40 and 28 GeV for the two different helicities. Since the natural phase space for backward and forward pion decay is much different, special care had to be taken in the design of the two beam configurations to ensure closely similar distributions of momenta.

A total of 5×10^9 muons were used for the experiment, giving a sensitivity varying from $<1\%$ near $Q^2 = 1$ GeV^2 to $\sim 4\%$ at $Q^2 \sim 10$ GeV^2. No effect was observed and an upper limit of $(-4 \pm 6) \times 10^{-3} \, Q^2$ was placed on the asymmetry. This corresponds to a limit on the strength of a supposed V-A interaction of the muon of $(+6 \pm 10) \, G_F$.

7.3 The SLAC ed Experiment

Following the series of deep inelastic scattering experiments performed at SLAC and referred to in other sections of this review, a program of experiments (157) using a polarized beam and polarized targets developed. These experiments yielded as a by-product a limit of $A^- < 5 \times 10^{-3}$ (95% C.L.) (160) on the asymmetry A^-. A strong limitation was imposed on these experiments by the inability of the targets to maintain their polarization in an intense beam.

Using a standard deuterium target Prescott et al (156) chose to maximize the beam intensity by using a source (161) that produced high yields but with $\sim 40\%$ polarization. The mean Q^2 of the experiment was in the range 1–2 GeV^2 and the asymmetry to be measured was therefore $\sim 10^{-4}$. That they achieved this and more is testified by the commonly applied prefix "classic" when describing the experiment. The resultant value for $\sin^2 \theta_W = 0.224 \pm 0.020$, to which the measurement of A^- is particularly sensitive, remains one of the strongest contributions to our knowledge of this parameter (54).

7.4 The BCDMS Measurement

Given the CERN muon beam with its comparatively high intrinsic polarization (158), the natural first attempt to measure a weak interference was of the asymmetry B. In this case the forward pion decay gives the highest possible beam yields and the fact that both beams, positive and negative, have the same configuration means that beam phase space differences are minimized. Nevertheless embarkation on the measurement of an effect $\lesssim 1\%$ with an apparatus > 50 meters long requires some optimism.

The experiment (53) was run in 1980 with beams of ± 200 and ± 120 GeV; the apparatus has been described in Section 3.3. A total of $\sim 3 \times 10^6$ events were recorded of which the 1.7×10^6 events at ± 200 GeV had $Q^2 > 40$ GeV^2. The beam sign was changed twice during each of eight, 12-day, data-taking periods. At each change of beam sign the spectrometer field was carefully cycled to remain on the equivalent part of the hysteresis curve of the iron core magnet modules. Similarly the critical beam elements were reversed in a controlled way and the fields were checked with an extensive system of Hall probes.

A major advantage in the experimental design is the azimuthal symmetry of the detector about the beam axis. This reduces, to second order, any effect due to beam phase space differences between positive and negative beams. Despite this advantage the analysis of the data had to use to the maximum the high degree of redundancy of both trigger hodoscopes and chamber planes in such a way as to check the time stability of the efficiencies of all elements.

It is not possible here to discuss the study of all the possible systematic effects, so we confine ourselves to one example to illustrate the level of care required. When a muon exits from a large number of radiation lengths of iron, it is accompanied in $\sim 20\%$ of the cases by a soft charged particle in general not collinear with the muon. Such particles can therefore be detected in the wire chambers and cause a slight mismeasurement of the muon track at that point. Any systematic tendency for such spurious "hits" to be to on one side or the other of the muon track can therefore distort the momentum measurement. If these hits were due to one of an electron-positron pair exiting from the iron they might be symmetrical under a change of field direction in the iron. On the other hand, if they are a result of electrons knocked on by the muon, then they would systematically distort the μ^+ to higher momentum and the μ^- to lower momentum. The track fit residuals had therefore to be studied carefully to search for, and establish a limit on the size of, this (and many other) possible effects.

The measured raw result for B as a function of $g(y)Q^2$ is shown in Figure 31 for the 200-GeV data. The expected theoretical curve is also shown for world standard parameter values. It is noticeable that, contrary to Equation 16, neither data nor curve, which agree very well, are linear in $g(y)Q^2$. This is due to higher order electromagnetic and weak (162) contributions to the asymmetry, the radiative corrections mentioned in Section 2.2. For example the two-photon-exchange term can contribute to this measured asymmetry. The model dependence in these calculations gives a significant contribution to the final systematic errors. Nevertheless, as can be seen in Figure 31, a significant asymmetry is observed in the raw data.

Corrected for radiative effects the data for both energies are displayed in Figure 32, again compared to the standard model predictions. By fitting the slopes of the data the results obtained are:

$b = (-0.147 \pm 0.037) \times 10^{-3}$ GeV^{-2} : 200 GeV

standard model ($\sin^2 \theta_W = 0.23$) $b = -0.151 \times 10^{-3}$ GeV^{-2} with $|\lambda| = 0.81$;

$b = (-0.174 \pm 0.075) \times 10^{-3}$ GeV^{-2} : 120 GeV

standard model ($\sin^2 \theta_W = 0.23$) $b = -0.153 \times 10^{-3}$ GeV^{-2} with $|\lambda| = 0.66$.

These are in a good agreement with the standard model of electroweak interactions.

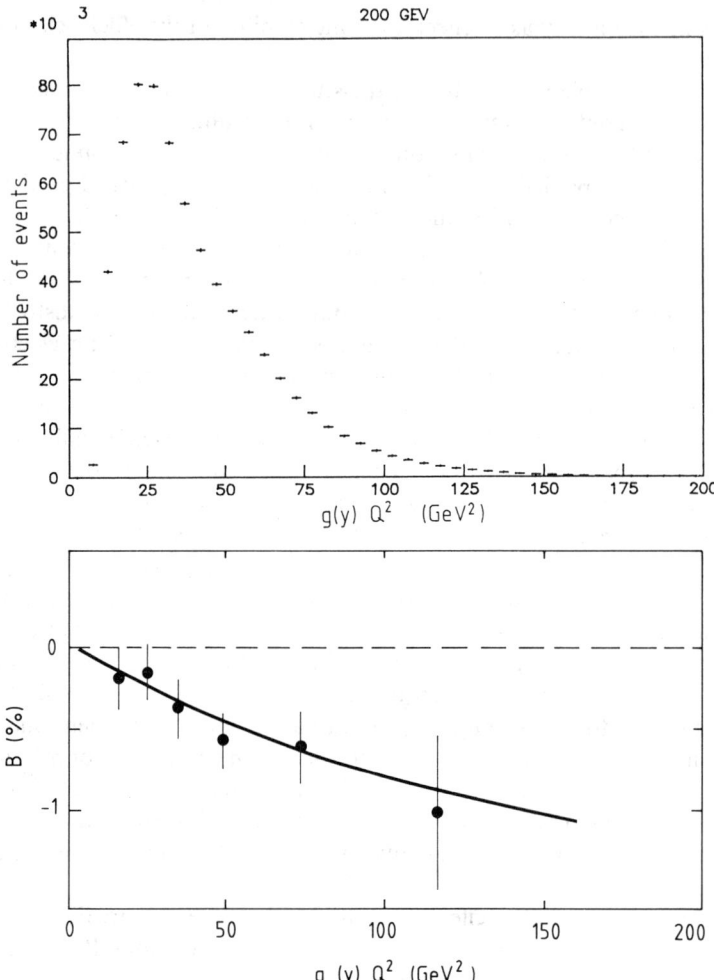

Figure 31 Measured asymmetry B at level of raw data as a function of $g(y)Q^2$ for the BCDMS 200-GeV data. Also shown is the distribution of events in this variable. The solid line indicates the standard model prediction.

If the measurement of $\sin^2 \theta_W = 0.23$ is taken from other experiments, the result can be expressed as a measurement of the right-handed weak coupling of the muon

$I_3^R = 0.00 \pm 0.06$ (stat.) ± 0.04 (syst.).

Alternatively, assuming $I_3^R = 0$, one can determine the Weinberg angle results:

$\sin^2 \theta_W = 0.23 \pm 0.07$ (stat.) ± 0.04 (syst.),

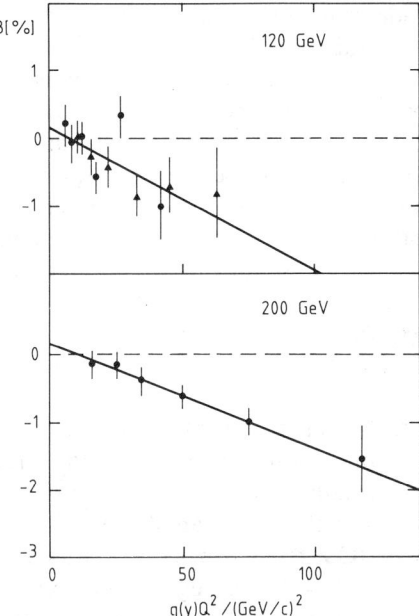

Figure 32 B at 120 and 200 GeV corrected for higher order effects as a function of $g(y)Q^2$. The solid line indicates the standard model prediction.

which is competitive with measurements using neutrinos or with the electron-deuteron experiment described earlier.

Measurements of the weak electromagnetic interference in $e^+e^- \to \mu^+\mu^-$ yield complementary information. Using the value for $a_\mu = -0.55 \pm 0.07$ obtained from measurements at the e^+e^- storage ring, PETRA, results in $v_\mu = -0.12 \pm 0.14$ (stat.) ± 0.08 (syst.). This also can be reinterpreted as a measurement of the Weinberg angle with $\sin^2 \theta_W = 0.19 \pm 0.07 \pm 0.04$.

From all these comments it seems that the standard electroweak model survives this measurement, which completes the interlinkings of lepton and quark coupling within the Sakurai tetrahedron. As is often the case with experiments of this type, one is at least as impressed by the skill, ingenuity, and perseverance of the experimental group as by the physics interpretation of the result.

8. CONCLUSIONS

The latest generation of muon experiments at CERN and Fermilab have covered a broad physics program reaching high standards with respect to both statistical and systematical uncertainties of their data. They have proven to be an excellent means of studying the structure functions of nucleons over a wide range of Q^2 and of investigating fine details, for example deviations of the point-like structure of heavy nuclei from that observed with free nucleons. Detailed confrontations of the measured Q^2 evolution with expectations from perturbative QCD have yielded smaller numerical values of the mass scale parameter Λ than previously assumed. The two CERN experiments are presently working on a completion of their structure function analyses, which will result in even more precise data sets. Extensive investigations of multimuon final states have led to some understanding of the mechanism of J/ψ and open charm production. The experiment of the BCDMS group has proven to be accurate enough to observe interference effects between the electromagnetic and the weak neutral current as predicted by electroweak theories and to determine $\sin^2 \theta_W$.

An aspect covered in an earlier article concerns the high statistics study of the dynamics of hadron production including the successful search for subtle phenomena originating only from higher order perturbative QCD effects. Here a new phase of the European Muon Collaboration experiment is in progress. The EMC extended their apparatus by the addition of a vertex detector with nearly 4π solid angle coverage in the center-of-mass system and by Čerenkov counters allowing particle identification in a wide momentum range. Further approved proposals of the CERN groups concern measurements with a polarized target of 1-m length and an extension of the structure function measurements with a variety of targets.

At Fermilab a new program with the high yield 750-GeV muon beam will start after the beginning of the TEVATRON operation. The construction of a new open geometry spectrometer consisting of a forward and a vertex detector is also approved, which will enable the study of a wide range of physics subjects, for example structure functions and aspects of the hadronic final state (Experiment E665). Furthermore the continuation of the high luminosity measurements based on the BFP multimuon spectrometer (Experiment 640) is anticipated.

We are grateful to Alan Edwards and Uwe Pietrzyk who provided us with some of the figures. We would also like to take this opportunity to express our appreciation to our many colleagues in all of the BFP, BCDMS, and EMC collaborations who have contributed in large measure to the pleasure we feel in having been able to participate in the physics described in this report.

Literature Cited

1. Anderson, G. D., Nedermeyer, S. H. 1936. *Phys. Rev.* 50:263, 1937. *Phys. Rev.* 51:884, 1938. *Phys. Rev.* 54:88
2. Harari, H. 1978. *Phys. Rep.* 42:235; Fritsch, H., Minkowski, P. 1981. *Phys. Rep.* 73:67
3. Scheck, F. 1978. *Phys. Rep.* 44:187
4. Combley, F., Picasso, E. 1974. *Phys. Rep.* C14:1; Farley, F. J. M., Picasso, E. 1979. *Ann. Rev. Nucl. Part. Sci.* 29:243
5. Francis, W. R., Kirk, T. W. 1979. *Phys. Rep.* C54:307
6. Renton, P., Williams, W. S. C. 1981. *Ann. Rev. Nucl. Part. Sci.* 31:193
7. West, G. 1975. *Phys. Rep.* 18:263
8. Hofstadter, R. 1963. *Nuclear and Nucleon Structure.* New York: Benjamin
9. Donnelly, T. W., Walecka, J. D. 1975. *Ann. Rev. Nucl. Sci.* 25:329
10. Hand, L. N., Miller, D. G., Wilson, R. 1963. *Rev. Mod. Phys.* 35:335; Rutherglen, J. G. 1969. *Proc. 4th Int. Symp. on Electron and Photon Interactions at High Energies,* ed. D. W. Braben. Liverpool: Daresbury Nucl. Phys. Lab. 163 pp.
11. Bjorken, J. D., Paschos, E. H. 1969. *Phys. Rev.* 185:1975
12. Feynman, R. P. 1972. *Photon-Hadron Interactions.* New York: Benjamin
13. Close, F. E. 1979. *An Introduction to Quarks and Partons.* New York: Academic
14. Politzer, H. D. 1973. *Phys. Rev. Lett.* 30:1346
15. Gross, D. J., Wilczek, F. 1973. *Phys. Rev. Lett.* 30:1343
16. Reya, E. 1981. *Phys. Rep.* 69:195
17. Bloom, E. D., et al. 1969. *Phys. Rev. Lett.* 23:930; Breidenbach, M., et al. 1969. *Phys. Rev. Lett.* 23:935
18. Friedman, J. I., Kendal, H. W. 1972. *Ann. Rev. Nucl. Sci.* 22:203
19. Taylor, R. E. 1975. *Proc. 1975 Int. Symp. on Lepton and Photon Interactions at High Energies.* Stanford Univ. 679 pp.
20. Fox, D. J., et al. 1974. *Phys. Rev. Lett.* 33:1504; Watanabe, Y., et al. 1975. *Phys. Rev. Lett.* 35:898
21. Chang, C., et al. 1975. *Phys. Rev. Lett.* 35:901
22. Anderson, H. L., et al. 1976. *Phys. Rev. Lett.* 37:4, 1034; 1977. *Phys. Rev. Lett.* 38:1450; Gordon, B. A., et al. 1978. *Phys. Rev. Lett.* 41:615
23. Gordon, B. A., et al. 1979. *Phys. Rev.* D20:2645
24. Buras, A. J. 1981. *Proc. 1981 Int. Symp. on Lepton and Photon Interactions at High Energies,* ed. W. Pfeil, Phys. Inst., Univ. Bonn. 636 pp.
25. Drees, J. 1981. See Ref. 24, p. 474.
26. Smadja, G. 1981. See Ref. 24, p. 444
27. Abramowicz, H., et al. 1982. *Z. Phys.* C12:289
28. Bergsma, F., et al. 1983. *Phys. Lett.* 123B:269
29. EMC. Aubert, J. J., et al. 1981. *Phys. Lett.* 105B:315
30. BCDMS. Bollini, D., et al. 1981. *Phys. Lett.* 104B:403
31. EMC. Aubert, J. J., et al. 1981. *Phys. Lett.* 105B:322
32. EMC. Aubert, J. J., et al. 1982. *Phys. Lett.* 114B:291
33. EMC. Aubert, J. J., et al. 1983. *Phys. Lett.* 123B:115
34. EMC. Aubert, J. J., et al. 1983. *Phys. Lett.* 123B:275
35. Altarelli, G., Parisi, G. 1977. *Nucl. Phys.* B126:298
36. BFP. Clark, A. R., et al. 1979. *Phys. Rev. Lett.* 43:187
37. BFP. Clark, A. R., et al. 1980. *Phys. Rev. Lett.* 45:2092
38. Markiewicz, T. 1981. PhD Thesis, Lawrence Berkeley Lab.
39. EMC. Aubert, J. J., et al. 1980. *Phys. Lett.* 89B:267
40. EMC. Aubert, J. J., et al. 1980. Preprint CERN-EP/80-84 (unpublished)
41. EMC. Aubert, J. J., et al. 1983. *Nucl. Phys.* B213:1
42. BFP. Clark, E. R., et al. 1980. *Phys. Rev. Lett.* 45:1465
43. BFP. Gollin, G. D., et al. 1981. *Phys. Rev.* D24:559
44. EMC. Aubert, J. J., et al. 1980. *Phys. Lett.* 94B:96
45. EMC. Aubert, J. J., et al. 1980. *Phys. Lett.* 94B:101
46. EMC. Aubert, J. J., et al. 1982. *Phys. Lett.* 110B:73
47. EMC. Aubert, J. J., et al. 1983. *Nucl. Phys.* B213:31
48. Phillips, R. J. N. 1980. *Proc. of 20th Int. Conf. on High Energy Physics, Wisconsin,* ed. L. Durand, L. G. Pondrom. New York: AIP (1981). 1470 pp.
49. Strovink, M. 1981. See Ref. 24, p. 594
50. BFP. Clark, A. R., et al. 1981. *Phys. Rev. Lett.* 46:299
51. BFP. Smith, W. H., et al. 1982. *Phys. Rev.* D25:2762
52. EMC. Aubert, J. J., et al. 1982. *Phys. Lett.* 106B:419
53. BCDMS. Argento, A., et al. 1982. *Phys. Lett.* 120B:245
54. Niebergall, F. 1982. *Proc. of Int. Conf. Neutrino '82, Balatonfüred,* ed. A. Frenkel, L. Jenik, Budapest

55. Drell, S. D., Walecka, J. D. 1964. *Ann. Phys.* 28:18
56. Gilman, F. J. 1972. *Phys. Rep.* 4C, Ch. 8–13
57. Mo, L. W., Tsai, Y. S. 1969. *Rev. Mod. Phys.* 41:205; Tsai, Y. S. 1971. *Rep. SLAC-PUB-848*
58. Akhundov, A. A., Bardin, D. Yu., Shumeiko, N. M. 1977. *Jad. Fiz.* 26:1251
59. EMC. Aubert, J. J., et al. 1981. *Z. Phys.* C10:101
60. Rochester, L. S., et al. 1976. *Phys. Rev. Lett.* 36:1284
61. Faucher, D. L., et al. 1976. *Phys. Rev. Lett.* 37:1323
62. Chen, K. W. 1975. *Proc. of the EPS Int. Conf. on High Energy Physics, Palermo*, ed. N. Zichichi. Bologna: Editrice Compositori. 458 pp.
63. Ballam, J., et al. 1974. *Phys. Rev.* D10:765; 1975. *Phys. Lett.* 56B:193
64. del Papa, C., et al. 1976. *Phys. Rev.* D13:2934; 1977. *Phys. Rev.* D15:2425
65. Clifft, R. W., Doble, N. 1974. *Rep. CERN/SPSC/74-78*
66. BFP. Clark, A. R., et al. 1979. In *Proc. 1979 Int. Symp. on Lepton and Photon Interactions at High Energies*, ed. T. W. Kirk. Batavia, Ill: Fermi Natl. Accel. Lab. Presented by M. Strovink. 135 pp.
67. BCDMS. Bollini, D., et al. 1983. *Nucl. Instrum. Methods* 204:333
68. BCDMS. Bollini, D., et al. 1982. *Nucl. Phys.* B199:27
69. EMC. Allkofer, O. C., et al. 1981. *Nucl. Instrum. Methods* 179:445
70. EMC. Aubert, J. J., et al. 1980. *Phys. Lett.* 95B:306; 1981. *Phys. Lett.* 100B:433; 103B:388; 1982. *Phys. Lett.* 114B:373; 119B:233
71. EMC. Aubert, J. J., et al. 1983. *Phys. Lett.* 121B:87
72. EMC. Albanese, J. P., et al. 1982. *Preprint CERN-EP/82-160*. To be published in *Nucl. Instrum. Methods*
73. Mount, R. P. 1981. *Nucl. Instrum. Methods* 187:401
74. Navarria, F. L., Zupancic, C., Feltesse, J. 1982. *Preprint CERN-EP/82-196*. To be published in *Nucl. Instrum. Methods*
75. Atwood, W. B. 1979. *Rep. SLAC-PUB-2428*
76. Bjorken, J. D. 1969. *Phys. Rev.* 179:1547
77. Georgi, H., Politzer, H. D. 1976. *Phys. Rev.* D14:1829
78. Bodek, A., et al. 1979. *Phys. Rev.* D20:1471
79. Mestayer, M. D., et al. 1983. *Phys. Rev.* D27:285
80. Abbott, L. F., Atwood, W. B., Barnett, R. M. 1980. *Phys. Rev.* D22:582
81. Abramowicz, H., et al. 1981. *Phys. Lett.* 107B:141
82. Abramowicz, H., et al. 1983. *Z. Phys.* C17:283
83. Ball, R. C., et al. 1979. *Phys. Rev. Lett.* 42:866
84. Eisele, F. 1982. *Proc. of 21st Int. Conf. on High Energy Physics, Paris*, ed. P. Petiau, M. Porneuf. *J. Phys. Coll.* C-3, Suppl. 12
85. Farrar, G. R., Jackson, D. R. 1975. *Phys. Rev. Lett.* 35:1416
86. Allen, P., et al. 1981. *Phys. Lett.* 103B:71; Giles, R. 1981. D.Phil. Thesis, Oxford Univ.
87. Field, R. D., Feynman, R. P. 1977. *Phys. Rev.* D15:2590
88. Glück, M., Hoffmann, E., Reya, E. 1982. *Z. Phys.* C13:119
89. Donnachie, A., Landshoff, P. V. 1980. *Phys. Lett.* 95B:437; Close, F. E., Roberts, R. G. 1981. *Z. Phys.* C8:57
90. Bodek, A., Ritchie, J. L. 1981. *Phys. Rev.* D23:1070; D24:1400
91. Frankfurt, L. L., Strikman, M. I. 1981. *Nucl. Phys.* B181:22
92. Berlad, G., Dar, A., Eilam, G. 1980. *Phys. Rev.* D22:1547
93. Eickmeyer, J., et al. 1976. *Phys. Rev. Lett.* 36:289; Bailey, J., et al. 1979. *Nucl. Phys.* B151:367; Stein, S., et al. 1975. *Phys. Rev.* D12:1884; May, M., et al. 1971. *Phys. Rev. Lett.* 35:407; Miller, M. et al. 1981. *Phys. Rev.* D24:1
94. Goodman, M. S., et al. 1981. *Phys. Rev. Lett.* 47:293
95. Altemus, S., et al. 1980. *Phys. Rev. Lett.* 44:965
96. Noble, J. V. 1981. *Phys. Rev. Lett.* 46:412
97. Høgaasen, H., Sorba, P., Viollier, R. 1980. *Z. Phys.* C4:131
98. Bergström, L., Fredriksson, S. 1980. *Rev. Mod. Phys.* 52:675
99. Pirner, H. J., Vary, J. P. 1981. *Phys. Rev. Lett.* 46:1376
100. Namiki, M., Okano, K., Oshimo, N. 1982. *Phys. Rev.* C25:2157
101. Godbole, R. M., Sarma, K. V. 1982. *Phys. Rev.* D25:120
102. Frankfurt, L. L., Strikman, M. I. 1981. *Phys. Rep.* 76C:215
103. Jaffe, R. L. 1983. *Phys. Rev. Lett.* 50:228
104. Llewellyn-Smith, C. H. Private communication
105. Bodek, A., et al. 1983. *Phys. Rev. Lett.* 50:1431
106. Sakurai, J. J. 1960. *Ann. Phys.* 11:1; Bauer, T. H., Spital, R. D., Yennie, D. R., Pipkin, M. M. 1978. *Rev. Mod. Phys.* 50:261
107. Jones, L. M., Wyld, H. W. 1978. *Phys. Rev.* D17:759; Novikov, V. A., Shifman, M. A., Vainshtein, A. I.,

Zakharov, V. I. 1978. *Nucl. Phys.* B136:125; Shifman, M. A., Vainshtein, A. I., Zakharov, V. I. 1978. *Nucl. Phys.* B136:157; Leveille, J. P., Weiler, T. 1979. *Phys. Lett.* 86B:377; *Nucl. Phys.* B147:147
108. Fritsch, H. 1979. *Phys. Lett.* 67B:217; Fritsch, H., Streng, K. H. 1978. *Phys. Lett.* 72B:385
109. Duke, D. W., Owens, J. F. 1980. *Phys. Lett.* 96B:184; 1981. *Phys. Rev.* D23:1671; Tajima, T., Watanabe, T. 1981. *Phys. Rev.* D23:1517
110. Berger, E. L., Jones, D. 1981. *Phys. Rev.* D23:1521
111. Baier, R., Rückl, R. 1982. *Nucl. Phys.* B201:1
112. Baier, R., Rückl, R. 1982. *Preprint MPI-PAE/P. Th. 58/82*
113. Halzen, F. 1982. See Ref. 84
114. Leveille, J., Weiler, T. 1981. *Phys. Rev.* D24:1789; Weiler, T. 1980. *Phys. Rev. Lett.* 44:304
115. Duke, D. W., Owens, J. F. 1981. *Phys. Rev.* D24:1403
116. Körner, J. G., Cleymans, J., Kuroda, M., Gounaris, G. J. 1982. *Phys. Lett.* 114B:195
117. Binkley, M., et al. 1982. *Phys. Rev. Lett.* 48:73; 1983. *Phys. Rev. Lett.* 50:302
118. Chang, C., Chen, K. W., Van Ginneken, A. 1977. *Phys. Rev. Lett.* 39:519
119. Bauer, D., et al. 1979. *Phys. Rev. Lett.* 43:1551
120. Bjorken, J. D. 1978. *Phys. Rev.* D17:171; Kartvelishvili, V. G., Likhoded, A. K., Petrov, V. A. 1978. *Phys. Lett.* 78B:615; Bowler, M. G. 1981. *Z. Phys.* C11:169
121. Brodsky, S. J., Hoyer, P., Peterson, C., Sakai, N. 1980. *Phys. Lett.* 93B:451; Brodsky, S. J., Peterson, C., Sakai, N. 1981. *Phys. Rev.* D23:2745
122. Gavai, R. V., Roy, D. P. 1981. *Z. Phys.* C10:333
123. Roy, D. P. 1981. *Phys. Rev. Lett.* 47:213; Godbold, S. J., Roy, D. P. 1981. *Tata Inst. TIFR/TH/82-5*; Brodsky, S. J., Petersen, C. 1982. *Rep. SLAC-PUB-2888*
124. SLAC. Abe, K., et al. 1981. See Ref. 24; WA4. Aston, D., et al. 1980. *Phys. Lett.* 94B:113; CIF. Atiya, M. S., et al. 1979. *Phys. Rev. Lett.* 43:414
125. Buras, A. J., Gaemers, K. J. F. 1978. *Nucl. Phys.* B132:249
126. Owens, J. F., Reya, E. 1978. *Phys. Rev.* D17:3003. Hwa, R. C., Zahir, S. M. 1981. *Phys. Rev.* D23:2539
127. BFP. Clark, A. R., et al. 1980. *Phys. Rev. Lett.* 45:686
128. Politzer, H. D. 1974. *Phys. Rep.* 14C:129
129. Gross, D., Wilczek, F. 1973. *Phys. Rev.* D8:3633; 1974. *Phys. Rev.* D9:980
130. Buras, A. J. 1980. *Rev. Mod. Phys.* 52:199
131. Buras, A. J. 1981. *Phys. Scr.* 23:863
132. Bardeen, W. A., Buras, A. J., Duke, D. W., Muta, T. 1978. *Phys. Rev.* D18:3998
133. Floratos, E. G., Ross, D. A., Sachrajda, C. T. 1979. *Nucl. Phys.* B152:493
134. Curci, G., Furmanski, W., Petronzio, R. 1980. *Nucl. Phys.* B175:27; Floratos, E. G., Lacaze, R., Kounnas, C. 1981. *Phys. Lett.* 98B:89; Herrod, R. T., Wada, S., Webber, B. R. 1981. *Z. Phys.* C9:351
135. Gonzales-Arroyo, A., Lopez, C., Yndurain, F. J. 1979. *Nucl. Phys.* B153:161; B159:512
136. Gonzales-Arroyo, A., Lopez, C. 1980. *Nucl. Phys.* B166:429
137. Abbott, L. F., Barnett, R. M. 1980. *Ann. Phys.* 125:276
138. Bialas, A., Buras, A. J. 1980. *Phys. Rev.* D21:1825
139. Barnett, R. M. 1982. *Phys. Rev. Lett.* 48:1657
140. Nachtmann, O. 1974. *Nucl. Phys.* B78:455
141. Barbieri, R., Ellis, J., Gaillard, M. K., Ross, G. G. 1976. *Nucl. Phys.* B117:50
142. Gunion, J., Nason, P., Blankenbecler, R. 1982. *Phys. Lett.* 117B:353
143. Abbott, L. F., Berger, E. L., Blankenbecler, R., Kane, G. L. 1979. *Phys. Lett.* 88B:157
144. Curci, G., Greco, M. 1980. *Phys. Lett.* 92B:175; Amati, D., et al. 1980. *Nucl. Phys.* B173:429; Ciafaloni, M. 1980. *Phys. Lett.* 95B:113
145. Pennington, M. R., Ross, G. G., Roberts, R. G. 1982. *Rep. RL-82-033*
146. Fadeev, N. G., Savin, I. A., Sanadze, V. V., Skachkov, N. B. 1982. *Phys. Lett.* 117B:349
147. Anderson, H. L., Matis, H. S., Myrianthopoulos, L. C. 1978. *Phys. Rev. Lett.* 40:1061
148. Bosetti, P. C., et al. 1978. *Nucl. Phys.* B142:1
149. de Groot, J. G. H., et al. 1979. *Phys. Lett.* 82B:292, 456
150. Ball, R. C., et al. 1980. *Rep. MSU-CSL-80*
151. Morfin, J. G., et al. 1981. *Phys. Lett.* 107B:450
152. Bosetti, P. C., et al. 1982. *Nucl. Phys.* B203:362
153. Weinberg, S. 1967. *Phys. Rev. Lett.* 19:1264; Salam, A. 1969. *Elementary Particle Theory*, ed. N. Svartholm. Stockholm: Alunquist & Wikselly; Glashow, S. L. 1961. *Nucl. Phys.*

22:579; Glashow, S. L., Illiopoulos, J., Maiani, L. 1970. *Phys. Rev.* D2:1285
154. Klein, M. 1979. In *Ecole Int. Phys. Particules Élémentaires*, Kupari, Dubrovnik. Strasbourg: Cent. Res. Nucl.
155. Nikolaev, N. N., Shifman, M. A., Shmatikov, M. Zh. 1973. *Pisma Zh. Eksp. Teor. Fiz.* 18:70; Derman, E. 1973. *Phys. Rev.* D7:2755; Berman, S. M., Primack, J. R. 1974. *Phys. Rev.* D9:2171; D10:3895 (erratum)
156. Prescott, C. Y., et al. 1978. *Phys. Lett.* 77B:347; *Phys. Lett.* 84B:524
157. Hughes, V. W. 1979. *High Energy Physics with Polarised Beams and Polarised Targets*, ed. G. H. Thomas, p. 171. Argonne Natl. Lab. (1978)
158. Bollini, D., et al. 1981. *Nuovo Cimento* 63A:441; Doble, N. 1975. *Rep. CERN/ Lab II/EA/75-117* (unpublished)
159. Bushnin, Yu. B., et al. 1976. *Phys. Lett.* 64B:102
160. Alguard, M. J., et al. 1976. *Phys. Rev. Lett.* 37:1258, 1261; 1978. *Phys. Rev. Lett.* 41:70
161. Garwin, E. L., Pierce, D. T., Siegmann, H. C. 1975. *Rep. SLAC-PUB-1576*
162. Bardin, D. Yu, Christova, P. Ch., Fedorenko, O. M. 1982. *Nucl. Phys.* B197:1
163. Davier, M. 1982. See Ref. 84

CHANNELING RADIATION

J. U. Andersen and E. Bonderup

Institute of Physics, University of Aarhus, DK-8000 Aarhus C, Denmark

R. H. Pantell

Electrical Engineering Department, Stanford University, Stanford, California 94305 USA

CONTENTS

1. INTRODUCTION 453
   1.1 Channeling 454
   1.2 Characteristics of Channeling Radiation 458
2. THEORY 462
   2.1 Classical Description 462
   2.2 Quantal Description 468
3. EXPERIMENTS 479
   3.1 Electrons 479
   3.2 Positrons 489
4. APPLICATIONS OF CHANNELING RADIATION 498
   4.1 Channeling Properties 498
   4.2 Crystal Properties 499
   4.3 Radiation Source 501

1. INTRODUCTION

As often happens with new physical phenomena, several independent developments have led to the discovery of channeling radiation (e.g. 1–4). The idea that the oscillatory motion of channeled charged particles should lead to emission of radiation is not surprising, but since the oscillation frequencies ω_0 are low, corresponding to energies $\hbar\omega_0$ of a few eV only, the observation of the radiation seemed difficult (1). A turning point was therefore the realization that relativistic effects will shift the photon energy into the keV or even the MeV region. In a series of publications by Kumakhov et al (5–7), the radiation was treated from both a classical and a quantal point of view; channeling radiation is sometimes referred to as "Kumakhov radiation" (8).

As an introduction, we briefly describe the relevant features of channeling and the main characteristics of channeling radiation. The detailed theoretical discussion in Section 2 is divided into a classical and a quantal description, and the comparison with experimental results for electrons and positrons is the subject of Section 3. The emphasis here is on experiments with MeV-particle beams, reflecting to some extent the research experience of the authors. The applications of channeling radiation are still to be explored in detail, but some of the possibilities are indicated in the last section.

1.1 *Channeling*

Channeling (9) was discovered in the early 1960s in measurements and calculations of ion ranges in crystals, and a comprehensive theoretical description was given by Lindhard in 1965 (10). For incidence nearly parallel to a major crystal axis or plane, the ion trajectories are governed by correlated soft collisions with atoms in such a way that hard collisions are avoided. In this section we review some of the basic concepts and results of the description of channeling.

CONTINUUM POTENTIALS   As shown by Lindhard, a classical picture may be applied in the analysis of how an ion trajectory is governed by correlated collisions; this is illustrated in Figure 1 for a string of atoms with spacing d. Since the deflection angles are very small, they may be evaluated in the impulse approximation from the momentum transfer in a single collision. The momentum transfer is nearly perpendicular to the direction of motion, which, in turn, almost coincides with the z direction,

$$\delta \mathbf{p} \simeq -\frac{1}{v}\int_{-\infty}^{\infty} dz\, \nabla_r V_a(\mathbf{r}, z).\qquad 1.$$

The projectile-atom interaction potential is $V_a(\mathbf{R})$, and the vector $\mathbf{R}$ is

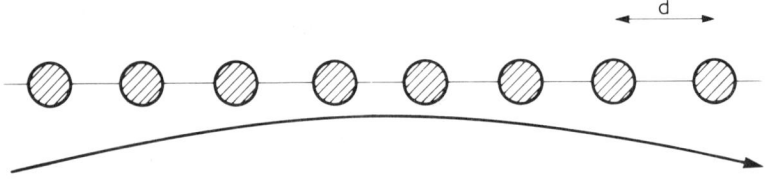

Figure 1   Governing of ion trajectory by correlated collisions with atoms in a string.

decomposed into $\mathbf{R} = (\mathbf{r}, z)$, where $\mathbf{r}$ is the projectile distance from the row when it passes the atom (the impact parameter).

According to Equation 1, the scattering by a single atom is determined by the integrated transverse force, which may be obtained from the continuum string potential,

$$U(\mathbf{r}) = \frac{1}{d}\int_{-\infty}^{\infty} dz\ V_a(\mathbf{r}, z), \qquad 2.$$

corresponding to a charge distribution smeared in the z direction. The gradient of $U(\mathbf{r})$ evaluated at the impact parameter may be replaced by an average of $\nabla U[\mathbf{r}(z)]$ over the distance $\delta z = d$ if $\mathbf{r}(z)$ varies little over this distance; the trajectory is then determined as motion in the two-dimensional potential $U(\mathbf{r})$. Similarly, for small angles of incidence with crystal planes, the motion is in effect governed by a continuum planar potential,

$$V(x) = Nd_p \int_{-\infty}^{\infty} dy\ dz\ V_a(x, y, z), \qquad 3.$$

where x is the coordinate perpendicular to the planes, and Nd_p is the density of atoms in the planes, N being the atomic density of the crystal and d_p the spacing of planes.

In the description of ion channeling, interaction potentials of the Thomas-Fermi type with simple scaling properties are useful (10, 11), but when the projectile charge is small enough to be considered a perturbation, more accurate Hartree-Fock calculations of atomic potentials may be applied (12). The expressions in Equations 2 and 3 give the potentials from a single string or plane, respectively. To include the contribution from other strings or planes, they should be replaced by one- or two-dimensional averages of the crystal potential, which to a first approximation is a sum of atomic potentials. The further correction for thermal vibrations of lattice atoms may in this approximation be included through convolution with the Gaussian distribution of atomic displacements. We denote the resulting continuum potentials by $U_T(\mathbf{r})$ and $V_T(x)$, respectively.

TRANSVERSE ENERGY AND CRITICAL ANGLES In the continuum approximation, there is a complete separation between longitudinal motion with constant momentum and transverse motion governed by an effective Hamiltonian, which in the axial case is given by

$$H_\perp(\mathbf{r}, \mathbf{p}_\perp) = \frac{p_\perp^2}{2M_1} + U_T(\mathbf{r}). \qquad 4.$$

For relativistic particles, the projectile mass M_1 should be interpreted as the relativistic mass (13). The transverse momentum is $\mathbf{p}_\perp = (p_x, p_y)$, and if the angle of motion relative to the axis is φ, its magnitude is $p_\perp = p \sin \varphi \simeq p\varphi$. The transverse kinetic energy may therefore be written as $E_\perp^{kin} = \frac{1}{2}pv\varphi^2$, and the total transverse energy, which is conserved, then becomes

$$E_\perp = \tfrac{1}{2}pv\varphi^2 + U_T(\mathbf{r}). \qquad 5.$$

For a positive particle approaching a string at an angle ψ at large distances, the transverse energy is $E_\perp \simeq \frac{1}{2}pv\psi^2$. Penetration to the center of the string is forbidden if this energy is lower than the continuum potential at $r \simeq 0$. This condition leads to a critical angle for channeling phenomena. Since the continuum-potential barrier is of order $(Z_1 e)(Z_2 e)/d$, where $Z_1 e$ and $Z_2 e$ are the charges of projectile and target atom, respectively, the critical angle is approximately equal to Lindhard's characteristic angle ψ_1, given by (10, 13),

$$\psi_1 = \left(\frac{2Z_1 Z_2 e^2}{\frac{1}{2}pvd}\right)^{1/2}. \qquad 6.$$

For planes, a similar argument applies, and the critical angle is (14) of order

$$\psi_p = \left(\frac{2Z_1 Z_2 e^2}{\frac{1}{2}pvd}\right)^{1/2} \left(\frac{Ca}{d}\right)^{1/2}, \qquad 7.$$

where the distance d is related to the planar spacing d_p through $Nd^2 d_p = 1$. The parameter Ca is a characteristic length in the Thomas-Fermi description of atomic potentials, $Ca \simeq 1.5 Z_2^{-1/3} a_0$, where $a_0 = 0.53$ Å is the Bohr radius.

Conservation of longitudinal and transverse energy is an idealization. The projectile loses energy by inelastic processes, and, owing to thermal fluctuations of atomic scattering and scattering by electrons, the transverse energy gradually changes ("dechanneling"). Such effects need to be considered in a quantitative description.

ELECTRONS AND POSITRONS, QUANTAL DESCRIPTION   Let us now turn to the light particles, electrons and positrons, which are of main interest in connection with radiation. The difference in sign of the charges leads to qualitative differences in the channeling phenomena. For electrons, the potential minima are at the center of atomic strings or planes, and channeled electrons have an increased probability for hard collisions with atoms. This implies that channeling states for electrons are much less stable than for positrons.

In the planar case, the oscillatory motion of channeled electrons and positrons is otherwise rather similar, but the difference in shape of the potential minimum is important for the spectrum of oscillation frequencies. For positrons, the potential is nearly harmonic and the frequency depends only weakly on amplitude, but for electrons this dependence is strong and leads to a broad range of oscillation frequencies.

The difference between positive and negative particles is most dramatic for the axial case. While channeled positrons move about freely between strings in the transverse plane, electrons may be localized to one atomic string. The continuum potential is rotationally symmetric at small distances from a string, and hence angular momentum with respect to the string is conserved for channeled electrons.

While classical pictures of trajectories are applicable in the description of ion channeling, the directional effects for low-energy positrons and electrons may be dominated by quantal interference effects, as is well known from electron microscopy. A quantal description may be obtained (13) by quantization of the transverse Hamiltonian, and the transverse energies and states are solutions to the corresponding stationary wave equation. Owing to the periodicity of the lattice potential, the eigenstates are Bloch waves, and the eigenvalues form a band structure in one or two dimensions (15). The bound states correspond to the lower, very narrow bands and may for electrons be obtained as solutions to the wave equation with the potential from a single string or plane.

A classical description is approached for large quantum numbers, and, owing to the relativistic increase of the effective mass in the Hamiltonian, the number v of bound channeling states increases with projectile energy where $E = \gamma mc^2$; γ is related to the particle velocity by $\gamma = [1-(v/c)^2]^{-1/2}$. Semiclassical estimates of v may be derived from the available volume in transverse phase space, and for planar channeling one obtains (15) for electrons and positrons, respectively,

$$v_p^- \simeq \gamma^{1/2}\left(\frac{4a_0}{d_p}\right)(Nd_p^3)^{1/2} \qquad 8.$$

$$v_p^+ \simeq \gamma^{1/2}Z_2^{1/3}(Nd_p^3)^{1/2}. \qquad 9.$$

For beam energies in the low MeV region, $\gamma \lesssim 10^2$, the number of quantum states is thus fairly small in the planar case, being somewhat higher for positrons, $v_p^+/v_p^- \simeq Z_2^{1/3} \simeq 2\text{--}4$.

For axial channeling of electrons, this type of estimate gives a number of bound states,

$$v_s^- \simeq \gamma\left(\frac{4a_0}{d}\right)Z_2^{1/3}, \qquad 10.$$

which increases more rapidly with γ. Axial-channeling states of positrons are not localized, but the number of states per unit cell, with transverse energy below the barrier for penetration into strings, is large even for small γ.

1.2 Characteristics of Channeling Radiation

Classically, the oscillatory motion of channeled electrons and positrons leads to emission of electromagnetic radiation, as in an undulator (16). Within a quantal picture, channeling radiation is associated with spontaneous transitions between bound eigenstates of the transverse Hamiltonian. We now briefly describe the main characteristics of the radiation: frequency range, linewidth, and intensity and polarization.

PHOTON FREQUENCY   As a first example, we consider a positron oscillating classically with frequency ω_0 between adjacent planes. For energies below the GeV region, the motion is nonrelativistic in the "rest system" R, where the translatory longitudinal motion has been transformed away, and this leads to a particularly simple description. (The superscript R is used throughout the text to indicate quantities measured in the particle's rest frame.) In R, the particle oscillates at frequency $\gamma\omega_0$ owing to time dilatation, and it radiates at this frequency (and higher harmonics) with the angular distribution characteristic of a dipole. A Doppler transformation to the laboratory gives the photon frequency

$$\omega = \frac{\gamma\omega_0}{\gamma(1-\beta\cos\vartheta)} = \frac{\omega_0}{1-\beta\cos\vartheta} \simeq \frac{2\gamma^2\omega_0}{1+\gamma^2\vartheta^2}, \qquad 11.$$

where ϑ is the laboratory angle between the beam and the direction of observation of the radiation.

Also within a quantal picture, the description is particularly simple in R. The situation is analogous to an atom undergoing a spontaneous transition, and the energy of the emitted photon $\hbar\omega^R$ is equal to the change in (transverse) energy $|\Delta E_\perp^R|$. To evaluate this change, we consider the transformation to R of the transverse Hamiltonian given by Equation 4 for the axial case. The kinetic energy is multiplied by γ due to the decrease in mass. The continuum potential is increased by the same factor since the (transverse) momentum transfer in a collision with a single atom is invariant under the transformation, and since the rate of collisions is higher in R by the factor γ owing to the contraction of the distance between scattering centers (Figure 2). We therefore have the simple relation $|\Delta E_\perp^R|$ = $\gamma|\Delta E_\perp|$, and the photon frequency may be obtained from Equation 11 through a replacement of ω_0 by $|\Delta E_\perp|/\hbar$.

The spacing between levels in the transverse potential in the laboratory usually decreases with increasing particle energy, owing to the appearance of the relativistic mass in the Hamiltonian. As an example, for a harmonic potential, the energy $\hbar\omega_0$ is proportional to $\gamma^{-1/2}$, the frequency being proportional to the square root of the ratio of the force constant and the mass. Still, the maximum photon energy, corresponding to emission in the forward direction, $\vartheta = 0$, always increases strongly with γ; for example it increases proportionally to $\gamma^{3/2}$ for the harmonic oscillator. As a result, transitions within a transverse potential of depth of the order of a Rydberg lead to photons in the keV and MeV regions for MeV and GeV projectiles, respectively.

A related type of radiation, also associated with correlated deflections, is coherent bremsstrahlung (17, 18), which is due to particles traversing planes at small angles at regular time intervals. The resulting photon frequencies are typically an order of magnitude higher than for channeling radiation.

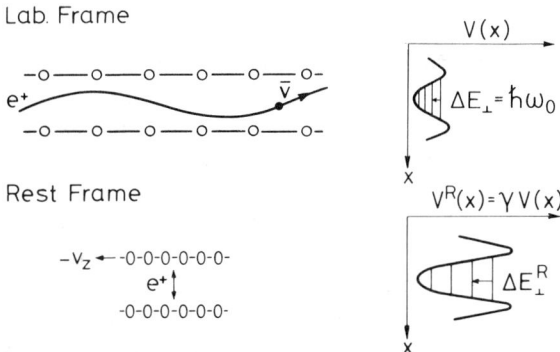

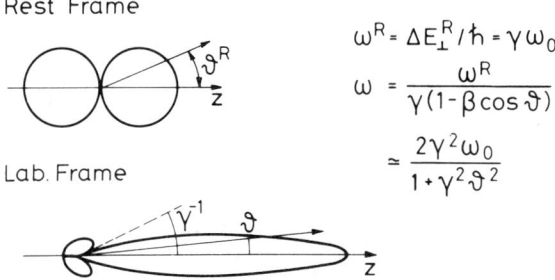

Figure 2 Illustration of Lorentz transformations between the laboratory and the rest frame for a radiating planar-channeled positron with kinetic energy 1 MeV, corresponding to $\gamma = 3$.

LINEWIDTH A number of phenomena contribute to the width of discrete lines from channeling radiation. One factor limiting the length for coherent emission is the crystal thickness. A relativistic particle radiating over a length L generates a full width at half maximum (FWHM), given by (see also Equation 50)

$$\text{FWHM} \simeq 3.5\pi\gamma^2 \frac{\hbar c}{L}. \qquad 12.$$

Another process limiting the coherence length is scattering of the channeled particle due to thermal displacements of crystal atoms, to defects, or, to a lesser extent, to excitation of target electrons. If the probability that a particle continues to radiate coherently decays exponentially with distance, and the e^{-1} coherence length is denoted by l, the corresponding linewidth is

$$\text{FWHM} = 2\gamma^2 \frac{\hbar c}{l}. \qquad 13.$$

For electrons, scattering is often the dominant contribution to line broadening for both planes and axes. A typical value is 15% linewidth at 60-keV photon energy, with $\gamma = 111$ for electron planar channeling. Substitution into Equation 13 gives $l = 0.52$ μm.

As is observed from Equation 11, the angular resolution of the photon detector influences the linewidth owing to the Doppler effect. The broadening is toward lower frequencies only, and therefore also leads to a shift of the line center. To avoid significant broadening, one must collimate the photon beam to angles much smaller than γ^{-1}. This is normally done in experiments with MeV electrons and positrons, but in the GeV region the angles are exceedingly small, and only photon yields integrated over angles have been measured. In this case, even a single frequency ω_0 gives rise to a broad photon spectrum, $0 \leq \omega \lesssim 2\gamma^2\omega_0$.

A Doppler broadening also results if the particles are moving at different angles relative to the direction of a fixed observation point. In axial channeling the transverse motion is constrained, but in the planar case the particles can move in the direction parallel to the planes. This motion can result either from scattering in the crystal or from emittance of the incident beam. Energy spread of the particle beam also causes broadening according to Equation 11, but this contribution to the linewidth is generally appreciably smaller than that from other factors.

For transverse states near or above the top of the potential well, the eigenfunctions are not limited to a single unit cell, and the energy becomes a function of a Bloch vector. If states with differing values of this pseudo-momentum are populated, either due to insufficient collimation of the incident beam or due to scattering, these band-structure effects give rise to a broadening of the photon lines.

A special type of broadening appears for positrons in the planar case, where the potential is nearly harmonic. A slight anharmonicity results in splitting of the radiation into close-lying lines, which fuse into a single line because of the broadening mechanisms discussed previously and because of the limited resolution of the detector. The width of this line is then determined mainly by anharmonicity.

INTENSITY AND POLARIZATION   The enhancement of channeling radiation and coherent bremsstrahlung over normal bremsstrahlung is determined by several factors such as photon collimation, line broadening, and angular spread and energy of the particle beam. Like normal bremsstrahlung, channeling radiation and coherent bremsstrahlung are emitted mainly in the forward direction, within angles $\sim \gamma^{-1}$, but in contrast to incoherent bremsstrahlung, the frequency and the direction of the photons are connected (Equation 11). As a result, the enhancement in the frequency spectrum is increased by photon collimation. A variation of the beam direction leads to changes in the population of the various transverse states, and therefore the intensity of the emitted channeling radiation and coherent bremsstrahlung depends on the angular range of the beam, determined by collimation and by multiple scattering. In contrast to the strong increase with γ of the radiation energy, the enhancement usually increases slowly with beam energy. In the experiments reported in Section 3, enhancements by typically an order of magnitude are found, both for MeV particles, where photons are observed in the forward direction, and for GeV particles, where the photon spectrum is integrated over angles.

The time structure of the radiation intensity is determined by the time dependence of the particle beam. With a linear accelerator as the source of relativistic projectiles, the microstructure of the bunched beam may be ten picoseconds long, and this leads to channeling x rays of picosecond duration.

For projectile energies below the GeV region, the radiation intensity is proportional to the square of the classical dipole moment or of a matrix element of the dipole operator. In the planar case, this vector is perpendicular to the planes, and the radiation is linearly polarized in the plane spanned by the dipole and the photon momentum.

ORIGIN OF DIFFERENT TYPES OF RADIATION   Channeling radiation results from acceleration of a charged particle in collisions with atoms in a crystal. From this perspective, channeling radiation has the same origin as ordinary bremsstrahlung and coherent bremsstrahlung. Ordinary bremsstrahlung is the radiation that occurs when a charged particle is accelerated by a single atom. In a quantal picture, the momentum of the atomic recoil is used to conserve energy and momentum between the initial and final states. Since the atomic momentum is continuous, the bremsstrahlung spectrum is also continuous. The electron (or positron) is nearly free, and thus its initial and final energy-momentum surfaces are approximately as for a free particle.

In coherent bremsstrahlung, the electron (or positron) follows a path, which brings it into periodic contact with the atoms in the crystal. Classically, the radiation frequency is given by this periodicity, but in a quantal picture it is determined by energy-momentum balance. The recoil of the crystal lattice is not continuous but restricted to reciprocal lattice vectors, and the energy of the emitted photon is fixed in the electron frame when the angle between the velocity and a reciprocal lattice vector is specified. Once again, the energy-momentum surface is approximately that of a free particle.

When a particle is trapped within a potential well, as in channeling, a multiplicity of energy-momentum surfaces are generated that differ from the surface for a free particle. For the lowest-lying states, the energy is independent of transverse quasi-momentum and, in the rest frame, the photon energy from a transition between such states is therefore independent of the direction of particle incidence.

In terms of the photon spectrum observed in a well-defined direction in the laboratory, the progression from normal bremsstrahlung to coherent bremsstrahlung, to channeling radiation, starts as a continuum, is then restricted by the range of angles between the particle motion and a reciprocal lattice vector, and finally becomes a set of fixed values.

2. THEORY

2.1 *Classical Description*

In this section, we outline the description of channeling radiation obtained within classical electrodynamics (7). As discussed in the Introduction, a classical picture should be useful for ultrarelativistic electrons and positrons. Channeling is due to correlated soft collisions with atoms, and channeling radiation results from constructive interference of the electromagnetic radiation from these deflections. The power spectrum may be evaluated from the trajectory $\mathbf{R}(t)$, determined from the continuum potentials. We concentrate here on planar-channeled positrons, which oscillate in a nearly harmonic potential between adjacent planes. Other types of channeling motion are not characterized by a single frequency, and the classical radiation spectrum is in general broad and structureless.

For projectiles also, which are not channeled but traverse planes at a small angle, the scattering by atoms in a plane is correlated, and the trajectory is determined by the continuum potential. The characteristic frequency now corresponds to the periodic crossing of planes. The associated radiation, known as coherent bremsstrahlung, has been studied for many years but has usually been treated by quantal perturbation theory (17). Within the present framework, channeling radiation and coherent bremsstrahlung appear as very similar phenomena, originating in the correlated scattering of particles, which are either bound or free in their transverse motion (19). In addition to these types of radiation, the spectrum contains incoherent bremsstrahlung, which is associated with uncorrelated deflections in collisions with impact parameters of the order of or smaller than the vibrational amplitude of atoms. The intensity of this component is modified by channeling, normally reduced for positrons and enhanced for electrons.

EQUATIONS OF MOTION   In the planar case, the trajectory may be decomposed into a uniform translation and a term periodic in time. For particles bound to a single plane (electrons) or to a pair of planes (positrons), the transverse motion $x(t)$ is purely oscillatory, whereas the x motion for projectiles with sufficient transverse energy to traverse planes also contains a translatory component. In both cases, the transverse motion is determined from the continuum potential $V(x)$ through the equation

$$\frac{d}{dt}(\gamma m \dot{x}) = -\frac{dV}{dx}[x(t)]. \qquad 14.$$

(For brevity, the label T on the potential is omitted.) Combining with the relation expressing energy conservation,

$$\frac{d}{dt}(\gamma m c^2) = -\frac{d}{dt} V[x(t)], \qquad 15.$$

we obtain

$$\gamma(t) m \ddot{x}(t) = -[1-(\dot{x}/c)^2]\, dV/dx. \qquad 16.$$

When γ is approximated by a constant, and the quantity $(\dot{x}/c)^2$ is neglected, we retrieve the conservation law discussed in the Introduction,

$$\tfrac{1}{2}\gamma m (\dot{x})^2 + V[x(t)] = E_\perp. \qquad 17.$$

The neglected terms are of relative order $E_\perp/\gamma m c^2$.

Although always totally insignificant for the transverse motion, the time dependence of γ, given by Equation 15, is responsible for the deviation of the longitudinal motion from a uniform translation. As the particle energy is increased into the GeV region, this deviation begins to be significant for the radiation.

RADIATION SPECTRUM For a particle with charge $\pm e$ moving along a trajectory $\mathbf{R}(t)$ with velocity $c\boldsymbol{\beta}(t)$, the energy radiated per unit frequency and unit solid angle is given (20) by

$$\frac{d^2 I}{d\omega\, d\Omega} = \frac{e^2 \omega^2}{4\pi^2 c} \left| \mathbf{n} \times \int dt\, \boldsymbol{\beta}(t) \exp\{i\omega[t - \mathbf{n} \cdot \mathbf{R}(t)/c]\} \right|^2. \qquad 18.$$

Here, the unit vector $\mathbf{n}$ indicates the direction of observation with polar angles (ϑ, φ) and differential solid angle $d\Omega$.

For a channeled positron moving in the z direction with average velocity $c\langle \beta_z \rangle$ and oscillating between planes with frequency $\omega_0(E_\perp)$, Equation 18 leads to the following relation between frequency and direction of the radiation, in the limit of many oscillations during the time T spent in the crystal,

$$\omega(1 - \cos\vartheta \langle \beta_z \rangle) = n\omega_0(E_\perp), \qquad n = 1, 2, 3, \ldots. \qquad 19.$$

When multiplied by $\gamma_z \equiv (1 - \langle \beta_z \rangle^2)^{-1/2}$, this formula may be recognized as the Doppler relation transforming frequencies between the laboratory and a frame moving with velocity $c\langle \beta_z \rangle$ in the z direction.

In this "average rest frame" R, the system can radiate only with frequencies that are integral multiples of the characteristic orbital frequency $\gamma_z \omega_0(E_\perp)$. In the relativistic limit $\gamma_z \gg 1$ considered here, the (transverse) electric and magnetic fields in R are of equal strength. Seen from this frame of reference, there is a longitudinal oscillation caused by the magnetic field, with an amplitude smaller than that of the transverse oscillation by a factor of order of the transverse velocity in units of c, β_x^R. When the motion in R is nonrelativistic, we may therefore in the discussion of radiation replace the actual "figure-eight" orbit by a one-dimensional oscillation in the x direction. Furthermore, it is convenient in this case to evaluate the radiation spectrum in R and then transform it to the laboratory. The reason is that, in R, the dipole approximation may be applied in Equation 18, i.e. the part of the exponential containing the trajectory may be approximated by unity, the exponent $\gamma_z \omega_0 x/c$ being of order β_x^R. Since the planar transverse motion of a positron is not too far from being harmonic, we expect a strong predominance of the lowest frequency in the radiation spectrum ($n = 1$) until the motion in R becomes relativistic at laboratory energies well up in the GeV region. Note that the lowest harmonic of the longitudinal oscillation corresponds to $n = 2$ in Equation 19.

As an indication of the results to be expected for not too high projectile energies, we therefore present the spectrum obtained in the dipole approximation with a harmonic potential (2, 7),

$$dI/d(\omega/\omega_m) = 3TI_{ch}\omega/\omega_m[1-2\omega/\omega_m+2(\omega/\omega_m)^2], \qquad 20.$$

$$\omega \leq \omega_m \simeq 2\gamma^2\omega_0,$$

where

$$I_{ch} = x_m^2 e^2 \omega_0^4 \gamma^4/3c^3. \qquad 21.$$

The maximum frequency ω_m corresponds to emission in the forward direction and is obtained from Equation 19; I_{ch} denotes the total power in channeling radiation, proportional to the square of the particle transverse amplitude x_m. The corresponding power for normal bremsstrahlung within the same frequency range $0 \leq \omega \leq \omega_m$ is given (18, 20, 21) by

$$I_b \simeq \frac{16}{3}\alpha^4 Z_2^2 \log(199Z_2^{-1/3})e^2 a_0^2 N\omega_m, \qquad 22.$$

where α denotes the fine-structure constant. The average enhancement factor I_{ch}/I_b is approximately proportional to $\gamma^{1/2}Z_2^{-2/3}$. For a 5-GeV positron channeled along a major plane in silicon, the enhancement is of the order of 20 and higher by a factor of 3 at the maximum frequency.

Let us briefly mention that the spin magnetic moment of the projectile does not lead to any significant modification of our discussion. In R, the moment is equal to $e\hbar/mc$, and the resulting relative change in the radiation intensity is of order of the square of the fine-structure constant.

The spectral form of the coherent bremsstrahlung from positrons traversing planes is also given by Equation 20 when the motion is nonrelativistic in the average rest frame, which moves at a small angle to the planes. Only the values of the frequency ω_0 and the amplitude x_m are changed. The frequency is given in terms of the average velocity perpendicular to planes,

$$\omega_0 = 2\pi c \langle \beta_x \rangle/d_p. \qquad 23.$$

When the angle ψ to the planes is large compared to the characteristic angle ψ_p for planar channeling (Equation 7), the transverse energy is well above the planar barrier, and the potential may be considered a perturbation. The frequency ω_0 is then linear in the angle ψ and much larger than the oscillation frequencies for bound states, and the amplitude $x_m^{(1)}$ of the first harmonic of the periodic perturbation $x(t)$ is given by the relation

$$(\gamma m)\omega_0^2 x_m^{(1)} = 2V_1 2\pi/d_p. \qquad 24.$$

Here V_1 is the coefficient corresponding to wavelength d_p in a Fourier expansion on complex exponentials of the continuum potential, assumed to be symmetric. The total intensity I_{cb} of the radiation, as determined from a combination of Equations 21 and 24, may be compared with the corresponding quantity for channeling radiation from a particle with transverse energy equal to the planar barrier, and for a harmonic potential we obtain $I_{cb}/I_{ch} = 4/\pi^2$.

MAGIC ENERGY   We now turn to the general case, where the transverse motion in the frame R may be relativistic. In the Doppler expression (Equation 19) connecting R to the laboratory, the small quantity $1-\langle \beta_z \rangle$ appears in a critical manner. To evaluate this average value, we apply the exact relation $\gamma^2(t) = [1-\beta_z^2(t)]^{-1}\{1+[p_x(t)/mc]^2\}$, where $p_x(t)$ denotes the transverse momentum. In this equation, the time variation of γ^2 is negligible, and we obtain

$$1-\langle \beta_z \rangle \simeq \frac{1}{2\gamma^2}[1+\langle p_x^2 \rangle/(mc)^2]. \qquad 25.$$

With this expression inserted into Equation 19 we have for photon emission around the forward direction

$$\omega = n\frac{2\gamma^2 \omega_0(E_\perp)}{1+2\gamma\langle E_\perp^{kin}\rangle/(mc^2)+(\gamma\vartheta)^2}, \qquad \vartheta \ll 1, \qquad 26.$$

where we have introduced the average transverse kinetic energy in the laboratory. The correction for transverse motion may be interpreted as an average Doppler correction, analogous to the third term in the denominator, since it may be written as $\langle (\gamma\beta_x)^2 \rangle$, and $\beta_x(t)$ is the angle of motion to the plane. Once again we observe that the deviation of the z motion from a uniform translation with velocity βc becomes significant for the radiation when the motion in the rest system with energy $\gamma E_\perp$ becomes relativistic. In addition, the formula shows that the high-energy part of the spectrum is confined to angles $\vartheta \lesssim \gamma^{-1}$. For GeV particles, these angles are so small that, in the channeling measurements performed so far, only photon yields integrated over solid angle have been obtained.

Even with a well-collimated particle beam, a channeling experiment includes projectiles with different transverse energies owing to the (uniform) distribution in the coordinate x at the front surface of the crystal. For the positrons contributing significantly to the radiation, the anharmonicity of the planar potential leads to an oscillation frequency ω_0, which is an increasing function of transverse energy. As is observed from Equation 26, the resulting increase in radiation frequency with increasing transverse energy competes with the decrease originating in the coupling between longitudinal and transverse motion. In a narrow region of γ values, the two effects nearly cancel, and it so happens that this occurs at a projectile energy of a few GeV, which is low enough that the radiation spectrum is dominated by the peak from the first harmonic. Close to a value "γ magic," one therefore observes a particularly abrupt decrease in the intensity at the high-energy side of the strong peak in $dI/d\omega$ (23).

The coherent radiation from the projectiles with free transverse motion, also contained in the beam, does not change the peak structure of the observed spectrum. The intensities of radiation from free and bound particles are similar, but the spectrum resulting from a group of free particles with a distribution in transverse energy is much broader owing to the strong variation of the oscillation frequency in Equation 23 with transverse energy.

MULTIPLE SCATTERING, LINE BROADENING, AND LINESHIFT As indicated in Section 1.2, multiple scattering leads to line broadening and a shift in average frequency in the geometry with forward photon collimation. Even in experiments with integration over angles of emission, there turns out to be an effect, which is particularly important close to the magic energy, where the frequency variations due to anharmonicity and to transverse motion nearly cancel. As discussed below, scattering parallel to the plane reduces the sharpness of the cut-off, owing to a limitation of the coherence length for emission, and it also leads to a small reduction of the cut-off frequency ω_m.

An upper limit to the coherence length is always set by the target thickness L, and the associated broadening may be expressed approximately (cf Equation 12) as

$$\text{FWHM}/\hbar\omega_m \simeq \lambda_0/L, \qquad 27.$$

where λ_0 is the wavelength of the particle oscillation. At GeV energies, this broadening can be made small since the distance, over which dechanneling may be neglected, is approximately proportional to γ (Equations 7, 28, and 29), whereas λ_0 increases as $\gamma^{1/2}$ only. However, since the incoherence due to multiple scattering sets in rather abruptly, Equation 27 may still be used to estimate the broadening, with L replaced by an effective coherence length l_c.

To estimate l_c, we consider the radiation formula (Equation 18) for emission in the average direction of motion (z direction). Because of scattering in the y direction, the z velocity is fluctuating, $\delta\beta_z \simeq (\cos\beta_y - 1)\beta_z \simeq -\frac{1}{2}\beta_y^2$. This introduces fluctuations in the phase of the exponential. Incoherence sets in when these fluctuations approach unity, and this leads to a coherence length determined by $l_c \omega \delta\beta_z/c \approx 1$. The magnitude of the fluctuations in β_y^2 is similar to the average value, which increases linearly with depth,

$$\langle \beta_y^2 \rangle \simeq 2Dz. \qquad 28.$$

For particles with large oscillation amplitudes, which yield most of the radiation, the scattering parallel to the plane is nearly as in a random medium, and the diffusion constant D is given by

$$D \simeq \frac{2\pi N Z_2^2 e^4}{(\gamma mc^2)^2} \log, \qquad 29.$$

where the logarithmic factor log is usually about 3 (24).

Apart from an uncertain numerical factor of the order of unity, these considerations lead to the estimate

$$l_c \simeq \left(\frac{\lambda_m}{2D}\right)^{1/2}, \qquad 30.$$

where λ_m is the radiation wavelength, $\lambda_m = 2\pi c/\omega_m$. The numerical factor here is chosen to reproduce the result of detailed calculations based on a diffusion equation for β_y (J. Golovchenko and O. Pedersen, private communication). The broadening obtained from Equations 30 and 27 is given by

$$\text{FWHM}/\hbar\omega_m \simeq 2\gamma^2 (2\lambda_m D)^{1/2}, \qquad 31.$$

and the shift analogous to the correction for x motion in Equation 26 becomes $\delta(\hbar\omega_m) \simeq -\text{FWHM}/4$ when $\langle \beta_y^2 \rangle$ is evaluated for $z = l_c/2$. The reason for the shift is partly that the average direction of motion of the individual particle deviates from the z direction, partly that the particle velocity fluctuates around this average direction. Only the latter effect remains after an average over photon directions, and in experiments without photon collimation the shift of ω_m is smaller than given above, by a factor estimated to be about two. The relative broadening and shift increase slowly with energy, approximately as $\gamma^{1/4}$.

The development of a detailed description of these effects is still at a preliminary stage, but we have included a qualitative discussion because they turn out to be important for the understanding of discrepancies between theory and experiments for GeV positrons, as discussed in Section 3.2. In addition, the sharpness of the cut-off at ω_m is very important for potential applications of the radiation.

2.2 Quantal Description

Within a quantal picture, the behavior of spin-$\frac{1}{2}$ particles such as electrons and positrons is described by the Dirac equation. However, as mentioned in the previous section and shown explicitly by Kumakhov and others (6, 25, 26), spin effects are not important for the transverse motion of channeled particles and the associated radiation. We therefore apply the simpler Klein-Gordon equation, describing relativistic spinless particles, to introduce the basic separation into longitudinal and transverse motion (13, 15).

TRANSVERSE WAVE EQUATION The stationary Klein-Gordon equation expresses the relativistic relation between momentum and energy for a particle of charge e and rest mass m,

$$\left\{c^2\left[-i\hbar\nabla - \frac{e}{c}\mathbf{A}(\mathbf{R})\right]^2 + m^2c^4\right\}\psi(\mathbf{R},\ldots)$$
$$= [E - H_r - H_l - V(\mathbf{R},\ldots)]^2 \psi(\mathbf{R},\ldots). \qquad 32.$$

Here E is the energy of the entire system, and H_r and H_l denote the Hamiltonians of the radiation field and the lattice, which interact with the projectile through the potentials $\bar{A}$ and V. Only the coordinates $\mathbf{R}$ of the particle are shown explicitly. One may obtain the Klein-Gordon equation from the Dirac equation by first transforming this equation into an eigenvalue problem for the square of the energy and then neglecting the (small) spin terms (27). In contrast to these terms, the coupling to lattice degrees of freedom is important since the lines of channeling radiation are broadened by thermal and electronic scattering, which limits the coherence length for emission of radiation.

From the wave function ψ we separate out a factor $\exp(iKz)$, corresponding to uniform motion in the channeling direction with energy E,

$$\psi(\mathbf{r}, z, \ldots) = \exp(iKz) w(\mathbf{r}, z, \ldots), \qquad \mathbf{R} = (\mathbf{r}, z)$$

with

$$E^2 = (\hbar K)^2 c^2 + m^2 c^4. \qquad 33.$$

In the equation for w, obtained through insertion of Equation 33 into Equation 32, we neglect terms $\partial^2/\partial z^2$ and $(V + H_r + H_l)^2$, where the initial energy for the lattice and the radiation field is taken to be zero. This approximation is based on the assumption of scattering through small angles only (13) and amounts to omission of terms of relative magnitude $(k_\perp/K)^2$, where $\hbar k_\perp$ is a typical transverse momentum. In addition, excitation energies for the lattice and the radiation field are assumed to be very small. In the Coulomb gauge, $\text{div}\,\mathbf{A} = 0$, we then obtain after division by $2E = 2\gamma mc^2$

$$i\hbar v \frac{\partial}{\partial z} w = \left[-\frac{\hbar^2}{2\gamma m}\Delta_\mathbf{r} + V(\mathbf{r}, z, \ldots) + H_r + H_l + H_{e,r}\right] w$$
$$H_{e,r} = -e\left(\beta A_z + \frac{1}{\gamma mc}\mathbf{A}\cdot\mathbf{p}\right), \qquad 34.$$

where $v = \beta c$ is given by the relation $\hbar K = \gamma mv$. As usual, contributions of order A^2 have been neglected in the term $H_{e,r}$. This equation has the form of a time-dependent Schrödinger equation.

The coupling of the particle to the degrees of freedom of the radiation field and of the lattice may be treated as perturbations, and to zeroth order we neglect the term $H_{e,r}$ and replace the potential by the thermally averaged lattice potential $V_T(\mathbf{r}, z)$, which is the expectation value of $V(\mathbf{r}, z, \ldots)$ in the ground state for the target electrons and in the lattice vibrational states weighted by Boltzmann factors.

The final simplification is obtained when $V_T(\mathbf{R})$ is replaced by the continuum approximation. The solutions of Equation 34 are then product functions,

$$w = u(\mathbf{r}) |\text{lattice}\rangle |\text{radiation}\rangle \exp(-i\varepsilon z/\hbar v), \qquad 35.$$

where the energy ε is the sum of contributions from the lattice, from the radiation field, and from the transverse motion of the projectile described by a "stationary" wave equation,

$$\left[-\frac{\hbar^2}{2\gamma m}\Delta_\mathbf{r} + U_T(\mathbf{r}) \right] u(\mathbf{r}) = E_\perp u(\mathbf{r}). \qquad 36.$$

In the planar case, the continuum potential V_T depends on one coordinate only.

EMISSION OF RADIATION We now consider radiative transitions between the eigenstates determined by Equation 36. The transition probabilities may be obtained from a standard perturbation treatment of the coupling term $H_{e,r}$ in Equation 34 and at first we neglect the lattice degrees of freedom. If for $z = 0$, the transverse state of the particle is $u_i(\mathbf{r})$, and the radiation field is in its ground state $|0\rangle$, the amplitude for finding the particle at depth $z = L$ in a state $u_f(\mathbf{r})$ together with a photon of momentum $\hbar\kappa$ and energy $\hbar\omega$ is given by

$$\langle u_f, \kappa | w(z = L, \ldots) \rangle \simeq \frac{1}{i\hbar v} \int_0^L dz \left\langle u_f(\mathbf{r}), \kappa \right|$$

$$-e\left(\beta A_z + \frac{1}{\gamma mc}\mathbf{A}\cdot\mathbf{p}\right)\left|u_i(\mathbf{r}), 0\right\rangle \qquad 37.$$

$$\times \exp[i(E_{\perp f} + \hbar\omega - E_{\perp i})z/\hbar v].$$

Consider for simplicity emission in the z direction. The photon-polarization vector $\boldsymbol{\varepsilon}_A$ is then in the xy plane, and the relevant matrix element of the radiation field, quantized within a volume L_0^3, is given (21) by

$$\langle \kappa | \mathbf{A} | 0 \rangle = \boldsymbol{\varepsilon}_A c \sqrt{2\pi/L_0^3} \sqrt{\hbar/\omega} \, \exp(-i\kappa z). \qquad 38.$$

With the connection $\kappa = n_r \omega/c$ between photon momentum and frequency, where n_r is the index of refraction, the integral in Equation 37 approaches a delta function for large L, which determines the photon energy:

$$\hbar\omega = \frac{(E_{\perp i}-E_{\perp f})}{(1-n_r\beta)} \simeq (1+\beta)\gamma^2 \frac{E_{\perp i}-E_{\perp f}}{1+2\gamma^2\delta}. \qquad 39.$$

The correction for refraction is important only for low beam energies since $\delta \equiv (1-n_r) \propto \omega^{-2}$ for not too low frequencies ω (28). With $\delta = 0$, Equation 39 corresponds to Equation 19 in the limit of a time-independent value of $(1-\beta)$ considered here. This equation also contains the generalization to other photon directions.

To obtain the number of photons dN_{ph} emitted within a solid angle $d\Omega$ around the forward direction, we multiply the square of the expression in Equation 37 by a photon phase-space factor $L_0^3\omega^2\,d\omega\,d\Omega/(2\pi c)^3$ and integrate over frequency. The result is

$$dN_{ph} = 2(1+\beta^{-1})\alpha\omega(L/c)(mc)^{-2}|\langle u_f|\mathbf{p}_\perp \cdot \boldsymbol{\varepsilon}_A|u_i\rangle|^2 \frac{d\Omega}{4\pi}. \qquad 40.$$

Here α denotes the fine-structure constant. Discussions of the polarization, and also of the generalization of the intensity formula to other angles of emission, may be found in the literature (6, 25, 26, 28).

AXIAL-CHANNELING RADIATION We now discuss the solution of the transverse wave equation (Equation 36). Methods known from atomic, molecular, and solid-state physics may be applied, and the analysis is much simplified by the fact that here we are dealing with a genuine one-particle problem, i.e. we do not have the complication of exchange and correlation. The coupling to lattice degrees of freedom can be treated as a perturbation (see below).

For electrons, the axial-channeling states of low transverse energy are localized in the vicinity of atomic strings and may therefore be determined with the potential from a single string. This isolated-string approximation is analogous to the description of a two-dimensional atom (28–30). The rotational symmetry of the potential leads to eigenstates, which in polar coordinates (r, φ) may be expressed as

$$u(\mathbf{r}) = R_l(r)\frac{1}{\sqrt{2\pi}}\exp(\pm il\varphi), \qquad l = 0, 1, 2, \ldots, \qquad 41.$$

where the radial wave function is obtained as a solution of the eigenvalue problem

$$\left\{-\left(\frac{d^2}{dr^2}+\frac{1}{r}\frac{d}{dr}\right)+\frac{l^2}{r^2}+\frac{2m}{\hbar^2}[\gamma U_T(r)-\gamma E_\perp]\right\}R_l(r) = 0. \qquad 42.$$

As seen from Equation 41, the energy levels are doubly degenerate for $l \neq 0$.

The selection rule for the matrix element in Equation 40, which is proportional to a dipole matrix element, is $\Delta l = \pm 1$, as for atomic transitions.

In Equation 42, the factor γ has been combined with the potential and the transverse energy. This may be interpreted as a transformation to the rest system for the case of nonrelativistic transverse motion considered here ($\gamma_z \simeq \gamma$), and the eigenvalue $\gamma E_\perp$ is the transverse energy in the rest system. The potential γU_T in Equation 42 depends on projectile energy and atomic spacing only through the ratio γ/d (cf Equation 2) and this leads to a scaling rule for radiation spectra obtained for electron incidence along different axes in a given crystal (29).

An example of a single-string potential $\gamma U_T(r)$ is shown in Figure 3 at the left-hand side, together with the corresponding energy levels. The potential is derived from atomic scattering factors, parameterized by Doyle & Turner

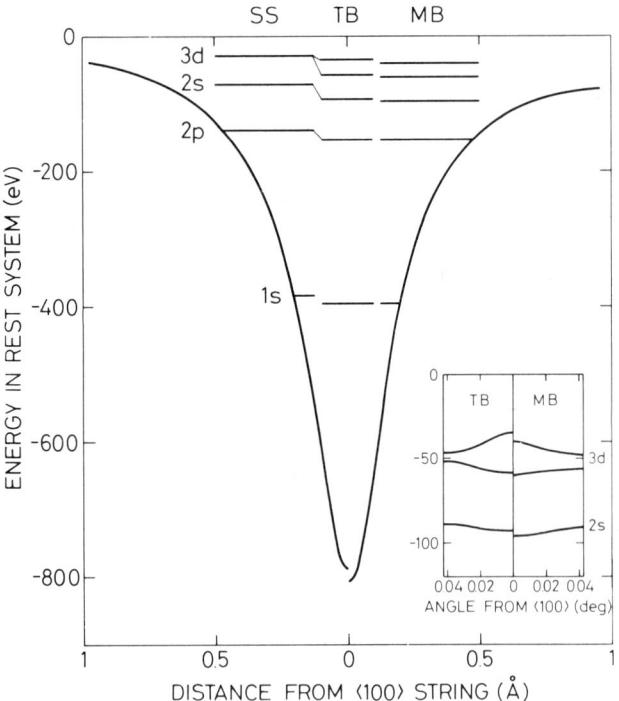

Figure 3 Continuum potentials $\gamma U_T(\mathbf{r})$ in a rest system for 4-MeV electrons channeled along the ⟨100⟩ direction in Si: (*Left*) the single-string potential, (*Right*) the average crystal potential plotted in the direction toward a nearest-neighbor string. Transverse-energy levels are from single-string (SS), tight-binding (TB), and many-beam (MB) calculations. The insert shows band structures corresponding to a tilt in the direction of a nearest-neighbor string. (From 28.)

(12), and it is given by

$$U_T(r) = -\frac{e^2}{a_0} \frac{2a_0^2}{d} \sum_{i=1}^{4} \frac{a_i}{B_i + \rho_2^2} \exp[-r^2/(B_i + \rho_2^2)], \qquad 43.$$

where a_i and $b_i = (2\pi)^2 B_i$ are the coefficients tabulated by Doyle & Turner, and ρ_2^2 is the two-dimensional mean square thermal displacement. The Doyle-Turner atomic potentials are based on relativistic Hartree-Fock calculations, but since the coefficients have been adjusted to give accurate Fourier transforms only for reciprocal distances less than 8π Å^{-1}, they are not valid at small distances. Corrections are necessary when the vibrational amplitude is small, $\rho_2 \lesssim (8\pi)^{-1}$ Å, and also for the evaluation of incoherent scattering (Equation 52).

At the right-hand side of the figure is shown the potential obtained when the contribution from neighboring strings is included. The energy levels determined from an expansion in plane waves (many-beam calculations, discussed below) are compared to the results from a tight-binding approximation, where the difference between the two potentials is treated as a perturbation. As shown in the insert, this perturbation splits the levels and for the upper levels the finite overlap between states in neighboring-string potentials leads to a dependence on the transverse wave vector $\mathbf{k}_\perp$, related to the incidence angle as in Equation 49 below.

PLANAR-CHANNELING RADIATION AND COHERENT BREMSSTRAHLUNG   In the planar case, the transverse states are obtained from the one-dimensional Schrödinger equation,

$$\left\{-\frac{\hbar^2}{2m}\frac{d^2}{dx^2} + [\gamma V_T(x) - \gamma E_\perp]\right\} u(x) = 0. \qquad 44.$$

Since the planar potential is periodic with period d_p, it may be expanded in a Fourier series, and the eigenfunctions become Bloch waves (15, 22),

$$V_T(x) = \sum_j V_j^T \exp(ijgx), \qquad j = 0, \pm 1, \ldots$$
$$u_{n,k_x}(x) = \exp(ik_x x)\sum_j C_j^{(n)}(k_x) \exp(ijgx), \qquad g = 2\pi/d_p. \qquad 45.$$

For a fixed value of k_x chosen, for example, within the first Brillouin zone, $|k_x| \leq g/2$, the band index n distinguishes the solutions of Equation 44, which assumes the matrix form (many-beam equation),

$$\left[\frac{\hbar^2}{2m}(k_x + jg)^2 - \gamma E_\perp^{(n)}(k_x)\right]C_j^{(n)} + \sum_l \gamma V_{j-l}^T C_l^{(n)} = 0. \qquad 46.$$

For a cubic lattice, the Doyle-Turner expression for the Fourier component

V_j^T for electrons is given by

$$V_j^T = -2\pi N a_0^2 \frac{e^2}{a_0} \sum_{i=1}^{4} a_i \exp\left[-\frac{1}{4}(B_i + 2\rho_1^2)(jg)^2\right], \qquad 47.$$

where ρ_1^2 is the one-dimensional mean square vibrational amplitude. This corresponds to a single-plane potential,

$$V_T(x) = -2\sqrt{\pi} N d_p a_0^2 \frac{e^2}{a_0} \sum_{i=1}^{4} \frac{a_i}{(B_i + 2\rho_1^2)^{1/2}} \exp[-x^2/(B_i + 2\rho_1^2)]. \qquad 48.$$

Figure 4 shows this potential, corrected for the contribution from neighboring planes, for a {111} plane in nickel. There is a significant temperature dependence of the energy levels, in particular for the ground state $n = 0$. Also shown are the squares of the wave functions for the two lowest-lying states, with even and odd parity, respectively. The alternating parity leads to the selection rule Δn odd for dipole transitions, and the matrix element in Equation 40 strongly favors transitions with $\Delta n = 1$. Illustrations of potentials and energy levels for positrons are shown in Section 3.2.

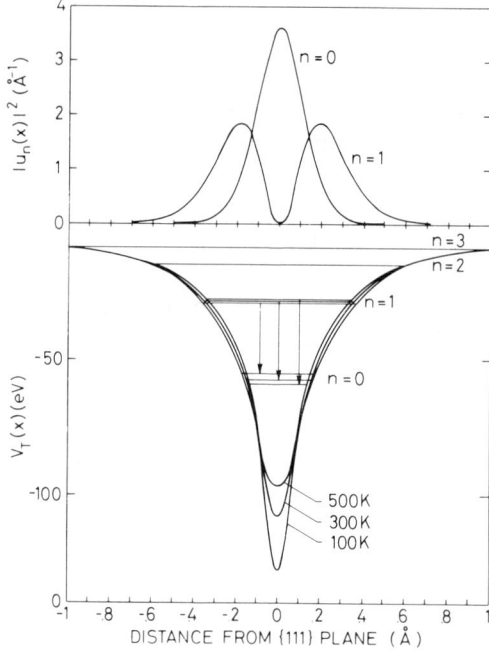

Figure 4 The {111}-planar potential in Ni, with energy levels of bound states for three crystal temperatures, and the density distribution across the plane for the two lowest levels at room temperature. (From 31.)

An example of a complete band-structure calculation is given in Figure 5 in the rest frame for 4-MeV electrons channeled along various planes in silicon. A beam incident at an angle ϑ to a plane populates only transverse states with a reduced value of the Bloch vector k_x given by

$$\gamma\beta mc\vartheta = \hbar(\pm k_x + jg), \qquad 49.$$

where j is an integer, and this relation has been used to relate k_x to the angle given as the abscissa in Figure 5 and in the insert in Figure 3. The transverse energy depends only weakly on angle for the low-lying bands; in the opposite limit of high transverse energies, the states are nearly plane waves, and the dispersion relation approaches that for a free particle. In this limit, the selection rule for dipole transitions is Δn even, $\Delta n = 2$ being the strongest, because neighboring levels correspond to opposite directions of transverse momentum in an extended-zone picture. For states with energies

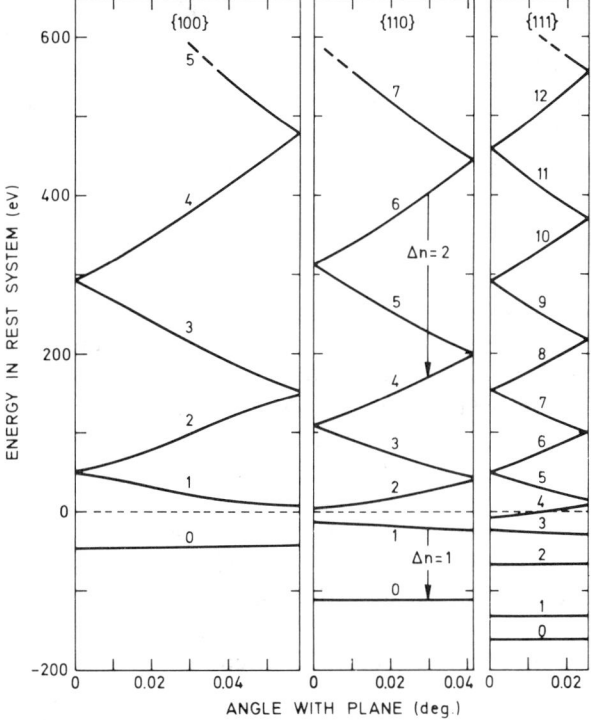

Figure 5 Energy bands for 4-MeV electrons moving along three different planes in Si, obtained from solution of Equation 46 with the Doyle-Turner potential given by Equation 47. The abscissa is the angle ϑ related to the transverse wave vector k_x through Equation 49. (From 22.)

close to the potential barrier between planes, there is no strong selection rule, and also radiative transitions to bound states occur.

The Δn even transitions between high-lying levels correspond to the well-known phenomenon of coherent bremsstrahlung. The transverse wave functions are plane waves, only slightly perturbed by the planar potential. Using instead plane waves as basis states, one may associate the radiation with the process of momentum transfer $\hbar(\Delta n/2)g$ from the lattice to the projectile. Apart from a small term independent of ϑ, the frequency calculated from Equation 39 is identical to the classical value given by Equation 23. The relation between the two pictures of coherent bremsstrahlung, as spontaneous transitions between eigenstates or as transitions between plane waves induced by Fourier components of the crystal potential, was discussed in detail in (22).

SINGLE SCATTERING AND LINEWIDTH As mentioned in the Introduction, a number of factors may contribute to the linewidth. We discuss in some detail here the limitation of the coherence length due to single scattering, which is usually the main source of line broadening in spectra from MeV electrons (31). The decay of states by radiation is negligible and the line broadening is analogous to collisional broadening of atomic transitions.

At first neglecting scattering events, we obtain a photon yield proportional to the square of the expression in Equation 37:

$$I(\omega) \propto \left(\frac{\sin x}{x}\right)^2, \quad x = \Delta\omega(1-n_r\beta)L/2v, \qquad 50.$$

where $\Delta\omega$ denotes the deviation from the line center given by Equation 39. For $\gamma \gg 1$ and $n_r = 1$, the width is then given by Equation 12. Single scattering leads to an exponential decay of the radiation matrix element, and, when the crystal thickness is much larger than the mean free path l for scattering leading to incoherence, $L \gg 2\pi l$, then the photon spectrum obtained from Equation 37 is a Lorentzian with the width given in Equation 13.

Classically, a scattering event results in incoherence when the change in frequency of the radiation exceeds the width of the line. We may distinguish interband from intraband scattering according to whether the state of motion perpendicular to the axis or plane (or more precisely, the band index) is changed or not. For MeV electrons, different transitions are usually sufficiently separated in energy that interband scattering always leads to incoherence. For intraband scattering, the situation is more complicated. The frequency change following an energy loss $\delta(\gamma mc^2)$ is very small, $\delta\omega/\omega = 2\delta\gamma/\gamma$, and also the Doppler shift due to a deflection is usually negligible: Deflection angles from thermal scattering are of order

$\vartheta \approx \hbar/\gamma mc\rho$ and, according to Equation 26, the corresponding shift is $|\delta\omega/\omega| \approx (\alpha a_0/\rho)^2 \simeq 10^{-2}$. The deflection angles from electronic scattering are typically even smaller. For this reason, single scattering may be neglected as a source of incoherence in the classical treatment. However, in a quantal picture, there is a finite difference between the particle states before and after emission of a photon, and therefore the amplitudes for scattering may be different in the two states. Since any lattice excitation confined to one of the states will lead to incoherence, intraband scattering will also contribute to some extent.

The scattering is due to the difference δV between the interaction potential and the thermally averaged continuum potential applied in the evaluation of transverse states. We concentrate here on the planar case, but a similar description is valid for axes. The total perturbation may conveniently be split into two parts,

$$\delta V = \delta V_1 + \delta V_2 \equiv [V(\mathbf{r}, z, \xi) - V_T(\mathbf{r}, z)] + [V_T(\mathbf{r}, z) - V_T(x)], \qquad 51.$$

where δV_1 is responsible for the excitation of lattice degrees of freedom ξ, and δV_2 leads to so-called nonsystematic reflections from transfers $\hbar\mathbf{g}$, where $\mathbf{g}$ is a reciprocal lattice vector not perpendicular to the plane. Such reflections turn out to be important only for electrons moving at a small angle to a strong axis in the plane. We may note in passing that the two terms in Equation 51 also lead to radiation. The first term contributes a background of incoherent bremsstrahlung, while the second produces coherent bremsstrahlung, which again is significant only for motion at a small angle to the axis.

The inelastic scattering caused by δV_1 may be separated into electronic and thermal scattering, which may be treated independently. Thermal scattering dominates and, from Reference (31), we quote a formula for its contribution to the reciprocal length for damping of the square of the radiation matrix element:

$$l_T^{-1} = \frac{N d_p}{(\hbar v)^2} \int dy \, \{\langle (\Gamma_a - \Gamma_{a,T})^2 \rangle_f - (\langle \Gamma_a - \Gamma_{a,T} \rangle_f)^2$$

$$+ \langle (\Gamma_a - \Gamma_{a,T})^2 \rangle_i - (\langle \Gamma_a - \Gamma_{a,T} \rangle_i)^2$$

$$+ (\langle \Gamma_a - \Gamma_{a,T} \rangle_f - \langle \Gamma_a - \Gamma_{a,T} \rangle_i)^2 \}_T. \qquad 52.$$

Here $\Gamma_a(x, y)$ denotes the integrated atomic potential,

$$\Gamma_a(x, y) = \int dz \, V_a(x, y, z), \qquad 53.$$

and, similarly, $\Gamma_{a,T}$ is the integral of the thermally averaged potential $V_{a,T}$.

The expectation values refer to the initial and final transverse states i and f, and the symbol $\{\ \}_T$ indicates a thermal average. The formula applies for uncorrelated vibrations, but correlations may easily be included (31).

As becomes apparent upon insertion of a complete set of intermediate states, the first two lines of Equation 52 represent the total interband scattering out of states f and i, resulting from the perturbation $V_a - V_{a,T}$. The intraband scattering represented by the last term in Equation 52 may be interpreted as phonon excitation induced by the difference between the expectation values of the perturbation in the initial and final states. Without this contribution, the formula (52) is equivalent to the result obtained by Bazylev & Goloviznin (32). If, on the other hand, intraband scattering is included without restrictions, the formula corresponds to application of the imaginary potential used in electron microscopy (30, 33). This amounts to retaining only the first term in lines one and two in the integrand. The different contributions are illustrated in Section 3.1 (Figure 9).

A proper treatment of intraband scattering is of particular importance for the electronic contribution to incoherence. In the expression analogous to Equation 52 an excitation with an x component of the momentum transfer $\hbar q_x$ is represented by an operator $\exp(-iq_x x)$; in the intraband term, this leads to a Fourier component of the difference between the spatial distributions for the projectile in the initial and final states. This Fourier component nearly vanishes for the small momentum transfers associated with the plasmon excitations, which dominate the scattering cross section.

MULTIPLE SCATTERING AND LINE INTENSITY   Scattering processes also affect the total intensity of a line, which is proportional to the population P_i of the initial band i. For a beam incident at an angle ϑ to a plane or axis, the population $P_i(\vartheta, 0)$ at the surface, $z = 0$, is determined by a Fourier component of the transverse wave function (Equation 57). For a fixed value of ϑ, the depth dependence of $P_i(\vartheta, z)$ is then obtained from the coupled differential equations,

$$\frac{dP_n}{dz} = \sum_{n'} W_{n,n'}(P_{n'} - P_n), \qquad 54.$$

where $W_{n,n'}$ denotes the rate of transitions between bands n and n'. Expressions for $W_{n,n'}$ in terms of the perturbations may be written in analogy to Equation 52, in which the first line represents the thermal contribution to the total scattering out of band f, $\sum_{n \neq f} W_{n,f}$.

There is a qualitative difference between multiple scattering at MeV and GeV energies. The change in transverse energy, $\delta E_\perp \approx \frac{1}{2} p v \vartheta^2$, corresponding to a typical deflection by thermal scattering, $\vartheta \approx \hbar/\rho \gamma mc$, may be compared to a planar potential barrier of the order of a Rydberg, $\delta E_\perp/Ry \simeq (a_0/\rho)^2 \gamma^{-1} \simeq 10^2 \gamma^{-1}$. For energies in the low MeV region, scattering angles are in this sense large, and there is a direct communication between channeling states and free states, while at GeV energies a diffusion picture applies as in classical treatments of dechanneling (9, 10). Furthermore, the situation is quite different for electrons and positrons, the scattering being enhanced and strongly suppressed, respectively. The situation is simplest for MeV electrons since a statistical equilibrium between channeling states and free states is quickly established, and the depth dependence of the populations may then be estimated from random multiple scattering (Equation 29).

Line intensities, which are independent of scattering (and of beam collimation) may be obtained through integration over incidence direction. This leads to a uniform population of the bands at the surface, expressed in the planar case by

$$(2\vartheta_B)^{-1} \int d\vartheta \, P_n(\vartheta) = 1, \qquad 55.$$

where ϑ_B denotes the Bragg angle. According to Equation 54, such a uniform population is stable in depth.

Finally, we may mention that scattering results in a stronger influence of the finite crystal thickness L on linewidths since the radiation following an incoherent-scattering event is emitted within a distance shorter than L. This makes the analysis of linewidths rather complex for intermediate thicknesses (31).

3. EXPERIMENTS

3.1 *Electrons*

For electrons, the spectrum of channeling radiation is broad, even in the rest frame, since the planar and axial continuum potentials are far from harmonic. An enhancement of the low-energy radiation has been observed for GeV electrons (34), where a classical description applies. A recent observation, in quantitative agreement with theory, is shown at the end of the following section. Here we discuss only experiments at lower energies, where quantization of transverse energy gives rise to structure in the spectrum.

A typical experimental arrangement is shown in Figure 6. The highly collimated electron beam is deflected into the target chamber, and after passage through the target, it is bent into a Faraday cup. In this geometry, the radiation from the target may be detected in the forward direction since the background of radiation from collimators is avoided. In contrast to measurements at GeV energies, it is easy to collimate the photons within angles much smaller than $1/\gamma$, and the observed photon energies are then related to the radiation frequencies ω^R in the rest frame through the forward Doppler transformation,

$$\hbar\omega = (1+\beta)\gamma\hbar\omega^R.$$ 56.

Thus a line structure of the spectrum in the rest frame is preserved.

PLANAR-CHANNELING RADIATION The first observation of a line spectrum was for planar channeling of 28- and 56-MeV electrons in silicon (35). The line energies vary with beam energy approximately as $\gamma^{1.7}$ (36), a somewhat stronger dependence on γ than the proportionality to $\gamma^{1.5}$ for a harmonic potential.

A more recent spectrum (37) is shown in Figure 7. The lines corresponding to $\Delta n = 1$ transitions have been fitted by Lorentzians superimposed on a smooth background. The line energies and widths are in Table 1 compared with calculations based on the many-beam equation (Equation 46) and the expression in Equation 52 for the contribution to the width from thermal scattering. This is clearly the dominant source of line broadening, and the results would be consistent with an additional contribution of 2–4 keV from electronic scattering.

Even sharper and relatively more intense lines have been observed (38) for channeling in diamond, and measured line energies and widths are compared with calculations in Table 1. As for silicon, the agreement is excellent for the energies, but the differences between the measured widths and the calculated values for thermal scattering are much larger for diamond. Since the electronic contribution should be similar in the two cases, this might indicate significant scattering by crystal defects, which are

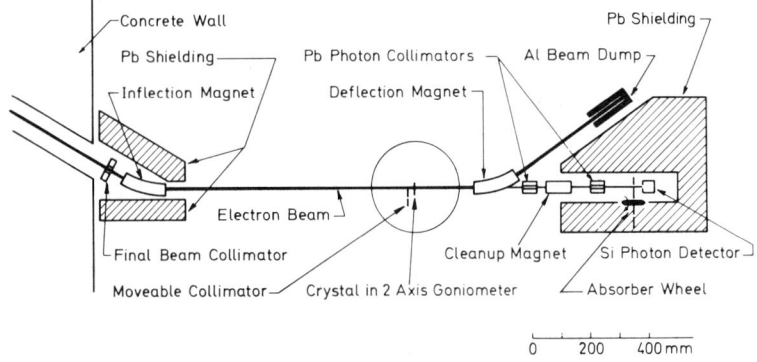

Figure 6 Experimental arrangement (see text) (22, 28, 31).

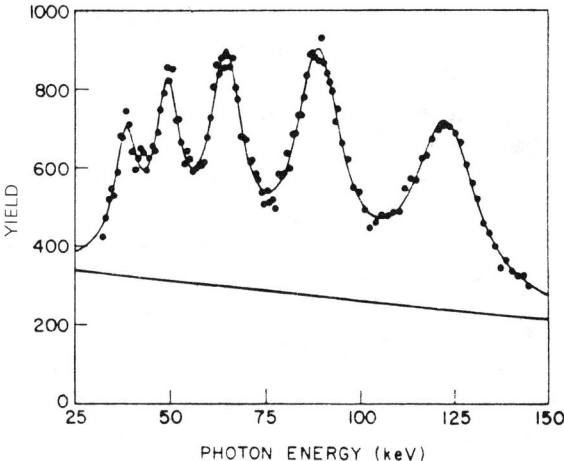

Figure 7 Photon spectrum obtained from 54-MeV electrons channeled along {110} planes of a 17-μm thick Si crystal (37).

Table 1 Characteristics for line radiation from 54-MeV electrons in bound {110} states in silicon and diamond[a]

Transition $n \rightarrow n'$	Photon energy (keV)		Linewidth (keV)	
	Experimental	Calculated	Experimental	Calculated
Silicon				
$1 \rightarrow 0$	122.5	125.3	19.5	15.2
$2 \rightarrow 1$	88.8	89.0	14.4	10.9
$3 \rightarrow 2$	64.3	64.5	10.5	7.3
$4 \rightarrow 3$	49.1	49.0	7.8	5.3
$5 \rightarrow 4$	38.3	38.4	7.0	4.0
Diamond				
$1 \rightarrow 0$	161.8	161.6	12.1	4.3
$2 \rightarrow 1$	104.4	104.5	8.7	2.5
$3 \rightarrow 2$	78.4	78.3	7.6	1.8
$4 \rightarrow 3$	58.0	58.1[b]	8.8	1.2

[a] Experimental results from (37) for silicon and (38) for diamond. Calculations based on Equations 46 and 52, with Doyle-Turner potential coefficients, modified in the evaluation of widths to represent also large momentum transfers (E. Lægsgaard, private communication).

[b] Band structure contribution is ±4.0 keV.

usually present even in diamonds of high quality. A contribution of 4 keV to the linewidth corresponds to a mean free path of ~2 μm.

A detailed study of the temperature dependence has been made for 4-MeV electrons in nickel (31), where thermal scattering is the dominant source of incoherence. The {111} planar potential and energy levels and wave functions of bound states were illustrated in Figure 4, and the radiation spectra observed for three different target temperatures are shown in Figure 8. The dominant line at about 4 keV corresponds to the $1 \to 0$ transition, and the measured line energies are close to the calculated values. A similar strong dependence of line energies on temperature has been observed for 54-MeV electrons in silicon (39).

Also the calculated linewidths are, as shown in Figure 9, in good agreement with the measured values, which have been corrected for finite crystal thickness. The dashed lines correspond to extreme assumptions about the contribution to incoherence from thermal intraband scattering: The upper curve includes all intraband scattering, while the lower one does

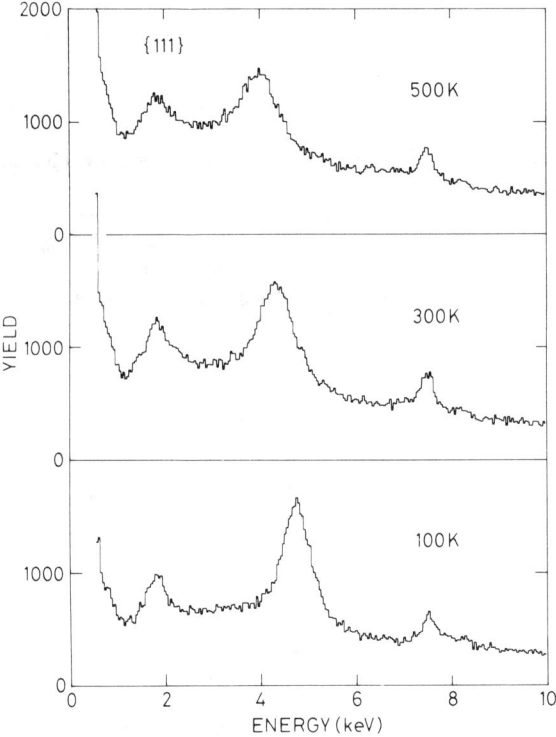

Figure 8 Radiation spectra for 4-MeV electrons channeled along {111} planes in a 0.5-μm Ni crystal at three different temperatures (31).

not include any. As expressed in Equation 52, intraband scattering contributes to the extent that it is different in the initial and final states, and this leads to the solid line. When evaluated with the same prescription, the contribution from electronic scattering is ~50 eV only, which should be added to the solid curves in Figure 9. If included indiscriminately, electronic scattering would increase the theoretical estimates in Figure 9 by about a factor of two.

Line broadening due to finite target thickness was observed in a study of planar-channeling radiation from 4-MeV electrons in silicon (22). Measurements of the $1 \to 0$ line for a {110} plane are shown in Figure 10, and the linewidths can be accounted for by the influence of the finite target thickness alone (Equation 12).

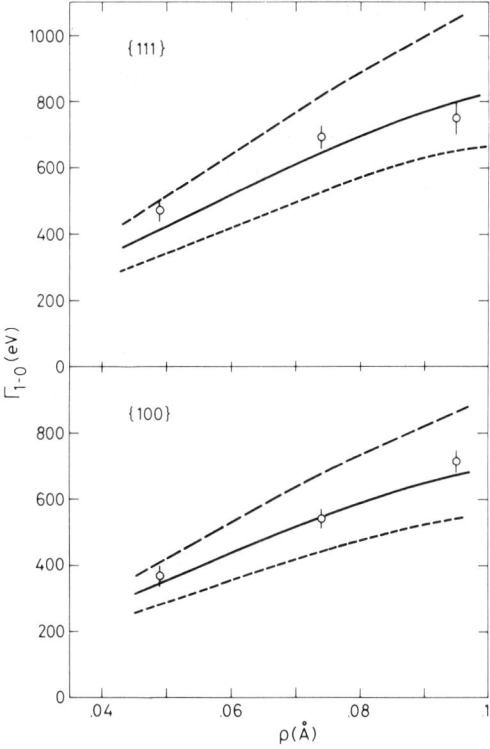

Figure 9 Linewidths (FWHM) obtained from fits to spectra of the type shown in Figure 8, corrected for detector resolution and the effect of finite target thickness. The curves are from calculations of the contribution from thermal scattering. The solid curves correspond to Equation 52, whereas the intraband scattering is neglected for the short-dashed curves and included without restrictions for the long-dashed curves. The additional contribution from electronic scattering is estimated to be ~50 eV. (From 31.)

The intensity and energy of the line were measured for the 0.5-μm sample as functions of the incidence angle to the plane, and the data are in Figure 11 compared with results of a many-beam calculation. The intensity is proportional to the population of the initial state, which for electrons with momentum $\hbar K$ incident at an angle ϑ is given by

$$P_1(\vartheta) = \frac{1}{d_p} \left| \int_{-d_p/2}^{d_p/2} dx \, u_1(x) \exp(iK\vartheta x) \right|^2, \qquad 57.$$

where $u_1(x)$ is the $n = 1$ wave function, normalized within one period of the planar potential. The variation of the intensity, shown in Figure 11, therefore reflects the square of the $n = 1$ wave function in transverse-momentum space. In the measurements for nickel discussed above, the angular dependence of the intensity was smeared considerably by multiple scattering, but the modification could be accounted for by application of Equation 54.

The incidence angle determines the pseudo-momentum or the reduced k_x value according to Equation 49. The $n = 1$ state in the $\{110\}$-planar potential is not completely localized, and Figure 11 includes a comparison of the measured and calculated variation of the $1 \rightarrow 0$ line energy. The

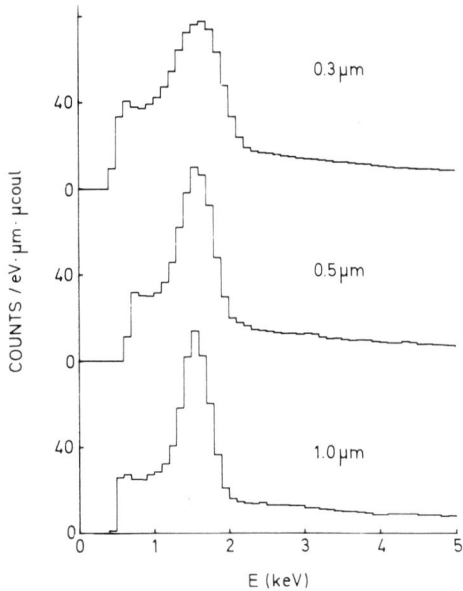

Figure 10 Photon spectra for 4-MeV electrons incident at an angle of 0.06° from a $\{110\}$ plane in Si crystals of different thickness L. The detector resolution was 280 eV, and the width from finite thickness is ~ 160 eV/L (μm) (Equation 12). (From 22.)

agreement is reasonable except for a region around $\vartheta = 0$, where the measured energy is strongly influenced by multiple scattering and angular divergence of the incident beam. When this is the case, the variation of the energy with k_x contributes to the linewidth.

BAND STRUCTURE AND COHERENT BREMSSTRAHLUNG Observation of radiative transitions between transverse energy levels may be used as a spectroscopic tool to map out the band structure, which was illustrated in Figure 5 for 4-MeV electrons along three different planes in silicon. Spectra obtained for different angles of incidence to a {110} plane are shown in Figure 12. The spectrum is simplest at the largest angle, where it consists of a single line of coherent bremsstrahlung (corresponding to a $\Delta n = 2$ transition between free states) and a background of incoherent bremsstrahlung, which close to and below the Si-K absorption edge at 1.8 keV is modified by absorption in target and detector and by Si-K x rays. At the smallest angle the spectrum is dominated by a $1 \to 0$ transition, and at the

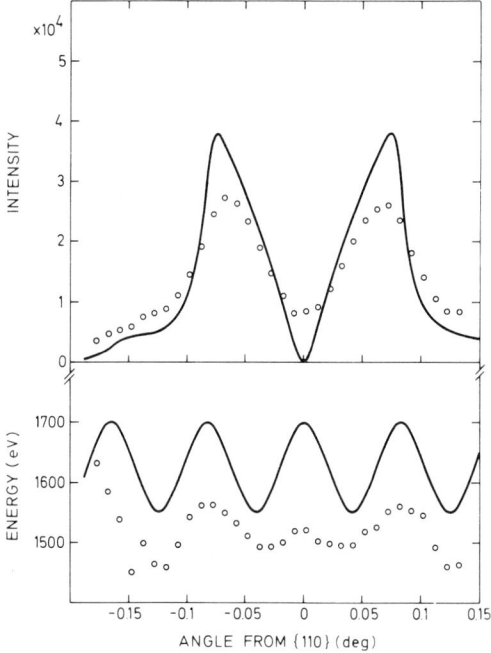

Figure 11 Variation with incidence angle to a {110} plane of the intensity of the $1 \to 0$ line for 4-MeV electrons in a 0.5-μm thick Si crystal. The variation of the line energy is shown in the lower part of the figure. The curves represent the results of calculations, based on Equations 40, 46, 47, 49, and 57. Part of the discrepancy in energy is caused by an error in the beam energy (-1%). (From 22.)

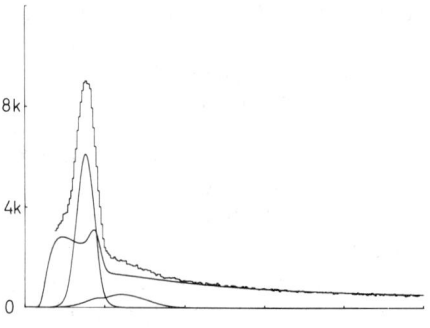
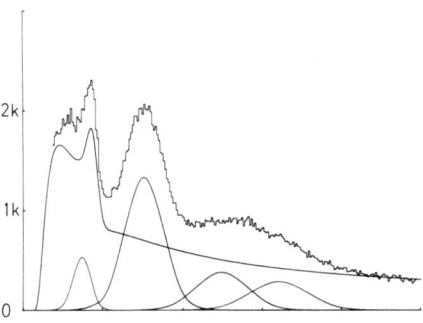
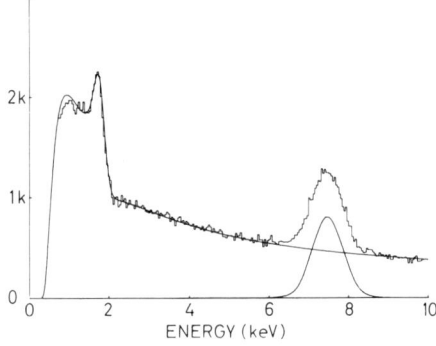

Figure 12 Photon spectra from 4-MeV electrons incident on a 0.5-μm thick Si crystal at angles 0.06°, 0.23°, and 0.50° to a {110} plane. The spectra have been fitted by lines of coherent radiation and a background of incoherent bremsstrahlung, modified by absorption, and by a line of Si-K x rays. At the smallest angle, the $1 \to 0$ line dominates, while at the largest angle, only a $\Delta n = 2$ line of coherent bremsstrahlung is visible. For the intermediate angle, the upper two broad lines are from free-to-bound transitions. (From 22).

intermediate angle, all types of transition are present, free-to-free, free-to-bound, and bound-to-bound.

The measured line energies are in Figure 13 compared to the transition energies calculated from the level structure in Figure 5. At the larger angles ϑ to the plane, only one band is significantly populated, and in the rest system, the transverse energy is approximately given by

$$E_\perp^R(\vartheta) \simeq \frac{\hbar^2(K\vartheta)^2}{2m} + \gamma V_0, \qquad 58.$$

where V_0 is the average crystal potential. The $\Delta n = 2$ transition energy is then determined by the change in $E_\perp^R$ due to transfer of a momentum $-\hbar \mathbf{g}$:

$$\hbar \omega^R \simeq \frac{\hbar^2 \mathbf{K} \cdot \mathbf{g}}{m} - \frac{\hbar^2 g^2}{2m}, \qquad 59.$$

which with $\mathbf{K} \cdot \mathbf{g} = Kg\vartheta$ corresponds to a linear dependence on ϑ, with a small off-set at $\vartheta = 0$. Also, the energies of transitions induced by higher harmonics of the potential, with $\Delta n = 4$ and $\Delta n = 6$, are straight lines as

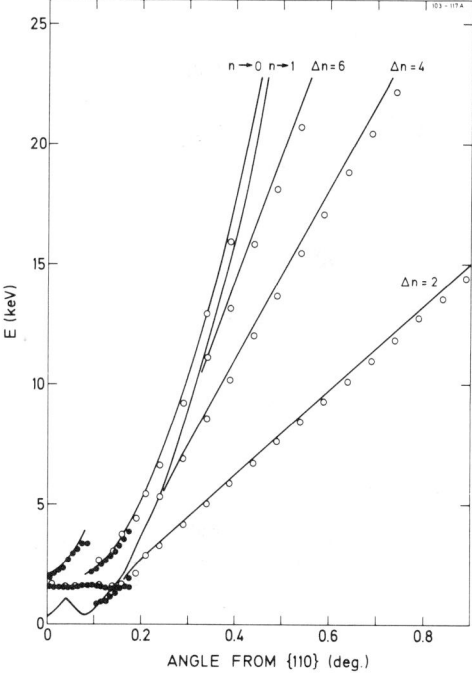

Figure 13 Line energies from spectra similar to those in Figure 12 are compared to calculated values for transitions between the levels shown in Figure 5 for 4-MeV electrons along a {110} plane in Si. (From 22.)

functions of ϑ, with slope proportional to Δn and off-set proportional to Δn^2. The free-to-bound transitions have a parabolic shape, according to Equation 58. These qualitative features are confirmed by the experiment, and when the small error in beam energy (-1%) is corrected for there is quantitative agreement between theory and experiment.

Coherent bremsstrahlung from electrons of a few MeV has been observed in other recent studies (40, 41). In experiments with fairly thick targets, the transition energies (Equation 59) are smeared by multiple scattering, and the radiation from free states gives rise to a continuous background under the peaks of channeling radiation (cf Figure 7).

AXIAL-CHANNELING RADIATION According to Equations 8 and 10, the number of localized channeling states is larger for the axial than for the planar case. As an example, the formulas predict for 56-MeV electrons in silicon that the number of quantum states in the $\{110\}$ and the $\langle 110 \rangle$ potentials should be $v_p^- = 7$ and $v_s^- = 147$. It is not surprising, therefore, that no distinct axial lines were observed in the experiment (35) on channeling radiation from 56-MeV electrons in silicon.

A spectrum is shown in Figure 14 for axial channeling of electrons in silicon at the much lower energy of 4 MeV, where the number of bound states is only about ten (28, 42). At the larger angles to the $\langle 111 \rangle$ axis, the radiation is mainly incoherent bremsstrahlung, but at the smaller angles, the spectrum is dominated by peaks of channeling radiation. The angular dependence of the intensity reflects the Fourier transform of the initial state, as in Equation 57 and for angular momentum $l \neq 0$, the intensity should

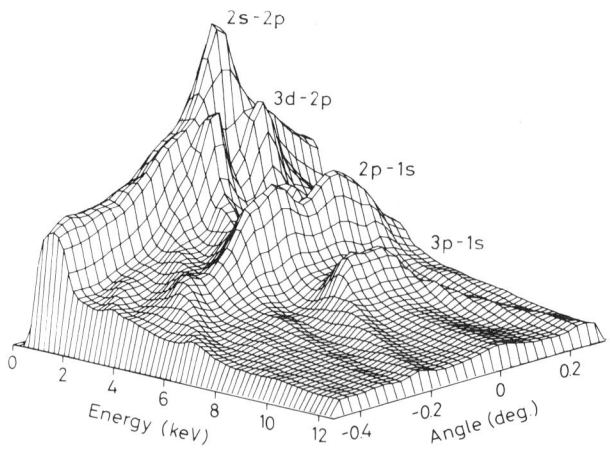

Figure 14 Experimental photon spectra vs angle of incidence to a $\langle 111 \rangle$ axis for 4-MeV electrons in a 0.5-μm thick Si crystal. The data were smoothed slightly to reduce statistical fluctuations. (From 28.)

vanish at zero angle of incidence. This is consistent with the assignments since all lines except 2s-2p have minima at $\vartheta = 0$. As for the planar case discussed above, the spectra also contain radiation due to transitions from free states. Most pronounced in the figure is the parabolic ridge merging with the 3p-1s transition at small angles, which may be interpreted as transitions from free states to the 1s state. The linewidths of transitions between bound states were found to agree reasonably well (within 15%) with calculations of the type discussed for planes (Equation 52). For the line energies, a more precise comparison with calculations is possible, and in general, the measurements agree with calculations with a Doyle-Turner potential to within a few percent.

Results for different axes and beam energies are in Figure 15 compared to calculations within the single-string approximation, where the radiation energy in the rest system $\hbar\omega^R$ in a given target material is a function of the parameter γ/d. The data are all from recent measurements (28), but apart from a correction for the 3p-1s line they agree with the measurements of Cue et al (29), where the γ/d scaling was first suggested and investigated. In energy measurements with a gas counter (42), the highest energies in Figure 15 were underestimated by $\sim 10\text{-}15\%$ owing to the rapid decrease of the detector efficiency above a few keV. On the other hand, in the experiments reported by Watson & Koehler (41) the line energies were in some cases higher by 10-20% than the results predicted in Figure 15. We believe that this may be explained by a small error in the measured value of the angle between the electron beam and the direction to the photon counter, which

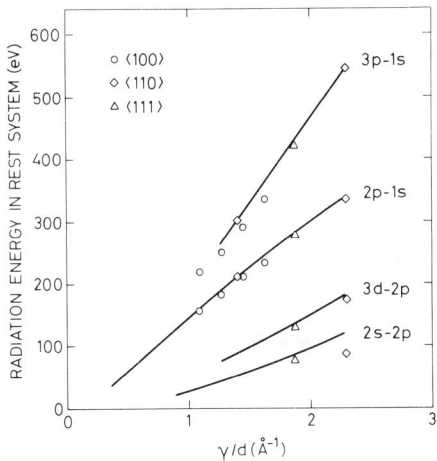

Figure 15 Comparison between transition energies in the rest frame, calculated within the single-string approximation with a Doyle-Turner potential (*solid curves*), and measurements for varying beam energy in three axial directions with different average atomic spacing d in silicon. (From 28.)

was about 3° and not zero, as in the setup used in the other measurements (Figure 6). As given by Equation 11, the line energy depends sensitively on this angle.

Axial-channeling radiation has also been measured for diamond (43), where the lines are sharper, just as for the planar radiation. For a direction with close-lying pairs of strings, transitions between molecular-type states were observed. For axial channeling in gold, a much heavier material, only featureless enhancement has been seen, in agreement with estimates of linewidths due to thermal scattering. This scattering limits the observation of a line spectrum of axial-channeling radiation to low-Z materials and to not too high electron energies, up to perhaps a few tens of MeV for the lightest materials (28).

3.2 Positrons

Channeling radiation has been observed from positrons at a few tens of MeV in silicon (44), diamond (45), germanium (46), and lithium fluoride (47), and for GeV positrons in diamond (48), silicon, and germanium (23, 49, 50).

MeV ENERGIES For monatomic crystals, the planar potentials are similar to those drawn for diamond in Figure 16. The potential is approximately harmonic, with nearly equally spaced energy eigenvalues, but the spacing increases somewhat with transverse energy. Line broadening occurs for the higher states because of extension of the wave function into the region between the neighboring planes (Bloch-wave broadening). The radiation intensity increases linearly with the quantum number n, but the coherence length decreases because of larger overlap with the planes of vibrating atoms. Therefore transitions between states near the top of the well produce broader spectral lines with lower peak intensity and higher integrated yield than transitions from states near the bottom of the well.

Figure 16 shows the thermally averaged Doyle-Turner continuum potential for the $\{100\}$, $\{110\}$, and the $\{111\}$ planes in diamond and the corresponding transverse energies for 54-MeV positrons. In the $\{111\}$ case, there are two sets of planes with separations in the ratio of three to one; in the other two cases, the planes are equally spaced.

The experimental results shown in Figure 17 were obtained with 54-MeV positrons incident parallel to planes (to within ~ 0.5 mrad) in a 23-μm thick type-II natural diamond. Each spectrum is the difference between spectra for the channeling direction and for a random orientation, both observed in the forward direction. The vertical lines indicate the calculated energies (Equation 39) obtained from the $\Delta n = 1$ transitions in Figure 16. The heights are proportional to the intensities of the transitions, calculated for equal populations of the transverse states. Dashed lines represent tran-

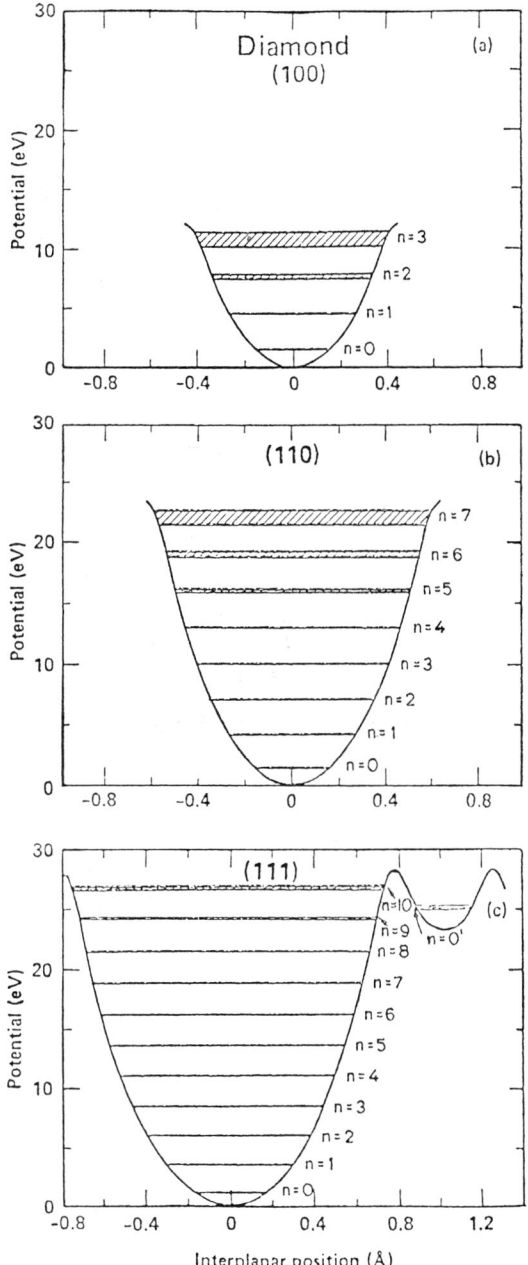

Figure 16 Planar Doyle-Turner potentials for positrons in diamond at room temperature, and transverse-energy levels at 54 MeV ($\gamma = 107$). (From 45.)

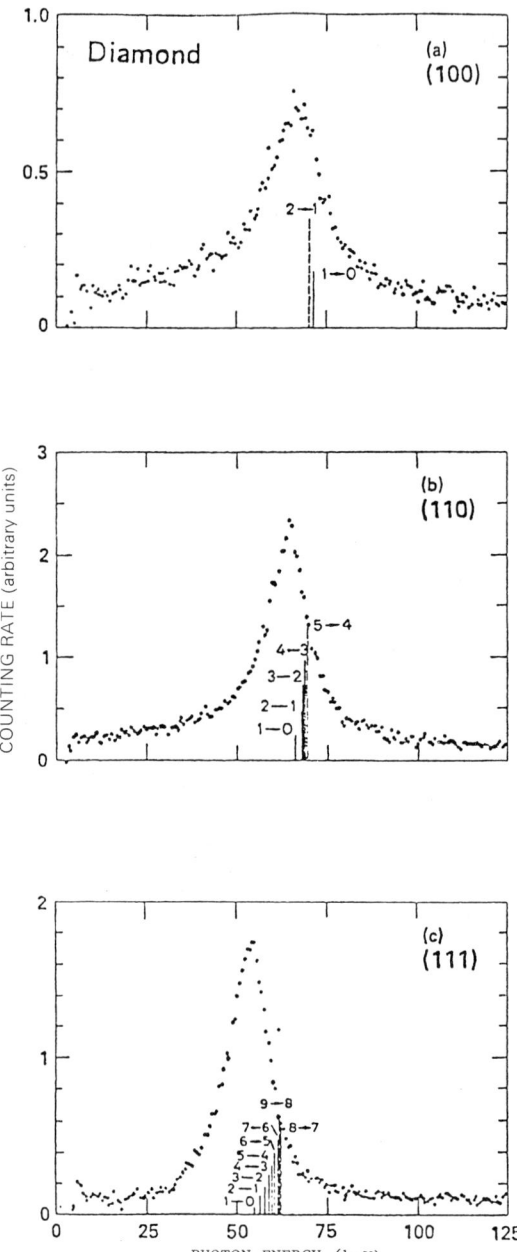

Figure 17 Measured planar-channeling radiation from 54-MeV positrons in diamond. Photon energies obtained from the levels in Figure 16 are indicated by vertical lines, with heights proportional to the intensities calculated for equal populations of the levels. (From 45.)

sitions near the top of the well and thus are likely to correspond to broader lines with a lower peak intensity.

The calculated photon energies are 5–10% higher than the measured ones. A conceivable explanation would be that, in the calculations, one did not take into account the redistribution of bonding electrons. However, a modification of the potentials based on experimental x-ray-scattering data did not significantly reduce the differences. Also the corrections for angular spread of the beam and for limited photon collimation are too small to explain the discrepancy.

Table 2 summarizes the experimental data obtained from planar-channeled 54-MeV positrons in silicon, diamond, and germanium. The calculated photon energies are averages over values corresponding to the various $\Delta n = 1$ transitions, with appropriate weightings for linewidths and transition strengths.

Potential functions are more complicated in binary crystals such as LiF (47). Along directions with all Miller indices odd such as $\{111\}$, the Li$^+$ and F$^-$ ions lie in different planes. The other planes contain both lithium and fluorine ions, and the potentials are similar to those in a monatomic crystal with $Z_2 = 6$. Figure 18 shows the calculated potentials for the $\{110\}$, $\{100\}$, and $\{111\}$ planes in LiF and the corresponding transverse energies for 54-MeV positrons. As expected, Figures 18a and 18b show a single well, whereas in Figure 18c there is a minor peak at $x = 0$, the position of the plane of Li$^+$ ions and a major peak at $x = 1.2$ Å, the position of the plane of F$^-$ ions.

Figure 19 gives the measured channeling radiation spectra, with the background subtracted, for 54-MeV positrons in a 25-μm thick LiF crystal. The data shown in Figures 19a and 19b are very similar to the results for the monatomic diamond, with a single line broadened primarily by the

Table 2 Planar-channeling radiation from 54-MeV positrons in silicon, diamond, and germanium

Crystal (Reference)	Plane	Photon energy (keV)		Linewidth (keV) (FWHM)
		Experimental	Calculated	Experimental
Si (44)	$\{100\}$	42.8 ± 0.7	44.2	9 ± 2
	$\{110\}$	38.8 ± 0.3	41.6	9.5 ± 0.3
	$\{111\}$	32.7 ± 0.7	34.5	9 ± 2
C (45)	$\{100\}$	66.6 ± 1	71.6	17 ± 1
	$\{110\}$	65.3 ± 1	69.2	12 ± 1
	$\{111\}$	54.5 ± 0.3	58.4	14 ± 1
Ge (46)	$\{110\}$	48.1 ± 0.8	52.2	18 ± 2

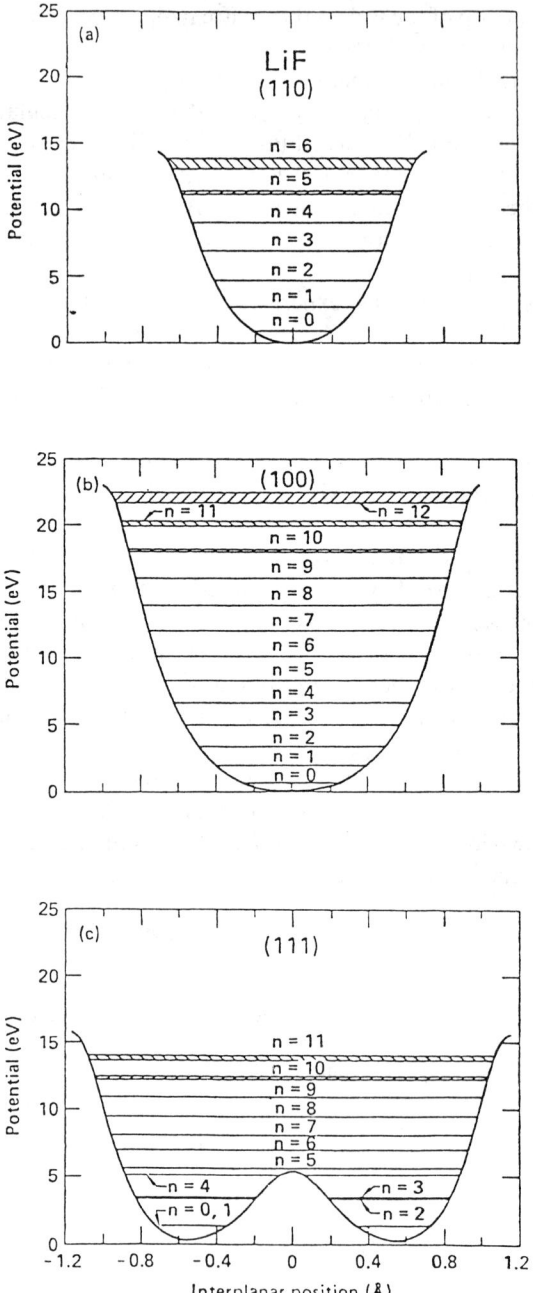

Figure 18 Planar potentials for LiF at room temperature, and transverse-energy levels for 54-MeV positrons ($\gamma = 107$). (From 47.)

anharmonicity of the potential well. The spectrum in Figure 19c, however, contains several lines because of the increased complexity of the potential for {111} planes. The major peak near 30 keV can be attributed to $\Delta n = 1$ transitions from the $n = 6$ to $n = 9$ levels. A second peak near 45 keV appears to arise from the $2 \rightarrow 1$ transition and $\Delta n = 3$ transitions between levels lying deeply within the well. A third peak at about 85 keV is probably from $\Delta n = 3$ transitions between higher levels.

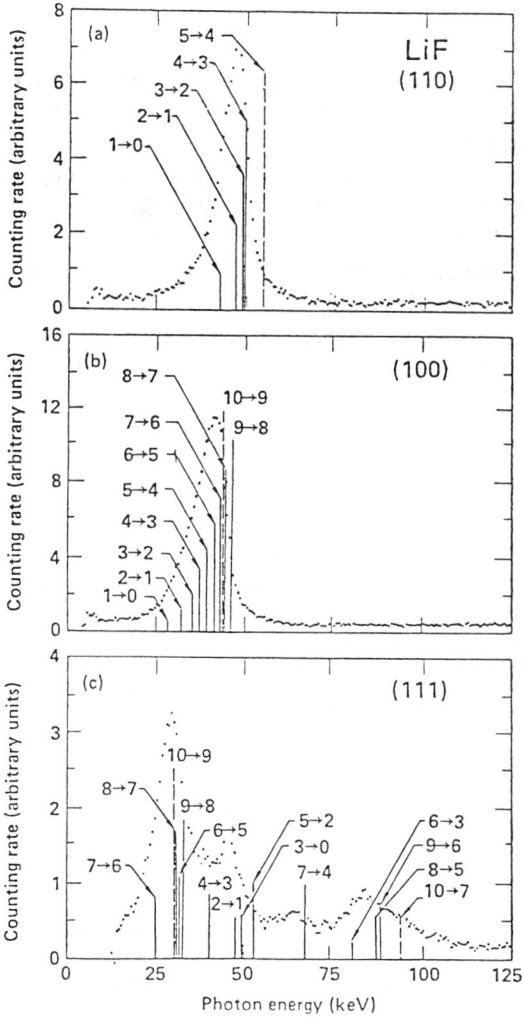

Figure 19 Spectra of planar-channeling radiation from 54-MeV positrons in LiF. The energies and relative intensities, indicated by vertical lines, are calculated as in Figure 17. The dashed lines correspond to transitions between levels near the barrier and are somewhat uncertain owing to band-structure broadening. (From 47.)

Table 3 Planar-channeling radiation from 54-MeV positrons in LiF (47)

	Photon energy (keV)		Linewidth (keV) (FWHM)
Plane	Experimental	Calculated	Experimental
{110}	46.8 ± 1	48.9	8.9 ± 2
{100}	41.8 ± 1	41.4	10.7 ± 2
{111}	29.9 ± 1	32.0	10.1 ± 2

Table 3 summarizes the theoretical and experimental results for the three different planes in LiF. For the {111} plane, only the major peak at $\simeq 30$ keV is included.

From the measurements of the intensity of channeling radiation from 54-MeV positrons incident along a {100} plane in LiF crystals of thickness 25, 125, and 150 μm, it was estimated that the population of bound states decays to half its initial value within a distance of $\simeq 150$ μm.

The results presented thus far have all been for planar channeling, but emission has also been measured from axial-channeled positrons, and Figure 20 shows the forward-directed spectrum for $\langle 100 \rangle$ axial-channeled

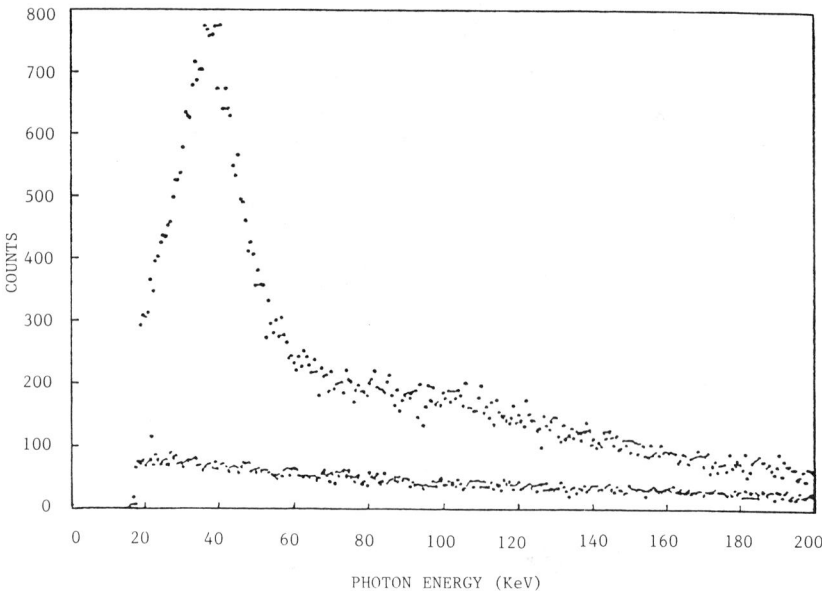

Figure 20 Spectrum of channeling radiation from 54-MeV positrons channeled along a $\langle 100 \rangle$ axis in Si. The lower curve is a spectrum for a "random" direction of incidence for the same number of positrons through the crystal. (From 44.)

54-MeV positrons in silicon. The observed intensity at the maximum of the peak at 40 keV is an order of magnitude higher than normal bremsstrahlung. Axial-channeling states of positive particles are not localized, and therefore the transverse energies form broad bands. The number of states per unit cell is high, even for low values of γ, and a classical description should therefore be applicable to axial-channeling radiation from positrons.

GeV ENERGIES Channeling radiation from GeV positrons has been observed by a few groups (23, 48–50), and we shall here give an example of a comparison between very recent data and classical calculations with the thermally averaged Doyle-Turner potential (50 and references therein; also E. Uggerhøj and O. Pedersen, private communication). In Figure 21a are shown measured and calculated photon spectra, integrated over emission angle, from 7-GeV positrons penetrating a 105-μm thick silicon crystal close to a $\{110\}$-planar direction. The distribution in angle of incidence was

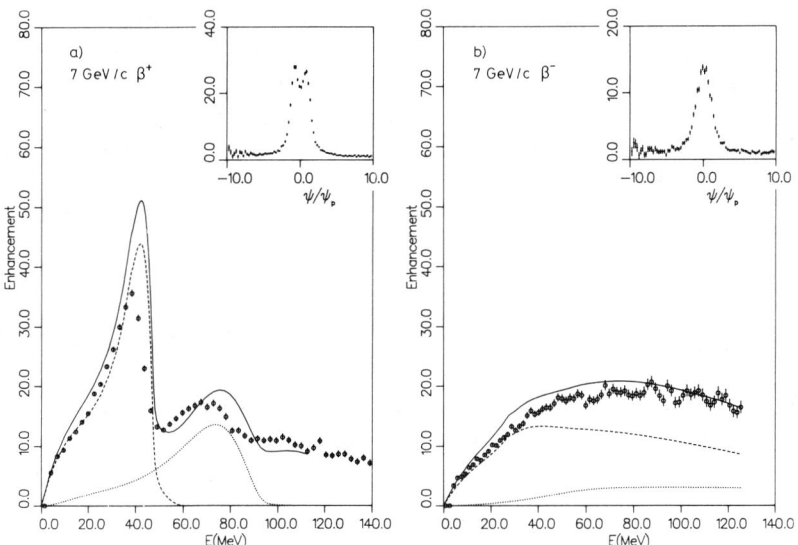

Figure 21 (a) Comparison of calculated and measured spectra of channeling radiation from 7-GeV positrons incident close to a $\{110\}$ plane in a 105-μm thick Si crystal. Broken and dotted curves give the contributions from the first and second harmonic of the motion of the channeled positrons. Multiple scattering is not included. A background measured with the target removed has been subtracted from the experimental spectrum. The insert shows the measured peak height as a function of projectile angle of incidence ψ to the plane. Values are given as enhancement factors over normal bremsstrahlung, i.e., in units of I_b/ω_m, with I_b defined in Equation 22. (b) As in (a) but for electrons. (From 50 and also E. Uggerhøj and O. Pedersen, private communication.)

uniform within the range from zero to the critical angle ψ_p. Both the measured points and the calculated curves represent absolute values for the power distribution expressed in units of the theoretical value for the corresponding amorphous material.

The beam contains both channeled particles and particles with free transverse motion, and the solid line shows the total spectrum from all projectiles. The broken and dotted curves refer to the channeled fraction only and give the partial contributions from the first and second harmonic of the oscillatory motion. Since the energy of 7 GeV is close to the magic value, where effects from anharmonicity and relativistic transverse motion nearly cancel, the cut-off in the photon spectrum at the high-energy side of the first harmonic should be sharper than in similar spectra at higher and lower projectile energies. This was confirmed by the experiments.

In all cases studied, the calculated cut-off frequency is sharper and a little higher than the experimental value. As indicated in Section 2.1, this discrepancy may be due to multiple scattering, which was neglected in the calculation. The coherence length l_c obtained for random multiple scattering, from Equations 29 and 30, is ~ 40 μm. Since this is much less than the target thickness, we may apply the estimate in Equation 31, which gives a broadening of $\sim 25\%$. This is of the right magnitude to explain the difference between theory and experiment in Figure 21a.

Figure 21b shows a similar comparison between calculated and measured spectra for electrons of the same energy. Here also, large enhancement factors are observed, and, as expected, the spectrum is broad and structureless. Broadening and shift from multiple scattering are therefore difficult to observe.

4. APPLICATIONS OF CHANNELING RADIATION

We discuss three possible applications: the study of aspects of channeling, the investigation of crystal properties, and channeling radiation as a source of x rays and γ rays. The first application has already been developed in detail, the second has been demonstrated, and the third is still only a proposal.

4.1 *Channeling Properties*

Radiation spectroscopy is a powerful tool in the study of energy levels and wave functions for electrons penetrating crystals. The results obtained may be valuable in other applications of electron transmission, for example in electron microscopy. Of special interest is the information on incoherent scattering, which may be derived from linewidths and from the variation of line intensities with incidence angle and with target thickness. Through observations of channeling radiation, the theoretical description can be tested in detail. In this connection the qualitative distinction between coherence length and occupation length is important. The latter may be strongly influenced by scattering into the state (feeding), which does not influence the coherence length. Also, the concept of an occupation length may be questionable when the depth dependence of the occupation is far from exponential, for example, for channeling of MeV electrons, where the occupation of a state is expected to vary asymptotically as one divided by the square root of the depth. For positrons, on the other hand, the decrease in population of channeling states will be close to exponential since the scattering back from free states is weak, and the reduction in scattering due to channeling results in very long occupation lengths, as mentioned in Section 3.2 with regard to LiF.

4.2 *Crystal Properties*

When the channeling behavior is understood, several interesting types of information on crystal properties may be derived from observations of channeling radiation. In an analysis of the first observation of planar-channeling radiation from MeV electrons, the planar potential was reconstructed (36). For the {110} planes in silicon, a potential function for electrons was assumed to be of the form

$$V(x) = \begin{cases} A_0 \exp(-b|x|) & \rho_1 \leq |x| \\ A_1 x^2 + B & \rho_1 \geq |x| \end{cases}, \qquad 60.$$

where x is the distance from the plane, and ρ_1 is the one-dimensional RMS vibrational amplitude. Without the modification in the lower line of Equation 60, the energy levels can be evaluated analytically, and the modification for small $|x|$ was then included to first order as a perturbation. Matching at $|x| = \rho_1$ leaves three free parameters in Equation 60, and they were adjusted to reproduce the observed line energies. Figure 22 shows the potential derived in this manner, compared to the Molière approximation to the Thomas-Fermi potential.

While this analysis clearly demonstrates the possibility of deriving information on potentials from line energies, procedures analogous to those applied in x-ray studies of electron densities are more promising. The starting point is the crystal potential corresponding to the thermally averaged charge density of free atoms, which may be represented by the Doyle-Turner approximation (Equation 47). As seen in Table 1, the measured line energies are then reproduced within experimental uncertainties, except perhaps for the $1 \to 0$ transition in silicon and the $4 \to 3$ transition in diamond. It has been demonstrated recently, however, that evidence for charge displacement into bonds can be obtained from

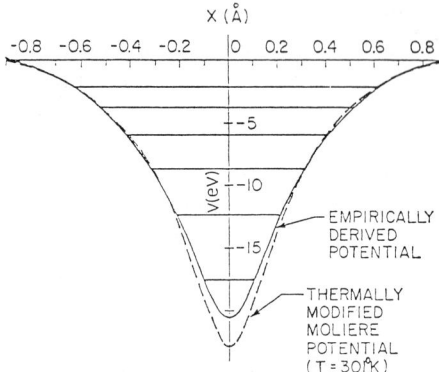

Figure 22 Planar potential derived from line spectrum compared to thermally averaged Molière potential. (From 36.).

observation of channeling radiation (43). Also the magnitude of thermal vibrations and the correlation between vibrations of neighboring atoms have been investigated (31).

It is not yet clear how competitive this new method is in relation to the well-established techniques based on x-ray or neutron diffraction. These techniques single out the interaction with either electrons or nuclei in the crystal, which may be an advantage. However, in the analysis of x-ray data, the vibrational amplitude is still an important parameter, which is poorly determined from such data alone, and some of the most accurate form factors of charge distributions in crystals have in fact been determined through measurements of "critical voltages" for transmission of high-energy electrons (51).

Compared to ions, high-energy electrons (and positrons) have the advantage of being a weak probe in the sense that inelastic scattering can be treated as a perturbation. This allows a more accurate description. In addition, complications from defect production should generally be much smaller.

Defects and impurities in crystals can also be investigated by channeling radiation. Emission from two diamonds was measured, one containing nitrogen in the form of platelets consisting of a double layer of nitrogen precipitated in a {100} plane.[1] Transmission electron microscopy showed platelet diameters varying from 40 to 200 Å, with an average separation of ~500 Å. Figure 23 shows the distortion of the {100} planes resulting from the presence of a platelet. The distortion of {110} and {111} planes is similar to that of a stacking fault, with an abrupt translation of the planes when they cross the platelet.

---

[1] Park, H., Pantell, R. H., Swent, R. L., Kephart, J. O., Berman, B. L., et al., in preparation.

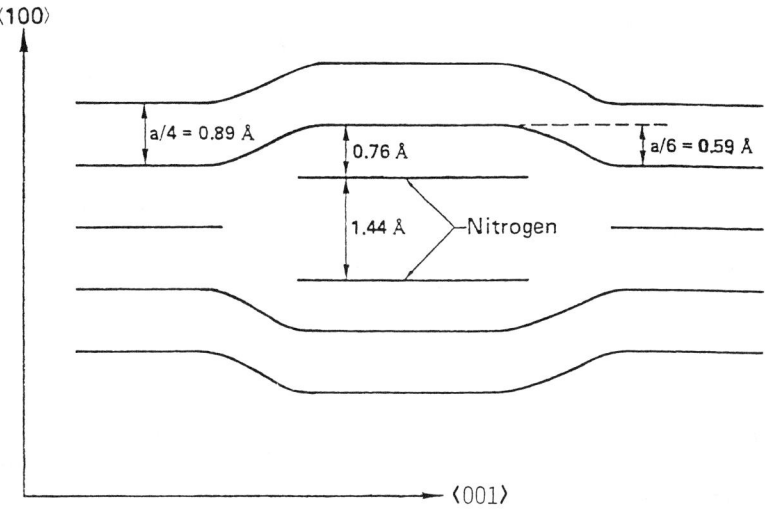

NITROGEN PLATELETS IN DIAMOND

Figure 23 Model geometry of nitrogen platelets in diamond.

For both positrons and electrons at 54 MeV, significant changes in radiation intensity, linewidth, and frequency were observed. Perhaps the most interesting result was obtained for positrons channeled by {111} planes, where the linewidth was reduced by more than 35% for the diamond with platelets. The explanation is probably that, owing to the displacement of the planes, states near the top or the bottom of the potential well are depleted, and this reduces the linewidth caused by anharmonicity (cf Figure 17). Normally, one would expect a line broadening due to scattering by defects, and this was indeed observed for axial channeling of 4-MeV electrons in the same type of diamond, in conjunction with the work reported in Reference (43). The mean free path for scattering by platelets, deduced from the line broadening, was consistent with estimates from electron microscopy and from proton dechanneling (52). It therefore appears that information on defect concentrations and in some cases also on defect configurations may be obtained from observation of channeling radiation.

4.3 *Radiation Source*

Finally, we consider the potential application of channeling radiation as a source of x rays or γ rays. The radiation is usually linearly polarized, is forward-peaked with a cone angle equal to γ^{-1} (Figure 2), and has a relatively narrow bandwidth (typically 10–20%). The peak of the spectrum is adjustable, and the pulses may be of picosecond duration.

With regard to this last characteristic, commercially available streak cameras can resolve 20-picosecond pulses at 20-keV photon energy. This is just the type of radiation that can be produced by channeling with a linear accelerator as the particle source. It may therefore be possible to investigate ultrafast relaxation processes in the x-ray region with channeling radiation.

To determine the intensity of such a source, it is necessary to consider the maximum current density that could be used to generate the x rays. Three factors determine this maximum: The current available, heating of the crystal, and the rate of defect formation.

While the first two factors are easily estimated, the influence of defect formation on the radiation is uncertain. An experiment was performed with 54-MeV electrons with 2-μA average current through an area of 3-mm diameter on a 17-μm thick silicon crystal for a period of three hours. It was estimated that, without annealing effects, this introduced defects to a concentration of about one part in 10^4. Channeling radiation was measured before and after the three-hour period, and there was no discernible degradation of the emission. Based on this observation, an average electron current of 100 μA in a 5-mm diameter beam appears reasonable. This beam can be produced by a linac, the temperature rises to $\sim 600°C$, and the defect formation rate is about one part in 10^3 per hour but will be strongly reduced by annealing. Since heating increases the linewidth, it may be better to operate at a somewhat reduced current.

For electrons at 54 MeV channeled along $\{110\}$ planes in silicon, the wavelength is 0.1 Å for the $1 \to 0$ transition. The radiation is linearly polarized and contained within a 10-mrad cone angle. If the intensity measured with average beam currents of 10^{-10}–10^{-11} A is extrapolated to a beam of 100 μA, one obtains $\sim 10^{10}$ photons per second within a 10% bandwidth, which is comparable to emission from a storage-ring synchrotron source.

The long mean free paths for thermal scattering in low-Z materials suggest the use of diamond as a target, but it is difficult to obtain defect-free samples of a reasonable size. LiF is a good crystal to use at low beam currents, but defect formation would severely limit the running time at higher beam currents.

The spectrum of radiation from planar-channeled positrons, consisting of a single fairly narrow line with a reduced background of normal bremsstrahlung, is in principle better suited for applications, but it is difficult to obtain high-quality, high-current beams. Radiation from GeV positrons may, however, turn out to be useful. Close to the "magic" energy, the line broadening by anharmonicity is cancelled by the effect of relativistic transverse motion, and there is a sharp cut-off in the radiation spectrum.

The photon energy at the cut-off is typically 30 MeV, and yields of the order of one photon per positron are produced in transmission through a 1-mm thick crystal.

Literature Cited

1. Vorobiev, A. A., Kaplin, V. V., Vorobiev, S. A. 1975. *Nucl. Instrum. Methods* 127:265–68
2. Kumakhov, M. A. 1976. *Phys. Lett.* A57:17–18
3. Baryshevskii, V. G., Dubovskaya, I. Ya. 1977. *Phys. Status Solidi B* 82:403–12
4. Terhune, R. W., Pantell, R. H. 1977. *Appl. Phys. Lett.* 30:265–68
5. Kumakhov, M. A. 1977. *Zh. Eksp. Teor. Fiz.* 72:1489–1503 (transl. *JETP* 45:781–89)
6. Kumakhov, M. A., Wedell, R. 1977. *Phys. Status Solidi B* 84:581–93
7. Kumakhov, M. A., Trikalinos, Ch. G. 1980. *Phys. Status Solidi B* 99:449–62
8. Wedell, R. 1980. *Phys. Status Solidi B* 99:11–49
9. Gemmel, S. D. 1974. *Rev. Mod. Phys.* 46:129–227
10. Lindhard, J. 1965. *K. Dan. Vidensk. Selsk. Mat. Fys. Medd.* 34 (14):1–64
11. Appleton, B. R., Erginsoy, C., Gibson, W. M. 1967. *Phys. Rev.* 161:330–49
12. Doyle, P. A., Turner, P. S. 1968. *Acta Crystallogr. Sect. A* 24:390–99
13. Lervig, Ph., Lindhard, J., Nielsen, V. 1967. *Nucl. Phys. A* 96:481–504
14. Andersen, J. U. 1967. *K. Dan. Vidensk. Selsk. Mat. Fys. Medd.* 36 (7):1–26
15. Andersen, J. U., Andersen, S. K., Augustyniak, W. M. 1977. *K. Dan. Vidensk. Selsk. Mat. Fys. Medd.* 30 (10):1–58
16. Alikhanyan, A. I., Esin, S. K., Ispiryan, K. A., Kankanyan, S. A., Korkhmazyan, N. A., et al. 1972. *Pis'ma Zh. Eksp. Teor. Fiz.* 15:142–46 (transl. *JETP Lett.* 15:98–100)
17. Überall, H. 1956. *Phys. Rev.* 103:1055–67
18. Ter-Mikaelian, M. L. 1972. *High-Energy Electromagnetic Processes in Condensed Media.* New York: Wiley. 457 pp.
19. Andersen, J. U. 1980. *Nucl. Instrum. Methods* 170:1–5
20. Jackson, J. D. 1962. *Classical Electrodynamics.* New York, London: Wiley. 641 pp.
21. Heitler, W. 1954. *The Quantum Theory of Radiation.* Oxford: Clarendon. 430 pp.
22. Andersen, J. U., Eriksen, K. R., Lægsgaard, E. 1981. *Phys. Scr.* 24:588–600
23. Atkinson, M., Bak, J. F., Bussey, P. J.,
 Christensen, P., Ellison, J. A., et al. 1982. *Phys. Lett. B* 110:162–66
24. Andersen, S. K., Fich, O., Nielsen, H., Schiøtt, H. E., Uggerhøj, E., et al. 1980. *Nucl. Phys. B* 167:1–40
25. Sáenz, A. W., Überall, H., Nagl, A. 1981. *Nucl. Phys. A* 372:90–108
26. Bird, D. M., Buxton, B. F. 1982. *Proc. R. Soc. London Ser. A* 379:459–79
27. Baym, G. 1969. *Lectures on Quantum Mechanics.* New York, Amsterdam: Benjamin. 594 pp.
28. Andersen, J. U., Bonderup, E., Lægsgaard, E., Marsh, B. B., Sørensen, A. H. 1982. *Nucl. Instrum. Methods* 194:209–24
29. Cue, N., Bonderup, E., Marsh, B. B., Bakhru, H., Benenson, R. E., et al. 1980. *Phys. Lett. A* 80:26–28
30. Komaki, K., Fujimoto, F., Ootuka, A. 1982. *Nucl. Instrum. Methods* 194:243–46
31. Andersen, J. U., Bonderup, E., Lægsgaard, E., Sørensen, A. H. 1983. *Phys. Scr.* In press
32. Bazylev, V. A., Goloviznin, V. V. 1982. *Radiat. Eff.* 60:101–9
33. Dederichs, P. M. 1972. *Solid State Phys.* 27:135–236
34. Adan'yants, A. O., Vartanov, Yu. A., Vartapetyan, G. A., Kumakhov, M. A., Trikalinos, Kh., Yaralov, V. Ya. 1979. *Pis'ma Zh. Eksp. Teor. Fiz.* 29:554–56 (transl. *JETP Lett.* 29:505–7)
35. Swent, R. L., Pantell, R. H., Alguard, M. J., Berman, B. L., Bloom, S. D., Datz, S. 1979. *Phys. Rev. Lett.* 43:1723–26
36. Pantell, R. H., Swent, R. L. 1979. *Appl. Phys. Lett.* 35:910–12
37. Berman, B. L., Bloom, S. D., Datz, S., Alguard, M. J., Swent, R. L., Pantell, R. H. 1981. *Phys. Lett. A* 82:459–61
38. Gouanere, M., Sillou, D., Spighel, M., Cue, N., Gaillard, M., et al. 1982. *Nucl. Instrum. Methods* 194:225–28
39. Swent, R. L., Pantell, R. H., Datz, S., Alvarez, R. 1982. *Nucl. Instrum. Methods* 194:235–37
40. Watson, J. E., Koehler, J. S. 1981. *Phys. Rev. A* 24:861–63
41. Watson, J. E., Koehler, J. S. 1982. *Phys. Rev. B* 25:3079–90
42. Andersen, J. U., Lægsgaard, E. 1980. *Phys. Rev. Lett.* 44:1079–82
43. Andersen, J. U., Datz, S., Lægsgaard,

E., Sellschop, J. P. F., Sørensen, A. H. 1982. *Phys. Rev. Lett.* 49:215–18
44. Alguard, M. J., Swent, R. L., Pantell, R. H., Berman, B. L., Bloom, S. D., Datz, S. 1979. *Phys. Rev. Lett.* 42:1148–51
45. Datz, S., Fearick, R. W., Park, H., Pantell, R. H., Swent, R. L., et al. 1983. *Phys. Lett. A* 96:314–18
46. Park, H., Swent, R. L., Kephart, J. O., Pantell, R. H., Berman, B. L., et al. 1983. *Phys. Lett. A* 96:45–48
47. Berman, B. L., Datz, S., Fearick, R. W., Kephart, J. O., Pantell, R. H., et al. 1982. *Phys. Rev. Lett.* 49:474–77
48. Miroshnichenko, I. I., Murray, J. D., Avakyan, R. O., Figut, T. Kh. 1979. *Pis'ma Zh. Eksp. Teor. Fiz.* 29:786–90 (transl. *JETP Lett.* 29:722–26)
49. Filatova, N. A., Golovatyuk, V. M., Isakov, A. N., Ivanchenko, I. M., Kadyrov, R. B., et al. 1982. *Phys. Rev. Lett.* 48:488–92; *Nucl. Instrum. Methods* 194:239–41
50. Uggerhøj, E. 1983. *Phys. Scr.* In press
51. Smart, D. J., Humphreys, C. J. 1980. *Inst. Phys. Conf. Ser.* No. 52C:211–14
52. Fearick, R. W., Sellschop, J. P. F. 1980. *Nucl. Instrum. Methods* 168:195–202

COSMIC-RAY RECORD IN SOLAR SYSTEM MATTER[1]

R. C. Reedy

Nuclear Chemistry Group, Los Alamos National Laboratory, Los Alamos, New Mexico 87545

J. R. Arnold

Department of Chemistry, University of California, San Diego, La Jolla, California 92093

D. Lal

Physical Research Laboratory, Navrangpura, Ahmedabad 380 009 India; and Scripps Institution of Oceanography, La Jolla, California 92093

CONTENTS

1. INTRODUCTION 506
   1.1 *Nature of the Cosmic Rays and their Interactions* 506
   1.2 *Records in Solar System Matter* 507
   1.3 *Outlook* 507
2. NATURE OF THE COSMIC RAYS 508
   2.1 *Galactic Cosmic Rays* 509
   2.2 *Solar Cosmic Rays* 511
3. COSMIC-RAY INTERACTIONS WITH MATTER 513
   3.1 *Nuclide Production* 514
   3.2 *Track Formation* 517
4. HISTORY OF THE TARGETS 518
   4.1 *Earth* 518
   4.2 *Moon* 521
   4.3 *Meteorites* 522
   4.4 *Cosmic Spherules and Dust* 524
5. HISTORY OF THE COSMIC RAYS 525
   5.1 *Solar Cosmic Rays* 525
   5.2 *Heavy (VH and VVH) Nuclei* 528
   5.3 *Galactic Cosmic Rays* 529
6. CONCLUSIONS 533

[1] The US Government has the right to retain a nonexclusive, royalty-free license in and to any copyright covering this paper.

1. INTRODUCTION

The nuclei in the cosmic rays have enough energy ($E \approx 1$ MeV to many GeV) to penetrate into and interact with matter in the solar system. Some of these interactions leave reaction products (such as nuclides, chemical effects, or atomic displacements) that persist for long periods of time. These reaction products can be used to study the nature of the cosmic rays and their interactions. Our emphasis in this article is on the use of these records to study the history both of the cosmic rays and of the bombarded targets, such as the earth, the moon, and meteorites. These cosmic-ray records have been used for a wide variety of investigations, many of which are discussed here, such as the nature of the ancient sun (1) and the origin of meteorites.

We consider only two types of energetic particles in the earth's environment: the galactic cosmic rays (GCR), which come from outside the solar system, and the solar cosmic rays (SCR), emitted irregularly by major flares on the sun. Our discussion excludes the trapped particles in the earth's magnetosphere, because we are unaware of any material that records an extended past bombardment in that region. Also, no mention will be made of the low-energy particles in the solar wind, although they do leave some records in extraterrestrial matter (2, 3). Likewise, we do not consider the role of the cosmic rays in the early history of the universe, such as in the formation of certain rare light isotopes like ^{9}Be (4).

1.1 Nature of the Cosmic Rays and their Interactions

The nuclei in both types of cosmic rays consist mainly of protons, $\sim 10\%$ alpha particles, and $\sim 1\%$ heavier nuclei (atomic number $Z = 3$ to ~ 90). The GCR particles have high energies but low fluxes, whereas the SCR particles have lower energies but higher fluxes (Table 1). The energy and charge of a particle control which mechanism, nuclear reaction or ionization energy loss, dominates its interaction in matter (5, 6). The SCR particles and the very heavy nuclei ($Z \geq 20$) in the GCR are mainly stopped in the top few centimeters of solid matter, but produce a fairly high density of records there. The lighter particles in the GCR are much more penetrating and induce nuclear reactions.

We consider two types of records left by these interactions. Nuclear reactions produce a variety of radioactive and stable nuclides (e.g. ^{53}Mn and noble gases) that can be detected and identified as having been produced by cosmic-ray particles. The paths travelled by individual nuclei with $Z \geq 20$ and with certain energies contain enough radiation damage that they can be etched by chemicals and be made visible as tracks (7). The ranges of depth in which these records are produced in solid matter are summarized in Table 1.

Table 1 Energies, mean fluxes, and interaction depths of the two types of cosmic-ray particles

Radiation	Energies (MeV nucleon^{-1})	Mean flux (particles cm^{-2} s^{-1})	Effective depth (cm)
Solar cosmic rays			
Protons and helium nuclei	5–100	~100	0–2
Iron-group and heavier nuclei	1–50	~1	0–0.1
Galactic cosmic rays			
Protons and helium nuclei	100–3000	3	0–100
Iron-group and heavier nuclei	~100	0.03	0–10

1.2 Records in Solar System Matter

The records of cosmic-ray interactions have been studied in terrestrial samples and in meteorites, cosmic dust, and lunar samples. Examples of material analyzed are ocean sediments, lunar rocks, the wood of tree rings, and glass parts of the Surveyor III spacecraft returned by the Apollo 12 astronauts. Radionuclides of various half-lives allow us to study the record over specific time periods. Tracks and stable cosmic-ray-produced nuclides extend the history further into the past, including records that were produced shortly after the formation of the solar system.

These records can be used both to study the history of the targets and to determine the nature of the cosmic rays in the past. This is unavoidably a bootstrap process, involving studying one with assumptions about the other. The availability of a suite of terrestrial and extraterrestrial materials, coupled with significant technological developments in studying the effects of charged-particle irradiation, and the big differences in the nature and the interactions of these two types of cosmic-ray particles have made it possible in the last decade to achieve significant progress in both fields.

1.3 Outlook

In the classical period of studying the cosmic-ray record, beginning in the 1950s and extending through the early studies of lunar samples, two things were accomplished. Methods were developed for the detection of a wide range of cosmic-ray-induced changes. Models were developed using these data to demonstrate that these effects were occurring in the distant past as well as now, and that the mean intensities of GCR and SCR, as recorded in the materials, have been roughly constant. We expected alterations in the mean flux of solar particles, and in the GCR flux, to be tied to the 11-year solar cycle. Near each sunspot maximum the GCR flux decreases and there is an increase in the mean intensity of particles emitted by solar flares.

Our ideas are changing. The simple picture of a clock-like 11-year solar cycle must now be modified to include periods like the Maunder Minimum from A.D. 1645–1715 (8), when solar activity was very low for many decades. There may also have been extended periods of high solar activity. We also are looking for other changes in the fluxes of the cosmic rays in the past. Complex exposure histories of the targets are now expected and frequently visible in their cosmic-ray records.

We now can make more precise measurements of the cosmic-ray record in much smaller samples. The use of accelerators as ultrasensitive mass spectrometers (9) is one of the several new or improved methods of detection currently being used to study the cosmic-ray record. Analyses of a wider variety of reaction products (e.g. tracks, many radionuclides, and noble gases) are now often made on aliquots of a sample; this provides a much larger data base for interpreting the cosmic-ray record.

We are also analyzing the record in new types of samples. Cosmic dust collected in the stratosphere is a new class of material now becoming available for the study of the cosmic-ray record (10, 11). The thousands of meteorites found on several ice fields in Antarctica are greatly expanding and diversifying our collection of extraterrestrial matter (12, 13). Their comparatively long terrestrial residence times provide new windows on the past. Deep sea sediments, and the cosmic spherules they contain, are examples of suites of samples containing a cosmic-ray record that can now be studied by sensitive new methods.

2. NATURE OF THE COSMIC RAYS

The energetic nuclei in the solar system have a wide range of energies and compositions (Table 1). The nuclei in both the GCR and SCR are mainly protons and alpha particles (with a ratio of protons to α particles of 10–20), with about 1% heavier nuclei. The magnetic field of the earth causes large latitudinal variations in the incident fluxes and energies of the cosmic-ray particles (14). The temporal and spatial distributions of these particles in the solar system are strongly influenced by interplanetary magnetic fields that are controlled by the sun (15).

Earth-based observations have established that the intensities of both types of cosmic rays are correlated with solar phenomena (15). These and other observations in the vicinity of the earth at 1 astronomical unit (AU) from the sun were the basis for our initial understanding of the flux variations of these particles in the solar system. During the 1970s, experiments on board many spacecraft provided results that have expanded the earlier observations and modified the initial theories. Important data have been obtained by spacecraft (Pioneers 10 and 11, Voyager) which carried instruments far beyond 1 AU (16), although these missions have all been confined to regions of space close to the ecliptic. Very valuable measurements will be made by experiments on future spacecraft, which will explore regions of the solar system far from the ecliptic.

2.1 Galactic Cosmic Rays

The initial sources of the GCR particles and the mechanisms for their acceleration are not well known, but probably include supernovae (discrete sources), the interstellar medium (diffuse sources), or both (17, 126). These particles diffuse or are transported to the solar system, during which time various interactions, including acceleration, may occur (18). Finally, the interplanetary magnetic fields modulate the spectrum of GCR particles as these particles enter the inner parts of the solar system. Although changes in the sources, acceleration, or interstellar propagation of these particles can change their fluxes in the solar system, solar modulation is the dominant source of the observed GCR variability (Figure 1). Near the earth, the fluxes of GCR particles with $E < 1$ GeV nucleon^{-1} are modulated by an order of magnitude during a solar cycle. At $E > 5$–10 GeV nucleon^{-1}, the spectrum of GCR particles is not influenced much by solar activity and its shape can be described roughly by a power law in energy, $E^{-2.5}$.

The modulation is due to the interactions of incoming GCR particles with the interplanetary magnetic fields convected outward by the highly conductive solar-wind plasma, which results in scattering, diffusion, and energy losses (19). It is effective within a zone called the heliosphere, extending out to ~ 50 AU (16, 18). Recent measurements of the fluxes of GCR particles using experiments on balloons and satellites have revealed that the modulation processes are more complex than a simple anticorrelation with solar activity (20, 21). The GCR fluxes were highest during July 1977 (20), even higher than the previous solar minimum in 1965, and are shown in Figure 1 as a cross at 200 MeV nucleon^{-1}, the peak flux. The modulation dynamics also depend on a number of recently recognized solar features, such as the polar coronal holes (22).

An important question in examining the cosmic-ray record is how the GCR spectra varied during sustained periods of very low or very high solar activity. During long periods of essentially no solar activity, such as the Maunder Minimum, A.D. 1645–1715 (8), the interplanetary disturbances would have been essentially absent and the GCR particles would not be hindered from reaching the inner solar system. The local interstellar spectrum (23) is estimated (24) to be similar to the one that is in the inner solar system during such periods of very low solar activity.

While model calculations are needed to predict GCR fluxes during periods of unusual solar activity or at high heliolatitudes, there are data on

GCR flux variations near the ecliptic that extend to distances ~ 30 AU from the sun. The Pioneer 10 and 11 spacecraft have measured out to and beyond 18 AU the fluxes and gradients for several cosmic-ray particles and various energy ranges. For protons with $E > 60$ MeV, there is a heliocentric radial gradient of 2–3% per AU (25–27). Pioneer 11 reached a heliographic latitude of 16°; there was little latitude variation (0 ± 1.5% per 10°) of the fluxes of protons with $E > 80$ MeV (25).

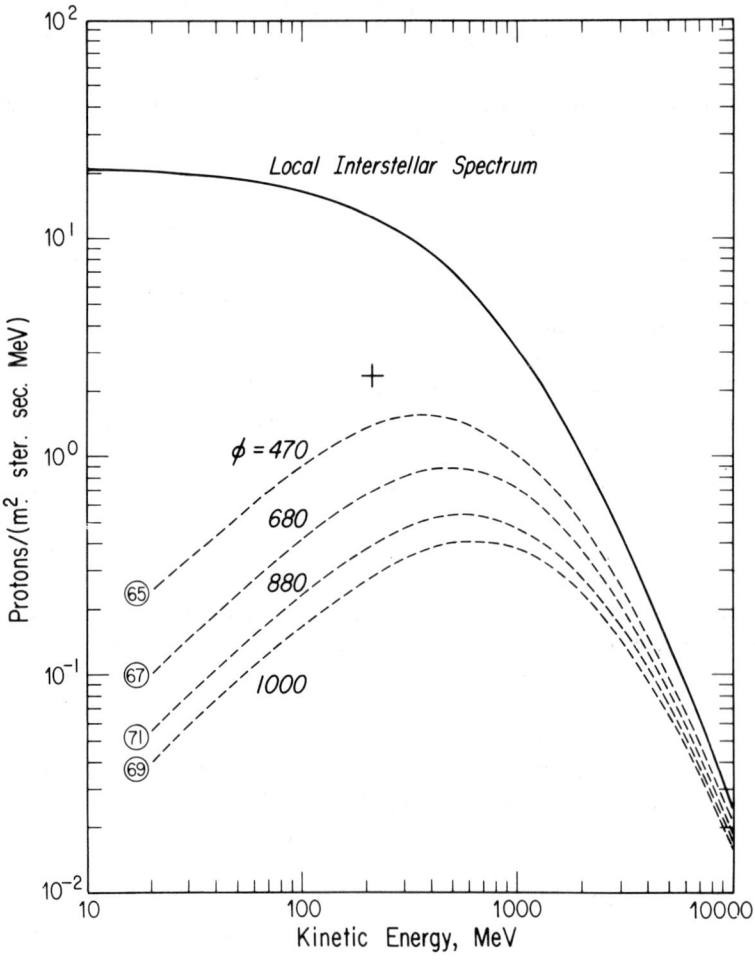

Figure 1 Fluxes of GCR protons near the earth and in local interstellar space. The curves for 1965, 1967, 1969, and 1971 are calculated fits to satellite data (23). The modulation coefficients, ϕ (in MeV nucleon^{-1}), are given for each curve. (Curves and calculations courtesy of M. Garcia-Munoz and J. A. Simpson.) The cross shows the peak proton flux observed in 1977 (20).

2.2 Solar Cosmic Rays

At a distance of 1 AU from the sun, particles emitted by solar flares are an important source of nuclei with $E < 300$ MeV, as seen in Figure 2, which compares GCR-proton fluxes with the time-averaged solar-proton flux. More than 100 SCR events have been observed since 1942 (15). The fluence of protons with $E > 10$ MeV has ranged from $< 10^5$ to almost 10^{11} protons cm^{-2} (6, 28). A few large flares produce most of the SCR particles emitted during an 11-year solar cycle, and few particles are observed when solar

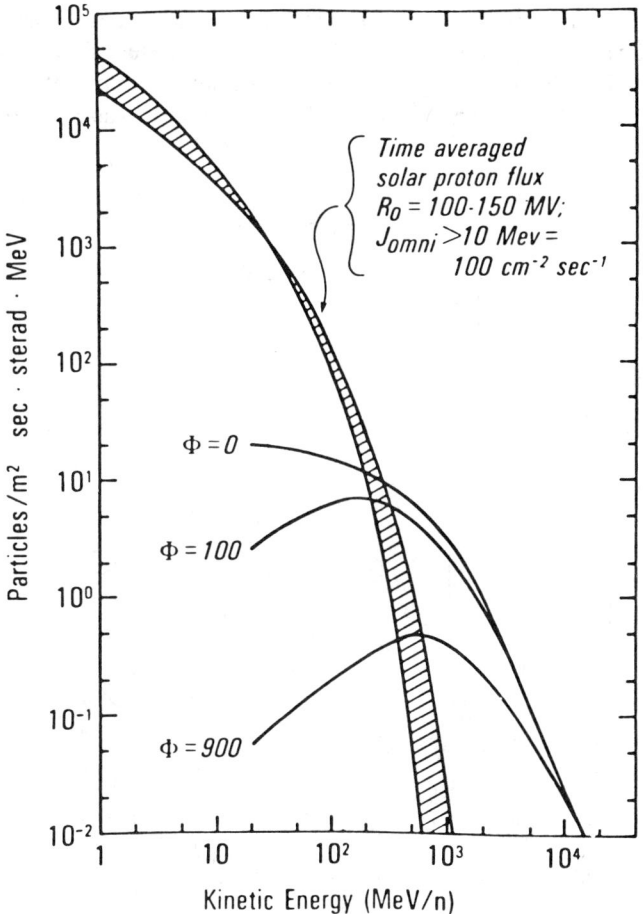

Figure 2 The long-term average fluxes of solar protons determined from lunar data are much larger at low energies than the GCR-proton fluxes for different modulation levels ($\phi = 0$ refers to no modulation; $\phi = 900$ is typical of the GCR fluxes during solar maximum.) [Modified version of Figure 8 of (24).]

activity is low (Figure 3). Individual events last typically for a couple of days and have time-averaged fluxes that are orders of magnitude higher than the total GCR flux.

The energy spectrum of solar particles is soft, with many particles of $E > 10$ MeV, but few with $E > 100$ MeV. In general, the spectrum of SCR particles can be represented fairly well as a power law, $E^{-\gamma}$, where E is the kinetic energy per nucleon. For proton energies between 20 and 80 MeV, the values for γ typically range from 2 to 4, with an average value of 2.9 at the time of maximum proton intensity (29). At lower proton energies, the value of γ generally is lower than that for $E = 20$–80 MeV, while the energy spectrum for energies above ~ 100 MeV usually is steeper. From 1967 to 1972, the values of γ for protons and alpha particles were similar, with the alpha particles having a slightly higher value, and the flux ratio of alpha particles to protons for energies of 3–8 MeV nucleon^{-1} was about 0.03 (30).

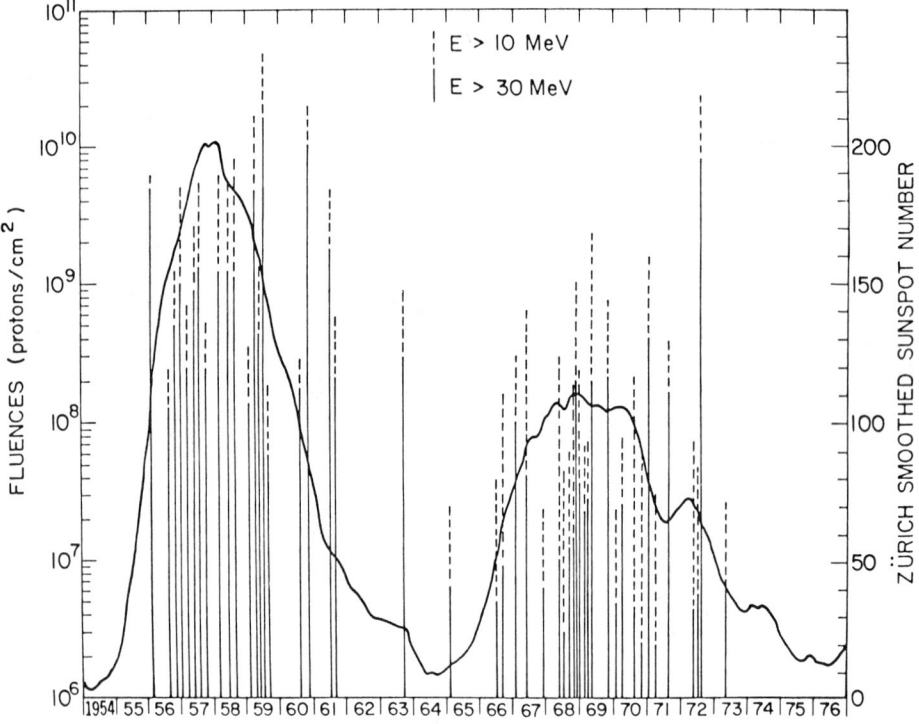

Figure 3 Zurich-smoothed sunspot number (*continuous solid curve*) and the omnidirectional integral fluxes of protons above 10 and 30 MeV emitted by solar flares from 1954 to 1976. Fluences for several flares occurring close to each other have been combined. [Based on the compilation and results of (28).]

The spectral shapes of SCR particles varied widely from event to event, and with the particle fluxes. An alternate spectral shape for solar protons with energies above about 10 MeV is $dJ/dR = k' \exp(-R/R_0)$ where R is rigidity, the momentum per unit charge (pc/Ze), and R_0 typically has values of 50–150 MV (6).

The calculations for the transport of solar particles predict that the momentary flux of solar particles father away from the earth will behave as $R^{-3.7}$ in the early phases, but the integrated particle fluence obeys the inverse square law (R^{-2}), where R is the distance from the sun (31). Analyses of SCR events observed by detectors on board Pioneers 10 and 11 showed that the decrease in the flux with distance from the sun varied about this calculated ratio (32). The interplanetary acceleration of solar particles to $E > 10$ MeV occurs quite rarely (15), but this mechanism is a source of much lower energy particles (32).

3. COSMIC-RAY INTERACTIONS WITH MATTER

The energy, charge, and mass of an energetic particle and the mineralogy or chemistry of the target mainly determine which interaction processes are important and which cosmogenic (cosmic-ray-produced) products are formed. Energetic nuclear particles mainly interact with matter in two ways: ionization energy losses and nuclear scattering or reactions (5). All charged nuclei continuously lose energy by ionizing atoms as they pass through matter. The resulting radiation damage can accumulate in matter and be detected as thermoluminescence (TL). The paths travelled by individual nuclei with $Z \gtrsim 20$ and with $E \approx 0.1$ to 1 MeV nucleon^{-1} can be etched by certain chemicals and made visible as tracks (7). A nuclear reaction between an incident particle and a target nucleus involves the formation of new, secondary particles (e.g. neutrons, pions, and gamma rays) and of a residual nucleus that usually is different from the initial one. Low-energy and high-Z nuclei are rapidly slowed by ionization energy losses. High-energy, low-Z particles lose energy more slowly and usually induce nuclear reactions before stopping. The numbers of secondary particles and of residual nuclei that can be produced by nuclear reactions depend on the energy of the incident projectile.

Because of the variety of the cosmic-ray particles and of their modes of interaction, the effective depths of the interactions and their products vary considerably (Table 1). This diversity of product types and their effective production depths is very useful in studying the record of cosmic-ray interactions. Below a depth of ~ 1000 g cm^{-2} (a few meters in solid matter or the thickness of the earth's atmosphere) there are few cosmic-ray particles (mainly muons) because most of them have been removed by nuclear reactions or stopped by ionization energy losses. It is so hard to detect the few interactions that do occur at such depths that these parts of objects are considered unexposed to cosmic rays. Only near the surfaces are the cosmic-ray records readily determined.

Collisions with large meteoroids can destroy parts of an extraterrestrial object that were exposed to cosmic-ray particles and can expose new surfaces previously shielded from the cosmic rays. Erosion rates due to microcratering are ~ 1 mm per 10^6 years for most exposed rocks. Erosion rapidly removes the top 100 μm that contain the tracks made by heavy nuclei in the solar cosmic rays. The production profiles for nuclides produced by SCR particles also are affected by erosion for time periods $\geq 10^6$ years. The simultaneous study of implanted solar-wind ions, solar and galactic heavy-nuclei tracks, and nuclides produced by GCR and SCR nuclei can reveal the changes in the irradiation geometry because each effect has a characteristic depth dependence.

3.1 Nuclide Production

When matter is exposed to cosmic rays, a large variety of cosmogenic stable and radioactive nuclides are made. The particles inducing the nuclear reactions are the primary ones, mainly protons and alpha particles, and the secondary ones made in matter by the primary particles, especially neutrons and pions. Their energies range from a few eV to many GeV. The more frequently studied cosmogenic radionuclides have half-lives that range from days to million of years and are made from the common elements in extraterrestrial matter or in the earth's atmosphere (Table 2). The activity of a cosmogenic radionuclide starts near zero for a freshly exposed sample and will approach its production rate (assumed to be constant) after the sample has been exposed to cosmic rays for several half-lives. Activities of long-lived radionuclides that are below their equilibrium levels can be used to determine relatively short exposure ages. Stable cosmogenic noble-gas nuclides, such as ^{21}Ne, are readily detected by mass spectrometry and often are used to determine a sample's integral exposure to cosmic rays. In iron meteorites, stable cosmogenic isotopes of certain elements (e.g. Cr, Ca, and K) have also been measured.

The relatively low-energy solar protons and alpha particles are usually stopped by ionization energy losses very near the surface of the material. The SCR particles that induce nuclear reactions produce few secondary particles. The product nucleus is close in mass to the target nucleus. Nuclear reactions induced by SCR-produced secondary neutrons are thus unimportant (33). The fluxes of SCR particles as a function of depth can be calculated accurately from ionization-energy-loss relations, so production rates of a nuclide as a function of depth can be calculated well if the cross sections for its formation are known (5).

The activities of SCR-produced nuclides decrease rapidly with increasing depth (Figure 4), the steepness of the profile roughly being inversely proportional to the threshold energy of the major reaction. The ^{56}Co made by the low-energy ^{56}Fe(p,n)^{56}Co reaction has a very steep profile (5). Because solar alpha particles have relatively short ranges, the depth-activity profiles of the nuclides they produce (e.g. ^{59}Ni from iron) are extremely steep (34). For most nuclides produced in lunar samples by SCR particles, solar-alpha-particle-induced reactions are relatively unimportant.

Because the high-energy GCR particles have ranges in matter that are much longer than their interaction lengths, most react before they are stopped and produce many secondary particles, especially neutrons and pions. The cascade that develops from GCR-particle interactions produces a population of particles with many low- and medium-energy particles. The fluxes of the high-energy GCR particles decrease roughly exponentially

Table 2 Cosmic-ray produced radionuclides frequently measured in terrestrial or extraterrestrial matter

Radionuclide	Half-life[a] (years)	Main targets[b]	Particles[c]
^{3}H	12.323	O, Mg, Si	GCR, SCR
^{10}Be	1.6×10^6	O, Mg, Si, (N)	GCR
^{14}C	5730	O, Mg, Si, (N)	GCR, SCR
^{22}Na	2.602	Mg, Al, Si	SCR, GCR
^{26}Al	7.16×10^5	Al, Si, (Ar)	SCR, GCR
^{32}Si	105[d]	(Ar)	GCR
^{36}Cl	3.0×10^5	Ca, Fe, (Ar)	GCR
^{37}Ar	35.0 days	Ca, Fe	GCR, SCR
^{39}Ar	269	K, Ca, Fe	GCR
^{40}K	1.28×10^9	Fe	GCR
^{46}Sc	83.82 days	Fe, Ti	GCR
^{48}V	15.97 days	Fe, Ti	GCR, SCR
^{53}Mn	3.7×10^6	Fe	SCR, GCR
^{54}Mn	312.2 days	Fe	SCR, GCR
	2.7	Fe	SCR, GCR
^{56}Co	78.76 days	Fe	SCR
^{59}Ni	7.6×10^{4e}	Fe, Ni	GCR, SCR
^{60}Co	5.272	Co, Ni	GCR
^{81}Kr	2.1×10^5	Sr, Zr	GCR, SCR
^{129}I	1.6×10^7	Te, Ba, La, Ce	GCR

[a] The half-lives, in years unless otherwise stated, are from (121).
[b] The target elements from which most production occurs; targets listed in parentheses are for production in the earth's atmosphere.
[c] The type(s) of cosmic-ray particles producing significant amounts of the radionuclides; GCR = galactic cosmic rays, SCR = solar cosmic rays.
[d] The average of new measurements (123, 124).
[e] The new half-life (125).

with depth. The fluxes of secondary neutrons increase with depth near the surface, but then level out and decrease exponentially with depth (5, 35). In the moon's surface, there are about 13 neutrons $cm^{-2} s^{-1}$ produced with energies below 10 MeV (36). The flux of GCR particles varies with solar activity (being highest at periods of solar minimum), but the shapes of GCR production rates versus depth do not change much over a solar cycle (33).

Neutrons slowed to energies of keV or eV produce nuclides by neutron capture reactions. Such a reaction can produce a very high yield of a

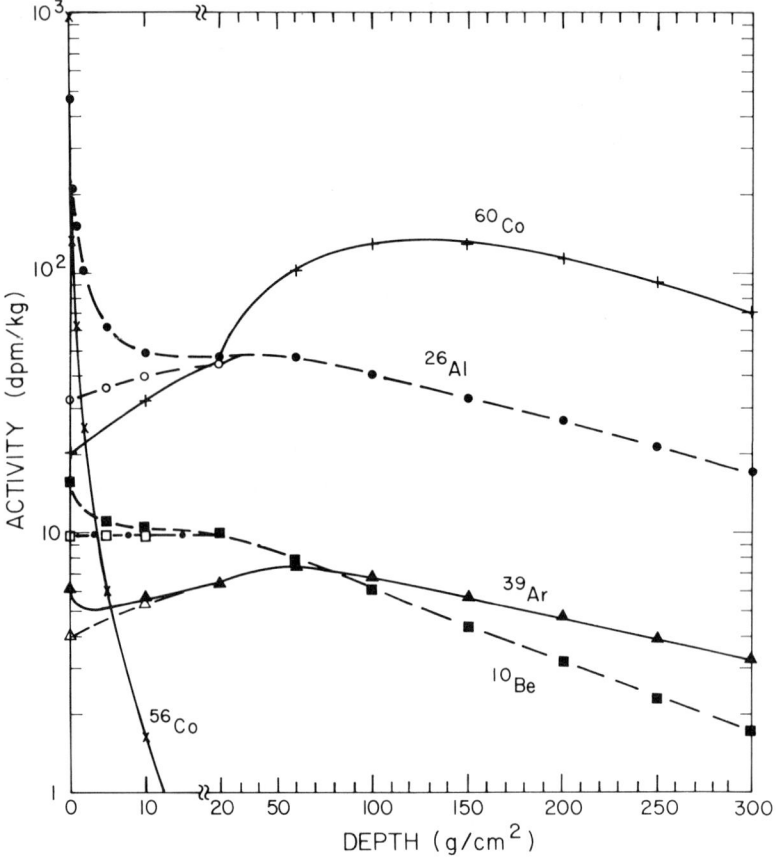

Figure 4 Predicted production rates of various radionuclides as a function of depth in the moon. The scale for depths changes at 20 g cm^{-2}. The curve for the production of ^{60}Co by the ^{59}Co(n,γ) reaction is from (120) and is in units of dpm (g Co)$^{-1}$. The other curves were calculated using the Reedy-Arnold model (5) and the chemical composition of lunar rock 12002. The curve for ^{56}Co represents only production by solar protons via the ^{56}Fe(p,n)^{56}Co reaction. For ^{26}Al, ^{39}Ar, and ^{10}Be, the solid symbols represent production by both solar protons and GCR particles and the open symbols are for GCR production only.

product nuclide if the capturing nucleus has a large (n,γ) cross section (e.g. ^{158}Gd and ^{60}Co made by neutron capture reactions with ^{157}Gd and ^{59}Co, respectively), and if the parent body is large enough (~ 100 g cm^{-2} or more) to slow neutrons to energies of the order of eV (37). Most GCR-induced reactions involve more energetic particles ($E > 1$ MeV) and the emission of one or more nucleons, and the products often are referred to as spallogenic nuclides. The reactions induced by high-energy particles produce many product nuclides, each in relatively low yield. Usually the largest production rates for cosmogenic nuclides involve low-energy reactions, like (n,2n), because both the fluxes of particles and the reaction cross sections are fairly large.

The fluxes of GCR particles as a function of depth in an extraterrestrial object are not precisely known, in particular for secondary neutrons. Many GCR-induced reactions have not had their cross sections measured, especially those induced by energetic neutrons. However, the general nature of the cascade caused by GCR particles and of the distribution of cosmogenic nuclides is known from theoretical calculations (33, 35), accelerator bombardments of thick targets (38, 39), measurements in the earth's atmosphere (40), and studies of cosmogenic nuclides in extraterrestrial objects (5, 6). Because both the attenuation of primary GCR particles and the build-up of secondary particles occur gradually as a function of depth, the depth-activity profile of a GCR-produced nuclide varies slowly (Figure 4) and depends on the excitation functions of the reactions producing it (5). A fairly high-energy product like ^{10}Be has a GCR depth-activity curve that is flat from the surface to a depth of about 10 g cm^{-2} and that decreases at greater depths, whereas a low-energy neutron-produced nuclide like ^{39}Ar has a profile in which its activity increases by about a factor of two from the surface to about 50 g cm^{-2} and then decreases. Below a depth of ~ 100 g cm^{-2} (~ 300 g cm^{-2} for neutron capture reactions), the production rate of GCR-produced nuclides decreases with depth with an e-folding length of ~ 200 g cm^{-2} (5, 35). The big difference in the depth-activity profiles for the production of nuclides by SCR and GCR particles (see Figure 4) allows these two components usually to be resolved from measured profiles, even for nuclides like ^{26}Al, which are readily made by both SCR and GCR particles.

3.2 Track Formation

Although nuclear reactions can occur in all the constituent phases of meteorites and lunar samples, observations of solid-state damage due to charged-particle irradiation are usually limited to the crystalline dielectric phases of the minerals present, usually olivines, pyroxenes, and feldspars. A variety of techniques are employed to observe solid-state damage due to solar-wind ions, heavy cosmic-ray nuclei (usually $Z > 20$), and fission fragments (6, 7). We are here primarily concerned with tracks due to cosmic-ray nuclei of $Z \geq 20$. From their observed abundances, these nuclei are conveniently subdivided into two groups, those with $20 < Z \leq 28$, termed the iron or VH (very heavy) group of nuclei, and those with $Z \geq 30$, termed the VVH (or very very heavy) group. The VVH nuclei (mainly $30 \leq Z \leq 40$) are less abundant than the VH nuclei by a factor of about 500 to 700 for energies above ~ 500 MeV nucleon^{-1}, similar to solar photospheric and universal abundance ratios (41). Abundances of nuclei with $Z > 40$ are much lower, about 5×10^{-5} nuclei per iron nucleus (41).

In dielectrics, when the primary ionization exceeds a certain critical value, the damaged material is appreciably altered chemically so that the trails of charged particles can be enlarged by suitable chemical treatment and seen with an optical microscope (7). The chemically developed (etched) holes are called tracks. In natural minerals, this ionization threshold is exceeded only by nuclei with $Z > 20$. As Z becomes higher, the range of energies at which ionization rates are above the critical value is greater. The VH and VVH group nuclei form chemically etchable tracks (~ 10–60 μm long) near the end of their range, where their ionization rates are the largest. The determination of Z for the nucleus forming a track can be made usually within ± 2 charge units (7, 41). The identification of charge group (VH or VVH) can be made with certainty.

Profiles of track densities as a function of depth have been determined in many meteorites and lunar samples with simple exposure histories and in a number of man-made materials exposed in space, including a glass filter from the Surveyor III camera returned by the Apollo 12 astronauts. In the track production profile for the moon (Figure 5), the tracks in the top ~ 1 mm are made mainly by heavy SCR nuclei with energies below ~ 10 MeV nucleon^{-1}; the deeper tracks are made by GCR nuclei ($E \geq 100$ MeV nucleon^{-1}) after they have slowed to $E \approx 2$ MeV nucleon^{-1}.

4. HISTORY OF THE TARGETS

There are records of the cosmic-ray bombardment of the earth and of extraterrestrial materials, including lunar rocks and soil, meteorites of various classes, interplanetary dust collected in the stratosphere, and cosmic spherules found in deep-sea sediments. For these materials, the study of cosmogenic nuclides and nuclear tracks gives significant historical information.

4.1 *Earth*

The earth is very active geologically and has had a number of climatic changes in the last few million years. Ice ages, sea-level changes, and

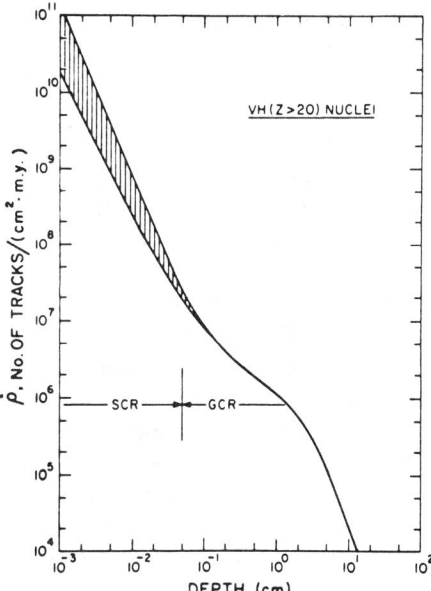

Figure 5 Predicted rates for the production of tracks (1 m.y. = 10^6 years) as a function of depth in a lunar rock (density of 3.4 g cm^{-3}) directly exposed to cosmic-ray particles. The shaded region reflects uncertainties in the fluxes of low-energy VH nuclei in the SCR. [Based on results summarized in (41).]

magnetic variations and polarity reversals have occurred. Cosmogenic radionuclides have provided records of these changes and other terrestrial processes, such as sedimentation and manganese module growth. The most famous cosmogenic radionuclide is ^{14}C (42), whose archeological uses are many but outside the scope of this paper. It has been used as an atmospheric tracer, especially for CO_2, and to study climatic variables that control the mixing of CO_2 among surface reservoirs (43).

On the shortest time scales accessible to ^{14}C, tens to hundreds of years, correlations are observed among solar-activity indices, climatic variables (especially temperature), and ^{14}C activity. When sunspots are few, ^{14}C activity is higher, and the climate is generally colder (8, 44–46). The higher ^{14}C content can be understood as a consequence of weak solar fields allowing more GCR protons to reach the earth. It is less clear how solar activity affects climate.

In the 10^3–10^4-year range, the dominant effect seems to be variation in the earth's magnetic field (Figure 6). The main dipole field was apparently substantially weaker one to two half-lives ago, so that more cosmic rays reached the earth's atmosphere, and the ^{14}C production rate was higher. The effect is modelled as a sinusoidal variation, with an amplitude in ^{14}C of about 10% and a period of about 10^4 years (44). However, only the past half-cycle is subject to detailed check, so the true variation curve need not be close to the model at times further back than 10^4 years.

Longer-lived cosmogenic radionuclides enter the terrestrial environment in at least three ways. They are carried in by meteoritic material bombarded in space. This is the main source of nuclides produced from iron-group targets in meteorites, the most important of which is ^{53}Mn. They also are produced by spallation reactions in the atmosphere. ^{10}Be is made from nitrogen and oxygen in this way (14); bombardment of atmospheric Ar is the main source of ^{26}Al (47). Finally, a small amount of production takes place in surface rocks and soil by reactions of neutrons and muons.

Highly-sensitive methods of detection, such as activation analysis for ^{53}Mn (48, 49) and the use of accelerators for high-energy ion counting (9), have stimulated renewed interest in these nuclides. Raisbeck and co-workers (50–52) in a series of papers have reported measurements of the ^{10}Be content of sea water, ice cores, and sediments. We expect this work to lead to a useful model of distribution of ^{10}Be in natural waters and sediments, which will permit this nuclide to play a useful role in geochronology and geochemistry. The chemical similarity of Be and Al in

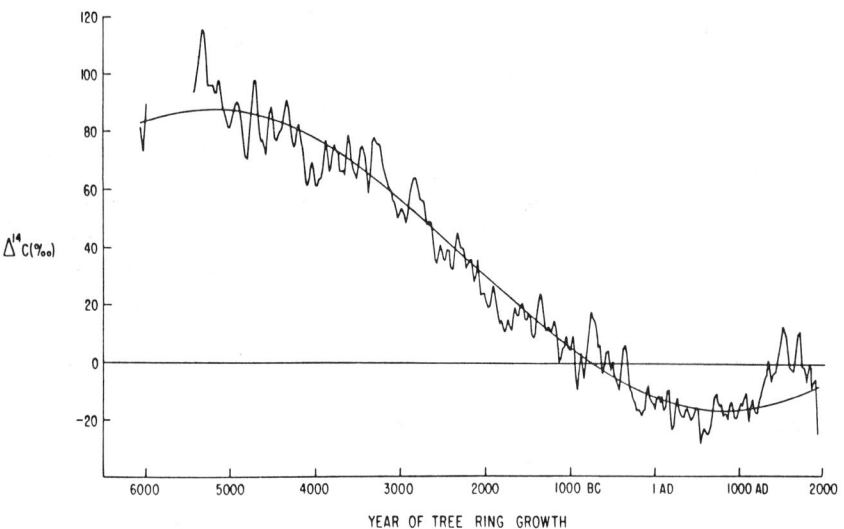

Figure 6 Best-fit sine curve and spline functions drawn through the experimental Δ^{14}C values in parts per thousand measured by the La Jolla Radiocarbon Laboratory for dated tree rings (45). The 10^4-year sine curve represents the slow variation caused by the changing geomagnetic field. The high-frequency fluctuations (Suess wiggles) are computer-drawn spline functions through the experimental Δ^{14}C variations (like the data in Figure 9). [Figure from (45) courtesy of H. E. Suess.]

the laboratory suggests that ^{10}Be and ^{26}Al made in the atmosphere should be distributed in the same way. The measurement of the ^{26}Al/^{10}Be ratio can provide another parameter and reduce the need for a well-supported geochemical model. Raisbeck et al (52a) also have reported ^{26}Al measurements using an accelerator. A potential application of ^{53}Mn is in the understanding of the special processes that lead to the deposition of Mn-rich nodules and crusts in the deep sea. Such measurements have not yet been done because of the low ^{53}Mn/^{55}Mn ratio in such samples.

4.2 Moon

The moon inhabits the same region of space as the earth, but it has no atmosphere and relatively weak magnetic fields. The cosmic rays produce all their effects in solid matter that is very stable by terrestrial standards. While there is evidence for other transport processes, gardening (mixing and transport of the soil) by meteoritic impact seems to be dominant at all scales of distance and time accessible to our study. The radionuclides ^{53}Mn and ^{26}Al, produced abundantly by solar protons at depths less than a few centimeters, are ideally suited for gardening studies on a time scale of 10^6–10^7 years.

Our collections contain basically two kinds of samples. Rocks collected on the surface have been there for $\geq 10^6$ years, with interesting exceptions. Most rocks studied seem to have had complex surface histories (53, 54), in which tumbling, fragmentation, burial and re-exposure have all played a role. On time scales as long as 10^7 years, a simple one-stage bombardment history on the moon's surface is improbable for rocks small enough to be collected. One well-characterized event, formation of the South Ray crater at the Apollo 16 site 2.0×10^6 years ago (55), ejected pristine material from below the zone penetrated by cosmic rays, including some rocks that since then have had simple bombardment histories. The dominant alteration effect in such rocks is erosion, produced mainly by sandblasting by micrometeorites, at a rate of ~ 1 mm per 10^6 years (6). The rate is dependent on the hardness of the rock.

We also have many samples of the lunar regolith (soil). The most instructive of these are cores, ranging in length from about 20 cm to 3 m for the long drill stems. Their histories are also complex, usually involving one or a few large cratering or depositional events, superimposed on a quasi-continuum of smaller disturbances. The nuclear track data (6, 56) make clear the local heterogeneity of each layer, except for rare materials like the Apollo 17 orange glass. The lunar-core track results also suggest several episodic enhancements in meteorite bombardment rates (57) during the last $\sim 10^9$ years. The data on SCR-produced ^{53}Mn (Figure 7) and ^{26}Al show that disturbances have occurred, on a time scale of 10^6–10^7 years, to a depth

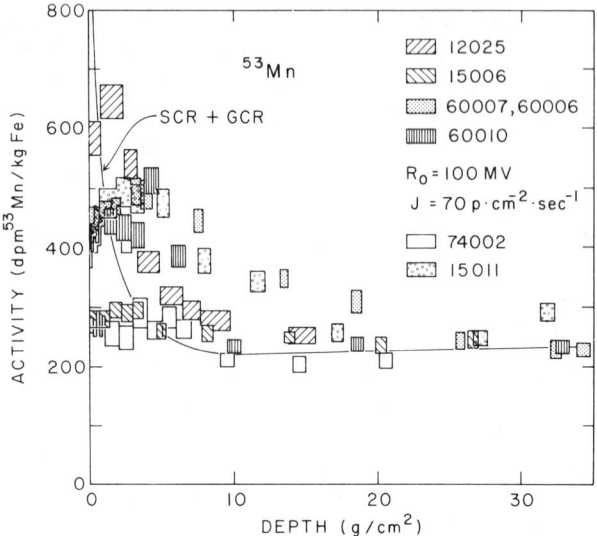

Figure 7 Measured ^{53}Mn activities for samples from six lunar cores. The solid curve is the predicted depth-activity profile for no gardening of the lunar soil and is based on theoretical calculations normalized to lunar rock data. [The R_0 and J values are the average spectral shape and flux above 10 MeV of the solar protons (91).] All six cores show considerable effects due to gardening processes on the moon's surface to depths of 10–20 g cm^{-2} (5–10 cm). [From (59) and K. Nishiizumi, unpublished.]

that varies from a few cm to more than 10 cm. The deeper layers are undisturbed on this time scale. Monte Carlo models have been developed that give a fairly satisfactory match with such observations (58, 59). Tracks and stable cosmogenic nuclides, such as ^{158}Gd and xenon isotopes, have indicated that deeper samples in lunar cores have had complex histories on both small and large depth scales (56).

4.3 Meteorites

Most classes of meteorites originate in the asteroid belt between Mars and Jupiter (60). The cosmic-ray record of meteorites is the main source of information concerning the recent histories of these small solar system objects. Cosmogenic noble gases, radionuclides (especially ^{53}Mn and ^{26}Al), and tracks have been measured in many meteorites. The rates for producing cosmogenic nuclides or tracks as a function of composition and shielding conditions (pre-atmospheric size and sample location) are known fairly well, and are used often to determine the length of time that a meteorite was exposed to cosmic-ray particles.

The exposure (or bombardment) age of a meteorite is most precisely measured with a radioactive-stable pair of nuclides, such as ^{39}Ar-^{38}Ar. The activity of the radionuclide is used to determine the production rate of the stable nuclide. The best pair is ^{81}Kr-^{83}Kr, but there are other good ones. For long-lived iron meteorites, the measurement of ^{40}K along with stable cosmogenic K isotopes is of great value (61). More commonly, a production rate for ^{21}Ne or some other nuclide is used. Recently, measured ^{22}Ne/^{21}Ne isotopic ratios in meteoritic samples have been used to correct ^{21}Ne and other production rates for the shielding conditions during the exposure of the sample (62, 63). Nuclear tracks and activities of long-lived radionuclides also provide information on the exposure history of meteorites. There have been uncertainties about the absolute exposure ages (64–66), but the ^{21}Ne data yield a good set of relative ages.

Chondritic meteorites have bombardment ages which range up to a few tens of millions of years (Reference 67 or Figure 8), while iron meteorites show long ages (61). The longest bombardment age so far, about 2×10^9 years on the ^{40}K–^{41}K scale, is that of the iron meteorite Deep Springs (61). There are statistically significant groups of H-group chondrites at 4–8 $\times 10^6$ years (Figure 8), of coarse octahedrites (irons) at 5–6 $\times 10^8$ years, and of other meteorite classes. These groupings, associated with specific meteorite types, appear to record individual events. Our picture is that meteorites spent their earlier histories contained in bodies that were large compared to the characteristic length of GCR penetration, a few meters. The start of the cosmic-ray exposure was a collision between two objects, which produced disruption or at least fragments of meteoritic size. These were brought into earth-crossing orbits by planetary perturbations (68, 69).

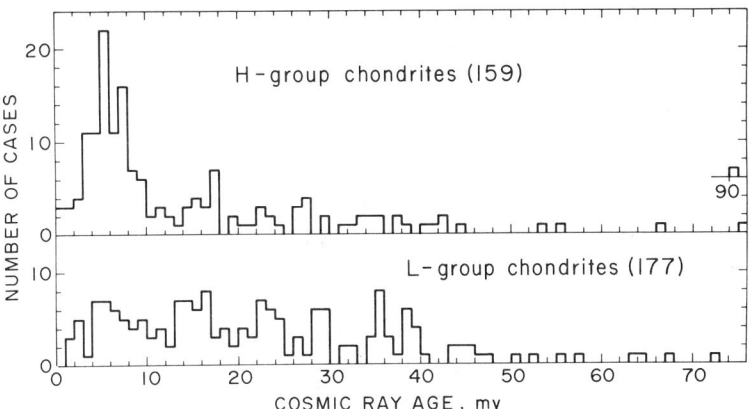

Figure 8 Cosmic-ray exposure ages (in 10^6 years) of the major types of stone meteorites calculated from measured concentrations of cosmogenic ^{21}Ne and ^{22}Ne/^{21}Ne ratios (for shielding corrections) taken from various sources. [From J. Smith and K. Marti, unpublished.]

The observed time scales and age groupings are essential data for planetary perturbation models.

Some meteorites have been small objects for a relatively short time. An example is the chondrite Farmington (70, 71), with a bombardment age of 5 × 10^4 years. In such a case there is little time for perturbation of the original orbit. The orbit of such a short-lived object, if it could be determined, would be of major importance.

Multistage bombardments in space were noted by Chang & Wänke (72) for large iron meteorites. For such objects, meter-sized or larger with bombardment ages $\sim 5 \times 10^8$ years, two or more stages of collisional breakup seem to be the rule rather than the exception. This is consistent with our models (60). Nishiizumi (73) has reported two cases, one very striking, of two-stage bombardment of chondrites in space. There has not yet been a critical evaluation of the implications for meteorite origin.

There is evidence for pre-compaction irradiation, the exposure to high-energy radiation preceding the assembly of the meteorite as a solid rock. Certain gas-rich meteorites seem to have been exposed to the solar wind before assembly and there are nuclear track records of MeV particles in some of these (74, 75). The best case for cosmogenic nuclide production prior to compaction in a meteorite is provided by inclusions (xenoliths) in the chondrite St. Mesmin (76).

The number of meteorites available for research is being multiplied, already by a factor of about three, by the collection of many objects found on "blue ice" regions of the Antarctic ice sheet (12, 13). Our special interest in them is that they are "old." From the evidence of cosmogenic nuclides, their time of fall on earth is nearly always $>3 \times 10^4$ years ago (77) and generally in the range $1-7 \times 10^5$ years. Some iron meteorites have been on the earth longer; Tamarugal, with a terrestrial age of $\sim 2 \times 10^6$ years (72), is the oldest we know. Such objects in principle give us a window on the cosmic-ray flux in past eras.

4.4 Cosmic Spherules and Dust

The work of Brownlee (78) has recently helped to focus attention on the "cosmic spherules" found in deep-sea sediments. These objects have long been known from their composition to be of extraterrestrial origin and are comparatively abundant ($\sim 10^{-7}$ parts by mass) in deep-sea sediments because of the slow rate of sedimentation of terrestrial materials there. The renewed interest comes from the realization that these spherules must include materials from the great majority of meteoroids that never reach the earth's surface (79). The main uncertainty about this material is whether it consists mainly of ablation droplets from large meteoroids, or of small meteoroids which have undergone partial or total fusion. There is structural and isotopic evidence (D. Brownlee, private communication) to support the latter view. A few recent determinations (80) of cosmogenic ^{53}Mn in deep-sea spherules suggest that these spherules were ablation droplets (but see also 80a). Because deep-sea sediment cores give a continuous record over tens to hundreds of millions of years, a detailed study of the intensity and nature of incoming extraterrestrial material over time is possible (79, 81). There is evidence (J. Czajkowski, private communication) of preservation of spherules in Jurassic sediments, more than 1.5×10^8 years old.

First Brownlee and co-workers (82, 83) and now NASA (11) have been collecting in the stratosphere fluffy, unaltered particles seemingly of cometary origin. These particles whose survival was predicted by Öpik (84), are often chondritic in composition, but much less dense and crystalline than the carbonaceous chondrites that have been collected on the earth's surface. The total amount collected is as yet only in the microgram range. Solar-wind noble gases have been detected in them (85, 86), verifying their extraterrestrial origin, but the small amount of material collected so far precludes measuring any cosmic-ray effects. Perhaps nuclear tracks will be first (87).

5. HISTORY OF THE COSMIC RAYS

Many of the targets bombarded by the cosmic rays have relatively simple exposure histories and their cosmic-ray records can be used to study the irradiating particles. Concentrations of radioactive and stable cosmogenic nuclides provide data for determining fluxes of SCR and GCR particles; track densities provide information on the heavy nuclei ($Z > 20$) in the cosmic rays. These cosmic-ray records not only extend our knowledge on the present nature of the particles and their interactions, but are our main sources of information about their histories over various time periods in the past.

Some meteorites and lunar breccias contain evidences of irradiation by cosmic rays (87a) shortly after the solar system was formed. However, in most samples such ancient records were never made or were erased by a variety of processes. The cosmic-ray histories presented here are mainly for the last 10^7 years for the SCR and $\sim 10^9$ years for the GCR. From most samples, only average cosmic-ray fluxes or properties can be determined for various time periods in the past. The derived mean cosmic-ray fluxes are within a factor of two of those observed during the last few decades.

5.1 Solar Cosmic Rays

Direct measurements of the fluxes of SCR particles by particle detectors in satellites have been made only since about 1960. Indirect measurements using absorption of radio waves in the ionosphere over the polar caps or using neutron monitors or ionization chambers go back to 1952 or 1936, respectively (15). The SCR record in meteorites usually is removed by ablation during passage through the earth's atmosphere, although some of it should be preserved in certain samples (88). However, the lunar samples have provided an excellent medium in which to study the record of SCR particles in the past.

The depth-activity profiles of many SCR-produced radionuclides were clearly evident in the surface layers of the returned lunar rocks. Generally the compositions and cross sections used to unfold the profiles of solar-proton-produced radionuclides in lunar samples and the corrections for GCR production of these nuclides are reasonably well known (89). The activities of short-lived radionuclides (e.g. ^{56}Co) produced in lunar rocks by solar protons were in good agreement with those calculated from satellite-measured proton fluxes for the solar flares occurring before each mission (28, 90). The SCR-produced activities measured for ^{22}Na, ^{55}Fe, and ^{3}H were made mainly by protons emitted prior to direct satellite measurements and, along with the relative intensities inferred from indirect measurements, were used to determine solar-proton fluxes for the solar cycle that occurred from 1954 to 1964 (28). The fluxes adopted from satellite measurements or determined from lunar radioactivities for these last two solar cycles are given in Table 3.

Several long-lived radionuclides have been used to study SCR fluxes in the past: ^{14}C, ^{59}Ni, ^{81}Kr, ^{26}Al, and ^{53}Mn. Usually they are measured in lunar rocks with long exposure ages, so each radionuclide can tell us about fluxes one to two half-lives before the present. Several rocks have short exposure ages, determined from their records of GCR particles, which can be used to study other intervals. These nuclides are produced mainly by

Table 3 Average solar-proton fluxes over various time periods as determined from lunar radioactivity measurements

			Fluxes (protons cm^{-2} s^{-1})			
Time period	Date source	Reference	$E > 10$ MeV	$E > 30$ MeV	$E > 60$ MeV	$E > 100$ MeV
1965–1976	SPME[a]		90	30	8	
1954–1964	^{22}Na, ^{55}Fe		380	140	60	26
10^4 years	^{14}C			70	26	9
3×10^5 years	^{81}Kr				18	9
10^6 years	^{26}Al		70	25	9	3
5×10^6 years	^{53}Mn		70	25	9	3

[a] Direct measurements by the Solar Proton Monitor Experiment (122).

solar protons, except ^{59}Ni, which is made mainly by solar alpha particles. Based on lunar ^{59}Ni activities and on satellite measurements of solar alpha particles, Lanzerotti et al (34) concluded that long-term and current solar-alpha-particle fluxes are comparable to within a factor of four.

Over the last 0.5–10 million years, ^{26}Al and ^{53}Mn activities indicate relatively little change in the average fluxes of solar protons. The solar-proton fluxes given in Table 3 for these radionuclides are our preferred values, based on measurements made by Kohl et al (91) using several lunar rocks with a variety of exposure ages. These fluxes agree with those derived from lunar rock measurements by the Battelle group, (91a) and are consistent with the record in lunar cores (59). Bhandari et al (92) measured ^{26}Al activities in four rocks with exposure ages ranging from 0.5 to 3.7 million years and concluded that the average solar-proton fluxes during the last 0.5 to 1.5 million years have varied less than $\pm 25\%$. However, their reported fluxes are much higher than those adopted here, and Potdar (92a) has further results consistent with theirs. The disagreement is well outside experimental error, and requires resolution. In any case, the fluxes measured during the last few decades and the last few million years are similar, which indicates that current solar activity is not atypical of what it has been in the past.

The activities of ^{14}C and ^{81}Kr indicate that SCR fluxes averaged over the last 10^4 and 10^5 years were considerably higher than they were over the last 10^6 years (see Table 3). The ^{14}C measurements for six depths in a lunar rock gave solar-proton fluxes about three times those determined from ^{26}Al and ^{53}Mn data (93). Concentration-depth profiles of ^{81}Kr have been measured mass-spectrometrically in two lunar rocks and imply that solar-proton fluxes above 60 MeV (the threshold energy for the main reactions producing ^{81}Kr) were considerably higher than those determined for the last 10^6 years (89, 94). However, there are some uncertainties in the ^{14}C- and ^{81}Kr-deduced fluxes, including the cross sections used to unfold the measured profiles (89).

While the average fluxes and spectral shapes of solar protons during the last $\sim 10^7$ years are similar to those observed recently, much less is known about the nature and distribution of individual SCR events in the past. Giant solar-proton events having fluxes several orders of magnitude greater than the largest ones observed during 1956–1972 have been postulated as causing extinction catastrophes in the fossil record (95). The tree ring record of ^{14}C (see below) can be used to conclude that no flares more than ten times greater than those observed since 1956 have occurred since ~ 5000 B.C. The lunar data on ^{26}Al and ^{53}Mn indicate that, during the last $\sim 10^7$ years, there have been very few solar flares that are orders of magnitude larger than those observed during the last two solar cycles (96).

5.2 Heavy (VH and VVH) Nuclei

The studies of tracks made by VVH and VH nuclei in meteorites and in lunar samples have provided long-term average values of relative fluxes and energy spectra of nuclei for $Z = 20$–92 and $E = 0.5$–2000 MeV nucleon^{-1} (6, 97, 98). The determination of absolute time-averaged fluxes of VH and VVH nuclei in the past is not yet possible because of the lack of an independent measure of time, but ratios of heavy nuclei to protons in the SCR have been obtained (41, 98). At low energies (< 10 MeV nucleon^{-1}), the duration of irradiation is erosion controlled; in the high-energy region (> 100 MeV nucleon^{-1}), fragmentation of rocks, which changes the exposure geometry, becomes important. The track profiles for rocks with multiple exposure histories are flatter than production profiles. Fortunately, it is possible to select rocks having a predominantly single-stage irradiation.

The flux and spectrum of iron-group nuclei with $E = 100$–200 MeV nucleon^{-1} are similar to these observed today. These tracks were formed during different intervals over the last 10^9 years. For nuclei with $Z \geq 30$, the available data primarily refer to the VVH/VH abundance ratios in the energy region 100–1000 MeV nucleon^{-1}. The ratio is found to be $1.5 (\pm 0.2) \times 10^{-3}$, in good agreement with the abundance ratios in the sun or primitive chondrites (98). The relative abundances of heavier nuclei in the four charge groups, 52–62, 63–75, 76–83, and 90–96, have also been found to match well with the solar and cosmic abundances.

Below 50 MeV nucleon^{-1}, the VVH/VH ratio increases as one goes to lower energies. For heavy nuclei, ~ 40 MeV nucleon^{-1} is the dividing line above which the flux is due mainly to GCR and below to SCR (98). This ratio increases down to energies ~ 1 MeV nucleon^{-1}, reaching $\sim 2 \times 10^{-2}$, about ten times higher than the solar ratio. Preferential acceleration of heavier nuclei at low energy also has been observed in recent SCR events (99).

High densities of tracks have been observed in grains inside the "gas-rich" meteorites (74, 75). These grains also have high concentrations of solar-wind-implanted noble gases (100, 101). They appear to have been irradiated $\sim 4 \times 10^9$ years ago while in a regolith similar to that found now on the moon's surface (102). Carbonaceous chondrites also have grains with tracks made during the early history of the solar system (97). The ratios of VVH/VH nuclei irradiating these meteoritic grains have varied, but within factors of 2–3.

The SCR tracks in many grains of lunar soil also were made a long time ago. Analyses of track densities and gradients in grains from lunar cores show no evidence for $\sim 10^3$–10^4-year periods with very high fluxes of heavy

SCR nuclei during the last $\sim 10^9$ years (103). The shape of the energy spectrum of VH nuclei has remained remarkably similar for the epochs for which data are available.

5.3 Galactic Cosmic Rays

Most GCR particles have such high energies that their interactions with matter are very complex, and so it is hard to unfold their records to obtain absolute fluxes. Most studies of GCR flux variations compare activities of radionuclides with different half-lives in meteoritic or lunar samples. Ratios of measured activities to those predicted by various models or to the concentrations of stable cosmogenic nuclides are used to look for variations in average GCR-particle fluxes over the mean-lives of various radionuclides. Activities of ^{10}Be and ^{14}C in terrestrial samples are used to study GCR flux variations over shorter time intervals.

The orbits that most meteorites had before colliding with the earth are not known. So far, there are only three meteorites whose orbits are known accurately from several photographs of their fireball trajectories. These meteorites, Pribram, Lost City, and Innisfree, had low orbital inclinations, and thus were exposed not far from the earth's orbital plane. The activities of long-lived radionuclides in these and other meteorites agree well, which indicates that almost all meteorites have been exposed to similar fluxes of GCR particles. The Malakal chondrite has an unusually high activity of ^{26}Al, which might be the consequence of an irradiation by a high flux of cosmic-ray particles before $\sim 2 \times 10^6$ years ago (104). Its ^{53}Mn content, however, is normal (73). High activity ratios of short-lived (^{22}Na, ^{54}Mn) to long-lived (^{26}Al, ^{53}Mn) radionuclides in the Dhajala chondrite have been interpreted as being due to higher GCR fluxes at heliographic latitudes of 15–40°S than within $\pm 15°$ of the ecliptic plane during solar minimum periods (105).

The activities of ^{22}Na, ^{46}Sc, and ^{54}Mn have been measured in 24 meteorites that fell between 1967 and 1978 (106). These activities, after correcting for sample chemistry and shielding effects, varied by factors of two or more, and the variations correlated with the sunspot cycle, with maximum activities being produced at solar minimum periods. These results indicate that production rates for cosmogenic nuclides in meteorites can vary by up to factors of three between solar minimum and solar maximum periods.

Production rates for radionuclides with half-lives from 16 days (^{48}V) to 3.7×10^6 years (^{53}Mn) were measured in the Aroos iron meteorite and compared with calculated activities for iron meteorites (40). The ratios of observed to calculated activities varied, but did not show any systematic trend with half-life. These and other results for radioactivities in meteorites and lunar samples indicate that the fluxes of energetic (above about 500 MeV) GCR particles have varied less than about 25–50% during the last few million years and are similar to present fluxes.

Many studies of long-term GCR flux variations have used iron meteorites or metallic (FeNi) phases of meteorites because they are chemically simple targets. Most of the radionuclides produced from iron have reaction threshold energies above 100 MeV so that secondary particles are relatively unimportant and results from accelerator bombardments can be used more easily to predict production-rate ratios. Forman et al (107, 108) examined ^{37}Ar and ^{39}Ar activities in metal from many meteorites. They found the ^{39}Ar activities were 10–18% higher than expected from the ^{37}Ar activities, as would be expected for long periods of reduced solar modulation during the last 500 years. Measurements of ^{39}Ar activities in meteorites that fell several centuries ago would help confirm the presence of higher GCR fluxes during the Maunder Minimum (108). Activity ratios of ^{39}Ar to ^{36}Cl measured in iron meteorites are within 10% of the production ratios measured in iron targets bombarded by high-energy protons (109); thus the flux of GCR particles during the last ~ 500 years is similar to the average flux over the last $\sim 5 \times 10^5$ years.

Other long-lived radionuclides, such as ^{26}Al, ^{53}Mn, and ^{40}K, are produced by such different reactions that their calculated production ratios are somewhat model sensitive. As discussed above, a pair of radioactive and stable nuclides that are produced by similar reactions can be used to obtain exposure ages. For iron meteorites, the ^{39}Ar/^{38}Ar, ^{36}Cl/^{36}Ar, and ^{26}Al/^{21}Ne ages usually are similar, but ^{40}K/^{41}K ages are usually about 50% higher (110, 111). It is hard to explain these age differences by meteorite orbital changes $\sim 10^6$ to 10^7 years ago and space erosion (112). Thus, the flux of the cosmic rays to which iron meteorites were exposed during the past $\sim 10^6$ years appears to have been $\sim 50\%$ more intense than the average for the last 10^9 years.

An important new result is the measurement of 1.6×10^7-year ^{129}I in three meteorites and a lunar surface rock (127). The 10^7–10^8-year time scale now accessible is an interesting one for possible GCR variations.

Among stony meteorites, production rates of many cosmogenic nuclides can vary considerably because of differences in chemistry, shape, and size (e.g. 62, 63). Corrections for chemistry and shielding conditions were applied by several investigators in determining ^{21}Ne production rates in many chondrites from measured radionuclide activities and neon concentrations (65, 66, 113). The ^{21}Ne production rates determined with measured activities of ^{22}Na, ^{81}Kr, ^{10}Be, and ^{53}Mn agreed fairly well, but were 0.6–0.7 of those obtained with ^{26}Al data. Other measurements have given results consistent with these observations (e.g. 62, 64). This large discrepancy in the production rate for ^{21}Ne based on ^{26}Al activities could have been caused by a higher flux of cosmic-ray particles sometime during the last few million years (65, 66). However, it is very difficult to make any model for GCR flux variations fit all of the results, especially to make the ^{81}Kr, ^{10}Be, and ^{53}Mn results agree with each other, but to differ significantly from the results for ^{26}Al, which has an intermediate half-life (113). Because the reactions producing these nuclides have much lower threshold energies than those for producing most nuclides from iron, the flux variations that could produce such changes in ^{21}Ne production rates are not inconsistent with the results discussed above for the fluxes of high-energy GCR particles. Other explanations for this unusual set of results might include ^{21}Ne produced before the last cosmic-ray exposure (65), biases in the data selection (66), and inadequacies in making shielding corrections.

In addition to GCR flux variations, there are other causes for different cosmogenic nuclide production rates and apparent exposure ages. For example, shielding changes due to multiple collisions or other causes could have occurred, especially in meteorites with long exposure ages. Some of our basic data, such as cross sections or half-lives, may be in error. These sources must be considered and eliminated before concluding that GCR flux variations caused production rate changes in meteorites.

A GCR flux change could be either solar or nonsolar in origin. The movement of the solar system through the galaxy or in and out of the galactic plane could cause long-term flux changes (114). Rare external events (like nearby supernovae) or solar variations (like the Maunder Minimum) can cause short-term GCR flux changes. Such rapid fluctuations would be difficult to detect in extraterrestrial samples, but are observable in terrestrial samples.

The most interesting cosmogenic radionuclides in terrestrial samples are those like ^{10}Be, ^{14}C, ^{32}Si, and ^{36}Cl, which are mainly made in the earth's atmosphere and which have half-lives that are long compared to the time scales for the removal from the atmosphere (14). These radionuclides produce differential records in organic matter, marine sediments, or glaciers that can be dated by independent techniques (e.g. dendrochronology, geophysical events like magnetic reversals, or natural radioactivity). Most other radionuclides in the terrestrial environment give GCR flux records that are more integral in nature, like those in meteorites, because they are mainly produced in interplanetary space (^{53}Mn) or because they mainly reside in the atmosphere (^{39}Ar or ^{81}Kr).

The activities of ^{10}Be have been measured in a number of cores taken in deep-sea sediments. Sections of these cores have been dated by their paleomagnetic stratigraphies or by thorium isotope methods, and other chemical and physical properties of the cores were determined, so that ^{10}Be concentrations could be converted to production rates (115, 116). The ^{10}Be contents of these sediments vary, mainly because of changes in sedimentation rates but also possibly because of climatic changes or reversals or other variations in the earth's magnetic field. The inferred global ^{10}Be production rates for the last 2.5×10^6 years have varied by less than $\pm 30\%$ when averaged over periods of 10^5 years and less than $\pm 10\%$ for periods longer than 2×10^5 years. Similar results involving ^{10}Be have been obtained for large manganese nodules (117). Studies of large-diameter sediment cores from the equatorial Pacific Ocean indicate that, during the last 10^6 years, the global ^{10}Be production rate could have changed once by as much as $30 \pm 7\%$, averaged over $\sim 10^5$ years, and had several smaller excursions with amplitudes of $<20\%$ (P. Sharma and B. L. K. Somayajulu, private communication).

The terrestrial ^{14}C data provide the sharpest limits available on short-term spikes, or sudden shifts, in the particle flux in the inner solar system, because atmospheric $^{14}CO_2$ is a small part of the total ^{14}C reservoir and a

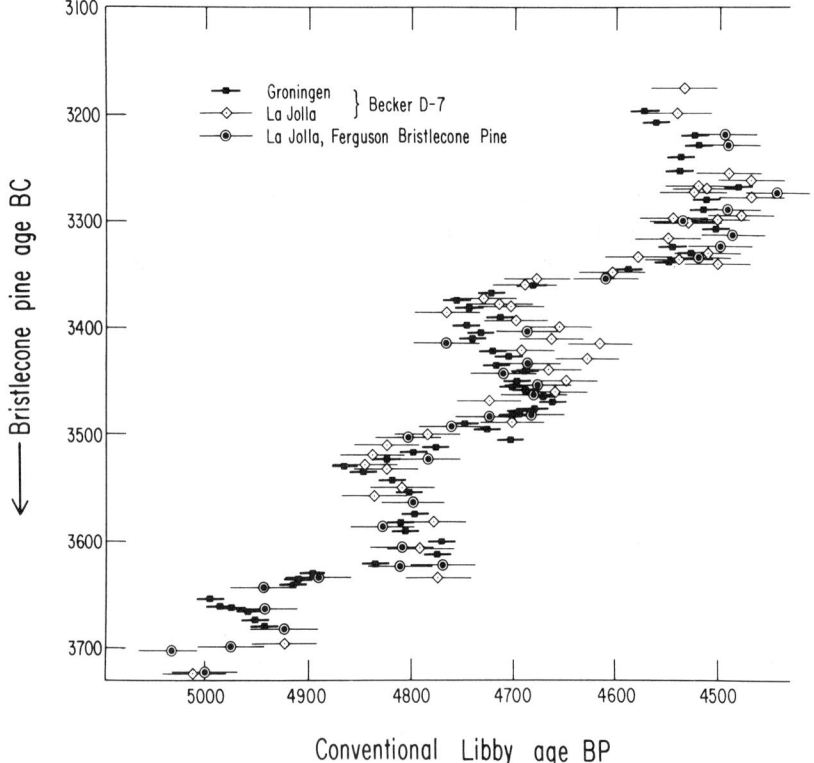

Figure 9 The experimental ^{14}C data with wiggles for the period 4500–5000 B.P. (before present). Note the excellent agreement between the La Jolla and Groningen data. Power spectrum analyses of such ^{14}C data for tree rings show a predominant 200-year wiggle period. [Based on Figure 3 of (45); figure courtesy of H. E. Suess.]

rapid fluctuation in production is well displayed in the record (43, 118). As discussed above and shown in Figures 6 and 9, the measured ^{14}C activities in dated tree rings can be resolved into two components: one slowly varying because of geomagnetic field changes and one oscillating rapidly. The rapid variations (called "Suess wiggles") have amplitudes of 1–2% and a prominent 200-year periodicity (44). These Suess wiggles are believed to provide evidence for periods of unusual solar activity (8, 44, 119). The amplitudes agree with those predicted for changes in the GCR fluxes due to extremes in solar modulation and having the observed frequencies (24). A larger change in the low-energy (<1 GeV) proton flux is not ruled out, because such protons only reach the earth's atmosphere in the polar regions.

6. CONCLUSIONS

The most definite evidence now available for time variations of the cosmic rays near the earth is provided by studies of SCR products in lunar samples. The ^{14}C profiles in lunar rocks require a rather high proton flux on a scale of 10^4 years; limited ^{81}Kr data suggest something similar for as long as a few hundred thousand years. Over the longer periods, represented by ^{26}Al and ^{53}Mn, the flux seems to have been lower, on the average "more like the present" if we knew how to define a present-day average.

As for the GCR intensity in space 1–3 AU from the sun, we have several lines of evidence. The best is from ^{14}C variations in terrestrial wood, which indicate a strong solar modulation effect leading to a large change in the global production of ^{14}C with an average period of about 200 years. These variations exceed in magnitude, but are similar to, solar-modulation effects observed during the last few decades. The last such period of unusual solar activity was 1645–1715 (the Maunder Minimum); it apparently also caused enhanced production of ^{39}Ar in meteorites.

On a longer time scale the classical result is that there has been no change in the time-averaged GCR flux near the earth within some error of about 30–40%. Variations might well be expected either from changes in solar modulation on longer time scales, or in the sun's location in the galaxy (in or out of spiral arms). On a time scale of 10^5–10^7 years, we now see effects that may be due to such changes, but we must eliminate some other possibilities to be sure.

On a 10^9-year scale, data on cosmogenic ^{40}K in iron meteorites require an increase in cosmic-ray flux toward the present, the average over $\sim 10^9$ years being about one third lower than that of the last 10^6 years. We cannot yet be sure whether this is a chapter in the history of meteorites or in that of the cosmic rays.

Track and noble-gas studies indicate that cosmic rays were present in the solar system near its beginning, with energy and charge spectra much like they are now.

Studies of SCR products in lunar samples allow us to measure rates of meteorite impact, regolith turnover (gardening), and rock fragmentation on the moon's surface. Gardening due to impact occurs to depths on the order of 10 cm in a few million years. Surfaces of lunar rocks are eroded at rates of ~ 1 mm per million years.

Meteorites are broken out of larger bodies, again by impact processes, on time scales of $\sim 10^5-10^9$ years. Some meteorites have been in unusual orbits and some differences might have been seen in the GCR fluxes there.

New developments, especially more sensitive techniques for measuring cosmogenic radionuclides, are expected to increase greatly the range and precision of the data available to us. The study of ^{10}Be in terrestrial samples, such as ice cores, may provide a detailed look at time variations on a range of time scales. The use of two or more isotopes (for example ^{10}Be and ^{26}Al) can remove uncertainties in interpretation. Deep-sea sediments, and the cosmic spherules they contain, and also Antarctic meteorites, will provide important windows on the past.

ACKNOWLEDGMENTS

We thank M. Honda, K. Marti, K. Nishiizumi, H. E. Suess, M. Garcia-Munoz, and J. A. Simpson for valuable discussions and for unpublished results and figures used in this paper. One author (R. C. R.) thanks F. Begemann, L. Schultz, and H. Wänke for their hospitality during his stay at the Max-Planck-Institut, Mainz, Federal Republic of Germany, when he did part of this synthesis. This work was supported in part by NASA Grant 7027, NASA work order 14,084, and the Max-Planck Gesellschaft.

Literature Cited[a]

1. Pepin, R. O., Eddy, J. A., Merrill, R. B., eds. 1980. *The Ancient Sun*. New York: Pergamon, 581 pp.
2. Bibring, J. P., Borg, J., Burlingame, A. L., Langevin, Y., Maurette, M., et al. 1975. *PLSC 6th*, pp. 3471–93

[a] The following abbreviations are used: *PLSC* (and *PLPSC*) refer to the *Proceedings of the Lunar (and Planetary) Science Conference*, the first 12 of which were published in New York by Pergamon (except the 2nd and 3rd, which were published in Cambridge by MIT Press) and which were supplements of *Geochim. Cosmochim. Acta*; and the 13th of which was published in Washington by the American Geophysical Union as a supplement to *J. Geophys. Res.*

3. Pepin, R. O. 1980. See Ref. 1, pp. 411–21
4. Reeves, H. 1974. *Ann. Rev. Astron. Astrophys.* 12:437–69
5. Reedy, R. C., Arnold, J. R. 1972. *J. Geophys. Res.* 77:537–55
6. Lal, D. 1972. *Space Sci. Rev.* 14:3–102
7. Fleischer, R. L., Price, P. B., Walker, R. M. 1975. *Nuclear Tracks in Solids*. Berkeley: Univ. Calif. Press. 605 pp.
8. Eddy, J. A. 1976. *Science* 192:1189–1202
9. Litherland, A. E. 1980. *Ann. Rev. Nucl. Part. Sci.* 30:437–73
10. Fraundorf, P., McKeegan, K. D., Sandford, S. A., Swan, P., Walker, R. M. 1982. *PLSC 13th*, A403–8
11. Mackinnon, I. D. R., McKay, D. S.,

Nace, G., Isaacs, A. M. 1982. *PLPSC 13th*, A413–21
12. Cassidy, W. A., Rancitelli, L. A. 1982. *Am. Sci.* 70:156–64
13. Yanai, K., Cassidy, W. A., Funaki M., Glass, B. P. 1978. *PLPSC 9th*, pp. 977–87
14. Lal, D., Peters, B. 1967. *Handb. Phys.* 46(2):551–612
15. Pomerantz, M. A., Duggal, S. P. 1974. *Rev. Geophys. Space Phys.* 12:343–61
16. Simpson, J. A. 1978. *Astronaut. Aeronaut.* 16(7/8):96–105
17. Lingenfelter, R. E. 1979. *Proc. 16th Int. Cosmic Ray Conf., Kyoto*, 14:135–45. Kyoto: Int. Union Pure Appl. Phys.
18. Simpson, J. A. 1983. *Ann. Rev. Nucl. Part. Sci.* 33:323–81
19. Moraal, H. 1976. *Space Sci. Rev.* 19:845–920
20. Webber, W. R., Yushak, S. M. 1979. See Ref. 17, 14:383–88
21. Garcia-Munoz, M., Mason, G. M., Simpson, J. A. 1977. *Astrophys J.* 213:263–68
22. Hundhausen, A. J., Sime, D. G., Hansen, R. T., Hansen, S. F. 1980. *Science* 207:761–63
23. Garcia-Munoz, M., Mason, G. M., Simpson, J. A. 1975. *Astrophys. J.* 202:265–75
24. Castagnoli, G., Lal, D. 1980. *Radiocarbon* 22:133–58
25. Van Allen, J. A. 1980. *Astrophys. J.* 238:763–67
26. Webber, W. R., Lockwood, J. A. 1981. *J. Geophys. Res.* 86:11458–62
27. McKibben, R. B., Pyle, K. R., Simpson, J. A. 1982. *Astrophys. J.* 254:L23–27
28. Reedy, R. C. 1977. *PLSC 8th*, pp. 825–39
29. Van Hollebeke, M. A. I., Ma Sung, L. S., McDonald, F. B. 1975. *Solar Phys.* 41:189–223
30. Stroscio, M. A., Katz, L., Yates, G. K., Sellers, B., Hanser, F. A. 1976. *J. Geophys. Res.* 81:283–86
31. Haffner, J. W. 1972. *Proc. 1971 Natl. Symp. Natural and Man-made Radiation in Space*, NASA Rep. TM X-2440, ed. E. A. Warman, pp. 336–44. Washington, DC: NASA
32. Zwickl, R. D., Webber, W. R. 1977. *Solar Phys.* 54:457–504
33. Armstrong, T. W., Alsmiller, R. G. Jr. 1971. *PLSC 2nd*, pp. 1729–45
34. Lanzerotti, L. J., Reedy, R. C., Arnold, J. R. 1973. *Science* 179:1232–34
35. Lingenfelter, R. E., Canfield, E. H., Hampel, V. E. 1972. *Earth Planet. Sci. Lett.* 16:355–69
36. Woolum, D. S., Burnett, D. S., Furst, M., Weiss, J. R. 1975. *Moon* 12:231–50
37. Eberhardt, P., Geiss, J., Lutz, H. 1963. *Earth Science and Meteoritics*, pp. 143–68. Amsterdam: North-Holland
38. Kohman, T. P., Bender, M. L. 1967. *High-Energy Nuclear Reactions in Astrophysics*, pp. 169–245. New York: Benjamin
39. Honda, M. 1962. *J. Geophys. Res.* 67:4847–58
40. Arnold, J. R., Honda, M., Lal, D. 1961. *J. Geophys. Res.* 66:3519–31
41. Lal, D. 1977. *Philos. Trans. R. Soc. London Ser. A* 285:69–95
42. Libby, W. F. 1955. *Radiocarbon Dating*. Chicago: Univ. Chicago Press. 175 pp. 2nd ed.
43. Siegenthaler, U., Oeschger, H. 1978. *Science* 199:388–95
44. Suess, H. E. 1980. *Radiocarbon* 22:200–9
45. Suess, H. E. 1980. *Endeavour* 4:113–17
46. Eddy, J. A. 1977. *Clim. Change* 1:173–90
47. Tanaka, S., Sakamoto, K., Takagi, J., Tsuchimoto, M. 1968. *Science* 160:1348–49
48. Millard, H. T. 1965. *Science* 147:503–4
49. Nishiizumi, K., Murrell, M. T., Arnold, J. R., Elmore, D., Ferraro, R. D., et al. 1981. *Earth Planet. Sci. Lett.* 52:31–38
50. Raisbeck, G. M., Yiou, F., Fruneau, M., Loiseaux, J. M., Lieuvin, M. 1978. *Nature* 275:731–33
51. Raisbeck, G. M., Yiou, F., Fruneau, M., Loiseaux, J. M., Lieuvin, M., et al. 1979. *Geophys. Res. Lett.* 6:717–19
52. Raisbeck, G. M., Yiou, F., Fruneau, M., Loiseaux, J. M., Lieuvin, M., et al. 1980. *Earth Planet. Sci. Lett.* 51:275–78
52a. Raisbeck, G. M., Yiou, F., Klein, J., Middleton, R. 1983. *Nature* 301:690–92
53. Hörz, F., Gibbons, R. V., Gault, D. E., Hartung, J. B., Brownlee, D. E. 1975. *PLSC 6th*, pp. 3495–3508
54. Wahlen, M., Honda, M., Imamura, M., Fruchter, J. S., Finkel, R. C., et al. 1972. *PLSC 3rd*, pp. 1719–32
55. Behrmann, C., Crozaz, G., Drozd, R., Hohenberg, C., Ralston, C., et al. 1973. *PLSC 4th*, pp. 1957–74
56. Walker, R. M. 1975. *Ann. Rev. Earth Planet. Sci.* 3:99–128
57. Goswami, J. N., Lal, D. 1979. *PLPSC 10th*, pp. 1253–67
58. Langevin, Y., Arnold, J. R. 1977. *Ann. Rev. Earth Planet. Sci.* 5:449–89
59. Langevin, Y., Arnold, J. R., Nishiizumi, K. 1982. *J. Geophys. Res.* 87:6681–91
60. Wasson, J. T., Wetherill, G. W. 1979. *Asteroids*, pp. 926–74. Tucson: Univ. Ariz. Press
61. Voshage, H., Feldmann, H. 1979. *Earth Planet. Sci. Lett.* 45:293–308

62. Cressy, P. J. Jr., Bogard, D. D. 1976. *Geochim. Cosmichim. Acta* 40:749–62
63. Herzog, G. F., Cressy, P. J. Jr. 1974. *Geochim. Cosmochim. Acta* 38:1827–41
64. Finkel, R. C., Kohl, C. P., Marti, K., Martinek, B., Rancitelli, L. 1978. *Geochim. Cosmochim. Acta* 42:241–50
65. Nishiizumi, K., Regnier, S., Marti, K. 1980. *Earth Planet. Sci. Lett.* 50:156–70
66. Müller, O., Hampel, W., Kirsten, T., Herzog, G. F. 1981. *Geochim. Cosmochim. Acta* 45:447–60
67. Crabb, J., Schultz, L. 1981. *Geochim. Cosmochim. Acta* 45:2151–60
68. Öpik, E. J. 1976. *Interplanetary Encounters*. Amsterdam: Elsevier. 155 pp.
69. Arnold, J. R. 1965. *Astrophys. J.* 141:1536–47
70. Kirsten, T., Krankowsky, D., Zahringer, J. 1963. *Geochim. Cosmochim. Acta* 27:13–42
71. Anders, E. 1962. *Science* 138:431–33
72. Chang, C., Wänke, H. 1969. *Meteorite Research*, ed. P. Millman, pp. 397–406. Dordrecht: Reidel
73. Nishiizumi, K. 1978. *Earth Planet. Sci. Lett.* 41:91–100
74. Lal, D., Rajan, R. S. 1969. *Nature* 223:269–71
75. Pellas, P., Poupeau, G., Lorin, J. C., Reeves, H., Audouze, J. 1969. *Nature* 223:272–74
76. Schultz, L., Signer, P. 1977. *Earth Planet. Sci. Lett.* 36:363–71
77. Fireman, E. L. 1980. *PLPSC 11th*, pp. 1215–21
78. Brownlee, D. E. 1978. *Cosmic Dust*, pp. 295–336. Chichester: Wiley
79. Murrell, M. T., Davis, P. A. Jr., Nishiizumi, K., Millard, H. T. Jr. 1980. *Geochim. Cosmochim. Acta* 44:2067–74
80. Nishiizumi, K., Arnold, J. R. 1982. See Ref. 91a, pp. 594–95
80a. Raisbeck, G. M., Yiou, F., Klein, J., Middleton, R., Yamakoshi, Y., Brownlee, D. 1983. *Lunar and Planetary Science XIV*, pp. 622–23. Houston: Lunar Planet. Inst.
81. Parkin, D. W., Sullivan, R. A. L., Andrews, J. N. 1980. *Philos. Trans. R. Soc. London Ser. A* 297:495–518
82. Brownlee, D. E. 1978. *Protostars and Planets*, ed. T. Gehrels, pp. 134–50. Tucson: Univ. Ariz. Press
83. Brownlee, D. E., Tomandl, D. A., Olszewski, E. 1977. *PLSC 8th*, pp. 149–60
84. Öpik, E. 1937. *Tartu Obs. Publ.* 29:51
85. Rajan, R. S., Brownlee, D. E., Tomandl, D., Hodge, P. W., Farrar, H. IV, et al. 1977. *Nature* 267:133–35
86. Hudson, B., Flynn, G. J., Fraundorf, P., Hohenberg, C. M., Shirck, J. 1981. *Science* 211:383–86
87. Frundorf, P., Lyons, T., Schubert, P. 1982. *PLPSC 13th*, A409–12
87a. Goswami, J. N., Hutcheon, I. D., Macdougall, J. D. 1976. *PLSC, 7th*, pp. 543–62
88. Michel, R., Brinkmann, G., Stuck, R. 1982. *Earth Planet. Sci. Lett.* 59:33–48
89. Reedy, R. C. 1980. See Ref. 1, pp. 365–86
90. Finkel, R. C., Arnold, J. R., Imamura, M., Reedy, R. C., Fruchter, J. S., et al. 1971. *PLSC 2nd*, pp. 1773–89
91. Kohl, C. P., Murrell, M. T., Russ, G. P. III, Arnold, J. R. 1978. *PLSC 9th*, pp. 2299–2310
91a. Fruchter, J. S., Evans, J. C., Reeves, J., Perkins, R. 1982. *Lunar and Planetary Science XIII*, pp. 243–44. Houston: Lunar Planet. Inst.
92. Bhandari, N., Bhattacharya, S. K., Padia, J. T. 1976. *PLSC 7th*, pp. 513–23
92a. Potdar, M. B. 1981. *Nuclear Interactions of the Solar and Galactic Cosmic Rays with Interplanetary Materials*. PhD thesis. Gujarat Univ, Physical Res. Lab., Ahmedabad, India
93. Boeckl, R. S. 1972. *Earth Planet. Sci. Lett.* 16:269–72
94. Yaniv, A., Marti, K., Reedy, R. C. 1980. *Lunar and Planetary Science XI*, pp. 1291–93. Houston: Lunar Planet. Inst.
95. Reid, G. C., Isaksen, I. S. A., Holzer, T. E., Crutzen, P. J. 1976. *Nature* 259:177–79
96. Lingenfelter, R. E., Hudson, H. S. 1980. See Ref. 1, pp. 69–79
97. Goswami, J. N., Lal, D., Macdougall, J. D. 1980. See Ref. 1, pp. 347–64
98. Lal, D. 1974. *Philos. Trans. R. Soc. London Ser. A* 277:395–411
99. Shirk, E. K. 1974. *Astrophys. J.* 190:695–702
100. Eberhardt, P., Geiss, J., Grögler, N. 1965. *J. Geophys. Res.* 70:4375–78
101. Wänke, H. 1965. *Z. Naturforsch. Teil A* 20:946–49
102. Housen, K. R., Wilkening, L. L., Chapman, C. R., Greenberg, R. 1979. *Icarus* 39:317–51
103. Crozaz, G. 1980. See Ref. 1, pp. 331–46
104. Cressy, P. J. Jr., Rancitelli, L. A. 1974. *Earth Planet. Sci. Lett.* 22:275–83
105. Bhandari, N., Bhattacharya, S. K., Somayajulu, B. L. K. 1978. *Earth Planet. Sci. Lett.* 40:194–202
106. Evans, J. C., Reeves, J. H., Rancitelli, L. A., Bogard, D. D. 1982. *J. Geophys. Res.* 87:5577–91
107. Forman, M. A., Schaeffer, O. A.,

Schaeffer, G. A. 1978. *Geophys. Res. Lett.* 5:219–22
108. Forman, M. A., Schaeffer, O. A. 1980. See Ref. 1, pp. 279–92
109. Schaeffer, O. A., Heymann, D. 1965. *J. Geophys. Res.* 70:215–24
110. Voshage, H. 1962. *Z. Naturforsch.* 17a:422–32
111. Hampel, W., Schaeffer, O. A. 1979. *Earth Planet. Sci. Lett.* 42:348–58
112. Schaeffer, O. A., Nagel, K., Fechtig, H., Neukum, G. 1981. *Planet. Space Sci.* 29:1109–18
113. Moniot, R. K., Kruse, T. H., Savin, W., Tuniz, C., Milazzo, T., 1982. See Ref. 91a, pp. 536–37
114. Forman, M. A., Schaeffer, O. A. 1979. *Rev. Geophys. Space Phys.* 17:552–60
115. Somayajulu, B. L. K. 1977. *Geochim. Cosmochim. Acta* 41:909–13
116. Tanaka, S., Inoue, T. 1979. *Earth Planet. Sci. Lett.* 45:181–87
117. Sharma, P., Somayajulu, B. L. K. 1982. *Earth Planet. Sci. Lett.* 59:235–44
118. Arnold, J. R., Anderson, E. C. 1957. *Tellus* 9:28–32
119. Stuiver, M., Quay, P. D. 1980. *Science* 207:11–19
120. Wahlen, M., Finkel, R. C., Imamura, M., Kohl, C. P., Arnold, J. R. 1973. *Earth Planet. Sci. Lett.* 19:315–20
121. Seelmann-Eggebert, W., Pfennig, G., Munzel, H., Klewe-Nebenius, H. 1981. *Nuklidkarte.* Karlsruhe: Kernforschungszentrum. 5th ed.
122. Bostrom, C. O., Williams, D. J., Arens, J. F., Kohl, J. W. 1967–1973. *Solar Geophys. Data*, pp. 282–341. Boulder, Colo: US Dept. Commerce, NOAA Environ. Data Serv.
123. Elmore, D., Anantaraman, N., Fulbright, H. W., Gove, H. E., Hans, H. S., et al. 1980. *Phys. Rev. Lett.* 45:589–92
124. Kutschera, W., Henning, W., Paul, M., Smither, R. K., Stephenson, E. J., et al. 1980. *Phys. Rev. Lett.* 45:582–96
125. Nishiizumi, K., Gensho, R., Honda, M. 1981. *Radiochim. Acta* 29:113–16
126. Trimble, V. 1983. *Rev. Mod. Phys.* 55:511–63
127. Nishiizumi, K., Elmore, D., Honda, M., Arnold, J. R., Gove, H. E. 1983. *Nature.* In press

MEASUREMENT OF CHARMED PARTICLE LIFETIMES

Ronald A. Sidwell, Neville W. Reay, and Noel R. Stanton

Department of Physics, Ohio State University, 174 West 18th Avenue, Columbus, Ohio 43210

CONTENTS

1. INTRODUCTION 539
    1.1 *Plan of This Review* 540
    1.2 *Weakly Decaying Charmed Particles* 540
    1.3 *Early Results* 542
2. DESCRIPTION OF RECENT EXPERIMENTS 545
    2.1 *E531* 545
    2.2 *WA58* 548
    2.3 *NA16* 548
    2.4 *NA18* 549
    2.5 *BC72* 550
    2.6 *NA1* 551
    2.7 *Mark II* 552
3. EFFECTS OF BACKGROUND AND SCANNING LOSSES 553
    3.1 *Decay Time Resolution* 554
    3.2 *Systematic Effects Due to Scanning Biases* 555
4. RESULTS 558
    4.1 *Calculating Mean Lifetimes* 558
    4.2 D^0 *Lifetime Measurements* 559
    4.3 Λ_c^+ *Lifetime* 560
    4.4 $F^{\pm}$ *Lifetime* 561
    4.5 $D^{\pm}$ *Lifetime* 563
5. COMPARISON OF LIFETIMES AND SEMILEPTONIC DECAY RATES 564
    5.1 *Ratio* $\tau(D^+)/\tau(D^0)$ 565
    5.2 *Semileptonic Partial Widths* 565
    5.3 *Nonleptonic Enhancement?* 566
6. SUMMARY 566

1. INTRODUCTION

The experimental study of the lifetimes of charmed particles has progressed enormously in the last five years. In 1978 the available fragmentary results were in conflict not only with each other, but with the expectations that $\tau_{\text{charm}} \approx \text{few} \times 10^{-13}$ seconds and that the cross section in hadron collisions should be of order tens of microbarns. Recent hybrid experiments, which consist of fine-grained vertex detectors coupled to downstream spectrometers, have now located several hundred examples of charmed decays. The measurements of lifetimes are in good agreement with one another, and differing lifetimes for the various species of states can be distinguished.

1.1 Plan of This Review

We first briefly review the classification of the charmed particle states, and theoretical expectations for the lifetimes of the lowest-lying particles. We then discuss the experimental techniques used to measure the lifetimes, with some emphasis on the inherent difficulties. Finally, we review and summarize the available results. Our references to the theoretical literature are abridged, but we try to present as complete a picture as possible of the experimental situation. Much of the data is available only in preliminary form: in preprints, conference proceedings, or in rapporteurs' summaries.

1.2 Weakly Decaying Charmed Particles

The strongly interacting elementary particle with quantum number charm is the fourth quark to be discovered. Like the other quarks, it has so far only been found bonded to an antiquark (charmed meson) or in a three-quark system (charmed baryon). The charmed mesons are expected to have quark compositions of $(c\bar{u})$, $(c\bar{d})$, $(c\bar{s})$, and $(c\bar{c})$, where c, u, d, s denote the charm, up, down, and strange quarks. (A table of their properties can be found on p. 338 of Ref. 28.) The lightest such family, the pseudoscalar mesons ($J^P = 0^-$), have all been observed; they are named the D^0, D^+, F^+ and η_c. The D^0, D^+, and F^+ exhibit explicit charm, that is they have quantum number charm = 1. The more massive vector particles ($J^P = 1^-$) include the J/ψ discovered in 1974 and the D^{*+} and D^{*0}. The spectroscopy of the charmed mesons found in e^+e^- interactions was reviewed recently by Goldhaber & Wiss (28).

The only firmly established charmed baryon at present is the Λ_c^+(cud), which is expected to be one of a large family of spin-$\frac{1}{2}$ baryons, several of which may be long-lived.

Decay of the lowest-lying particles with explicit charm, $C = \pm 1$, must proceed through the change in flavor of a constituent quark, which is only observed to happen via the charged current electroweak interaction. The expected meson transitions are classified by the decay products into leptonic, semileptonic, and nonleptonic final states. The estimated widths for the leptonic decay mode are very small (except possibly for $F \to \tau \nu_\tau$) (27), and none has yet been observed. The Λ_c^+ baryon can only decay semileptonically or into all-hadron final states.

The semileptonic decay of charmed particles is viewed as proceeding through the beta decay of the charmed quark, $c \to s + \ell^+ + \nu_\ell$, or $c \to d + \ell^+ + \nu_\ell$, as illustrated in Figure 1a. The partial width for this diagram is approximately given by scaling from muon decay:

$$\Gamma_{SL}(c \to s\ell^+ \nu) \simeq \left(\frac{m_c}{m_\mu}\right)^5 \Gamma(\mu \to e\bar{\nu}_e \nu_\mu)$$

where m_c is the mass of the charmed quark. For $m_c \simeq 1.75 \text{ GeV}/c^2$, $\Gamma_{SL} \simeq 5 \times 10^{11} \text{ s}^{-1}$. The W^+ in Figure 1a can also couple with equal strength to a quark-antiquark (colored) pair as shown in Figure 1b, and this diagram produces a nonleptonic final state. In the simplest form of the spectator quark model for charmed particle decays, one ignores the presence of other quarks in the initial state and

$$\Gamma_{all} = \Gamma_{SL} \times 2 + \Gamma(c \to s u \bar{d}) \times 3 \simeq 5\Gamma_{SL} \simeq 10^{+12}\text{–}10^{13} \text{ s}^{-1}$$

where the multiplying factors are for two kinds of leptons and three colors of quarks. Thus, in the simplest model Γ_{SL} is the same for all charmed particles, as is the total width.

Particle proper decay times t_0 are related to decay distance l by $l = pct_0/m$ where $p = $ momentum, $m = $ mass, and $c = $ speed of light. Thus for a charmed particle with momentum 10 GeV/c, mass about 2 GeV/c^2, and reciprocal width 10^{-13} seconds, the mean distance traveled before decaying is 150 μm. Since decay times are distributed exponentially and particles are usually produced with a broad range of energies, flight paths are typically tens of micrometers to occasionally as much as several

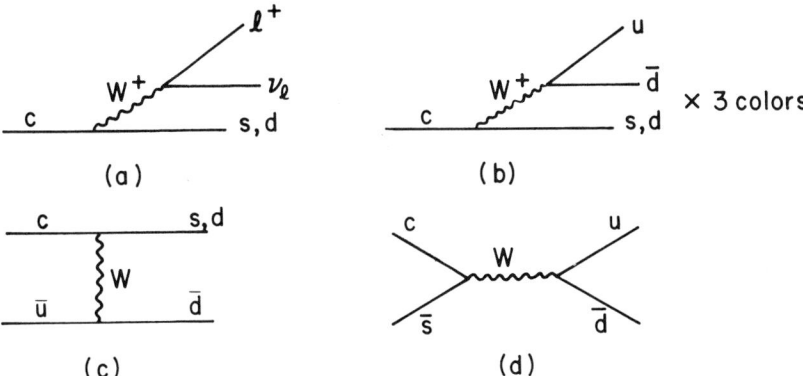

Figure 1 Flavor-changing diagrams for the charmed quark: (a) and (b) are sometimes known as W^+ radiation; (c) and (d) are W exchange and annihilation.

centimeters. This range of distances is not well matched to the conventional visual detectors available several years ago. Most bubble chamber experiments had bubble sizes of several hundred μm, so that only the longest decays would be resolved. Nuclear research emulsions can resolve decays shorter than 10 μm; however, the finding of events under the microscope can have a low success rate without guidance from an external particle detector, and the usual methods by which one finds decays of charged and neutral particles are inherently different. Charged particle decays can be found efficiently in emulsion by following tracks from the point of emission (primary interaction) to the decay point, for distances up to several centimeters. Neutral decays are conventionally searched for by carefully examining the volume of emulsion downstream of the primary interaction. This process, called volume scanning, is usually carried out at magnifications of $100 \times$ to $250 \times$. It is time consuming and becomes less successful as the emulsion scanner moves his field of view away from the primary interaction. The efficiency for finding decays, nearly 100% within 300 μm of the primary, drops to 20% or less at distances greater than one mm, and is difficult to calibrate. The interested reader can find a longer discussion of the relative merits of bubble chambers and emulsions and the methods for recognizing decays in these two media in Ref. 34. In Section 2.1 we describe a new emulsion technique (scan-back) which solves some of the difficulties with this medium.

As well as finding events, the experimenter must determine decay times by measuring p and m. An external-momentum-analyzing spectrometer containing neutral particle identification and charged particle tagging (i.e. able to tag e, μ, π, k, p) is highly desirable. The lifetime $\tau = 1/\Gamma$ is then found by performing a maximum likelihood fit for an ensemble of particles of a given type (such as Λ_c^+), or by fitting the decay time spectrum dN/dt. In either case the efficiency for finding a particle with a decay time t must be known.

The expectation that the ground-state charmed particles are long-lived, with decay lengths in the range of $10-10^3$ μm, led to a considerable research effort to observe the decays directly; at first in bare emulsion exposures and in existing bubble chambers, more recently in a variety of detectors coupled to powerful spectrometers. Before discussing the latter experiments (post 1978) it is interesting to look at the early results, those reported prior to 1979. We recommend the rapporteur reviews by K. Niu (36) and R. Diebold (21).

1.3 Early Results

The first observation of candidates for charmed particle decays came from the exposure of emulsion chambers to cosmic rays (35). By 1975, 16 examples had been found, in some cases by reanalyzing old experiments dating back to 1952. The results were summarized in 1975 by Hoshino et al (29) and are listed in Table 1.

The next five experiments in the table come from bare emulsion exposures to accelerator beams and gave contradictory results both for the lifetime and production rate. This discrepancy is now largely understood: the charm decay search regions in the high statistics experiments with null result were too short to have seen the neutral decays found by K. Niu and his group (25), who use follow-back techniques to look for decays producing showers in the analyzer portion of their emulsion chambers. Very short single-decay candidates seen by Diambrini-Palazzi (20) and Komer et al (31) have not been confirmed by other groups.

To our knowledge, the first successful hybrid emulsion experiment, in which a downstream spark chamber spectrometer was used to localize interactions in the emulsion target, was Fermilab E247 (17). On average 0.7 cm^3 of emulsion volume was scanned around each vertex prediction, yielding 37 neutrino interactions and one charmed candidate.

The first hybrid setup with full momentum analysis of charged secondaries was experiment WA17 in which 32 liters of Ilford emulsion were exposed to a neutrino beam in front of BEBC (the Big European Bubble Chamber) at CERN. The success rate for location by volume scanning of charged current events was 32%; 169 such neutrino interactions were found. Events were localized by reconstruction of bubble chamber tracks to ± 1 mm transverse to the beam direction, and ± 8 mm in the direction parallel to the beam. The major limitations of the experiment were poor emulsion quality, the lack of neutral particle identification, and the inability to follow tracks back into the emulsion pellicles.

In the E564 experiment (10) the emulsion stack was placed inside the 15-foot Fermilab bubble chamber. One candidate for F decay was found.

Charmed particle searches using large conventional bubble chambers as the detection medium were summarized by Reeder at the 1979 Lepton-Photon Conference (39). Five candidates with observable decay lengths were found in more than 300 dilepton events. The resulting lifetime limits were $1.5 < \tau^0 < 7 \times 10^{-13}$ s and $2.0 < \tau^\pm < 6 \times 10^{-12}$ s.

It is difficult to draw quantitative conclusions from these first experiments. Because of incomplete reconstruction of most events it was hard to distinguish among particle species, and, in the case of the bubble chambers, between charged and neutral topologies. The minimum distance at which decays could be seen in chambers with 500-μm bubbles is 5–10 mm, so that results from this technique should be perhaps regarded as upper limits. Scanning biases were possible in emulsion because of differing techniques and search regions for neutral and charged particles.

Table 1 Early searches for charmed particle decay

Reference	Technique	Beam	Comments	Yield/interaction	σ_{tot} (μb/nucleon)	τ ($\times 10^{-13}$ s)
(29) (compilation)	emulsions	cosmic rays (multi-TeV)	require pair production	2–5%	~1000	~4 neutral decay 10–20 charged
(19)	emulsion	300-GeV p	$l < 150$ μm	0/60,000	<1.5	
(16)	emulsion	300, 400-GeV p	10–600 μm	0/16,000	<7	
(31)	emulsion	400-GeV p	$l < 1,000$ μm	9 single/1,120	~120	0.2
(25)	emulsion chambers	400-GeV p	follow-back and direct search	2 pairs/1,637	30 ± 20 (assume A^1)	~5
(see 21)	emulsion	60-GeV π- 70-GeV p	$l < 100$ μm	4 single/20,000	5 ± 3 (semileptonic)	0.6
(20)	emulsion	20–80-GeV γ	$l < 3,000$	2 single/482	<1	0.02–0.05
(17) (E247)	emulsion/spark chamber	wide band ν	2,000 neutral	1/37		~6
(7) (WA17)	emulsion/BEBC	wide band ν	≤2,000 neutrals ≤5,000 charged	8/214		$\tau^0 = 0.6^{+0.6}_{-0.3}$ $\tau^\pm = 2.5^{+2.1}_{-1.1}$

2. DESCRIPTION OF RECENT EXPERIMENTS

All of the experiments we discuss consist of a relatively fine-grained vertex detector followed by a downstream spectrometer. The seven groups that had presented results by Fall 1982 are listed under their experiment number in Table 2. The organization is by type of vertex detector: emulsion experiments, bubble chambers, and electronic vertex detectors. Most of these collaborations are still analyzing data and/or recording new events so that the event totals in the last columns will be growing in the future. Because of losses due to incomplete reconstruction and cuts to define scanning volume, the number of events used for lifetime fits is generally smaller than the charmed candidate total.

2.1 *E531*

The E531 group exposed 23 liters of Fuji emulsion in 1978–1979, and 35 liters in 1980–1981 to the wide-band neutrino beam at Fermilab. The emulsion targets were comprised half of conventional pellicles of pure emulsion, and half of thickly coated plastic sheets (70 μm of polystyrene coated on each side by 330 μm of Fuji emulsion), organized in modules of $\frac{1}{2}$–1 liter each.

The technique of event finding by scan-back of spectrometer tracks into emulsion was first tried by members of the E531 collaboration using the Fermilab test beam (26). In this scan-back procedure charged tracks reconstructed in an external spectrometer are searched for at the downstream edge of the emulsion. Track candidates in the spectrometer and in the emulsion are considered linked up when measured angles and positions agree within limits determined by the experimental resolution. Once the track is located in the emulsion it can then be followed upstream to its point of origin, which may be an interaction, gamma conversion, or decay. An important advantage of this method is that the efficiency for event finding is independent of the topology of the event being searched for. Thus, for example, there is no bias against finding low multiplicity or neutral decays.

Even in this first test it was found that event-finding efficiency was much higher (68% vs 38%) than that obtained in conventional volume scanning, and scanning time per event was eight times shorter (2 hours against 16 hours). The technique has been further improved so that the finding efficiency is 95% in E531 for events in the fiducial volume. The scan-back technique is used for finding charmed decays for 60% of the events from the first run and all of the events taken in 1981. Calibration is done in part using electrons found in the spectrometer from γ conversions. A comparison of

Table 2 Recent experiments measuring charmed particle lifetimes

Experiment	Beam	Vertex detector	Charged hadron identification	γ Detection	Hadron calorimeter	Interactions found ($\times 10^3$)	Approx. charm yield	Events used in τ determination			
								D^0	D^+	F^+	Λc^+
E531	Fermilab ν wide band	emulsion	time of flight (TOF)	yes	yes	3.7	>120	48	11	4	8
WA58	CERN Photon 20–70 GeV	emulsion	Čerenkov	yes	none	7	60	18	8		8
NA16	CERN π^-, p 350 GeV	LEBC	ISIS	yes	none	340	77	16	15	2	4
NA18	CERN π^- 340 GeV	BIBC	none	none	none	95	21	9	7		
BC72/73	SLAC Photon 20 GeV	SLAC 40″ B.C.	Čerenkov	yes	none	490	59	13	14		
NA1	CERN Photon 70 GeV	silicon wafers	Čerenkov	yes	none	1000	>100		98		
Mark II	SLAC e^+e^- 29 GeV C.M.	drift chambers	TOF	yes	yes	17 pb^{-1}		7			

predicted and measured electron pair distributions is shown in Figure 2a, where number of events is plotted against the separation between production vertex and conversion point. The fall-off of events at large distance is due to the finite target thickness. The relative scanning efficiency, defined as the ratio of found to predicted events, is plotted in Figure 2b. In the region from 0 to 10 mm where most of the charmed events occur, the efficiency is $93 \pm 9\%$. Other calibration has been done from the scan-back of tracks known to come from charm decay and from primary interactions. Again the efficiency is high and independent of distance.

Searching for charm decays also is performed using conventional techniques: a volume scan for neutral events out to 1 mm, and track following for at least 6 mm, so that a double scan is done for the majority of events.

The E531 spectrometer permits charged particle tagging by time-of-

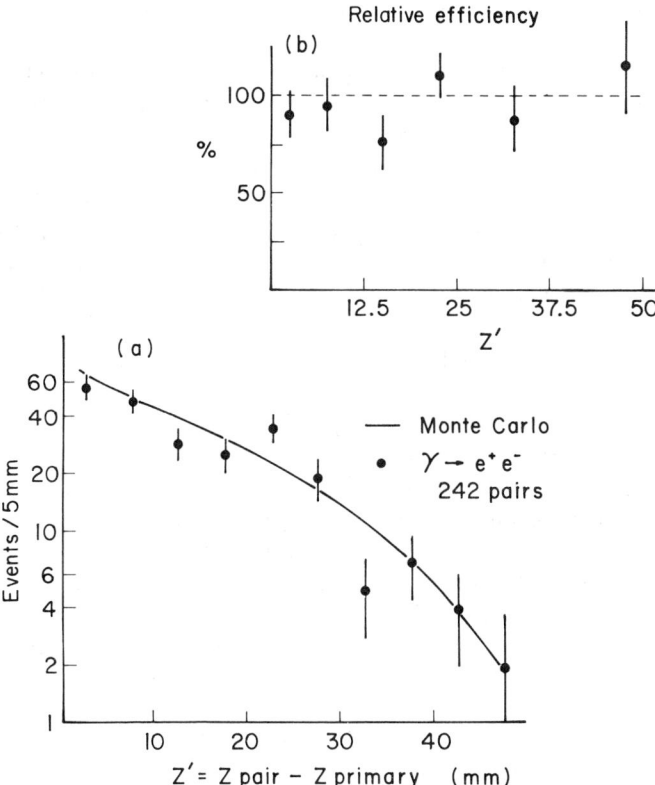

Figure 2 (a) Comparison of predicted and measured distribution of γ conversions. The predicted spectrum (MC) is normalized to the total number of events found. (b) Unfolded efficiency for finding electron pairs by scan-back.

flight, position and energy measurement of electromagnetic showers using a lead converter and lead glass; it also permits tagging of muons and neutral hadrons with a steel, scintillator, and proportional tube calorimeter. Because the spectrometer is designed with a very wide aperture, most secondary decay products are detected.

Analysis of data from the 1979 run is complete and mostly published (42–44). The analysis of data taken in 1981 is two thirds finished. Only D^0 events from this new data are included in the results presented below. The average detection efficiency (summed over both runs) is 73% for D^0 decaying into two or more charged tracks (plus neutrals), 69% for single-prong decays (where the missing P_T is > 300 MeV/c) and 85% for multiple-prong charged decays.

2.2 WA58

The WA58 collaboration exposed single-emulsion pellicles 0.6 mm thick at an angle of 5° with respect to the CERN tagged photon beam. The effective emulsion thickness is thus 6 mm. Event finding is done by volume scanning 45 mm^3 centered on the vertex prediction from the OMEGA spectrometer. The success rate is 50%. Interactions are also found that are not matched to the spectrometer. Charmed decay candidates are found first by following charged tracks to the edge of the emulsion (typically 3 mm). If a decay candidate is not found, a volume scan for neutrals is done for 550 μm, otherwise the volume scan continues to $l_{max} = 1000$ μm. At present 60 decays (27 pairs, 6 single events) have been found (G. Diambrini-Palazzi, private communication), of which 34 are used for lifetime determinations. Some partial results have been published. [Adamovich (2) includes references to earlier work.]

2.3 NA16

NA16 (LEBC–EHS Collaboration) employs the small hydrogen bubble chamber LEBC (Lexan Bubble Chamber) built specifically for the study of charm decays (15). Small bubble size (45 μm) and high density (80 "blobs" per cm) are achieved by running at higher pressure and temperature than is done conventionally and by using a short delay (300 μs) between the interaction and the firing of the flash lamp. The bubble chamber is coupled to the European Hybrid Spectrometer (EHS) for charged particle momentum analysis and π^0 reconstruction. Information is available from the dE/dx chamber ISIS for identifying some tracks.

Decays are looked for in a double scan of the film. Events hidden in the confusion region near the primary vertex are detected from a change in ionization or from a secondary track not pointing to this vertex (thus having large impact parameter).

After subtraction of events that are unreconstructed or that are

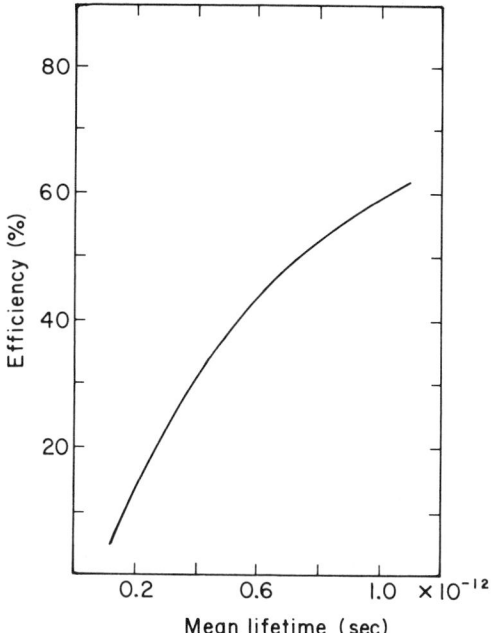

Figure 3 Estimated efficiency for detection of charmed mesons as a function of mean lifetime in LEBC.

consistent with a photon conversion, strange particle decay, or Σ^+ Dalitz decay, 60 identified constrained charm decays (from 51 events) were found. This sample is further refined to 52 decays (23 neutral, 28 charged, one ambiguous) by rejecting short decays that are underconstrained. The final filtering made prior to a maximum likelihood fit to determine lifetimes is a cut to eliminate events for which the decay length is smaller than the computed minimum detectable length l_{min}.

The maximum detectable length is determined by the size of LEBC (20 cm along the beam), the depth of focus, and a 0.06 cm cut on the transverse distance of the decay point to the beam direction.

The estimated scanning efficiency as a function of mean lifetime for D mesons is plotted in Figure 3. It varies from $\sim 10\%$ at 10^{-13} s to 65% at 10^{-12} s (8) for an assumed bubble size of 50 μm.

Final results for NA16 used in this article were published in 1983 (4). The collaboration is pursuing charmed lifetime measurements in a new experiment at CERN, NA27.

2.4 *NA18*

The NA18 group uses the BIBC (Berne Infinitesimal Bubble Chamber) vertex detector. It is a small (6.5 cm long) freon chamber operated at CERN in a mode similar to LEBC. Because of the heavy liquid fill, the bubble density achieved is higher, 300 cm^{-1} (38). The downstream spectrometer employs a streamer chamber for charged track reconstruction. No particle identification is available nor is there a π^0 detector, so that events are considered for analysis only if the decay products are all charged and the event is completely constrained; 21 such events have been found. The efficiency for finding $D\bar{D}$ pairs is estimated to be 11%. Background in the final data selection is higher than for the other bubble chamber experiments because of the freon target. It is 1.4 events in the 9-event D^0 sample, and 2.2 events in the 7-event $D^{\pm}$ sample (11a).

2.5 BC72

In this experiment the SLAC 40-inch hydrogen bubble chamber is modified for studying charmed decays by running hot (29 K) and by adding a single high resolution camera whose flash is triggered earlier than the three conventional cameras. The resolution obtained is 70 bubbles cm^{-1} averaging a 55-μm diameter. An unusual nearly monochromatic photon beam with a spectrum peaking at 20 GeV and a full width at half maximum of 2 GeV was chosen for this experiment. The downstream spectrometer is the SLAC Hybrid Facility, which contains two Čerenkov counters for particle identification and an electromagnetic shower detector comprised of hodoscopes and lead glass blocks.

The hadronic interactions have been double scanned for decay candidates, which are recorded if the decay point is visible or if there is at least one track observed with an impact distance greater than one track width. A sketch of an observed pair decay is shown in Figure 4 to clarify these cuts. The (positive) charged particle decays into three prongs after 0.86 mm, and the neutral charmed particle after 1.8 mm. [A photograph of this event can be found in (1).]

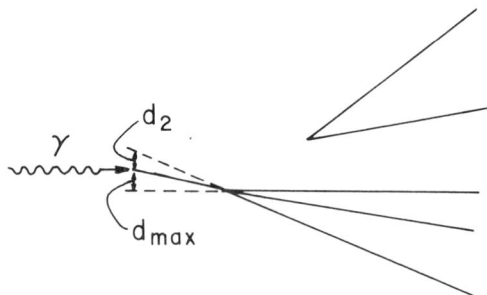

Figure 4 Sketch of charmed pair decay in the SLAC bubble chamber.

Events passing the following cuts are considered for lifetime analysis:

1. fiducial volume,
2. maximum impact distance $d_{max} > 110$ μm,
3. second track from same decay has $d_2 > 40$ μm,
4. decay length > 500 μm, and < 10 mm.

At the time of the 1982 Paris Conference, 29 events with 31 decays survived these cuts (30); earlier in the year 23 of the events were published in *Physical Review Letters* (1). The breakdown of the events is 13 neutral, 5 positive, 9 negative, and 4 ambiguous. Eleven of the nonambiguous events can be completely reconstructed. For the unconstrained events the momentum is estimated on an event-by-event basis (the nearly monochromatic incident energy serves as a constraint). Lifetimes are obtained by comparing the means of various experimental distributions such as average lifetime, maximum impact parameter d_{max}, projected total length L, and projected effective length with those obtained by generating Monte Carlo events with the same cuts. The results obtained are consistent to 20% with different models for charmed particle production.

2.6 NA1

Perhaps the most novel detector is that built by the FRAMM collaboration for use in experiment NA1 at CERN. The target consists of forty 300-μm thick silicon wafers, 14 mm in diameter, spaced 100 μm apart (Figure 5). Charged particles leave the equivalent of 90 keV in each Si wafer compared to an RMS noise level of 30 keV (14). The pulse height signature of an event consisting of an interaction of the photon beam followed by two charmed decays is illustrated in the lower part of the figure. The spectrometer deployed contains four magnets for charged track momentum analysis ($\Delta p/p \simeq 0.5$–1.5%) and five sets of photon detectors using either lead-scintillator hodoscopes or lead glass absorbers. Two Čerenkov counters separate pions and kaons from 5 to 21 GeV/c.

The analysis of $D^{\pm}$ mesons, unlike that of the preceding experiments, requires that the charmed event first be selected by reconstruction in the spectrometer; the path length traveled by the D meson is then measured in the silicon target (5). The pulse height steps in the target must follow this pattern:

1. the interaction is in the target;
2. the pulse height after the interaction is level for more than four wafers and corresponds to at least two minimum ionizing particles;
3. a step in pulse height of $\Delta n = 2, 4,$ or 6 is seen, and is again stable for at least four Si wafers.

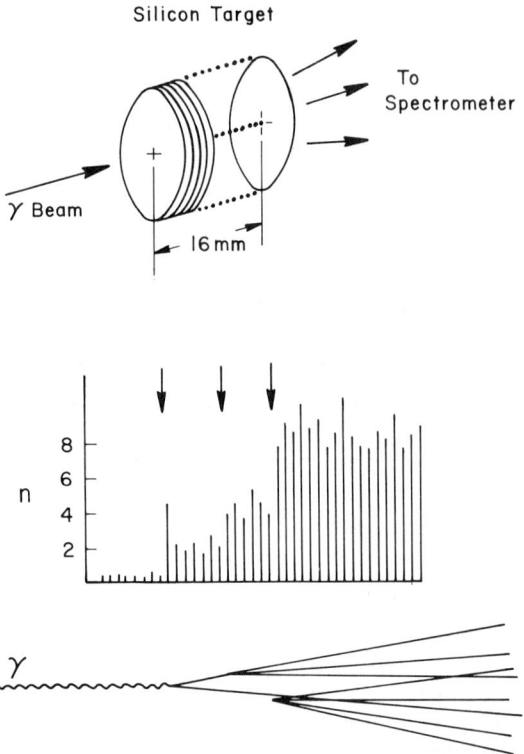

Figure 5 The silicon wafer target used in the NA1 experiment. The middle part of the figure shows the pattern of pulse height steps in the target that would be left by the decay of two charmed particles.

There were 86 events showing this pulse height pattern, 12 of which showed two steps in ionization.

Since the association of decay lengths in the target and D's reconstructed in the spectrometer is ambiguous, the average Lorentz boost ($\gamma = E_{tot}/2M_D$) is attributed to both charmed particles. Because the analysis requires coherent charmed pair production, this approximation is believed to contribute little to the error in lifetime (5).

Additional data has recently been taken by some members of the NA1 collaboration using a new target with finer segmentation (G. Bellini, private communication). Results are not yet available.

2.7 *Mark II*

Mark II results for the D^0 lifetime were presented by J. Jaros at the 1982 Paris Conference and at the SLAC Summer Institute (29a). The spectrometer employed at PEP is well known and is not described here. The method of analysis employed is to select D^0's coming from $D^{*\pm}$ decays at high momentum and to reconstruct the effective length of the decays using drift chambers. The average error in path length is 700 μm, compared to an average path length of 500 μm for the seven events selected. A similar analysis is used to determine the lifetime of the τ lepton (23) from 126 events.

3. EFFECTS OF BACKGROUND AND SCANNING LOSSES

In order to determine particle lifetimes it is necessary to have a well-measured data sample for which the background and scanning biases are known.

Backgrounds for the D^0 and D^+ have been estimated by the E531 and NA16 groups (see Table 3) for events in the final data selection; it is found to be generally small, < 3%, except possibly in the case of feed-through of F^+ and Λ_c^+ decays into the D^+ category. This $D^+/F^+/\Lambda_c^+$ ambiguity arises because no experiment has complete particle identification, and it is unfortunately true that if $M(K\pi\pi) = M_{D^+}$, then $M(KK\pi) \simeq M_{F^+}$ and $M(Kp\pi) \simeq M_{\Lambda_c^+}$ for many events. In the case of the F and Λ_c^+ samples both experiments retain only those events that are uniquely fit or where particle identification precludes any other hypothesis. Ambiguity between charged and neutral decays is nonexistent for the emulsion experiments. For BC72, 3 of 23 published events are charged/neutral ambiguous, as is one of 52 decays, which otherwise passes all cuts for NA16. In both analyses these events are eliminated from further consideration. In the silicon target of

Table 3 Comparison of backgrounds by source: E531 and NA16

Background	E531 (%)	NA16 (%)
D^0 background		
γ conversion	< 0.5	nil
K^0, Λ^0 decay	< 0.2	< 3 (charged and neutral)
neutral interactions	< 1	< 1
D^+ background		
strange particle decay	< 0.2	< 3
charged interactions	≤ 0.9 ($p > 4$ GeV/c)	< 1
F^+, Λ_c^+ feed-through	0^{+3}_{-0} events	< 5 (3c fits)

NA1 such neutral/charged ambiguities are impossible to distinguish on an event-by-event basis and constitute an estimated 20% background under the D^+.

3.1 Decay Time Resolution

Error in decay time measurements can arise from resolution effects in measuring decay lengths or from inadequate event fitting. Decay lengths in emulsion are measured to a few micrometers, while in the LEBC chamber 0.1 mm is achieved (15). The Mark II experiment is unique in that the measuring resolution is larger than the average observed decay path. In principle no net bias is introduced into the decay time measurements as long as the e^+e^- beam positions are well known and event statistics are large enough to determine the mean decay path.

The subject of event fitting is rather complicated. Most charm particle decays involve the emission of one or more neutral secondaries (π^0, K^0, Λ^0, ν, etc). Many of these neutral secondaries cannot be observed in the spectrometer; in addition, charged tracks from decays at wide angles are often not momentum analyzed. Most experiments thus find that as many as half or more of their decay candidates are unconstrained. Additionally, because of the combinations possible, events fits using neutral vertices are not always unique: in particular a π^0 may be wrongly assigned. The NA16–LEBC group has considered the effect of spurious fits using a π^0 and find any effect on lifetime determination is small, < 10%, and introduces no systematic bias.

It is desirable that unconstrained events not be excluded from lifetime determinations, because of the effort and cost required to obtain charmed decays and because omitting such events may introduce biases. At least two methods for fitting such events are possible, although little used. One technique is the observation of excited states of the D^0, D^+, F, and Λ_c^+. Thus far this constraint has been used only by E531, for the decay chain $D^{*+} \to D^0\pi^+$. Results for all $D^0\pi^+$ or $\bar{D}^0 \pi^-$ mass combinations from 48 events are plotted in Figure 6: 13 events consistent with D* decay are

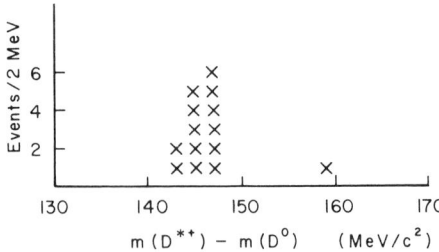

Figure 6 Histograms of $D^0\pi^+$ and $\bar{D}^0\pi^-$ masses (E531), after subtraction of the D^0 mass.

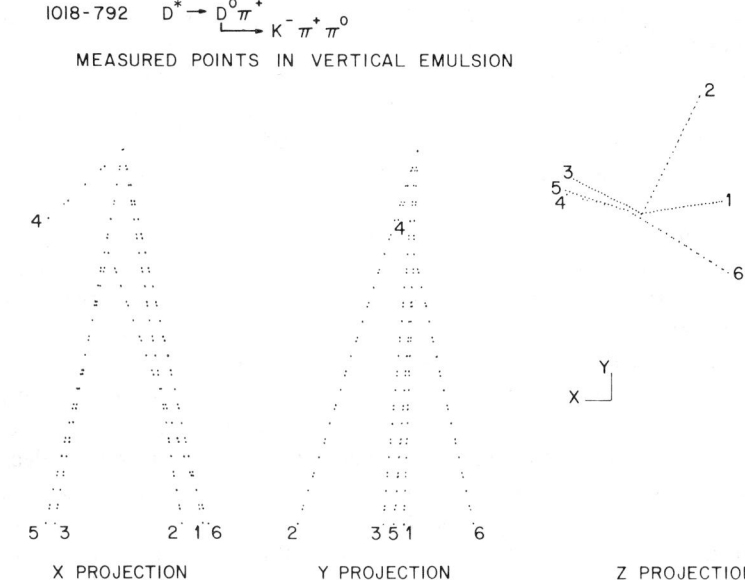

Figure 7 Emulsion measurements of a D^0 decay. All six tracks are reconstructed in the E531 spectrometer.

found; the background under the peak is much less than one event. Measurements in emulsion of one of the D* candidates is shown in Figure 7 in x, y, and z projection (z is along the beam direction). Use of the D* and other excited states should prove to be a useful tool in event fitting, once the spectroscopy of more of these vector mesons is known.

Another potentially useful constraint comes from the relation between impact parameter and decay time, which is expected to be linear for relativistic decays (34). The scatter plot of average impact parameter ($\bar{d}$) versus decay time is shown in Figure 8 for the events found to be consistent with D* decay in Figure 6. The spread in the ratio of $\bar{d}$ to decay time is $\pm 26\%$ if one uses the average impact parameter; it is 50% worse and skewed if one uses only the maximum. The range of Lorentz boosts is from $\gamma = 2.3$ to 11.8 with a mean value $\bar{\gamma} = 6.5$. The dependence of the ratio on prong count and other factors remains to be determined. The maximum impact parameter is sometimes used by experiments to determine decay times: a better choice is the average impact for unreconstructed decays.

3.2 Systematic Effects Due to Scanning Biases

The effect of an unknown scanning bias is to systematically increase or decrease the lifetime estimate of a sample, depending upon whether the observed decays are depleted at short or long distances (Figure 9). The presence of systematic biases can be tested by varying minimum (or maximum) cuts on decay path.

In addition to cuts on minimum decay distance $l_{min} > 0.5$ mm, the bubble chamber experiments are insensitive to events with small impact parameter (typically $d_{max} \geq 100$ μm). These cuts are not trivial; they would, for example, eliminate 63% of the D^0 events found by E531, all of their Λ_c^+, and three of four F decays.

Scanning cutoffs that are too short in relation to the average decay path lead to a situation where only a lower limit can be quoted or where two standard-deviations (SD) limits become very large, even when the scanning efficiency up to the cutoff is well known. In Figure 10 we show the likelihood function for two samples with 16 and 18 events. For the first sample $l_{max} \simeq 20$ mm ($t_{max} \simeq 30\tau$). For the second sample we cut at $l_{max} \simeq 400$ μm ($t_{max} \simeq 0.7\tau$). [See (22) for a discussion of maximum likelihood.] In the present example the effect of the short cut-off on decay distance is to increase the 1-SD upper limit from 1.35τ to 1.5τ. The 2-SD upper limit grows even more, from 1.8τ to 2.7τ. In the limit as the number of events decreases or the cut-off becomes even shorter there is no upper limit at all. The minimum useful span of distances over which decays are

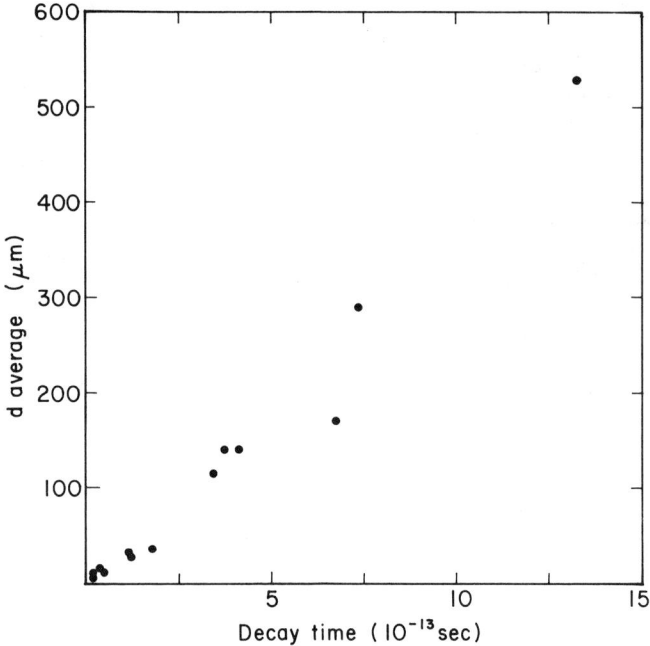

Figure 8 Average impact parameter versus decay time for events in Figure 6 consistent with a D* hypothesis.

CHARMED PARTICLE LIFETIMES 557

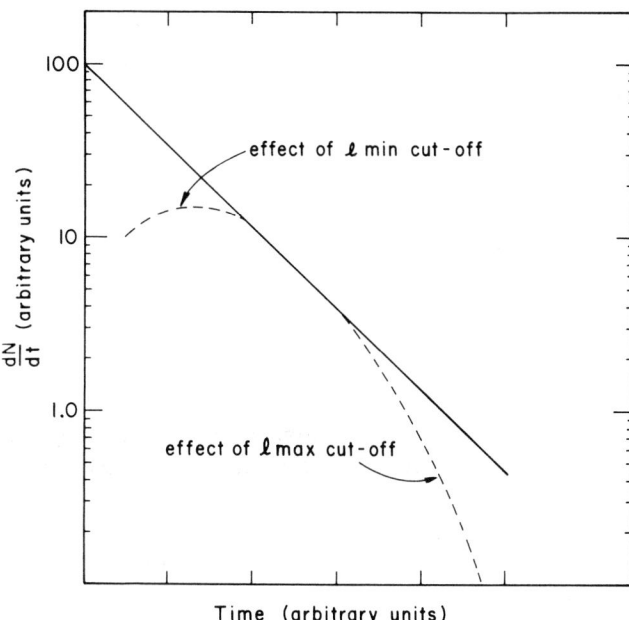

Figure 9 Effect of hypothetical scanning distance cuts on an exponential decay time distribution.

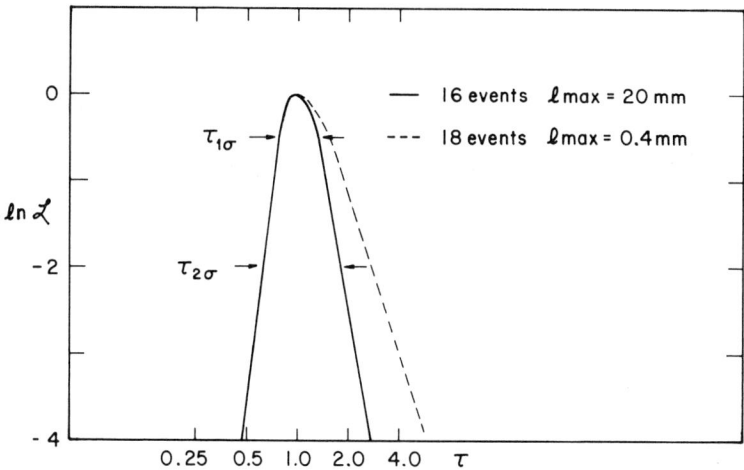

Figure 10 Comparison of likelihood functions for two event samples with differing decay path cuts.

observed is a matter of taste and experimental technique, but $l_{max} > 2\beta\gamma c\tau$ seems desirable. By this criterion, the search decay region for both neutral and charged events is too short in WA58.

4. RESULTS

In summarizing results for the charmed lifetimes the order presented is D^0, Λ_c^+, F^+, and D^+. D^+ measurements are last because of possible feedthrough from other event types.

4.1 Calculating Mean Lifetimes

Traditional methods of calculating a weighted mean do not give a satisfactory answer for lifetime measurements with asymmetric errors. The procedure we have adopted for combining results is suggested by consideration of what one expects from performing a maximum likelihood fit to decay time data.

In the absence of biases the best estimate of τ is the mean decay time

$$\tau = t_{av} = \sum t_i/N$$

where the total number of decays with decay time t_i is N. The values of τ corresponding to 1 SD are given (approximately) by

$$\tau_{1SD} = \{(1 \pm \sqrt{1/N})/(1-1/N)\}\tau.$$

Alternatively,

$$N_{eff} = 1/\{\tau/t_{1SD} - 1\}^2.$$

With real data the 1-SD limits are larger than those derived above. We define an effective event number N_{eff} using the upper and lower limits quoted and use this as the weight for a result. N_{eff} is always smaller than the observed number of decays. The weighted mean is then

$$\tau_{mean} = \frac{\sum N_{eff} \tau_{measured}}{\sum N_{eff}},$$

where the sum is over all experiments. We calculate upper and lower 1-SD values of the lifetime from

$$\tau_{1SD} = [(1 + \sqrt{1/\sum N_{eff}})/(1-1/\sum N_{eff})]\tau_{mean}.$$

We have checked this prescription for computing the weighted mean against samples of events drawn from our own data (E531) and find the mean so computed to agree to 1% with results from maximum likelihood fits. The computed errors are in agreement to within 5% of those obtained from fitting.

4.2 D^0 Lifetime Measurements

Published results for the D^0 lifetime so far are available in part for E531 (43), BC72 (1), WA58 (2), NA16 (3, 4), and NA18 (11a).

Additional events were presented at the 1982 Paris conference from the above groups and Mark II. Since these later data more than double the available statistics, we have used them in computing the mean D^0 lifetime. The results available in late 1982 are listed as part of Figure 11. Data from the second run of E531 are treated as a separate entry since a number of changes were made both to the spectrometer and to scanning procedures. The weighted mean D^0 lifetime from all the available data is $3.9^{+0.5}_{-0.4} \times 10^{-13}$ s. The consistency of the measurements can be seen in Figure 11; WA58 and BC72 are low and high with respect to the mean, but in both cases by less than 2 SD by our estimate.

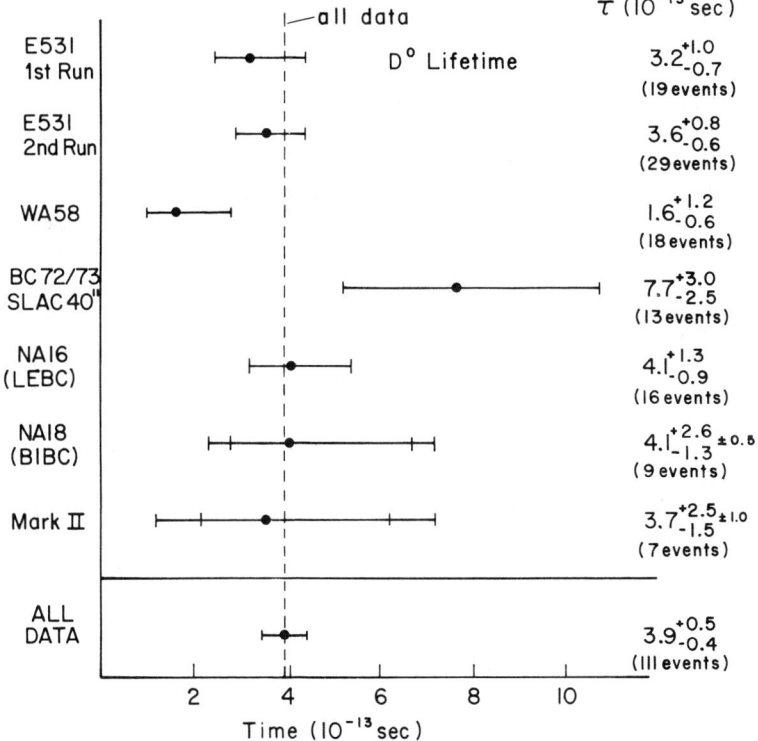

Figure 11 D^0 lifetime measurements. The effect of systematic errors in the NA18 and Mark II results is shown by the outside set of error bars.

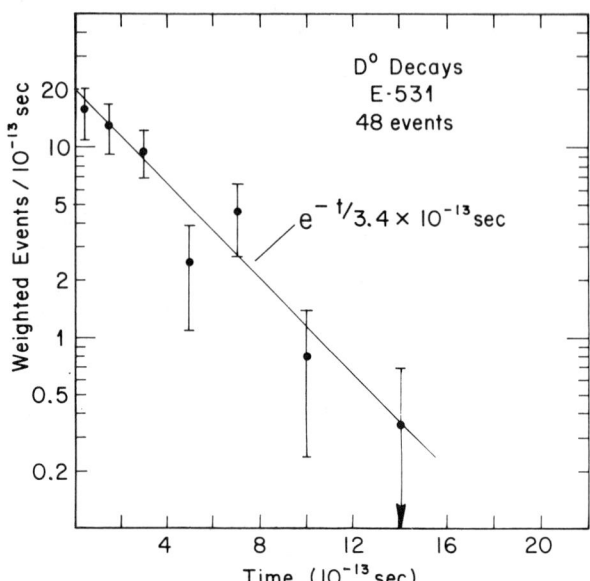

Figure 12 Differential time distribution for D^0 decays from E531. The events are individually weighted by the reciprocal of the scanning efficiency.

A plot of dN/dt using all the events published is not practical because of the differing scanning efficiencies and decay path cutoffs, but can be done for the 48 E531 events (Figure 12). The data are in good agreement with the hypothesis of a single-parent D^0 population.

4.3 Λ_c^+ Lifetime

Very few Λ_c^+ decays have been reported. At the 1982 Paris Conference (30) the total was 21 events of which six had been published: four from E531 (42), one by WA17 (9), and one by WA58 (2). The events found by the emulsion experiments are all very short ($\bar{l} = 164$ μm for E531, 202 μm for WA58) and have an average boost of $\beta\gamma < 3$, so that Λ_c^+ production in νN and γN interactions is target-like. This production mechanism might explain why no events are reported by BC72/73, which is also a γN exposure and which has l_{min} cuts that are substantial compared to 200 μm. NA16 reported observation of five events; four had $l < l_{min}$ and were observed while being measured for another decay candidate.

A puzzling difference exists in event topology seen by E531 (2 one-prong, 6 three-prong decays) and WA58 (7 one-prong and only 1 three-prong). Part of the difference may arise from differing acceptance criteria for one-prong decays in the two experiments. E531 requires, perhaps too conserva-

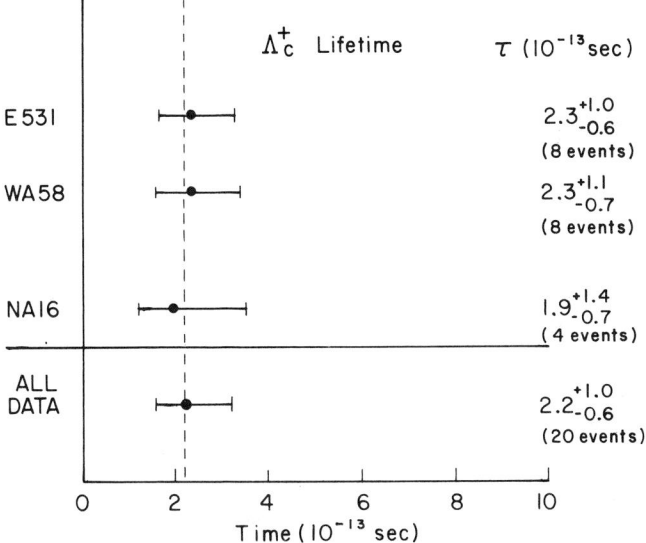

Figure 13 Comparison of Λ_c^+ lifetime measurements.

tively, that the missing P_T be > 300 MeV/c before single-prong decays are classified as charm decay candidates.

Including all experiments with more than one event yields a lifetime of $2.2^{+1.0}_{-0.6} \times 10^{-13}$ s (see Figure 13). An event from WA17 that had a decay time of 7.3×10^{-13} s is consistent with this value of τ (9). The statistics of the Λ_c samples are small and it is possible that longer-lived decays as yet unobserved will change the mean lifetime in an unpredictable way. To allow for this effect we have increased the errors on the Λ_c lifetime by 50% over the calculated result.

4.4 $F^{\pm}$ Lifetime

The first visual observation of the F is from E531 and was reported by Prentice (37) at the 1979 Lepton-Photon Conference. The decay seen was $F^- \to \pi^-\pi^-\pi^+\pi^0$, with $t = 3.6 \times 10^{-13}$ s. Since then additional decays have been reported by E564 (10), E531 (42), and by the NA16–LEBC collaboration (4). Results presented at Paris by NA18 and NA1 are considered preliminary by these groups.

Because of the difficulties in establishing a sample free of contamination from D and Λ_c, the number of clean examples of F decay remains small, and many detailed questions remain about its mass, decay modes, and other properties. The experimental situation was recently reviewed by Trilling (41).

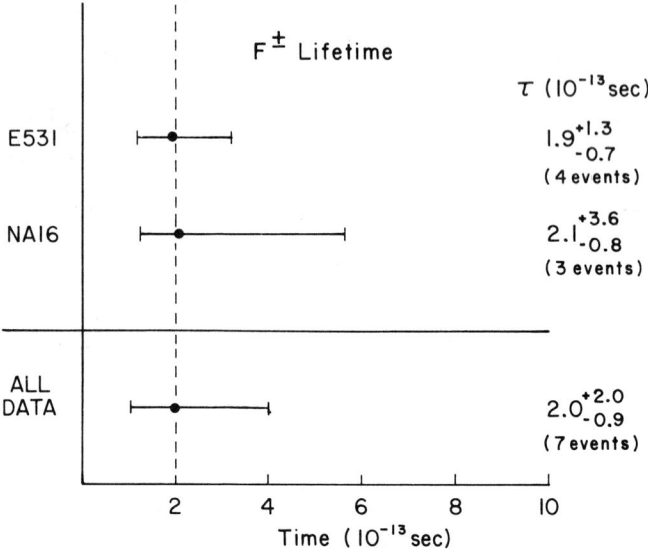

Figure 14 Comparison of $F^{\pm}$ lifetime measurements.

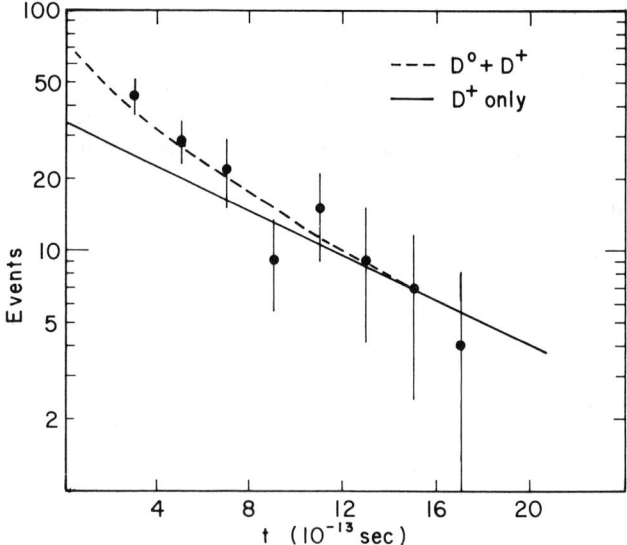

Figure 15 Decay time distribution for D^+ candidates from NA1. The dashed curve is a fit to all the data including an estimated 20% D^0 contamination; the full curve shows the D^+ contribution.

The mean lifetime is $2.0^{+2.0}_{-0.9} \times 10^{-13}$ s based on the seven events of E531 and NA16. The errors have again been scaled up by 50% because of the small statistics (Figure 14). The average mass of the E531 events is 2044 ± 30 MeV, and the E564 event had 2017 ± 25 MeV. The NA16 group has not reported the masses of their events.

4.5 $D^{\pm}$ Lifetime

The highest statistics determination of the D^+ lifetime comes from the NA1 experiment using a silicon wafer target described in Section 2.6. The experimenters estimate a 10% contamination of the 98 D^+ candidates owing to photon conversions and hadron interactions and a 20% background from D^0, assuming $\tau(D^0) = 2.5 \times 10^{-13}$ s. The fit to the decay time spectrum including D^0 is indicated by the dashed curve in Figure 15, while the solid curve gives the D^+ contribution. The D^+ lifetime measurement error includes an estimate of systematic effects. Use of the current value of $\tau(D^0) = 3.9 \times 10^{-13}$ s would by our estimate raise the D^0

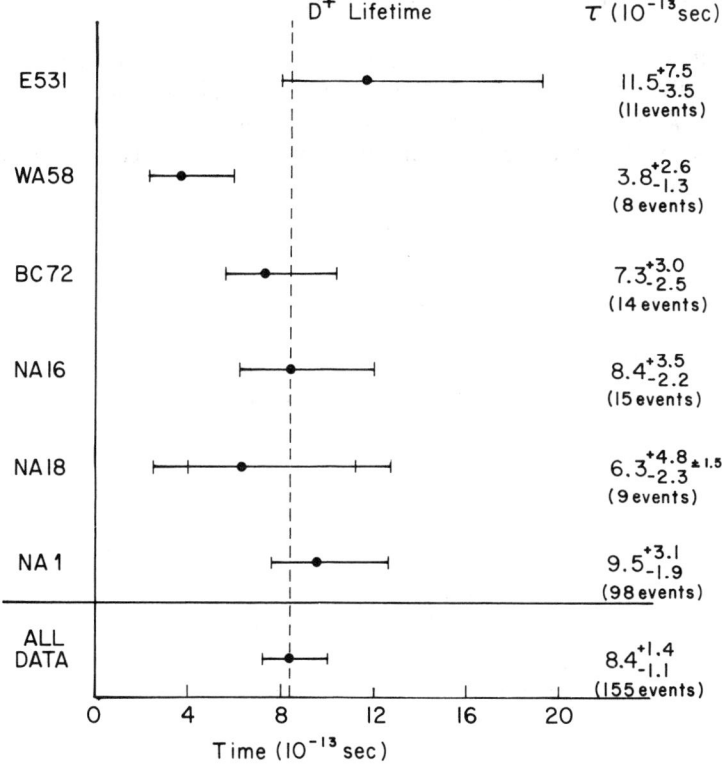

Figure 16 Comparison of $D^{\pm}$ lifetime measurements.

background and slightly increase the fitted D^+ lifetime from NA1, but by much less than 1 SD.

Results from all experiments are summarized in the bar graph in Figure 16. Except for the result from NA1, all of the measurements are as yet unpublished. Partial results based on smaller statistics can be found in the literature for E531 (42), WA58 (2), and BC72 (1). Estimated backgrounds are $\leq 10\%$ for E531, BC72, and NA16 and 30% for NA18 and NA1. Background in the WA58 sample has not been calculated.

The D^+ lifetime based on all the measurements is

$$\tau(D^+) = 8.4^{+1.4}_{-1.1} \times 10^{-13} \text{ s}.$$

The consistency of the experiments is quite good, with $\chi^2/\text{D.F.} < 1$.

We note that the remaining D^0, F, Λ_c background under the D^+ means its lifetime may be somewhat longer than that quoted by all the groups.

5. COMPARISON OF LIFETIMES AND SEMILEPTONIC DECAY RATES

The lifetimes we have computed from existing data are summarized pictorially in Figure 17. The D^0 and D^+ are now quite well measured with about 100 events in each channel. The number of Λ_c^+ and $F^\pm$ events is still quite small but the lifetimes for these two particles seem to be more consistent with $\tau(D^0)$ than with $\tau(D^+)$.

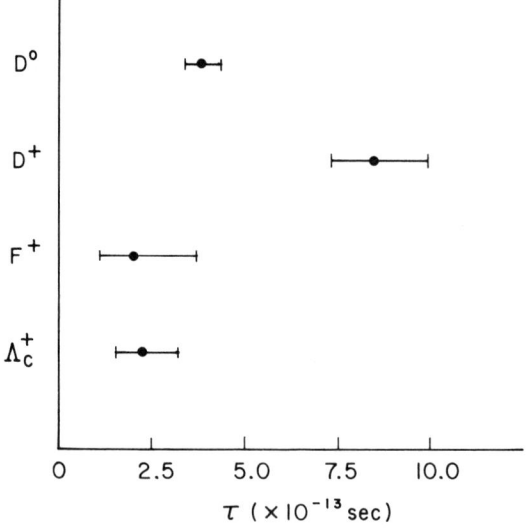

Figure 17 D^0, D^+, F, Λ_c^+ measured lifetimes.

5.1 Ratio $\tau(D^+)/\tau(D^0)$

The ratio $R = \tau(D^+)/\tau(D^0)$ is not directly measured in any experiment so in calculating it we use the mean lifetimes rather than the values of R reported by several of the experiments. The value of R obtained from the lifetime measurements, $R = 2.1 \pm 0.4$, is in good agreement with that inferred from D^+ and D^0 semileptonic branching rates measured by Mark II (40) but is a factor of 2 smaller than the lower limit obtained by the DELCO collaboration (11).

5.2 Semileptonic Partial Widths

It is important to test the theoretical expectation that Γ_{SL} is the same for all charmed particle decays. A common value of Γ_{SL} is expected, regardless of the correctness of its calculation.

Γ_{SL} is obtained experimentally by taking the ratio of the inclusive semileptonic branching rate (B.R.) to the lifetime. Thus for the D^0, Γ_{SL} = B.R. $(D^0 \to e^+ + \text{anything})/\tau(D^0)$. One expects, since charmed particles are quite massive, that $\Gamma_{SL}(D^0 \to e + \text{anything}) = \Gamma_{SL}(D^0 \to \mu + \text{anything})$, apart from a very small correction due to the differing muon and electron masses. The inclusive semileptonic branching rates measured for the D^0, D^+, and Λ_c^+ are listed in Table 4. From these rates, and from the lifetimes determined in this review, one obtains the partial widths (in units of 10^{11} s^{-1})

$$\Gamma_{SL}(D^0) = 1.4 \pm 0.8,$$

$$\Gamma_{SL}(D^+) = 2.3 \pm 0.6,$$

and

$$\Gamma_{SL}(\Lambda_c^+) = 2.0 \pm 1.1.$$

The error in Γ is dominated in all cases by the imprecision of the branching rate measurements. The calculations of Cabibbo et al (18) and Ali &

Table 4 Measured branching ratios

Decay	Branching ratio (%)	Reference
$D^0 \to e^+ x^-$	5.5 ± 3.7	(40) (Mark II)
	< 4	(11) (DELCO)
	$5.1^{+4.8}_{-1.4}$	NA16–LEBC (preprint)
	mean = 5.3 ± 3	(Mark II and NA16 only)
$D^+ \to e^+ x^0$	19^{+4}_{-3}	(41)
$\Lambda_c^+ \to e^+ x^0$	4.5 ± 1.7	(45) (Mark II)

Pietarinen (6), which are first order in gluon corrections, yield $\Gamma_{SL} = 2.0 \pm 0.5 \times 10^{11}$ s^{-1}, in good agreement with the measurements.

5.3 *Nonleptonic Enhancement?*

Better measurements of the D^0 and D^+ lifetimes and branching rates are highly desirable, but the rough agreement in Γ_{SL} strongly suggests that the cause of lifetime differences between D^+ and the other particles lies in the nonleptonic decay channels. This result is not unexpected; the K^0, K^+ lifetime difference is the result of "nonleptonic enhancement."

A complete catalog of phenomenological explanations for lifetimes differing from the prediction of the simplest spectator model is beyond the scope of this review. A promising approach is the consideration of weak interactions among the initial-state quarks, such as the processes shown in Figure 1c and 1d, which leads to the expectation that D^0, F, and Λ_c^+ decays to hadrons may be enhanced over those of the D^+. [See, for example, (12, 13, 24, 32) or the rapporteur review by Maiani (33) at Paris.] Crucial tests of the phenomenology still await better measurements of charmed particle lifetimes and particular branching modes.

6. SUMMARY

The experimental study of charmed particle lifetimes has progressed enormously in the last five years, from chaos to the point where measurements are consistent within rather large errors. If surprises still await us, they must be uncovered by the rapidly improving precision of the new results appearing during the next few years.

New experimental techniques continue to be developed to increase the yield of well-measured decays. As examples we cite the silicon wafer target and the high resolution bubble chambers, which in turn may be supplanted by yet more sophisticated solid-state devices such as silicon microstrip detectors, and by holographic photography of bubbles. Photographic emulsion continues to remain the detector with highest spatial resolution; with the development of computer-aided microscopes and measuring machines and techniques for pinpointing interesting tracks, this technique will remain competitive for at least another generation of experiments.

In the near future we expect that collaborations will obtain thousands of charm decay events. In addition to measuring lifetimes precisely, these new experiments will measure production cross sections and decay modes for charm in a relatively bias-free and model-independent way. Exciting new physics will be explored. Certainly new ground-state charmed particles remain to be discovered, especially in the baryon sector, and the spectroscopy of excited states remains to be elucidated. The all-leptonic decay $F \to \tau\nu_\tau$ should be seen, yielding an improved limit for the tau neutrino mass as well as a clean measurement of the F form factor. $D^0 - \bar{D}^0$ mixing will be searched for with a sensitivity at least ten times better than can be achieved at present. And finally, the study of weakly decaying beauty particles will (we hope) commence, with as yet unforeseen results.

ACKNOWLEDGMENTS

We thank Pat Kimball for typing this manuscript and Robin Sidwell for drawing the figures. This work has been supported in part by the US Department of Energy through contract ACO2-76ER 01545.

Literature Cited

1. Abe, K., Bacon, T. C., Ballam, J., et al. 1982. *Phys. Rev. Lett.* 48:1526–29
2. Adamovich, M. I., Alexandrov, Y. A., Bolta, J. M., et al. 1981. *Phys. Lett.* 99B:271–76
3. Adeva, B., Aguilar-Benitez, M., Allison, W., et al. 1981. *Phys. Lett.* 102B:285–90
4. Aguilar-Benitez, M., Allison, W. W., Bagnaia, P., et al. 1983. *Phys. Lett.* 122B:312–16
5. Albini, E., Amendolia, S. R., Baldini Celio, R., et al. 1982. *Phys. Lett.* 110B:339–43
6. Ali, A., Pietarinen, E. 1979. *Nucl. Phys.* B154:519–34
7. Allasia, D., Angelini, C., Bagnaia, P., et al. 1980. *Nucl. Phys.* B176:13–36
8. Allison, W., Bettini, A., Bizzarri, R., et al. 1980. *Phys. Lett.* 93B:509–16
9. Angelini, C., Bagnaia, P., Baroni, G., et al. 1979. *Phys. Lett.* 84B:150–55
10. Ammar, R., Coppage, D., Davis, R., et al. 1980. *Phys. Lett.* 94B:118–22
11. Bacino, W., Ferguson, T., Nodulman, L., et al. 1980. *Phys. Rev. Lett.* 45:329–32
11a. Badertscher, A., Hahn, B., Hugentobler, E., et al. 1983. *Phys. Lett.* 123B:471–76
12. Bander, M., Silverman, D., Soni, A. 1980. *Phys. Rev. Lett.* 44:7–9
13. Barger, V., Leveille, J. P., Stevenson, P. M. 1980. *Phys. Rev.* D22:693–97
14. Bellini, G., D'Angelo, P., Manfredi, P. F., et al. 1982. *Nucl. Instrum. Methods* 196:351–60
15. Benichou, J. L., Herve, A., Leutz, H., et al. 1981. *Nucl. Instrum. Methods* 190:487–502
16. Bozzoli, W., Campanini, R., Capiluppi, P., et al. 1977. *Lett. Nuovo Cimento* 19:32–36
17. Burhop, E. H. S., Davis, D. H., Tovee, D. N., et al. 1976. *Phys. Lett.* 65B:299–304
18. Cabibbo, N., Corbo, G., Maiani, L. 1979. *Nucl. Phys.* B155:93–103
19. Coremans-Bertrand, G., Sacton, J., Breslin, A., et al. 1976. *Phys. Lett.* 65B:480–82
20. Diambrini-Palazzi, G. 1978. *Proc. 19th Int. Conf. High Energy Phys., Tokyo*, pp. 297–99. Tokyo: Phys. Soc. Jpn.
21. Diebold, R. 1978. See Ref. 20, pp. 666–89
22. Eadie, W. T., Drijard, D., James, F. E., et al. 1971. *Statistical Methods in Experimental Physics*. New York: North-Holland
23. Feldman, G. J., Trilling, G. H., Abrams, G. S., et al. 1982. *Phys. Rev. Lett.* 48:66–69
24. Fritzsch, H., Minkowski, P. 1980. *Phys. Lett.* 90B:455–59
25. Fuchi, H., Hoshino, K., Kuramata, S., et al. 1979. *Phys. Lett.* 85B:135–41
26. Fuchi, H., Hoshino, K., Kuramata, S., et al. 1979. *J. Phys. Soc. Jpn.* 47:687–94
27. Gaillard, M. K., Lee, B. W., Rosner, J. L. 1975. *Rev. Mod. Phys.* 47:277–310
28. Goldhaber, G., Wiss, J. E. 1980. *Ann. Rev. Nucl. Part. Sci.* 30:337–81
29. Hoshino, K., Kuramata, S., Niu, K., et al. 1975. *14th Int. Cosmic Ray Conf., Munich*, 7:2442–47
29a. Jaros, J. 1983. *Proc. SLAC Summer Inst. (1982)*, pp. 25–32. Stanford, Calif: Stanford Univ.
30. Kalmus, G. 1982. *21st Int. Conf. High Energy Phys., J. Phys.* 43:C3–431–69
31. Komer, A. A., Orlova, G. I., Salmanova, N. A., et al. 1978. *Pis'ma Zh. Eksp. Teor. Fiz.* 28:490–94
32. Körner, J. G., Kramer, G., Willrodt, J. 1979. *Z. Phys.* C 2:117–35
33. Maiani, L. 1982. See Ref. 30, pp. 631–57

34. Mulvey, J. H. 1981. *Proc. SLAC Summer Inst.*, pp. 573–99. Stanford, Calif: Stanford Univ.
35. Niu, K., Mikumo, E., Maeda, Y. 1971. *Prog. Theor. Phys.* 46:1644–46
36. Niu, K. 1978. See Ref. 20, pp. 447–50
37. Prentice, J. D. 1979. *Proc. 1979 Int. Symp. Lepton Photon Int.*, pp. 563–68. Batavia, Ill: Fermilab
38. Ramseyer, E., Hahn, B., Hugentobler, E. 1982. *Nucl. Instrum. Methods* 201:335–40
39. Reeder, D. D. 1979. See Ref. 37, pp. 553–62
40. Schindler, R. H., Alam, M. S., Boyarski, A. M., et al. 1981. *Phys. Rev.* D24:78–97
41. Trilling, G. H. 1981. *Phys. Rep.* 75:57–124
42. Ushida, N., Kondo, T., Fujioka, G., et al. 1980. *Phys. Rev. Lett.* 45:1053–56
43. Ushida, N., Kondo, T., Fujioka, G., et al. 1982. *Phys. Rev. Lett.* 48:844–47
44. Ushida, N., Kondo, T., Fujioka, G., et al. 1983. *Phys. Lett.* 121B:287–91, 292–96
45. Vella, E., Trilling, G., Abrams, G. S., et al. 1982. *Phys. Rev. Lett.* 48:1515–18

INELASTIC ELECTRON SCATTERING FROM NUCLEI

J. Heisenberg

Department of Physics, University of New Hampshire, Durham, New Hampshire 03824

H. P. Blok

Department of Physics, Vrije Universiteit, Amsterdam, The Netherlands

CONTENTS

1. INTRODUCTION 569
2. FORMALISM 573
   2.1 Quantities Measured in Electron Scattering 573
   2.2 Determination of the Densities 577
   2.3 Spectroscopy 581
   2.4 Measurements at Low Momentum Transfer 585
3. NUCLEAR STRUCTURE RESULTS 587
   3.1 The Rotational Model 589
   3.2 Small-Amplitude Vibrations 592
   3.3 Single-Particle Transition Densities 595
   3.4 Single-Particle Transitions and Effective Charges 597
   3.5 Transverse Excitations 601
   3.6 Influence of Non-nucleonic Degrees of Freedom 604
4. SUMMARY 607

1. INTRODUCTION

Electrons as a probe of nuclear structure have a special place in the zoo of nuclear probes. This is because the interaction of electrons with nucleons, the electromagnetic interaction, is well known and relatively weak. Hence, in practice electron scattering can be described with sufficient accuracy in the Distorted Wave Born Approximation (DWBA); the only terms not known a priori in this description are the matrix elements of the nuclear charge and current operators between the initial and final nuclear wave functions, the so-called transition charge and current densities. If the calculated cross sections are found to lie outside the experimental uncertainty, one must conclude that the calculated charge or current is not quite correct. The discrepancies cannot be blamed on the reaction mechanism, influence of two-step processes, for instance, or other effects outside of the nuclear structure calculation as sometimes is the case in hadron scattering. It is this (sometimes discouraging) high precision that gives electron scattering its special role.

In electron scattering experiments an electron with energy E_i is scattered from a target nucleus and the scattered electron of energy E_f is detected at an angle θ. Excitation of the nucleus makes itself visible in the energy loss $\hbar\omega = E_i - E_f$, so a measurement of the energy spectrum of the scattered electrons determines the excitation spectrum of the nucleus in question. By varying E_i and/or θ, one can measure the cross section for excited states as a function of the momentum transfer q, where $\mathbf{q} = \mathbf{k}_i - \mathbf{k}_f$. This is the key to the determination of the densities as the cross section at a certain q is basically the Fourier transform of the transition charge and current densities for that state. (See Section 2.1. There we also discuss the role of the continuity equation, which gives a relation between the charge and the current density.) This means that if one has data for a sufficient range of q values, one can reconstruct these densities (see Section 2.2).

In the past, most inelastic electron scattering experiments were restricted by low beam energies. The data provided much useful information on spins and parities, and on moments of the nuclear transition charge and current densities, such as electromagnetic transition probabilities and transition radii (see sections 2.3 and 2.4 and References 1–3), but the lack of data at high q made it impossible to determine the densities themselves. With the availability of new accelerators and experimental techniques, which combine the full range of beam energies (50–500 MeV) with a high resolution ($\Delta E/E \sim 1 \times 10^{-4}$), it has become possible to map in a detailed way the radial behavior of the transition charge and current densities for many nuclear levels of different character, ranging from collective to single-particle states. This means that nuclear structure calculations can be checked with high precision for a variety of cases in which the nuclear many-body system can present itself.

In recent years, transition densities have been measured in increasing numbers and with increasing precision. In this review we did not attempt to be complete in our presentation of the data because such an attempt would necessarily be outdated by the time of publication. Instead, we tried to select some typical examples of what one learns from these experiments. A few years back, the bare fact that densities could be measured with such precision caused excitement. Now, since the technique is well established, the focus has changed from how we construct the picture to what we see on the picture. For that reason we see as one of the important tasks of a review like this to show the theoretical developments stimulated by the precise results from electron scattering.

For a long time most electron scattering results were interpreted in terms of macroscopic models, i.e. models in which the behavior of the nuclear system is described in terms of collective variables like the radius and the skin thickness of the charge distribution and changes thereof when the nucleus is excited. These models have had considerable success, partly because in many cases strongly excited levels were studied, e.g. the first 2^+ and 3^- states that are collective in nature. Further, most of the early data did not extend to high enough q to allow an extraction of the details of the transition densities in the nuclear interior. Therefore the macroscopic model evidently was able to describe the gross features of the measured transition (charge) density in terms of a surface peaked shape.

This situation is changing now. With the availability of more precision data, also for weakly excited levels, one observes structures in the nuclear interior for many densities that cannot be described by macroscopic models. As examples standing on this border line, we discuss in Section 3.1 the determination of different multipoles of the intrinsic charge distribution of the deformed nucleus ^{154}Gd and in Section 3.2 the excitation of members of the β- and γ-vibrational bands in the same nucleus. As in spherical nuclei, one finds in these cases also that the general behavior of the measured densities can be correctly described by this macroscopic treatment of the vibration, but upon closer inspection discrepancies are observed, requiring a microscopic treatment in which the nucleus is described in terms of the motion of individual particles.

This microscopic description has its own problems. In a quantitative comparison of measured transition *charge* densities with results of microscopic models, one often finds that too much structure is predicted in the nuclear interior and not enough strength near the nuclear surface. (In macroscopic models this last point does not occur as the strength of the deformation is usually chosen so as to reproduce the measured density in the surface region.) This effect has been known in the measurement of electromagnetic transition strengths [$B(E\lambda)$ values] for a long time and has led to the introduction of the concept of effective charge. This effective charge is supposed to include, apart from the normal charge, the contributions of transitions that are outside the model space used in the calculations. The results of electron scattering show that this extra polarization charge is concentrated near the nuclear surface and, in general, does not follow the shape of the charge density calculated from the valence wave functions (see Sections 3.3 and 3.4).

More refined calculations, such as Random Phase Approximation (RPA) calculations, should be able to describe this so-called core polarization. In this field considerable progress has been made in recent years. A necessary ingredient in such calculations is the residual interaction and specifically its strength and radial dependence. It seems that precise electron scattering results for several states of different multipolarity can give information on the q dependence, i.e. on the radial behavior of this residual interaction.

The situation is much less clear for the transition *current* densities (see Section 3.5). There is steadily increasing evidence that the strength of transverse excitations is smaller than given by even refined calculations. The origin of this quenching is not clear at present. Speculations include nuclear structure effects or non-nucleonic degrees of freedom. It might be due to a more diffuse pairing distribution in the ground state of the nuclei studied than is assumed in the calculations (ground-state correlations) or to a fragmentation of the strength through mixing with more complicated configurations. A determination of partial occupations of orbits in the ground state will probably come from (e, e'p) experiments. In some cases, however, (e, e') data may give some information on this. An example in ^{90}Zr is given in Section 3.5. Fragmentation of strength can, in principle, also be investigated experimentally, but it requires finding small fragments at high excitation energy where the level density is high. These are possible explanations in terms of nucleons only.

The question has been raised especially in the case of M1 transitions, if part of the anomaly is due to non-nucleonic degrees of freedom, such as virtual excitation of a nucleon into a $\Delta(1232)$ (see Section 3.6). This question can be asked more generally. Ever since Yukawa introduced the meson as mediator for the nuclear force it has been a goal in nuclear physics to identify the interplay between nucleonic and non-nucleonic degrees of freedom in the nucleus. (At present there are attempts to describe both in terms of quarks.) It is well known that this interplay appears, for example, in the form of various types of meson exchange currents (MEC), which are usually a small correction to the dominant nucleonic current. In order to get a handle on those exchange effects one has to have precise experiments for cases where MEC effects are expected to be non-negligible, e.g. at high q or in certain forbidden transitions. In addition, one has to improve the precision of the nuclear structure calculation to the level of accuracy where these exchange effects become important. The latter point also requires high precision data for many levels where MEC effects are expected to be small in order to check the nucleonic calculations.

This situation seems to be typical for the present status of nuclear physics. The high precision of the experimental data and the improved accuracy in the numerical calculations, which last but not least has been made possible by the development in the computer technology in recent years, have both stimulated a more precise description of the nuclear system. It is difficult to say whether the remaining discrepancies are due to failures in the approximations or whether there is a need for new concepts. In either case, possible new developments in our understanding of the many-body system are fostered by the modern electron scattering results.

2. FORMALISM

Because the electron waves are considerably distorted through the Coulomb field of a large-Z nucleus, all practical calculations have to be performed in DWBA. However, the underlying physics is much more transparent in the plane wave formalism and the distortions do not introduce basic changes. For that reason, we present that formalism.

2.1 *Quantities Measured in Electron Scattering*

The Plane Wave Born Approximation (PWBA) cross section has been derived often (1, 4, 5). It is usually written in the following approximate form:

$$\frac{d\sigma}{d\Omega} = 4\pi\sigma'_M f_{rec} \left[\sum_{\lambda=0}^{\infty} |F^C_\lambda(q)|^2 + (\tfrac{1}{2} + tg^2\theta/2) \sum_{\lambda=1}^{\infty} \{|F^E_\lambda(q)|^2 + |F^M_\lambda(q)|^2\} \right]. \qquad 1.$$

Here the Mott cross section for unit charge σ'_M is given by

$$\sigma'_M = \frac{\alpha^2(\hbar c)^2 \cos^2\theta/2}{4E_i^2 \sin^4\theta/2}. \qquad 2.$$

The recoil factor f_{rec} is given as

$$f_{rec} = \left(1 + \frac{2E_i \sin^2\theta/2}{Mc^2}\right)^{-1}, \qquad 3.$$

with M being the mass of the target nucleus.

In the derivation of Equation 1, one-photon exchange is assumed. Further, apart from using plane waves for the incoming and outgoing electron, the approximations $m_e = 0$ and $\hbar\omega = E_i - E_f \ll \hbar c q$ have been made. The nuclear structure enters into the cross section only through the longitudinal form factor F^C and the transverse form factors F^E and F^M. These form factors are functions of the momentum transfer q only, where $q = 2(\sqrt{E_i E_f}/\hbar c)\sin\theta/2$. They can be expressed as Fourier-Bessel transforms of the nuclear charge and current transition densities:

$$F_\lambda^C(q) = \frac{\hat{J}_f}{\hat{J}_i} \int_0^\infty \rho_\lambda(r) j_\lambda(qr) r^2 \, dr$$

$$F_\lambda^E(q) = \frac{\hat{J}_f}{\hat{J}_i \lambda} \int_0^\infty \{\sqrt{\lambda+1} J_{\lambda,\lambda-1}(r) j_{\lambda-1}(qr) - \sqrt{\lambda} J_{\lambda,\lambda+1}(r) j_{\lambda+1}(qr)\} r^2 \, dr$$

$$F_\lambda^M(q) = \frac{\hat{J}_f}{\hat{J}_i} \int_0^\infty J_{\lambda\lambda}(r) j_\lambda(qr) r^2 \, dr. \qquad 4.$$

$$\hat{J} = \sqrt{2J+1}$$

These transition densities are defined as the reduced matrix elements[1] of the charge or current operator between the initial and final nuclear states:

$$\rho_\lambda(r) = \int \langle \psi_f \| \rho_{op}(\mathbf{r}) Y_\lambda(\hat{r}) \| \psi_i \rangle \, d\hat{r}$$

$$J_{\lambda\lambda'}(r) = \frac{i}{c} \int \langle \psi_f \| \mathbf{J}_{op}(\mathbf{r}) \cdot \mathbf{Y}_{\lambda\lambda'1}(\hat{r}) \| \psi_i \rangle \, d\hat{r} \qquad 5.$$

where Y_λ are the usual spherical harmonics and $\mathbf{Y}_{\lambda\lambda'1}$ the vector spherical harmonics.

In the DWBA, Equations 1 and 4 are no longer valid, but the cross section still depends on the densities as given by Equation 5. The changes introduced by calculating the cross section in DWBA can be understood in a qualitative way through the Eikonal approximation: through the Coulomb attraction the electrons are accelerated upon approaching the nucleus and the electron wave is focussed onto the nucleus. As a consequence, an experiment actually samples the form factor at a larger momentum transfer than given by the asymptotic values of the kinematic variables. Therefore, if one wants to compare the experimental results with some theoretical form factor calculated in PWBA, the data measured for a certain q must be plotted at q_{eff}, which is slightly larger than the asymptotic q:

$$q_{eff} = q \left(1 + 1.5 \frac{Z\alpha\hbar c}{ER_{eq}} \right), \qquad 6.$$

where R_{eq} is the equivalent radius of a hard sphere. In practice one uses $R_{eq} = 1.12 \times A^{1/3}$.

[1] The Wigner-Eckart theorem is used in the form:
$$\langle J_f M_f | T_{\lambda\mu} | J_i M_i \rangle = \langle J_i M_i \lambda\mu | J_f M_f \rangle \langle J_f \| T_\lambda \| J_i \rangle.$$

The focussing of the electron wave onto the nucleus increases the cross section and causes a smearing of the square of the form factor, which mainly fills in the minima one would get with plane waves. The DWBA calculations include all these effects and no corrections need to be applied in plotting data over a DWBA calculation. Nevertheless one quite often plots both the data and the DWBA calculation as a function of q_{eff}, especially if data were obtained at different values of both E_i and θ.

The form (Equation 5) for the nuclear transition charge or transition current is a consequence of the fact that the nucleus has a well-defined spin before and after the scattering in addition to the approximation that we have only a "one-photon exchange." We thus have to know the nuclear charge or current only at the interaction point given by the variable r. This also holds for exchange currents even though the latter, in terms of nucleons, are due to two-body operators. It is only in the explicit calculation of the exchange current from the nuclear wave function that the two-body nature matters.

The selection rules for the various terms are such that for natural-parity or "electric" transitions where $\pi_i \pi_f = (-1)^\lambda$ the magnetic form factor F_λ^M is zero. Similarly for unnatural-partity or "magnetic" transitions where $\pi_i \pi_f = (-1)^{\lambda+1}$, F_λ^C and F_λ^E are zero. If we scatter from a spin-zero nucleus, only the term with $\lambda = J_f$ in each of the sums of Equation 1 remains, and in the second sum either F^E or F^M is zero, depending on the nature of the transition. So only a single multipolarity contributes to the cross-section measurement of a level and the cross section is determined by the three densities ρ_λ, $J_{\lambda,\lambda-1}$, and $J_{\lambda,\lambda+1}$ for electric transitions and just $J_{\lambda,\lambda}$ in the case of a magnetic transition.

The continuity equation $\mathbf{V}\mathbf{J} + (1/c)(\partial \rho / \partial t) = 0$, which as a statement of charge conservation is assumed to hold quite universally, connects the currents $J_{\lambda,\lambda-1}$ and $J_{\lambda,\lambda+1}$ to the transition charge $\rho_\lambda(r)$:

$$\hat{\lambda}\frac{\omega}{c}\rho_\lambda(r) = \sqrt{\lambda}\left(\frac{d}{dr} - \frac{\lambda-1}{r}\right)J_{\lambda,\lambda-1}(r) - \sqrt{\lambda+1}\left(\frac{d}{dr} + \frac{\lambda+2}{r}\right)J_{\lambda,\lambda+1}(r). \quad 7.$$

Thus, in the case of natural-parity transitions one has only two instead of three independent densities. This can be used to eliminate one of the terms in Equation 4. It is most convenient to eliminate $J_{\lambda,\lambda-1}(r)$ so that the transverse electric form factor becomes

$$F_\lambda^E(q) = -\sqrt{\frac{\lambda+1}{\lambda}} \frac{\omega}{qc} F_\lambda^C(q) - \frac{\hat{\lambda}}{\sqrt{\lambda}} \frac{\hat{J}_f}{\hat{J}_i} \int J_{\lambda,\lambda+1}(r) j_{\lambda+1}(qr) r^2 \, dr. \quad 8.$$

So, if one determines F_λ^C and F_λ^E from a combination of forward and backward angle measurements, both $\rho_\lambda(r)$ and $J_{\lambda,\lambda+1}(r)$ and therefore also

$J_{\lambda,\lambda-1}(r)$ (using Equation 7) can be determined. The essential point is that the charges and currents in nature that one measures obey the continuity equation. This is because the continuity equation, as a statement about the (relativistic) charge and current of the nucleus, follows from the gauge invariance of the probing electromagnetic field and of its coupling to the particle field (6).

Problems arise only if one wants to calculate the charge and current densities from theoretical wave functions, as in most cases these are (a) nonrelativistic and (b) contain nucleon degrees of freedom only. Both of these limitations are less severe for the charge densities than the current densities. The relativistic corrections to the charge operator, e.g. the relativistic spin-orbit term, are of order $(v/c)^2$. [There are other types of corrections of the same order (8) that eventually also should be included. At the present level of accuracy of the experiments such corrections start to be noticeable.] Further, the non-nucleonic degrees of freedom, i.e. the fact that nucleons are surrounded by a meson cloud, can presumably be accounted for by folding the point charge distribution with the nucleon charge distribution.

For current densities, the situation is considerably more severe. First of all, the usual Schrödinger form of the current operator does not obey the continuity equation if one introduces velocity-dependent terms such as a spin-orbit potential in the potential used to calculate the single-particle wave functions. (This is essentially a relativistic effect that survives in the Schrödinger nonrelativistic limit for bound Dirac particles.) Further, there are additional relativistic corrections and meson exchange currents. These cannot be approximated by folding with the nucleon charge distribution, as the masses of nucleons and mesons, and therefore their contributions to the current, are different.

For these reasons, a comparison between experiment and theory for the purpose of checking wave functions should preferentially be done by comparing charge densities, where the uncertainties in the operator are smallest. In the comparison of calculated current densities $J_{\lambda,\lambda-1}(r)$ or $J_{\lambda,\lambda+1}(r)$ with experimental results, one may at present learn more about the adequacy of the current operator used than about the wave functions. In any case, in comparisons of currents, one can not avoid dealing with mesonic and relativistic corrections, as they may be quite significant.

There are cases where one may want to calculate cross sections directly from the theoretical densities. In that case, one should for the reasons mentioned, express the cross section as much as possible in terms of the charge density as this can be calculated most reliably. This is implied by using, for example, Equation 8 instead of Equation 4 for F_λ^E.

2.2 Determination of the Densities

The measured form factors are Fourier-Bessel (FB) transforms of the densities. This FB transform can be inverted and the densities can be written as integrals over the form factors:

$$\rho_\lambda(r) = \frac{2}{\pi} \frac{\hat{J}_i}{\hat{J}_f} \int_0^\infty F_\lambda^C(q) j_\lambda(qr) q^2 \, dq$$

$$J_{\lambda,\lambda+1}(r) = -\frac{2}{\pi} \frac{\hat{J}_i}{\hat{J}_f} \frac{\sqrt{\lambda}}{\hat{\lambda}} \int_0^\infty \left[F_\lambda^E(q) + \sqrt{\frac{\lambda+1}{\lambda}} \frac{\omega}{qc} F_\lambda^C(q) \right] j_{\lambda+1}(qr) q^2 \, dq$$

$$J_{\lambda\lambda}(r) = \frac{2}{\pi} \frac{\hat{J}_i}{\hat{J}_f} \int_0^\infty F_\lambda^M(q) j_\lambda(qr) q^2 \, dq. \qquad 9.$$

This indicates that for a complete inversion the form factor should be known up to infinite momentum transfer. While this is obviously impossible, we find that nature helps. For several cases, the form factor has been measured over a large range in momentum transfer. In those cases it has been observed that the measured form factors, which fell off with a certain rate as a function of q, start to drop much more rapidly once q gets above twice the Fermi momentum of a nucleon in the nucleus, which is approximately 1.4 fm^{-1}. This has been observed on the 2^+ level in ^{58}Ni (9), and the 3^- (10) and 5^- levels in ^{208}Pb (11). In Figure 1 we give the form factors for the lowest two 5^- levels in ^{208}Pb as an example. These form factors verify that the densities contain very little strength in the high Fourier components, which allows us to truncate this integral of Equation 9 typically at about 3.5 fm^{-1} in heavy nuclei and at about 4.0 fm^{-1} in light nuclei. However, another consequence of this observation is that, unless we measure to that limit, we may miss a substantial part of the nuclear structure. This was demonstrated in the study of the analysis of the 3^- level in ^{208}Pb (see 12).

But the problems are not only at high momentum transfer. Quite often it is particularly difficult at low momentum transfer to measure the inelastic form factor because of the large background from the radiation tail of the elastic scattering. In such cases, additional information may be obtained from other experiments that use the electromagnetic interaction. In particular, the measurement of the $B(E\lambda)$ or $B(M\lambda)$ helps in the determination of the transition charge density ρ_λ or the transition current density $J_{\lambda,\lambda}$, respectively. These quantities are given by:

$$B(E\lambda) = \left[\frac{\hat{J}_f}{\hat{J}_i} \int_0^\infty \rho_\lambda(r) r^{\lambda+2} \, dr \right]^2 \qquad 10.$$

and

$$B(M\lambda) = \frac{\lambda}{\lambda+1}\left[\frac{\hat{J}_f}{\hat{J}_i}\int_0^\infty J_{\lambda\lambda}(r)r^{\lambda+2}\,dr\right]^2. \qquad 11.$$

They correspond to the electron scattering form factors F_λ^C and F_λ^M, respectively, near $q = 0$ and thus help particularly in cases of insufficient data at low momentum transfer.

For certain levels (mostly E2 transitions in strongly deformed nuclei) information on the transition charge density is known from muonic x-ray transitions. In these cases the sampling function for the transition density is different from that in Equation 9 and depends on the radial wave function of

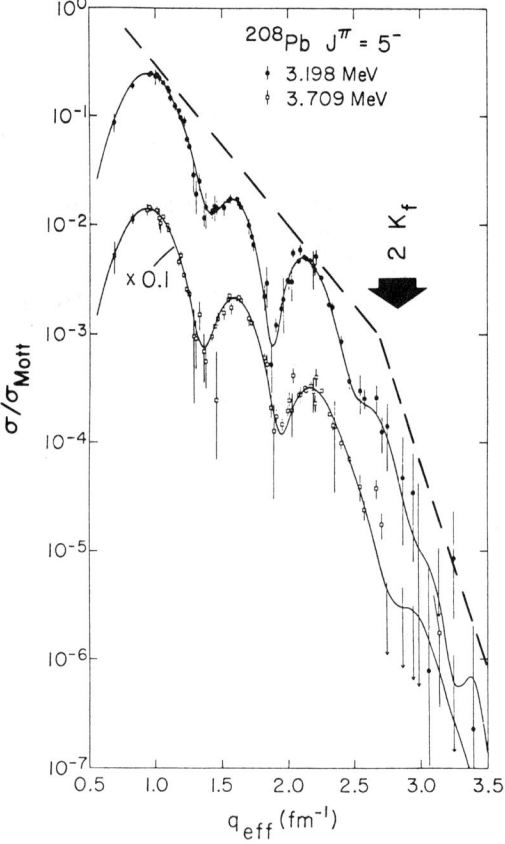

Figure 1 Measured cross sections with fit for the lowest two E5 transitions in ^{208}Pb.

the muon in the various orbits. The equivalent integral has the form

$$\int_0^\infty \rho_\lambda(r) \, e^{-\alpha r} \, r^{\lambda+2} \, dr, \qquad 12.$$

where α has to be determined for each nucleus from the calculation of the muon wave function and can be found in tables (e.g. 13). Such data correspond to points at fairly small momentum transfer (14) and play a role in the analysis similar to that of the $B(E\lambda)$ value.

So far we have discussed the reconstruction of the radial densities from the form factors only in the framework of the PWBA. For general cases where the DWBA has to be used, the technique of reconstructing the densities is somewhat more complicated but well established. The densities for r greater than a cut-off radius R_c are assumed to be zero, while for $r < R_c$ they are expanded in a FB series as

$$\rho_\lambda(r) = \sum_\mu A_\mu q_\mu^{\lambda-1} j_\lambda(q_\mu^{\lambda-1} r)$$

$$J_{\lambda,\lambda+1}(r) = \frac{\hat{\lambda}}{\sqrt{\lambda+1}} \frac{\hbar\omega}{hc} \sum_\mu B_\mu j_{\lambda+1}(q_\mu^\lambda r) \qquad 13.$$

$$J_{\lambda\lambda}(r) = \sum_\mu C_\mu j_\lambda(q_\mu^\lambda r).$$

Here the q_μ^λ is the μth zero of the spherical Bessel function of order λ. The current density, $J_{\lambda,\lambda-1}(r)$, is then given by

$$J_{\lambda,\lambda-1}(r) = \frac{\hat{\lambda}}{\sqrt{\lambda}} \frac{\hbar\omega}{hc} \left[\sum_\mu A_\mu j_{\lambda-1}(q_\mu^{\lambda-1} r) - \sum_\mu B_\mu j_{\lambda-1}(q_\mu^\lambda r) \right]. \qquad 14.$$

In the analysis of the experimental data extending to a certain q_{max}, the coefficients A_μ and B_μ, or C_μ are fitted to the data. Thereafter the coefficients with index μ for which $q_\mu > q_{max}$, which are not sensitive to the data, are chosen in such a way that the form factors drop as rapidly as possible. We refer to (12) for further details.

In the case of natural-parity transitions it has been customary, instead of determining the coefficients A and B at the same time, to proceed in an iterative way. In order to be sensitive to the current and to the charge density separately, one needs data in the forward and backward direction for the same momentum transfer. One then first fits the forward scattering data to obtain a zero-order transition charge density. Using this transition charge, one can calculate the backward cross sections and then fit the current density $J_{\lambda,\lambda+1}$ to the excess cross section in the backward scattering data. This current is then used to correct the forward scattering data for contributions from the scattering due to $J_{\lambda,\lambda+1}$. These corrected data can now be used in fitting the charge density again in next order. This iterative procedure converges quite rapidly and replaces the Rosenbluth separation (see 3) that cannot be used in the presence of Coulomb distortions.

With this procedure, one obtains in a model-independent way a density that gives a best fit to the data. Further, one also gets an error envelope giving the uncertainty of the densities at each value of r. One has to keep in mind that these uncertainties are strongly correlated, and thus not every density within the error band will fit the data within the experimental uncertainty. Secondly, the uncertainty from the form factor beyond the maximum measured momentum transfer q_{max} enters into the error band even though it cannot be treated in the same way as a statistical uncertainty. Of course, it is always possible to check densities given by some model calculation by calculating cross sections and comparing these with the measured ones. As a matter of fact, the ultimate check on the quality of theoretical predictions should always be performed in that way.

In the case of negligible transition current $J_{\lambda,\lambda+1}(r)$ one can present recalculated data. These are defined as

$$\left[\frac{d\sigma}{d\Omega}(E, q_{eff})\right]_{Rec} = \left[\frac{d\sigma}{d\Omega}(E_{ex}, q_{eff})\right] \exp \frac{\left[\frac{d\sigma}{d\Omega}(E, q_{eff})\right]_{DWBA}}{\left[\frac{d\sigma}{d\Omega}(E_{ex}, q_{eff})\right]_{DWBA}}. \qquad 15.$$

This recalculation is slightly model dependent, i.e. it depends on the shape of $\rho(r)$; thus it is only fully correct if the correct density is used for that recalculation. Therefore, one has to do this recalculation a few times as the fitted density comes closer to the final density.

Unfortunately it is not always possible to extract densities without ambiguity. Quite often the information provided by the experiment is insufficient for a full reconstruction, for instance because the data do not cover a large enough range in q or because the errors in the cross sections are too large. This is particularly so in the determination of the currents. In those cases one has to resort to a nuclear model to tell which configurations might be important in the transition. One then adjusts the amplitudes of these configurations in the transition and possibly the radial shape of the single-particle orbits and/or the g_ℓ and g_s factors occurring in the transition operators, in order to describe the experimental data. As one does not know how much model dependence is introduced by choosing certain configurations, densities obtained in that way have to be viewed with caution.

2.3 Spectroscopy

In this section we discuss electron scattering as a tool to determine spin and parity, possibly of previously unknown nuclear levels. As with other nuclear probes, this is a useful technique if proper care is taken in the interpretation. For that reason we discuss several possible difficulties.

If one observes a longitudinal form factor, we can conclude that the excitation of the state proceeds via a natural-parity transition. The opposite is unfortunately not true. From the absence of a longitudinal form factor it does not follow that one has an unnatural-parity transition, since the longitudinal form factor can be too small to be measured. This will occur, for instance, in the excitation of almost pure neutron states such as the 10^+ or 12^+ states in ^{208}Pb. Even though the transitions to these states are of natural parity they are not accompanied by a Coulomb form factor (15).

If one starts with a 0^+ nucleus, the spin of an excited state is determined by the multipolarity of the transition to that state. For strongly excited states with longitudinal form factors we may assume that such states are collective and therefore their transition charge densities are peaked at the nuclear surface. This feature is predicted by most models, as is shown in Section 3. The shape of the form factor is then approximately given by $j_\lambda(qR)$, where R is the nuclear radius. Since R is usually quite well known, one can use the position of the first maximum in the form factor to determine the multipolarity of the transition. This method was used successfully in identifying the first two 10^+ states in ^{208}Pb by Friedrich et al (16). We show in Figure 2 the form factors for the lowest $2^+, 4^+, 6^+$, and 8^+ states in ^{90}Zr (17). It can be seen that in a regular pattern the first maximum in the form factor appears at larger and larger momentum transfer for increasing multipolarity.

For weaker transitions or transverse form factors this method can be used as some indication for the multipolarity of the transition but it is not reliable. Figure 3 shows the form factors of the lowest 5^-, the third 5^-, and the lowest 10^+ state in ^{208}Pb as measured at 90°. It is clear that the form factor of the third 5^- state could easily be mistaken as being due to an E10 transition.

Similarly confusing can be the identification of the magnetic excitations. In Figure 4 we show the form factors of the first four 6^- levels in ^{208}Pb, two of which have been measured while the other two were predicted using the wave functions of Heusler & von Brentano (19), which agree quite well for the two observed 6^- levels. Obviously in such a situation the form factor cannot be taken as a strong signature of the multipolarity of the transition.

The problem in these cases is that electrons probe the nuclear interior as well as the surface region. If the transition density is not just peaked at the surface, the contribution from the interior interferes with that from the surface, and this clouds the picture. This situation does not exist as much with strongly interacting probes. Because the projectile is strongly absorbed in nuclear matter, the interior region affects the cross sections much less. For that reason the identification of the multipolarity is usually more reliable with strongly interacting probes.

Excited 0^+ and 1^- states may have unusual form factors because of selection rules, which refer to the photon point ($q = k$). For E0 transitions, an exact selection rule states that $B(E0) = 0$, which means that the integral of $\rho(r)r^2$ must vanish, so a nonvanishing density must have at least one

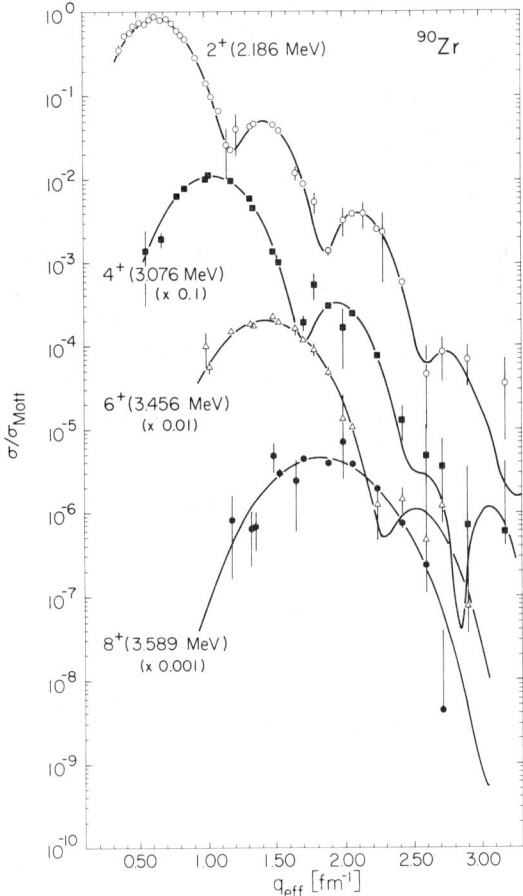

Figure 2 Measured cross sections and fit for the lowest $J^\pi = 2^+, 4^+, 6^+$, and 8^+ states in ^{90}Zr.

node. As a consequence, the lowest order term in the expansion of the form factor in powers of the momentum transfer vanishes, causing the form factor to behave like the one for an E2 transition (see also Section 2.4). Figure 5 shows the form factor for the 7.6-MeV 0^+ state in ^{12}C from the experiment of Crannell et al (20), together with a fit to the data and the extracted transition charge density. No transverse form factors exist for E0 transitions.

In the case of E1 transitions one has a similar situation. Low-lying 1^- states are normally excited via isoscalar transitions from the ground state, which means that proton and neutron excitations contribute equally to the transition. In that case the $B(E1)$ value is strongly reduced [in self-conjugate

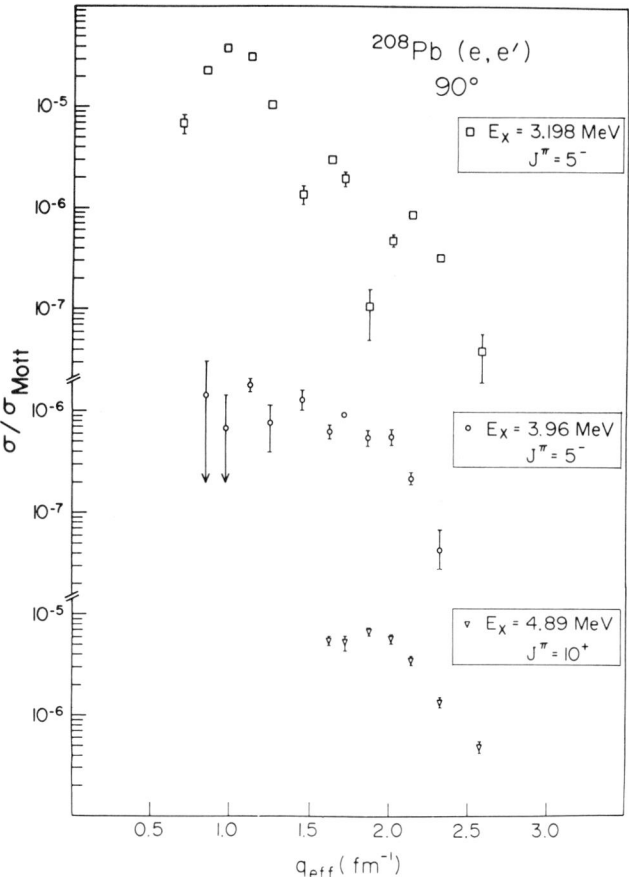

Figure 3 Measured cross sections (18) taken at 90° for three levels in ^{208}Pb as indicated.

nuclei it is even zero, see (21)]. Therefore, the form factors for these transitions show in most cases atypical shapes. Figure 6 shows the form factor of the 4.84-MeV 1^- state in ^{208}Pb plus the transition charge density for that level (22). Again, the suppression of the form factor at low momentum transfer is an indication of the selection rule at the photon point.

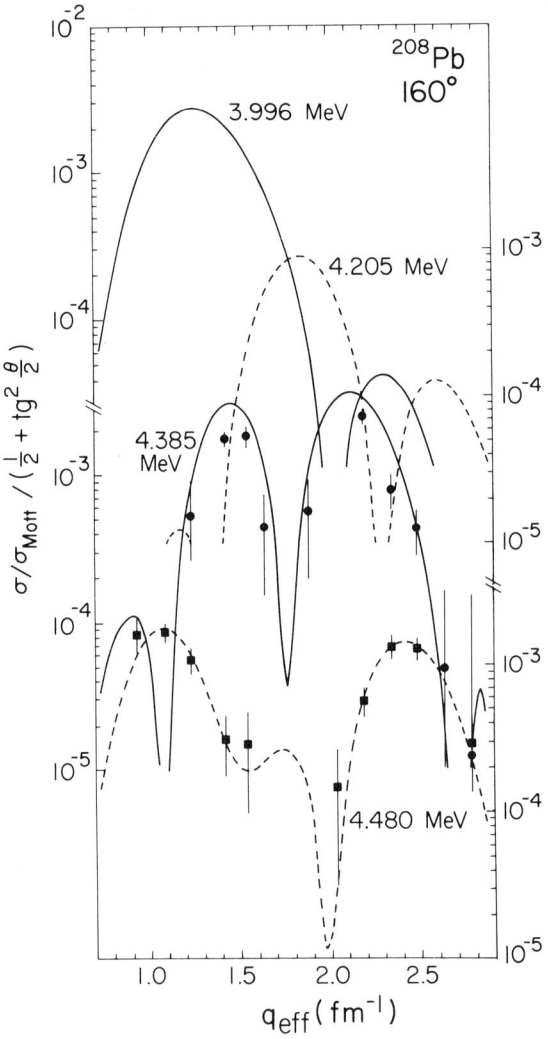

Figure 4 Predicted cross sections for the four lowest 6^- levels in ^{208}Pb. The data are from (18).

Quite often nuclear structure considerations suggest a certain character for a state, e.g. stretched particle-hole excitations. In such cases comparison of the measured form factor with the predicted one may permit an assignment. Thus, in spite of these occasional ambiguous cases, electron scattering form factors have been used successfully in the identification of many states.

2.4 Measurements at Low Momentum Transfer

Quite often form factors have been measured only at low momentum transfer. From such data one can extract the multipolarity of the transition, the transition probability, and possibly a transition radius. This is apparent if one calculates the Coulomb form factor squared in the small-q limit, where the spherical Bessel function can be expanded:

$$F^c_\lambda(q) \approx \frac{q^\lambda}{(2\lambda+1)!!} \sqrt{B(E\lambda)} \left[1 - \frac{q^2}{2(2\lambda+3)} R^2_{tr} + \cdots \right]. \qquad 16.$$

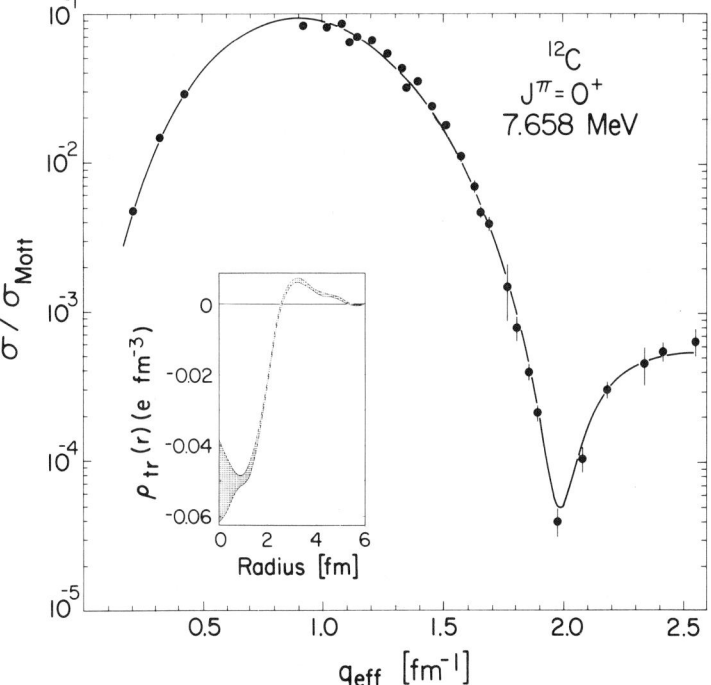

Figure 5 Measured cross sections and fit for the 7.65-MeV $J^\pi = 0^+$ state in ^{12}C. The insert shows the transition charge density that is the result of the best fit.

Here the transition radius is defined as

$$R_{tr}^2 = \frac{\int \rho_\lambda(r) r^{\lambda+4} \, dr}{\int \rho_\lambda(r) r^{\lambda+2} \, dr}.\qquad 17.$$

A similar form with $B(E\lambda)$ replaced by $B(M\lambda)$ and the transition radius R_{tr} by the equivalent quantity for $J_{\lambda\lambda}$ is obtained in the small-q expansion of the magnetic form factor. Therefore, the quantity $q^{-\lambda} F_\lambda^c(q)$ plotted against q^2 will give a straight line at small q, if λ is chosen properly. From the extrapolation of this line to $q = 0$ and the slope, one can determine $B(E\lambda)$ and R_{tr}^2. We refer to (3) for a detailed discussion of this subject. It should be remembered that Equation 16 is only valid in PWBA. This makes it very hard to apply such an analysis to heavy nuclei where the Coulomb distortions are large (especially because measurements at low q are usually done at low energy).

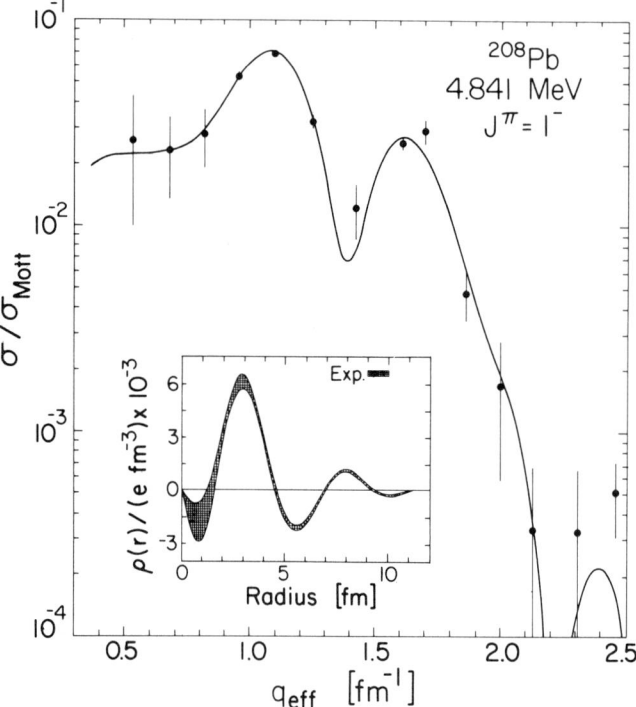

Figure 6 Measured cross sections and fit for the isoscalar E1 excitation to the 4.841-MeV state in ^{208}Pb. The insert shows the transition charge distribution that is the result of the fit.

The small-q expansion of the transverse electric form factor leads to an interesting result. Since the second part of this form factor is folded with the spherical Bessel function of order $\lambda + 1$ (see Equation 8), it drops at low q faster than the first term. So at small q the transverse electric form factor is directly related to the Coulomb form factor. This is known as Siegert's theorem (23). It is apparent from our formulation that if $J_{\lambda,\lambda+1}(r)$ is vanishingly small, Siegert's theorem holds for all momentum transfers. It is this simple form at small momentum transfer that makes it preferable to eliminate $J_{\lambda,\lambda-1}(r)$ by means of the continuity equation rather than eliminating any other term.

In considering the reliability of the $B(E\lambda)$ or $B(M\lambda)$ values extracted from such an analysis, one has to keep in mind that these are extrapolated quantities and that the corrections for DWBA effects are not model independent. If there have been discrepancies in the past between the value extrapolated from electron scattering and the value measured directly at the photon point, it has probably been because not enough terms were considered in the low-q expansion, or because the effect of distortions were not properly taken care of. Certainly in those cases where the full transition density was determined, no discrepancies have been found between the value extracted from electron scattering data and the direct measurement at the photon point.

For those cases where the $B(E\lambda)$ value vanishes because of selection rules or the nuclear dynamics, the first term in the small-q expansion vanishes. In these cases the form factors assume quite different shapes causing possible misinterpretations of the multipolarity. Such a suppression of the form factor at low q is actually not an uncommon situation since quite often most of the strength for the transition probability (low q) is concentrated in a single collective state whereas for the form factors at higher q this is not the case.

3. NUCLEAR STRUCTURE RESULTS

In the past, the most frequently used model has been the Tassie model (24). In this model the nucleus is described as an incompressible irrotational fluid, which leads to the result

$$\rho_\lambda(r) = Nr^{\lambda-1} \frac{\partial \rho_0(r)}{\partial r}. \qquad 18.$$

Here ρ_0 is the ground-state charge distribution. However, in most applications the form of ρ_0 was adjusted to fit the inelastic data. In this form it is only a convenient parameterization for a surface peaked shape but not a nuclear model. In addition, more recent experiments show that transition charge densities of even very collective states like the octupole vibration in ^{208}Pb (see Figure 7) show a significant structure in the nuclear interior that cannot be explained by macroscopic models.

Such a structure must be attributed to the participation of orbits with nodes in the collective motion. In the case of this octupole vibration in ^{208}Pb the structure arises partly from the $3s_{1/2}$ proton orbit, which is the last orbit filled in the ground state. This orbit extends into the far interior of the nucleus. As this orbit participates in the vibration, the transition charge density also reaches into the interior.

This example demonstrates that one needs microscopic descriptions if one wants to understand such structure in the interior. Figure 7 also shows several RPA predictions as outlined in (10). In spite of the remaining small discrepancies, such calculations reproduce the interior structure as well as the strength at the surface.

Presently we have two areas where we can make meaningful comparisons between theory and experiment. These are the collective levels in spherical nuclei such as ^{208}Pb where in the calculation all particle-hole excitations up

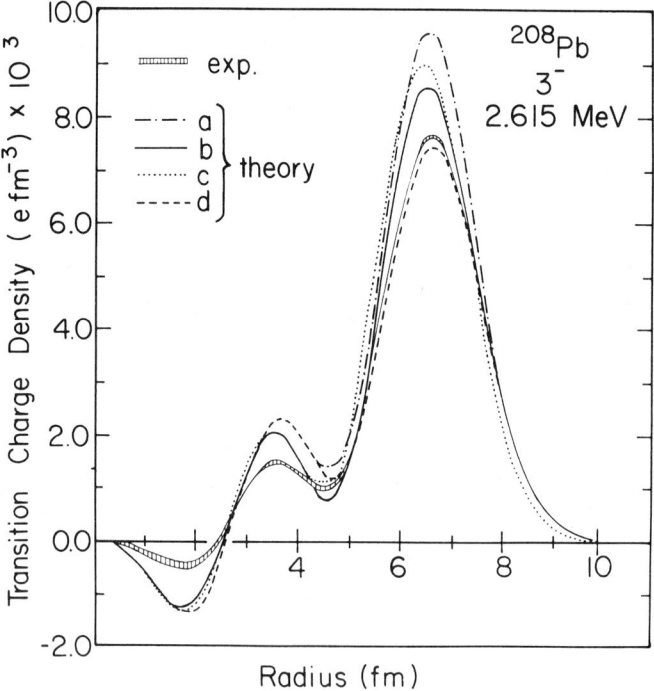

Figure 7 Transition charge density of the octupole vibration in ^{208}Pb. The theoretical predictions come from RPA calculations as explained in (10).

to >100 MeV can be included, or the ground-state rotational band of strongly deformed nuclei discussed in the next section. Away from the closed shells and outside the strongly deformed region considerable difficulties in the interpretation of electron scattering results remain, especially in the interpretation of the currents.

3.1 *The Rotational Model*

In the discussion of experimental results we start with the measurement of form factors for the members of the ground-state rotational band of strongly deformed nuclei since they have yielded probably the most direct and interpretable information so far. In the region of strongly deformed nuclei, the Bohr-Mottelson picture provides a model that allows association of the measured transition charge densities for the various members of the rotational band with the various multipoles of the intrinsic deformed charge distribution. This permits a full reconstruction of the intrinsic shape, which can be compared to the results of nuclear structure calculations such as Hartree-Fock calculations. Thus, the impressive power of elastic electron scattering to determine with high precision the shape of the nuclear ground state (25a,b,c) can be extended into the deformed region through the inclusion of the results of inelastic electron scattering.

The model of rotational motion is well established. According to this model, strongly deformed nuclei show a rotational band built on the ground state or any intrinsic excited state with strong E2 transitions between the members of the band.

If we assume the wave functions to be of the standard form (26)

$$|IMK\rangle = \sqrt{\frac{2I+1}{8\pi(1+\delta_{K0})}} [D^I_{MK}(\Omega)\chi_K(r) + (-1)^{I+K} D^I_{M-K}(\Omega)\chi_{-K}(r)], \quad 19.$$

transition matrix elements for the one-body operator O_λ of multipolarity λ can be written as

$$\langle I_2 K_2 \| O_\lambda \| I_1 K_1 \rangle = \hat{I}_1 [\langle I_1 K_1 \lambda (K_2-K_1) | I_2 K_2 \rangle \langle K_2 | O_{\lambda(K_2-K_1)} | K_1 \rangle$$
$$+ (-1)^{I_1+K_1} \langle I_1 -K_1 \lambda(K_2+K_1)|I_2 K_2\rangle \langle K_2|O_{\lambda(K_2+K_1)}|-K_1\rangle]. \quad 20.$$

The strong intraband E2 transitions then are a measure of the intrinsic quadrupole moment, i.e. the quadrupole moment of the deformed charge distribution in a body-fixed frame. This statement is an approximation assuming a stable deformation and assuming that the overlap of the rotated wave function with the original wave function is a δ function (27). For the purpose of this paper we assume that this is a good approximation for the nuclei discussed.

The multipole operator that yields the transition charge density can be transformed in the same way so that we obtain

$$\rho_\lambda(r) = \frac{1}{\hat{\lambda}} \int d\Omega \left[\langle \chi_{K_f} | \rho^{op} | \chi_{K_i} \rangle Y_{\lambda, K_f - K_i}(\Omega) \right.$$
$$\left. + (-1)^{I_i + K_i} \langle \chi_{K_f} | \rho^{op} | \chi_{-K_i} \rangle Y_{\lambda, K_f + K_i}(\Omega) \right]. \quad 21.$$

For the ground-state rotational band of even-even nuclei, where $K_i = K_f = 0$ and $\chi_f = \chi_i$, the matrix element for ρ^{op} reduces to the deformed charge distribution $\rho_i(r,\Omega)$ in the intrinsic frame, so the transition charge densities can be written as

$$\rho_\lambda(r) = \frac{1}{\hat{\lambda}} \int \rho_i(r,\Omega) Y_{\lambda 0}(\Omega) \, d\Omega. \quad 22.$$

Assuming that all $\rho_\lambda(r)$ have been measured in some experiment, we can reconstruct the intrinsic deformed charge distribution as

$$\rho(r,\Omega) = \sum_\lambda \hat{\lambda} \rho_\lambda(r) Y_{\lambda 0}(\Omega). \quad 23.$$

Several such reconstructions have been published (28). Here we present in Figure 8 the reconstructed density for the nucleus ^{154}Gd in the form of a density contour plot.

Even though such a contour plot gives a suggestive picture of the shape of a nucleus, it is not convenient for a quantitative comparison with HF

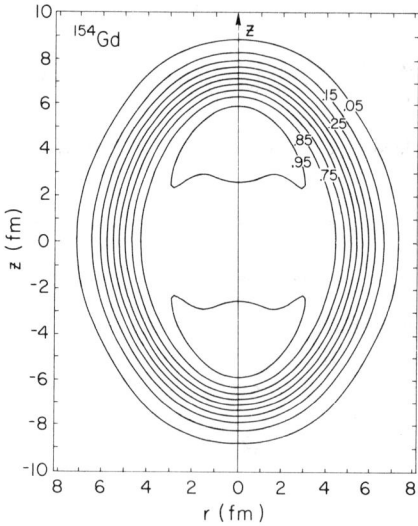

Figure 8 Lines of constant density in a cross section of the nucleus ^{154}Gd.

predictions. For that purpose, it is better to plot the multipole densities directly and compare them with the HF prediction. This is done in Figure 9.

It may be questionable how good the assumption of a stable ground-state deformation is for this nucleus. Nevertheless, we find the kind of agreement that is usual in such comparisons: the spherical part $\rho_0(r)$ and the quadrupole part $\rho_2(r)$ are quite well reproduced even in the interior. (For most strongly deformed nuclei, the agreement in ρ_2 is even considerably better than in this example.) This is reasonable because these calculations are fully microscopic. However, the comparison in the higher order transition densities gets progressively worse. This is not such a surprise. The small transition densities of high multipolarity are not very collective. Therefore, the inaccuracies in the microscopic treatment, which average out in the more collective densities, become visible in the less collective ones.

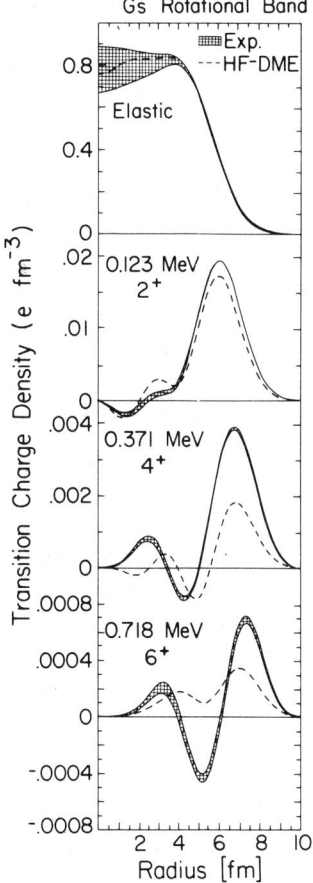

Figure 9 Transition charge densities for the various members of the ground-state rotational band in ^{154}Gd. While the uncertainty shown in ρ_0 accounts for the incompleteness error, the other error bands do not include this uncertainty. The dashed curves show a HF prediction.

3.2 Small-Amplitude Vibrations

The Bohr-Mottelson picture also includes the small-amplitude oscillations of the nuclear surface around the stable deformation. In order to get a form for the charge operator, one has to assume a certain behavior of the surface oscillation. We assume that the surface thickness of the nucleus, i.e. the density gradient, stays the same and that only the position of the surface, as given by, for example, the half density radius R, is changing. With this assumption, the charge density is a function of $r - R$.

We can parameterize the half density radius in the standard form:

$$R(\Omega) = R_0 \left[1 + \sum_{\lambda\mu} (\beta_{\lambda\mu} + \alpha_{\lambda\mu}) Y_{\lambda\mu}(\Omega) - \frac{1}{4\pi} (\beta_{\lambda\mu} + \alpha_{\lambda\mu})^2 \right]. \qquad 24.$$

The second sum, quadratic in the deformation parameters is needed to preserve the volume as one deforms the nucleus. If one considers deformations for odd-order λ, one has to add a term that corrects for the shift in the center of mass that accompanies such a deformation.

In this notation the $\beta_{\lambda\mu}$ describe the static deformations assumed to be nonzero only for λ = even and $\mu = 0$, while the $\alpha_{\lambda\mu}$ describe the vibrations. Equation 24 includes surface vibrations of spherical nuclei when we set $\beta_{\lambda\mu} = 0$. For a strongly deformed nucleus and quadrupole deformation, one distinguishes the β vibration, which is due to the variable α_{20}, from the γ vibrations due to α_{22} and α_{2-2}, which are equal. Since these amplitudes are assumed to be small, we can ignore second-order terms in them. This is not true for the static deformations, which can be quite large.

In quantizing this oscillatory motion, one replaces the deformation parameters by the phonon creation and annihilation operators $b^\dagger$ and b:

$$\alpha_{\lambda\mu} = \sqrt{\frac{\hbar}{2B\omega}} (b_{\lambda\mu} + (-1)^\mu b^\dagger_{\lambda-\mu}). \qquad 25.$$

Here ω is the frequency and $\sqrt{\hbar/2B\omega}$ the amplitude of the oscillation. Thus, for each intrinsic excitation mode one has two constants that need to be determined from the experiment. These are taken from the excitation energy of the band head and from the strength of the strongest transition to the band. With this choice of the two constants, not only is the transition strength to all members of the rotational band built on the vibration fixed, but also the radial shape of the transition charge densities. This is not trivial since these densities can assume quite a variety of different shapes.

To obtain the transition charge density, one has to replace $\alpha_{\lambda\mu}$ in the expression for the charge by the phonon creation and annihilation operators and then take matrix elements between the various states. The transition charge density between the ground state and the one-phonon state now becomes

$$\rho_\lambda(r) = -1/\hat{\lambda} R_0 \sqrt{\frac{\hbar}{2B\omega}} \int \frac{\partial \rho_i(r,\Omega)}{\partial r}$$

$$\times \left(Y^*_{\lambda,\Delta K}(\Omega) - \frac{1}{2\pi} \beta_{\lambda,\Delta K} \right) Y_{\lambda,\Delta K}(\Omega) \, d\Omega. \qquad 26.$$

Let us first discuss the β and γ vibrations in a qualitative way. For the β vibration the nucleus is changing its deformation. Since the volume is conserved, the nucleus is elongated at the poles more than it is reduced at the waist. Therefore, the amplitude of the vibration is largest at the long axis of the nucleus and the transition charge reaches its maximum where r is equal to the long axis. The γ vibration is a deformation of the nucleus at the waist. The density at the poles is not changing at all. So the transition charge for the γ vibration has a maximum where r is equal to the short axis.

This corresponds exactly to what is observed by Hersman et al (29a,b) in ^{154}Gd. Figure 10 shows the densities for the 2^+ states of the ground-state rotational band, the β band and the γ band. The dashed line gives the prediction using Equation 26 with the experimentally determined shape of the nuclear ground state, based on the assumption of the rotational model previously discussed.

It is clear that our derivation of the charge operator works only at the nuclear surface as we assumed the gradient of the density to remain constant. One cannot expect, therefore, such a description to give a realistic density in the interior of the nucleus. In addition, the experimental data for ^{154}Gd do not range far enough in momentum transfer to determine reliably the densities in the nuclear interior (the incompleteness uncertainty is not included in this figure). Therefore, this experiment verifies the geometric interpretation of the vibrations at the surface but it does not show the limitations of this macroscopic description in the interior. A complete understanding of the measured transition charge densities requires a detailed microscopic calculation. While such calculations for deformed nuclei have only been done for a few cases, vibrations in spherical nuclei have been investigated quite extensively (see 12).

Some words should be said on the interacting boson model (IBM) (30). The IBM is an algebraic model in which states of nuclei are described not in a geometrical way, as is done by Bohr and Mottelson, but by building them from bosons, which can be interpreted as pairs of nucleons coupled to $L = 0$ (s bosons) or $L = 2$ (d bosons). Using this framework, both vibrational and rotational nuclei can be described in a uniform way.

In the IMB, E2 excitations are described as transitions of an s into a d boson (or reversed) or as the recoupling of d bosons. The "single-boson" transition charge densities $\alpha(r)$ and $\beta(r)$ (for s → d and d → d, respectively) accompanying these basic transitions are not predicted by the model, as the model itself has no interpretation in r space. While there is an effort underway to derive the boson parameters microscopically starting from the underlying fermions, these presently have to be taken from the experiment. Again, inelastic electron scattering provides a unique tool for studying these quantities. Globally one finds (29, 31) that $\alpha(r)$ is surface peaked and resembles the one-phonon density, whereas $\beta(r)$ resembles the second derivative of the ground-state charge distribution similar to a two-phonon density.

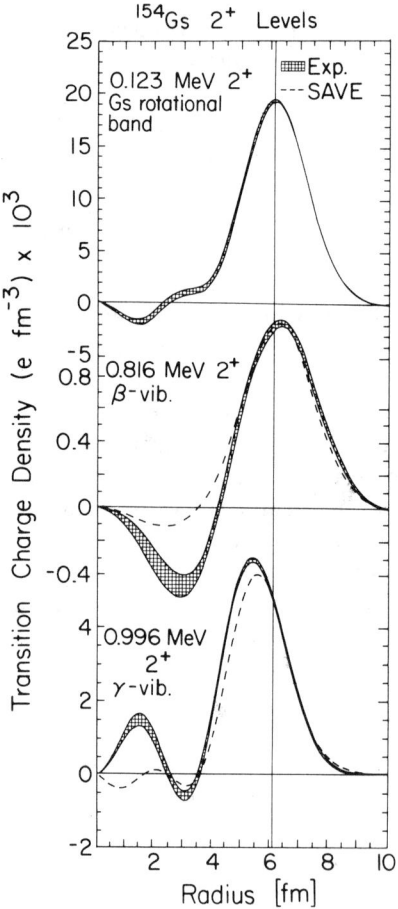

Figure 10 E2 transition charge densities for the ground-state rotation, the β vibration, and the γ vibration in ^{154}Gd. The vertical line indicates the position of the maximum of the ground-state ρ_2. The dashed line gives the predictions with the geometric model.

One consequence of this model is that any third 2^+ state should have a density that is a linear combination of the other two E2 densities, where the IBM predicts the mixing ratios. This prediction was tested for the nucleus ^{154}Gd by Hersman et al (29a,b) and was found to fail.

3.3 Single-Particle Transition Densities

If we describe the nucleus through the motion of individual nucleons, the total transition density can be expressed as

$$\rho_\lambda(r) = \sum_{a,b} \rho_\lambda^{ab}(r) S_{ab,\lambda} \frac{1}{\hat{J}_f} \qquad 27.$$

and similarly for the $J_{\lambda,\lambda'}(r)$. That means that the transition densities are sums over single-particle densities weighted by the spectroscopic amplitude or density matrix element:[2]

$$S_{ab,\lambda} = \frac{\hat{J}_f}{\hat{\lambda}} \langle \psi_f \| [a_a^+ \otimes \tilde{a}_b]_\lambda \| \psi_i \rangle. \qquad 28.$$

Here a and b denote single-particle orbitals with quantum numbers n, l, and j, while a^+ and $\tilde{a}$ are particle and hole creation operators. The sum over a and b runs over both protons and neutrons.

The values of $S_{ab,\lambda}$ must be provided by the nuclear structure calculation. Care should be taken that in calculating these, the same conventions are used as in the calculation of the single-particle densities, e.g. (ls)j coupling, no i^l in the single-particle wave functions, and $u_{nlj}(r) > 0$ near the origin. For pure single-particle transitions $\psi_i = |b\rangle$ and $\psi_f = |a\rangle$ we have $S_{ab,\lambda} = 1$. The same value is found if $\psi_i = |0\rangle$ and $\psi_f = (a_a^+ \otimes a_b)_\lambda |0\rangle$.

Assuming the nonrelativistic form of the charge and current operators, several authors (1, 5) calculated the single-particle densities. Using the conventions given above, one can write the transition charge density as

$$\rho_\lambda^{ab}(r) = eC_{ab\lambda} u_a(r) u_b(r) \qquad 29.$$

with

$$C_{ab\lambda} = (-1)^{\lambda + j_a - \frac{1}{2}} \frac{\hat{j}_a \hat{j}_b}{\sqrt{4\pi}} \langle j_a \tfrac{1}{2} j_b - \tfrac{1}{2} | \lambda 0 \rangle.$$

The transition current densities have two contributions. The first comes from the charge of the nucleons, while the second stems from the

---

[2] The factor $\hat{J}_f$ in Equations 27 and 28 is due to the form of the Wigner-Eckart theorem we use plus the choice $S_{ab,\lambda} = 1$ for a pure single-particle transition.

magnetization of the nucleons. They are given by

$$J^{ab,C}_{\lambda,\lambda-1} = \frac{e\hbar}{2mc} C_{ab\lambda} \frac{1}{\sqrt{\lambda\hat{\lambda}}} \left\{ -\lambda [u'_a u_b - u_a u'_b] \right.$$
$$\left. + [\ell_b(\ell_b+1) - \ell_a(\ell_a+1)] \frac{u_a u_b}{r} \right\} \qquad 30.$$

$$J^{ab,M}_{\lambda,\lambda-1} = \mu \frac{e\hbar}{2mc} C_{ab\lambda} \frac{1}{\sqrt{\lambda\hat{\lambda}}} (\chi_b - \chi_a) \left(\frac{d}{dr} + \frac{\lambda+1}{r} \right) u_a u_b$$

$$J^{ab,C}_{\lambda,\lambda+1} = \frac{e\hbar}{2mc} C_{ab\lambda} \frac{1}{\sqrt{\lambda+1\hat{\lambda}}} \left\{ (\lambda+1)[u'_a u_b - u_a u'_b] \right.$$
$$\left. + [\ell_b(\ell_b+1) - \ell_a(\ell_a+1)] \frac{u_a u_b}{r} \right\} \qquad 31.$$

$$J^{ab,M}_{\lambda,\lambda+1} = \mu \frac{e\hbar}{2mc} C_{ab\lambda} \frac{1}{\sqrt{\lambda+1\hat{\lambda}}} (\chi_b - \chi_a) \left(\frac{d}{dr} - \frac{\lambda}{r} \right) u_a u_b$$

$$J^{ab,C}_{\lambda\lambda} = \frac{e\hbar}{2mc} C_{ab\lambda} \frac{1}{\sqrt{\lambda(\lambda+1)}} (\chi_a + \chi_b - \lambda)(\chi_a + \chi_b + \lambda + 1) \frac{u_a u_b}{r}$$
$$J^{ab,M}_{\lambda\lambda} = \mu \frac{e\hbar}{2mc} C_{ab\lambda} \frac{1}{\sqrt{\lambda(\lambda+1)}} \left[\frac{\lambda(\lambda+1)}{r} + (\chi_a + \chi_b)\left(\frac{d}{dr} + \frac{1}{r} \right) \right] u_a u_b, \qquad 32.$$

with $\chi = (l-j)(2j+1)$, $e\hbar/2mc$ is the nuclear magneton, and μ is the magnetic moment of the active nucleon in nuclear magnetons. The prime indicates the derivative with respect to r. There are higher order corrections due to relativistic effects and exchange currents (32a–d). Some approximate forms for the exchange currents have been given in (33). The way to derive these higher order terms is not yet fully established and the explicit calculation is quite cumbersome. For that reason we do not discuss these calculations, but instead refer to (32a–d).

There are several effects that we have to discuss at this point. First, the single-particle densities have to be folded with the nucleon charge or magnetization distribution. In momentum transfer space this means that the form factors have to be multiplied by the respective nucleon form factors. This correction can be applied most conveniently in the Fourier-Bessel expansion. Assuming that a density $\rho_\lambda(r)$ is given as

$$\rho_\lambda(r) = \sum A_\mu j_\lambda(q_\mu r), \qquad (r \leq R_c), \qquad 33.$$

the convoluted density is given by

$$\rho_\lambda(r)^{\text{conv}} = \sum A_\mu F_n(q_\mu) j_\lambda(q_\mu r), \qquad (r \leq R_c), \qquad 34.$$

where $F_n(q)$ is the nucleon form factor. The form factors that apply are the neutron and proton charge form factors for the transition charge density and for the convection current term in the transition current densities, and the magnetic form factor for the terms due to the magnetization as given, for example, in (34). Because of the smallness of the neutron charge form factor, contributions from the neutron to the convection current are rather small and in most cases can be neglected.

Quite often the single-particle wave functions have been calculated by means of Hartree-Fock. In that case, one has to be aware that the calculated ground-state wave function includes a residual center-of-mass motion. Therefore, the single-particle wave functions are not yet in the rest frame of the nucleus. Only in the approximation of harmonic oscillator wave functions can this residual center-of-mass motion be removed exactly. Since it is usually a small correction anyway, one can apply this correction in the harmonic oscillator approximation. The transformation has been given in (4) and again can be incorporated by means of the Fourier-Bessel expansion. Assuming a single-particle density as given by Equation 34, one can express the equivalent density, where the CM motion has been removed, by

$$\rho_\lambda(r)^{\text{conv,cmc}} = \sum A_\mu F_n(q_\mu) \exp\left(\frac{q_\mu^2 b^2}{4A}\right) j_\lambda(q_\mu r) \qquad r \leq R_c. \qquad 35.$$

Here b is the harmonic oscillator length and A is the mass number.

3.4 Single-Particle Transitions and Effective Charges

As previously mentioned, the details of measured charge densities, even for collective states, cannot be described by macroscopic models. In this section we discuss a microscopic description of the measured charge densities. Current densities are discussed in the next section.

If one compares measured transition charge densities with those calculated (as described in Section 3.3) one finds quite often that qualitatively the structure of the measured density is described, but that the calculated density is too small in the nuclear surface region. This applies both to collective states and to single-particle states. Such an effect is well known from γ-ray studies, where generally the measured $B(E\lambda)$ values are (much) larger than the ones calculated with, for instance, shell model wave functions.

This is most clearly observed in transitions between levels that are neutron single-particle or single-hole states. A typical example is ^{207}Pb, where pick-up reactions on ^{208}Pb have shown that the low-lying levels originate from the removal of a neutron from one of the orbits close to the Fermi level (35). In spite of this character of being almost pure neutron-hole states, one observes electric transitions between those states (36). Such observations have led to the introduction of the widely used concept of effective charge by giving protons and neutrons charges of $e + \Delta e_p$, and Δe_n, respectively. The extra charge Δe is needed because in the model calculations the model space always is truncated: the motion of only a certain number of nucleons (valence nucleons) in a certain number of shells (valence shells) is included in the model and the rest of the nucleons are supposed to form an inert core. The fact that transitions from such core nucleons to other orbits will, in reality, contribute leads to an underestimation of the measured transition probability.

Physically the extra charge Δe can be viewed as being due to the polarizing action of a valence nucleon on the core, hence the name core polarization and/or induced polarization charge. This polarizing action should be calculated explicitly. In practice, however, the values of Δe_p and Δe_n have almost always been adjusted to reproduce the measured values of the transition rates.

While the introduction of these numbers may be sufficient for the description of electromagnetic transition probabilities, which are also numbers (in practice most analyses have been limited to E2 transitions), electron scattering data provide additional information by determining not only the transition strengths for levels with other multipolarities, but also the radial structure of the induced charge. The results thereof show that the description of this effect by a single number is insufficient and demonstrate the need for explicit calculations of the induced charge as a radial function.

The first experiment that demonstrated this deficiency was on the excitation of the neutron-hole states in ^{207}Pb (37). This experiment showed that the shape of the measured transition charge density does not follow the dominant neutron transition, but instead is a surface peaked density similar to that observed for the low-lying collective states of the core nucleus ^{208}Pb.

A similar result was obtained in the measurement of the transition charge density of the proton single-particle transition in ^{89}Y from the $2p_{1/2}$ orbit to the $1g_{9/2}$ orbit (38). The experiment showed (see Figure 11) that the node in the measured transition charge density is not at the place of the node of the $\pi 2p_{1/2}$ wave function. (Remember that according to Equation 29 the charge density for a pure single-particle transition is the product of the radial wave functions.) This demonstrates that it is not possible to account for the core polarization by simply rescaling the single-particle density. Instead, it was found also for this proton transition that the core polarization corresponds to a surface peaked density.

The results obtained for the single-proton-hole states in ^{89}Y (39) are very similar (see Figure 12). In all three cases the spectroscopic amplitudes are less than one because of pairing correlations. The dashed line gives the shell

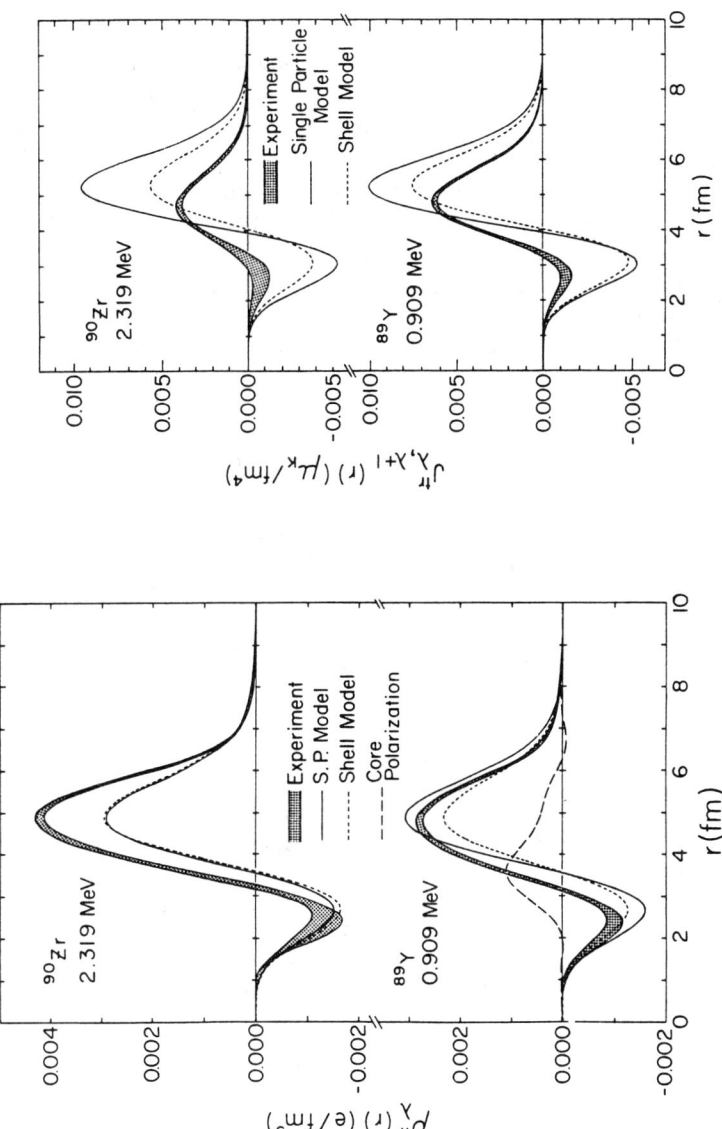

Figure 11 Transition charge and current densities for the lowest E5 transitions in ^{89}Y and ^{90}Zr.

model prediction corrected for this spectroscopic amplitude while the solid line is a result of a (SMRPA) calculation that includes the effects of core polarization by coupling the valence particle transition to the core particle-hole excitations (40).

The explicit calculation of core polarization involves the inclusion of a rather large configuration space. For that reason it has been proposed that the core polarization be calculated as a coupling between the single-particle transition and the giant multipole resonances (41). In this approach the giant resonance is used as a doorway state that accumulates all the strength.

This model has been tested by Papanicolas et al (42) for ^{206}Pb and ^{204}Pb. The structure of the lowest 2^+ excitation in these nuclei is dominated by the coupling of different neutron-hole pairs. The shapes of the transition charge densities for these levels, shown in Figure 13, are very similar to the shape of the one for the lowest 2^+ level in ^{208}Pb. The transition charge density of the

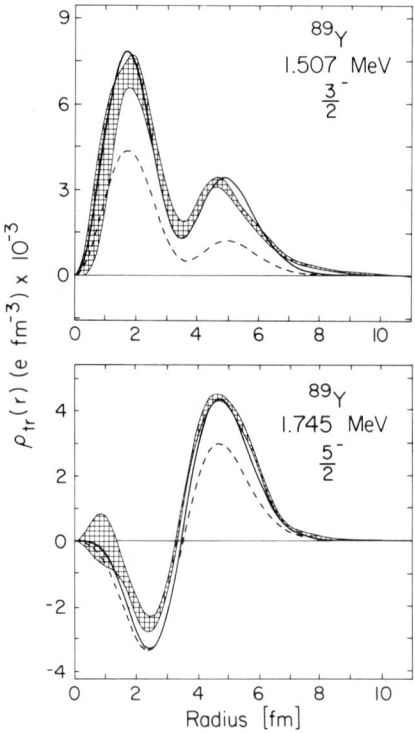

Figure 12 Transition charge densities for the $2p_{3/2}$ and $1f_{5/2}$ proton-hole states in ^{89}Y. The dashed line gives the shell model prediction, the solid line shows a prediction that includes core polarization.

giant quadrupole resonance has not been measured. However, calculations using sum rule techniques, which are supposed to be quite reliable, predict a transition charge density (also shown in Figure 13) that differs considerably in shape from the one for the low-lying collective states. Therefore, the experimental result demonstrates that in this case the dominant core polarization effect is the coupling to the low-lying particle-hole excitations rather than to the high-lying giant quadrupole resonance.

Since core polarization is a substantial fraction of the total density, shell model calculations in a limited valence space without the explicit inclusion of core polarization are not very useful for comparison with inelastic scattering experiments. This is particularly so for the heavier nuclei where even the $0\hbar\omega$ valence space is too large for a full shell model calculation.

3.5 Transverse Excitations

So far we have discussed almost exclusively the transition charge density. Considerable information on nuclear structure can come also from the

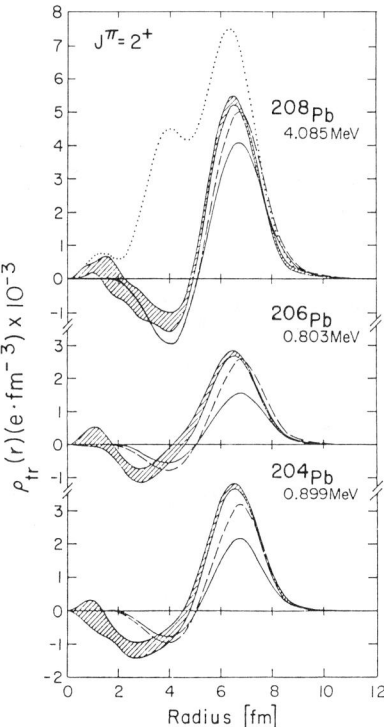

Figure 13 Transition charge densities for the lowest E2 transitions in ^{208}Pb, ^{206}Pb, and ^{204}Pb.

measurement of transition current densities in nuclei (the term current here includes both the contribution from the convection current and the one from the magnetization). However, while transition charge densities have been measured for many transitions, the measurement of the detailed shape of transition current densities is fairly new. While this statement applies for unnatural-parity transitions, which are determined by the current density $J_{\lambda,\lambda}(r)$, it is especially true for natural-parity transitions, in which $J_{\lambda,\lambda+1}(r)$ plays a role. (Even though, for practical reasons, one should distinguish between the currents in these two types of transitions, the nuclear structure information one obtains is to a large extent the same.) In the latter case, the reason is simply that the contribution from the transverse electric current to the cross section is usually very small compared to the longitudinal part, and it often requires measurements at 180° to see any effect of it. This is particularly so for levels that are very collective. By collective we mean that in a microscopic description of the excitation of such a state, many particle-hole transitions play a role, none of which is dominant. For such transitions, RPA calculations show that the current densities of the separate transitions tend to cancel each other. Further, in a particular particle-hole transition, the process in which a nucleon moves from the hole orbit to the particle orbit (the "forward-going" amplitude) can be accompanied (because of ground-state correlations, i.e. the presence of a pairing distribution) by the process in which the nucleon moves from the particle orbit to the hole orbit (the "backward-going" amplitude). These two processes act constructively for the transition charge density, but destructively for the current density (52).

In the case of positive-parity transitions of even multipolarity (e.g. E2 transitions), away from closed shells such transitions also get appreciable contributions from the pair break-up transitions $(j^2)_0 \to (j^2)_J$, for which the microscopic theory predicts the absence of transition currents (compare Equations 30–32). So, in general, the transverse electric current $J_{\lambda,\lambda+1}(r)$ is small except when an excitation is dominated by a few particle-hole transitions.

Generally, transverse magnetic excitations appear to be weak. So far no collective magnetic transitions have been found, contrary to the situation in electric transitions (remember that electric transitions are defined as natural-parity transitions and that these have both a charge form factor and a transverse electric form factor). The reason is that the spin-independent interaction between nucleons is attractive, which makes the lowest natural-parity state of a given multipolarity collective. The spin-dependent interaction, however, which plays a role for the unnatural-parity states, is repulsive, which makes the collective state lie at high excitation energy where it will be severely fragmented. So again, the larger currents are observed in cases where a few particle-hole transitions are dominant.

The high spin transitions of natural or unnatural parity are most prominent. Examples are the high spin states in the Pb isotopes (15, 43, 44); the E7, M7, or M10 transitions in ^{90}Zr and ^{88}Sr (17, 45); M8 transitions in the Ni region (46a–d); M8 and E5 transitions in ^{48}Ca (47); M6 transitions in the s-d shell (48a,b); and M4 transitions in the p-shell nuclei (49a,b).

In all cases of magnetic excitations in this list, electron scattering was used to identify the states and to determine the transition strength. This strength was found to be only 30–50% of the strength predicted for a pure particle-hole state according to Equation 32. The reasons for this reduction are not yet fully understood.

Many mechanisms can lead to quenching. As mentioned above, core polarization leads in the case of unnatural-parity transitions to a reduction in strength due to a repulsive interaction. Hamamoto et al (50a,b) attributed the reduction of the strength for the high spin states in ^{208}Pb to core polarization. Mixing of a pure particle-hole configuration with particle-hole vibration coupled states will lead to fragmentation of the strength. Krewald et al (51) calculated this effect in addition to core polarization using a q-dependent interaction based on π and ρ exchange. They found that both effects contribute about equally to the observed reduction. There is little experimental evidence favoring one mechanism or the other so far. The mixing mechanism leads to fragments at higher excitation energies (see 51), which might be found in very accurate experiments. However, in ^{58}Ni where many fragments have been found, the total strength is still only less than 50% of the predicted one (46a–d). Core polarization will change the q dependence of the form factor and could be used as signature but this occurs mainly beyond the first maximum. As high spin transitions have their first maximum already at fairly high q, in most cases only the first maximum has been measured. Moreover, other effects (see Section 3.6) may contribute in that region.

Ground-state correlations can also cause quenching (without changing the q dependence of the form factor) owing to a destructive interference of the forward- and backward-going amplitudes. As pointed out by Rowe (52), the dynamics of pairing is such that in even-even nuclei these amplitudes interfere constructively for the charge but destructively for the currents; in odd-A nuclei the reverse occurs for transitions between one-quasi-particle states. Away from closed shells, this can lead to a considerable reduction, but for a nucleus like ^{208}Pb it is usually assumed that all shells are either completely filled or empty, although some calculations indicate that this approximation may not be good enough. In principle, one-nucleon transfer reactions could give information on the filling of shells but the uncertainties in the extraction of absolute spectroscopic factors (which are directly related to the filling of an orbital) are usually too large to be of quantitative help. The best information on shell filling will probably come from the (e, e'p) reaction, which by now can also be studied in heavy nuclei (53).

In the experiments on ^{90}Zr it was possible to get some information on the filling of the proton $2p_{1/2}$ and $1g_{9/2}$ shells from a comparison of the measured E5 charge and current densities for the $0^+ \to 5^-$ (2.32-MeV) transition in ^{90}Zr and for the $1/2^- \to 9/2^+$ (0.91-MeV) transition in ^{89}Y (54). Both transitions are almost pure proton $2p_{1/2} \to 1g_{9/2}$ transitions (plus some core polarization, which is assumed to be the same for both cases) but, because in the ground state of ^{90}Zr both the $2p_{1/2}$ and the $1g_{9/2}$ orbitals are partly filled, the 5^- state in ^{90}Zr, which has a $\pi p_{1/2} g_{9/2}$ character, can be excited both by the transition $\pi 2p_{1/2} \to 1g_{9/2}$ and the transition $\pi 1g_{9/2} \to 2p_{1/2}$. Denoting the amplitudes for these two transitions by A and $\bar{A}$, respectively, one can write the transition charge density in ^{90}Zr as

$$\rho(r) = (A + \bar{A})\rho_{2p_{1/2} \to 1g_{9/2}}$$

and the current as

$$J(r) = (A - \bar{A})J_{2p_{1/2} \to 1g_{9/2}}.$$

Similar equations hold for ^{89}Y with amplitudes B and $\bar{B}$. In ^{89}Y the amplitude $\bar{B}$ is quite small. Using the value $\bar{B}/B = -0.069$ calculated by Hofstra & Allaart (57) and taking the ratio of the densities in ^{90}Zr and ^{89}Y (see Figure 11), one finds $(A - \bar{A})/(A + \bar{A}) = 0.48 \pm 0.07$, which yields $\bar{A}/A = 0.35 \pm 0.05$.[3] This number is independent of a possible quenching of the current as long as the quenching is the same in ^{89}Y and ^{90}Zr. Model calculations give for this ratio 0.09–RPA (55), 0.22–SM (56), and 0.29–Broken Pair (57). The RPA calculations do not include the pairing distribution explicitly. It is clear that, away from closed shells, RPA calculations lead to unrealistic results.

3.6 Influence of Non-nucleonic Degrees of Freedom

In the introduction, we referred to the possibility that effects outside the standard nucleonic type of description for the nucleus, i.e. those due to the exchange of mesons, may be visible in some experimental data. We gave some arguments that transverse excitations would be the best place to look for the effects of such meson exchange currents (MEC), which include π and

---

[3] The numbers given here differ slightly from those quoted in (54) since in that analysis the amplitude $\bar{B}$ had not been taken into account.

ρ exchange, nucleon-antinucleon, and Δ-hole terms. There are higher order corrections and relativistic effects. We refer the reader to (32a–d) for details.

From the experimental side, the most striking feature is that all transverse excitations, both the transverse electric and the magnetic, seem to be quenched. This was discussed in Section 3.5 together with some possible quenching mechanisms. MEC probably will also have some effects but generally these are estimated to be of the order of 10% (33, 58). Further, excess cross section is often observed at values of q greater than 2 fm^{-1}. At such values of q the normal nuclear structure form factor drops rapidly since one goes beyond twice the Fermi momentum in the nucleus. As MEC are two-body operators on the nucleon level, they will not drop as fast, which may give a qualitative explanation of the observed behavior. However, short-range correlations between the nucleons, which are mediated also by the exchange of mesons, introduce additional high-q strength in the standard nuclear form factors. The calculation of these effects is quite complex. Probably the best documented case is electrodisintegration from the deuteron where the MEC have a large influence at high q (59). For heavier nuclear systems few measurements exist for $q > 2.5$ fm^{-1}. A recent example is the M7 multipole in magnetic elastic scattering from ^{51}V (58).

Of special interest are M1 transitions, as the observed quenching at low q values has been attributed to the influence of Δ-hole effects (60a–e). This is a non-nucleonic form of core polarization, in which a nucleon that moves into another orbit is excited into a $\Delta(1232)$ under the influence of the Δ-N interaction, thereby creating a Δ-hole state. Even if the energy of such Δ-hole states is about 300 MeV, this process may lead to non-negligible contributions, even at $q = 0$, since the Δ-N interaction is relatively strong and since all nucleons can contribute because the Pauli principle does not block Δ-hole states. This is in contrast to nuclear core polarization where in the case of M1 transitions at small q only transitions between spin-orbit partners can contribute to the form factor (the other ones are n- or l-forbidden). So, if these are included in the model space, nuclear core polarization has no effect at low q, which removes one source of uncertainty in the analysis.

Strong M1 transitions have been measured in ^{12}C (61a,b) and in ^{48}Ca (62a,b) that are due to the $1p_{3/2} \to 1p_{1/2}$ and the $\nu 1f_{7/2} \to 1f_{5/2}$ transitions, respectively. For other nuclei in this region the M1 strength is strongly fractionated. The search for M1 transitions in heavy nuclei has not been very successful. In addition, the form factors for states that could be 1^+ often cannot be followed over a large range in momentum transfer because of background due to the high level density.

The heaviest nucleus with a well-documented M1 transition is ^{88}Sr (63).

The form factor for the 1^+ state at 3.48 MeV was measured over a large range of momentum transfer (Figure 14). This M1 transition is mainly due to the proton $2p_{3/2} \to 2p_{1/2}$ transition, but the strength is reduced by ground-state correlations. A sum rule analysis of the combined results of pick-up and stripping reactions on ^{88}Sr (64) and projected BCS calculations (65) both give a reduction factor of 0.25–0.35. With this reduction factor the high-q points of the form factor, where core polarization is small, are correctly described, but the data at low q [and also the B(M1) value, which has been measured independently (66)] are overestimated by a factor of about two. This discrepancy is similar to what is observed for the 10.23-MeV state in ^{48}Ca. The reduction has been attributed to Δ-hole polarization and calculated reduction factors are in reasonable agreement with both experiments (67).

The inclusion of Δ-hole polarization also changes the shape of the form factor in ^{88}Sr in the region between $q = 0.5$ and 1.5 fm^{-1}. However, this is also the region, where nuclear core polarization has an effect (at small q, contributions are small for the reason mentioned, and at high q they are small because the residual interaction is weak). The influence of MEC still has to be calculated.

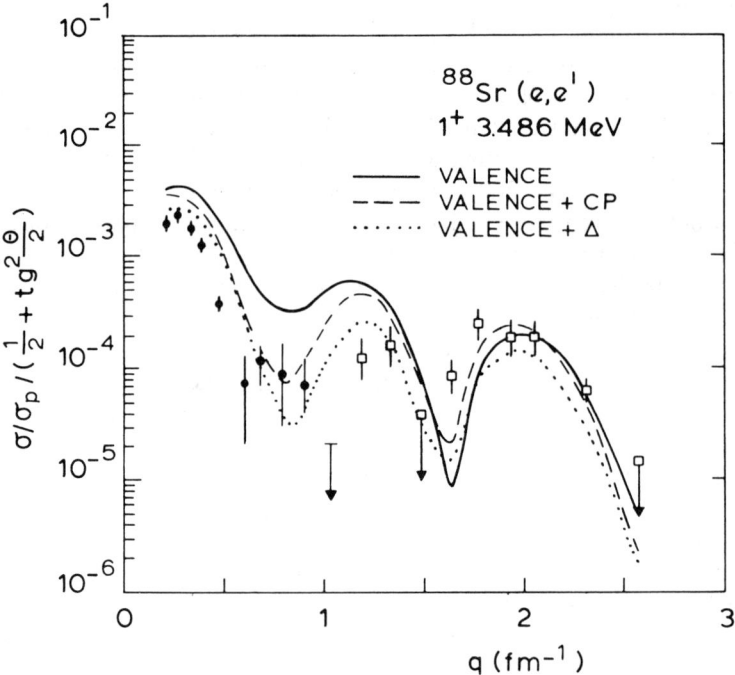

Figure 14 Measured and calculated form factors for the 1^+ level in ^{88}Sr.

Altogether there are indications that non-nucleonic degrees of freedom such as Δ-hole polarization are needed to explain the reduction of M1 strength at low q. The reduction seems to be too large to be explained by reasonable values for ground-state correlations, but the evidence is certainly not yet conclusive. Particularly when one observes that magnetic transitions of higher multipolarity seem to be quenched almost as much. For these the influence of Δ-hole polarization is rather small as the Δ-N interaction is weak for higher q.

4. SUMMARY

In our review we have concentrated on the relation between electron scattering and nuclear structure. It is the precision of the probe that challenges the quantitative aspect of nuclear structure calculations. We have shown that in two areas we can make quantitative comparisons. These are the ground-state rotational band of strongly deformed nuclei and the collective vibrations of spherical nuclei. Apart from these two areas, a quantitative calculation of core polarization seems to be the missing link to allow a meaningful comparison of theoretical predictions with electron scattering results. Considerable progress has been made in this area and it seems that we have come much closer to a situation where electron scattering can be used to map out the effective residual interaction.

There are other areas where electron scattering results have considerable influence on the advancement of our understanding. For instance, electron scattering results from $T = 0$ nuclei are almost always used as calibration points for the interaction of other probes with nuclei. This way of calibrating the isoscalar interaction of α particles with nuclei was used by Martens & Bernstein (68). The need for the density dependence of the two-nucleon effective interaction in understanding medium energy proton scattering experiments was demonstrated by Kelly et al (69), again on the basis of electron scattering results.

There are many other areas of electron scattering that we have not touched upon in this review such as coincidence experiments, excitations into the continuum, or the use of polarized electrons. In addition, one has to keep in mind that electron scattering measures mostly a one-body density. For that reason one sees the effects of the two-body densities and, for example, the correlation function only in an indirect way. Thus, in spite of our advances in the understanding of one-body densities, very little is known about many-body densities. Still we have shown that modern electron scattering results have an impact on virtually every nuclear model and thus have led and continue to lead us to considerable improvements in our understanding of nuclei.

ACKNOWLEDGMENTS

We would like to thank our colleagues H. Crannell and W. Hersman for letting us use their data prior to publication. Special thanks are also given to W. Hersman for his help in the preparation of the figures and for a careful reading of the manuscript. Much of the research presented was funded by the US-DOE contract no. AS03-79ER10338.

Literature Cited

1. de Forest, T., Walecka, J. D. 1966. *Adv. Phys.* 15:57
2. Donnelly, T. W., Walecka, J. D. 1975. *Ann. Rev. Nucl. Sci.* 25:329
3. Theissen, H. 1972. *Springer Tracts in Modern Physics*, Vol. 65. New York: Springer
4. Überall, H. 1971. *Electron Scattering from Complex Nuclei*. New York: Academic
5. Lee, H. C. 1975. *Atomic Energy of Canada*, Rep. AECL-4839. Ontario
6. Bjorken, J. D., Drell, S. D. 1964. *Relativistic Quantum Mechanics*. New York: McGraw-Hill
7. Bertozzi, W., Heisenberg, J., Friar, J., Negele, J. W. 1972. *Phys. Lett.* 41B:408
8. Friar, J. L. 1979. *Proc. Int. Conf. on Nucl. Phys. with Electromagnetic Interactions*, Mainz. Berlin: Springer-Verlag. 445 pp.
9. Frois, B., Turck-Chieze, S., Bellicard, J. B., Huet, M., Leconte, P., et al. 1983. *Phys. Lett.* 122B:347
10. Goutte, D., Bellicard, J. B., Cavedon, J. M., Frois, B., Huet, M., et al. 1980. *Phys. Rev. Lett.* 45:1618
11. Papanicolas, C. N., Heisenberg, J., Lichtenstadt, J., McCarthy, J. S., Goutte, D., et al. 1982. *Phys. Lett.* 108B:279
12. Heisenberg, J. 1981. *Adv. Nucl. Phys.* 12:61. (The phases in this reference differ from the presently used ones mainly by a substitution of $J_{\lambda,\lambda+1}$ by $-J_{\lambda,\lambda+1}$. In addition, some misprints were corrected.)
13. Engfer, R., Schneuwly, H., Vuilleumier, J. L., Walter, H. K., Zehnder, A. 1974. *At. Data Tables* 14:509–97
14. Reuter, W., Shera, E. B., Wohlfahrt, H. D., Tanaka, Y. 1983. *Phys. Lett.* 124B:293
15. Lichtenstadt, J., Papanicolas, C. N., Sargent, C. P., Heisenberg, J., McCarthy, J. S. 1980. *Phys. Rev. Lett.* 44:858
16. Friedrich, J., Voegler, N., Euteneur, H. 1976. *Phys. Lett.* 64B:269
17. Heisenberg, J., Dawson, J., Robb, J., Schwentker, O., Lichtenstadt, J., et al. To be published
18. Heisenberg, J., Lichtenstadt, J., Papanicolas, C. N., McCarthy, J. S. 1982. *Phys. Rev.* C25:2292
19. Heusler, A., von Brentano, P. 1973. *Ann. Phys.* 75:381
20. Crannell, H., O'Brien, J. T., Sober, D. I., Kowalski, S., Williamson, C. F., et al. To be published
21. de Shalit, A., Feshbach, H. 1974. *Theoretical Nuclear Physics*, 1:706–9. New York: Wiley
22. Heisenberg, J., Lichtenstadt, J., McCarthy, J. S., Papanicolas, C. N. To be published
23. Siegert, A. J. F. 1937. *Phys. Rev.* 52:787
24. Tassie, L. J. 1956. *Aust. J. Phys.* 9:407
25a. Cavedon, J. M. 1980. PhD thesis. Orsay
25b. Frois, B., Bellicard, J. B., Cavedon, J. M., Huet, M., Leconte, Ph., et al. 1977. *Phys. Rev. Lett.* 38:152
25c. Sick, I., Bellicard, J. B., Bernheim, M., Frois, B., Huet, M., Leconte, Ph., et al. 1975. *Phys. Rev. Lett.* 35:910
26. Rogers, J. D. 1965. *Ann. Rev. Nucl. Sci.* 15:241
27. Zaringhalam, A., Negele, J. W. 1977. *Nucl. Phys.* A288:417
28. Cardman, L. S., Dowell, D. H., Gulbranson, R. L., Ravenhall, D. G., Mercer, R. L. 1977. *Phys. Rev.* C18:1388
29a. Hersman, F. W. 1982. PhD thesis, Cambridge, Mass: MIT
29b. Hersman, F. W., Bertozzi, W., Buti, T. N., Finn, J. M., Hyde, C., et al. 1983. *Phys. Lett.* In press
30. Arima, A., Iachello, F. 1981. *Ann. Rev. Nucl. Part. Sci.* 31:75
31. Moinester, M. A., Alster, J., Azuelos, G., Dieperink, A. E. L. 1982. *Nucl. Phys.* A383:264
32a. Friar, J. L. 1977. *Ann. Phys.* 104:380
32b. Friar, J. L. 1977. *Phys. Lett.* 69B:51
32c. Friar, J. L. 1980. In *Electron and Pion Interactions with Nuclei at Intermediate Energies*, ed. W. Bertozzi, S. Costa, C. Schaerf, pp. 175–201. Chur, Switzerland: Harwood Academic
32d. Dubach, J. 1979. *Phys. Lett.* 81B:124

33. Dehesa, J. S., Krewald, S., Donnelly, T. W. To be published
34. Simon, G. G., Schmitt, Ch., Borkowski, F., Walther, V. H. 1980. *Nucl. Phys.* A333:381
35. Smith, S. M., Roos, P., Moazed, C., Bernstein, A. M. 1971. *Nucl. Phys.* A173:32
36. Körner, H. J., Auerbach, K., Braunsfurth, J., Gerdau, E. 1966. *Nucl. Phys.* 86:395
37. Papanicolas, C. N., Heisenberg, J., Lichtenstadt, J., McCarthy, J. S., Courtemanche, A. 1978. *Phys. Rev. Lett.* 41:537
38. Schwentker, O., Dawson, J., McCaffrey, S., Robb, J., Heisenberg, J., et al. 1982. *Phys. Lett.* 112B:40
39. Schwentker, O., Dawson, J., Robb, J., Heisenberg, J., Lichtenstadt, J., et al. 1983. *Phys. Rev. Lett.* 50:15
40. Heisenberg, J., Dawson, J. To be published
41. Bohr, A., Mottelson, B. R. 1975. *Nuclear Structure.* Ch. 6.5, Vol. 2. New York: Benjamin
42. Papanicolas, C. N., Heisenberg, J., Lichtenstadt, J., McCarthy, J. S., Goutte, D., et al. To be published
43. Lichtenstadt, J., Heisenberg, J., Papanicolas, C. N., Sargent, C. P., Courtemanche, A. N., et al. 1979. *Phys. Rev.* C20:497
44. Papanicolas, C. N., Heisenberg, J., Lichtenstadt, J., McCarthy, J. S. 1981. *Phys. Lett.* 99B:96
45. van der Bijl, L. T. 1982. PhD thesis, Vrije Univ., Amsterdam
46a. Lindgren, R. A., Williamson, C. F., Kowalski, S., et al. 1978. *Phys. Rev. Lett.* 40:504
46b. Lindgren, R. A., Flanz, J. B., Gerace, W. J., Hicks, R. S., Hotta, A., et al. 1978. *Phys. Rev. Lett.* 41:1705
46c. Lindgren, R. A., Flanz, J. B., Hicks, R. S., Parker, B., Peterson, G. A., et al. 1981. *Phys. Rev. Lett.* 46:706
46d. Lindgren, R. A., Plum, M. A., Gerace, W. J., Hicks, R. S., Parker, B., et al. 1981. *Phys. Rev. Lett.* 47:1266
47. Wise, J. 1982. PhD thesis, Univ. Virginia, Charlottesville
48a. Zarek, H., Pich, B. O., Drake, T. E., Rowe, D. J., Bertozzi, W., et al. 1977. *Phys. Rev. Lett.* 38:750
48b. Donnelly, T. W., Walecka, J. D., Walker, G. G., Sick, I. 1970. *Phys. Lett.* 32B:545
49a. Donnelly, T. W., Walecka, J. D., Sick, I., Hughes, E. B. 1968. *Phys. Rev. Lett.* 21:1196
49b. Sick, I., Hughes, E. B., Donnelly, T. W., Walecka, J. D., Walker, G. E. 1969. *Phys. Rev. Lett.* 23:1117
50a. Hamamoto, I., Bertsch, G. F., Lichtenstadt, J. 1980. *Phys. Lett.* 93B:213
50b. Hamamoto, I., Lichtenstadt, J., Bertsch, G. F. 1980. *Phys. Lett.* 96B:249
51. Krewald, S., Speth, J. 1980. *Phys. Rev. Lett.* 45:417
52. Rowe, D. J. 1970. *Nuclear Collective Motion.* London: Methuen. 200 pp.
53. *NIKHEF BULLETIN.* August 1982, No. 7
54. Heisenberg, J., Dawson, J., Schwentker, O., Blok, H. P. 1982. *Proc. Telluride Conf. on Spin Excitations in Nuclei.* March 1982. To be published
55. Heisenberg, J., Krewald, S. 1982. Unpublished.
56. Gazzaly, M. M., Hintz, N. M., Franey, M. A., Dubach, J., et al. 1983. *Phys. Rev.* C28:294
57. Hofstra, P., Allaart, K. 1979. *Z. Phys.* A292:159
58. Platchkov, S. K., Cavedon, J. M., Clemens, J. C., Frois, B., Goutte, D., et al. To be submitted to *Phys. Lett.*
59. Arenhövel, H. 1981. *Z. Phys.* A302:25
60a. Oset, E., Rho, M. 1979. *Phys. Rev. Lett.* 42:47
60b. Mukhopadhyay, H., Toki, H., Weise, W. 1979. *Phys. Lett.* 84B:35
60c. Knüpfer, W., Dillig, M., Richter, A. 1980. *Phys. Lett.* 95B:349
60d. Toki, H., Weise, W. 1980. *Phys. Lett.* 97B:12
60e. Bohr, A., Mottelson, B. 1981. *Phys. Lett.* 100B:10
61a. Flanz, J. B., Hides, R. S., Lindgren, R. A., Peterson, G. A., Dubach, J., et al. 1979. *Phys. Rev. Lett.* 43:1922
61b. Delorme, J., Figureau, A., Giraud, N. 1980. *Phys. Lett.* 91B:328
62a. Steffen, W., Gräf, H. D., Gross, W., Meuer, D., Richter, A., et al. 1980. *Phys. Lett.* 95B:23
62b. Steffen, W., Gräf, H. D., Richter, A., Harting, A., Weise, W., et al. To be published
63. van der Bijl, L. T., Blok, H. P., Frey, R., Meuer, D., Richter, A., et al. 1982. *Z. Phys.* A305:231
64. Comfort, J. R., Duray, J. R., Braithwaite, W. J. 1979. *Phys. Rev.* C8:1354
65. Akkermans, J. N. L., Allaart, K. 1981. *Z. Phys.* A304:245
66. Metzger, F. R. 1971. *Nucl. Phys.* A173:141
67. Weise, W. 1982. *Int. Conf. Nucl. Structure.* Amsterdam: North Holland
68. Martens, E. J., Bernstein, A. M. 1968. *Nucl. Phys.* A117:24
69. Kelly, J., Bertozzi, W., Buti, T. N., Hersman, F. W., Hyde, C., et al. 1980. *Phys. Rev. Lett.* 45:2012

INTERNAL SPIN STRUCTURE OF THE NUCLEON

Vernon W. Hughes

Physics Department, Yale University, New Haven, Connecticut 06520

Julius Kuti

Institute for Theoretical Physics, University of California, Santa Barbara, California 93106

CONTENTS

1. INTRODUCTION 611
2. THEORY 615
3. EXPERIMENT 624
4. COMPARISON OF THEORY AND EXPERIMENT 631
5. FUTURE PROSPECTS 638

1. INTRODUCTION

The study of the structure of the proton and neutron through deep inelastic scattering, initially with electrons but subsequently with muons and neutrinos as well, has played a central role in establishing the quark-parton theory of the composition of hadrons and of quantum chromodynamics (QCD). One important aspect of these theoretical and experimental developments is the two spin-dependent structure functions, which are independent of the two spin-averaged structure functions and define the internal spin structure of the nucleon.

Since both quarks and gluons possess spin and the forces between them are spin dependent, we can expect important information on these forces and on nucleon structure to be obtained through the study of the spin-dependent aspects of the nucleon wave function, as has been the case before in atomic and nuclear physics. The deep inelastic polarization experiment probes the spin distribution of quark constituents inside the nucleon. The theoretical description is expected to come from QCD.

Quantum chromodynamics—the theory of colored quarks and gluons and their interactions—is the presently accepted theory of strong interactions (1–4). Just as for the spin-independent structure functions, QCD predicts certain general sum rules and scaling behavior for the spin-dependent structure functions, whose verification can provide important confirmation of theory.

The hadronic currents of the electromagnetic and weak interactions are built from bilinear expressions of local quark fields providing a mathematical and physical relation between electromagnetic and weak hadronic phenomena. In the years before QCD emerged as the leading candidate for the theory of strong interactions, the quark-parton model had an elegant formulation in terms of the algebra of hadron currents (5–8). It was assumed that the product of two local hadronic currents with light-like separation between them was calculable as though the quark fields obeyed the free Dirac equation on the light cone. The scaling prediction for the spin-averaged and spin-dependent structure functions was an immediate consequence of this assumption (quark light-cone algebra). The scaling functions had a direct interpretation in terms of quark and gluon distributions inside the nucleon. The sum rules of the quark-parton model were elegant consequences of the quark light-cone algebra.

Quantum chromodynamics provides a theoretical explanation and a physical picture of why the interaction between quarks and gluons would vanish asymptotically at very short distances. This perturbative QCD supports the assumptions of the quark light-cone algebra and the quark-parton model. Even more powerfully, as a theory should predict new things, it describes quantitatively the logarithmic rate at which quarks become asymptotically free at short distances (9, 10).

The logarithmic corrections to the scaling functions and sum rules are the most important new predictions of perturbative QCD. In particular, the scaling behavior of spin-dependent structure functions and the corrections to it are accurately predicted in QCD together with some important sum rules of the scaling functions.

The calculation of the quark spin distributions inside the nucleon is a difficult nonperturbative calculation. As yet it has not been possible to calculate spin-dependent (or spin-independent) structure functions from the basic equations of QCD. However, models of nucleon structure have been developed consistent with the general picture of QCD, which make quantitative predictions of the structure functions. In atomic and nuclear physics it has often been found that spin-dependent observables provide particularly sensitive tests of a system's wave function, and we may anticipate that for the nucleon also spin-dependent quantities will be illuminating. We may remark that an enormous theoretical literature on the spin-dependent structure functions of the nucleon has developed since about 1970.

Information about the spin-dependent structure functions of the nucleon requires the measurement of spin-dependent asymmetries in the deep inelastic scattering of high energy polarized electrons (or muons) from polarized nucleons (11). The measured asymmetry A compares the cross sections for the case of parallel and antiparallel spins of the colliding particles and is given by

$$A = \frac{d\sigma^{\uparrow\downarrow} - d\sigma^{\uparrow\uparrow}}{d\sigma^{\uparrow\downarrow} + d\sigma^{\uparrow\uparrow}} \qquad 1.$$

where $d\sigma^{\uparrow\uparrow}$ is the cross section when the spins of electron and proton are parallel and along the direction of motion of the incident electron; $d\sigma^{\uparrow\downarrow}$ is the cross section for antiparallel spins. The measurements are made for inclusive scattering so that the momentum and scattering angle of the outgoing electron is observed but the final state of the proton is not detected. Since the technologies of polarized electron sources and polarized nucleon targets are difficult and also limiting with respect to intensity and polarization, relatively few data are available on the spin-dependent structure functions as compared to the data that have been obtained on spin-independent structure functions. Indeed, experiments at SLAC on polarized e-p scattering of longitudinally polarized electrons by longitudinally polarized protons provide our only experimental information to date.

The theory of polarized e-p scattering in leading order of the fine structure constant α is based on the exchange of a single virtual photon between the electron and the nucleon. Since the hadronic final state is not detected in the experiment, the spin-dependent differential cross section of the scattering is proportional to the antisymmetric part of the hadronic tensor amplitude $W_{\mu\nu}(Q^2, \nu)$, which is completely determined in terms of two spin-dependent structure functions $G_1(Q^2, \nu)$ and $G_2(Q^2, \nu)$. The terms Q^2 and ν are the relativistic invariants of four-momentum transfer to the nucleon and of energy loss by the electron, respectively.

The spin-dependent structure functions appear in the numerator of virtual photon-nucleon asymmetries $A_1(Q^2, \nu)$ and $A_2(Q^2, \nu)$. The asymmetry A_1 is the quantity measured thus far in polarized e-p scattering and it is given by

$$A_1 = \frac{\sigma_{1/2} - \sigma_{3/2}}{\sigma_{1/2} + \sigma_{3/2}} \qquad 2.$$

where $\sigma_{1/2}$ and $\sigma_{3/2}$ are the virtual photon-nucleon absorption cross sections when the projection of total angular momentum of virtual photon plus nucleon along the virtual photon direction is 1/2 and 3/2, respectively.

There is an important general sum rule related to the spin-dependent structure functions—the Bjorken polarization sum rule—which is given in its scaling form by the integral (12–17)

$$2\int_0^1 [g_1^p(x) - g_1^n(x)]\,dx = \frac{1}{3}\left|\frac{g_A}{g_V}\right|, \qquad 3.$$

where g_A/g_V is the ratio of axial to vector weak coupling constants in nucleon beta decay. The limiting scaling function as derived from $G_1(Q^2, \nu)$ is designated by $g_1(x)$. It has a remarkable and simple physical interpretation (18) in terms of quark spin distributions in the quark-parton model,

$$g_1(x) = \frac{1}{2}\sum_i e_i^2 [f_i^\uparrow(x) - f_i^\downarrow(x)], \qquad 4.$$

where $f_i^\uparrow(x)$ and $f_i^\downarrow(x)$ are interpreted as the probabilities of finding a quark parton of type i in a fast moving longitudinally polarized nucleon with a fraction x of the nucleon total momentum and with spin projected along or opposite to the nucleon spin, respectively. The charge e_i of the quark is measured in electron charge units. The Bjorken polarization sum rule involves the difference of scaling functions for protons and neutrons. The variable x is also the scaling variable $x = Q^2/2M\nu$ in the deep inelastic limit where x is fixed and $Q^2, \nu \to \infty$. The quark-parton picture is valid only in this limit (19–21).

The remarkable relation in Bjorken's sum rule between scaling functions in polarized deep inelastic electron scattering and the nucleon beta decay, which is a low energy phenomenon, is explained by the free light-cone behavior of the quark fields. The hadronic tensor amplitude is given by the commutator of the hadronic part of the electromagnetic current sandwiched between identical proton states. The two currents in the commutator are separated by light-like distances for the dominant contribution to deep inelastic scattering. The commutator is calculable there in terms of free quark fields and reduced to an axial vector current with calculable SU(3) flavor structure. This axial vector matrix element relates to nucleon beta decay for the proton-neutron difference on the right-hand side.

The Bjorken sum rule is a rigorous consequence of QCD in the limit $Q^2 \to \infty$. QCD also predicts the leading correction to the sum rule for finite Q^2 in terms of the strong interaction coupling constant α_s (22–25).

The quark-parton theory of nucleon structure makes a general qualitative prediction that one of the spin-dependent asymmetries A_1 for the proton is positive. Current phenomenological models of nucleon structure based on the quark-parton model predict values of the spin-dependent nucleon structure functions. The measured large positive asymmetry is against the early simple bootstrap picture of the nucleon. In the bootstrap model the nucleon was the bound state of a nucleon and a pion in P-orbital wave. This angular momentum structure would produce a large negative asymmetry A_1 ruled out by the Yale-SLAC data (26). Although the structure functions cannot be calculated yet from the basic QCD Hamiltonian, perturbative QCD predicts that at large x near to one, $A_1 \simeq 1$ for both proton and neutron, i.e. a quark (either u or d) that carries a large fraction of the momentum of the nucleon also carries the nucleon's spin (27–29).

The data obtained on polarized e-p scattering (11) determine the spin-dependent structure function A_1 of the proton over the deep inelastic kinematic range $0.1 < x < 0.7$ and $1 < Q^2 < 10$ $(\text{GeV}/c)^2$ with accuracy of 15–30%. There are also some data for elastic scattering and for resonance-region scattering (27).

The experimental data do confirm the Bjorken polarization sum rule at the parton-quark level (31) under the assumption that the neutron contributes a negligible amount; the data also verify the scaling behavior of A_1^p within their limited accuracy. Furthermore the data successfully distinguish the phenomenological models of the internal spin structure of the proton, as well as support the prediction of perturbative QCD that $A_1^p(x) \to 1$ as $x \to 1$.

We should also mention here an interesting connection between polarized electron-proton scattering and the hyperfine splitting in atomic hydrogen. The hyperfine splitting of the hydrogen ground state may receive an appreciable contribution from the proton's polarizability by the orbiting electron (12, 32–34). This contribution is directly calculable from the spin-dependent structure functions $G_1(Q^2, v)$ and $G_2(Q^2, v)$. This information would provide an exceptional bridge between quark physics of short distances and high precision atomic physics.

The possibilities of obtaining additional experimental information on the spin-dependent structure functions both with polarized electron and polarized muon beams are excellent. Additional data will allow exciting comparisons to be made with the many theoretical predictions. Perhaps the most outstanding quantitative prediction at present is the Bjorken polarization sum rule. Its reliable measurement would require data on polarized neutron (polarized deuteron) scattering which remains the most important experiment project for the future in the field of polarized electron (muon)-nucleon scattering.

2. THEORY

We review here the theoretical significance of inelastic scattering of polarized electrons from polarized nucleons (35, 36). The Feynman diagram that describes the scattering process in leading order of the electromagnetic

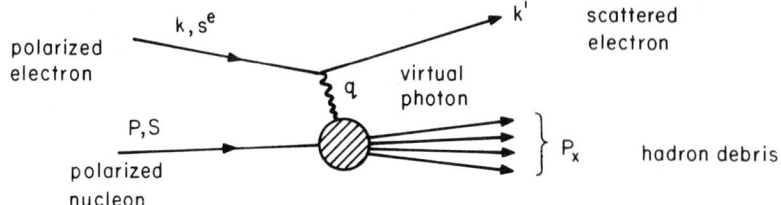

Figure 1 The Feynman diagram of spin-dependent deep inelastic electron-nucleon scattering in the one-photon exchange approximation.

coupling is shown in Figure 1. The polarized electron emits a virtual photon of four-momentum q which is absorbed by the polarized nucleon target. The nucleon explodes into some hadron debris that is not detected in the experiment.

The incoming electron is characterized by its four-momentum k and polarization four-vector S^e. The polarization state of the scattered electron is summed in the experiment. The target nucleon has four-momentum P and its polarization four-vector is S. The polarization vector S is reduced to the nucleon's spin three-vector $\mathbf{S}$, $S = (0, \mathbf{S})$ in the target rest frame.

The relativistic invariants,

$$-q^2 = Q^2 = -(k-k')^2 = 4EE' \sin^2 \frac{\theta}{2},$$

$$v = \frac{P \cdot Q}{M} = E - E', \qquad 5.$$

$$W^2 = (P+q)^2 = M^2 + 2Mv - Q^2,$$

are well-known from spin-averaged deep inelastic lepton-nucleon scattering. In Equation 5, M designates the nucleon mass; E and E' are the incoming and outgoing electron energies, respectively. The laboratory scattering angle of the electron is θ. The deep inelastic limit of the scattering process is defined by the kinematic conditions $Q^2 \gg M^2$ and $W \gg M$.

The inelastic cross section in the laboratory is given by

$$\frac{d^2 \sigma^{S_e S}}{d\Omega \, dE'} = \frac{4\alpha^2}{Q^4} \frac{E'}{E} L_{\mu\nu} W^{\mu\nu}, \qquad 6.$$

where α is the fine structure constant and $d\Omega$ designates the differential solid angle of the scattered electron with respect to the beam. The lepton current is described by the tensor $L_{\mu\nu}$, which is split into the sum of a polarization-independent symmetric part and a polarization-dependent antisymmetric part,

$$L_{\mu\nu} = k'_\mu k_\nu + k'_\nu k_\mu - g_{\mu\nu} k' \cdot k + im_e \varepsilon_{\mu\nu\alpha\beta} S_e^\alpha q^\beta, \qquad 7.$$

after summation over the polarization of the scattered electron. The metric is defined by $g_{00} = -g_{11} = -g_{22} = -g_{33} = +1$, and $\varepsilon_{\mu\nu\alpha\beta}$ designates the completely antisymmetric tensor in the four indices μ, ν, α, and β with $\varepsilon_{\mu\nu\alpha\beta} = +1$ for even permutations of 1234. For longitudinally polarized electrons the polarization four-vector S^e in the laboratory system is $S^e = 1/m_e(k, 0, 0, E)$ so that the electron mass m_e in the spin-dependent part of $L_{\mu\nu}$ cancels.

The tensor for the longitudinally polarized lepton is characterized by the parameter ε,

$$\varepsilon = \left[1 + 2\left(1 + \frac{v^2}{Q^2}\right)\tan^2\frac{\theta}{2}\right]^{-1}, \qquad 8.$$

which is the ratio of the longitudinally to the transversely polarized virtual photon fluxes.

The hadronic tensor amplitude $W_{\mu\nu}$ also has a decomposition into the sum of of a symmetric spin-independent part and an antisymmetric spin-dependent part,

$$W_{\mu\nu} = W_{\mu\nu}^S + iW_{\mu\nu}^A = \frac{1}{2\pi}\int d^4x\, e^{iqx}\langle P, S|[J_\mu(x), J_\nu(0)]|P, S\rangle, \qquad 9.$$

where $J_\mu(x)$ designates the hadronic electromagnetic current. The spin-independent symmetric part defines the well-known spin-independent structure functions $W_1(Q^2, v)$ and $W_2(Q^2, v)$,

$$W_S^{\mu\nu} = \left(-g^{\mu\nu} + \frac{q^\mu q^\nu}{q^2}\right)W_1 + \frac{1}{M^2}\left(P^\mu - \frac{P\cdot q}{q^2}q^\mu\right)\left(P^\nu - \frac{P\cdot q}{q^2}q^\nu\right)W_2. \qquad 10.$$

The antisymmetric tensor amplitude defines the spin-dependent structure functions $G_1(Q^2, v)$ and $G_2(Q^2, v)$,

$$W_A^{\mu\nu} = M\varepsilon^{\mu\nu\alpha\beta}q_\alpha S_\beta G_1 + \frac{1}{M}\varepsilon^{\mu\nu\alpha\beta}q_\alpha[(P\cdot q)S_\beta - (S\cdot q)P_\beta]G_2. \qquad 11.$$

The measured asymmetry A in the Yale-SLAC experiments compares the cross sections for the case of parallel and antiparallel spin of the colliding particles, and is defined by

$$A = \frac{d\sigma^{\uparrow\downarrow} - d\sigma^{\uparrow\uparrow}}{d\sigma^{\uparrow\downarrow} + d\sigma^{\uparrow\uparrow}}, \qquad 12.$$

where $d\sigma^{\uparrow\uparrow}$ is the cross section when the spins of electron and proton are parallel and along the direction of motion of the incident electron; $d\sigma^{\uparrow\downarrow}$ is the cross section for antiparallel spins.

The spin dependence of the differential cross sections in terms of the spin-dependent structure function is given by

$$\frac{d^2\sigma^{\uparrow\downarrow}}{d\Omega\,dE'} - \frac{d^2\sigma^{\uparrow\uparrow}}{d\Omega\,dE'} = \frac{4\alpha^2}{Q^2}\frac{E'}{E}[M\cdot G_1(E+E'\cos\theta) - Q^2 G_2]. \qquad 13.$$

It is also useful to define virtual photon-nucleon asymmetries, A_1 and A_2, given by

$$A_1 = \frac{\sigma_{1/2} - \sigma_{3/2}}{\sigma_{1/2} + \sigma_{3/2}}, \qquad 14.$$

$$A_2 = \frac{\sigma_{\mathrm{TL}}}{\sigma_{\mathrm{T}}}, \qquad 15.$$

where $\sigma_{1/2}$ and $\sigma_{3/2}$ are the total virtual photoabsorption cross sections when the projection of total angular momentum of virtual photon plus nucleon along the virtual photon direction is 1/2 and 3/2, respectively. The cross section σ_{T} is defined by the relation $\sigma_{\mathrm{T}} = 1/2(\sigma_{1/2} + \sigma_{3/2})$, and σ_{TL}, which may be negative, is a term arising from the interference between transverse and longitudinal photon-nucleon amplitudes.

The asymmetry A in Equation 12 relates to A_1 and A_2 by

$$A = D(A_1 + \eta A_2), \qquad 16.$$

where

$$D = \frac{1 - (E'/E)\varepsilon}{1 + \varepsilon R} \qquad 17.$$

can be regarded as a kinematic depolarization factor of the virtual photon. The quantity $R = \sigma_{\mathrm{L}}/\sigma_{\mathrm{T}}$ is the ratio of longitudinal and transverse virtual photoabsorption cross sections. The kinematic factor η is given by

$$\eta = \varepsilon(Q^2)^{1/2}/(E - E'\varepsilon). \qquad 18.$$

There are some rigorous positivity limits (15, 16, 37, 38) imposed on A_1 and A_2:

$$|A_1| \leq 1, \qquad |A_2| \leq \sqrt{R} \qquad 19.$$

The asymmetries A_1 and A_2 can be written in terms of spin-dependent structure functions

$$A_1 = \frac{M\nu G_1 - Q^2 G_2}{W_1} \qquad 20.$$

$$A_2 = \frac{\sqrt{Q^2}}{W_1}(MG_1 + \nu G_2). \qquad 21.$$

We turn now to the most important properties of the spin-dependent structure functions. One can take the scaling limit where $v, Q^2 \to \infty$ with $x = Q^2/2Mv$ fixed. In that limit the light-cone behavior of the electromagnetic current commutator in Equation (9) dominates the hadronic tensor amplitude. Since $J_\mu(x)$ is built from local quark fields, the scaling limit is governed by the quark light-cone algebra of the hadronic currents (15–17).

It follows from the leading singularities of the quark light-cone commutators through Fourier transformation that

$$M^2 v G_1(Q^2, v) \to g_1(x) \qquad 22.$$

and

$$Mv^2 G_2(Q^2, v) \to g_2(x) \qquad 23.$$

in the scaling limit. This scaling behavior of the spin-dependent structure functions is a rigorous consequence of the short-distance properties of QCD, since the quark light-cone algebra itself follows from the short distance part of QCD. Equations 22 and 23 are actually modified by logarithmic corrections in Q^2 in the rigorous derivation from QCD. Those corrections will appear manifestly in important sum rules for the spin-dependent structure functions.

The scaling functions $g_1(x)$ and $g_2(x)$ have direct physical interpretations in terms of quark partons (18, 21, 39). The following relations are valid in the quark-parton model.

$$g_1(x) = \frac{1}{2} \sum_i e_i^2 [f_i^\uparrow(x) - f_i^\downarrow(x)], \qquad 24.$$

$$g_1(x) + g_2(x) = \frac{1}{2Mx} \sum_i e_i^2 m_i [f_i^{T\uparrow}(x) - f_i^{T\downarrow}(x)]. \qquad 25.$$

Here $f_i^\uparrow(x)$ and $f_i^\downarrow(x)$ are interpreted as the probabilities of finding a quark parton of type i in a longitudinally polarized nucleon with a fraction x of the nucleon total momentum and with spin projected along or opposite to the nucleon spin, respectively. The charge of the ith quark parton is measured in electron charge units and designated by e_i. The sum runs over all quark types.

The other quark spin distribution functions $f_i^{T\uparrow}(x)$ and $f_i^{T\downarrow}(x)$ have a slightly more complicated physical interpretation. They refer to a situation where the nucleon and the quark partons are polarized in the transverse direction.

The second relation in Equation 25 relates to the asymmetry A_2 as

$$\frac{v}{\sqrt{Q^2}} A_2 \xrightarrow[\substack{v, Q^2 \to \infty \\ x \text{ fixed}}]{} \frac{g_1(x) + g_2(x)}{F_1(x)} \qquad 26.$$

in the above discussed scaling limit, where

$$F_1(x) = \frac{1}{2} \sum_i e_i^2 [f_i^\uparrow(x) + f_i^\downarrow(x)]. \qquad 27.$$

It is shown that, in the approximation in which the quark partons are on their mass shell without transverse momenta, the second structure function $g_2(x)$ vanishes. The asymmetry A_2 is proportional to the quark masses m_i through Equations 25 and 26. Therefore, A_2 would vanish for massless quarks.

The asymmetry A_1 is the dominant term in deep inelastic scattering with longitudinal polarization. It has a very simple and transparent physical interpretation in the scaling limit,

$$A_1(x) = \frac{\sum_i e_i^2 [f_i^\uparrow(x) - f_i^\downarrow(x)]}{\sum_i e_i^2 [f_i^\uparrow(x) + f_i^\downarrow(x)]} \qquad 28.$$

if $R \to 0$ in the same limit as predicted from the quark-parton picture, or, more generally, from the short-distance behavior of QCD.

The most important sum rule for spin-dependent deep inelastic electron-nucleon scattering was derived first by Bjorken from the commutator algebra of hadronic currents with underlying quark fields. At that time the theoretical scaling behavior of the spin-dependent structure functions was not known yet. Knowing now the scaling behavior of $G_1(Q^2, v)$, the sum rule for the difference of $g_1(x)$ for protons and neutrons is

$$2 \int_0^1 [g_1^p(x) - g_1^n(x)] \, dx = \frac{1}{3} \left| \frac{g_A}{g_V} \right| = 0.418 \pm 0.002 \qquad 29.$$

where g_A/g_V is the ratio of axial to vector weak coupling constants in nucleon beta decay.

The scaling form of Bjorken's sum rule in Equation 29 can be rewritten in a form

$$\int_0^1 \frac{dx}{x} \left[\frac{A_1^p(x) F_2^p}{1 + R^p} - \frac{A_1^n(x) F_2^n(x)}{1 + R^n} \right] = \frac{1}{3} \left| \frac{g_A}{g_V} \right| \qquad 30.$$

that is convenient for the experimental analysis. The scaling function $F_2(x)$ is the scaling limit of vW_2. The sum rule is valid in the scaling limit where R

= 0, but is convenient to keep R in Equation 30 for comparison with finite Q^2 data where $R \neq 0$ is known from the spin-independent experiment.

Separate sum rules were derived (40) from the quark light-cone algebra for protons and neutrons under the assumption that the net spin polarization of strange sea quarks is zero:

$$2 \int_0^1 g_1^p(x) \, dx = \int_0^1 \frac{dx}{x} \frac{A_1^p(x) F_2^p(x)}{1+R^p} = \left|\frac{g_A}{g_V}\right| \frac{0.89}{3} = 0.372 \pm 0.002, \qquad 31.$$

and

$$2 \int_0^1 g_1^n(x) \, dx = \int \frac{dx}{x} \frac{A_1^n(x) F_2^n(x)}{1+R^n} = \left|\frac{g_A}{g_V}\right| \frac{(-0.11)}{3} = -0.046 \pm 0.0002. \qquad 32.$$

There exist data now to test the sum rule in Equation 31. The neutron sum rule should be tested in future experiments.

The scaling form of the Bjorken sum rule is not strictly valid in quantum chromodynamics. The quark-gluon coupling constant is approaching zero with increasing Q^2 in QCD. Therefore the quark partons become free only asymptotically when $Q^2 \to \infty$ and the structure functions receive logarithmic Q^2 corrections from quark-gluon and gluon-gluon interactions. The running coupling constant has the form (4, 8)

$$\alpha_s(Q^2) = \frac{1}{c \ln(Q^2/\Lambda^2)} \qquad 33.$$

where double logarithmic corrections are neglected. The constant c in Equation 33 is calculable theoretically,

$$c = \frac{33 - 2f}{12\pi} \qquad 34.$$

where f is the number of quark flavors. The scale parameter Λ sets the strength of quark-gluon coupling in QCD. With careful definition of the neglected corrections (69) in Equation 33, the experimental value of Λ is somewhere around 100 MeV from deep inelastic scattering and e^+e^- annihilation.

The effect of asymptotic freedom on the Bjorken sum rule is an additive Q^2 dependent correction (22):

$$\int_0^1 dx \, [g_1^p(x) - g_1^n(x)] = \frac{1}{6} \left|\frac{g_A}{g_V}\right| \left(1 - \frac{\alpha_s(Q^2)}{\pi}\right), \qquad 35.$$

where the running coupling constant $\alpha_s(Q^2)$ is given in Equation 33.

A sum rule also for the second spin-dependent structure function G_2

follows from angular momentum conservation. The sum rule for protons or neutrons (17, 18, 41) is

$$\int_0^1 g_2(x)\,dx = 0 \qquad 36.$$

in the scaling limit. There are some subtleties about the convergence of the integral in Equation 36 that we do not discuss here (42).

Scaling of the structure functions, the Bjorken sum rule, and asymptotic freedom corrections to the scaling limit are accurate predictions from the short-distance behavior of QCD. The calculation of the quark spin distributions inside the nucleon is a more difficult task. It would require the nonperturbative calculation of the nucleon structure in terms of quark and gluon wavefunctions. There is an exciting new development in QCD which treats that difficult problem from the numerical point of view using stochastic methods and computer simulation (43). Though the first nonperturbative results are very interesting they do not yield enough information yet on the quark and gluon distributions inside the nucleon. We have to turn, therefore, to somewhat more model-dependent descriptions.

We estimate first the asymmetry A_1 in a very simple quark model where the SU(6) quarks are identified with the partons scattering the incoming polarized electrons. The static wave function of the proton with spin up is given by

$$|\text{proton}\uparrow\rangle = \frac{1}{\sqrt{18}}(2|u^\uparrow d^\downarrow u^\uparrow\rangle + 2|u^\uparrow u^\uparrow d^\downarrow\rangle + 2|d^\downarrow u^\uparrow u^\uparrow\rangle$$
$$- |u^\uparrow u^\downarrow d^\uparrow\rangle - |u^\uparrow d^\uparrow u^\downarrow\rangle - |u^\downarrow d^\uparrow u^\uparrow\rangle \qquad 37.$$
$$- |d^\uparrow u^\downarrow u^\uparrow\rangle - |d^\uparrow u^\uparrow u^\downarrow\rangle - |u^\downarrow u^\uparrow d^\uparrow\rangle).$$

The probability of interaction with a $u^\uparrow$ quark (spin up) is 5/9, and with a $u^\downarrow$ quark it is 1/9 according to the spin wave function in Equation 37. The corresponding probabilities for $d^\uparrow$ and $d^\downarrow$ quarks are 1/9 and 2/9, respectively.

The asymmetry A_1 is given by simple counting (12). Only quarks with spin antiparallel to the spin of the virtual photon contribute, and the cross section is proportional to (charge)^2. Quarks with spin $+1/2$ contribute to $\sigma_{1/2}$ and we find $\sigma_{1/2} = 21/81$. Quarks with spin $-1/2$ contribute to $\sigma_{3/2}$ with $\sigma_{3/2} = 9/81$. This leads to an asymmetry $A_1 = 5/9$. From isospin symmetry the prediction for the neutron is $A_1 = 0$.

This simple picture has to be modified in various aspects. In the limit $x \to 0$ the electron is scattered by quarks in the $q\bar{q}$ "sea" decoupled from the proton's spin. We expect the asymmetry A_1 to vanish in this limit. In the other extreme limit $x \to 1$, one quark carries the proton's momentum. With its spin parallel to the proton's it gives $A_1 = 1$ in that limit.

A simple quark-parton model was proposed with some modifications to incorporate the expected qualitative features of quark and gluon distributions inside the nucleon (18, 39, 44–50). This simple quark-parton model picture which has emerged during the years describes the observed asymmetry $A_1(x)$ both in shape and magnitude.

The quark-parton models of spin-dependent structure functions emerged from the unification of the spin-isospin wave function of the SU(6) quark model and of Feynman's parton picture. The calculation of structure functions in the MIT bag model proceeds in a different fashion (51). Relativistic quarks are confined inside the nucleon by vacuum pressure and the structure functions are calculated in the rest frame of the nucleon from the light-cone behavior of the hadronic currents.

There was also some attempt to describe the spin-dependent structure functions in terms of direct channel resonance excitations (52–54). General sum rules were proposed as a spin-dependent electroproduction test of relativistic constituent quarks (55).

There is a very interesting side application of polarized deep inelastic electron-proton scattering for the hyperfine splitting in atomic hydrogen. This is an exceptional bridge between the usually disconnected fields of high energy quark physics and high precision atomic physics.

One finds a contribution δ_{pol} to the hyperfine splitting of the hydrogen ground state (56) from the Feynman diagram of Figure 2. This diagram has an unknown part in it that is the spin-dependent Compton scattering amplitude of a virtual photon on the proton. The diagram with a loop integral describes the proton polarizability contribution to the hyperfine splitting. Through dispersion relations the virtual Compton amplitude in Figure 2 relates to the spin-dependent structure functions G_1 and G_2.

Our partial knowledge of the properties of the structure functions G_1 and

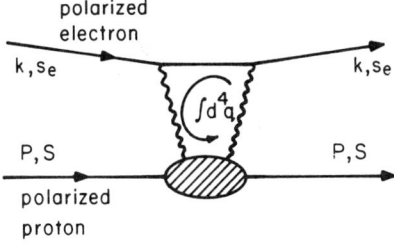

Figure 2 The Feynman diagram of the proton's polarizability contribution to the hyperfine splitting of the hydrogen ground state.

G_2, together with rigorous inequalities, puts limits on the polarization contribution (33, 34)

$$-4 \text{ ppm} \leq \delta_{\text{pol}} \leq 4 \text{ ppm}. \qquad 38.$$

Further information on G_1 and G_2 would make it feasible to actually calculate δ_{pol} completing this nice bridge between high energy quark physics and low energy, high precision atomic physics.

3. EXPERIMENT

As discussed in Section 2, determination of the spin-dependent structure functions of the nucleon requires the use of a high energy polarized electron (or muon) beam and of a polarized proton (or neutron) target to measure spin-dependent asymmetries in the inclusive scattering $e^- + p \rightarrow e^- + X$ (unobserved). Thus far, data have been obtained only for the scattering of longitudinally polarized electrons by longitudinally polarized protons (Equation 16).

3.1 Polarized Electron Beam

There are many possible schemes for producing polarized electrons (57). The polarized electron source that was developed at Yale and used in the high energy e-p polarization experiment at SLAC was an atomic beam source based on the photoionization of electron spin polarized, lithium-6 atoms (58, 59). A diagram of the energy levels and magnetic moments for the ground $^2S_{1/2}$ state of ^{6}Li in a magnetic field is shown in Figure 3, and a

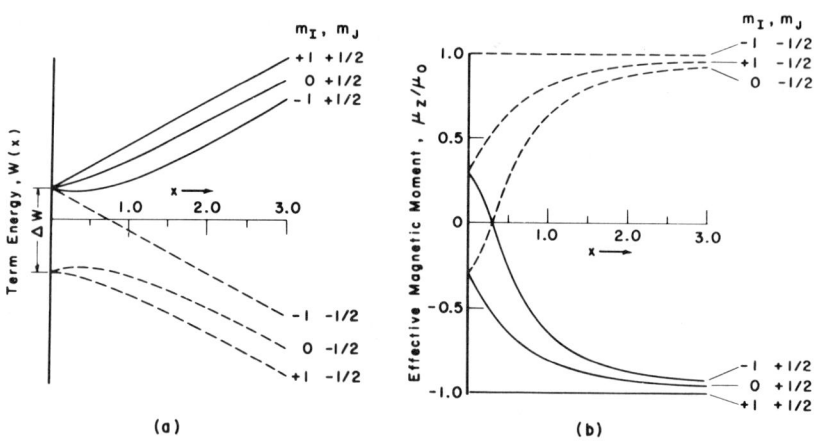

Figure 3 Energy levels and magnetic moments of ^{6}Li (nuclear spin $I = 1$) in the ground $^2S_{1/2}$ atomic state as a function of magnetic field H, where m_J, m_I are the electronic and nuclear magnetic quantum numbers, and $x = (g_J - g_I)\mu_0 H/\Delta W$ in which g_I and g_J are the nuclear and electronic g values, μ_0 is the Bohr magneton, and $\Delta W = h\Delta\nu$ is the hyperfine structure interval.

schematic diagram of the source in Figure 4. An intense atomic beam of ^{6}Li atoms in its ground state is formed by heating ^{6}Li in an oven with an orifice and by collimating the resulting flux of atoms. The sixpole magnet with its strong inhomogeneous magnetic field transmits only atoms with electronic magnetic quantum number $m_J = +1/2$, which then pass adiabatically into the ionization region where there is a longitudinal magnetic field of about 200 G provided by the polarizing coil. Intense light in the UV range from 1700 to 2300 Å is produced by a vortex-stabilized argon flash lamp and is focussed onto the polarized ^{6}Li atomic beam. The resulting photoelectrons are extracted from the ionization region with a kinetic energy of about 70 keV, and then transported either into the accelerator or into the polarization analyzer that employs double Mott scattering. The electron beam is longitudinally polarized either parallel or antiparallel to the beam direction depending on the direction of the current in the polarizing coil. The operating characteristics of the resulting high energy polarized electron beam are given in Table 1.

The intensity of the high energy polarized electron beam is less by a factor of about ~400 than that of an unpolarized electron beam at SLAC, but because of radiation damage of the polarized target (see below) its intensity is appropriate for polarized e-p scattering.

The electron polarization from the source at 70 keV was measured to an accuracy of 5% by double Mott scattering in which the first scattering at 90° converts the longitudinal polarization to transverse polarization.

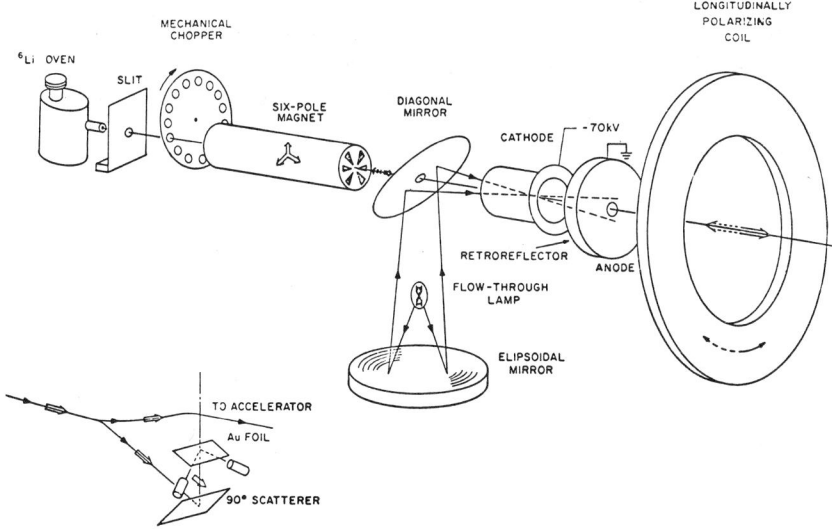

Figure 4 Schematic diagram of PEGGY I showing the principal components of the lithium atomic beam, the UV optics, the ionization region electron optics, and the double Mott scattering polarization analysis.

Table 1 Operating characteristics of polarized electron beam

Characteristic	Value
Pulse length	1.5 μs
Repetition rate	180 pps
Average intensity at GeV energies	5×10^8 e⁻ per pulse
Pulse-to-pulse intensity variation	<5%
Polarization	0.80 ± 0.03
Polarization reversal time	3 s
Intensity difference upon reversal	<5%

After acceleration in the linear accelerator to GeV energies, the longitudinal polarization was measured by elastic electron-electron scattering (Möller scattering) from a magnetized iron foil (60). Möller scattering is an attractive method for measuring the electron polarization at high energy because the cross section and analyzing power are large and the process is purely quantum electrodynamic. Figure 5 shows the Möller asymmetry and laboratory cross section at the representative beam energy of 9.71 GeV. The longitudinal beam polarization P measured as a function of beam energy E is shown in Figure 6. The variation of P with E is caused by the $g-2$ precession of the spin relative to the momentum in the beam switchyard, where the beam from the accelerator is bent by 24.5° into the experimental area. The accuracy with which the longitudinal electron polarization has been determined in the experiments with polarized electrons is about 4%. The error is due to the statistical counting error in the measured asymmetry, uncertainty in the background subtraction, and uncertainty about the electron spin magnetism of the magnetized iron foils.

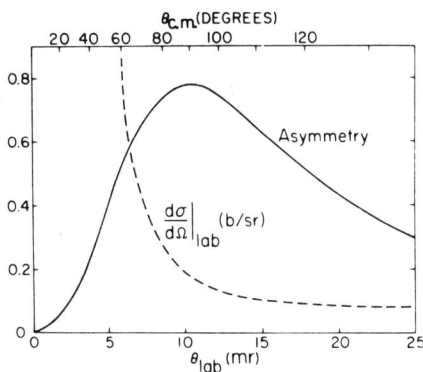

Figure 5 The Möller asymmetry and laboratory cross section plotted versus laboratory angle for the incident electron energy of 9.712 GeV.

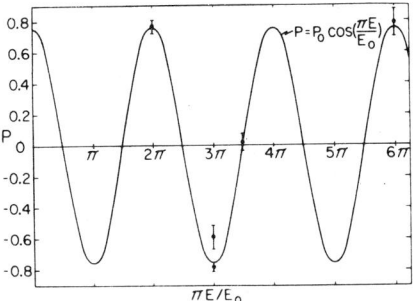

Figure 6 The longitudinal component, P, of the beam polarization plotted versus $\pi E/E_0$, the angle through which the spin precesses relative to the momentum during the 24.5° bend into the experimental area. E is the beam energy and $E_0 = 3.237$ GeV. The curve shown is a best fit to the data and has an amplitude $P_0 = 0.76 \pm 0.03$. P_0 is the only free parameter.

The polarized proton target used is based on the well-known method of dynamic nuclear orientation using a hydrocarbon (butanol) sample with a paramagnetic dopant (porphyrexide) (61–64). Its principal special feature has been the large energy dissipation in the target due to the relatively intense electron beam and the associated damage to the target. Techniques for rastering the beam over the target area in order to minimize the effect of radiation and provide uniform target polarization, for annealing radiation damage, and for rapid changing of target material have been developed. The target operates at 1 K and used a 50-kG magnetic field. A drawing of the target assembly is shown in Figure 7 and the operating characteristics of the target are given in Table 2.

The asymmetries measured in polarized e-p scattering are small. If we designate the experimental counting rate asymmetry by Δ, then we can write

$$\Delta = \frac{N(+) - N(-)}{N(+) + N(-)} = P_e P_p F A \qquad 39.$$

in which $N(+)$ and $N(-)$ designate the scattered electron counts from a polarized proton target of electrons with $+$ and $-$ helicity, respectively. P_e is the polarization of the electron beam, P_p is the polarization of free protons (i.e. those associated with hydrogen) in the target, F is the fraction of scattering events from the target that arise from the free (polarizable) protons, and A is the intrinsic asymmetry associated with polarized e-p scattering. Measurements have shown that $A \simeq 0.2$, in agreement with an early prediction from the Bjorken sum rule. Practically realizable values for P_e, P_p, and F are $P_e \simeq 0.8$m $P_p \simeq 0.6$, and $F \simeq 0.1$. Hence the predicted size of Δ is about 0.01.

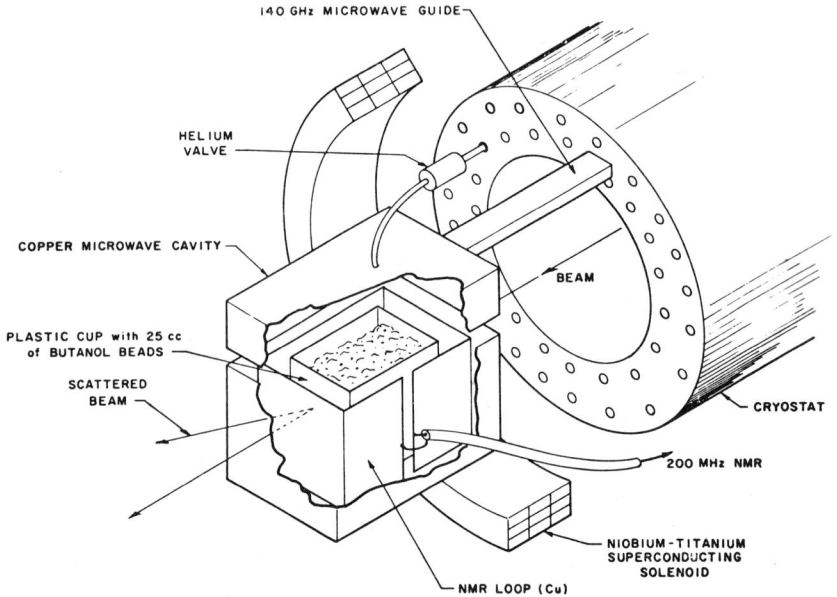

Figure 7 Schematic diagram of the Yale-SLAC polarized proton target operating at 1 K and 50 kG.

The small values of Δ to be measured require careful measurement and control of the intensity, position, and direction of the incident electron beam in order to assure that no false asymmetries are associated with reversal of the helicity of the electron beam.

The electron beam from the accelerator was momentum-analyzed by a transport system whose absolute momentum calibration was $\sim 0.1\%$ and a momentum slit in the transport system limited the beam spread to $\pm 0.375\%$. The electron beam charge per pulse was monitored with two precision toroidal charge monitors. Several sensitive microwave beam

Table 2 Operating characteristics of polarized proton target

Characteristic	Value
Magnetic field (superconducting)	50 kG
Temperature	1 K
Target material	25 cm^3 of butanol-porphyrexide
Maximum polarization, P_p	0.75
Depolarizing dose $(1/e)$	3×10^{14} e$^-$ cm^{-2}
Polarizing time $(1/e)$	~ 4 min
Anneal or target change time	~ 45 min

position monitors located along the beam line from the accelerator output to the target, together with a computer-controlled feedback system including steering magnets and accelerator klystons, controlled the position, angle, and energy of the electron beam incident on the target so that false asymmetries were maintained below 10^{-4} and hence were negligibly small (11, 65).

The scattered electrons were detected and their momentum and scattering angle were measured. In the first experiment (E80) the SLAC 8-GeV spectrometer was used (26, 31, 63). Electron identification was achieved with a gas threshold Čerenkov counter, a 3.25-radiation-length-thick, lead-glass, counter array that sampled the buildup of the electromagnetic shower, and a lead-Isolite shower counter. Less than one pion in 10^3 was misidentified as an electron by this system. An on-line XDS 9300 computer monitored the experiment and recorded data on magnetic tape.

In the second experiment (E130) (11, 65) a new larger acceptance spectrometer was used (see Figure 8). It utilizes two large dipole magnets (B201 and B81) and a detector system consisting of a 1-m diameter × 4-m long N_2 gas Čerenkov counter, a 4000-wire PWC system, a hodoscope, and a segmented lead-glass shower counter. The spectrometer may cover momenta up to 18 GeV/c, and its acceptance $\int d\Omega \, dp/p$ is 0.3 msr with the total momentum acceptance $\Delta p/p$ being about 50%. The momentum resolution $\delta p/p$ of the spectrometer is better than $\pm 1\%$.

The kinematic points at which data have been obtained are shown in Figure 9 and include elastic, resonance region, and deep inelastic points.

The intrinsic e-p asymmetry A is obtained from the measured values of Δ using Equation 39. The measured values of P_e and P_p are used. The quantity F is taken to be the ratio of the number of free protons to the total number of nucleons, corrected for the measured ratio of the neutron to proton scattering cross sections at the kinematic point (66). Radiative corrections to the values of A so obtained are then made using the extensive data

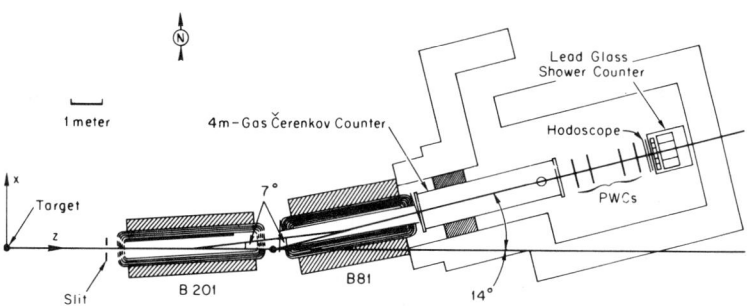

Figure 8 Spectrometer used in SLAC E130 experiment.

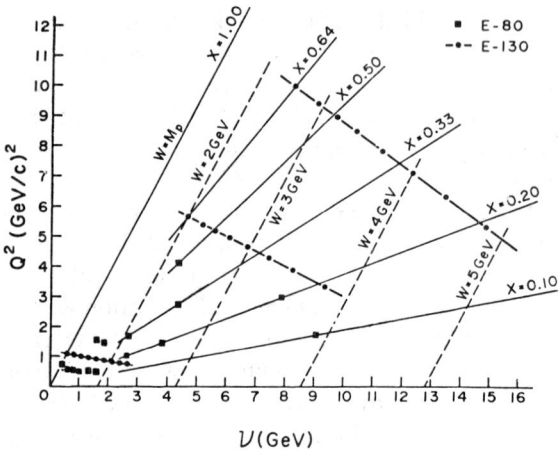

Figure 9 Kinematic points measured.

available on the spin-averaged cross sections, measured values of A, and the calculated values of A for elastic scattering. The equivalent radiator method in the peaking approximation was employed in a self-consistent type of calculation (31, 67, 68). The radiative corrections contribute only a small change in the A values but do increase the statistical errors by about 20–70%. Table 3 shows some typical data.

Exploratory measurements were made of the asymmetry in resonance region scattering (30). Data were obtained at $Q^2 = 0.5$ and 1.5 $(\text{GeV}/c)^2$ in the missing-mass range $W = 1.1$ to 1.9 GeV (see Figure 10).

Table 3 Data sample from SLAC E130 ($E = 22.659$ GeV; $\theta = 10°$)

x	Q^2	v	W	Δ	A^a	A^b	A/D^b
0.19	5.32	14.93	4.85	0.030(0.009)	0.439(0.137)	0.461(0.163)	0.69(0.24)
0.25	6.32	13.47	4.45	0.017(0.003)	0.248(0.049)	0.263(0.056)	0.44(0.09)
0.31	7.14	12.28	4.09	0.020(0.003)	0.279(0.047)	0.289(0.053)	0.53(0.10)
0.37	7.83	11.28	3.77	0.026(0.004)	0.358(0.054)	0.366(0.060)	0.74(0.12)
0.43	8.41	10.43	3.46	0.021(0.005)	0.288(0.064)	0.294(0.070)	0.65(0.16)
0.49	8.92	9.70	3.18	0.020(0.006)	0.257(0.085)	0.261(0.094)	0.62(0.22)
0.55	9.35	9.06	2.92	0.018(0.009)	0.288(0.118)	0.231(0.124)	0.60(0.32)
0.64	9.91	8.25	2.54	0.017(0.010)	0.212(0.126)	0.214(0.131)	0.61(0.38)

[a] Measured values without radiative correction (total errors in parentheses).
[b] Relatively corrected values (total errors in parentheses).

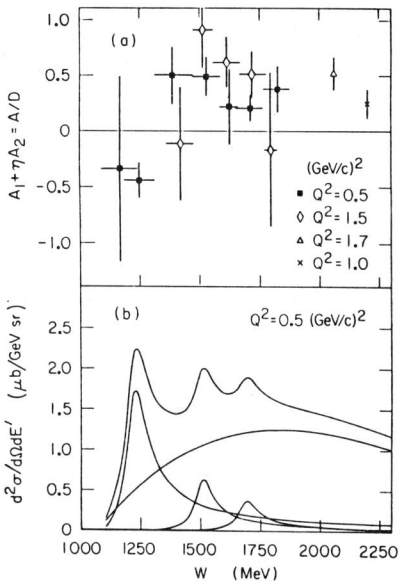

Figure 10 (a) Asymmetry vs missing-mass W. (b) Differential cross section vs W. Also shown is a decomposition into individual resonances and the background.

4. COMPARISON OF THEORY AND EXPERIMENT

For elastic scattering the theoretical value of the asymmetry in the one-photon exchange approximation (68) is given by

$$A = \frac{\tau G_M}{G_E}\left\{\frac{2M}{E} + \frac{G_M}{G_E}\left[\frac{2\tau M}{E} + 2(1+\tau)\tan^2\frac{\theta}{2}\right]\right\}$$
$$\times \left\{1 + \tau\left(\frac{G_M}{G_E}\right)^2\left[1 + 2(1+\tau)\tan^2\frac{\theta}{2}\right]\right\}^{-1} \quad 40.$$

in which $\tau = Q^2/4M^2$, $q^2 = -Q^2 = -4EE'\sin^2(\theta/2)$ is the square of the four-momentum of the virtual photon, M is the proton mass, E' is the scattered electron energy, and G_E and G_M are the electric and magnetic elastic form factors of the proton. The electron mass has been neglected. The measurement of A for elastic scattering was chosen primarily to test the validity of the experimental method. Alternatively, we can regard the measurement as a test of Equation 41 and as a determination of the sign of G_E/G_M. The experimental value of A is 0.103 ± 0.015, in reasonable agreement with the theoretical value $A = 0.112 \pm 0.001$ (63).

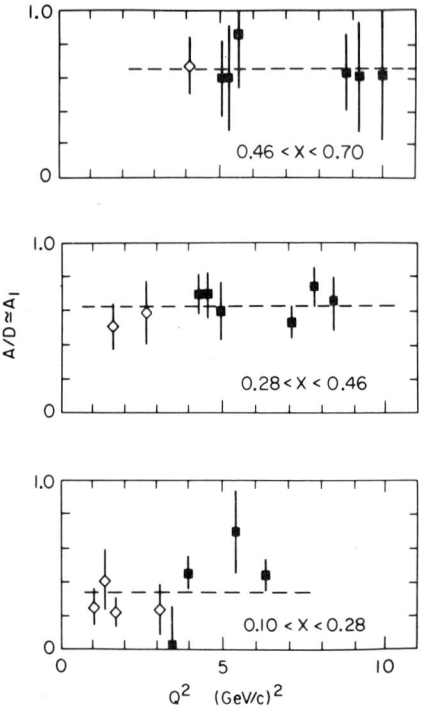

Figure 11 Radiatively corrected values of $A/D \simeq A_1$ obtained in SLAC E80 (*open diamonds*) and SLAC E130 (*closed squares*).

For deep inelastic scattering Figure 11 shows values of $A/D \simeq A_1$ obtained from experiments E80 and E130 plotted vs Q^2 in three intervals of x (26, 36, 65). The error bars include statistical and systematic errors. To test scaling of A_1 the values A/D have been divided by $\sqrt{x}$ (which described well the x dependence of the Q^2-combined data) and least-squares straight lines have been fit in the region $Q^2 > 2$ GeV^2. The assumption of scaling (zero slope) gives χ^2/DOF of 0.43/5, 2.4/5, and 5/3, and confidence levels of 99%, 80%, and 18%, for the top, middle, and bottom boxes, respectively. We therefore conclude that scaling of A_1 holds within errors. The Q^2-combined values of A/D are shown in Figure 12. The data are best described by $A/D = (0.94 \pm 0.08)\sqrt{x}$ (with $\chi^2/\text{DOF} = 9.5/11$).

The data permit a test of the Ellis-Jaffe sum rule (40) for the proton

$$S_{\text{BJ}}^{\text{p}} = 2\int_0^1 g_1^{\text{p}} \, dx = \int_0^1 \frac{dx}{x} \frac{A_1^{\text{p}} F_2^{\text{p}}}{1+R^{\text{p}}} \frac{(0.89)}{3} = 0.372 \pm 0.002 \qquad 41.$$

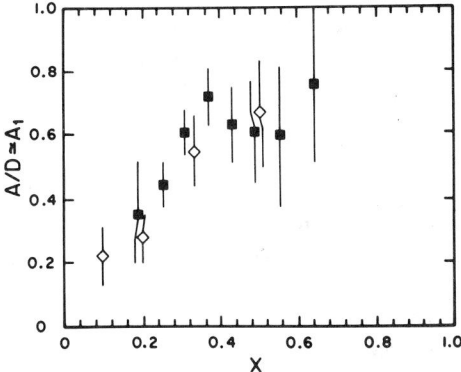

Figure 12 Measured values of A/D vs x. Points were obtained from Figure 11 data assuming A/D values are independent.

and of the Bjorken sum rule (12, 14)

$$S_{Bj} = 2\int_0^1 (g_1^p - g_1^n)\,dx = \int_0^1 \frac{dx}{x}\left(\frac{A_1^p F_2^p}{1+R^p} - \frac{A_1^n F_2^n}{1+R^n}\right)$$

$$= \frac{1}{3}\left|\frac{g_A}{g_V}\right| = 0.418 \pm 0.002 \qquad 42.$$

if A_1^n is approximated by zero. The integrand $A_1^p F_2^p/(1+R^p)$ is plotted in Figure 13 using $F_2^p(x, Q^2)$ from available data and the value $R = 0.25 \pm 0.10$ from the SLAC e-p data. The smooth curve in the region $0.1 < x < 0.64$ is obtained from our fit $A_1 = 0.94\sqrt{x}$ and F_2^p evaluated at $Q^2 = 4\,(\text{GeV}/c)^2$ (which is the mean Q^2 value of our data). The integral under this curve in the data region $0.1 < x < 0.64$ is 0.189 ± 0.016, which saturates 45% of the

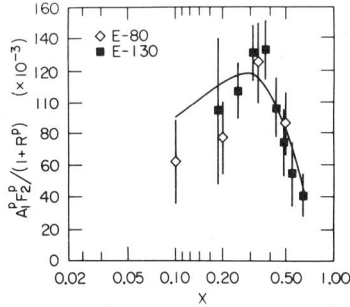

Figure 13 Experimental values of $A_1^p F_2^p/(1+R^p)$. F_2^p and R are from unpolarized data. The smooth curve is obtained using $A_1^p(x) = 0.94\sqrt{x}$.

Bjorken sum rule. The integral over the full x range using the Regge theory prediction $A_1 \propto x^{1.14}$ for $x < 0.01$ and our fit $A_1 = 0.94\sqrt{x}$ for $x < 0.1$ gives

$$2 \int_0^1 g_1^p(x) \, dx = 0.33 \pm 0.10.$$

In conclusion, our result is consistent with the Ellis-Jaffe sum rule for the proton. This implies that our results are also consistent with the Bjorken sum rule provided that the neutron contribution is as small as suggested by the Ellis-Jaffe sum rule for the neutron.

Comparison of our data on A_1^p with theoretical values provides a major test for our understanding of nucleon structure. The generally accepted theory of quantum chromodynamics involving quarks and gluons has not yet been successfully applied from its own first principles to calculate either spin-independent or spin-dependent structure functions. However, perturbative QCD does make some important predictions about nucleon structure functions including A_1 for x near 1, which is the high momentum tail of the wave function. The models of nucleon structure picture the proton as consisting of three valence quarks, two u quarks and a d quark, together with gluons and a sea of quark-antiquark pairs. They picture the neutron as two d quarks and a u quark together with gluons and the sea. The early models assumed SU(6) symmetry for the wave function. However, experimental data on F_2^n/F_2^p (69) and on A_1^p at large x required that SU(6) symmetry breaking be introduced. The important and unsymmetrical aspect of the wave function for the proton (neutron) near $x = 1$, which is predicted by perturbative QCD, is the occurrence with high probability of a single u (d) quark with large x and a diquark with isotopic spin $I = 0$ and spin component $S_z = 0$ (27–29). Of the various models for the proton wave function that are intended to represent the nonperturbative QCD solution, perhaps the most basic is the MIT bag model (70), which incorporates confinement.

A comparison of our data on $A_1^p(x)$ with various model predictions is shown in Figure 14. We should remark that some earlier nonquark models of the proton predicted negative values for A_1, but all quark models predict that A_1 is positive. Hence the earliest data indicating that A_1 is positive provided a crucial test of the quark model (26). In the quark model A_1 can be written (71)

$$A_1(x) = \frac{\sum_i e_i^2 [f_i^\uparrow - f_i^\downarrow]}{\sum_i e_i^2 [f_i^\uparrow + f_i^\downarrow]} \qquad 43.$$

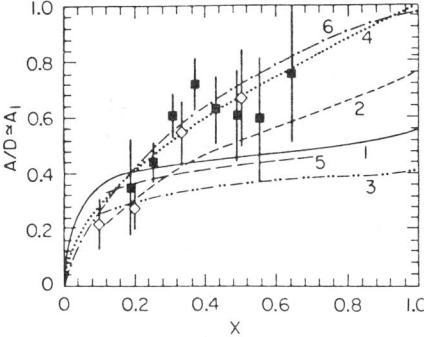

Figure 14 Experimental values A_1 compared with theories. 1. Symmetrical valence-quark model (18). 2. Current quarks (44). 3. Orbital angular momentum (39, 49). 4. Unsymmetrical model (45, 46). 5. MIT bag model (51, 70). 6. Source theory (72).

in which the sum is over the quarks i, e_i is the charge of the ith quark, and $f_i^\uparrow$ ($f_i^\downarrow$) is the probability for quark i to have its spin parallel (antiparallel) to the target nucleon spin. A_1 clearly provides a measure of the probability that the quark spins are aligned with the nucleon spin. Our data are consistent only with the Carlitz-Kaur (45, 46), the Schwinger (72), and possibly the Close (44) models of A_1. Our confidence levels in these models are 70%, 70%, and 3%, respectively. Curve 4 of Figure 14 provides an unsymmetrical model of the quark distributions involving SU(6) breaking, Regge theory at small x, the Melosh transformation (73), and agreement, with the Bjorken sum rule. Curve 6 is based on Schwinger's source theory (72), which is not a quark model.

For resonance region scattering, the measured asymmetries A/D are predominantly large and positive throughout the entire range in missing mass W except in the region of the $\Delta(1232$ MeV$)$ resonance, where A/D is expected to be negative because of magnetic dipole excitation. In principle the measured asymmetry values can be predicted from a multipole analysis of complete but unpolarized electroproduction data (74). Figure 15 displays the predictions based on a multipole analysis of single-pion electroproduction data only, which accounts for about half of the differential cross section. The agreement between these predictions and the data is rather good, and hence indicates that the net asymmetry contributed by channels other than single-pion production cannot be very different from the measured asymmetries. Figure 16 indicates that scaling applies for the resonance region data except at the $\Delta(1232)$ point, and hence that the spin-dependent behavior is also consistent with a global duality mechanism in analogy to the unpolarized case.

It is well known in the theory of atomic hyperfine structure that a

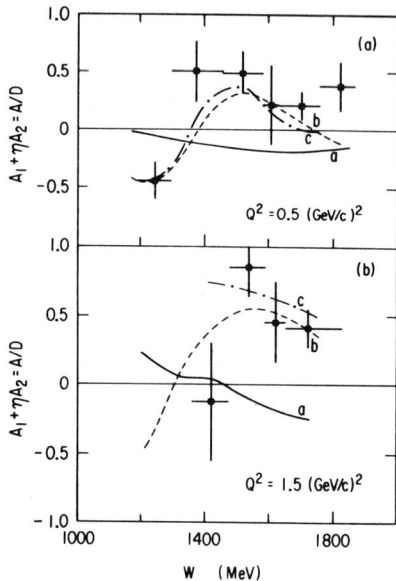

Figure 15 (a) Asymmetry data is $Q^2 \simeq 0.5$ (GeV/c)^2 compared with a multipole analysis performed by Devenish & Lyth (74): curve a, Born term alone; curve b, Born terms plus $\Delta(1232)$; and curve c, Born terms plus all resonances. (b) Same for $Q^2 \simeq 1.5$ (GeV/c)^2.

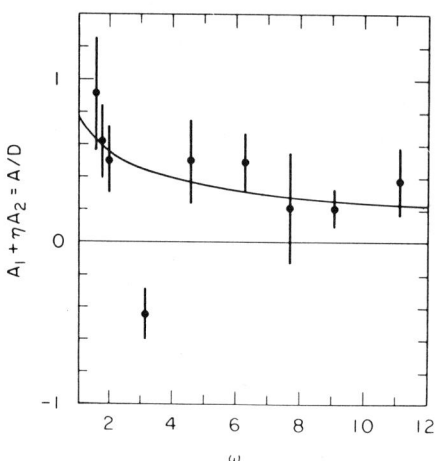

Figure 16 Asymmetry vs scaling variable ω. The curve $0.7\omega^{-1/2}$ is a fit to deep inelastic data ($W > 2$ GeV). The data points are the resonance region results ($W < 2$ GeV).

significant contribution to the hyperfine splitting interval Δv in hydrogen arises from the spin-dependent polarizability of the proton. Figure 17 gives the experimental and theoretical values for Δv. The contribution of the spin-dependent polarizability is designated $\delta_p(\text{pol})$. The principal theoretical uncertainty in Δv is due to $\delta_p(\text{pol})$, for which a positivity bound $|\delta_p(\text{pol})| \lesssim 4$ ppm has been calculated (33, 34). The quantity $\delta_p(\text{pol})$ can be expressed in terms of the spin-dependent structure functions G_1 and G_2, which are measured in polarized e-p scattering. The greatest contribution to $\delta_p(\text{pol})$ comes from the small-Q^2 region, including the proton resonances. The experimental data available to date support the theoretical estimates of $\delta_p(\text{pol})$. However, considerably more data are needed to provide a significant quantitative determination of $\delta_p(\text{pol})$.

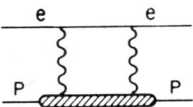

$\Delta v_{\text{expt.}} = 1\,420\,405\,751.766\,7(10)$ Hz

$\Delta v_{\text{theory}} = \Delta v_F (1 + \delta_{\text{QED}} + \delta_p)$ $\Delta v_F =$ Fermi value; $\delta_{\text{QED}} =$ QED corrections

$\delta_p =$ Proton recoil and structure term

$\delta_p = \delta_p(\text{rigid}) + \delta_p(\text{polarizability}) = -34.6(9) \times 10^{-6} + \delta_p(\text{pol})$

$$\delta_p(\text{pol}) = \frac{\alpha}{\pi} \frac{m_e}{M} \frac{1}{2(1+\mu_A)} \int_0^\infty \frac{d(-q^2)}{(-q^2)} [\Delta_1(q^2) + \Delta_2(q^2)]$$

$$\Delta_1(q^2) = \frac{9}{4}[F_2(q^2)]^2 + 5M^3 \int_{v_I(q^2)}^\infty \frac{dv}{v} \beta_1\left(\frac{v^2}{-q^2}\right) G_1(v, q^2)$$

$$\Delta_2(q^2) = 3M^2 \int_{v_I(q^2)}^\infty \frac{dv}{v^2} \beta_2\left(\frac{v^2}{-q^2}\right) q^2 G_2(v, q^2)$$

$\beta_1(x) \equiv \frac{4}{5}(-3x + 2x^2 + 2(2-x)\sqrt{x(1+x)})$; $\beta_2(x) \equiv 4x(1 + 2x - 2\sqrt{x(1+x)})$

$F_2(q^2) =$ Pauli form factor; $F_2(0) = \mu_A$; $v_I(q^2) = m_\pi + \frac{(m_\pi^2 - q^2)}{2M}$

Figure 17 Hyperfine structure interval Δv in hydrogen. The Feynman diagram and the expression given for $\delta_p(\text{pol})$ indicate the contribution to Δv of the spin-dependent polarizability of the proton.

5. FUTURE PROSPECTS

On the experimental side only a modest amount of data on polarized e-p scattering have been obtained, primarily because of the difficulty of the experiment. However, major advances in our knowledge are certainly possible both with polarized electron and with polarized muon scattering from polarized nucleon targets.

Thus far essentially only the spin-dependent structure function A_1^p has been measured. Actually the intrinsic e-p asymmetry $A = D(A_1 + \eta A_2)$ has been determined, but for the kinematics of scattering longitudinally polarized electrons by longitudinally polarized protons the kinematic term η is relatively small so that $A \simeq DA_1$. Determination of A_2 can be done with the protons polarized transverse to the incident longitudinally polarized electron beam. For this case the measured asymmetry $A_T = d(\zeta A_1 - A_2)$ and the kinematic factor ζ is relatively small so that, with measurements of both A and A_T, both A_1 and A_2 can be determined.

The spin-dependent structure functions for the neutron, A_1^N and A_2^N, can also be determined using a polarized deuteron target with both longitudinal and transverse polarizations, as well as a polarized proton target. In first approximation an appropriate subtraction of proton from deuteron data provides the information on the neutron. Effects of the deuteron binding and of the admixture of $^3D_1$ state in the deuteron wave function are estimated to be small but must also be considered.

A proposal (75) was submitted to SLAC (E138) to do these measurements using recent advances in the technology of polarized electron beams and polarized targets. For many years the material used in a polarized target has been a hydrocarbon chemically doped with a paramagnetic substance. Recently it has been found (76–78) that NH_3 and ND_3 after irradiation in a high energy electron (or parton) beam can be successfully dynamically polarized. The resulting polarized target has a significantly higher fraction of polarizable protons than a hydrocarbon and also has ~ 30 times the resistance to radiation. With the irradiated NH_3 it is then advantageous to use the higher intensity GaAs polarized electron source, despite its lower polarization as compared to the atomic ^6Li source.

Within less than two years the CERN European Muon Collaboration plan to measure polarized μ-p scattering using a longitudinally polarized muon beam with energies up to about 250 GeV and a longitudinally polarized proton target 1 m in length. The kinematic range for the scattering will extend up to $Q^2 \simeq 60$ $(GeV/c)^2$ with $0.1 \lesssim x \lesssim 0.7$. The muon beam at Fermilab from the Tevatron with energies up to 700 GeV can also be considered for polarized μ-p scattering experiments. Extensive measurement of asymmetries in the resonance region can be done with lower energy polarized e^- beam in the several GeV range, e.g. with SLAC or Bonn beams or with the planned 4-GeV CW National Accelerator in the US.

Knowledge of the neutron spin-dependent asymmetry A_1^N is necessary for a test of the Bjorken polarization sum rule. Several theoretical predictions of A_1^N have been made on the basis of models of nucleon structure and are shown in Figure 18. The SU(6) relativistic symmetric valence-quark model of the neutron predicts that $A_1^N = 0$ for all x. The other models also predict that A_1^N is small for $x < 0.5$. The unsymmetrical model, which fits well the measured A_1^p values, incorporates the perturbative QCD prediction that $A_1^N \to 1$ as $x \to 1$, when the single d or u quark with high x carries the entire spin component of the nucleon.

The asymmetry A_2 arises from an interference between amplitudes for absorption of virtual longitudinal and transverse photons by the proton. There is a sum rule related to A_2 (41). In the scaling limit simple parton models predict that A_2 becomes zero, and there is a positivity bound $|A_2|$

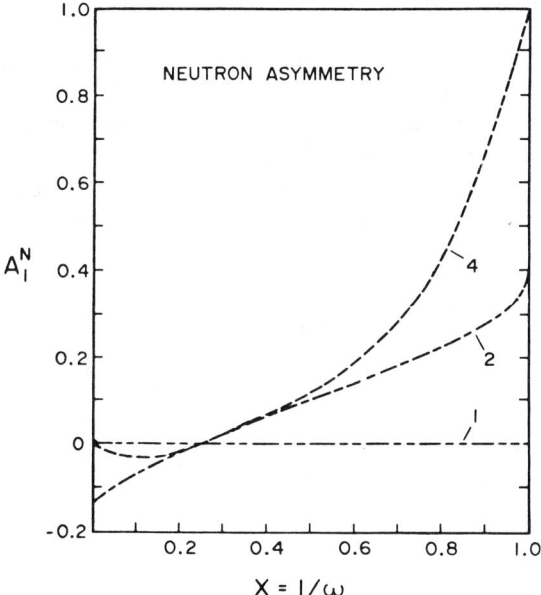

Figure 18 Theoretical predictions for A_1 (neutron). The models are as follows: 1. a relativistic symmetric valence-quark model of the neutron (18), 2. a model incorporating the Melosh transformation, which distinguishes between constituent and current quarks; 4. an unsymmetrical model in which the entire spin of the neutron is carried by a single quark in the limit of $x = 1$.

$< R^{1/2}$. Physically A_2 arises from quark masses and quark transverse momenta. Figure 19 shows various theoretical predictions for A_2 for the kinematics of the proposed E138 experiment. The positivity limit of $|A_2| < R^{1/2}$ is 0.5, since the best current value of R in this kinematic range is $R = 0.25 \pm 0.10$. Parenthetically, this large experimental value for R, which is expected theoretically to be zero in the scaling limit poses a problem for QCD theory, which may be related to higher-twist terms; the comparison of theory and experiment for A_2 can be expected to pose a similar problem. In addition Figure 19 shows the prediction of the MIT bag model and a prediction given from $g_2(x) = 0$ that is consistent with SU(6). Data on A_2 are important for comparison with these and other theories for A_2. In addition, data on A_2 are important to the experimental determination of A_1, since we measure $A/D = A_1 + \eta A_2$, and we only obtain a value of A_1 provided ηA_2 is sufficiently small. With the positivity bound for A_2, the value of ηA_2 for E80-E130 data is between 0.2 and 0.8 times the experimental one-standard-deviation error in the determination of A/D.

A major experiment is planned to be run at CERN by the European Muon Collaboration in 1984 (79) to measure the asymmetry in deep inelastic scattering of longitudinally polarized muons by longitudinally polarized protons. A significant test of the predicted scaling of the spin-dependent structure function A_1^p will be obtained from this CERN data at high Q^2 $[Q^2 \leq 60$ $(\text{GeV}/c)^2]$ and the SLAC data with $2 < Q^2 < 10$ $(\text{GeV}/c)^2$. Indeed the proton data to be obtained in the proposed experiment will be more precise than that obtained in E130 (by a factor of 2 to 3), and hence useful for the test of scaling. Figure 20 gives a theoretical prediction of scaling violation using $\Lambda = 0.1$ and 0.4 GeV.

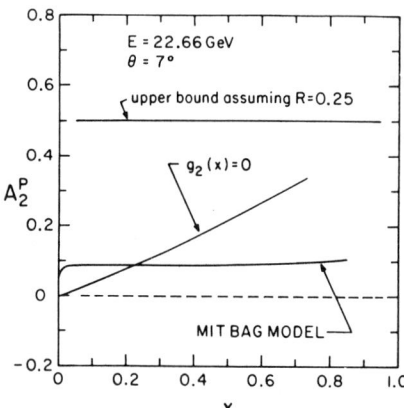

Figure 19 Theoretical predictions for A_2 (proton) for the kinematics of the E138 proposal.

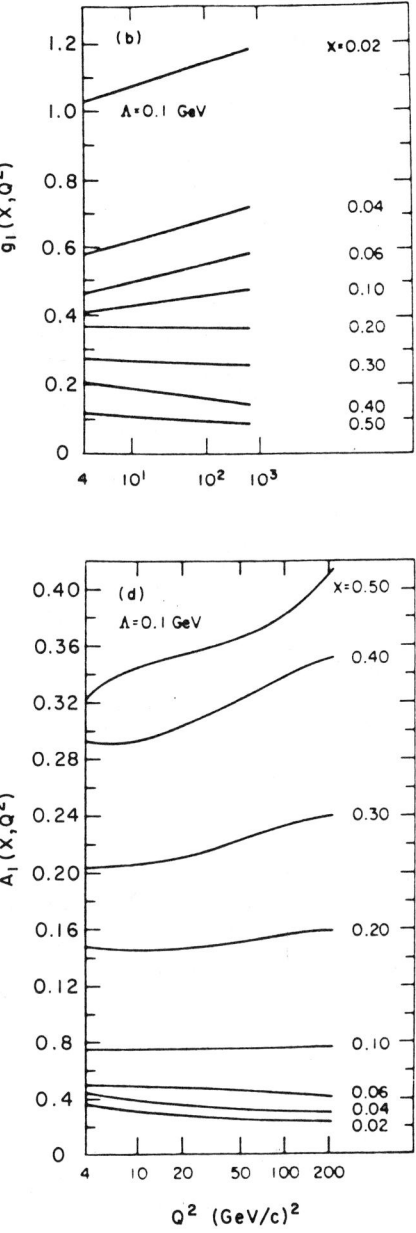

Figure 20 (*Top*) Scaling violation of $g_1(x, Q^2)$ (23) for two values of Λ, obtained by QCD and a broken SU(6) model (44) at $Q_0 = 2$ GeV. (*Bottom*) Scaling violations of $A_1(x, Q^2)$ obtained from (*top*) and known values of $F_1(x, Q^2)$.

Finally we emphasize that knowledge of the internal spin structure of the nucleon, apart from its importance to our understanding of nucleon structure, is essential to the interpretation of spin-dependent high energy phenomena involving hadrons (80–83). These include hadron-hadron scattering, the polarized Drell-Yan process, and production of polarized W or Z vector bosons in collisions of polarized protons in a high energy storage ring.

In conclusion, we wish to emphasize that the polarization data in deep inelastic electron (muon) scattering from polarized nucleons has proven to be very important for our understanding of the nucleon structure. During the years a consistent and remarkably simple picture of the nucleon has been emerging from experimental information and from QCD as the theoretical guidance to organize the observations. Accordingly, the nucleon as an extended object (with form factors and resonance excitations) is best described in terms of three permanently bound constituent quarks. The bound constituent quarks appear to deep inelastic probes as an almost free parton distribution with three valence quarks and a cloud of quark-antiquark pairs and gluons. The spin structure of the nucleon from the wave function of the constituent quarks is inherited in the spin distribution of partons. The aim of deep inelastic polarization experiments is to explore and understand this parton spin distribution in our general investigation of the nucleon structure. The first polarization experiments were remarkably successful. They clearly established the quark spin connection. Further polarization experiments on the neutron would improve significantly our present understanding of the nucleon structure.

ACKNOWLEDGMENT

This work was supported in part by the US Department of Energy under Contract No. DE-AC02-76ERO3075. One of us (JK) would like to acknowledge the kind hospitality extended to him from the Institute for Theoretical Physics of the University of California, Santa Barbara.

Literature Cited

1. Altarelli, G. 1982. *Phys. Rep.* 81(1): 1–127
2. Itzykson, C., Zuber, J. B. 1980. *Introduction to Quantum Field Theory*. New York: McGraw-Hill. 705 pp.
3. Marciano, W., Pagels, H. 1978. *Phys. Rep.* 36C: 137
4. Wilczek, F. 1982. *Ann. Rev. Nucl. Part. Sci.* 32: 177
5. Gell-Mann, M. 1964. *Phys. Lett.* 8: 217
6. Adler, S. L., Dashen, R. F. 1968. *Current Algebra and Applications to Particle Physics*. New York: Benjamin. 394 pp.
7. Wilson, K. G. 1971. *Phys. Rev.* D3: 1818
8. Fritzsch, H., Gell-Mann, M. 1972. *XVI Int. Conf. on High Energy Physics, Proc.*, Vol. II, p. 135
9. Gross, D. J., Wilczek, F. 1973. *Phys. Rev. Lett.* 30: 1343

10. Politzer, H. D. 1973. *Phys. Rev. Lett.* 30:1346
11. Hughes, V. W., et al. 1981. In *High Energy Physics with Polarized Targets*, ed. C. Joseph, J. Soffer, p. 331. Lausanne: Berkhauser Verlag.
12. Bjorken, J. D. 1966. *Phys. Rev.* 148:1467
13. Galfi, L., et al. 1970. *Phys. Lett.* 31B:465
14. Bjorken, J. D. 1970. *Phys. Rev.* D1:465
15. Kuti, J. 1972. *Proc. Neutrino '72*, ed. G. Marx, A. Frenkel, Vol. 2, p. 101
16. Galfi, L., et al. 1972. *Acta Phys. Hungarica* 21:85
17. Hey, A. J. G., Mandula, J. E. 1972. *Phys. Rev.* D5:2610
18. Kuti, J., Weisskopf, V. W. 1971. *Phys. Rev.* D4:3418
19. Feynman, R. P. 1969. *Phys. Rev. Lett.* 23:1415
20. Bjorken, J. D., Paschos, E. A. 1969. *Phys. Rev.* 185:1975
21. Feynman, R. P. 1972. *Photon-Hadron Interactions*. Reading: Benjamin
22. Kodaira, J., et al. 1979. *Phys. Rev.* D20:627
23. Darrigol, O., Hayot, F. 1978. *Nucl. Phys.* B141:391
24. Matsuda, S., Uematsu, T. 1980. *Nucl. Phys.* B168:181
25. Kodaira, J. 1980. *Nucl. Phys.* B169:181
26. Alguard, M. J., et al. 1976. *Phys. Rev. Lett.* 37:1261
27. Feynman, R. P. 1972. *Proc. Neutrino '72*, Vol. II, p. 75
28. Brodsky, S. J. 1981. See Ref. 11, p. 169
29. Farrar, G., Jackson, D. 1975. *Phys. Rev. Lett.* 35:1416
30. Baum, G., et al. 1981. *Phys. Rev. Lett.* 45:2000
31. Alguard, M. J., et al. 1978. *Phys. Rev. Lett.* 41:70
32. Iddings, C. K. 1965. *Phys. Rev.* 138B:446
33. Gnädig, P., Kuti, J. 1972. *Phys. Lett.* 42B:241
34. de Rafael, E. 1971. *Phys. Lett.* 37B:201
35. Gilman, F. 1973. *Proc. SLAC Summer Inst. on Part. Phys.*, Vol. I, p. 71
36. Hey, A. J. G. 1974. *Daresbury Lect. Note Ser.*, No. 13
37. Doncel, M. G., de Rafael, E. 1971. *Nuovo Cimento* 4A:363
38. Gnädig, P., Niedermayer, F. 1973. *Nucl. Phys.* B55:612
39. Sehgal, L. M. 1974. *Phys. Rev.* D10:1663
40. Ellis, J., Jaffe, R. 1974. *Phys. Rev.* D9:1444
41. Burkhardt, H., Cottingham, W. N. 1970. *Ann. Phys.* 56:453
42. Heimann, R. L. 1973. *Nucl. Phys.* B64:429
43. Kuti, J. 1982. In *Unified Theories of Elementary Particles*, ed. P. Breitenlohner, H. P. Dürr, p. 90. Berlin: Springer Verlag
44. Close, F. E. 1974. *Nucl. Phys.* B80:269
45. Carlitz, R., Kaur, J. 1976. *Phys. Rev. Lett.* 38:673
46. Kaur, J. 1977. *Nucl. Phys.* B128:219
47. Altarelli, G., et al. 1974. *Nucl. Phys.* B69:531
48. Gourdin, M. 1972. *Nucl. Phys.* B38:418
49. Look, G. W., Fischbach, F. 1977. *Phys. Rev.* D16:211
50. Joshipura, A. S., Roy, P. 1980. *Phys. Lett.* B92:348
51. Hughes, R. J. 1977. *Phys. Rev.* D16:622
52. Domokos, G., et al. 1971. *Phys. Rev.* D3:1191
53. Close, F., et al. 1974. *Nucl. Phys.* B77:281
54. Le Yaovanc, A., Oliver, L., Pène, O., Raynal, J. C. 1975. *Phys. Rev.* D11:680
55. Wandzura, S., Wilczek, F. 1977. *Phys. Lett.* B72:195
56. Brodsky, S. J., Drell, S. D. 1970. *Ann. Rev. Nucl. Sci.* 20:147
57. Kessler, J. 1976. *Polarized Electrons*. Berlin: Springer Verlag
58. Hughes, V. W., et al. 1971. *Phys. Rev.* A5:192
59. Alguard, M. J., et al. 1979. *Nucl. Instrum. Methods* 163:29
60. Cooper, P. S., et al. 1975. *Phys. Rev. Lett.* 34:1589
61. Abragam, A., Goldman, M. 1978. *Rep. Prog. Phys.* 41:395
62. Borghini, M. 1971. *Proc. 2nd Int. Conf. on Polarized Targets*, ed. G. Shapiro, p. 1. Berkeley: Univ. Calif. Press
63. Alguard, M. J., et al. 1976. *Phys. Rev. Lett.* 37:1258
64. Ash, W. W. 1976. *High Energy Physics with Polarized Beams and Targets*, ed. M. L. Marshak, p. 485. New York: Am. Inst. Phys.
65. Baum, G., et al. 1983. *Phys. Rev. Lett.* In press
66. Bodek, A., et al. 1979. *Phys. Rev.* D20:1471
67. Schuler, K. P. 1979. *High Energy Physics with Polarized Beams and Polarized Targets*, ed. G. H. Thomas, p. 217. New York: LAIP
68. Dombey, N. 1969. *Rev. Mod. Phys.* 41:236
69. Buras, A. J., Gaemers, K. H. F. 1978. *Nucl. Phys.* B132:249
70. Jaffe, R. L. 1975. *Phys. Rev.* D11:1953
71. Close, F. E. 1979. *An Introduction to Quarks and Partons*. London: Academic
72. Schwinger, J. 1977. *Nucl. Phys.* B123:223

73. Melosh, H. J. 1974. *Phys. Rev.* D9:1095
74. Devenish, R. C. E., Lyth, D. H. 1975. *Nucl. Phys.* B93:109
75. Hughes, V. W., Schuler, P. K. 1982. *SLAC Proposal E138*
76. Seely, M. L., et al. 1982. *High Energy Spin Physics—1982*, ed. G. Bunce, *AIP Conf. Proc. No. 95*, p. 526
77. Niinikoski, R. O., Rieubland, J.-M., 1979. *Phys. Lett.* 72A:141
78. Hartel, U., et al. 1982. See Ref. 11, p. 453
79. Gabathuler, E. 1974. *CERN EMC Proposal*
80. Baldracchini, F., et al. 1980. *Int. Cent. for Theor. Phys. Trieste Tech. Rep.*
81. Hidaka, K., et al. 1980. *Phys. Rev.* D21:1316
82. Craigie, N., et al. 1980. *Phys. Lett.* B96:381
83. Sivers, D. 1978. *High Energy Physics with Polarized Beams and Polarized Targets*, ed. H. G. Thomas, *AIP Conf. Proc. No. 51*, p. 505

GRAND UNIFIED THEORIES AND THE ORIGIN OF THE BARYON ASYMMETRY

Edward W. Kolb

Theoretical Division, Los Alamos National Laboratory, Los Alamos, New Mexico 87545

Michael S. Turner

Astronomy and Astrophysics Center, The University of Chicago, Chicago, Illinois 60637

CONTENTS

1. INTRODUCTION	646
1.1 *Overview*	646
1.2 *The Standard Cosmology*	647
2. AN ASYMMETRIC UNIVERSE	650
2.1 *Evidence for a Baryon Asymmetry*	650
2.2 *The Tragedy of a Symmetric Cosmology*	652
3. BARYOGENESIS—THE QUALITATIVE PICTURE	653
3.1 *The Necessary Ingredients*	653
3.2 *The Standard Scenario: Out-of-Equilibrium Decay*	655
4. BARYON NUMBER VIOLATION IN GRAND UNIFIED MODELS	658
5. CP VIOLATION IN GRAND UNIFIED MODELS	660
5.1 *Origin and Classification of CP Violation*	660
5.2 *A Simple Model Calculation*	662
6. BOLTZMANN EQUATIONS FOR BARYOGENESIS	665
6.1 *A Simple Model for Baryogenesis*	665
6.2 *Analytic Solutions*	669
6.3 *Damping of Preexisting Asymmetries*	673
7. SPECIFIC MODELS	674
7.1 $SU(5)$	674
7.2 $SO(10)$	679
7.3 *Realistic Boltzmann Equations*	681

0163-8998/83/1201-0645$02.00

8. BEYOND THE STANDARD SCENARIO	683
8.1 *Primordial Black Holes*	683
8.2 *Supersymmetric Models*	684
8.3 *Monopole Catalysis*	686
8.4 *Low-Temperature Scenarios*	687
8.5 *Initial Conditions*	689
8.6 *The Lepton Number*	691
9. SUMMARY AND CONCLUDING REMARKS	692

1. INTRODUCTION

1.1 *Overview*

The hot big-bang cosmology is a highly successful model for the evolution of the Universe; it readily accounts for the Hubble expansion of the Universe, the cosmic microwave background, and the cosmic abundances of ^{4}He, ^{2}H, and possibly ^{7}Li. Until very recently, however, it shed no light on the fact that we live in a Universe in which there is a preponderance of matter over antimatter. But for a tiny effect in the K^0-$\bar{K}^0$ system, the laws of physics do not seem to differentiate between matter and antimatter—so why should the Universe be so asymmetric? This matter-antimatter asymmetry (usually called the baryon asymmetry) is quantified by a dimensionless number, "the baryon number" of the Universe. This number B is roughly the ratio of the average baryon number density to the average photon number density (more precisely, it is the ratio of the baryon number density to the entropy density, cf Equation 12). Since the amount of antimatter in the Universe appears to be negligible, the average baryon number density may be obtained from the mass density contributed by matter (baryons), while the photon number density is very closely given by the number density of photons in the cosmic microwave background radiation. This parameterization is extremely useful since B remains constant during most of the history of the Universe. The value of B today is about 10^{-10}. Why does the baryon number take on this curiously small value? That the net baryon number differs from zero is crucial; if it did not, then matter-antimatter annihilations would render the present Universe essentially devoid of both matter and antimatter.

The most startling prediction of grand unified theories (GUTs) is proton decay—the instability of matter. However, the most impressive achievement of GUTs to date may be the attractive framework they provide for explaining the origin of the baryon asymmetry in an initially symmetric Universe. Baryogenesis links together physics at the distance scale of 10^{-28} cm, the tiny violation of matter-antimatter symmetry in the laws of physics, and the predominance of matter in our observable Universe (present size $\approx 10^{28}$ cm).

In this article we review the great progress made over the past few years in relating the baryon asymmetry to reactions predicted by GUTs to have taken place about 10^{-34} s after the bang. Since no one complete and unified model has emerged as *the* Grand Unifed Theory, it is not possible to make an exact prediction for the magnitude and/or sign of the baryon asymmetry. Therefore, we concentrate on the common features that different models have in the context of baryogenesis.

In the remainder of this section we discuss microphysics in the standard cosmology. In Section 2 we briefly review the observational evidence for the existence of a baryon asymmetry. The GUTs scenario for baryogenesis is presented qualitatively in Section 3, and the three key ingredients—baryon number violation, C and CP violation, and departure from thermal equilibrium—are discussed further in Sections 4, 5, and 6. In Section 7 the SU(5) and SO(10) models are considered in more detail, and in Section 8 some simple variants of the standard scenario are discussed. Section 9 contains a brief summary and some concluding remarks.

1.2   *The Standard Cosmology*

The hot big-bang model, the so-called standard cosmology, is based upon the isotropic and homogeneous Friedmann-Robertson-Walker (FRW) cosmology. Here we briefly summarize the relevant aspects of the model for baryogenesis. For a more detailed discussion of the model we refer the reader to Weinberg (1) and Wagoner (2). Throughout we shall use units where $\hbar = c = k_B = 1$. In this system 1 GeV = $(2.0 \times 10^{-14}$ cm$)^{-1}$ = $(6.6 \times 10^{-25}$ s$)^{-1}$ = 1.2×10^{13} K, and $G = m_{pl}^{-2}$ where $m_{pl} = 1.2 \times 10^{19}$ GeV (the Planck mass).

The line element in the FRW cosmology is,

$$ds^2 = -dt^2 + R(t)^2 \left[\frac{dr^2}{1-kr^2} + r^2(d\theta^2 + \sin^2\theta \, d\phi^2) \right], \qquad 1.$$

where $k = 1, 0, -1$ specifies the positively curved and finite, the flat and infinite, and the negatively curved and infinite models, respectively. The proper spatial separation between two points that are co-moving with the expansion (i.e. have fixed coordinates r, θ, ϕ) grows with the cosmic scale factor $R(t)$. The evolution of $R(t)$ is given by

$$H^2 = (\dot{R}/R)^2 = \frac{8\pi G \rho}{3} - \frac{k}{R^2}, \qquad 2.$$

where ρ is the total energy density and the overdot denotes time derivative. The expansion rate H sets the timescale of the expansion: $R(t)$ e-folds in a time of the order of H^{-1}. In the FRW model the stress-energy tensor of the cosmic fluid (T^μ_ν) must take on the diagonal form, diag($-\rho, p, p, p$), and the equation of motion for the cosmic fluid is

$$d(\rho R^3) + p\, d(R^3) = 0. \qquad 3.$$

Today the energy density contributed by nonrelativistic matter (baryons and possibly other nonrelativistic particles such as massive neutrinos) is about three or four orders of magnitude greater than that contributed by relativistic particles (photons, very light neutrino species, etc). (That this ratio is only 10^3 or 10^4 may seem a bit paradoxical since the average energy of a photon in the microwave background is about 10^{-3} eV and the rest energy of a nucleon is about 10^9 eV; however, it must be remembered that there are some 10^9 photons per nucleon.) From Equation 3 it follows that the energy density contributed by nonrelativistic matter ($p = 0$) decreases as $R^{-3}(t)$, while that of relativistic matter ($p = \rho/3$) decreases as $R^{-4}(t)$, and, because of this, when the scale factor of the Universe was a factor of 10^4 or more smaller than it is today (corresponding to temperatures greater than about 10 eV and times earlier than about 10^{10} s after the bang) the energy density contributed by relativistic particles dominated the energy density of the Universe. Thus during its early history (which is the epoch of interest for baryogenesis) the Universe was radiation-dominated.

The number density of, and the energy density contributed by, a very relativistic species ($m \ll T$) that is in thermal equilibrium is

$$n = g \frac{\zeta(3)}{\pi^2} T^3 \times \begin{cases} 1 & \text{Bose-Einstein} \\ (3/4) & \text{Fermi-Dirac} \\ \zeta(3)^{-1} & \text{Maxwell-Boltzmann} \end{cases} \qquad 4.$$

$$\rho = g \frac{\pi^2}{30} T^4 \times \begin{cases} 1 & \text{Bose-Einstein} \\ 7/8 & \text{Fermi-Dirac} \\ 90/\pi^4 & \text{Maxwell-Boltzmann} \end{cases}, \qquad 5.$$

where $\zeta(3) = 1.202$ and g is the total number of degrees of freedom of the species (e.g. 2 for a photon, 4 for $e^\pm$, 12 for $q\bar{q}$, $q = u, d, c, \ldots$). The total energy density contributed by all relativistic species is

$$\rho = g_* \frac{\pi^2}{30} T^4, \quad g_* = \sum_{\text{bosons}} g + 7/8 \sum_{\text{fermions}} g. \qquad 6.$$

If all species are in equilibrium then it follows that the total entropy ($\propto sR^3$) remains constant, where

$$s = \frac{\rho + p}{T} = \frac{2\pi^2}{45} g_* T^3 \qquad 7.$$

is the entropy density. Note that s is proportional to the number density of relativistic particles present. So long as g_* is constant (g_* changes whenever $T \approx$ mass of a species), T decreases as $R(t)^{-1}$. Whenever g_* is approximately constant and the Universe is radiation-dominated, it follows that $R(t) \propto t^{1/2}$, and

$$\dot{R}/R = -\dot{T}/T = (4\pi^3/45)^{1/2} g_*^{1/2} T^2/m_{pl} \qquad 8a.$$

$$t = (45/16\pi^3)^{1/2} g_*^{-1/2} m_{pl}/T^2 \approx 2.4 \times 10^{-6} g_*^{-1/2} (T/\text{GeV})^{-2} \text{ s.} \qquad 8b.$$

The asymptotic freedom exhibited by non-Abelian gauge theories justifies the weakly interacting gas approximation tacitly made in Equations 4 and 5. Quantum corrections to general relativity should be small for times later than $t_{pl} \cong 10^{-43}$ s ($T \lesssim m_{pl} \approx 10^{19}$ GeV). Thus it is not unreasonable (although it may eventually prove to be wrong!) to expect Equations 8 to be valid for 10^{19} GeV $\gtrsim T \gtrsim 10$ eV.

What does thermal equilibrium mean in the expanding Universe? In a truly rigorous sense it cannot be defined. Operationally, of course, it can be defined. If the rates Γ for the reactions driving the cosmic fluid to equilibrium are rapid compared to the expansion rate H, then those reactions can bring the Universe to a state of equilibrium in a time short compared to the timescale on which T is changing, H^{-1} ($H = \dot{R}/R = |\dot{T}/T|$). If for the important reactions $\Gamma \gg H$, then we expect the Universe to pass through a succession of nearly thermal equilibrium states.

In thermal equilibrium the phase space number density of a particle species is

$$f(p) = g\{\exp[(\mu + E)/T] + \theta\}^{-1}, \qquad 9.$$

where $\theta = 1$ (Fermi-Dirac), -1 (Bose-Einstein), 0 (Maxwell-Boltzmann), and μ is the chemical potential. For a species with $\mu, m \ll T$, the number and energy densities are given by Equations 4 and 5, while for a species with $m \gg \mu, T$ they are

$$n \approx g(mT/2\pi)^{3/2} \exp(-\mu/T - m/T), \qquad 10.$$

$$\rho \approx mn. \qquad 11.$$

In addition, for each reaction $a + b \leftrightarrow c + d$, we have $\mu_a + \mu_b = \mu_c + \mu_d$, i.e. all species are in *chemical* equilibrium.

In the expanding Universe it often happens that not all reaction rates Γ are much faster than the expansion rate H. A typical example is the case where reactions that change the energy of a particle of species a are occurring rapidly (e.g. $a + e^- \rightarrow a + e^-$), but reactions that change the number of particles of species a are not occurring rapidly (e.g. $a + \bar{a} \rightarrow 2\gamma$). Here we expect the phase space density to be of the form expressed by

Equation 9 but we do not expect $\mu_a + \mu_b = \mu_c + \mu_d$ (e.g. $\mu_a + \mu_{\bar{a}} = 2\mu_\gamma = 0$) for the reactions that change the number of a's. Species a is said to be in *kinetic* equilibrium, but not in *chemical* equilibrium. The distribution of a's in momentum space is determined by the temperature, but the number density (which is determined by μ_a) is not.

If the Universe were always in thermal equilibrium, then its state could be completely specified by the temperature and all the conserved quantum numbers (e.g. charge, color)—easy to describe, but very uninteresting. If, on the other hand, it were nowhere near thermal equilibrium, then the evolution of its state would necessitate integrating the complete set of Boltzmann equations for all species—a very formidable task! However, if kinetic equilibrium is maintained, and all but a few species are in chemical equilibrium, then the state of the Universe can be obtained by evolving a manageable set of Boltzmann equations. One such example is the evolution of the baryon asymmetry.

One expects a departure from thermal equilibrium whenever a rate crucial for maintaining chemical or kinetic equilibrium is less than the expansion rate ($\Gamma < H$). However, this is not always true. Consider two noninteracting, massless species (a and b), which at some epoch ($R = R_0$, $T = T_0$) are in equilibrium, with chemical potentials related by $\mu_b = C\mu_a$. Owing to the expansion, each species will have its number density diluted by $(R/R_0)^3$, and each particle will have its momentum redshifted by R_0/R. It is simple to show that at a later epoch each species will have a distribution again given by Equation 9, but with $T' = T_0(R_0/R)$, $\mu' = \mu(R_0/R)$, and $\mu'_b = C\mu'_a$. In the absence of interactions, massless species initially in thermal equilibrium remain in thermal equilibrium. Witness the microwave background for which the epoch of last scattering was $R \approx 10^{-3} R_{\text{today}}$, and which today has a blackbody spectrum to within the observational uncertainties.

2. AN ASYMMETRIC UNIVERSE

Here we briefly summarize the evidence for the baryon asymmetry and the seemingly insurmountable problems that render baryon symmetric cosmologies untenable. For a more detailed discussion of these two issues we refer the reader to Steigman's review of the subject (3). For a review of recent attempts to reconcile a symmetric Universe with both baryogenesis and the observational constraints, we refer the reader to Stecker (4, 4a).

2.1 *Evidence for a Baryon Asymmetry*

Within the solar system we can be very confident that there are no concentrations of antimatter (e.g. antiplanets). If there were, solar wind particles striking such objects would be the strongest γ-ray sources in the sky. Also, NASA has yet to lose a space probe because it annihilated with antimatter in the solar system.

Cosmic rays more energetic than O(0.1 GeV) are generally believed to be of "extrasolar" origin, and thereby provide us with samples of material from throughout the galaxy (and possibly beyond). The ratio of antiprotons to protons in the cosmic rays is about 3×10^{-4} (5–7), and the ratio of anti-^{4}He to ^{4}He is less than 10^{-5} (5). Antiprotons are expected to be produced as cosmic-ray secondaries (e.g. $p+p \to 3p+\bar{p}$) at about the 10^{-4} level. At present both the spectrum and total flux of cosmic-ray antiprotons are at variance with the simplest model of their production as secondaries. A number of alternative scenarios for their origin have been proposed including the possibility that the detected $\bar{p}$ are cosmic rays from distant antimatter galaxies (8). Although the origin of these $\bar{p}$ remains to be resolved, it is clear that they do not provide evidence for an appreciable quantity of antimatter in our galaxy. [For a recent review of the cosmic-ray antiproton puzzle and possible solutions, we refer the reader to Gaisser (9).]

The existence of both matter and antimatter galaxies in a cluster of galaxies containing intracluster gas would lead to a significant γ-ray flux from decays of π^0s produced by nucleon-antinucleon annihilations. Using the observed γ-ray background flux as a constraint, Steigman (3) argues that clusters like the Virgo cluster, which is at a distance ≈ 20 Mpc ($\approx 10^{26}$ cm) and contains several hundred galaxies, must not contain both matter and antimatter galaxies.

Based upon the above-mentioned arguments, we can say that if there exist equal quantities of matter and antimatter in the Universe, then we can be absolutely certain they are separated on mass scales greater than $1\ M_\odot$, and reasonably certain they are separated on scales greater than $(1-100)\ M_{\text{galaxy}} \approx 10^{12}-10^{14}\ M_\odot$. As discussed below, this fact is virtually impossible to reconcile with a symmetric cosmology.

It has often been pointed out that we derive most of our direct knowledge of the large-scale Universe from photons, and since the photon is a self-conjugate particle we obtain no clue as to whether the source is made of matter or antimatter. Neutrinos, on the other hand, can in principle reveal information about the matter-antimatter composition of their source. Large neutrino detectors such as DUMAND may someday provide direct information about the matter-antimatter composition of the Universe on the largest scales (10, 11, 11a).

Baryons account for only a tiny fraction of the particles in the Universe, the 3K-microwave photons being the most abundant species (yet detected). The number density of 3K photons is $n_\gamma = 400\ (T/2.7\ \text{K})^3\ \text{cm}^{-3}$. The baryon density is not nearly as well determined. Contributions to the present mass density of the Universe are usually expressed as a fraction of the closure density, $\rho_c = 3H_0^2/8\pi G$, where H_0 is the present value of the Hubble parameter. Luminous matter (baryons in stars) contribute at least 0.01 of closure density ($\Omega_{\text{lum}} > 0.01$), while observations indicate that Ω_{tot} (and Ω_{baryon}) must be $< O(2)$ (12). These direct determinations place the baryon-to-photon ratio $\eta \equiv n_b/n_\gamma$ in the range 3×10^{-11} to 6×10^{-8}. [Hereafter, we denote the present baryon-to-photon ratio by η, and reserve B for the baryon-to-entropy ratio.] The yields of big-bang nucleosynthesis depend directly on η, and the production of amounts of D, ^{3}He, ^{4}He, and ^{7}Li that are consistent with their present measured abundances restricts η to the narrow range $(4-7) \times 10^{-10}$ (13, 14).

Since today it appears that $n_b \gg n_{\bar{b}}$, η is also the ratio of net baryon number to photons. The number of photons in the Universe has not remained constant, but has increased at various epochs when particle species have annihilated (e.g. $e^\pm$ pairs at $T \approx 0.5$ MeV). Assuming the expansion has been isentropic (i.e. no significant entropy production), the entropy ($\propto sR^3$) has remained constant. The "known entropy" is presently about equally divided between the 3K photons and the three cosmic neutrino backgrounds (e, μ, τ). Taking this to be the present entropy, the ratio of baryon number to entropy is

$$B \equiv n_B/s \approx (1/7)\eta \approx (6-10) \times 10^{-11}, \qquad 12.$$

where $n_B \equiv n_b - n_{\bar{b}}$ and η is taken to be in the range $(4-7) \times 10^{-10}$. So long as the expansion is isentropic and baryon number is at least effectively conserved this ratio remains constant and is what we refer to as the baryon number of the Universe.

Although the matter-antimatter asymmetry appears to be "large" today (in the sense that $n_B \approx n_b \gg n_{\bar{b}}$), the fact that B is small implies that at very early times the asymmetry was "small" ($n_B \ll n_b$). To see this, let us assume for simplicity that nucleons are the fundamental baryons. Earlier than 10^{-6} s after the bang the temperature was greater than the mass of a nucleon. Thus nucleons and antinucleons should have been about as abundant as photons, $n_N \approx n_{\bar{N}} \approx n_\gamma$. The entropy density s is $\approx g_* n_\gamma \approx g_* n_N \approx O(10^2) n_N$. The constancy of $n_B/s \approx O(10^{-10})$ requires that for $t < 10^{-6}$ s, $(n_N - n_{\bar{N}})/n_N (\approx 10^2 n_B/s) \approx O(10^{-8})$. During its earliest epoch, the Universe was nearly (but not quite) baryon symmetric.

2.2 *The Tragedy of a Symmetric Cosmology*

Suppose that the Universe were initially locally baryon symmetric. Earlier than 10^{-6} s after the bang nucleons and antinucleons were about as abundant as photons. For $T < 1$ GeV the equilibrium abundance of nucleons and antinucleons is $(n_N/n_\gamma)_{\text{EQ}} \approx (m_N/T)^3 \exp(-m_N/T)$, and as the Universe cooled the number of nucleons and antinucleons would decrease, tracking the equilibrium abundance as long as the annihilation rate $\Gamma_{\rm ann} \approx n_{\rm N}(\sigma v)_{\rm ann} \approx n_{\rm N} m_\pi^{-2}$ was greater than the expansion rate H. At a temperature $T \approx {\rm O}(20\,{\rm MeV})$, annihilations freeze out ($\Gamma_{\rm ann} \approx H$), nucleons and antinucleons being so rare they can no longer find each other to annihilate. Because of the incompleteness of the annihilations, residual nucleon and antinucleon to photon ratios $n_{\bar{\rm N}}/n_\gamma = n_{\rm N}/n_\gamma \cong 10^{-18}$ are "frozen in." Even if the matter and antimatter could subsequently be separated, $n_{\rm N}/n_\gamma$ is a factor of 10^8 too small. To avoid this annihilation catastrophe, matter and antimatter must be separated on large scales before $t \approx 3 \times 10^{-3}\,{\rm s}\,(T \approx 20\,{\rm MeV})$.

STATISTICAL FLUCTUATIONS One possible mechanism for doing this is statistical (Poisson) fluctuations. The co-moving volume that encompasses our galaxy today contains $\sim 10^{12}\,M_\odot \approx 10^{69}$ baryons and $\sim 10^{79}$ photons. Earlier than 10^{-6} s after the bang this same co-moving volume contained $\sim 10^{79}$ photons and $\sim 10^{79}$ baryons and antibaryons. In order to avoid the annihilation catastrophe, this volume would need an excess of baryons over antibaryons of $\sim 10^{69}$, but from statistical fluctuations one would expect $N_{\rm b} - N_{\bar{\rm b}} \approx {\rm O}(N_{\rm b}^{1/2}) \sim 3 \times 10^{39}$—a mere $29\frac{1}{2}$ orders of magnitude too small!

CAUSALITY CONSTRAINTS Clearly, statistical fluctuations are of no help, so consider a hypothetical interaction that separates matter and antimatter. In the standard cosmology the distance over which light signals (and hence causal effects) could have propagated since the bang (the horizon distance) is finite and $\sim 2t$. When $T \approx 20\,{\rm MeV}$ ($t \approx 3 \times 10^{-3}$ s) causally coherent regions contained only about $10^{-5}\,M_\odot$. Thus, in the standard cosmology causal processes could have only separated matter and antimatter into lumps of mass $\lesssim 10^{-5}\,M_\odot \ll M_{\rm galaxy} \approx 10^{12}\,M_\odot$. The inflationary scenarios discussed in Section 8.5 can alter the causality constraints. However, a symmetric Universe model will have to contend with the domain wall problem discussed in Section 5.

It should be clear that the two observations, $n_{\rm b} \gg n_{\bar{\rm b}}$ on scales at least as large as $10^{12}\,M_\odot$ and $n_{\rm b}/n_\gamma \approx (4\text{–}7) \times 10^{-10}$, effectively render all baryon-symmetric cosmologies untenable. A viable pre-GUT cosmology needed to have as an initial condition a tiny baryon number, $n_{\rm B}/s \approx (6\text{–}10) \times 10^{-11}$—a very curious initial condition at that!

3. BARYOGENESIS—THE QUALITATIVE PICTURE

3.1 *The Necessary Ingredients*

More than a decade ago Sakharov (15) and others (16, 17) suggested that an initially baryon-symmetric Universe might dynamically evolve a baryon excess of ${\rm O}(10^{-10})$, which after baryon-antibaryon annihilations destroyed essentially all of the antibaryons would leave the one baryon per 10^{10} photons that we observe today. In his 1967 paper Sakharov outlined the three ingredients necessary for baryogenesis: (a) B-nonconserving interactions; (b) a violation of both C and CP; (c) a departure from thermal equilibrium.

It is clear that B (baryon number) must be violated if the Universe begins baryon symmetric and then evolves a net B. In 1967 there was no motivation for B nonconservation. After all, the proton lifetime is more than 35 orders of magnitude longer than that of any unstable elementary particle—pretty good evidence for B conservation (18–22). Of course, grand unification provides just such motivation [see Section 4 or the reviews in (23, 24)], and proton decay experiments are likely to verify B nonconservation in the next decade if the proton lifetime is $\lesssim 10^{33}$ years (25).

Under C (charge conjugation) and CP (charge conjugation combined with parity), the B of a state changes sign. Thus a state that is either C or CP invariant must have $B = 0$. If the Universe begins with equal amounts of matter and antimatter, and without a preferred direction (as in the standard cosmology), then its initial state is both C and CP invariant. Unless both C and CP are violated, the Universe will remain C and CP invariant as it evolves, and thus cannot develop a net baryon number even if B is not conserved. Both C and CP violations are needed to provide an arrow to specify that an excess of matter be produced. C is maximally violated in the weak interactions, and both C and CP are violated in the K^0-$\bar{K}^0$ system (26, 27). Although a fundamental understanding of CP violation is still lacking at present, GUTs can accommodate CP violation (this is discussed in Section 5). It would be very surprising if CP violation only occurred in the K^0-$\bar{K}^0$ system and not elsewhere in the theory also (including the B-nonconserving sector). In fact, without miraculous cancellations the CP violation in the neutral kaon system will give rise to CP violation in the B-nonconserving sector at some level.

The necessity of a departure from thermal equilibrium is a bit more subtle. It has been shown (2, 28–30, 30a) that CPT and unitary alone are sufficient to guarantee that equilibrium particle distributions are given by Equation 9: $f(p) = [\exp(\mu/T + E/T) \pm 1]^{-1}$. In equilibrium, processes like $\gamma + \gamma \rightleftarrows b + \bar{b}$ imply that $\mu_b = -\mu_{\bar{b}}$, while processes like (but not literally) $\gamma + \gamma \rightleftarrows b + b$ require that $\mu_b = 0$. Since $E^2 = p^2 + m^2$ and $m_b = m_{\bar{b}}$ by CPT, it follows that in thermal equilibrium, $n_b \equiv n_{\bar{b}}$.

As we discussed in Section 1.2, because the temperature of the Universe is changing on a characteristic timescale H^{-1}, thermal equilibrium can only be maintained if the rates for reactions that drive the Universe to equilibrium are much greater than H. Departures from equilibrium have occurred often during the history of the Universe. For example, because the rate for γ + matter → γ' + matter' is $\ll H$ today, matter and radiation are not in equilibrium (thank God!).

3.2 The Standard Scenario: Out-of-Equilibrium Decay

The basic idea of baryogenesis has been discussed by many authors (31–38). The model that incorporates the three ingredients discussed above and that has become the "standard scenario" is the so-called out-of-equilibrium decay scenario (32, 38). We now describe the scenario in some detail, though only qualitatively.

Denote by "X" a superheavy ($\gtrsim 10^{14}$ GeV) boson whose interactions violate B conservation. X might be a gauge or a Higgs boson (see Section 4 for further discussion of these bosons). [Scenarios in which the X particle is a superheavy fermion have been suggested in (39, 39a, 40).] Let its coupling strength to fermions be $\alpha^{1/2}$, and its mass be M. From dimensional considerations its decay rate $\Gamma_D = \tau^{-1}$ should be

$$\Gamma_D \approx \alpha M. \qquad 13.$$

At the Planck time ($\sim 10^{-43}$ s) we will assume that the Universe is baryon symmetric ($n_B/s = 0$), with all fundamental particle species (fermions, gauge and Higgs bosons) present with equilibrium distributions. (Nonequilibrium initial conditions are discussed in Section 8.6.) At this epoch $T \sim g_*^{-1/4} m_{pl} \sim 3 \times 10^{18}$ GeV $\gg M$. (Here we have taken $g_* \sim O(100)$; in minimal SU(5) $g_* \approx 160$.) So X, $\bar{X}$ bosons are very relativistic and up to statistical factors as abundant as photons: $n_X = n_{\bar{X}} \approx n_\gamma$. Nothing of importance occurs until $T \approx M$.

For $T < M$ the equilibrium abundance of X, $\bar{X}$ bosons relative to photons is

$$X_{EQ} \approx (M/T)^{3/2} \exp(-M/T), \qquad 14.$$

where $X \equiv n_X/n_\gamma$ is just the number of X, $\bar{X}$ bosons per co-moving volume. In order for X, $\bar{X}$ bosons to maintain an equilibrium abundance as T falls below M, they must be able to diminish in number rapidly compared to $H = |\dot{T}/T|$. The most important process in this regard is decay; other processes (e.g. annihilation) are higher order in α. If $\Gamma_D \gg H$ for $T = M$, then X, $\bar{X}$ bosons can adjust their abundance (by decay) rapidly enough so that X "tracks" the equilibrium value. In this case thermal equilibrium is maintained and no asymmetry is expected to evolve.

More interesting is the case where $\Gamma_D < H \approx 1.66\, g_*^{1/2} T^2/m_{pl}$ when $T = M$, or equivalently $M > g_*^{-1/2} \alpha\, 10^{19}$ GeV. In this case X, $\bar{X}$ bosons are *not* decaying on the expansion timescale ($\tau > t$) and so remain as abundant as photons ($X = 1$) for $T \lesssim M$; hence they are overabundant relative to their equilibrium number. This overabundance is the departure from thermal equilibrium. Much later, when $T \ll M, \Gamma_D \approx H$ (i.e. $t \approx \tau$), and X, $\bar{X}$ bosons begin to decrease in number as a result of decays. To a good approximation they decay freely since the fraction of fermion pairs with sufficient center-of-mass (cm) energy to produce an X or $\bar{X}$ is $\sim \exp(-M/T) \ll 1$, greatly suppressing inverse decay processes ($\Gamma_{ID} \approx \exp(-M/T)\Gamma_D \ll H$). Figure 1 summarizes the time evolution of X; Figure 2 shows the relationship of the various rates (Γ_D, Γ_{ID}, and H) as a function of M/T ($\sim t^{1/2}$).

Now consider the decay of X and $\bar{X}$ bosons: suppose X decays to channels 1 and 2 with baryon numbers B_1 and B_2, and branching ratios r and $(1-r)$. Denote the corresponding quantities for $\bar{X}$ by $-B_1$, $-B_2$, $\bar{r}$, and $(1-\bar{r})$ [e.g. $1 = (\bar{q}\bar{q})$, $2 = (q\ell)$, $B_1 = -2/3$, and $B_2 = 1/3$]. The mean net baryon number of the decay products of the X and $\bar{X}$ are, respectively, $B_X = rB_1 + (1-r)B_2$ and $B_{\bar{X}} = -\bar{r}B_1 - (1-\bar{r})B_2$. Hence the decay of an X, $\bar{X}$ pair on average produces a baryon number ε,

$$\varepsilon \equiv B_X + B_{\bar{X}} = (r-\bar{r})(B_1 - B_2). \qquad 15.$$

If $B_1 = B_2$, $\varepsilon = 0$. In this case X could have been assigned a baryon number B_1, and B would not be violated by X, $\bar{X}$ bosons.

It is simple to show that $r = \bar{r}$ unless both C and CP are violated. Let $\bar{X}$ = the charge conjugate of X, and $r_\uparrow$, $r_\downarrow$, $\bar{r}_\uparrow$, $\bar{r}_\downarrow$ denote the respective branching ratios in the upward and downward directions (for simplicity we have reduced the angular degree of freedom to up and down). The quantities r and $\bar{r}$ are branching ratios averaged over angle: $r = (r_\uparrow + r_\downarrow)/2$, $\bar{r} = (\bar{r}_\uparrow + \bar{r}_\downarrow)/2$ and $\varepsilon = (r_\uparrow - \bar{r}_\uparrow + r_\downarrow - \bar{r}_\downarrow)/2$. If C is conserved, $r_\uparrow = \bar{r}_\uparrow$ and $r_\downarrow = \bar{r}_\downarrow$, and $\varepsilon = 0$. If CP is conserved $r_\uparrow = \bar{r}_\downarrow$ and $r_\downarrow = \bar{r}_\uparrow$, and once again $\varepsilon = 0$.

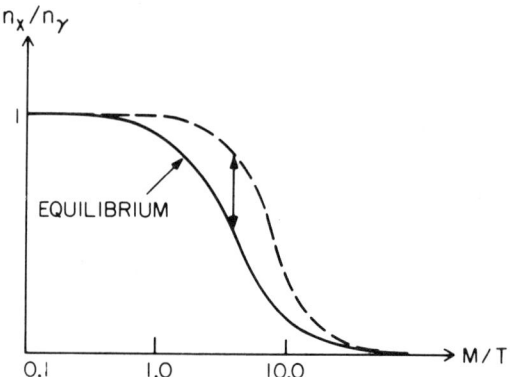

Figure 1 The abundance of X bosons relative to photons as a function of $z \equiv M/T$. The broken curve shows the actual abundance, while the solid curve shows the equilibrium abundance. The arrow indicates the departure from equilibrium.

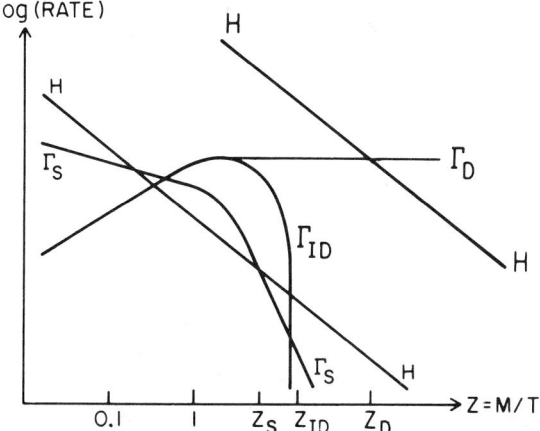

Figure 2 The log of important rates as a function of $z \equiv M/T$. H is the expansion rate, Γ_D the decay rate, Γ_{ID} the inverse decay rate, and Γ_S the $2 \leftrightarrow 2$ scattering rate. The upper line marked H corresponds to the case where $K \ll 1$. For $z \approx 1$ all reaction rates are ineffective ($\Gamma < H$). The X bosons decay when $z \approx z_D$ ($\Gamma_D = H$). The lower line marked H corresponds to the case where $K_c > K > 1$. For $z \approx 1$ all reaction rates are effective ($\Gamma > H$); inverse decays (ID) and $2 \leftrightarrow 2$ scatterings (S) freeze out ($\Gamma = H$) at $z \approx z_{ID}$ and z_S, respectively.

When the X, X̄ bosons decay ($T \ll M$, $t \approx \tau$) $n_X = n_{\bar{X}} \approx n_\gamma$. Therefore, the net baryon number density produced is $n_B \approx \varepsilon n_\gamma$. The entropy density $s \approx g_* n_\gamma$, and so the baryon asymmetry produced is $n_B/s \approx \varepsilon/g_* \approx 10^{-2}\,\varepsilon$.

Recall that the condition for a departure from equilibrium to occur is $K \equiv (\Gamma_D/H)|_{T=M} < 1$ or $M > g_*^{-1/2} \alpha m_{pl}$. If X is a gauge boson, then $\alpha \approx 1/45$, and so M must be $\gtrsim 10^{16}$ GeV. If X is a Higgs boson, then α is essentially arbitrary, although $\alpha \approx (m_f/M_W)^2\, \alpha_{gauge} \approx 10^{-3}\text{–}10^{-6}$ if the X is in the same representation as the light Higgs bosons responsible for giving mass to the fermions (here m_f = fermion mass, M_W = mass of the W boson ≈ 80 GeV). It is apparently easier for Higgs bosons to satisfy this mass condition than it is for gauge bosons. If $M > g_*^{-1/2} \alpha m_{pl}$, then only a modest C, CP-violation ($\varepsilon \approx 10^{-8}$) is necessary to explain $n_B/s \approx (6\text{–}10) \times 10^{-11}$. As we discuss in Section 7, ε is expected to be larger for a Higgs boson than for a gauge boson. For both these reasons a Higgs boson is a more likely candidate for producing the baryon asymmetry.

In Section 6 we discuss this scenario more quantitatively. For $K \lesssim 1$ this simple qualitative picture is borne out. For $K \gtrsim 1$ there is a surprising result; the asymmetry produced does not fall off rapidly with increasing K until $K \gg 1$ (30, 41–44). It is not that the nonequilibrium condition is not necessary; rather, it is more difficult to maintain thermal equilibrium in the expanding Universe than a comparison of reaction rates with the expansion rate would suggest.

4. BARYON NUMBER VIOLATION IN GRAND UNIFIED MODELS

One of the three crucial ingredients for generating a baryon asymmetry is baryon number violation. At present there is no experimental evidence to suggest that baryon number is not conserved. While the observation of proton (or neutron) decay seems the most likely possibility for direct confirmation of baryon number violation, one might argue that since baryon number violation is necessary to produce the baryons in the Universe, the very existence of physicists searching for proton decay is already *indirect* evidence for baryon number nonconservation.

The fact that the proton lifetime is greater than 10^{30} years (18–22), many orders of magnitude longer than the lifetime of any known unstable particle, once seemed to suggest that baryon number is exactly conserved. However, theorists are now willing, indeed eager, to abandon the once-cherished belief in baryon number conservation. Part of the motivation comes from the observation that baryon number conservation corresponds to a *global* (i.e. ungauged) symmetry (45). In this modern era of gauge theories, global symmetries are regarded as unwanted and ugly. Even more damaging to the idea of baryon number conservation is the remarkable discovery of 't Hooft (46) that even in the standard electroweak interactions, which were thought exactly to conserve baryon number, there are nonperturbative effects that violate conservation of baryon number. Although the nonperturbative effects are small, corresponding to a proton lifetime well in excess of 10^{100} years, their existence leads one to expect that baryon number conservation is not sacred, and may be violated in the perturbative sector as well.

Grand unified theories are theories that unify the strong and electroweak interactions. In these models quarks, antiquarks, and leptons (or antileptons) are usually members of the same irreducible representations of the gauge group. Thus there exist gauge bosons that mediate transitions between quarks and antiquarks or leptons, thereby violating baryon number conservation. In theories where there is a "desert" (i.e. no new physics) between the weak scale and the unification scale, the unification scale is predicted to be about 10^{14} GeV. This scale sets the mass of the baryon-number-violating bosons. Such an enormous mass for the particles responsible for mediating baryon number violation is necessary to account for the fact that the proton lifetime is greater than 10^{30} years. If boson exchange is responsible for proton decay, we may estimate the proton lifetime as

$$\tau_p \approx M^4 g^{-4} m_p^{-5} \approx 10^{30} \text{ yr} \left(\frac{M}{10^{15} \text{ GeV}} \frac{1}{g}\right)^4, \qquad 16.$$

where g is the coupling strength of the boson to fermions, M is its mass, and m_p is the proton mass. If the baryon-number-violating boson is a gauge particle and g a typical gauge coupling, then M must be greater than about 10^{14} GeV. If instead it is a Higgs boson, then g is a typical Yukawa coupling to light fermions, and M could be as light as 10^{10} GeV. In the simplest GUT, SU(5), the unification scale is predicted to be few $\times$ 10^{14} GeV, while in more complicated models the unification scale can be 10^{15}–10^{17} GeV. Since the proton lifetime scales as the *fourth* power of the unification scale, it will be in excess of 10^{33} years, if the scale is greater than about 10^{16} GeV. The background from cosmic-ray neutrino interactions seems to result in a practical experimental upper limit of 10^{33} years for terrestrial proton decay experiments (25), and so the early Universe may well turn out to be our only laboratory for studying baryon number nonconservation.

Remarkably enough there are only five different kinds of bosons that can mediate B-nonconserving interactions between the known fermions (47). Classified according to their transformation properties under $SU(3)_C \otimes SU(2)_L \otimes U(1)_Y$ they are:

Vector bosons: (3, 2, 5/6), (3, 2, −1/6)
Scalar bosons: (3, 1, 1/3), (3, 1, 4/3), (3, 3, 1/3),

where (i, j, k) indicates their $SU(3)_C$ and $SU(2)_L$ multiplicities and their hypercharge Y (normalized to the electric charge, Q, and weak isospin, I_3, by $Q = I_3 - Y$) respectively. The (3, 2, 5/6) vector and (3, 1, 1/3) scalar bosons are present in the minimal SU(5) model (see Section 7.1). These bosons violate baryon number conservation in the following sense. Assign a baryon number of $1/3$ ($-1/3$) to all quarks (antiquarks) and 0 to all leptons. The baryon number of all other particles is determined by their interactions with quarks and leptons. It is not possible to assign a baryon number to the bosons discussed above and have B conserved in all reactions. For instance, consider the decays of the (3, 2, 5/6) vector bosons. The charge $-4/3$ species can decay (in a real or virtual process) to two channels: (a) $\bar{u}_L \bar{u}_R$ ($B = -2/3$) and (b) $d_L e_R$ ($B = 1/3$); thus it is not possible consistently to assign it a definite baryon number. Since this boson couples to two states differing by one unit of baryon number, it can mediate processes with $\Delta B = \pm 1$.

The cross section for baryon-number-violating scattering of light fermions ($m_f \ll M$) mediated by the s-channel exchange of a boson of mass M is approximately

$$\sigma \propto \frac{g^4 s}{(s+M^2)^2} \approx \frac{g^4 s}{M^4} \qquad (s \ll M^2), \qquad 17.$$

where s is the square of the center-of-mass energy. The longevity of the proton implies that $M/g \gtrsim 10^{15}$ GeV, and so at presently available energies

($s \lesssim 10^6$ GeV^2) the cross section for such reactions is exceedingly tiny, $\sigma \lesssim 10^{-80}$ cm^2!

However, the situation is very different in the early Universe. If the universe was ever at temperatures greater than M, cross sections for baryon-number-violating reactions are not suppressed relative to other interactions (e.g. electroweak or strong), as they are at low energies. At temperatures T greater than the mass of the light fermions, $\langle s \rangle \approx 10T^2$. If $\langle s \rangle \gtrsim M^2$, then the cross section for baryon-number-violating reactions (and all other interactions) is

$$\sigma \propto \frac{g^4 s}{(s+M^2)^2} \approx \frac{g^4}{s} \quad (s \gg M^2). \qquad 18.$$

Temperatures $T \gtrsim M$ correspond to times since the bang earlier than (cf Equation 8b)

$$t \approx (45/16\pi^3)^{1/2} g_*^{-1/2} m_{\rm pl}/M^2 \approx 10^{-35} \text{ s}. \qquad 19.$$

Therefore in the *very* early Universe, baryon-number-violating reactions should have been important, and, if the idea of baryogenesis is correct, they were necessary to insure the existence of matter in the Universe today.

5. CP VIOLATION IN GRAND UNIFIED MODELS

5.1 *Origin and Classification of* CP *Violation*

The generation of a baryon asymmetry requires interactions that not only violate B conservation, but also violate both C and CP conservation. In the standard scenario this occurs in the decay of a superheavy boson. C is violated in ordinary weak interactions, and is also violated in grand unified models based on the gauge group SU(5). In larger gauge groups, such as SO(10), C may be conserved, and as discussed in Section 7.2, C violation is an important consideration when calculating baryon number generation in SO(10) models. The occurrence of CP violation in most unified models is not required, but can be accommodated.

CP violation has only been observed in the K^0-$\bar{K}^0$ system. For CP-conserving interactions, it is well known that the coupling constants in the Lagrangian are relatively real. CP nonconservation is manifest by coupling constants in the Lagrangian that are relatively complex. Currently, there are two popular explanations for the origin of complex couplings: (*a*) the complexity of the fermion mass matrix gives rise to complex fermion-W couplings when the mass matrix is diagonalized [the so-called Kobayashi-Maskawa model (48)]; (*b*) CP violation is due to the exchange of Higgs bosons whose couplings to fermions are complex (49). At present, it is not possible to determine which of these mechanisms (if either) is responsible for CP violation in the neutral kaon system.

In general, CP violation may be intrinsic (the result of explicit complex couplings) or spontaneous (the result of a Higgs field obtaining a complex vacuum expectation value). In addition, CP violation may also be classified as being hard or soft. "Hard" CP violation results from terms of dimension 4 in the Lagrangian, while "soft" CP violation results from terms of dimension 3 or 2 in the Lagrangian. Spontaneous CP violation is necessarily soft, while intrinsic CP violation may be soft or hard.

In the standard big-bang model, CP violation must be intrinsic. If CP were spontaneous, then at high temperatures, the symmetry should be restored, and during the phase transition that breaks CP the Universe would break up into domains, each with a different sign of CP violation (i.e. $\langle \phi \rangle = \pm \phi_0$), and a characteristic size $\lesssim$ horizon distance ≈ 2 ct. [The horizon contains $\cong (t/s)^{1.5}$ $M_\odot$ of baryons.] Walls separate these domains, and the energy density in the domain walls would prevent the Universe from evolving to the Universe we observe today (50, 51). [It is possible to arrange the Higgs potential so that the symmetry is not restored at high temperatures, thus avoiding this problem (51a, b).]

In nonstandard big-bang models, CP violation can be spontaneous, and yet the problem of CP domains could be avoided (4, 52). For instance, in the inflationary Universe scenarios (53–55), a single horizon volume, and thus a "domain" of constant sign of $\langle \phi \rangle$, easily expands to a size that encompasses our entire observable Universe. In this case, any domain walls would be well outside our horizon and their effects unobservable. Another possibility of reconciling spontaneous CP violation with the observed baryon asymmetry and lack of domain walls is to assume that the Universe is "cold," i.e. degenerate in some fermion species. This degeneracy prevents high-temperature restoration of the symmetry (56) and picks out a preferred vacuum with a definite sign of CP.

At present, there is no way to relate the magnitude or even the sign of CP violation required for baryogenesis to any experimentally observed parameter (57). A prevalent (and we think not unreasonable) attitude is to believe that the K^0-$\bar{K}^0$ system is not the only place where Nature has chosen to violate CP, and to suspect that it is also likely to be violated at very high energies, e.g. in the decays of superheavy bosons. Unfortunately, at present the CP violation in the superheavy boson system must be taken as a parameter, usually related to the imaginary part of some coupling constant.

It is interesting to note that there has been one attempt to relate the magnitude of the CP violation responsible for the baryon asymmetry to another low-energy parameter, the electric dipole moment of the neutron (58). Although the proposal is attractive since it predicts the electric dipole moment to be just below the present experimental upper limit, the analysis relies on a number of nontrivial assumptions. Nevertheless, it gives one hope that eventually, as *CP* violation becomes better understood, it may be possible to relate the *CP* violation responsible for the baryon number of our Universe to experimentally determined parameters.

It might be argued that baryogenesis is an exercise in trading one unexplained parameter, the baryon number, for another unknown parameter, the amount of *CP* violation. However, *CP* violation is observed in Nature, and there are reasonable physical models to account for its occurrence. This should be contrasted with the lack of models (other than baryogenesis) capable of explaining the baryon number of our Universe.

5.2   *A Simple Model Calculation*

In this section we use a simple model (59) to show how the complexity of coupling constants can result in the *CP* violation required to produce a baryon asymmetry. The mean net baryon number produced by the decay of an X boson and its antiparticle $\bar{X}$ is

$$\varepsilon_X = \sum_f B_f \frac{\Gamma(X \to f) - \Gamma(\bar{X} \to \bar{f})}{\Gamma_X}, \qquad 20.$$

where B_f is the baryon number of the final state f, $\Gamma(X \to f)$ is the partial width for the indicated decay mode, and Γ_X is the total X decay width.

Let us assume a Lagrangian of the form

$$\mathscr{L} = g_1 X i_2^\dagger i_1 + g_2 X i_4^\dagger i_3 + g_3 Y i_1^\dagger i_3 + g_4 Y i_2^\dagger i_4 + \text{h.c.}, \qquad 21.$$

where X and Y are two massive boson fields, and $i_{1 \to 4}$ are light particle fields. This interaction leads to the following decay modes for X and Y:

$$X \to \bar{i}_1 i_2 \text{ and } \bar{i}_3 i_4, \quad Y \to \bar{i}_3 i_1 \text{ and } \bar{i}_4 i_2.$$

In the Born approximation, the X and Y decay widths are given by the square of the tree-diagrams shown in Figure 3, e.g.

$$\Gamma(X \to \bar{i}_1 i_2) = I_X |g_1|^2,$$
$$\Gamma(\bar{X} \to i_1 \bar{i}_2) = I_{\bar{X}} |g_1^*|^2 \qquad 22.$$

where I_X accounts for the kinematic factors (phase space integrals, etc). Since the kinematic factors are the same for particles and antiparticles, e.g. $m_X = m_{\bar{X}}$ by *CPT*, it follows that $I_X = I_{\bar{X}}$. Clearly, in the Born approximation ε_X vanishes.

At next order in perturbation theory there are corrections in X decay due to Y exchange, shown in Figures 4a and 4b. The rate for $X \to i_2 \bar{i}_1$ receives contributions from the square of Figure 3a and the interference between Figures 3a and 4a, and can be written as

$$\Gamma(X \to \bar{i}_1 i_2) = I_X |g_1|^2 + I_{XY} g_1 g_2^* g_3 g_4^* + (I_{XY} g_1 g_2^* g_3 g_4^*)^*, \qquad 23.$$

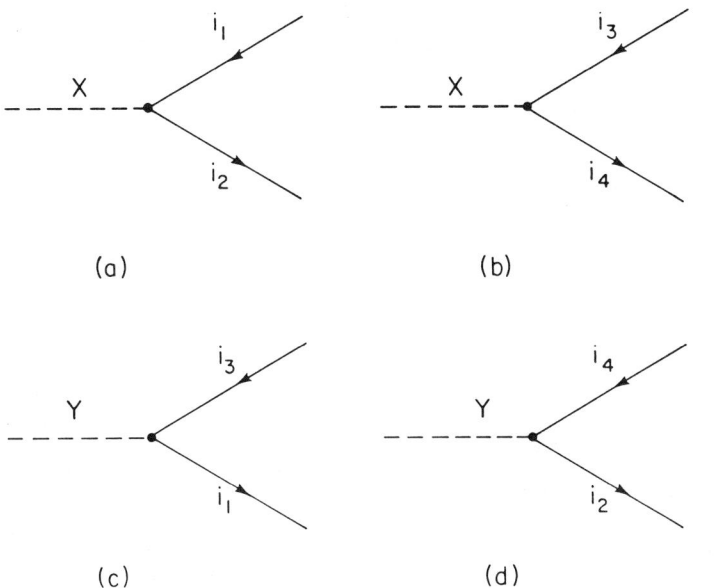

Figure 3 The diagrams for the decays $X \to \bar{i}_1 + i_2$ and $\bar{i}_3 + i_4$, and $Y \to i_1 + \bar{i}_3$ and $i_2 + \bar{i}_4$.

where I_{XY} includes the kinematic factors from the loop diagram. Although I_X is real, I_{XY} will be complex if the intermediate particles in the loop (the $\bar{i}_3 i_4$ intermediate state) are kinematically allowed to propagate on mass shell. This requires $m_X \geq m_3 + m_4$.

In a similar way the width for $\bar{X} \to i_1 \bar{i}_2$ can be written as

$$\Gamma(\bar{X} \to i_1 \bar{i}_2) = I_X |g_1^*|^2 + I_{XY} g_1^* g_2 g_3^* g_4 + (I_{XY} g_1^* g_2 g_3^* g_4)^*. \qquad 24.$$

We are interested in the difference between $X \to \bar{i}_1 i_2$ and $\bar{X} \to i_1 \bar{i}_2$, which is given by

$$\Gamma(X \to \bar{i}_1 i_2) - \Gamma(\bar{X} \to i_1 \bar{i}_2)$$
$$= 2I_{XY} \operatorname{Im}(g_1 g_2^* g_3 g_4^*) + 2I_{XY}^* \operatorname{Im}(g_1^* g_2 g_3^* g_4)$$
$$= 4 \operatorname{Im} I_{XY} \operatorname{Im}(g_1 g_2^* g_3 g_4^*). \qquad 25.$$

A similar calculation for the other X decay mode shows that

$$\Gamma(X \to \bar{i}_3 i_4) - \Gamma(\bar{X} \to i_3 \bar{i}_4) = -[\Gamma(X \to \bar{i}_1 i_2) - \Gamma(\bar{X} \to i_1 \bar{i}_2)].$$

Therefore it follows that

$$\varepsilon_X = \frac{4}{\Gamma_X} \operatorname{Im} I_{XY} \operatorname{Im}(g_1 g_2^* g_3 g_4^*) [(B_{i_4} - B_{i_3}) - (B_{i_2} - B_{i_1})]. \qquad 26.$$

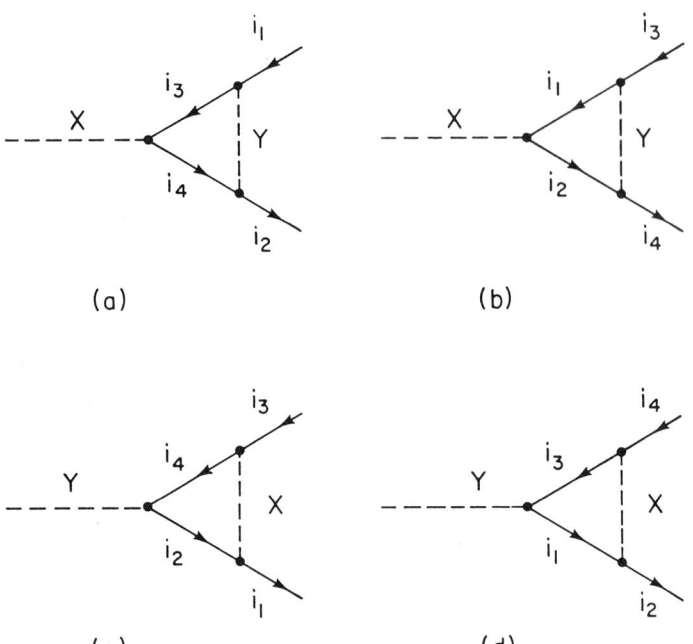

Figure 4 One-loop corrections to X and Y decay. The one-loop diagram for X (Y) decay with an X (Y) boson in the loop is not shown.

(At the same order there is also a contribution from the interference of the tree diagram with a one-loop diagram where an X is exchanged; however, it follows trivially that the product of the coupling constants involved is real.)

Equation 26 illustrates several results that are more general than our simple model. (*a*) Baryon number must be violated in X decay since ε_X is proportional to the difference between the baryon numbers of the two final states of X decay. (*b*) The particle exchanged in the loop (in this case, Y) must also violate baryon number (30, 47). (*c*) The X, Y bosons must be more massive than $m_3 + m_4$ and $m_1 + m_2$ in order for I_{XY} to have an imaginary part (47, 60). (*d*) Some of the coupling constants in the Lagrangian must be complex. It is also interesting to consider the sum of ε_X and ε_Y:

$$\varepsilon_X + \varepsilon_Y = 4 \left\{ \frac{\text{Im } I_{XY}}{\Gamma_X} - \frac{\text{Im } I_{YX}}{\Gamma_Y} \right\} \text{Im}(g_1 g_2^* g_3 g_4^*)$$
$$\times [(B_{i_4} - B_{i_3}) - (B_{i_2} - B_{i_1})]. \qquad 27.$$

In the limit that $m_X = m_Y$, $\varepsilon_X + \varepsilon_Y$ vanishes, since $\Gamma_X = \Gamma_Y$ and Im I_{XY} = Im I_{YX} when the masses are degenerate, and thus the asymmetry produced by the X will exactly cancel that produced by the Y.

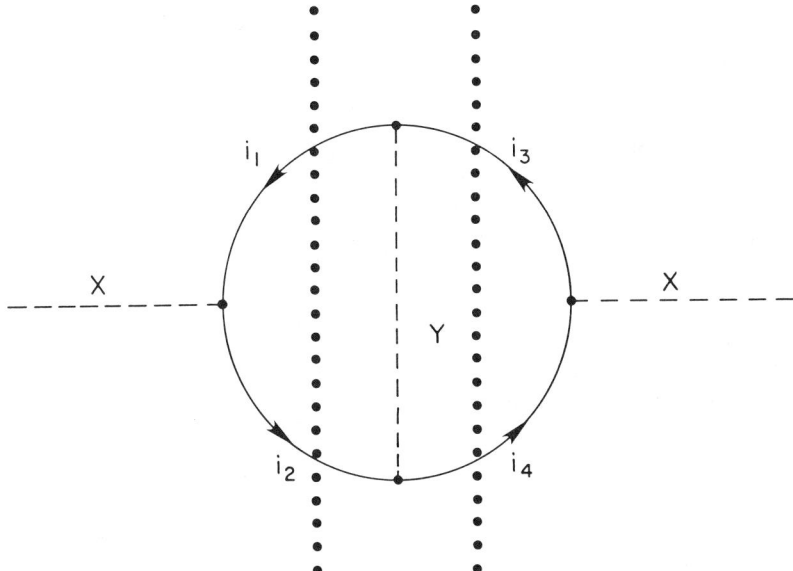

Figure 5 The double-cut diagram for X decay with one-loop corrections from Y exchange.

Finally, we note that evaluating the double-cut diagram of Figure 5 is equivalent to the method described above in calculating ε_X. For a more detailed discussion of CP violation in the context of baryogenesis we refer the interested reader to (30, 47, 59, 60).

6. BOLTZMANN EQUATIONS FOR BARYOGENESIS

For a given GUT a network of Boltzmann equations can be set up and numerically integrated to find the evolution of the baryon asymmetry. Such codes have been written for the SU(5) (42–44, 59) and SO(10) (59) models. [For earlier attempts, see (60a, 60b).] Essentially all of the results can be understood in terms of a highly simplified model, which we develop in this section. In Section 7.3 we briefly discuss how to set up the network of Boltzmann equations for a realistic unified model.

6.1 A Simple Model for Baryogenesis

In the simple model (30, 41) we have a massive boson X ($\equiv \bar{X}$), whose interactions violate B, C, and CP, and a massless species b ($\bar{b}$) with baryon number $+\frac{1}{2}$ ($-\frac{1}{2}$). We assume that both species obey Maxwell-Boltzmann statistics. This should be a reasonable approximation in the absence of a Bose condensate or degenerate fermions.

Baryon number is violated in this model by decays (D): $X \to bb$, $\bar{b}\bar{b}$; inverse decays (ID) bb, $\bar{b}\bar{b} \to X$; and $2 \leftrightarrow 2$ scattering processes (S) mediated by X: $bb \leftrightarrow \bar{b}\bar{b}$. Higher order processes need not be considered because they are suppressed by additional powers of a coupling constant. In this model we do not explicitly consider B-conserving reactions since they do not directly affect the evolution of the baryon asymmetry. We do assume, however, that these reactions are occurring rapidly enough ($\Gamma > H$) to maintain kinetic equilibrium for X, b, and $\bar{b}$. These thermalizing reactions include the interaction of X, b, and $\bar{b}$ with all the other species whose evolution we do not trace, e.g. $\gamma + b \to \gamma + b$, $X + \gamma \to X + \gamma$, etc. Ellis & Steigman (61) have shown that for $T \lesssim 10^{16}$ GeV the rate of these thermalizing interactions should be greater than H, justifying our assumption. Kinetic equilibrium allows the phase space density of b, $\bar{b}$, and X to be written as

$$f_b(p) = \exp[-(E-\mu)/T],$$
$$f_{\bar{b}}(p) = \exp[-(E+\mu)/T], \qquad 28.$$
$$f_X(p) = \exp[-(E+\mu_X)/T],$$

where $E^2 = p^2 + m^2$, and the chemical potentials of b and $\bar{b}$ are equal and opposite owing to reactions like $b + \bar{b} \leftrightarrow \gamma + \gamma$. The number density of species i (i = b, $\bar{b}$, X) is, as usual $n_i = \int f_i \, d^3 p_i / (2\pi)^3$.

We will take the absolute squares of the decay and inverse decay amplitudes for the X boson to be

$$|M(X \to bb)|^2 = |M(\bar{b}\bar{b} \to X)|^2 = \tfrac{1}{2}(1+\varepsilon) M_0^2,$$
$$|M(X \to \bar{b}\bar{b})|^2 = |M(bb \to X)|^2 = \tfrac{1}{2}(1-\varepsilon) M_0^2. \qquad 29.$$

Note that CP is violated since $|M(X \to bb)|^2 \neq |M(X \to \bar{b}\bar{b})|^2$, while CPT is conserved, since $|M(X \to bb)|^2 = |M(\bar{b}\bar{b} \to X)|^2$. The CP violation is proportional to the parameter ε. As before, ε is the mean net baryon number produced by the decay of an X boson.

The X number density evolves according to

$$\dot{n}_X + 3(\dot{R}/R) n_X = \Lambda_{12}^X [-f_X(p_X) + \tfrac{1}{2}(1+\varepsilon) f_{\bar{b}}(p_1) f_{\bar{b}}(p_2)$$
$$+ \tfrac{1}{2}(1-\varepsilon) f_b(p_1) f_b(p_2)] M_0^2$$
$$= \Lambda_{12}^X [-f_X(p_X) + f_X^{eq}(p_X)] M_0^2 + O(\varepsilon) + O(\mu/T), \qquad 30.$$

where the $\dot{R}/R$ term accounts for the dilution of Xs due to the expansion, $f_X^{eq}(p_X) = \exp[-(p_X^2 + m_X^2)^{1/2}/T]$ is the equilibrium phase space density of

Xs, and Λ is the phase space integral operator defined by

$$\Lambda^{i,j,k,\ldots}_{l,m,n,\ldots} = \int \frac{d^3p_i}{(2\pi)^3 2E_i} \int \frac{d^3p_j}{(2\pi)^3 2E_j} \cdots \int \frac{d^3p_l}{(2\pi)^3 2E_l} \int \frac{d^3p_m}{(2\pi)^3 2E_m} \cdots$$
$$\times (2\pi)^4 \delta^4(p_i + p_j + p_k + \cdots - p_l - p_m - p_n \cdots). \qquad 31.$$

The second equality in Equation 30 follows by using the

$\delta(E_X - E_1 - E_2)$ in Λ^X_{12}.

The phase space integral over $f_X(p_X)M_0^2$ gives the thermally averaged X decay width (Γ_D) times the number density of X bosons, so that Equation 30 can be written as,

$$\dot{n}_X + 3(\dot{R}/R)n_X = -\Gamma_D(n_X - n_X^{eq}). \qquad 32.$$

In a similar manner, one obtains the equation governing the number density of bs and $\bar{b}$s,

$$\dot{n}_b + 3(\dot{R}/R)n_b = \Lambda^X_{12}[-(1-\varepsilon)f_b(p_1)f_b(p_2) + (1+\varepsilon)f_X(p_X)]M_0^2$$
$$+ 2\Lambda^{34}_{12}[-f_b(p_1)f_b(p_2)|M'(bb \to \bar{b}\bar{b})|^2$$
$$+ f_{\bar{b}}(p_3)f_{\bar{b}}(p_4)|M'(\bar{b}\bar{b} \to bb)|^2], \qquad 33a.$$

$$\dot{n}_{\bar{b}} + 3(\dot{R}/R)n_{\bar{b}} = \Lambda^X_{12}[-(1+\varepsilon)f_{\bar{b}}(p_1)f_{\bar{b}}(p_2) + (1-\varepsilon)f_X(p_X)]M_0^2$$
$$+ 2\Lambda^{34}_{12}[-f_{\bar{b}}(p_1)f_{\bar{b}}(p_2)|M'(\bar{b}\bar{b} \to bb)|^2$$
$$+ f_b(p_3)f_b(p_4)|M'(bb \to \bar{b}\bar{b})|^2], \qquad 33b.$$

where $|M'(bb \to \bar{b}\bar{b})|^2$ and $|M'(\bar{b}\bar{b} \to bb)|^2$ are the squares of the matrix elements for $2 \leftrightarrow 2$ scattering, with the part due to real intermediate-state Xs removed (this part has already been taken into account in the D and ID terms). The first set of terms in each equation represents D and ID processes, while the second set represents $2 \leftrightarrow 2$ scattering processes. Subtracting these two equations we obtain the equation for the evolution of the baryon asymmetry, $n_B = \frac{1}{2}(n_b - n_{\bar{b}})$,

$$\dot{n}_B + 3(\dot{R}/R)n_B = \varepsilon\Gamma_D(n_X - n_X^{eq}) - n_B[n_X^{eq}\Gamma_D + 2n(\sigma v)_S] + O(\varepsilon^2 + n_B^2 + \varepsilon n_B),$$
$$34.$$

where $n = \int \exp(-p/T)\, d^3p/(2\pi)^3 = T^3/\pi^2$ is the equilibrium number density of a light species, and

$$(\sigma v)_S = \Lambda^{34}_{12} f_b(p_1)f_b(p_2)|M'(bb \to \bar{b}\bar{b})|^2/n^2 \qquad 35.$$

is the thermally averaged $2 \leftrightarrow 2$ scattering cross section (with the part due to real intermediate-state Xs removed). Unitarity and CPT have been used to relate the CP-violating parts of the squared matrix elements for D, ID, and $2 \leftrightarrow 2$ scattering; the $|M'(bb \to \bar{b}\bar{b})|^2 - |M'(\bar{b}\bar{b} \to bb)|^2$ terms account for the crucial $-\varepsilon \Gamma_D n_X^{eq}$ term. The three ingredients needed for baryogenesis are manifest in Equation 34. The source term for baryon number generation, $\varepsilon \Gamma_D (n_X - n_X^{eq})$, is only nonzero if B, C, and CP are violated, and n_X is not equal to n_X^{eq}.

Assuming that the expansion is isentropic, sR^3 remains constant, where the entropy density s is defined in Equation 7 and is $s \approx g_* n$. Using this fact, and transforming to the dimensionless variable $z \equiv M/T (z \sim t^{1/2})$, where M is the mass of the superheavy X boson, we can write Equations 32 and 34 as

$$\Delta' = -X'_{EQ} - zK\gamma_D \Delta,$$
$$B' = \varepsilon zK\gamma_D \Delta - zK\gamma_B B,$$ 36.

where prime indicates d/dz and

$$B \equiv n_B/s \approx n_B/(g_* n)$$

$$X \equiv n_X/s \approx n_X/(g_* n)$$

$$X_{EQ} \equiv n_X^{eq}/s \approx g_*^{-1} \begin{cases} 1, & z \ll 1 \\ (\pi/2)^{1/2} z^{3/2} e^{-z}, & z \gg 1 \end{cases}$$

$$\Delta = X - X_{EQ}$$

$$K \equiv \Gamma_D(z=1)/2H(M) \approx \alpha m_{pl}/3 g_*^{1/2} M$$

$$\gamma_D = \Gamma_D(z)/\Gamma_D(z=1) \approx \begin{cases} z/2, & z \ll 1 \\ 1, & z \gg 1 \end{cases}$$

$$\gamma_B = \Gamma_B(z)/\Gamma_D(z=1) = [g_* X_{EQ} \gamma_D + 2(\sigma v)_s n]/\Gamma_D(z=1)$$
$$\cong \begin{cases} z/2 + A\alpha z^{-1}, & z \ll 1 \\ z^{3/2} e^{-z} + A\alpha z^{-5}, & z \gg 1 \end{cases}.$$ 37.

Note, $K \equiv (\Gamma_D/2H)|_{z=1}$ as before measures the "effectiveness" of decays, i.e. the rate relative to expansion rate when $T = M$. The decay rate $\Gamma_D(z=1)$ has been used to normalize all reaction rates so that γ_D and γ_B are both dimensionless and are just equal to the decay rate and rate of B-nonconserving reactions relative to the decay rate at $T \approx M$.

The physics of these equations is quite simple. The departure from equilibrium Δ drives the production of a baryon asymmetry. Inverse decays and $2 \leftrightarrow 2$ scatterings, embodied in γ_B, tend to damp any baryon asymmetry. The changing value of X_{EQ} drives Δ; in the *absence* of the expansion $T' = X'_{EQ} = 0$, and any departure from equilibrium would decay away exponentially $\propto \exp(-\frac{1}{2}Kz^2)$. However, because of the expansion, this does not occur.

At high temperatures ($z \ll 1$) the equilibrium abundance of Xs is just that of the photons, and so $X_{EQ} \approx g_*^{-1}$. At low temperatures the equilibrium abundance of Xs is a Boltzmann factor less than that of the photons. Now consider the thermally averaged decay rate Γ_D. At low temperatures ($z \gtrsim 1$), it is just the decay rate for an X at rest, which is $O(\alpha M)$. At high temperatures ($z \ll 1$) Xs are relativistic, and so their decays are suppressed by a time dilatation factor, and $\Gamma_D \cong O(\alpha M z)$.

The quantity $\gamma_B \cong \Gamma_B/\alpha M$, which is the rate of B-nonconserving reactions, contains two terms. The first term is due to IDs. At high temperatures ($z \ll 1$) the ID process occurs at essentially the same rate as the decay process since $X_{EQ}(z \ll 1) = g_*^{-1}$. At low temperatures ($z \gtrsim 1$) the inverse process is suppressed since $X_{EQ} \propto z^{3/2} e^{-z}$; physically this occurs because only a small fraction of the bb or $\bar{b}\bar{b}$ pairs are energetic enough to produce an X boson. The second term in γ_B arises from $2 \leftrightarrow 2$ scatterings. At high temperatures ($z \ll 1$) $(\sigma v)_S \approx A\alpha^2/T^2$, while at low temperatures ($z \gg 1$) it is suppressed due to the exchange of the superheavy X boson, and $(\sigma v)_S \approx A\alpha^2 T^2/M^4$. Here A is a dimensionless constant proportional to the number of scattering channels and other dimensionless factors that arise in computing $(\sigma v)_S$. Of course, the α^2 reflects the fact that the $2 \leftrightarrow 2$ scatterings are a second-order process. To mimic SU(5) one needs to take $A \approx O(10^3)$. Since $n \approx T^3$, the $2 \leftrightarrow 2$ scattering term in γ_B is at high temperatures ($z \ll 1$) $\approx A\alpha z^{-1}$ and at low temperatures ($z \gg 1$), $\approx A\alpha z^{-5}$.

6.2 Analytic Solutions

It is straightforward to integrate (62) Equations 36 to obtain $\Delta(z)$ and $B(z)$,

$$\Delta(z) = \Delta_0 \exp\left[-\int_0^z z' K\gamma_D(z')\,dz'\right]$$
$$-\int_0^z X'_{EQ}(z') \exp\left[-\int_{z'}^z z'' K\gamma_D(z'')\,dz''\right] dz' \qquad 38a.$$

$$B(z) = B_0 \exp\left[-\int_0^z z' K\gamma_B(z')\,dz'\right]$$
$$+\varepsilon K \int_0^z z'\Delta(z')\gamma_D(z') \exp\left[-\int_{z'}^z z'' K\gamma_B(z'')\right] dz', \qquad 38b.$$

where Δ_0 and B_0 are the departure from equilibrium and baryon asymmetry respectively at $z = 0$.

By using Equations 37 for γ_D, γ_B, and X_{EQ}, the solution for $B(z)$ can be obtained in two limiting regimes: $K \ll 1$ and $K \gg 1$. First consider $K \ll 1$, i.e. $M \gg \alpha m_{pl}/3g_*^{1/2}$; in this case the integrating factors are approximately equal to 1, since the arguments of the exponentials are $\propto K^{-1}$. In this limit we have:

$$B(z) \approx B_0 + \varepsilon[X(0) - X(z)],$$
$$X(z) \approx X(0)\exp(-\tfrac{1}{2}Kz^2).$$
39.

Physically, the Xs decay at $z \cong O(K^{-1/2})$, producing a baryon asymmetry of $n_B/s \cong \varepsilon/g_*$ in agreement with the qualitative picture described in Section 3. The evolution of $B(z)$ is shown in Figure 6.

In the regime $K \gg 1$, $\Delta(z)$ is easily found,

$$\Delta(z) \approx -X'_{EQ}/[zK\gamma_D(z)] \approx X_{EQ}/zK, \quad z \gtrsim 1,$$
40.

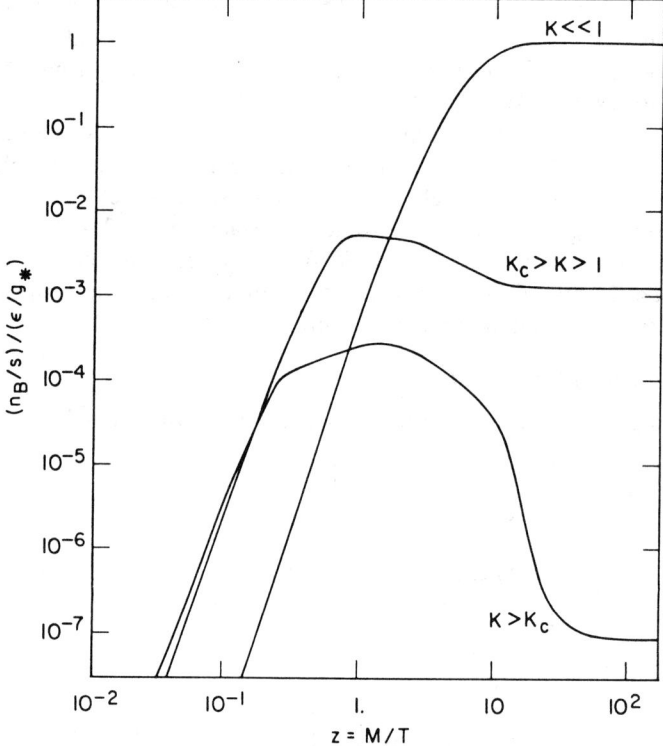

Figure 6 The evolution of n_B/s as a function of $z \equiv M/T$ ($z \sim t^{1/2}$). For $K \ll 1$ the X bosons decay out of equilibrium when $z \gg 1$. For $K_c > K > 1$, inverse decays (IDs) are effective until $z_f \approx O(10)$, n_B/s decreases slowly ($\propto z^{-1}$) and freezes out when $z = z_f$. For $K > K_c$, n_B/s decreases slowly at first because of IDs, then more rapidly because of $2 \leftrightarrow 2$ scatterings when they dominate the IDs, and finally becomes constant when $2 \leftrightarrow 2$ scatterings freeze out, $z_f \approx O(30)$.

that is, $\Delta(z)$ assumes the value that keeps $\Delta'(z)$ nearly zero. As noted earlier, $\Delta(z)$ does not decrease as $\exp(-\frac{1}{2}Kz^2)$ because of the X'_{EQ} term in Equation 36. However, any initial departure from equilibrium does.

Using the method of steepest descent, the integral for $B(\infty)$ can be evaluated,

$$B(\infty) \approx (\varepsilon/g_*)(\pi/2a^{1/2})z_f^{3/2} \exp\left[-z_f - \int_{z_f}^\infty zK\gamma_B(z)\,dz\right], \qquad 41.$$

where z_f is determined by: $z_f K \gamma_B(z_f) = 1$, and $a \equiv -(zK\gamma_B)'|_{z_f}$. The epoch $z \approx z_f$ corresponds to the time when the processes that diminish the baryon asymmetry (IDs and $2 \leftrightarrow 2$ scatterings) "freeze out," i.e. $(\Gamma_B/H)|_{z_f} \approx O(1)$. These processes are effective until $z \cong z_f$ and thereafter cease to be important, and so the asymmetry present at this epoch "freezes in."

If K is greater than 1, but not too large (we quantify this later), then during the epoch when B-nonconserving processes are important $(1 \lesssim z \lesssim z_f)$, the ID processes dominate the $2 \leftrightarrow 2$ scatterings and z_f is determined by when ID freezes out: $Kz_f^{5/2} \exp(-z_f) \approx 1$. Asymptotically the solution is $z_f \approx \ln K$; however, for $K \lesssim 10^6$, $z_f \approx 4.2(\ln K)^{0.6}$ is a better fit. Bringing everything together we have,

$$n_B/s = B(\infty) \approx (\varepsilon/g_*)K^{-1}z_f^{-1} \approx (\varepsilon/g_*)0.3K^{-1}(\ln K)^{-0.6}. \qquad 42.$$

The asymmetry produced only slowly decreases with increasing K, which is somewhat surprising. One might have expected that, if the decay rate Γ_D is greater than the expansion rate when $T \approx M(K \gtrsim 1)$, an equilibrium abundance of X bosons would be maintained and $B(\infty)$ would abruptly decrease for $K \gtrsim 1$. As we now discuss, the rapid decrease in $B(\infty)$ does occur when the $2 \leftrightarrow 2$ scatterings become important.

When $K \gg 1$ the $2 \leftrightarrow 2$ scatterings dominate γ_B at "freeze out." In this case z_f is determined by $AK\alpha z_f^{-4} = 1$, which implies $z_f \approx (AK\alpha)^{1/4}$, and it follows that

$$n_B/s = B(\infty) \approx (\varepsilon/g_*)(AK\alpha)^{1/2} \exp\left[-\frac{4}{3}(AK\alpha)^{1/4}\right]. \qquad 43.$$

That is, for very large K when the $2 \leftrightarrow 2$ scatterings dominate the final baryon asymmetry decreases exponentially with $K^{1/4}$.

To determine approximately how large K must be before the abrupt fall off of n_B/s with increasing K occurs, we can compare freeze out for ID processes $[z_f \approx 4.2(\ln K)^{0.6}]$ with freeze out for $2 \leftrightarrow 2$ scattering processes $[z_f \approx (AK\alpha)^{1/4}]$. The critical value of K is determined by

$$K_c(\ln K_c)^{-2.4} \approx 300/A\alpha. \qquad 44.$$

For $1 \lesssim K \lesssim K_c$, n_B/s is given by Equation 42; for $K \gtrsim K_c$, n_B/s given by Equation 43.

In Figure 6 the evolution of B is shown as a function of z ($\propto t^{1/2}$) for $1 \lesssim K \lesssim K_c$ and $K \gtrsim K_c$. In the first case, $B(z)$ decreases slowly until the ID processes freeze out (for $K \leq K_c$ the $2 \leftrightarrow 2$ scattering processes are not important) and thereafter remains constant. In the case of $K \gtrsim K_c$, $B(z)$ decreases more rapidly because of the $2 \leftrightarrow 2$ scattering processes; they also eventually freeze out, and $B(z)$ becomes constant.

Figure 7 summarizes the final asymmetry produced as a function of K. For $K \lesssim 1$, n_B/s is ε/g_*, independent of K. For $1 \lesssim K \lesssim K_c$ the ID processes are important and n_B/s decreases slowly with K. For $K \lesssim K_c$ the final asymmetry depends only upon K. Finally, for $K \gtrsim K_c$, n_B/s decreases exponentially with increasing K as a result of the $2 \leftrightarrow 2$ scattering processes. The final asymmetry depends only on $(A\alpha K)$. The crossover value of K, K_c, depends upon $A\alpha$ (see Equation 44).

For a GUT like SU(5) or SO(10), g_*, the effective number of species lighter than $O(M)$, is $O(100-300)$, and the gauge coupling is $O(1/45)$, so that for a gauge species $K \approx 7 \times 10^{15}$ GeV/M. For the XY gauge bosons of SU(5), $M \approx$ few $\times 10^{14}$ GeV and $A \approx$ few $\times 10^3$. Thus K_{XY} is $O(30)$, while K_c is $O(100)$. Such a species would strongly damp initial asymmetries (see Section 6.3), and since $K \leq K_c$, we find $B \approx \varepsilon/g_* K_{XY}^{-1}(\ln K_{XY})^{-0.6}$ $\approx 10^{-2} \varepsilon/g_*$.

The mass and coupling of a superheavy Higgs bosons are far less certain. For the triplet member of the 5-dimensional Higgs, the coupling to

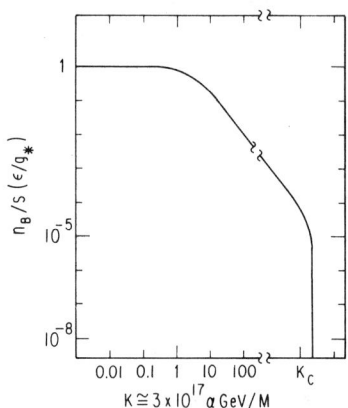

Figure 7 The final baryon asymmetry produced as a function of $K \approx 3 \times 10^{17} \alpha$ GeV/M. For $K \lesssim 1$, n_B/s is independent of K and $\approx (\varepsilon/g_*)$. For $K_c \gtrsim K \gtrsim 1$, n_B/s decreases slowly, $\propto K^{-1}$ $(\ln K)^{-0.6}$ (K_c is given by Equation 44). For $K \gtrsim K_c$, n_B/s decreases exponentially with increasing K.

fermions is set by fermion masses and $h \sim O(m_f/m_W)g$. The heaviest generation determines the effective values of h. For a top quark mass of O(30 GeV), $\alpha \approx h^2/4\pi \approx 10^{-1} \alpha_{\text{gauge}} \approx O(10^{-3})$. [To really "mock up" the full network of equations, one should average h^2 over generations, which is equivalent to dividing by the number of generations.] For $\alpha \approx 10^{-4}$–10^{-3}, K is $(3 \times 10^{13}$ GeV–3×10^{14} GeV$)/M$, while K_c is 10^5–few $\times 10^3$ (again taking $A \approx$ few $\times 10^3$). It seems likely that $K \leq 1$, so the Higgs triplet probably does not damp initial asymmetries, and produces an asymmetry $B \approx \varepsilon/g_*$. All of these results are borne out by the more detailed reaction network codes (30, 41–44).

6.3 Damping of Preexisting Asymmetries

IDs and $2 \leftrightarrow 2$ scatterings reduce any initial baryon asymmetry (42–44) by a factor (cf Equation 38b) of

$$\exp\left[-K \int_0^\infty z \gamma_B(z) \, dz\right] \approx \exp[-K(4+A\alpha)], \qquad 45.$$

where the first term in the exponent is due to IDs and the second to $2 \leftrightarrow 2$ scatterings. If K is greater than one the damping is substantial. In fact the "discovery" of a superheavy boson whose interactions violate B (e.g. by proton decay) for which K is greater than unity would seem to simply that our baryon asymmetry must have evolved dynamically, since any initial baryon excess would have been exponentially damped.[For the XY gauge bosons of SU(5), $K \approx 30$.] However, this is not necessarily the case; our simplified model was a bit too simple.

The model does not take into account quantum numbers that are conserved by the B-violating interactions of the X boson (63). For example in SU(5), $B-L$ and 5-ness are conserved by the XY bosons; $B-L$ is baryon minus lepton number and 5-ness is the property associated with a species in the 5-dimensional representation (see Section 7). If the initial baryon asymmetry has a projection onto such a conserved quantity (e.g. the initial conditions: $B \neq 0$ and $L = 0$ have a projection onto $B-L$), then that part of $B(0)$ cannot be damped. In the context of our simple model we can take this into account by writing: $B(0) = B_{\text{NT}}(0) + B_T(0)$, where the first term represents the part of $B(0)$ that is "nonthermalizing," and the second term represents the "thermalizing" part. The surviving part of the initial asymmetry is then

$$B_{\text{NT}}(0) + \exp[-K(4+A\alpha)]B_T(0). \qquad 46.$$

In general, the complete network of equations for the damping of initial baryon asymmetries (due to one superheavy species) can be diagonalized, and written as N equations of the same form as our simplified equations.

The nonthermalizing parts of B correspond to eigenmodes for which $\gamma_B = 0$.

7. SPECIFIC MODELS

In Sections 3 to 6 we outlined the recipe for generating a baryon asymmetry. Two of the ingredients, baryon number violation and C, CP violation are expected to be present in grand unified models; the third ingredient, a departure from equilibrium, is likely to be supplied by the expansion of the Universe. In Section 6 we constructed a very simple model to illustrate the interplay of the three ingredients. In this section we examine baryogenesis in context of "realistic" grand unified models. Although the many free parameters in such models preclude definite predictions, it is possible to make general statements about the compatibility of specific models with baryogenesis.

We discuss realistic models based upon SU(5) and SO(10). In SU(5) the choice of the Higgs representations present in the model is crucial. For some choices of the Higgs representations, a sufficient baryon asymmetry does not evolve; the minimal Higgs structure and three families of fermions are incompatible with baryogenesis. In SO(10), the pattern of symmetry breaking is very important. Some intermediate stages of symmetry breaking will prevent baryon number generation; models in which these symmetries persist to low energies are incompatible with baryogenesis. We also discuss how one constructs a network to compute the evolution of the baryon asymmetry in a realistic model.

7.1 SU(5)

SU(5) is the simplest candidate for unification of the strong and electroweak interactions; that is, it is the smallest group containing $SU(3)_C \otimes SU(2)_L \otimes U(1)_Y$ as a subgroup (64). The decomposition of an SU(5) representation by the embedding of $SU(3)_C \otimes SU(2)_L \otimes U(1)_Y$ in SU(5) will be denoted by (i, j, k), where i is the $SU(3)_C$ multiplicity, j is the $SU(2)_L$ multiplicity, and k is the $U(1)_Y$ charge.

In SU(5), the gauge fields transform as the 24-dimensional adjoint representation of SU(5); denoted as 24_V. Of the 24 gauge bosons, only the interactions of the XY gauge bosons $(3, 2, 5/6)$ and their antiparticles $(\bar{3}, 2, -5/6)$ violate baryon number conservation. The mass of these bosons is expected to be about 5×10^{14} GeV. They should be nearly degenerate in mass since the splitting of their masses only arises due to weak symmetry breaking. The 15 left-handed fermion fields in each generation are assigned to the reducible representation $\bar{5}_f + 10_f$

$$\bar{5}_f = \{D_{iL}^c, \nu_L, E_L\}; \quad 10_f = \{U_{iL}, D_{iL}, U_{iL}^c, E_L^c\} \qquad 47.$$

where i is the color index and the superscript c denotes charge conjugation. In the generic form above, U (D) refers to the charge $+2/3$ $(-1/3)$ quark, and E (ν) refers to the charged (neutral) lepton of a given generation.

The freedom in constructing an SU(5) model is in the Higgs sector. The 5_H, 10_H, 15_H, 45_H, and 50_H representations all appear in the decomposition of the products $5_f \otimes 5_f$, $\bar{5}_f \otimes 10_f$, and $10_f \otimes 10_f$, and so Higgs bosons in all of these representations can couple to fermions. In the simplest model only a single 5-dimensional Higgs representation, 5_H, couples to the fermions. Since the fermion representation is reducible, even with a single 5_H there are two independent Yukawa coupling matrices (in generation space), h_U and h_D.

We consider several SU(5) models with different choices of Higgs representations. The inherent freedom in the Higgs sector precludes a definite prediction of a baryon asymmetry. However, for some choices of Higgs representations, generation of a baryon asymmetry is impossible. Hence consideration of the baryon asymmetry gives us valuable information about a sector of grand unified theories (and gauge theories in general) that is poorly understood.

MINIMAL SU(5) In this model a single 5-dimensional Higgs representation (5_H) couples to fermions. The $SU(3)_C \otimes SU(2)_L \otimes U(1)_Y$ content of this Higgs multiplet is $(1, 2, -\frac{1}{2})$—the Weinberg-Salam doublet of light Higgs—and $(3, 1, 1/3)$—a color triplet of Higgs whose interactions violate B conservation; we denote the triplet by H_3. In the minimal model all B nonconservation arises from the interactions of the XY and H_3 bosons. The fermionic portion of the Lagrangian can be written in the schematic form:

$$\mathscr{L} = \frac{g}{\sqrt{2}} 24_V [(\bar{5}_f)_i (5_f)_i + \overline{(10_f)}_i (10_f)_i]$$
$$+ (h_U)_{ij} (10_f)_i^T (10_f)_j 5_H,$$
$$+ (h_D)_{ij} (\bar{5}_f)_i (10_f)_j \bar{5}_H, \qquad 48.$$

where g is the gauge coupling, and h_D and h_U are Yukawa coupling matrices in generation space.

In the minimal model the undetermined parameters relevant for baryogenesis are: the mass of the superheavy Higgs H_3, the number of generations, and the Yukawa coupling matrices. The mass of the gauge boson and its coupling strength are determined in this model (65–68) ($m_X \approx 5 \times 10^{14}$ GeV; $\alpha \equiv g^2/4\pi \approx 1/45$). The mass of the H_3 boson is presumably of the same order of magnitude; a generous range of uncertainty would be $10^{-2} \lesssim m_{H_3}/m_X \lesssim 10^2$. The Yukawa couplings of the H_3 are proportional to the fermion masses:

$$h_D = g \, \text{diag}(m_D^i)/\sqrt{2} m_W; \qquad h_U = g \, \text{diag}(m_U^i)/\sqrt{2} m_W \qquad 49.$$

(in the absence of generalized Cabibbo mixing), where m_D^i (m_U^i) are the charge 2/3 ($-1/3$) quark masses of the ith generation. [Throughout we ignore renormalization corrections to the fermion masses.] Clearly, the couplings to the heaviest family are the most important. Although the uncertainty in m_{H_3}, the number of families, and the fermion masses of all the

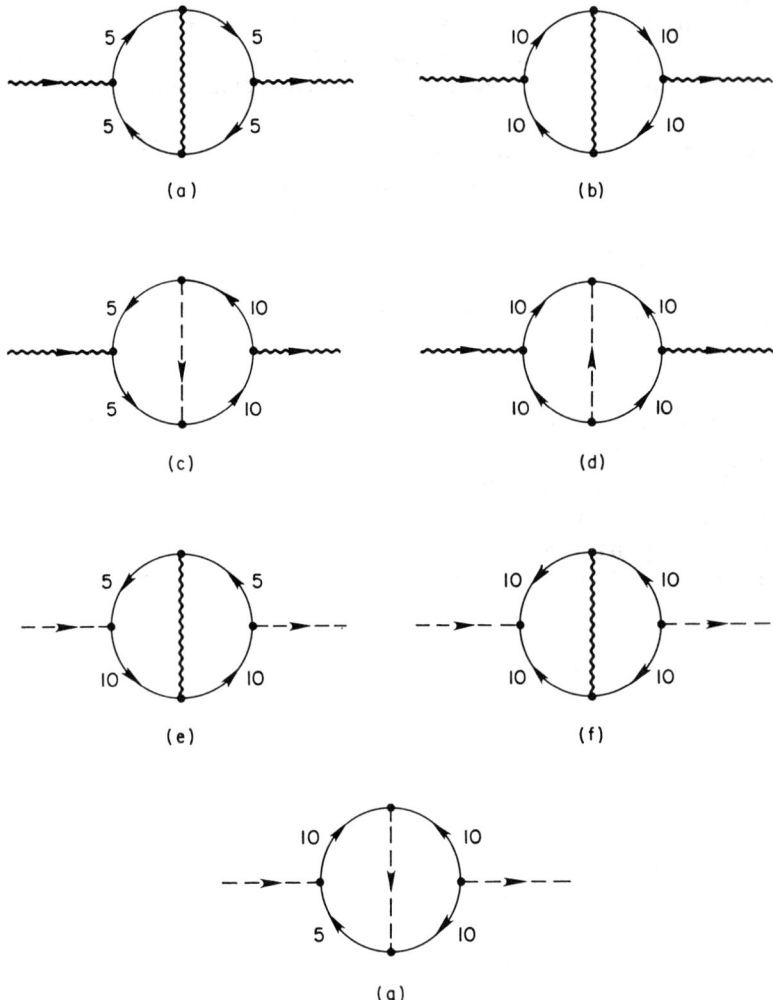

Figure 8 One-loop corrections to Higgs and gauge boson decays. The wavy line is the 24_V and the broken line is the 5_H. In the minimal SU(5) model none of these diagrams can give rise to the CP violation required for baryon number generation. Diagrams (*a*) and (*b*) can never give rise to CP violation since the gauge-fermion couplings are real.

families prevent a precise prediction of the baryon asymmetry, the real difficulty is the calculation of the C, CP violation. In the minimal model this requires detailed knowledge of the matrices h_U and h_D—family mixings and phases.

As discussed in Section 5, CP violation arises in loop-diagram corrections to XY and H_3 decay. Since the particles in the loops must violate baryon number, they must be XY and H_3 bosons. Candidate one-loop diagrams leading to CP violation are shown in Figure 8. Several groups (37, 47, 60, 69, 70) have pointed out that none of these diagrams leads to CP violation. For instance, consider Figure 8g:

$$\varepsilon_{8g} \propto \text{Im Tr } [(h_D)^\dagger(h_D)(h_U)^\dagger(h_U)], \qquad 50.$$

where the trace is over the suppressed family indices. The imaginary part of the trace, and hence ε, obviously vanishes. The same is true for all the other diagrams in Figure 8. It is not until three loops for H_3 bosons (four for XY bosons) that a class of diagrams can be found to yield a nonzero imaginary part. An example of such a diagram is shown in Figure 9. These diagrams result in

$$\varepsilon_{H_3} \approx \text{Im}[I] (h_U^i, h_D^i)^6 \sin \delta \, f(\theta_i), \qquad 51.$$

where I is the "phase space" integral (see Section 5), δ represents a complex phase in the Yukawa matrices, $(h_U^i, h_D^i)^6$ is a specific combination of Yukawa couplings (37, 69), and $f(\theta_i)$ is a specific combination of sines and cosines of generation mixing angles.

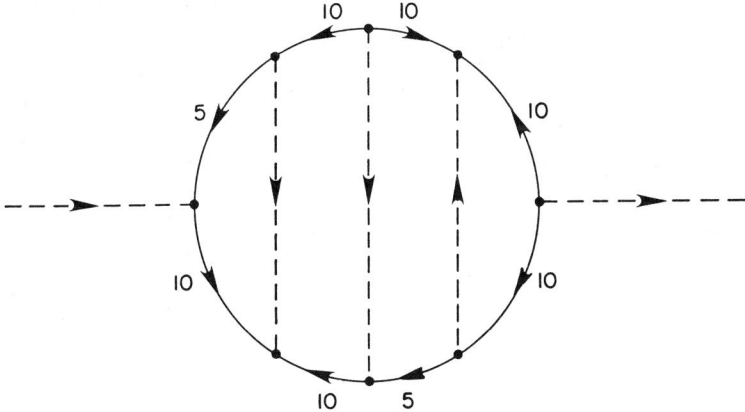

Figure 9 An example of a 3-loop correction to Higgs decay that gives rise to CP violation in minimal SU(5).

Approximating Im[I] by the available volume of phase space and setting $f(\theta_i) = 1$ we have

$$\varepsilon_H \approx 10^{-8}(m_b^4 m_t m_c/m_W^6)\sin\delta, \qquad 52.$$

with just the usual three families of fermions. Even with $m_t \approx m_W$, ε_H is at most $10^{-15}\sin\delta$—far too small to account for the observed asymmetry $n_B/s \approx (6-10) \times 10^{-11}$.

Introducing a fourth, heavy generation of fermions (t', b'; $m_{b'} \approx m_{t'} \approx m_W$) improves the situation a bit. In Equation 52 with $m_b \to m_{b'}$, $m_t \to m_{t'}$, $m_c \to m_t$, and for $m_{b'} \approx m_{t'} \approx m_t \approx m_W$, we have $\varepsilon_H \approx 10^{-8}\sin\delta$. Since the asymmetry produced can be as large as $\varepsilon/g_* \approx 10^{-2}\varepsilon$, it might be just barely possible to produce $n_B/s \approx 10^{-10}$–10^{-11} if $\sin\delta \approx O(1)$. However, since this scheme requires $m_f \approx m_W$, the Higgs coupling is about equal to the gauge coupling, and $K \approx 3 \times 10^{17}$ GeV $\alpha/M \approx 7 \times 10^{15}$ GeV/M. Thus for $M \lesssim 7 \times 10^{15}$ GeV, K is greater than 1 and the asymmetry produced is $n_B/s \approx (\varepsilon/g_*)K^{-1}\ln K^{-0.6}$, which is less than $10^{-2}\varepsilon$ (cf Equation 43). Thus, even this scheme does not seem likely to work (59, 71, 72). Needless to say, since ε_{XY} is only nonzero at four-loops, the vector bosons do not play a significant role in generating the baryon asymmetry.

NONMINIMAL SU(5) The fact that the Higgs structure of the minimal model appears to be incompatible with the observed baryon number of the Universe is strong motivation for expanding the Higgs sector. There are other motivations for a nonminimal Higgs structure. The minimal model makes the embarrassing prediction that: $m_d/m_s \approx m_e/m_\mu \approx 1/207$ (66).

The common feature of all the methods for getting CP violation at one-loop is an increase in the complexity of the Higgs sector. The simplest change is to add an additional 5-dimensional representation of Higgs ($5'_H$). CP violation will arise at one-loop, for instance, in diagrams like Figure 8g, where the exchanged Higgs is different from the decaying Higgs. This simple replication is automatically present in unified models that employ the Peccei-Quinn symmetry (73) to suppress strong CP violation. In such models, the Peccei-Quinn symmetry makes it necessary to have one 5_H that couples only to $\bar{5}_f 10_f$, and a second $5'_H$ that couples only to $10_f 10_f$ (74). In models with a second 5_H it is possible for ε to be nonzero at one-loop (47, 60, 69, 70, 70a, 75), and an adequate baryon asymmetry will evolve for a wide range of the other free parameters (Higgs masses, coupling constants, mixing angles, and phases) (59, 72).

A minor variation on this theme is to add a *different* (other than a 5-dimensional) second Higgs representation. One possibility is a 45-dimensional Higgs representation. The qualitative features of this choice are the same as with the addition of a second 5_H: CP violation will be present at the one-loop level, and an adequate baryon asymmetry can be generated for a range of choices of unknown parameters (59).

7.2 SO(10)

Although SU(5) is the simplest (and in that sense possibly the most attractive) model, the grand unified theory based on the gauge group SO(10) (76, 77) has several features that are very attractive from the viewpoint of particle physics (23, 24). SO(10) also differs from SU(5) in several ways that are of cosmological interest. These differences include the nonconservation of $B-L$, C symmetry of the theory, and richer possibilities for intermediate symmetries. These new features are also expected in even more complicated models (78), such as E_6 (79).

The usual 15 left-handed fermion fields (per family), together with a new field, the C partner of v_L, denoted as N_L^C, are placed in the 16-dimensional spinor representation of SO(10), 16_f. The gauge bosons are in the 45-dimensional adjoint representation, 45_V. Higgs bosons that couple to fermions must be in representations that occur in the product $16 \otimes 16$

$$16 \otimes 16 = (10+126)_S + (120)_A, \qquad 53.$$

where the S and A denote symmetric or antisymmetric product.

SO(10) is C symmetric, as is apparent since each particle and its C partner occur in 16_f. In addition, $B-L$ is a gauge symmetry of unbroken SO(10). The persistence of these symmetries depends upon the pattern of symmetry breaking. Possible symmetry-breaking patterns include (78)

$$SO(10) \to SU(5) \to \cdots \qquad 54a.$$

$$SO(10) \to SU(5) \otimes U(1) \to \cdots \qquad 54b.$$

$$SO(10) \to SU(3)_C \otimes SU(2)_L \otimes U(1)_Y \to \cdots \qquad 54c.$$

$$SO(10) \to SU(4) \otimes SU(2)_L \otimes SU(2)_R \to \cdots \qquad 54d.$$

$$SO(10) \to SU(4) \otimes SU(2)_L \otimes U(1)_R \to \cdots \qquad 54e.$$

$$SO(10) \to SU(3)_C \otimes SU(2)_L \otimes U(1)_R \otimes U(1) \to \cdots \qquad 54f.$$

$B-L$ remains unbroken so long as an $SU(2)_R$ or $U(1)_R$ factor is present. If $B-L$ is spontaneously broken before baryogenesis, then a nonzero value of $B-L$ may be generated in much the same way a baryon number is generated (59). If $B-L$ is an unbroken symmetry during baryogenesis, then a nonzero value of $B-L$ will not be produced until $B-L$ is broken. If the Universe begins with zero values of all quantum numbers, then the values of B and $B-L$ that evolve are the same order of magnitude. On the other hand, if there are initial asymmetries, it is possible for the final values of B

and $B-L$ to differ by many orders of magnitude. This point is discussed further in Section 8.6.

As we discussed in Section 3, the generation of an asymmetry in any quantum number that is odd under C requires C violation. Both baryon and lepton number are examples of such quantum numbers. SU(5) is not C symmetric, in fact the C operator cannot even be defined since the neutrino has no C partner. Since SO(10) is C symmetric, the C symmetry must be broken before a baryon asymmetry can be generated. A simple way to break C symmetry is to split the masses of particles that are C partners. To see why the masses of C partners must be split, consider the baryon number produced in the decays of the bosons $\chi\bar{\chi}$ and $\chi^c\bar{\chi}^c$ (note that in general, $\chi^c \neq \bar{\chi}$). In the limit of exact C symmetry, $\varepsilon_\chi = -\varepsilon_{\chi^c}$, so the baryon number produced in $\chi\bar{\chi}$ decays is exactly cancelled by the baryon number produced in $\chi^c\bar{\chi}^c$ decays. However, if the masses of particles in the intermediate states or in the decay products differ, or if $m_\chi \neq m_{\chi^c}$, then the kinematic factors in ε_χ and ε_{χ^c} will not be identical, the cancellation will not take place, and a net baryon number can be produced. This can be done either in the fermion sector by splitting the mass of the left-handed neutrino v_L, and its C partner N_L^C, or in the boson sector.

A nonzero mass for N_L^C allows a superheavy boson whose decays include neutrinos (either in the final or intermediate states) to produce a baryon asymmetry. In detail this occurs because the phase space volume for decay channels involving Ns and vs are different because of their unequal masses, and so in Equation 26 $\mathrm{Im}[I] \neq \mathrm{Im}[I^c]$. The C, CP violation, and hence the baryon number produced, is proportional to m_N^2/m_χ^2, where m_N and m_χ are right-handed neutrino and superheavy boson masses respectively. In such a scenario it is clear that m_N can be at most a few orders of magnitude less than m_χ. In these models the mass of the left-handed neutrino is often $O(m_\mathrm{f}^2/m_\mathrm{N})$ (m_f = fermion mass), and so compatibility with baryogenesis also implies light left-handed neutrinos (80, 80a). This scenario for breaking C symmetry is only possible for symmetry-breaking patterns described in Equations 54a–54c.

A second possibility involves splitting the masses of B-violating bosons that are C partners (59, 81, 81a). In the symmetry-breaking pattern SO(10) $\to$ SU(4) $\otimes$ SU(2)$_\mathrm{L}$ $\otimes$ SU(2)$_\mathrm{R}$, breaking C symmetry does not split the masses of baryon-number-violating C partners. The vector bosons in the 45$_\mathrm{V}$ adjoint representation have the embedding $(6, 2, 2)+(1, 2, 2)$ in the decomposition SO(10) $\supset$ SU(4) $\otimes$ SU(2)$_\mathrm{L}$ $\otimes$ SU(2)$_\mathrm{R}$, where (i, j, k) are the SU(4), SU(2)$_\mathrm{L}$, and SU(2)$_\mathrm{R}$ multiplicities respectively. All the baryon-number-violating vector bosons are in the $(6, 2, 2)$; they are the XY bosons of SU(5) and X' and Y' bosons, with electric charge $(-1/3, 2/3)$. Under charge conjugation $X \leftrightarrow \bar{X}$, $Y \leftrightarrow \bar{X}'$, $X' \leftrightarrow \bar{Y}$, and $Y' \leftrightarrow \bar{Y}'$.

Since all the C partners are in the same irreducible representation of $SU(4) \otimes SU(2)_L \otimes SU(2)_R$, they must be degenerate in mass, and thus cannot be responsible for generating a baryon number. The same is true for the baryon-number-violating bosons in the Higgs representations (10_H, 120_H, and 126_H)—the only representations that can couple directly to fermions. Because of the L−R symmetry the neutrino masses cannot be split either.

However, in $SO(10) \to SU(3)_C \otimes SU(2)_L \otimes U(1)_Y$, the (X, Y) doublet is split in mass from the (X′, Y′) doublet and baryon number generation is possible in the decays of vector bosons. If instead, SO(10) is broken only to $SU(4) \otimes SU(2)_L \otimes U(1)$, such a mass splitting does not occur, and vector bosons cannot produce a baryon asymmetry. Therefore if $SU(4) \otimes SU(2)_L \otimes U(1)$ symmetry persists to low energies, the $v-N$ mass splitting must be responsible for the baryon number and the baryon number is proportional to m_N^2/m_χ^2 (59, 80, 80a, 81). If $SU(3)_C \otimes SU(2)_L \otimes U(1)_Y$ is the relevant symmetry for baryon number generation, the baryon asymmetry may be traced to mass splittings in either the boson sector or the fermion sector ($v-N$).

As a final example, suppose that a $SU(4) \otimes SU(2)_L \otimes U(1)_R$ symmetry persists to low energies. The $U(1)_R$ factor prevents a mass for N_L^C, so baryon number generation requires splitting the mass of a baryon-number-violating boson from its C partner. Such a splitting does not occur for gauge bosons, nor for scalar bosons in the 16_H or 126_H. Only the (6, 3, 1)+(6, 1, 3) bosons in the 120_H both have baryon-number-violating interactions and are split in mass from their C partners. [Here (i, j, k) are the $SU(4)$, $SU(2)_L$, $U(1)_R$ multiplicities respectively.] Therefore if one wishes to have $SU(4) \otimes SU(2)_L \otimes U(1)_R$ symmetry persist to low energies, it is necessary to have a 120_H in the model to account for the baryon number of the Universe (59).

In discussing SO(10) we have tried to demonstrate that the requirement of generating a baryon asymmetry can potentially give valuable information about the scales and patterns of the intermediate stages of symmetry breaking. Though we have only discussed a few examples and only for SO(10), it is clear that an analogous analysis is possible for more complicated models. For a more detailed discussion of SO(10) the reader is referred to (59, 81).

7.3 *Realistic Boltzmann Equations*

In a simple model developed in Section 6 only two degrees of freedom are tracked: the baryon asymmetry and the abundance of X bosons. Even in the minimal SU(5) model with three generations there are nearly 100 degrees of freedom (color, flavor, generation, etc), which must, in principle, be tracked to follow the evolution of the baryon asymmetry. However, by making assumptions analogous to that of kinetic equilibrium, by taking advantage of all the conserved or nearly conserved quantum numbers, and by using the gauged and global symmetries of the model, we can reduce the degrees of freedom to a manageable number—seven in minimal SU(5). This reduction is more than just a convenience, for without it numerical errors in the integration of the network can dominate the evolution of the baryon asymmetry. For example, as long as C symmetry persists in SO(10), B must remain identically zero; while numerically it will typically have a value of order the accuracy of the integration scheme. Since the observed asymmetry is $O(10^{-10})$, a numerical accuracy of much better than this is required in order reliably to track the baryon asymmetry. Since the reduction follows in a similar manner in all models, we illustrate it in detail only for minimal SU(5).

At the temperatures of baryogenesis, $SU(2)_L \times U(1)_Y$ is an exact symmetry (having been restored by finite temperature effects), and the $W^\pm$ and Z bosons are massless. Thus there are long-range forces associated with weak isospin, color, and hypercharge (or equivalently electric charge). The Universe should be neutral in all quantum numbers associated with long-range forces. Therefore the Universe must be an isosinglet, color singlet, and have zero net charge associated with isospin, hypercharge, and the two commuting generators of $SU(3)_C$ (λ_3 and λ_8 for example). In addition to quantum numbers associated with unbroken gauge symmetries, there may be quantum numbers associated with global symmetries present in the model; for example, $B-L$ in minimal SU(5).

At temperatures less than about 10^{16} GeV, $2 \leftrightarrow 2$ scattering processes mediated by massless gauge bosons occur rapidly compared to the expansion rate ($\Gamma > H$). Not only should these reactions maintain kinetic equilibrium, but they should also result in all asymmetries (B, L, etc) being shared by all the species carrying that quantum number. For instance, a baryon asymmetry in left-handed up quarks will immediately be shared with left-handed down quarks (through W exchange), and with left-handed charm, strange, top, bottom,... quarks (through Cabibbo mixing).

In addition to the gauged and global symmetries, in most models there are additional "partly conserved" symmetries, i.e. symmetries that are conserved in some sector of the theory (63). For instance, if we consider only the fermion-gauge interactions in SU(5), we see that the Lagrangian is invariant under two global phase transformations: $5_f \to \exp(i\alpha) 5_f$ and $10_f \to \exp(i\beta) 10_f$. These two global symmetries of the vector interactions correspond to two conserved (by gauge interactions) quantum numbers: $\Pi_5 \equiv +1(-1)$ for each field in 5_f ($\bar{5}_f$), and $\Pi_{10} \equiv +1(-1)$ for each field in 10_f ($\overline{10}_f$) (63). The scalar interactions do not respect these global symmetries.

However, the linear combination of Π_5, Π_{10} and hypercharge corresponding to $B-L$ is conserved, as $B-L$ is an exact global symmetry in SU(5). During the epoch of baryogenesis hypercharge is conserved, so only Π_5 needs to be tracked. The usefulness of these partial symmetries lies in the fact that they are only violated by scalar interactions, and in some regimes the scalar interactions are ineffective ($\Gamma < H$) and can be neglected.

The use of conserved and partially conserved symmetries, and the assumption of kinetic equilibrium and asymmetry sharing allow the complete set of Boltzmann equations for minimal SU(5) to be written as 7 coupled, first-order differential equations for the following quantities: B, Π_5, v_-, X_+, X_-, S_+, and S_-. The quantities with a minus subscript refer to the excess number density of that particle species over its antiparticle; quantities with a plus subscript refer to the total number density of that particle species and its antiparticle. Note, $Y_+ = X_+$ and $Y_- = X_-$ are guaranteed by the $SU(2)_L$ symmetry of the interactions. Besides the extra degrees of freedom, the Boltzmann equations for these seven quantities will be analogous to the simplified set of equations developed in Section 6, with the required departure from equilibrium being provided by both $X_+ - X_{+EQ}$ and $S_+ - S_{+EQ}$, i.e. both the gauge and Higgs bosons can generate a baryon asymmetry. For more detail, the reader is referred to (59).

8. BEYOND THE STANDARD SCENARIO

8.1 *Primordial Black Holes*

In 1974 Hawking (82, 83) pointed out that black holes are black in more than name only. He showed that a black hole of mass m should emit a spectrum of radiation characteristic of a blackbody at a temperature $T \approx m_{pl}^2/m$, and in doing so should evaporate in a time $\tau \approx m^3 m_{pl}^{-4}$. For a one $M_\odot$ hole, the temperature is $T \cong 10^{-19}$ GeV and the lifetime is $\tau \approx 10^{71}$ s. For solar mass objects or larger, Hawking radiation is completely negligible. The Hawking process is significant, however, for black holes less massive than about 10^{15} g, since for these holes $T \gtrsim 0.1$ GeV and $\tau \lesssim 10^{17}$ s (age of the universe). There are no known astrophysical processes that would lead to the formation of such small holes in the present epoch; however, density inhomogeneities in the early Universe [if present with amplitude $\delta\rho/\rho \gtrsim O(1)$] could lead to the formation of primordial black holes (PBHs) (84).

In 1975 Hawking (85) suggested that, because of C, CP violation, evaporating PBHs might radiate an excess of baryons over antibaryons. In the context of GUTs the idea has been taken up again by many authors (32, 86–89c). There are two basic scenarios for baryogenesis by PBHs: 1. PBHs

radiate X bosons, which subsequently decay and produce the asymmetry;
2. PBHs directly radiate an excess of baryons over antibaryons.

SCENARIO 1 Only PBHs less massive than $m_* \cong m_{pl}^2 M^{-1}$ are hot enough ($T \gtrsim M$) to radiate X bosons of mass M. All PBHs will eventually be hot enough to radiate X bosons; however, PBHs initially more massive than m_* will radiate away most of their mass when $T \lesssim M$, into species other than X bosons, and hence will not contribute significantly to n_B/s. Only PBHs less massive than m_* are important for baryogenesis. In the standard scenario for baryogenesis there are two important timescales: $\tau_M \approx m_{pl} M^{-2}$, the age of the Universe when $T \approx M$, and $\tau_X \approx (\alpha M)^{-1}$, the lifetime of the X. In the PBH scenarios there is an additional timescale: $\tau \approx m^3 m_{pl}^{-4}$ the lifetime of a PBH of mass m. Note that $\tau_* \approx (m_{pl}/M)\tau_M \gtrsim \tau_M$; X bosons emitted by PBHs of mass m_* will always decay out of equilibrium (i.e. when $T < M$). In the usual scenario this only occurs when $\tau_X \gtrsim \tau_M$, which requires $M \gtrsim 3 \times 10^{17} \alpha$ GeV. All PBHs heavier than $(M/m_{pl})^{1/3} m_*$ will also produce X bosons that decay when $T \lesssim M$. Thus the usual condition on the boson mass for out-of-equilibrium decay can be relaxed. However, the baryon asymmetry produced depends crucially upon the initial mass spectrum of PBHs present, i.e. the initial degree of irregularity in the Universe. Any initial spectrum of PBHs is tightly constrained by the number of $\sim 10^{15}$-g PBHs it predicts, as these holes should be evaporating during the present epoch and contributing to the γ-ray background (90, 91). It is possible to produce the observed baryon asymmetry with a spectrum of PBHs consistent with these constraints.

SCENARIO 2 PBHs directly radiate an excess of baryons over antibaryons. All PBHs less massive than 10^{14} g are hot enough to radiate baryons, and all PBHs that evaporate after B-nonconserving processes became ineffective ($T \lesssim$ the grand scale $= M$), $m \gtrsim m_2 = (m_{pl}/M)^{2/3} m_{pl}$ will produce an excess that cannot get erased. Since the number of PBHs must fall with increasing PBH mass, PBHs of mass m_2 make the largest contribution to the baryon asymmetry. It is also possible to produce the observed asymmetry in this scenario.

8.2 *Supersymmetric Models*

There has long been interest in supersymmetry because of the intrinsic beauty of the Bose-Fermi symmetry in these theories and the possibility of using supergravity (local supersymmetry) to construct a quantum theory of gravity (92–94). Recently, there has been a flurry of interest in supersymmetric extensions of GUTs (95–98), because there is hope that these models can solve the "hierarchy problem" (the existence of two scales of spontaneous symmetry breaking separated by twelve or so orders of magnitude).

To date there does not exist a single compelling supersymmetric GUT. However, there are certain generic features that all such models have, and these features have been used to discuss in a semiquantitative way how supersymmetry affects baryogenesis (62, 99).

First, the number of particle species doubles; for every fermion (boson) there exists a supersymmetric bosonic (fermionic) partner. This doubles the number of light quark and lepton species (the supersymmetric partners of quarks and leptons are expected to have masses $\lesssim$ few TeV). Next, in the absence of a symmetry forbidding them, there exist dimension-5 (dim-5) operators in the effective, low-energy ($\lesssim$ the grand scale) theory that violate B conservation. These operators correspond to processes involving two scalar quarks and leptons and two ordinary quarks and leptons, which are mediated by the fermionic partner of the superheavy color triplet Higgs. [In SU(5) the superheavy color triplet Higgs and light Weinberg-Salam Higgs are members of the same 5-dimensional representation.] Finally, because of the additional light spin-0 and spin-$\frac{1}{2}$ particle species, the unification scale in a supersymmetric SU(5) model rises to O(10^{16} GeV), and the unified coupling constant is $\alpha \approx 1/25$ rather than 1/45 (100).

The main difference between baryogenesis in a supersymmetric GUT and a nonsupersymmetric GUT is the increased significance of $2 \leftrightarrow 2$ scatterings. This is because there are more scattering channels, a larger coupling constant, and most importantly there may be dim-5 scattering processes. To understand this quantitatively, recall that in our simple model in Section 6 we showed that the transition to the "scattering-dominated" regime (where n_B/s falls off exponentially with increasing $K \approx 3 \times 10^{17} \, \alpha/M$) occurs for $K \approx K_c$. With the usual (dim-6) $2 \leftrightarrow 2$ scattering processes we found (cf Equation 44) that

$$\frac{K_c}{(\ln K_c)^{2.4}} \approx \frac{300}{A\alpha}, \qquad 55.$$

where A is proportional to the number of scattering channels, and α is the coupling constant of the superheavy boson to light fermions. In the case of a dim-5 operator at low temperatures, $(\sigma v) \propto M^{-2}$ rather than T^2/M^4, and $z_f \approx (A\alpha K)^{1/2}$ rather than $(A\alpha K)^{1/4}$. The formula for K_c becomes

$$\frac{K_c}{(\ln K_c)^{1.2}} \approx \frac{18}{A\alpha}, \qquad 56.$$

and the final asymmetry in the scattering-dominated regime becomes

$$n_B/s \approx (\varepsilon/g_*) A\alpha K \, \exp[-2(A\alpha K)^{1/2}]. \qquad 57.$$

In a supersymmetric GUT the effective number of scattering channels A might increase by a factor of 10 owing to the addition of scalar quarks and leptons; changing $A\alpha$ from 10 to 100 decreases K_c from $O(10^3)$ to $O(100)$. The effect of dim-5 $2 \leftrightarrow 2$ scatterings is even more dramatic. For $A\alpha = 10$ changing to dim-5 $2 \leftrightarrow 2$ scatterings decreases K_c to $O(10)$, and for $A\alpha = 100$ it decreases K_c to less than $O(1)$.

For a given $K \gtrsim 1$, the effect of dim-5 scatterings is to decrease the baryon asymmetry that evolves by many orders of magnitude. Of course, by going to smaller K (larger M or smaller α) a reasonable asymmetry can still be generated. To be more specific, for $\alpha \approx 10^{-3}$ (the value expected from the coupling of light Higgs to fermions) M must be greater than about 3×10^{14} GeV in order for the observed asymmetry to evolve (unless dim-5 operators are forbidden by an additional symmetry).

To summarize, for a given superheavy mass a smaller baryon asymmetry evolves in a supersymmetric GUT, as a result of the increased importance of $2 \leftrightarrow 2$ scatterings. In this regard dim-5 operators are the most significant cause (62, 99). The baryon asymmetry also depends upon ε. Haber (101) has studied how supersymmetry affects CP violation, and finds that the situation is substantially the same as for nonsupersymmetric GUTs. There appear to be no additional cancellations that might tend to suppress ε; also see (101a).

8.3 Monopole Catalysis

Callan (102), Rubakov (103, 104), and Wilczek (105) have all discussed the fact that the 't Hooft–Polyakov (106, 107) monopoles predicted to exist in GUTs should catalyze B nonconservation. At the core of the monopole $(r \lesssim M_X^{-1})$ the full symmetry of the GUT is realized, and so on geometric grounds one would expect a cross section for this process $\sigma \sim M_X^{-2}$—which is utterly negligible since $M_X \gtrsim O(10^{14}$ GeV$)$. In rather remarkable and similar papers, Callan and Rubakov showed that, because of the singular nature of the potential between a monopole and a fermion $[V(r) \sim r^{-2}]$, the s-wave part of the fermion wavefunction (which has no centrifugal barrier) is "sucked into" the origin (where the full symmetry of the GUT is realized); this leads to a "hadronic" cross section for the catalysis process. Their calculations indicate that the cross sections for processes like $M + p \rightarrow M + \pi^0 + e^+$ should be: $(\sigma v) \approx \sigma_0 E^{-2} (E \gg \Lambda)$ and $(\sigma v) \approx \sigma_0 \Lambda^{-2} (E \ll \Lambda)$, where E is the incident energy of the fermion, $\Lambda \sim O(1$ GeV$)$ sets the scale of fermion masses, strong interaction physics, etc, and σ_0 is a dimensionless constant of order unity.

In order to investigate the effect of monopole catalysis on the evolution of n_B/s, consider the following simple model. 1. Particle species consisting of M—monopole, b—baryon, and ϕ—light particle that does not carry B number. 2. Interactions consisting of $b + M$ ($\bar{M}$) $\leftrightarrow \phi + M$ ($\bar{M}$), $\bar{b} + M$ ($\bar{M}$) $\leftrightarrow \bar{\phi} + M$ ($\bar{M}$). Assuming that the monopole is stable and $M - \bar{M}$

annihilations are unimportant, it follows that $n_M \gg (n_M)_{EQ}$, and in a manner analogous to Section 6 the following equations can be derived:

$$B(z)' = zK(r\gamma\varepsilon - r\gamma B), \qquad 58a.$$

$$B(\infty) = B(0) \exp\left[-\int_0^\infty rKz\gamma(z) \, dz\right]$$

$$+ (\varepsilon/g_*)rK \int_0^\infty z'\gamma(z') \exp\left[-\int_{z'}^\infty z''rK\gamma(z'') \, dz''\right] dz' \qquad 58b.$$

where $B = n_B/s$, $K = (\Gamma_B/H)|_{T \cong \Lambda} \approx m_{pl}/\Lambda$, $z = \Lambda/T$, $\Gamma_B \approx T^3 \langle \sigma v \rangle$, $\gamma = \Gamma_B(z)/\Gamma_B(1) \approx \sigma_0 z^{-1} (z \ll 1)$ and $\sigma_0 z^{-3} (z \gg 1)$, $r \equiv n_M/n_\gamma \approx 3 \times 10^{-24} h^2 (10^{16} \text{ GeV}/m_M) \Omega_M$, and ε measures the fractional difference between the cross section for $b + M \leftrightarrow \phi + M$ and $\bar{b} + \bar{M} \leftrightarrow \bar{\phi} + \bar{M}$ (m_M is the monopole mass, Ω_M is the present contribution of monopoles to closure density, and the Hubble parameter is $H = 100h$ km s^{-1} Mpc^{-1}).

Most of the damping due to monopole catalysis occurs for $T \approx \Lambda$, and $\int_0^\infty rKz\gamma(z) \, dz = 2rK\sigma_0 \approx 10^{-4} h^2 \Omega_M (10^{16} \text{ GeV}/m_M)\sigma_0$. Thus for $m_M \gtrsim 10^{12}$ GeV the damping of the baryon asymmetry by monopoles should be negligible. [This is the conclusion reached in (108).] Assuming $m_M \gtrsim 10^{12}$ GeV, the baryon asymmetry produced by monopole collisions is

$$(n_B/s) \approx 10^{-4} h^2 \Omega_M (10^{16} \text{ GeV}/m_M)\sigma_0 (\varepsilon/g_*). \qquad 59.$$

The baryon excess is also produced primarily when $T \approx \Lambda$, in low-energy collisions. As was discussed in Section 7, ε arises from loop diagrams involving the exchange of species whose interactions do not conserve B, and which therefore must necessarily be very heavy (mass $M \gtrsim 10^{14}$ GeV). Thus in low-energy collisions ε will be suppressed by powers of $(T/M) \approx (\Lambda/M) \approx O(10^{-14})$. Therefore the baryon asymmetry produced by monopole collisions should be negligible. We have not, however, considered nonperturbative contributions to ε. (The possibility that monopole catalysis might lead to a baryon asymmetry was suggested to us by C. Hill.)

8.4 Low-Temperature Scenarios

In the standard scenario the temperature at which the baryon asymmetry evolves is set by the mass M of the particle (denoted by "X") whose decays violate B, C, and CP conservation. In turn, the unification scale M_G sets the mass of the X: $M \approx M_G$. The quantity $K = (\Gamma_D/H)|_{T=M} \approx 3 \times 10^{17} \alpha$ GeV/M, which measures the effectiveness of decay processes, is an indicator of whether or not Xs will become overabundant when $T \lesssim M$ and provide the required departure from equilibrium. In the simplest GUTs $M \approx M_G \gtrsim O(10^{14} \text{ GeV})$, and K is at most a little larger than unity. In a low-energy scenario (10^{14} GeV $\gg M_G \sim 1\text{--}10^6$ TeV), K is naturally much greater than unity making baryogenesis difficult, if not impossible.

However, by suppressing decays, e.g. by making the coupling of the X to other particles very weak (α small), or by making its decay a three-body decay mediated by a massive boson, K can be order unity or even less if M is $\ll 10^{14}$ GeV. This alone may not be sufficient. If X carries a charge associated with a gauge symmetry that is unbroken at $T \approx M$ [e.g. for $M \gtrsim 1$ TeV, at least $SU(3)_C \otimes SU(2)_L \otimes U(1)_L$], then X-$\bar{\text{X}}$ annihilations will play an important role in maintaining an equilibrium abundance of Xs. The annihilation rate is $\Gamma_a \approx n_X (\sigma v)_{\text{ann}}$; for $T \lesssim M$, $\Gamma_a \approx (n_X/n_\gamma) T^3 \alpha^2/M^2$ since $(\sigma v)_{\text{ann}} \approx \alpha^2/M^2$. The effectiveness of annihilations, $\Gamma_a/H \approx (n_X/n_\gamma) z^{-1} (\alpha^2 m_{\text{pl}}/M)$, is large for $z \approx 1$ if $M \ll 10^{14}$ GeV (α = gauge coupling, $z \equiv M/T$). By including annihilations in the simplified set of equations used to analyze baryogenesis in Section 6, it is easy to show that

$$\Delta(z) \lesssim z^2 (M/\alpha^2 m_{\text{pl}}), \quad z \lesssim z_f$$

$$X(z_f) \lesssim X_{\text{EQ}}(z_f) \approx z_f^{3/2} \exp(-z_f),$$

$$n_B/s \lesssim (\varepsilon/g_*)(M/\alpha^2 m_{\text{pl}})[\ln(\alpha^2 m_{\text{pl}}/M)]^4, \qquad 60.$$

where $z_f \approx \ln(\alpha^2 m_{\text{pl}}/M) + \frac{1}{2}\ln[\ln(\alpha^2 m_{\text{pl}}/M)]$ is the epoch when annihilations freeze out ($\Gamma_a \approx H$); equality in Equations 60 holds only if decays are neglected. For $M \ll 10^{14}$ GeV the baryon asymmetry produced is very small. This difficulty can only be avoided if X transforms as a singlet under the gauge symmetry that remains unbroken at $T \approx M$.

Assuming that this is the case, then the only concern is suppressing the total decay rate Γ_D. A simple way to do this is to have the dominant decay mode be to a three-body final state, and be mediated by a heavy boson (in analogy to $\mu \rightarrow e\bar{\nu}\nu$). Then $\Gamma_D \approx \alpha_X^2 M^5/M_I^4$, where α_X is the coupling of the X to the intermediate boson and M_I is the mass of the intermediate boson. The condition for out-of-equilibrium decay ($K \lesssim 1$) is

$$M/M_I^{4/3} \lesssim g_*^{1/6} \alpha_X^{-2/3} m_{\text{pl}}^{-1/3}. \qquad 61.$$

The right-hand side of Equation 61 is 2×10^{-5} when $\alpha_X \approx 10^{-2}$ and all masses are measured in GeV.

Finally, there is the issue of CP violation in these scenarios. Recall $\varepsilon = (r - \bar{r})(B_1 - B_2)$, where r ($1 - r$) is the branching ratio to channel 1 (2) with baryon number B_1 (B_2). If $r \ll 1$ (as can occur in these low-energy scenarios) then the X can almost be assigned a baryon number ($= B_2$) so that its decays conserve B, and $\varepsilon \lesssim O(r)$ will be very small. Also recall from Section 5 that $(r - \bar{r})$ arises from the interference of the tree graph with loop graphs, which must have particles in the loops whose interactions violate B. Unless the masses of the particles in the loops M_{loop} are also $O(M)$, ε will be suppressed by powers of M/M_{loop}.

While it is difficult to generate a baryon asymmetry at low temperature, it is not impossible. A scenario based on $SU(2)_L \otimes SU(2)_R \otimes U(1)_{B-L}$ where $X = N_R$ (the right-handed neutrino) and $M = 10^4$ GeV has been discussed by Masiero & Mohapatra (40). Other low-temperature scenarios have also been considered (108a, 108b).

8.5 Initial Conditions

Thus far all our discussions of baryogenesis have been based upon the assumption of "standard initial conditions": (a) initial temperature $T_0 \sim 10^{19}$ GeV $\gg M$; and (b) equilibrium distributions for all particle species (with chemical potential $\mu \ll T_0$) at $z_0 = M/T_0$. Is there any reason to question (or believe for that matter!) the validity of these assumptions? Two-body collisions of light particles mediated by massless gauge bosons occur rapidly for $T \lesssim 10^{16}$ GeV (61), and at these temperatures should establish equilibrium distributions for light species. However, if $K \lesssim 1$, the ID production rate of Xs is never rapid ($\Gamma_{\text{ID}} < H$), and, unless Xs were initially present in equilibrium numbers, they will never be. [If Xs carry a charge associated with an unbroken gauge symmetry, inverse annihilations may help in this regard.] In the inflationary scenarios (53–55), the Universe undergoes a first-order phase transition during the spontaneous symmetry breaking (SSB) of the GUT, which results in supercooling and then eventually reheating to $T \approx O(M_G)$ (M_G is the scale of SSB). Clearly $z_0 \approx M/M_G$ is not much less than one in these scenarios; in fact, depending upon M, z_0 could be $\gtrsim 1$. Motivated by both the lack of compelling reasons for the standard initial conditions and the above-mentioned considerations, we review the effect of initial conditions on baryogenesis (109–113), using the simple model developed in Section 6.

ARBITRARY z_0, $0 \leq X(z_0) \leq$ FEW $X_{\text{EQ}}(z_0)$ Here we assume that all light species are present with equilibrium distributions, but that the initial abundance of the X bosons is not necessarily given by its equilibrium abundance. The only restriction we make here on $X(z_0)$ is that it is not much greater than $X_{\text{EQ}}(z_0)$. Using the simple model of Section 6, we summarize the results of (113).

1. $K \lesssim 1$—The ID and $2 \leftrightarrow 2$ scattering processes are never effective. Those X bosons present initially, and the few that are produced by inverse decays, eventually freely decay producing a baryon asymmetry that is not

subsequently damped since $K \lesssim 1$:

$$n_B/s \cong \varepsilon X(z_0) + (\varepsilon/g_*)K^2$$
$$\times \begin{cases} 1 & (z_0 \lesssim 1) \\ (1+z_0 K)^{-1} z_0^{5/2} \exp(-z_0)[z_0^{5/2} \exp(-z_0) + (A\alpha)z_0^{-4}] & (z_0 \gtrsim 1). \end{cases} \quad 62.$$

2. $K \gtrsim 1$—Both Ds and IDs (and possibly $2 \leftrightarrow 2$ scatterings) are occurring rapidly ($\Gamma > H$) for $1 \lesssim z_0 \lesssim z_f$. [Recall, $z_f \approx \max\{4.2(\ln K)^{0.6}, (A\alpha K)^{1/4}\}$.] For this reason, regardless of $X(z_0)$, if $z_0 \lesssim z_f$, $B(z)$ and $\Delta(z)$ rapidly approach their values in the standard scenario, and the final asymmetry is the same as that in the standard scenario. On the other hand, if $z_0 \gtrsim z_f$, IDs and $2 \leftrightarrow 2$ scatterings are never effective ($\Gamma < H$), and those Xs present initially and the few that are produced by IDs decay freely, producing an asymmetry that is also given by Equation 62.

From these results we can draw the following conclusions. With or without equilibrium numbers of Xs initially and $z_0 \ll 1$, the basic scenario works. Since $z_f \approx O(10)$, for $K \gtrsim 1$ if $z_0 \lesssim O(10)$ the initial conditions $[z_0, X(z_0)]$ are irrelevant. Regardless of K, for z_0 much greater than 10, the baryon excess that evolves decreases rapidly with increasing z_0.

INFLATIONARY UNIVERSE SCENARIOS The reheating process (conversion of vacuum energy to radiation) occurs as the oscillations (about the minimum of the potential) of the Higgs field responsible for SSB are damped by the creation of particles that couple to it (114, 115). The particle species coupling to this Higgs field directly all receive superheavy masses. Thus it is possible that after reheating the energy density of the Universe $[\rho(z_0) \approx M_G^4]$ is primarily in superheavy species, in particular, X bosons. In the case where initially $\rho(z_0) \approx \rho_X$, the simple model of Section 6 must be modified to take into account the entropy produced when the Xs decay into light species. In Table 1 we summarize the results of (113) for this scenario.

Table 1 Baryon asymmetry produced when $\rho(z_0) \approx M_G^4 \approx \rho_X$[a]

	$M > M_G$			$M_G > M$	
$K \lesssim g_*^{-1/2}(M_G/M)^2$	$K \gtrsim g_*^{-1/2}(M_G/M)^2$			$K \gtrsim 1$	$K \lesssim 1$
	$g_*^{1/4} \dfrac{M}{M_G} \gtrsim z_f$	$g_*^{1/4} \dfrac{M}{M_G} \lesssim z_f$			
$n_B/s \cong \varepsilon K^{1/2}$	$n_B/s \cong g_*^{-1/4}(M_G/M)\varepsilon$	standard result[b]		standard result[b]	$n_B/s \cong \varepsilon K^{1/2}$

[a] Such an initial condition is possible in the inflationary Universe scenario.
[b] By "standard result" we mean the value of n_B/s that evolves from standard initial conditions: initial temperature $\gg M$ and equilibrium abundances (see Section 6.2 and Figure 7).

Note that in some instances n_B/s is O(10–100) larger than the maximum value that can be attained in the standard scenario ($\approx \varepsilon/g_*$).

8.6 *The Lepton Number*

Although the present baryon number of the Universe is fairly well determined, the lepton number of the universe is not. A large lepton number (or equivalently $\mu_\nu \gtrsim T$) could well be hidden in the undetected neutrino background. A lepton number, L, defined as

$$L \equiv n_L/s = [n_{e^-} - n_{e^+} + \sum_i (n_{\nu_i} - n_{\bar{\nu}_i})]/s, \qquad 63.$$

as large as 10^4 is consistent with the limit on the mass density of the universe. A lepton number greater than O(1) would have several interesting cosmological effects. First, it would change significantly the predictions of primordial nucleosynthesis (117–119). A degenerate neutrino sea would also prevent the high-temperature restoration of spontaneously broken symmetries (56). Scenarios where neutrinos provide the dark matter of the universe would have to be modified if $L \gtrsim O(1)$ (120). Finally, a large neutrino chemical potential could make detection of the neutrino background radiation possible.

Just as the baryon number of the Universe may be understood in GUTs, the lepton number of the Universe may also be determined by GUT interactions. The possibility of generating a lepton number much greater than the baryon number depends crucially upon initial conditions (116).

Symmetric Initial Conditions: If we assume that initially all fermion species have zero chemical potentials, the initial B and L (hence $B-L$) are zero. In models such as SU(5) where $B-L$ is conserved, the L generated will be equal in magnitude to B, hence small today. In models such as SO(10) where $B-L$ is not conserved, a nonzero $B-L$ may be generated in the same manner as a nonzero B is generated (59). For some range of parameters L may be 100 times larger than B, which is still too small to be cosmologically interesting.

Degenerate Initial Conditions: Suppose that initially one or more fermion species is degenerate. As discussed in Section 7.3, the rapid ($\Gamma > H$) vector-mediated interactions will share the fermion asymmetries among all fermion species in a given irreducible representation. In SU(5) there are two irreducible fermion representations, $\bar{5}_f$ and 10_f. Any arbitrary fermion asymmetry can be expressed as a combination of the asymmetry in 5_f, Π_5, and the asymmetry in 10_f, Π_{10}. For instance $B = \Pi_5 + \Pi_{10}$, while $B-L = 3\Pi_5 + \Pi_{10}$. The quantum numbers Π_5 and Π_{10} are conserved by vector interactions, and in the limit that scalar interactions can be ignored (which is justified if $K_S \leq 1$, as is often the case) Π_5 and Π_{10} will be conserved quantum numbers. A large $B-L$, but a small B, is possible if initially both

Π_5 and Π_{10} are large, but $\Pi_5 \approx -\Pi_{10}$. An arbitrarily large L will eventually (before nucleosynthesis) relax to an $L \approx O(1)$ (121, 122), so in this scenario a small $B \approx 10^{-10}$ and a lepton number $L \approx O(1)$ are compatible.

There are two unseemly problems with this scenario. First, it is necessary to have two large numbers, Π_5 and Π_{10}, that almost exactly cancel. Second, there is no physical reason to expect either Π_5 or Π_{10} to be large. The first problem can be nicely addressed in SO(10) models, where $\Pi_5 + \Pi_{10} = 0$ is "natural" (59). The second objection cannot be addressed, other than to say that a large initial fermion degeneracy cannot be ruled out.

9. SUMMARY AND CONCLUDING REMARKS

The out-of-equilibrium decay scenario is a rather simple model for producing (and thereby explaining) the baryon number of the universe, $B \approx 10^{-10}$. The scenario works quite well for bosons more massive than about 10^{14} GeV and requires only a modest C, CP violation, $O(10^{-8})$, in the superheavy boson system. Because Higgs bosons couple more weakly to fermions than do gauge bosons, it is easier for their decays to occur out of equilibrium. In addition the C, CP violation in their decays generally occurs at one order lower in perturbation theory than for gauge bosons. For both of these reasons a Higgs boson seems to be the more likely candidate for producing the baryon asymmetry.

Baryogenesis is a very attractive framework for understanding the origin of the baryon asymmetry. It ties together two puzzles, the tiny CP violation that occurs in Nature, and the small baryon number of the Universe, which, of course, is necessary to guarantee the existence of matter in the Universe. In explaining the baryon number of the Universe, it reduces the number of parameters that must be specified as initial data in the standard big-bang cosmology. Although it is not yet possible to compute precisely the magnitude or even the sign of the baryon asymmetry, this is only because, at present, a complete unified model is still lacking. We hope we have demonstrated that, given a definite model, the baryon asymmetry can be computed in a straightforward manner.

As we discussed, if the unification scale is greater than about 10^{16} GeV, then the early Universe may be the only place where baryon number violation (and perhaps even unification) can be studied. Although baryogenesis is not yet on the same firm footing as big-bang nucleosynthesis, it has already been of use in constraining unified models. For instance, as we discussed in Section 7.1, baryogenesis and the minimal SU(5) model with only three families of fermions are not compatible. In Section 7.2, we showed that in order for an acceptable baryon asymmetry to evolve, the C symmetry of SO(10) must be broken at a scale not too disparate from the unification scale. Baryogenesis appears to be very difficult in low-energy unification scenarios and in supersymmetric unified models with dimension-5 operators that violate B conservation.

If indeed baryogenesis is an accurate accounting of the origin of baryon number of the Universe, then there are important cosmological implications as well. First, the existence of the baryon asymmetry implies that the Universe was once hotter than about 10^{14} GeV, and thereby extends our knowledge of its evolution back to within 10^{-35} s of the bang. In the standard cosmology the value of n_B/s is fixed only by parameters of the unified model, and is thus spatially constant. This favors adiabatic initial density fluctuations, as it rules out isothermal fluctuations (86, 123). [In nonstandard cosmological models it is possible to have both baryogenesis and isothermal perturbations (124, 125). Isothermal fluctuations are characterized by $\delta(n_B/s) \neq 0$, while adiabatic perturbations have $\delta(n_B/s) = 0$.] In the absence of very unusual initial conditions the baryon and lepton numbers of the Universe should be comparable, which suggests that the neutrino seas are not degenerate (126, 127). Finally, since the ratio of baryon number to entropy generated is at most $(\varepsilon/g_*) \lesssim O(10^{-4})$, a limit can be placed on the total entropy production occurring after baryogenesis—no more than a factor of about 10^6.

Although filling in all the details of baryogenesis awaits *the* Grand Unified Theory, baryogenesis may well be the most impressive achievement of unification to date.

ACKNOWLEDGMENTS

We would like to thank many people with whom we have collaborated in studying grand unified theories and the origin of the baryon asymmetry: J. N. Fry, J. A. Harvey, K. A. Olive, S. Raby, D. B. Reiss, G. Segrè, and S. Wolfram. This work was supported in part by the Department of Energy at The University of Chicago (through contract DE-ACO2 80ER 10773A003) and at Los Alamos National Laboratory.

Literature Cited

1. Weinberg, S. 1972. *Gravitation and Cosmology*, pp. 469–609. New York: Wiley
2. Wagoner, R. V. 1980. In *Physical Cosmology, Proc. Les Houches Summer Sch., XXXII*, ed. R. Balian, J. Auduoze, D. N. Schramm, pp. 395–442. New York/Amsterdam: North-Holland
3. Steigman, G. 1976. *Ann. Rev. Astron. Astrophys.* 14:339–72
4. Stecker, F. W. 1981. *Ann. NY Acad. Sci.* 375:69
4a. Stecker, F. W. 1982. In *Proc. 1981 Oxford Symp. on Progress in Cosmology*, ed. A. Wolfendale, p. 1. Dordrecht: Reidel
5. Buffington, A., Schindler, S. M. 1981. *Astrophys. J.* 247:L105
6. Golden, R. L., et al. 1979. *Phys. Rev. Lett.* 43:1196–1200
7. Bogomolov, E. A., et al. 1979. In *Proc. 16th Int. Cosmic Ray Conf.* 1:33, Kyoto
8. Stecker, F. W., Protheroe, R. J.,

Kazanas, D. 1981. In *Proc. 17th Int. Cosmic Ray Conf.*, Paris
9. Gaisser, T. K. 1982. In *The Birth of the Universe*, ed. J. Auduoze, J. Tran Thanh Van, p. 347. Gif sur Yvette: Editions Frontiers
10. Berezinsky, V. S., Ginzburg, V. L. 1981. *Mon. Not. R. Astron. Soc.* 194:3–14
11. Learned, J. G., Stecker, F. W. 1980. *Proc. Neutrino '79*, ed. A. Haatuft, C. Jarlskog. Bergen, Norway
11a. Brown, R. W., Stecker, F. W. 1982. *Phys. Rev.* D26:373
12. Peebles, P. J. E. 1980. See Ref. 2, pp. 213–70
13. Olive, K. A., Schramm, D. N., Steigman, G., Turner, M. S., Yang, J. 1981. *Astrophys. J.* 246:557
14. Yang, J., Turner, M. S., Steigman, G., Schramm, D. N., Olive, K. A. 1983. Univ. Chicago preprint. Submitted to *Astrophys. J.*
15. Sakharov, A. D. 1967. *Zh. Eksp. Teor. Fiz. Pis'ma* 5:32 (transl. *JETP Lett.* 5:24)
16. Parker, L. 1977. In *Asymptotic Structure of Space-Time*, ed. F. P. Esposito, L. Witten, pp. 118, 160. New York: Plenum
17. Weinberg, S. 1964. In *Lectures on Particles and Fields*, ed. S. Deser, K. Ford, p. 482. Englewood Cliffs, NJ: Prentice-Hall
18. Learned, J., Reines, F., Soni, A. 1979. *Phys. Rev. Lett.* 43:907–10
19. Cherry, M. L., et al. 1981. *Phys. Rev. Lett.* 47:1507
20. Krishnaswamy, M. R., et al. 1982. *Phys. Lett. B* 118:461
21. Battistoni, G., et al. 1982. *Phys. Lett. B* 118:461
22. Bartelt, J., et al. 1983. *Phys. Rev. Lett.* 50:651
23. Ramond, P. 1983. *Ann. Rev. Nucl. Part. Sci.* 33:31–66
24. Langacker, P. 1981. *Phys. Rep.* 72:185
25. Goldhaber, M., Sulak, L. 1981. *Comments Nucl. Part. Phys.* 10:215
26. Christenson, J. H., Cronin, J. W., Fitch, V. L., Turlay, R. 1965. *Phys. Rev. Lett.* 13:138; *Phys. Rev. B* 140:74
27. Kleinknecht, K. 1976. *Ann. Rev. Nucl. Sci.* 26:1–50
28. Aharony, A. 1973. In *Modern Developments in Thermodynamics*, pp. 95–114. New York: Wiley
29. Dolgov, A. D. 1979. *Pis'ma Zh. Eksp. Teor. Fiz.* 29:254–58
30. Kolb, E. W., Wolfram, S. 1980. *Nucl. Phys. B* 172:224–84
30a. Barr, S. M. 1979. *Phys. Rev. D* 19:3803–7
31. Yoshimura, M. 1978. *Phys. Rev. Lett.* 41:281–84; (E) 42:746
32. Toussaint, D., Treiman, S. B., Wilczek, F., Zee, A. 1979. *Phys. Rev. D* 19:1036–45
32a. Toussaint, D., Wilczek, F. 1979. *Phys. Lett. B* 81:238–40
33. Dimopoulos, S., Susskind, L. 1978. *Phys. Rev. D* 18:4500–9
34. Dimopoulos, S., Susskind, L. 1979. *Phys. Lett. B* 81:416–18
35. Papastamatiou, N. J., Parker, L. 1979. *Phys. Rev. D* 19:2283
36. Ignatiev, A. Yu., Krasnikov, N. V., Kuzmin, V. A., Tavkhelidze, A. N. 1978. *Phys. Lett. B* 76:486
37. Ellis, J., Gaillard, M. K., Nanopoulos, D. V. 1979. *Phys. Lett. B* 80:360–64; (E) 82:464
38. Weinberg, S. 1979. *Phys. Rev. Lett.* 42:850–53
39. Yanagida, T., Yoshimura, M. 1980. *Phys. Rev. Lett.* 45:71–74
39a. Harvey, J. A., Kolb, E. W., Reiss, D. B., Wolfram, S. 1981. *Nucl. Phys. B* 177:456–60
40. Masiero, A., Mohapatra, R. N. 1981. *Phys. Lett. B* 103:343
41. Kolb, E. W., Wolfram, S. 1980. *Phys. Lett. B* 91:217
42. Fry, J. N., Olive, K. A., Turner, M. S. 1980. *Phys. Rev. D* 22:2953
43. Fry, J. N., Olive, K. A., Turner, M. S. 1980. *Phys. Rev. D* 22:2977
44. Fry, J. N., Olive, K. A., Turner, M. S. 1980. *Phys. Rev. Lett.* 45:2074–77
45. Lee, T. D., Yang, C. N. 1955. *Phys. Rev.* 98:1501
46. 't Hooft, G. 1976. *Phys. Rev. Lett.* 37:8
47. Nanopoulos, D. V., Weinberg, S. 1979. *Phys. Rev. D* 20:2484–93
48. Kobayashi, M., Maskawa, T. 1973. *Prog. Theor. Phys.* 49:652
49. Weinberg, S. 1976. *Phys. Rev. Lett.* 37:657
50. Kobzarev, I., Okun, L., Zeldovich, Ya. B. 1974. *Phys. Lett. B* 50:340
51. Zeldovich, Ya. B., Kobzarev, I., Okun, L. 1974. *Zh. Eksp. Teor. Fiz.* 67:3
51a. Mohapatra, R. N., Senjanovic, G. 1980. *Phys. Rev. D* 21:3470
51b. Kuzmin, V. A., Tkachev, I. I., Shaposhnikov, M. E. 1981. *Pis'ma Zh. Eksp. Teor. Fiz.* 33:557–60 (transl. *JETP Lett.* 33:540–43)
52. Sato, K. 1981. *Phys. Lett. B* 99:66–70
53. Guth, A. H. 1981. *Phys. Rev. D* 23:347
54. Linde, A. 1982. *Phys. Lett. B* 108:389
55. Albrecht, A., Steinhardt, P. J. 1982. *Phys. Rev. Lett.* 48:1220
56. Linde, A. 1976. *Phys. Rev. D* 14:3345
57. Masiero, A., Mohapatra, R. N., Peccei, R. D. 1982. *Phys. Lett. B* 108:111
58. Ellis, J., Gaillard, M. K., Nanopoulos,

D. V., Rudaz, S. 1981. *Phys. Lett. B* 99:101
59. Harvey, J. A., Kolb, E. W., Reiss, D. B., Wolfram, S. 1982. *Nucl. Phys. B* 201:16–100
60. Yanagida, T., Yoshimura, M. 1980. *Nucl. Phys. B* 168:534–38
60a. Yoshimura, M. 1979. *Phys. Lett. B* 88:294
60b. Honda, M., Yoshimura, M. 1979. *Prog. Theor. Phys.* 62:1704
61. Ellis, J., Steigman, G. 1980. *Phys. Lett. B* 89:186
62. Fry, J. N., Turner, M. S. 1983. *Phys. Lett. B* 125:379–84
63. Treiman, S. B., Wilczek, F. 1980. *Phys. Lett. B* 95:222–26
64. Georgi, H., Glashow, S. L. 1974. *Phys. Rev. Lett.* 32:438
65. Georgi, H., Quinn, H., Weinberg, S. 1974. *Phys. Rev. Lett.* 33:451
66. Buras, A., Ellis, J., Gaillard, M. K., Nanopoulos, D. V. 1978. *Nucl. Phys. B* 135:66
67. Goldman, T., Ross, D. 1980. *Nucl. Phys. B* 171:273
68. Marciano, W. J. 1979. *Phys. Rev. D* 20:274
69. Barr, S. M., Segrè, G., Weldon, H. A. 1979. *Phys. Rev. D* 20:2494
70. Yildiz, A., Cox, P. 1980. *Phys. Rev. D* 21:906–9
70a. Ignatev, A. Yu., Kuzmin, V. A., Shaposhnikov, M. E. 1979. *Pis'ma Zh. Eksp. Teor. Fiz.* 30:726–30
71. Segrè, G., Turner, M. S. 1981. *Phys. Lett. B* 99:339
72. Harvey, J. A., Kolb, E. W., Reiss, D. B., Wolfram, S. 1981. *Phys. Rev. Lett.* 47:391–94
73. Peccei, R. D., Quinn, H. R. 1977. *Phys. Rev. Lett.* 38:1440
74. Wise, M. B., Georgi, H., Glashow, S. L. 1981. *Phys. Rev. Lett.* 47:402
75. Masiero, A., Yanagida, T. 1982. *Phys. Lett. B* 109:353
76. Georgi, H. 1975. In *Particles and Field*, ed. C. E. Carlson. New York: AIP. 575 pp.
77. Fritzsch, H., Minkowski, P. 1975. *Ann. Phys.* 94:193
78. Slansky, R. 1981. *Phys. Rep.* 79:128
79. Gürsey, F., Ramond, P., Sikivie, P. 1975. *Phys. Lett. B* 60:177
80. Yanagida, T., Yoshimura, M. 1981. *Phys. Rev. D* 23:2048
80a. Fukugita, M., Yanagida, T., Yoshimura, M. 1981. *Phys. Lett. B* 106:183
81. Haber, H. E., Segrè, G., Soni, S. K. 1982. *Phys. Rev. D* 25:1400
81a. Kuzmin, V. A., Shaposhnikov, M. E. 1980. *Phys. Lett. B* 92:115–17
82. Hawking, S. W. 1974. *Nature* 248:30
83. Hawking, S. W. 1975. *Commun. Math. Phys.* 43:199
84. Carr, B. J. 1975. *Astrophys. J.* 201:1
85. Hawking, S. W. 1975. Unpublished
86. Turner, M. S., Schramm, D. N. 1979. *Nature* 279:303–4
87. Turner, M. S. 1979. *Phys. Lett. B* 89:155
88. Barrow, J. D. 1980. *Mon. Not. R. Astron. Soc.* 192:427
89. Lindley, D. 1981. *Mon. Not. R. Astron. Soc.* 196:317–38
89a. Dolgov, A. D. 1980. *Zh. Eksp. Teor. Fiz.* 79:337–49
89b. Grillo, A. F. 1980. *Phys. Lett. B* 94:364
89c. Krauss, L. M. 1982. *Phys. Rev. Lett.* 49:1459
90. Page, D. N., Hawking, S. W. 1976. *Astrophys. J.* 206:1
91. Carr, B. J. 1976. *Astrophys. J.* 206:8–25
92. Gol'fand, Y. A., Likhtman, E. P. 1971. *JETP Lett.* 13:323
93. Volkov, D., Akulov, V. P. 1973. *Phys. Lett. B* 46:109
94. Wess, J., Zumino, B. 1974. *Nucl. Phys. B* 70:39
95. Dimopoulos, S., Raby, S. 1981. *Nucl. Phys. B* 192:353
96. Witten, E. 1981. *Nucl. Phys. B* 188:513
97. Dimopoulos, S., Raby, S., Wilczek, F. 1981. *Phys. Rev. D* 24:1681
98. Dine, M., Fischler, W., Srednicki, M. 1981. *Nucl. Phys. B* 189:575
99. Kolb, E. W., Raby, S. 1983. *Phys. Rev. D* 27:2990–96
100. Einhorn, M. B., Jones, D. R. T. 1982. *Nucl. Phys. B* 196:475
101. Haber, H. E. 1982. *Phys. Rev. D* 16:1317
101a. Nanopoulos, D. V., Tamvakis, K. 1982. *Phys. Lett. B* 114:235
102. Callan, C. G. 1982. *Phys. Rev. D* 25:2141; 26:2058
103. Rubakov, V. A. 1982. *Nucl. Phys. B* 203:311
104. Rubakov, V. A. 1981. *Pis'ma Zh. Eksp. Teor. Fiz.* 33:658 (transl. *JETP Lett.* 33:644)
105. Wilczek, F. 1982. *Phys. Rev. Lett.* 48:1146
106. 't Hooft, G. 1974. *Nucl. Phys. B* 79:276
107. Polyakov, A. 1974. *Pis'ma Zh. Eksp. Teor. Fiz.* 20:430 (transl. *JETP Lett.* 20:194)
108. Ellis, J., Nanopoulos, D. V., Olive, K. A. 1982. *Phys. Lett. B* 116:127
108a. Masiero, A., Senjanovic, G. 1982. *Phys. Lett. B* 108:191
108b. Masiero, A., Nieves, J. F., Yanagida, T. 1982. *Phys. Lett. B* 116:11
109. Hut, P., Klinkhamer, F. K. 1982. *Astron. Astrophys.* 106:245

110. Turner, M. S., Fry, J. N. 1981. *Phys. Rev. D* 24: 3341
111. Turner, M. S. 1982. *Phys. Rev. D* 25: 299
112. Dolgov, A. D., Linde, A. D. 1982. *Phys. Lett. B* 116: 329
113. Fry, J. N., Kolb, E. W., Turner, M. S. 1983. *Fermilab Preprint*
114. Albrecht, A., Steinhardt, P. J., Turner, M. S., Wilczek, F. 1982. *Phys. Rev. Lett.* 48: 1437
115. Abbott, L., Fahri, E., Wise, M. B. 1982. *Phys. Lett. B* 117: 29
116. Harvey, J. A., Kolb, E. W. 1981. *Phys. Rev. D* 24: 2090–99
117. Wagoner, R. F., Fowler, W. A., Hoyle, F. 1967. *Astrophys. J.* 148: 3
118. Yahil, A., Beaudet, G. 1976. *Astrophys. J.* 206: 261
119. David, Y., Reeves, H. 1980. See Ref. 2, pp. 443–64
120. Freese, K., Kolb, E. W., Turner, M. S. 1983. *Phys. Rev. D* 27: 1689–95
121. Langacker, P., Segrè, G., Soni, S. 1982. *Phys. Rev. D* 26: 3425
122. Fry, J. N., Hogan, C. J. 1982. *Phys. Rev. Lett.* 49: 1873
123. Lindley, D. 1981. *Nature* 291: 133–34
124. Barrow, J. D., Turner, M. S. 1981. *Nature* 291: 469
125. Bond, J. R., Kolb, E. W., Silk, J. 1982. *Astrophys. J.* 255: 341–60
126. Dimopoulos, S., Feinberg, G. 1979. *Phys. Rev. D* 20: 1283
127. Turner, M. S. 1981. *Phys. Lett. B* 98: 145–48

CUMULATIVE INDEXES

CONTRIBUTING AUTHORS VOLUMES 23–33

A

Alexander, J. M., 24:279–339
Allison, W. W. M., 30:253–98
Amaldi, U., 26:385–456
Andersen, J. U., 33:453–504
Appelquist, T., 28:387–499
Arad, B., 24:35–67
Arianer, J., 31:19–51
Arima, A., 31:75–105
Arnold, J. R., 33:505–37

B

Ballam, J., 27:75–138
Barnett, R. M., 28:387–499
Barschall, H. H., 28:207–37
Baym, G., 25:27–77
Bég, M. A. B., 24:379–449
Benczer-Koller, N., 30:53–84
Ben-David, G., 24:35–67
Berko, S., 30:543–81
Berry, H. G., 32:1–34
Bertrand, F. E., 26:457–509
Bienenstock, A., 28:33–113
Birkelund, J. R., 33:265–322
Blann, M., 25:123–66
Blok, H. P., 33:569–609
Bloom, E. D., 33:143–97
Bøggild, H., 24:451–513
Bohr, A., 23:363–93
Bonderup, E., 33:453–504
Brewer, J. H., 28:239–326
Brosco, G., 23:75–122
Bucksbaum, P. H., 30:1–52

C

Cahill, T. A., 30:211–52
Cerny, J., 27:331–51
Chinowsky, W., 27:393–464
Clayton, R. N., 28:501–22
Cline, D., 27:209–78
Cobb, J. H., 30:253–98
Cole, F. T., 31:295–335
Commins, E. D., 30:1–52
Conversi, M., 23:75–122
Cormier, T. M., 32:271–308
Crowe, K. M., 28:239–326

D

Darriulat, P., 30:159–210
de Forest, T. Jr., 25:1–26
DeTar, C. E., 33:235–64
Diamond, R. M., 30:85–157
Dieperink, A. E. L., 25:1–26
Donnelly, T. W., 25:329–405
Donoghue, J. F., 33:235–64
Dover, C. B., 26:239–317
Drees, J., 33:385–452

E

Ellis, J., 32:443–97

F

Fabjan, C. W., 32:335–89
Farley, F. J. M., 29:243–82
Feld, M. S., 29:411–54
Ferbel, T., 24:451–513
Fernow, R. C., 31:107–44
Fick, D., 31:53–74
Fisk, H. E., 32:499–573
Fleury, A., 24:279–339
Flocard, H., 28:523–96
Fox, G. C., 23:219–313
Franzini, P., 33:1–29
Freedman, D. Z., 27:167–207
Freedman, M. S., 24:209–47
French, J. B., 32:35–64
Fry, W. F., 27:209–78
Fulbright, H. W., 29:161–202

G

Gaillard, M. K., 32:443–97
Gaisser, T. K., 30:475–542
Geller, R., 31:19–51
Gentry, R. V., 23:347–62
Gibson, W. M., 25:465–508
Girardi, G., 32:443–97
Glashausser, C., 29:33–68
Goeke, K., 32:65–115
Goldhaber, A. S., 28:161–205
Goldhaber, G., 30:337–81
Goodman, A. L., 26:239–317
Gottschalk, A., 29:283–312

Goulding, F. S., 23:45–74;
  25:167–240
Greenberg, O. W., 28:327–86
Grunder, H. A., 27:353–92
Guinn, V. P., 24:561–91

H

Hansen, P. G., 29:69–119
Hardy, J. C., 27:333–51
Harvey, B. G., 25:167–240
Hass, M., 30:53–84; 32:1–34
Hecht, K. T., 23:123–61
Heckman, H. H., 28:161–205
Heisenberg, J., 33:569–609
Herrmann, G., 32:117–47
Hirsch, R. L., 25:79–121
Hoffman, D. C., 24:151–207
Hoffman, M. M., 24:151–207
Hughes, V. W., 33:611–44
Huizenga, J. R., 27:465–547;
  33:265–322
Hung, P. Q., 31:375–438

I

Iachello, F., 31:75–105

J

Jacob, M., 26:385–456
Jaklevic, J. M., 23:45–74
Jackson, A. D., 33:105–41

K

Keefe, D., 32:391–441
Keller, O. L. Jr., 27:139–66
Kim, Y. E., 24:69–100
Kirsten, F. A., 25:509–54
Kleinknecht, K., 26:1–50
Kohaupt, R. D., 33:67–104
Kolb, E. W., 33:645–96
Koltun, D. S., 23:163–92
Kota, V. K. B., 32:35–64
Krisch, A. D., 31:107–44
Kuo, T. T. S., 24:101–50
Kuti, J., 33:611–44

L

Lach, J., 29:203–42
Lal, D., 33:505–37
Lande, K., 29:395–410
Lane, K., 28:387–499
Lattimer, J. M., 31:337–74
Lee-Franzini, J., 33:1–29
Lingenfelter, R. E., 32:235–69
Litherland, A. E., 30:437–73
Litt, J., 23:1–43
Ludlam, T., 32:335–89

M

Mahaux, C., 23:193–218; 29:1–31
Mark, J. C., 26:51–87
Matthiae, G., 26:385–456
Maurette, M., 26:319–50
McGrory, J. B., 30:383–436
Measday, D. F., 29:121–60
Meunier, R., 23:1–43
Meyerhof, W. E., 27:279–331
Miller, G. A., 29:121–60
Mills, F. E., 31:295–335
Montgomery, H. E., 33:383–452
Mottelson, B. R., 23:363–93
Müller, B., 26:351–83
Murnick, D. E., 29:411–54
Myers, W. D., 32:309–34

N

Neumann, R. D., 29:283–312

O

Oeschger, H., 25:423–63

P

Pantell, R. H., 33:453–504
Peck, C. W., 33:143–97
Pendleton, H. N., 30:543–81
Perl, M. L., 30:299–335
Pethick, C., 25:27–77
Picasso, E., 29:243–82
Pigford, T. H., 24:515–59
Pondrom, L., 29:203–42
Ponomarev, L. I., 23:395–430
Povh, B., 28:1–32
Primakoff, H., 31:145–92
Protopopescu, S. D., 29:339–93

Q

Quentin, P., 28:523–96
Quigg, C., 23:219–313

R

Ramaty, R., 32:235–69
Ramond, P., 33:31–66
Ramsey, N. F., 32:211–33
Reay, N. W., 33:539–68
Reedy, R. C., 33:505–37
Renton, P., 31:193–230
Rindi, A., 23:315–46
Rolfs, C., 28:115–59
Rosen, S. P., 31:145–92
Rosenfeld, A. H., 25:555–98

S

Sak, J., 30:53–84
Sakurai, J. J., 31:375–438
Samios, N. P., 29:339–93
Sanford, J. R., 26:151–98
Saudinos, J., 24:341–77
Scharff-Goldhaber, G., 26:239–317
Schramm, D. N., 27:37–74; 27:167–207
Schröder, W. U., 27:465–547
Schwitters, R. F., 26:89–149
Sciulli, F., 32:499–573
Seaborg, G. T., 27:139–66
Segrè, E., 31:1–18
Seki, R., 25:241–81
Selph, F. B., 27:353–92
Shifman, M. A., 33:199–233
Sidwell, R. A., 33:539–68
Silbar, R. R., 24:249–77
Simpson, J. A., 33:323–81
Sirlin, A., 24:379–449
Sivers, D., 32:149–75
Söding, P., 31:231–93
Sorba, P., 32:443–97
Speth, J., 32:65–115
Stanton, N. R., 33:539–68
Steigman, G., 29:313–37
Stephens, F. S., 30:85–157
Sternheim, M. M., 24:249–77
Strauch, K., 26:89–149
Swiatecki, W. J., 32:309–34

T

Taulbjerg, K., 27:279–331
Taylor, T. B., 25:406–21
Thomas, R. H., 23:315–46
Trautmann, N., 32:117–47
Trautvetter, H. P., 28:115–59
Tubbs, D. L., 27:167–207
Tubis, A., 24:69–100
Turner, M. S., 33:645–96

V

Vandenbosch, R., 27:1–35
Voss, G.-A., 33:67–104

W

Wagoner, R. V., 27:37–74
Wahlen, M., 25:423–63
Walecka, J. D., 25:329–405
Watt, R. D., 27:75–138
Weidenmüller, H. A., 29:1–31
Wetherill, G. W., 25:283–328
Wiegand, C. E., 25:241–81
Wilczek, F., 32:177–209
Wildenthal, B. H., 30:383–436
Wilkin, C., 24:341–77
Williams, W. S. C., 31:193–230
Winick, H., 28:33–113
Wiss, J. E., 30:337–81
Wolf, G., 31:231–93

Y

Yan, T.-M., 26:199–238
Yodh, G. B., 30:475–542
Yoshida, S., 24:1–33

CHAPTER TITLES, VOLUMES 23–33

PREFATORY CHAPTER

Fifty Years Up and Down a Strenuous and Scenic Trail	E. Segrè	31:1–18

ACCELERATORS

The Radiation Environment of High-Energy Accelerators	A. Rindi, T. H. Thomas	23:315–46
The Fermi National Accelerator Laboratory	J. R. Sanford	26:151–98
Heavy-Ion Accelerators	H. A. Grunder, F. B. Selph	27:353–92
Synchrotron Radiation Research	H. Winick, A. Bienenstock	28:33–113
Ultrasensitive Mass Spectrometry with Accelerators	A. E. Litherland	30:437–73
The Advanced Positive Heavy Ion Sources	J. Arianer, R. Geller	31:19–51
Increasing the Phase-Space Density of High Energy Particle Beams	F. T. Cole, F. E. Mills	31:295–335
Inertial Confinement Fusion	D. Keefe	32:391–441
Progress and Problems in Performance of e^+e^- Storage Rings	R. D. Kohaupt, G.-A. Voss	33:67–104

ASTROPHYSICS

Neutron Stars	G. Baym, C. Pethick	25:27–77
Element Production in the Early Universe	D. N. Schramm, R. V. Wagoner	27:37–74
The Weak Neutral Current and Its Effects in Stellar Collapse	D. Z. Freedman, D. N. Schramm, D. L. Tubbs	27:167–207
Experimental Nuclear Astrophysics	C. Rolfs, H. P. Trautvetter	28:115–59
Isotopic Anomalies in the Early Solar System	R. N. Clayton	28:501–22
Cosmology Confronts Particle Physics	G. Steigman	29:313–37
Experimental Neutrino Astrophysics	K. Lande	29:395–410
Baryon Number and Lepton Number Conservation Laws	H. Primakoff, S. P. Rosen	31:145–92
The Equation of State of Hot Dense Matter and Supernovae	J. M. Lattimer	31:337–74
Gamma-Ray Astronomy	R. Ramaty, R. E. Lingenfelter	32:235–69
Elemental and Isotopic Composition of the Galactic Cosmic Rays	J. A. Simpson	33:323–81
Cosmic-Ray Record in Solar System Matter	R. C. Reedy, J. R. Arnold, D. Lal	33:505–37
Grand Unified Theories and the Origin of the Baryon Asymmetry	E. W. Kolb, M. S. Turner	33:645–96

ATOMIC, MOLECULAR, AND SOLID STATE PHYSICS

Photon-Excited Energy-Dispersive X-Ray Fluorescence Analysis for Trace Elements	F. S. Goulding, J. M. Jaklevic	23:45–74
Molecular Structure Effects on Atomic and Nuclear Capture of Mesons	L. I. Ponomarev	23:395–430
Atomic Structure Effects in Nuclear Events	M. S. Freedman	24:209–47
Kaoniç and Other Exotic Atoms	R. Seki, C. E. Wiegand	25:241–81
K-Shell Ionization in Heavy-Ion Collisions	W. E. Meyerhof, K. Taulbjerg	27:279–331
Synchrotron Radiation Research	H. Winick, A. Bienenstock	28:33–113
Advances in Muon Spin Rotation	J. H. Brewer, K. M. Crowe	28:239–326
The Parity Non-Conserving Electron-Nucleon Interaction	E. D. Commins, P. H. Bucksbaum	30:1–52
Proton Microprobes and Particle-Induced X-Ray Analytical Systems	T. A. Cahill	30:211–52

699

Positronium	S. Berko, H. N. Pendleton	30:543–81
Channeling Radiation	J. U. Andersen, E. Bonderup, R. H. Pantell	33:453–504

BIOLOGY AND MEDICINE

Photon-Excited Energy-Dispersive X-Ray Fluorescence Analysis for Trace Elements	F. S. Goulding, J. M. Jaklevic	23:45–74
The Radiation Environment of High-Energy Accelerators	A. Rindi, R. H. Thomas	23:315–46
Environmental Aspects of Nuclear Energy Production	T. H. Pigford	24:515–59
Synchrotron Radiation Research	H. Winick, A. Bienenstock	28:33–113
Intense Sources of Fast Neutrons	H. H. Barschall	28:207–37
Diagnostic Techniques in Nuclear Medicine	R. D. Neumann, A. Gottschalk	29:283–313

CHEMISTRY

Molecular Structure Effects on Atomic and Nuclear Capture of Mesons	L. I. Ponomarev	23:395–430
Chemistry of the Transactinide Elements	O. L. Keller Jr., G. T. Seaborg	27:139–66
Advances in Muon Spin Rotation	J. H. Brewer, K. M. Crowe	28:239–326
Rapid Chemical Methods for Identification and Study of Short-Lived Nuclides	G. Herrmann, N. Trautmann	32:117–47

EARTH AND SPACE SCIENCES

Radioactive Halos	R. V. Gentry	23:347–62
Radiometric Chronology of the Early Solar System	G. W. Wetherill	25:283–328
Fossil Nuclear Reactors	M. Maurette	26:319–50
Isotopic Anomalies in the Early Solar System	R. N. Clayton	28:501–22
Cosmology Confronts Particle Physics	G. Steigman	29:313–37
Experimental Neutrino Astrophysics	K. Lande	29:395–410
Particle Collisions Above 10 TeV As Seen By Cosmic Rays	T. K. Gaisser, G. B. Yodh	30:475–542
Elemental and Isotopic Composition of the Galactic Cosmic Rays	J. A. Simpson	33:323–81
Cosmic-Ray Record in Solar System Matter	R. C. Reedy, J. R. Arnold, D. Lal	33:505–37

FISSION AND FUSION ENERGY

Environmental Aspects of Nuclear Energy Production	T. H. Pigford	24:515–59
Nuclear Safeguards	T. B. Taylor	24:407–21
Fossil Nuclear Reactors	M. Maurette	26:319–50
Inertial Confinement Fusion	D. Keefe	32:391–441

INSTRUMENTATION AND TECHNIQUES

Cerenkov Counter Technique in High-Energy Physics	J. Litt, R. Meunier	23:1–43
Photon-Excited Energy-Dispersive X-Ray Fluorescence Analysis for Trace Elements	F. S. Goulding, J. M. Jaklevic	23:45–74
Flash-Tube Hodoscope Chambers	M. Conversi, G. Brosco	23:75–122
The Radiation Environment of High-Energy Accelerators	A. Rindi, R. H. Thomas	23:315–46
Identification of Nuclear Particles	F. S. Goulding, B. G. Harvey	25:167–240
Low Level Counting Techniques	H. Oeschger, M. Wahlen	25:423–63
Blocking Measurements of Nuclear Decay Times	W. M. Gibson	25:465–508
Computer Interfacing for High-Energy Physics Experiments	F. A. Kirsten	25:509–54
The Particle Data Group: Growth and Operations—Eighteen Years of Particle Physics	A. H. Rosenfield	25:555–98
The Fermi National Accelerator Laboratory	J. R. Sanford	26:151–98

Hybrid Bubble-Chamber Systems	J. Ballam, R. D. Watt	27:75–138
Synchrotron Radiation Research	H. Winick, A. Bienenstock	28:33–113
Intense Sources of Fast Neutrons	H. H. Barschall	28:207–37
Advances in Muon Spin Rotation	J. H. Brewer, K. M. Crowe	28:239–326
Nuclei Far Away from the Line of Beta Stability: Studies by On-Line Mass Separation	P. G. Hansen	29:69–119
Hyperon Beams and Physics	J. Lach, L. Pondrom	29:203–42
The Muon (g-2) Experiments	F. J. M. Farley, E. Picasso	29:243–82
Diagnostic Techniques in Nuclear Medicine	R. D. Neumann, A. Gottschalk	29:283–312
Applications of Lasers to Nuclear Physics	D. E. Murnick, M. S. Feld	29:411–54
Transient Magnetic Fields at Swift Ions Traversing Ferromagnetic Media and Applications to Measurements of Nuclear Moments	N. Benczer-Koller, M. Hass, J. Sak	30:53–84
Proton Microprobes and Particle-Induced X-Ray Analytical Systems	T. A. Cahill	30:211–52
Relativistic Charged Particle Identification by Energy Loss	W. W. M. Allison, J. H. Cobb	30:253–98
Ultrasensitive Mass Spectrometry with Accelerators	A. E. Litherland	30:437–73
The Advanced Positive Heavy Ion Sources	J. Arianer, R. Geller	31:19–51
Increasing the Phase-Space Density of High Energy Particle Beams	F. T. Cole, F. E. Mills	31:295–335
Beam Foil Spectroscopy	H. G. Berry, M. Hass	32:1–34
Rapid Chemical Methods for Identification and Study of Short-Lived Nuclides	G. Herrmann, N. Trautmann	32:117–47
Electric-Dipole Moments of Particles	N. F. Ramsey	32:211–33
Calorimetry in High-Energy Physics	C. W. Fabjan, T. Ludlam	32:335–89
Inertial Confinement Fusion	D. Keefe	32:391–441
Progress and Problems in Performance of e^+e^- Storage Rings	R. D. Kohaupt, G.-A. Voss	33:67–104
Physics with the Crystal Ball Detector	E. D. Bloom, C. W. Peck	33:143–97
Channeling Radiation	J. U. Andersen, E. Bonderup, R. H. Pantell	33:453–504

NUCLEAR APPLICATIONS

Radioactive Halos	R. V. Gentry	23:347–62
Environmental Aspects of Nuclear Energy Production	T. H. Pigford	24:515–59
Applications of Nuclear Science in Crime Investigation	V. P. Guinn	24:561–91
Status and Future Directions of the World Program in Fusion Research and Development	R. L. Hirsch	25:79–121
Nuclear Safeguards	T. B. Taylor	25:407–21
Global Consequences of Nuclear Weaponry	J. C. Mark	26:51–87
Intense Sources of Fast Neutrons	H. H. Barschall	28:207–37
Diagnostic Techniques in Nuclear Medicine	R. D. Neumann, A. Gottschalk	29:283–312
Proton Microprobes and Particle-Induced X-Ray Analytical Systems	T. A. Cahill	30:211–52
Ultrasensitive Mass Spectrometry with Accelerators	A. E. Litherland	30:437–73
The Equation of State of Hot Dense Matter and Supernovae	J. M. Lattimer	31:337–74

NUCLEAR REACTION MECHANISMS—HEAVY PARTICLES

Reactions Between Medium and Heavy Nuclei and Heavy Ions of Less than 15 MeV/amu	A. Fleury, J. M. Alexander	24:279–339
Positron Creation in Superheavy Quasi-Molecules	B. Müller	26:351–83

CHAPTER TITLES

Heavy-Ion Accelerators	H. A. Grunder, F. B. Selph	27:353–92
Damped Heavy-Ion Collisions	W. U. Schröder, J. R. Huizenga	27:465–547
High Energy Interactions of Nuclei	A. S. Goldhaber, H. H. Heckman	28:161–205
Nuclei at High Angular Momentum	R. M. Diamond, F. S. Stephens	30:85–157
Polarization in Heavy Ion Reactions	D. Fick	31:53–74
Resonances in Heavy-Ion Nuclear Reactions	T. M. Cormier	32:271–308
Fusion Reactions Between Heavy Nuclei	J. R. Birkelund, J. R. Huizenga	33:265–322

NUCLEAR REACTION MECHANISMS—LIGHT PARTICLES

Meson-Nucleus Scattering at Medium Energies	M. M. Sternheim, R. R. Silbar	24:249–77
Proton-Nucleus Scattering at Medium Energies	J. Saudinos, C. Wilkin	24:341–77
Knock-Out Processes and Removal Energies	A. E. L. Dieperink, T. De Forest Jr.	25:1–26
Preequilibrium Decay	M. Blann	25:123–66
Excitation of Giant Multipole Resonances through Inelastic Scattering	F. E. Bertrand	26:457–509
Element Production in the Early Universe	D. N. Schramm, R. V. Wagoner	27:37–74
Experimental Nuclear Astrophysics	C. Rolfs, H. P. Trautvetter	28:115–59
High Energy Interactions of Nuclei	A. S. Goldhaber, H. H. Heckman	28:161–205
Intense Sources of Fast Neutrons	H. H. Barschall	28:207–37
Nuclear Physics with Polarized Beams	C. Glashausser	29:33–68
Hopes and Realities for the (p, π) Reaction	D. F. Measday, G. A. Miller	29:121–60
Alpha Transfer Reactions in Light Nuclei	H. W. Fulbright	29:161–202
Theory of Giant Resonances	K. Goeke, J. Speth	32:65–115
Rapid Chemical Methods for Identification and Study of Short-Lived Nuclides	G. Herrmann, N. Trautmann	32:117–47
Inelastic Electron Scattering from Nuclei	J. Heisenberg, H. P. Blok	33:569–609

NUCLEAR STRUCTURE

Symmetries in Nuclei	K. T. Hecht	23:123–61
Linear Relations Among Nuclear Energy Levels	D. S. Koltun	23:163–92
Intermediate Structure in Nuclear Reactions	C. Mahaux	23:193–218
The Many Facets of Nuclear Structure	A. Bohr, B. R. Mottelson	23:363–93
Resonance Fluorescence of Excited Nuclear Levels in the Energy Range 5–11 MeV	B. Arad, G. Ben-David	24:35–67
Shell-Model Effective Interactions and the Free Nucleon-Nucleon Interaction	T. T. S. Kuo	24:107–50
Post-Fission Phenomena	D. C. Hoffman, M. M. Hoffman	24:151–207
Knock-Out Processes and Removal Energies	A. E. L. Dieperink, T. de Forest Jr.	25:1–26
Preequilibrium Decay	M. Blann	25:123–66
Electron Scattering and Nuclear Structure	T. W. Donnelly, J. D. Walecka	25:329–405
Blocking Measurements of Nuclear Decay Times	W. M. Gibson	25:465–508
The Variable Moment of Inertia (VMI) Model and Theories of Nuclear Collective Motion	G. Scharff-Goldhaber, C. B. Dover, A. L. Goodman	26:239–317
Excitation of Giant Multipole Resonances through Inelastic Scattering	F. E. Bertrand	26:457–509
Spontaneously Fissioning Isomers	R. Vandenbosch	27:1–35
Delayed Proton Radioactivities	J. Cerny, J. C. Hardy	27:333–51
Hypernuclei	B. Povh	28:1–32
Self-Consistent Calculations of Nuclear Properties with Phenomenological Effective Forces	P. Quentin, H. Flocard	28:523–96
Nuclear Physics with Polarized Beams	G. Glashausser	29:33–68
Nuclei Far Away from the Line of Beta Stability: Studies by On-Line Mass Separation	P. G. Hansen	29:69–119
Hopes and Realities for the (p, π) Reaction	D. F. Measday, G. A. Miller	29:121–60
Alpha Transfer Reactions in Light Nuclei	H. W. Fulbright	29:161–202
Applications of Lasers to Nuclear Physics	D. E. Murnick, M. S. Feld	29:411–54

Transient Magnetic Fields at Swift Ions Traversing Ferromagnetic Media and Applications to Measurements of Nuclear Moments	N. Benczer-Koller, M. Haas, J. Sak	30:53–84
Nuclei at High Angular Momentum	R. M. Diamond, F. S. Stephens	30:85–157
Large-Scale Shell-Model Calculations	J. B. McGrory, B. H. Wildenthal	30:383–436
Polarization in Heavy Ion Reactions	D. Fick	31:53–74
The Interacting Boson Model	A. Arima, F. Iachello	31:75–105
Beam Foil Spectroscopy	H. G. Berry, M. Hass	32:1–34
Statistical Spectroscopy	J. B. French, V. K. B. Kota	32:35–64
Theory of Giant Resonances	K. Goeke, J. Speth	32:65–115
Resonances in Heavy-Ion Nuclear Reactions	T. M. Cormier	32:271–308
The Macroscopic Approach to Nuclear Masses and Deformations	W. D. Myers, W. J. Swiatecki	32:309–34
Fusion Reactions Between Heavy Nuclei	J. R. Birkelund, J. R. Huizenga	33:265–322
Inelastic Electron Scattering from Nuclei	J. Heisenberg, H. P. Blok	33:569–609

NUCLEAR THEORY

Symmetries in Nuclei	K. T. Hecht	23:123–61
Linear Relations Among Nuclear Energy Levels	D. S. Koltun	23:163–92
Intermediate Structure in Nuclear Reactions	C. Mahaux	23:193–218
The Many Facets of Nuclear Structure	A. Bohr, B. R. Mottelson	23:363–93
Time Description of Nuclear Reactions	S. Yoshida	24:1–33
The Theory of Three-Nucleon Systems	Y. E. Kim, A. Tubis	24:69–100
Shell-Model Effective Interactions and the Free Nucleon-Nucleon Interaction	T. T. S. Kuo	24:101–50
Post-Fission Phenomena	D. C. Hoffman, M. M. Hoffman	24:151–207
Meson-Nucleus Scattering at Medium Energies	M. M. Sternheim, R. R. Silbar	24:249–77
Proton-Nucleus Scattering at Medium Energies	J. Saudinos, C. Wilkin	24:341–77
Neutron Stars	G. Baym, C. Pethick	25:27–77
Electron Scattering and Nuclear Structure	T. W. Donnelly, J. D. Walecka	25:329–405
The Variable Moment of Inertia (VMI) Model and Theories of Nuclear Collective Motion	G. Scharff-Goldhaber, C. B. Dover, A. L. Goodman	26:239–317
Element Production in the Early Universe	D. N. Schramm, R. V. Wagoner	27:37–74
Self-Consistent Calculations of Nuclear Properties with Phenomenological Effective Forces	P. Quentin, H. Flocard	28:523–96
Recent Developments in Compound-Nucleus Theory	C. Mahaux, H. A. Weidenmüller	29:1–31
Large-Scale Shell-Model Calculations	J. B. McGrory, B. H. Wildenthal	30:383–436
The Interacting Boson Model	A. Arima, F. Iachello	31:75–105
The Equation of State of Hot Dense Matter and Supernovae	J. M. Lattimer	31:337–74
Statistical Spectroscopy	J. B. French, V. K. B. Kota	32:35–64
Theory of Giant Resonances	K. Goeke, J. Speth	32:65–115
The Macroscopic Approach to Nuclear Masses and Deformations	W. D. Myers, W. J. Swiatecki	32:309–34
Nuclear Matter Theory: A Status Report	A. D. Jackson	33:105–41
Grand Unified Theories and the Origin of the Baryon Asymmetry	E. W. Kolb, M. S. Turner	33:645–96

PARTICLE INTERACTIONS AT HIGH ENERGIES

Production Mechanisms of Two-to-Two Scattering Processes at Intermediate Energies	G. C. Fox, C. Quigg	23:219–313
Meson-Nucleus Scattering at Medium Energies	M. M. Sternheim, R. R. Silbar	24:249–77
Proton-Nucleus Scattering at Medium Energies	J. Saudinos, C. Wilkin	24:341–77
Inclusive Reactions	H. Bøggild, T. Ferbel	24:451–513
The Physics of e^+e^- Collisions	R. F. Schwitters, K. Strauch	26:89–149
The Fermi National Accelerator Laboratory	J. R. Sanford	26:151–98

The Parton Model	T.-M. Yan	26:199–238
Diffraction of Hadronic Waves	U. Amaldi, M. Jacob, G. Matthiae	26:385–456
Neutrino Scattering and New-Particle Production	D. Cline, W. F. Fry	27:209–78
Psionic Matter	W. Chinowsky	27:393–464
Charm and Beyond	T. Appelquist, R. M. Barnett, K. Lane	28:387–499
Hopes and Realities for the (p, π) Reaction	D. F. Measday, G. A. Miller	29:121–60
Hyperon Beams and Physics	J. Lach, L. Pondrom	29:203–42
Large Transverse Momentum Hadronic Processes	P. Darriulat	30:159–210
The Tau Lepton	M. L. Perl	30:299–335
Charmed Mesons Produced in e^+e^- Annihilation	G. Goldhaber, J. E. Wiss	30:337–81
Particle Collisions Above 10 TeV As Seen By Cosmic Rays	T. K. Gaisser, G. B. Yodh	30:475–542
High Energy Physics with Polarized Proton Beams	R. C. Fernow, A. D. Krisch	31:104–44
Hadron Production in Lepton-Nucleon Scattering	P. Renton, W. S. C. Williams	31:193–230
Experimental Evidence on QCD	P. Söding, G. Wolf	31:231–93
The Structure of Neutral Currents	P. Q. Hung, J. J. Sakurai	31:375–438
What Can We Count On? A Discussion of Constituent-Counting Rules for Interactions Involving Composite Systems	D. Sivers	32:149–75
Calorimetry in High-Energy Physics	C. W. Fabjan, T. Ludlam	32:335–89
Charged-Current Neutrino Interactions	H. E. Fisk, F. Sciulli	32:499–573
Upsilon Resonances	P. Franzini, J. Lee-Franzini	33:1–29
Progress and Problems in Performance of e^+e^- Storage Rings	R. D. Kohaupt, G.-A. Voss	33:67–104
Physics with the Crystal Ball Detector	E. D. Bloom, C. W. Peck	33:143–97
Sum Rule Approach to Heavy Quark Spectroscopy	M. A. Shifman	33:199–233
Muon Scattering	J. Drees, H. E. Montgomery	33:383–452
Cosmic-Ray Record in Solar System Matter	R. C. Reedy, J. R. Arnold, D. Lal	33:505–37
Measurement of Charmed Particle Lifetimes	R. A. Sidwell, N. W. Reay, N. R. Stanton	33:539–68
Internal Spin Structure of the Nucleon	V. W. Hughes, J. Kuti	33:611–44
Grand Unified Theories and the Origin of the Baryon Asymmetry	E. W. Kolb, M. S. Turner	33:645–96

PARTICLE SPECTROSCOPY

Kaonic and Other Exotic Atoms	R. Seki, C. E. Wiegand	25:241–81
The Particle Data Group: Growth and Operations—Eighteen Years of Particle Physics	A. H. Rosenfeld	25:555–98
The Physics of e^+e^- Collisions	R. F. Schwitters, K. Strauch	26:89–149
Psionic Matter	W. Chinowsky	27:393–464
Quarks	O. W. Greenberg	28:327–86
Charm and Beyond	T. Appelquist, R. M. Barnett, K. Lane	28:387–499
Light Hadronic Spectroscopy: Experimental and Quark Model Interpretations	S. D. Protopopescu, N. P. Samios	29:339–93
The Tau Lepton	M. L. Perl	30:299–335
Charmed Mesons Produced in e^+e^- Annihilation	G. Goldhaber, J. E. Wiss	30:337–81
Experimental Evidence on QCD	P. Söding, G. Wolf	31:231–93
Upsilon Resonances	P. Franzini, J. Lee-Franzini	33:1–29
Physics with the Crystal Ball Detector	E. D. Bloom, C. W. Peck	33:143–97
Sum Rule Approach to Heavy Quark Spectroscopy	M. A. Shifman	33:199–233
Bag Models of Hadrons	C. E. DeTar, J. F. Donoghue	33:235–64

Measurement of Charmed Particle Lifetimes	R. A. Sidwell, N. W. Reay, N. R. Stanton	33:539–68

PARTICLE THEORY

Production Mechansims of Two-to-Two Scattering Processes at Intermediate Energies	G. C. Fox, C. Quigg	23:219–313
Proton-Nucleus Scattering at Medium Energies	J. Saudinos, C. Wilkin	24:341–77
Gauge Theories of Weak Interactions (Circa 1973–74 C. E.)	M. A. B. Bég, A. Sirlin	24:379–449
The Parton Model	T.-M. Yan	26:199–238
Diffraction of Hadronic Waves	U. Amaldi, M. Jacob, G. Matthiae	26:385–456
Quarks	O. W. Greenberg	28:327–86
Charm and Beyond	T. Applequist, R. M. Barnett, K. Lane	28:387–499
Light Hadronic Spectroscopy: Experimental and Quark Model Interpretations	S. D. Protopopescu, N. P. Samios	29:339–93
Baryon Number and Lepton Number Conservation Laws	H. Primakoff, S. P. Rosen	31:145–92
Experimental Evidence on QCD	P. Söding, G. Wolf	31:231–93
The Structure of Neutral Currents	P. Q. Hung, J. J. Sakurai	31:375–438
What Can We Count On? A Discussion of Constituent-Counting Rules for Interactions Involving Composite Systems	D. Sivers	32:149–75
Quantum Chromodynamics: The Modern Theory of the Strong Interaction	F. Wilczek	32:177–209
Physics of Intermediate Vector Bosons	J. Ellis, M. K. Gaillard, G. Girardi, P. Sorba	32:443–97
Gauge Theories and Their Unification	P. Ramond	33:31–66
Sum Rule Approach to Heavy Quark Spectroscopy	M. A. Shifman	33:199–233
Bag Models of Hadrons	C. E. DeTar, J. F. Donoghue	33:235–64
Internal Spin Structure of the Nucleon	V. W. Hughes, J. Kuti	33:611–44
Grand Unified Theories and the Origin of the Baryon Asymmetry	E. W. Kolb, M. S. Turner	33:645–96

RADIATION EFFECTS

K-Shell Ionization in Heavy-Ion Collisions	W. E. Meyerhof, K. Taulbjerg	27:279–331
Synchrotron Radiation Research	H. Winick, A. Bienenstock	28:33–133
Calorimetry in High-Energy Physics	C. W. Fabjan, T. Ludlam	32:335–89
Channeling Radiation	J. U. Andersen, E. Bonderup, R. H. Pantell	33:453–504

WEAK AND ELECTROMAGNETIC INTERACTIONS

Atomic Structure Effects in Nuclear Events	M. S. Freedman	24:209–47
Gauge Theories of Weak Interactions (Circa 1973–74 C. E.)	M. A. B. Bég, A. Sirlin	24:379–449
CP Violation and K° Decays	K. Kleinknecht	26:1–50
The Weak Neutral Current and Its Effects in Stellar Collapse	D. Z. Freedman, D. N. Schramm, D. L. Tubbs	27:167–207
Neutrino Scattering and New-Particle Production	D. Cline, W. F. Fry	27:209–78
Hypernuclei	B. Povh	28:1–32
Charm and Beyond	T. Applequist, R. M. Barnett, K. Lane	28:387–499
Hyperon Beams and Physics	J. Lach, L. Pondrom	29:203–42
The Muon (g-2) Experiments	F. J. M. Farley, E. Picasso	29:243–82
Cosmology Confronts Particle Physics	G. Steigman	29:313–37
Experimental Neutrino Astrophysics	K. Lande	29:395–410
The Parity Non-Conserving Electron-Nucleon Interaction	E. D. Commins, P. H. Bucksbaum	30:1–52

The Tau Lepton	M. L. Perl	30:299–335
Charmed Mesons Produced in e^+e^- Annihilation	G. Goldhaber, J. E. Wiss	30:337–81
Positronium	S. Berko, H. N. Pendleton	30:543–81
Baryon Number and Lepton Number Conservation Laws	H. Primakoff, S. P. Rosen	31:145–92
The Structure of Neutral Currents	P. Q. Hung, J. J. Sakurai	31:375–438
Electric-Dipole Moments of Particles	N. F. Ramsey	32:211–33
Physics of Intermediate Vector Bosons	J. Ellis, M. K. Gaillard, G. Girardi, P. Sorba	32:443–97
Charged-Current Neutrino Interactions	H. E. Fisk, F. Sciulli	32:499–573
Muon Scattering	J. Drees, H. E. Montgomery	33:383–452
Measurement of Charmed Particle Lifetimes	R. A. Sidwell, N. W. Reay, N. R. Stanton	33:539–68
Inelastic Electron Scattering from Nuclei	J. Heisenberg, H. P. Blok	33:569–609
Internal Spin Structure of the Nucleon	V. W. Hughes, J. Kuti	33:611–44

A NONPROFIT SCIENTIFIC PUBLISHER
Annual Reviews Inc.
4139 EL CAMINO WAY • PALO ALTO, CA 94306 USA • (415) 493-4400

Please list the volumes you wish to order by volume number. If you wish a standing order (the latest volume sent to you automatically each year), indicate volume number to begin order. Volumes not yet published will be shipped in month and year indicated. All prices subject to change without notice. Prepayment required from individuals. Telephone orders charged to VISA, MasterCard, American Express, welcomed.

ANNUAL REVIEW SERIES

		Prices Postpaid per volume USA/elsewhere	Regular Order Please send: Vol. number	Standing Order Begin with: Vol. number
Annual Review of ANTHROPOLOGY				
Vols. 1-10	(1972-1981)	$20.00/$21.00		
Vol. 11	(1982)	$22.00/$25.00		
Vol. 12	(1983)	$27.00/$30.00		
Vol. 13	(avail. Oct. 1984)	$27.00/$30.00	Vol(s). _____	Vol. _____
Annual Review of ASTRONOMY AND ASTROPHYSICS				
Vols. 1-19	(1963-1981)	$20.00/$21.00		
Vol. 20	(1982)	$22.00/$25.00		
Vol. 21	(1983)	$44.00/$47.00		
Vol. 22	(avail. Sept. 1984)	$44.00/$47.00	Vol(s). _____	Vol. _____
Annual Review of BIOCHEMISTRY				
Vols. 29-50	(1960-1981)	$21.00/$22.00		
Vol. 51	(1982)	$23.00/$26.00		
Vol. 52	(1983)	$29.00/$32.00		
Vol. 53	(avail. July 1984)	$29.00/$32.00	Vol(s). _____	Vol. _____
Annual Review of BIOPHYSICS AND BIOENGINEERING				
Vols. 1-10	(1972-1981)	$20.00/$21.00		
Vol. 11	(1982)	$22.00/$25.00		
Vol. 12	(1983)	$47.00/$50.00		
Vol. 13	(avail. June 1984)	$47.00/$50.00	Vol(s). _____	Vol. _____
Annual Review of EARTH AND PLANETARY SCIENCES				
Vols. 1-9	(1973-1981)	$20.00/$21.00		
Vol. 10	(1982)	$22.00/$25.00		
Vol. 11	(1983)	$44.00/$47.00		
Vol. 12	(avail. May 1984)	$44.00/$47.00	Vol(s). _____	Vol. _____
Annual Review of ECOLOGY AND SYSTEMATICS				
Vols. 1-12	(1970-1981)	$20.00/$21.00		
Vol. 13	(1982)	$22.00/$25.00		
Vol. 14	(1983)	$27.00/$30.00		
Vol. 15	(avail. Nov. 1984)	$27.00/$30.00	Vol(s). _____	Vol. _____

SEE ORDERING INFORMATION ON PAGE 4.

		Prices Postpaid per volume USA/elsewhere	Regular Order Please send: Vol. number	Standing Order Begin with: Vol. number
Annual Review of ENERGY				
Vols. 1-6	(1976-1981)	$20.00/$21.00		
Vol. 7	(1982)	$22.00/$25.00		
Vol. 8	(1983)	$56.00/$59.00		
Vol. 9	(avail. Oct. 1984)	$56.00/$59.00	Vol(s). _____	Vol. _____
Annual Review of ENTOMOLOGY				
Vols. 7-16, 18-26	(1962-1971; 1973-1981)	$20.00/$21.00		
Vol. 27	(1982)	$22.00/$25.00		
Vol. 28	(1983)	$27.00/$30.00		
Vol. 29	(avail. Jan. 1984)	$27.00/$30.00	Vol(s). _____	Vol. _____
Annual Review of FLUID MECHANICS				
Vols. 1-13	(1969-1981)	$20.00/$21.00		
Vol. 14	(1982)	$22.00/$25.00		
Vol. 15	(1983)	$28.00/$31.00		
Vol. 16	(avail. Jan. 1984)	$28.00/$31.00	Vol(s). _____	Vol. _____
Annual Review of GENETICS				
Vols. 1-15	(1967-1981)	$20.00/$21.00		
Vol. 16	(1982)	$22.00/$25.00		
Vol. 17	(1983)	$27.00/$30.00		
Vol. 18	(avail. Dec. 1984)	$27.00/$30.00	Vol(s). _____	Vol. _____
Annual Review of IMMUNOLOGY				
Vol. 1	(1983)	$27.00/$30.00		
Vol. 2	(avail. April 1984)	$27.00/$30.00	Vol(s). _____	Vol. _____
Annual Review of MATERIALS SCIENCE				
Vols. 1-11	(1971-1981)	$20.00/$21.00		
Vol. 12	(1982)	$22.00/$25.00		
Vol. 13	(1983)	$64.00/$67.00		
Vol. 14	(avail. Aug. 1984)	$64.00/$67.00	Vol(s). _____	Vol. _____
Annual Review of MEDICINE: Selected Topics in the Clinical Sciences				
Vols. 1-3, 5-15	(1950-1952; 1954-1964)	$20.00/$21.00		
Vols. 17-32	(1966-1981)	$20.00/$21.00		
Vol. 33	(1982)	$22.00/$25.00		
Vol. 34	(1983)	$27.00/$30.00		
Vol. 35	(avail. April 1984)	$27.00/$30.00	Vol(s). _____	Vol. _____
Annual Review of MICROBIOLOGY				
Vols. 17-35	(1963-1981)	$20.00/$21.00		
Vol. 36	(1982)	$22.00/$25.00		
Vol. 37	(1983)	$27.00/$30.00		
Vol. 38	(avail. Oct. 1984)	$27.00/$30.00	Vol(s). _____	Vol. _____
Annual Review of NEUROSCIENCE				
Vols. 1-4	(1978-1981)	$20.00/$21.00		
Vol. 5	(1982)	$22.00/$25.00		
Vol. 6	(1983)	$27.00/$30.00		
Vol. 7	(avail. March 1984)	$27.00/$30.00	Vol(s). _____	Vol. _____
Annual Review of NUCLEAR AND PARTICLE SCIENCE				
Vols. 12-31	(1962-1981)	$22.50/$23.50		
Vol. 32	(1982)	$25.00/$28.00		
Vol. 33	(1983)	$30.00/$33.00		
Vol. 34	(avail. Dec. 1984)	$30.00/$33.00	Vol(s). _____	Vol. _____